Lecture Notes in Physics

The Lecture Notes in Physics

The series Lecture Notes in Physics (LNP), founded in 1969, reports new developments in physics research and teaching – quickly and informally, but with a high quality and the explicit aim to summarize and communicate current knowledge in an accessible way. Books published in this series are conceived as bridging material between advanced graduate textbooks and the forefront of research and to serve three purposes:

- to be a compact and modern up-to-date source of reference on a well-defined topic
- to serve as an accessible introduction to the field to postgraduate students and nonspecialist researchers from related areas
- to be a source of advanced teaching material for specialized seminars, courses and schools

Both monographs and multi-author volumes will be considered for publication. Edited volumes should, however, consist of a very limited number of contributions only. Proceedings will not be considered for LNP.

Volumes published in LNP are disseminated both in print and in electronic formats, the electronic archive being available at springerlink.com. The series content is indexed, abstracted and referenced by many abstracting and information services, bibliographic networks, subscription agencies, library networks, and consortia.

Proposals should be sent to a member of the Editorial Board, or directly to the managing editor at Springer:

Christian Caron
Springer Heidelberg
Physics Editorial Department I
Tiergartenstrasse 17
69121 Heidelberg / Germany
christian.caron@springer.com

For further volumes:
www.springer.com/series/5304

Donald Melrose

Quantum Plasmadynamics

Magnetized Plasmas

Donald Melrose
School of Physics
University of Sydney
Sydney, New South Wales
Australia

ISSN 0075-8450 ISSN 1616-6361 (electronic)
ISBN 978-1-4614-4044-4 ISBN 978-1-4614-4045-1 (eBook)
DOI 10.1007/978-1-4614-4045-1
Springer New York Heidelberg Dordrecht London

Library of Congress Control Number: 2012941608

Printed on acid-free paper

Springer is part of Springer Science+Business Media (www.springer.com)

Preface

The motivation and background for the two-volume book on Quantum Plasmadynamics (QPD) were explained in the Preface to volume 1 [1]. In brief, my objective in QPD is to synthesize quantum electrodynamics (QED) and the kinetic theory of plasmas. My interest in doing this has extended over more than four decades, as an ongoing but secondary research topic. I have used the development of QPD as a training tool for some of my students in theoretical physics. As a result, my collaborators have mostly been research students, who have written theses either partially or completely on problems in this field (including Wilson Sy and Ray Stoneham in the 1970s, Andrew Parle, Leith Hayes, Peter Robinson and Michelle Storey (née Allen) in the 1980s, Whayne Padden, Qinghuan Luo, Stephen Hardy and Malcolm Kennett in the 1990s, and Jock McOrist, Alex Judge and Matthew Verdon in the 2000s). Occasional international visitors (John Kirk, Jan Kuijpers, V.V. Zheleznyakov) were collaborators briefly in earlier years. Longer term collaborations have been with Jeanette Weise, since 1995, and with Mushtaq Ahmed for 2 years (2009–2010).

Volume 2 is essentially an extension of the theory in volume 1 from unmagnetized plasmas to magnetized plasma. Both volumes consist of two parts, with the first part concerned with a covariant reformulation of nonquantum plasma kinetic theory, and the second part concerned with the use of QED to calculate plasma processes and plasma responses. The writing of volume 2 has taken me longer than originally anticipated: most of the material in Chaps. 1–8 already existed in some form 4 years ago, when volume 1 was completed. A major part of the delay has been due to my desire to derive a completely general form for the response of a relativistic quantum electron gas for the magnetized case. The aim was to generalize the results in Chap. 9 of volume 1 to the magnetized case, resulting in Chap. 9 of this volume. These new results have been applied to only a few special cases, and there is much more to be explored relating to dispersion in relativistic quantum magnetized plasmas. I had originally intended to include a tenth chapter, analogous to Chap. 10 of volume 1, extending the theory to a magnetized neutrino plasma, based on [2], and to a magnetized boson plasma, based on [3,4], but decided to omit this material.

One specific problem that delayed the completion of this volume concerns the relation between QPD and quantum fluid theory (QFT), on which there has been a rapidly expanding literature over the past decade or so. I have included a short section (§ 1.5) on QFT in this volume. The problem concerns spin-dependent plasmas: I have been unable to identify how the correct relativistic quantum (QPD) result reproduces the quasi-classical QFT result (§ 9.6). There remains a need to justify (or otherwise) extensions of QFT to include spin in a magnetized electron gas.

Sydney, Australia Don Melrose

References

1. D.B. Melrose, *Quantum Plasmadynamics: Unmagnetized Plasmas* (Springer, New York, 2008)
2. D.B. Melrose, M.P. Kennett, Phys. Rev. D **58**, 093011 (1998)
3. N.S. Witte, R.L Dawe, K.C. Hines, J. Math. Phys. **28**, 1864 (1987)
4. N.S. Witte, V. Kowalenko, K.C. Hines, Phys. Rev. D **38**, 3667 (1988)

Contents

Chapter 1
Covariant Fluid Models for Magnetized Plasmas

Three applications of covariant fluid theory are discussed in this chapter: a covariant form of cold-plasma theory, covariant MHD theory, and a covariant form of quantum fluid theory.

In a cold plasma model each species in the plasma is described as a fluid. The fluid equations are used to calculate the response of each species, with the response of the plasma obtained by summing over the contributions of all species. Magnetohydrodynamics (MHD) is the conventional fluid description of a plasma. It differs from a cold plasma model in that the plasma is treated as a single fluid, rather than separate fluids for each species, and through the inclusion of a pressure. In quantum fluid theory, additional quantum effects are included in the fluid equation: quantum mechanical diffusion (the Bohm term), degeneracy and spin.

A covariant form of the fluid equations for a magnetized cold plasma requires a covariant description of a magnetostatic field. The Maxwell 4-tensor, $F^{\mu\nu}$, is used to construct basis 4-vectors in § 1.1. The covariant description of the orbit of a particle is identified, and used in § 1.2 to find the response of a cold plasma using both the forward-scattering and the fluid methods. The cold plasma model is generalized to include the effect of motions along the field lines in § 1.3. A covariant form of MHD theory for a relativistic plasma is introduced in § 1.4 and used to derive the properties of the MHD modes. Quantum fluid theory is discussed in § 1.5. SI units are used in introducing the theory, before reverting to natural units for the formal development of the theory. Except where indicated otherwise, formulae are in natural units.

1.1 Covariant Description of a Magnetostatic Field

A static electromagnetic field is said to be a magnetostatic field if there exists an inertial frame in which there is a magnetic field but no electric field. In this section, the Maxwell 4-tensor is written down for an arbitrary static electromagnetic field, and then specialized to a magnetostatic field. The Maxwell tensor is used to separate

D. Melrose, *Quantum Plasmadynamics: Magnetized Plasmas*, Lecture Notes in Physics 854, DOI 10.1007/978-1-4614-4045-1_1,

space-time into two 2-dimensional subspaces, one containing the time-axis and the direction of the magnetic field, and the other perpendicular to the magnetic field.

1.1.1 Maxwell 4-Tensor

The Maxwell tensor for any electromagnetic field, corresponding to an electric field, $\boldsymbol{E}$ and a magnetic field, $\boldsymbol{B}$, has components (SI units)

$$F^{\mu\nu}(x) = \begin{pmatrix} 0 & -E^1(x)/c & -E^2(x)/c & -E^3(x)/c \\ E^1(x)/c & 0 & -B^3(x) & B^2(x) \\ E^2(x)/c & B^3(x) & 0 & -B^1(x) \\ E^3(x)/c & -B^2(x) & B^1(x) & 0 \end{pmatrix}, \tag{1.1.1}$$

where argument x denotes the components of the 4-vector $x^\mu = [ct, \boldsymbol{x}]$. As in volume 1, the 4-tensor indices, μ, ν run over $0, 1, 2, 3$, and the metric tensor, $g^{\mu\nu}$, is diagonal, $(+1, -1, -1, -1)$, such that one has $x_\mu = [ct, -\boldsymbol{x}]$. Thus, in cartesian coordinates, the contravariant components with $\mu = 0, 1, 2, 3$ are $x^0 = ct$, $x^1 = x$, $x^2 = y$, $x^3 = z$, and the covariant components are $x_0 = ct$, $x_1 = -x$, $x_2 = -y$, $x_3 = -z$. The contravariant components $E^i(x)$, $B^i(x)$, with $i = 1, 2, 3$ correspond to the respective cartesian components of the corresponding 3-vectors.

The dual of the Maxwell tensor is defined by

$$\mathcal{F}^{\mu\nu}(x) = \frac{1}{2}\,\epsilon^{\mu\nu\alpha\beta}\,F_{\alpha\beta}(x), \tag{1.1.2}$$

with $\epsilon^{\mu\nu\alpha\beta}$ the completely antisymmetric tensor with $\epsilon^{0123} = +1$. The dual has components (SI units)

$$\mathcal{F}^{\mu\nu}(x) = \begin{pmatrix} 0 & -B^1(x) & -B^2(x) & -B^3(x) \\ B^1(x) & 0 & E^3(x)/c & -E^2(x)/c \\ B^2(x) & -E^3(x)/c & 0 & E^1(x)/c \\ B^3(x) & E^2(x)/c & -E^1(x)/c & 0 \end{pmatrix}. \tag{1.1.3}$$

Maxwell's equations in covariant form are

$$\partial_\mu F^{\mu\nu}(x) = \mu_0 J^\nu(x), \qquad \partial_\mu \mathcal{F}^{\mu\nu}(x) = 0, \tag{1.1.4}$$

with $\partial_\mu = \partial/\partial x^\mu$.

Classification of Static Fields

An arbitrary static homogeneous electromagnetic field may be classified as a magnetostatic field, an electrostatic field, or an electromagnetic wrench. This classification is based on the fact that there are two independent electromagnetic invariants (SI units)

$$S = -\tfrac{1}{4}F^{\mu\nu}F_{\mu\nu} = \tfrac{1}{2}\left(\boldsymbol{E}^2/c^2 - \boldsymbol{B}^2\right), \qquad P = -\tfrac{1}{4}F^{\mu\nu}\mathcal{F}_{\mu\nu} = \boldsymbol{E}\cdot\boldsymbol{B}/c, \tag{1.1.5}$$

where the argument x is omitted. An arbitrary static electromagnetic field is (a) a magnetostatic field for $S < 0$, $P = 0$, (b) an electrostatic field for $S > 0$, $P = 0$, or (c) an electromagnetic wrench for $P \neq 0$. This classification follows from the fact that in these three cases, one can always choose an inertial frame such that (a) there is no electric field and the magnetic field is along a chosen axis, (b) there is no magnetic field and the electric field is along a chosen axis, and (c) the electric and magnetic fields are parallel and along a chosen axis.

Magnetostatic Field

It is convenient to denote the magnetostatic field by $F_0^{\mu\nu}$, and to write

$$F_0^{\mu\nu} = Bf^{\mu\nu}, \qquad B = \left(\tfrac{1}{2}F_0^{\mu\nu}F_{0\mu\nu}\right)^{1/2}. \tag{1.1.6}$$

Equation (1.1.6) defines the invariant B, interpreted as the magnetic field strength, and introduces a dimensionless 4-tensor $f^{\mu\nu}$. A second dimensionless 4-tensor is the dual of $f^{\mu\nu}$:

$$\phi^{\mu\nu} = \tfrac{1}{2}\epsilon^{\mu\nu\alpha\beta}f_{\alpha\beta}, \qquad \mathcal{F}^{\mu\nu} = B\,\phi^{\mu\nu}. \tag{1.1.7}$$

The tensors $f^{\mu\nu}$ and $\phi^{\mu\nu}$ are both antisymmetric and they are orthogonal to each other:

$$f^{\mu\nu} = -f^{\nu\mu}, \qquad \phi^{\mu\nu} = -\phi^{\nu\mu}, \qquad f^{\mu\alpha}\phi_\alpha{}^\nu = 0. \tag{1.1.8}$$

Alternative Form for the Maxwell 4-Tensor

The most general form for a static electromagnetic field is an electromagnetic wrench. Such a field may be written in the form (SI units)

$$F^{\mu\nu} = Bf^{\mu\nu} + (E/c)\phi^{\mu\nu}, \qquad E/c, B = \left[(S^2+P^2)^{1/2} \pm S\right]^{1/2}. \tag{1.1.9}$$

The tensors $f^{\mu\nu}$, $\phi^{\mu\nu}$ have the forms (1.1.16) in the frame in which the electric and magnetic fields are parallel and along the 3 axis. For $E = 0$ (1.1.9) describes a magnetostatic field along the 3 axis, and for $B = 0$ (1.1.9) describes an electrostatic field along the 3 axis.

4-Tensor B^μ

The description of the magnetostatic field in terms of $F_0^{\mu\nu}$ applies in any frame. There is an alternative description of a magnetostatic field in terms of a 4-vector, B^μ, that is available if there is some preferred frame. In the presence of a medium there is a preferred frame: the rest frame of the medium. Let $\tilde{u}^\mu$ be the 4-velocity of the rest frame, with $\tilde{u}^2 = \tilde{u}^\mu \tilde{u}_\mu = 1$. One has $\tilde{u} = [1, \mathbf{0}]$ in the rest frame. The 4-tensor is

$$B^\mu = \mathcal{F}_0^{\mu\nu} \tilde{u}_\nu = B\, b^\mu. \tag{1.1.10}$$

One has $B^\mu B_\mu = -B^2$ and hence $b^\mu b_\mu = -1$. In the rest frame one may write $b^\mu = [0, \boldsymbol{b}]$, where $\boldsymbol{b}$ is a unit vector along the direction of the magnetic field.

1.1.2 Projection Tensors $g_\perp^{\mu\nu}$ and $g_\parallel^{\mu\nu}$

The tensors $f^{\mu\nu}$ and $\phi^{\mu\nu}$ allow one to construct projection tensors $g_\perp^{\mu\nu}$ and $g_\parallel^{\mu\nu}$:

$$g_\perp^{\mu\nu} = -f^\mu{}_\alpha f^{\alpha\nu}, \qquad g_\parallel^{\mu\nu} = \phi^\mu{}_\alpha \phi^{\alpha\nu}. \tag{1.1.11}$$

The tensors $g_\perp^{\mu\nu}$ and $g_\parallel^{\mu\nu}$ span the 2-dimensional perpendicular and time-parallel subspaces, respectively. They correspond to a separation of the metric tensor into metric tensors for these two subspaces:

$$g^{\mu\nu} = g_\parallel^{\mu\nu} + g_\perp^{\mu\nu}. \tag{1.1.12}$$

In a frame in which the magnetostatic field is along the 3-axis, $g_\parallel^{\mu\nu}$ is diagonal $(1, 0, 0, -1)$ and $g_\perp^{\mu\nu}$ is diagonal $(0, -1, -1, 0)$.

An alternative definition of $g_\parallel^{\mu\nu}$ assumes the existence of a frame in which (1.1.10) applies. This frame is described by its 4-velocity $\tilde{u}^\mu$, and by the 4-vector b^μ along the direction of the magnetic field. One has

$$g_\parallel^{\mu\nu} = \tilde{u}^\mu \tilde{u}^\nu - b^\mu b^\nu. \tag{1.1.13}$$

Despite appearances, $g_\parallel^{\mu\nu}$ is independent of $\tilde{u}$, that is, it is not dependent on the choice of a preferred frame.

Parallel and Perpendicular Invariant Components

The projection tensors allow one to separate any 4-vector, a^μ say, into a sum of two orthogonal 4-vectors, $a^\mu = a_\perp^\mu + a_\parallel^\mu$;

$$a_\perp^\mu = g_\perp^{\mu\nu} a_\nu, \qquad a_\parallel^\mu = g_\parallel^{\mu\nu} a_\nu. \tag{1.1.14}$$

Similarly, for a single 4-vector, a^μ, the invariant a^2 is written in the form $a^2 = (a^2)_\perp + (a^2)_\parallel$, and for two 4-vectors, a^μ and b^μ, the invariant ab is written in the form $ab = (ab)_\perp + (ab)_\parallel$. These separations are made by writing

$$(a^2)_\perp = g_\perp^{\mu\nu} a_\mu a_\nu, \qquad (ab)_\perp = g_\perp^{\mu\nu} a_\mu b_\nu,$$
$$(a^2)_\parallel = g_\parallel^{\mu\nu} a_\mu a_\nu, \qquad (ab)_\parallel = g_\parallel^{\mu\nu} a_\mu b_\nu. \tag{1.1.15}$$

The components of the wave 4-vector, $k^\mu = [\omega/c, \mathbf{k}]$, with $\mathbf{k} = (k_x, k_y, k_z)$, give $(k^2)_\perp = -k_\perp^2 = -(k_x^2 + k_y^2)$, $(k^2)_\parallel = \omega^2/c^2 - k_z^2$.

Components for B Along the 3-Axis

In any frame in which $F_0^{\mu\nu}$ corresponds to $\boldsymbol{E} = 0$, one is free to write $\boldsymbol{B} = B\boldsymbol{b}$, and to orient the axies such that $\boldsymbol{b} = (0, 0, 1)$ is along the 3-axis. The tensors $f^{\mu\nu}$ and $\phi^{\mu\nu}$ then have components

$$f^{\mu\nu} = \begin{pmatrix} 0 & 0 & 0 & 0 \\ 0 & 0 & -1 & 0 \\ 0 & 1 & 0 & 0 \\ 0 & 0 & 0 & 0 \end{pmatrix}, \qquad \phi^{\mu\nu} = \begin{pmatrix} 0 & 0 & 0 & -1 \\ 0 & 0 & 0 & 0 \\ 0 & 0 & 0 & 0 \\ 1 & 0 & 0 & 0 \end{pmatrix}. \tag{1.1.16}$$

The projection tensors (1.1.11) have components

$$g_\perp^{\mu\nu} = \begin{pmatrix} 0 & 0 & 0 & 0 \\ 0 & -1 & 0 & 0 \\ 0 & 0 & -1 & 0 \\ 0 & 0 & 0 & 0 \end{pmatrix}, \qquad g_\parallel^{\mu\nu} = \begin{pmatrix} 1 & 0 & 0 & 0 \\ 0 & 0 & 0 & 0 \\ 0 & 0 & 0 & 0 \\ 0 & 0 & 0 & -1 \end{pmatrix}. \tag{1.1.17}$$

1.1.3 Basis 4-Vectors

Consider the response of a medium described by the response tensor $\Pi^{\mu\nu}(k)$. The tensor indices can depend only on the available 4-vectors and 4-tensors. In a magnetized medium the available 4-tensors and 4-vectors are $f^{\mu\nu}$, k^μ, and the 4-velocity, $\tilde{u}^\mu$, of the rest frame of the medium. It is possible to construct several different sets of basis 4-vectors that may be used to represent $\Pi^{\mu\nu}(k)$. For the magnetized vacuum, $\tilde{u}^\mu$ is undefined, and one needs a set that does not involve it. A set that involves $\tilde{u}^\mu$ is more convenient for a material medium.

Basis 4-Vectors for the Magnetized Vacuum

One can construct four mutually orthogonal 4-vectors from k^μ and the Maxwell tensor for the magnetostatic field. A convenient choice is

$$k_\parallel^\mu = g_\parallel^{\mu\nu} k_\nu = (\omega/c, 0, 0, k_z), \qquad k_\perp^\mu = g_\perp^{\mu\nu} k_\nu = (0, k_\perp, 0, 0),$$
$$k_G^\mu = -f^{\mu\nu} k_\nu = (0, 0, k_\perp, 0), \qquad k_D^\mu = \phi^{\mu\nu} k_\nu = (k_z, 0, 0, \omega/c), \tag{1.1.18}$$

where the components apply in a frame in which the magnetic field is along the 3-axis, and $\boldsymbol{k}$ is in the 1–3 plane. In terms of this choice of basis 4-vectors, the 4-tensors (1.1.16) and (1.1.17) become

$$f^{\mu\nu} = -\frac{k_\perp^\mu k_G^\nu - k_G^\mu k_\perp^\nu}{k_\perp^2}, \qquad \phi^{\mu\nu} = -\frac{k_\parallel^\mu k_D^\nu - k_D^\mu k_\parallel^\nu}{(k^2)_\parallel},$$
$$g_\parallel^{\mu\nu} = \frac{k_\parallel^\mu k_\parallel^\nu - k_D^\mu k_D^\nu}{(k^2)_\parallel}, \qquad g_\perp^{\mu\nu} = -\frac{k_\perp^\mu k_\perp^\nu + k_G^\mu k_G^\nu}{k_\perp^2}, \tag{1.1.19}$$

with $(k^2)_\parallel = k_\parallel^\mu k_{\parallel\mu} = \omega^2 - k_z^2$, $(k^2)_\perp = k_\perp^\mu k_{\perp\mu} = -k_\perp^2$.

Writing the response 4-tensor in terms of invariant components along the 4-vectors (1.1.18) gives

$$\Pi^{\mu\nu}(k) = \sum_{A,B} \Pi_{AB}(k)\, k_A^\mu k_B^\nu, \qquad \Pi_{AB}(k) = \frac{k_A^\mu k_B^\nu}{k_A^2 k_B^2} \Pi_{\mu\nu}(k), \tag{1.1.20}$$

with $A, B = \parallel, \perp, D, G$. The charge-continuity and gauge-invariance relations, $k_\mu \Pi^{\mu\nu}(k) = 0$, $k_\nu \Pi^{\mu\nu}(k) = 0$, imply that the invariant components $\Pi^{AB}(k)$ in (1.1.20) satisfy

$$(k^2)_\parallel \Pi_{\parallel B}(k) = k_\perp^2 \Pi_{\perp B}(k), \qquad (k^2)_\parallel \Pi_{A\parallel}(k) = k_\perp^2 \Pi_{A\perp}(k), \tag{1.1.21}$$

with $A, B = \parallel, \perp, D, G$.

Shabad's Basis 4-Vectors

A related choice of basis 4-vectors, made by Shabad [24], is

$$b_1^\mu = k_G^\mu, \quad b_2^\mu = k_D^\mu, \quad b_3^\mu = k_\perp^\mu + k^\mu k_\perp^2/k^2, \quad b_4^\mu = k^\mu. \tag{1.1.22}$$

In a form analogous to (1.1.20), the response tensor,

$$\Pi^{\mu\nu}(k) = \sum_{A,B=1}^{3} \Pi_{AB}(k)\, b_A^\mu b_B^\mu, \tag{1.1.23}$$

involves only six invariant components, with the counterpart of (1.1.21) becoming $\Pi_{4B}(k) = 0 = \Pi_{A4}(k)$.

The set (1.1.22) of basic vectors b_A^μ with $A = 1, 2, 3$ may be replaced by any linear combination of them that preserves the orthogonality condition. In particular, one can choose a linear combination that separates longitudinal and transverse parts. Let this choice be denoted $b_{A'}^\mu$, with $A' = 1', 2', 3'$. One requires that two of these satisfy $\boldsymbol{b}_{A'} \cdot \boldsymbol{k} = 0$, and that the third has its 3-vector part parallel to $\boldsymbol{k}$. The combination $k_z b_2^\mu + \omega b_3^\mu$ has its 3-vector part proportional to $\boldsymbol{k}$. Thus the choice of (unnormalized) basis 4-vectors,

$$b_{1'}^\mu = b_1^\mu, \quad b_{2'}^\mu \propto \omega k_\perp^2 b_2^\mu - k^2 k_z b_3^\mu, \quad b_{3'}^\mu \propto k_z b_2^\mu + \omega b_3^\mu, \tag{1.1.24}$$

allows one to project onto the transverse plane, due to $b_{1'}^\mu, b_{2'}^\mu$ being transverse and $b_{3'}^\mu$ being longitudinal. (This procedure breaks down for the special case $k^2 = 0$.)

The normalization of these 4-vectors is problematic. Only one of the 4-vectors can be time-like, and this 4-vector is to be normalized to unity with the others three 4-vectors normalized to minus unity. However, which of them is time-like depends on the signs of the invariants k^2 and $(k^2)_\parallel = \omega^2 - k_z^2$. The choice (1.1.24) is not used in the following.

Basis 4-Vectors for a Magnetized Medium

In the presence of a medium, the 4-vector, $\tilde{u}^\mu$, of the rest frame of the medium may be chosen as one of the basis 4-vectors, and it is then the only time-like basis 4-vector. One needs three independent space-like 3-vectors. Given the unit 3-vectors $\boldsymbol{\kappa} = \boldsymbol{k}/|\boldsymbol{k}|$ and $\boldsymbol{b}$ along the wave 3-vector and the magnetostatic field, respectively, an orthonormal set of 3-vectors in the rest frame is $\kappa^\mu = [0, \boldsymbol{\kappa}]$, $a^\mu = [0, \boldsymbol{a}]$, $t^\mu = [0, \boldsymbol{t}]$, with

$$\boldsymbol{a} = -\boldsymbol{\kappa} \times \boldsymbol{b}/|\boldsymbol{\kappa} \times \boldsymbol{b}|, \qquad \boldsymbol{t} = -\boldsymbol{\kappa} \times \boldsymbol{a}. \tag{1.1.25}$$

A covariant form of these definitions is

$$\kappa^\mu = \frac{k^\mu - k\tilde{u}\,\tilde{u}^\mu}{[(k\tilde{u})^2 - k^2]^{1/2}}, \quad a^\mu = -\frac{b_1^\mu}{k_\perp}, \quad t^\mu = \epsilon^{\mu\alpha\beta\gamma}\tilde{u}_\alpha \kappa_\beta a_\tau, \tag{1.1.26}$$

with the fourth basis 4-vector being $\tilde{u}^\mu$.

To interpret these basis vectors further, it is helpful to express other quantities that appear in the rest frame in terms of invariants. Let the angle between the 3-vector $\boldsymbol{k}$ and $\boldsymbol{B}$ in the rest frame be θ. One has

$$\omega = k\tilde{u}, \qquad |\boldsymbol{k}|^2 = (k\tilde{u})^2 - k^2,$$

$$\sin^2\theta = \frac{k_\perp^2}{(k\tilde{u})^2 - k^2}, \qquad \cos^2\theta = \frac{(k\tilde{u})^2 - (k^2)_\parallel}{(k\tilde{u})^2 - k^2}. \tag{1.1.27}$$

Using these relations one may rewrite many of the formulae in the noncovariant theory in a covariant form.

The 4-vectors (1.1.26) may be written in terms of the angles (1.1.27) in the rest frame of the medium. With the 3-axis along the magnetic field and $\boldsymbol{k}$ in the 1–3 plane, one finds

$$\tilde{u}^\mu = (1,0,0,0), \qquad \kappa^\mu = (0, \sin\theta, 0, \cos\theta),$$

$$a^\mu = (0,0,1,0), \qquad t^\mu = (0, \cos\theta, 0, -\sin\theta). \tag{1.1.28}$$

The 4-vectors t^μ, a^μ span the 2-dimensional transverse space, and may be chosen as basis vectors for the representation of transverse polarization states.

An alternative choice of basis 4-vectors, closely related to the set (1.1.28), is the set

$$\tilde{u}^\mu = (1,0,0,0), \qquad e_1^\mu = (0,1,0,0),$$

$$e_2^\mu = (0,0,1,0), \qquad b^\mu = (0,0,0,1). \tag{1.1.29}$$

Covariant definitions of the last three of these correspond to $e_1^\mu = k_\perp^\mu/k_\perp$, $e_2^\mu = k_G^\mu/k_\perp$, $b^\mu = B^\mu/B$, with B^μ defined by (1.1.10).

1.1.4 Linear and Nonlinear Response Tensor

The response of a plasma is described in terms of the linear response tensor and a hierarchy of nonlinear response tensors. The definition of these response tensors follow from an expansion of the induced current in the plasma in powers of a fluctuating electromagnetic field. This is referred to as the weak-turbulence expansion. The linear response tensor is defined by the leading term in this expansion.

Weak-Turbulence Expansion

The expansion of the induced current density, $J^{\mu}_{\rm ind}(x)$, in powers of the electromagnetic field, described by $A^{\mu}(x)$, is made in terms of their Fourier transforms in time and space. The Fourier transform and its inverse for the induced current is

$$J^{\mu}_{\rm ind}(k) = \int d^4x\, e^{ikx}\, J^{\mu}_{\rm ind}(x), \qquad J^{\mu}_{\rm ind}(x) = \int \frac{d^4k}{(2\pi)^4}\, e^{-ikx}\, J^{\mu}_{\rm ind}(k). \quad (1.1.30)$$

The weak-turbulence expansion is

$$\begin{aligned} J^{\mu}_{\rm ind}(k) &= \Pi^{\mu}{}_{\nu}(k)A^{\nu}(k) + \int d\lambda^{(2)}\, \Pi^{(2)\mu}{}_{\nu\rho}(-k,k_1,k_2)A^{\nu}(k_1)A^{\rho}(k_2) \\ &\quad + \int d\lambda^{(3)}\, \Pi^{(3)\mu}{}_{\nu\rho\sigma}(-k,k_1,k_2,k_3)A^{\nu}(k_1)A^{\rho}(k_2)A^{\sigma}(k_3) + \cdots \\ &\quad + \int d\lambda^{(n)}\, \Pi^{(n)\mu}{}_{\nu_1\nu_2\ldots\nu_n}(-k,k_1,k_2,\ldots,k_n)\, A^{\nu_1}(k_1)A^{\nu_2}(k_2)\ldots A^{\nu_n}(k_n) \\ &\quad + \cdots, \end{aligned} \quad (1.1.31)$$

where the n-fold convolution integral is defined by

$$d\lambda^{(n)} = \frac{d^4k_1}{(2\pi)^4}\frac{d^4k_2}{(2\pi)^4}\cdots\frac{d^4k_n}{(2\pi)^4}(2\pi)^4\delta^4(k-k_1-k_2-\cdots-k_n), \quad (1.1.32)$$

with $\delta^4(k) = \delta(k^0)\delta(k^1)\delta(k^2)\delta(k^3)$. The expansion (1.1.31) defines the linear response tensor $\Pi^{\mu\nu}(k)$ and a hierarchy of nonlinear response tensors, of which only the quadratic response tensor $\Pi^{(2)\mu\nu\rho}(k_0,k_1,k_2)$, with $k_0+k_1+k_2=0$, and the cubic response tensor $\Pi^{(3)\mu\nu\rho\sigma}(k_0,k_1,k_2,k_3)$, with $k_0+k_1+k_2+k_3=0$, are usually considered when discussing specific weak-turbulence processes.

General Properties of $\Pi^{\mu\nu}(k)$

The linear response tensor satisfies several general conditions.

The reality condition on Fourier transforms is that reversing the sign of k^{μ} and complex conjugating has no net effect. Hence one has

$$\Pi^{\mu\nu}(-k) = \Pi^{*\mu\nu}(k), \quad (1.1.33)$$

where * denotes the complex conjugate. The Fourier transform of the charge-continuity condition, $\partial_{\mu}J^{\mu}_{\rm ind}(x) = 0$ implies the first of

$$k_{\mu}\Pi^{\mu\nu}(k) = 0, \qquad k_{\nu}\Pi^{\mu\nu}(k) = 0. \quad (1.1.34)$$

The second of (1.1.34) is the gauge-invariance condition. It is desirable that the description of the response be independent of the choice of gauge for $A^\nu(k)$, and an arbitrary gauge transformation involves adding a term proportional to k^ν to $A^\nu(k)$. Imposing the second of the conditions (1.1.34) ensures that $\Pi^{\mu\nu}(k)$ is independent of the choice of gauge.

A third general condition is that a separation of $\Pi^{\mu\nu}(k)$ into hermitian (H) and antihermitian (A) parts is equivalent to a separation into the time-reversible or non-dissipative part of the response, and the time-irreversible or dissipative part of the response, respectively. These two parts are

$$\Pi^{H\mu\nu}(k) = \tfrac{1}{2}[\Pi^{\mu\nu}(k) + \Pi^{*\nu\mu}(k)], \qquad \Pi^{A\mu\nu}(k) = \tfrac{1}{2}[\Pi^{\mu\nu}(k) - \Pi^{*\nu\mu}(k)], \tag{1.1.35}$$

respectively.

Onsager Relations

The Onsager relations follow from the time-reversal invariance properties of the equations of motion used in any calculation of the response tensor. The choice of basis 4-vectors (1.1.28) is particularly convenient for expressing the Onsager relations:

$$\Pi^{00}(\omega,-\boldsymbol{k})\,|_{-\boldsymbol{B}_0} = \Pi^{00}(\omega,\boldsymbol{k})\,|_{\boldsymbol{B}_0}, \qquad \Pi^{0i}(\omega,-\boldsymbol{k})\,|_{-\boldsymbol{B}_0} = -\Pi^{i0}(\omega,\boldsymbol{k})\,|_{\boldsymbol{B}_0},$$
$$\Pi^{ij}(\omega,-\boldsymbol{k})\,|_{-\boldsymbol{B}_0} = \Pi^{ji}(\omega,\boldsymbol{k})\,|_{\boldsymbol{B}_0}, \tag{1.1.36}$$

where the reversal of the sign of any external magnetostatic field is noted explicitly. With the choice of coordinate axes in (1.1.28), the Onsager relations (1.1.36) imply

$$\begin{aligned}
\Pi^{01}(\omega,\boldsymbol{k}) &= \Pi^{10}(\omega,\boldsymbol{k}), & \Pi^{02}(\omega,\boldsymbol{k}) &= -\Pi^{20}(\omega,\boldsymbol{k}),\\
\Pi^{03}(\omega,\boldsymbol{k}) &= \Pi^{30}(\omega,\boldsymbol{k}), & \Pi^{12}(\omega,\boldsymbol{k}) &= -\Pi^{21}(\omega,\boldsymbol{k}),\\
\Pi^{13}(\omega,\boldsymbol{k}) &= \Pi^{31}(\omega,\boldsymbol{k}), & \Pi^{23}(\omega,\boldsymbol{k}) &= -\Pi^{32}(\omega,\boldsymbol{k}).
\end{aligned} \tag{1.1.37}$$

In terms of the invariant components introduced in (1.1.20), the Onsager relations imply

$$\Pi^{AB}(k) = \Pi^{BA}(k), \quad A,B = \parallel,\perp,D; \quad \Pi^{AG}(k) = -\Pi^{GA}(k), \quad A = \parallel,\perp,D, \tag{1.1.38}$$

with $k_G^\mu = [0, k_\perp \boldsymbol{a}]$ in the frame defined by (1.1.28).

An important implication for the polarization vector of a wave in a magnetized medium is that its component along the $\boldsymbol{a}$-axis (the 2-axis or y-axis for $\mathbf{k}$ in the 1–3 plane) is out of phase with the other components. Thus, one can choose an overall phase factor such that the 2-component ($\boldsymbol{a}$-component) is imaginary and the 0-, 1-, 3-components ($\tilde{u}$-, $\boldsymbol{\kappa}$-, $\boldsymbol{t}$-components) are all real.

1.2 Covariant Cold Plasma Model

A covariant formulation of the cold plasma model is used in this section to calculate the linear response tensor for a cold magnetized plasma. Cold plasma theory generalizes the magnetoionic theory, also called the Appleton-Hartree theory, which was developed in the 1930s [4, 13] to describe radio wave propagation in the ionosphere. The magnetoionic theory describes the response of a cold electron gas, and the generalization to include the motion of the ions was made in the 1950s [26, 27].

Natural units ($\hbar = c = 1$) are used in this an subsequent sections, except where stated otherwise.

1.2.1 Fluid Description of a Cold Plasma

A cold plasma, in which thermal or other random motions are neglected, can be described using fluid equations, with one fluid for each species of particle. Consider particles of species α, with charge q_α and mass m_α. For simplicity in writing, the affix α, denoting the species, is suppressed for the present, with the charge and mass denoted by q and m, respectively.

The fluid equations in covariant form involve the proper number density, $n_{\rm pr}(x)$, and the fluid 4-velocity, $u^\mu(x)$. The equation of continuity is

$$\partial_\mu \left[n_{\rm pr}(x) u^\mu(x)\right] = 0. \tag{1.2.1}$$

The equation of motion for the fluid is

$$u^\sigma(x)\partial_\sigma u^\mu(x) = \frac{q}{m}\left[F_0^{\mu\nu} + \partial^\mu A^\nu(x) - \partial^\nu A^\mu(x)\right]u_\nu(x), \tag{1.2.2}$$

where the contributions of the static field $F_0^{\mu\nu}$ and of a fluctuating field $F^{\mu\nu}(x) = \partial^\mu A^\nu(x) - \partial^\nu A^\mu(x)$ are included separately in the 4-force on the right hand side. The operator $u^\sigma(x)\partial_\sigma$ in (1.2.2) may be interpreted as the total derivative $\partial/\partial\tau(x)$, where $\tau(x)$ is the proper time along the flow lines of the fluid. The derivative $\partial/\partial\tau(x)$ arises naturally in a covariant (Lagrangian or Hamiltonian) treatment of particle dynamics, and in a fluid (Eulerian) theory it is interpreted as the convective derivative $u^\sigma(x)\partial_\sigma$.

The response of the plasma is identified as the induced 4-current density:

$$J^\mu_{\rm ind}(x) = q n_{\rm pr}(x)\, u^\mu(x). \tag{1.2.3}$$

The objective is to expand $n_{\rm pr}(x)\, u^\mu(x)$ in powers of the perturbing electromagnetic field, described by $A^\mu(x)$, and to identify the response tensor from the Fourier transform of (1.2.3).

Fourier Transform of the Fluid Equations

The first steps in evaluating the induced current for a cold plasma are to Fourier transform the fluid Eqs. (1.2.1) and (1.2.2) and the induced current (1.2.3), and to carry out a perturbation expansion in powers of A^μ.

The Fourier transformed form of the continuity Eq. (1.2.1) is

$$\int d\lambda^{(2)}\, n_{\rm pr}(k_1)\, ku(k_2) = 0, \tag{1.2.4}$$

where the convolution integral is defined by (1.1.32) with $n = 2$. The Fourier transform of the equation of fluid motion (1.2.2) is

$$\int d\lambda^{(2)}\, k_2u(k_1)\, u^\mu(k_2) = i\frac{q}{m}F_0^{\mu\nu}u_\nu(k) - \frac{q}{m}\int d\lambda^{(2)}\, k_1u(k_2)\, G^{\mu\nu}\big(k_1, u(k_2)\big)A_\nu(k_1), \tag{1.2.5}$$

with

$$G^{\mu\nu}(k,u) = g^{\mu\nu} - \frac{k^\mu u^\nu}{ku}. \tag{1.2.6}$$

The induced 4-current (1.2.3) becomes

$$J^\mu_{\rm ind}(k) = \int d\lambda^{(2)}\, n_{\rm pr}(k_1)\, u^\mu(k_2). \tag{1.2.7}$$

The 4-current has contributions from each species, only one of which is included in (1.2.7).

Expansion in Powers of the 4-Potential

In the weak-turbulence expansion for a fluid, the proper number density and the fluid 4-velocity are expanded in powers of $A(k)$.

The expansion of the proper number density is

$$n_{\rm pr}(k) = \bar{n}_{\rm pr}\,(2\pi)^4\delta^4(k) + \sum_{N=1}^{\infty} n^{(N)}_{\rm pr}(k), \tag{1.2.8}$$

where the zeroth order proper number density is denoted $\bar{n}_{\rm pr}$. In the rest frame of a cold fluid, all the particle are at rest, and in this frame the proper number density is equal to the actual number density, $\bar{n}$ say. In the following, $\bar{n}$ is used to describe the number density of the fluid, where $\bar{n}$ is the actual number density in the rest frame of the fluid. Where no confusion should result, $\bar{n}$ is written as n. The expansion of the fluid 4-velocity gives

$$u^\mu(k) = \tilde{u}^\mu(2\pi)^4\delta^4(k) + \sum_{N=1}^{\infty} u^{(N)\mu}(k). \tag{1.2.9}$$

The zeroth order fluid 4-velocity is necessarily nonzero, being $\tilde{u}^\mu = (1,0,0,0)$ in the rest frame of the fluid. More generally, the fluid has a nonzero velocity along the direction of the magnetic field, and the associated 4-velocity may be interpreted as that of the rest frame of the fluid, and written

$$\tilde{u}^\mu = \tilde{u}^\mu_\parallel = \tilde{\gamma}(1,0,0,\tilde{v}_z), \qquad \tilde{\gamma} = \left(1-\tilde{v}_z^2\right)^{-1/2}. \tag{1.2.10}$$

Fluid flow (for a charged fluid) across the magnetic field is inconsistent with the assumption that there is no static electric field. The expansion of the induced 4-current (1.2.7) gives

$$J^\mu_{\rm ind}(k) = qn\tilde{u}^\mu(2\pi)^4\delta^4(k) + \sum_{N=1}^{\infty} J^{(N)\mu}(k). \tag{1.2.11}$$

The zeroth order current density, $qn\tilde{u}^\mu$, associated with a given fluid is nonzero, but is usually ignored. The justification for this is either that the contributions to the 4-current from different species sum to zero, or that the static fields generated by this current density are negligible.

1.2.2 *Linear Response Tensor: Cold Plasma*

The derivation of $\Pi^{\mu\nu}(k)$ using the fluid model involves only the linear terms in the expansion of the fluid equations. The space components, $\Pi^{ij}(k)$, are simply related to the components of the dielectric tensor $K^i{}_j(k)$, used in a conventional 3-tensor description.

First Order Current

The first order term in the expansion (1.2.11) of the current determines the linear response. On substituting (1.2.8) and (1.2.9) into (1.2.11), for $N = 1$ one has

$$J^{(1)\mu}(k) = q\left[n\,u^{(1)\mu}(k) + n^{(1)}(k)\,\tilde{u}^\mu\right], \tag{1.2.12}$$

where the subscript 'ind' is omitted. To first order, the equation of continuity (1.2.4) gives

$$k\tilde{u}\,n^{(1)}(k) = -n\,ku^{(1)}(k), \tag{1.2.13}$$

which determines the first order number density in terms of the first order fluid velocity. The first order fluid velocity is determined by the first order terms in (1.2.5):

$$k\tilde{u}\,u^{(1)\mu}(k) = i\frac{q}{m}F_0^{\mu\nu}u^{(1)}_\nu(k) - \frac{q}{m}\,k\tilde{u}\,G^{\mu\nu}(k,\tilde{u})A_\nu(k). \tag{1.2.14}$$

Let the solution of (1.2.14) be written

$$u^{(1)\mu}(k) = -\frac{q}{m}\tau^{\mu}{}_{\rho}(k\tilde{u})G^{\rho\nu}(k,\tilde{u})A_{\nu}(k). \tag{1.2.15}$$

An an explicit form for the tensor $\tau^{\mu\nu}(\omega)$ is derived below.

On inserting (1.2.13) and (1.2.15) into (1.2.12), one identifies the linear response tensor by writing $J^{(1)\mu}(k) = \Pi^{\mu\nu}(k)A_{\nu}(k)$. The resulting cold plasma response tensor is

$$\Pi^{\mu\nu}(k) = -\frac{q^2 n}{m\bar{\gamma}}\,G^{\alpha\mu}(k,\tilde{u})\tau_{\alpha\beta}(k\tilde{u})G^{\beta\nu}(k,\tilde{u}). \tag{1.2.16}$$

Note that, by hypothesis, any zeroth order motion is along the magnetic field lines, so that $\tilde{u}$ is equal to $\tilde{u}_{\parallel}$, implying that the response tensor (1.2.16) is independent of $k_{\perp}$.

1.2.3 Tensor $\tau^{\mu\nu}(\omega)$

The tensor $\tau^{\mu\nu}(\omega)$, introduced in (1.2.15), is constructed as follows. Write (1.2.14) in the form

$$[k\tilde{u}g^{\mu\nu} + i\epsilon\Omega_0 f^{\mu\nu}]\,u_{\nu}^{(1)}(k) = -\frac{q}{m}\,k\tilde{u}\,G^{\mu\nu}(k,\tilde{u})A_{\nu}(k), \tag{1.2.17}$$

with $F_0^{\mu\nu} = Bf^{\mu\nu}$, $\epsilon = -q/|q|$, $\Omega_0 = |q|B/m$. In the rest frame of the plasma, $k\tilde{u}$ is the frequency ω, and $\tau^{\mu\nu}(\omega)$ is defined as the inverse of the tensor $\omega g^{\mu\nu} + i\epsilon\Omega_0 f^{\mu\nu}$ on the left hand side of (1.2.17). Specifically, the definition is

$$[\omega g^{\mu\nu} + i\epsilon\Omega_0 f^{\mu\nu}]\,\tau^{\nu}{}_{\rho}(\omega) = \omega g^{\mu\rho}. \tag{1.2.18}$$

Solving (1.2.18) gives

$$\tau^{\mu\nu}(\omega) = g_{\parallel}^{\mu\nu} + \frac{\omega}{\omega^2 - \Omega_0^2}\left(\omega g_{\perp}^{\mu\nu} - i\epsilon\Omega_0 f^{\mu\nu}\right). \tag{1.2.19}$$

The matrix representation of $\tau^{\mu\nu}(\omega)$ is

$$\tau^{\mu\nu}(\omega) = \begin{pmatrix} 1 & 0 & 0 & 0 \\ 0 & -\dfrac{\omega^2}{\omega^2 - \Omega_0^2} & \dfrac{i\epsilon\Omega_0\omega}{\omega^2 - \Omega_0^2} & 0 \\ 0 & -\dfrac{i\epsilon\Omega_0\omega}{\omega^2 - \Omega_0^2} & -\dfrac{\omega^2}{\omega^2 - \Omega_0^2} & 0 \\ 0 & 0 & 0 & -1 \end{pmatrix}. \tag{1.2.20}$$

In the unmagnetized limit, $\Omega_0 \to 0$, $\tau^{\mu\nu}(\omega)$ reduces to $g^{\mu\nu}$.

Note the unconventional choice of the sign $\epsilon = -q/|q|$, which is positive, $\epsilon = +1$, for electrons and negative, $\epsilon = -1$, for positively charged ions. This choice is made for convenience with comparison with the relativistic quantum case, where $\epsilon = \pm 1$ is a quantum number labeling electron and positron states.

Alternative Forms for the $\Pi^{\mu\nu}(k)$

The explicit form (1.2.16) for the cold plasma response tensor is written in a concise notation, and a more explicit form is obtained by using the explicit form (1.2.6) for $G^{\mu\nu}(k,u)$ and (1.2.19) for $\tau^{\mu\nu}(\omega)$. This form is

$$\begin{aligned}\Pi^{\mu\nu}(k) &= -\frac{q^2 n}{m\bar{\gamma}}\Bigg\{g_{\parallel}^{\mu\nu} - \frac{k_{\parallel}^{\mu}\tilde{u}^{\nu} + \tilde{u}^{\mu}k_{\parallel}^{\nu}}{k\tilde{u}} + \frac{(k^2)_{\parallel}\tilde{u}^{\mu}\tilde{u}^{\nu}}{(k\tilde{u})^2} \\ &\quad + \frac{(k\tilde{u})^2}{(k\tilde{u})^2 - \Omega_0^2}\left[g_{\perp}^{\mu\nu} - \frac{k_{\perp}^{\mu}\tilde{u}^{\nu} + \tilde{u}^{\mu}k_{\perp}^{\nu}}{k\tilde{u}} - \frac{k_{\perp}^2\tilde{u}^{\mu}\tilde{u}^{\nu}}{(k\tilde{u})^2}\right] \\ &\quad - \frac{i\epsilon\Omega_0}{(k\tilde{u})^2 - \Omega_0^2}\left[k\tilde{u}f^{\mu\nu} + k_G^{\mu}\tilde{u}^{\nu} - \tilde{u}^{\mu}k_G^{\nu}\right]\Bigg\}, \end{aligned} \tag{1.2.21}$$

where $k_{\perp}^{\mu}$, k_G^{μ} are defined by (1.1.18). Note that because $\tilde{u}$ is restricted to the 0–3 plane, one has $(k\tilde{u})_{\perp} = 0$ and $k\tilde{u} = (k\tilde{u})_{\parallel}$.

An alternative way of writing (1.2.21) is in terms of the 4-vector $k_D^{\mu} = (k_z, 0, 0, \omega)$, rather than $\tilde{u}^{\mu}$. One has

$$(k^2)_{\parallel}\tilde{u}^{\mu} = \omega k_{\parallel}^{\mu} - k_z k_D^{\mu}. \tag{1.2.22}$$

The result (1.2.21) can be rewritten so that $\tilde{u}^{\mu}$ does not appear explicitly using $k\tilde{u} = \omega$,

$$g_{\parallel}^{\mu\nu} = \frac{k_{\parallel}^{\mu}k_{\parallel}^{\nu} - k_D^{\mu}k_D^{\nu}}{(k^2)_{\parallel}}, \tag{1.2.23}$$

so that the tensors in (1.2.21) become

$$\begin{aligned} g_{\parallel}^{\mu\nu} - \frac{k_{\parallel}^{\mu}\tilde{u}^{\nu} + \tilde{u}^{\mu}k_{\parallel}^{\nu}}{k\tilde{u}} + \frac{(k^2)_{\parallel}\tilde{u}^{\mu}\tilde{u}^{\nu}}{(k\tilde{u})^2} &= -\frac{k_D^{\mu}k_D^{\nu}}{\omega^2}, \\ k_{\perp}^{\mu}\tilde{u}^{\nu} + \tilde{u}^{\mu}k_{\perp} &= \omega\frac{k_{\perp}^{\mu}k_{\parallel}^{\nu} + k_{\parallel}^{\mu}k_{\perp}^{\nu}}{(k^2)_{\parallel}} - k_z\frac{k_{\perp}^{\mu}k_D^{\nu} + k_D^{\mu}k_{\perp}}{(k^2)_{\parallel}}, \\ \frac{\tilde{u}^{\mu}\tilde{u}^{\nu}}{(k\tilde{u})^2} &= \frac{\omega^2}{(k^2)_{\parallel}^2}(k_{\parallel}^{\mu}k_{\parallel}^{\nu} + k_D^{\mu}k_D^{\nu}) - \frac{k_D^{\mu}k_D^{\nu}}{(k^2)_{\parallel}} - \frac{\omega k_z}{(k^2)_{\parallel}^2}(k_{\parallel}^{\mu}k_D^{\nu} - k_D^{\mu}k_{\parallel}^{\nu}), \\ k_G^{\mu}\tilde{u}^{\nu} - \tilde{u}^{\mu}k_G^{\nu} &= \omega\frac{k_G^{\mu}k_{\parallel}^{\nu} - k_{\parallel}^{\mu}k_G^{\nu}}{(k^2)_{\parallel}} - k_z\frac{k_G^{\mu}k_D^{\nu} - k_D^{\mu}k_G^{\nu}}{(k^2)_{\parallel}}. \end{aligned} \tag{1.2.24}$$

The forms on the right hand sides of (1.2.24) appear naturally when the relativistic quantum form of the response tensor, derived in § 9.2, is use to rederive the cold-plasma form in § 9.4.

1.2.4 Cold Plasma Dielectric Tensor

The dielectric tensor, $K^i{}_j(k)$, is related to the mixed 3-tensor components $\Pi^i{}_j(k)$, which are numerically equal to $-\Pi^{ij}(k)$. The contravariant space components of (1.2.21), with $\tilde{u}^\mu \to \gamma(1,0,0,\beta)$, are

$$\begin{aligned}\Pi^{ij}(k) &= -\frac{q^2 n}{m\gamma}\left\{\left[-\frac{\omega^2}{\gamma^2(\omega-k_z\beta)^2}-\frac{k_\perp^2\beta^2}{(\omega-k_z\beta)^2-\Omega^2}\right]b^i b^j\right.\\ &\quad+\frac{(\omega-k_z\beta)^2 g_\perp^{ij}-(\omega-k_z\beta)(k_\perp^i b^j+k_\perp^j b^i)\beta}{(\omega-k_z\beta)^2-\Omega^2}\\ &\quad\left.-\frac{i\epsilon\Omega\left[(\omega-k_z\beta)f^{ij}+(k_G^i b^j-b^j k_G^i)\beta\right]}{(\omega-k_z\beta)^2-\Omega^2}\right\}, \qquad (1.2.25)\end{aligned}$$

where $\boldsymbol{b}$ is a unit vector along the 3-axis, $\boldsymbol{k}_\perp$ and $\boldsymbol{k}_G$ are vectors of magnitude $k_\perp$ along the 1- and 2-axes, respectively, with $g_\parallel^{ij} = -b^i b^j$, $g_\perp^{ij}$ diagonal $-1,-1,0$, f^{ij} nonzero only for $f^{12} = -f^{21} = -1$, and with $\Omega = \Omega_0/\gamma$. In the absence of a streaming motion, (1.2.25) simplifies to

$$\Pi^{ij}(k) = -\frac{q^2 n}{m}\tau^{ij}(\omega) = -\frac{q^2 n}{m}\left[\frac{\omega}{\omega^2-\Omega_0^2}\left(\omega g_\perp^{ij} - i\epsilon\Omega f^{ij}\right) - b^i b^j\right]. \qquad (1.2.26)$$

The dielectric tensor is

$$K^i{}_j(k) = \delta^i_j + \frac{1}{\varepsilon_0\omega^2}\Pi^i{}_j(k) = \delta^i_j - \frac{\omega_p^2}{\omega^2}\tau^i{}_j(\omega), \qquad (1.2.27)$$

with $\omega_p^2 = q^2 n/\varepsilon_0 m$, and with

$$\tau^i{}_j(\omega) = \begin{pmatrix} \dfrac{\omega^2}{\omega^2-\Omega_0^2} & -\dfrac{i\epsilon\Omega_0\omega}{\omega^2-\Omega_0^2} & 0 \\ \dfrac{i\epsilon\Omega_0\omega}{\omega^2-\Omega_0^2} & \dfrac{\omega^2}{\omega^2-\Omega_0^2} & 0 \\ 0 & 0 & 1 \end{pmatrix}. \qquad (1.2.28)$$

The cold plasma dielectric 3-tensor follows from (1.2.27) with (1.2.28) by adding a label, α say, for each species and summing over the species. This gives the standard form [26,27]

$$K^i{}_j = \begin{pmatrix} S & -iD & 0 \\ iD & S & 0 \\ 0 & 0 & P \end{pmatrix}, \tag{1.2.29}$$

with each of S, D, P involving a sum over the species

$$S = \frac{1}{2}(R_+ + R_-), \qquad D = \frac{1}{2}(R_+ - R_-), \qquad P = 1 - \sum_\alpha \frac{\omega_{p\alpha}^2}{\omega^2}, \tag{1.2.30}$$

with

$$R_\pm = 1 - \sum_\alpha \frac{\omega_{p\alpha}^2}{\omega^2} \frac{\omega}{\omega \mp \epsilon_\alpha \Omega_\alpha}, \tag{1.2.31}$$

where $-\epsilon_\alpha$ is the sign of the charge of species α. The dependence of $K^i{}_j$ and of $S, D, P, R_\pm$ on ω is implicit. The $\pm$ labeling of $R_\pm$ is chosen such that the resonance, specifically an infinite value at $\omega = \Omega_\alpha$, occurs in R_+ for charges that spiral in a right hand sense (electrons, $\epsilon_e = 1$) and in R_- for charges that spiral in a left hand sense (ions, $\epsilon_i = -1$).

No damping is included in the version of cold plasma theory discussed here; this is reflected in the dielectric tensor (1.2.29) being hermitian. It implies that the waves are undamped. Collisional damping can be included by including a frictional term in the fluid equations, but this is not done here.

Further quantities constructed from $K^i{}_j$ appear in the derivation of the dispersion equation and the construction of the polarization vectors. These include the trace and the longitudinal part of the tensor,

$$K_1 = K^s{}_s = 2S + P, \qquad K^L = -\kappa_i \kappa^j K^i{}_j = S \sin^2\theta + P\cos^2\theta, \tag{1.2.32}$$

the determinant of the tensor,

$$\det[K^i{}_j] = P(S^2 - D^2), \tag{1.2.33}$$

the square of the tensor,

$$K^i{}_s K^s{}_j = \begin{pmatrix} S^2 + D^2 & -iSD & 0 \\ iSD & S^2 + D^2 & 0 \\ 0 & 0 & P^2 \end{pmatrix}, \tag{1.2.34}$$

and the trace and the longitudinal part of the square

$$K_2 = K^r{}_s K^s{}_r = 2(S^2 + D^2) + P^2, \qquad K_2^L = (S^2 + D^2)\sin^2\theta + P^2\cos^2\theta. \tag{1.2.35}$$

Dielectric Tensor for a Cold Electron Gas

At frequencies much higher than the ion plasma frequency and the ion gyrofrequency, the contribution of the ions can be neglected. The dielectric tensor (1.2.29) then describes the response of a cold electron gas, with S, D, P given by

$$S = \frac{\omega^2 - \omega_p^2 - \Omega_e^2}{\omega^2 - \Omega_e^2}, \qquad D = -\frac{\epsilon\omega_p^2\Omega_e}{\omega(\omega^2 - \Omega_e^2)}, \qquad P = 1 - \frac{\omega_p^2}{\omega^2}, \tag{1.2.36}$$

with $\epsilon = 1$ for electrons. In the presence of an admixture of positrons, the parameter ϵ can be re-interpreted as the ratio of the difference to the sum of the number densities of the cold electrons and positrons.

The response tensor can be written in terms of two dimensionless magnetoionic parameters

$$X = \frac{\omega_p^2}{\omega^2}, \qquad Y = \frac{\Omega_e}{\omega}. \tag{1.2.37}$$

In terms of these parameters, S, D, P in (1.2.36) become

$$S = 1 - \frac{X}{1 - Y^2}, \qquad D = -\frac{\epsilon XY}{1 - Y^2}, \qquad P = 1 - X. \tag{1.2.38}$$

1.3 Inclusion of Streaming Motions

A covariant formulation of the cold plasma response facilitates the inclusion of streaming motions. The response 4-tensor may be calculated in the frame in which there is no streaming motion, and transformed to the frame where the plasma has the specified streaming motion. In the presence of a magnetic field this applies only to streaming motions along the magnetic field lines. This procedure is used here in three ways. First, it is applied to the plasma as a whole. Second, different streaming motions are introduced for different components in the plasma, allowing the existence of instabilities due to counterstreaming motions. Third, a multi-fluid model is used to write down the response tensor for an arbitrary one-dimensional (1D), strictly-parallel ($p_\perp = 0$) distribution of particles by summing over infinitesimal distributions with different streaming velocities.

1.3.1 Lorentz Transformation to Streaming Frame

Given the response tensor, $\Pi^{\mu\nu}(k)$, in one inertial frame, one can write it down in any other frame by making the appropriate Lorentz transformation. Let K and K' be two inertial frames, and let $L^{\mu'}{}_{\mu}$ and its (matrix) inverse $L^{\mu}{}_{\mu'}$ be the Lorentz transform matrices between K and K'. These matrices are defined such that any 4-vector with contravariant components a^μ in K and $a^{\mu'}$ in K' satisfies

$$a^{\mu'} = L^{\mu'}{}_{\mu}\, a^{\mu}, \qquad a^{\mu} = L^{\mu}{}_{\mu'}\, a^{\mu'}, \tag{1.3.1}$$

with

$$L^{\mu}{}_{\mu'}\, L^{\mu'}{}_{\nu} = \delta^{\mu}_{\nu}, \qquad L^{\mu'}{}_{\mu}\, L^{\mu}{}_{\nu'} = \delta^{\mu'}_{\nu'}. \tag{1.3.2}$$

Given $\Pi^{\mu\nu}(k)$ in K, the response tensor in K' is $\Pi^{\mu'\nu'}(L^{-1}k')$, with

$$(L^{-1}k')^{\mu} = L^{\mu}{}_{\mu'}\, k^{\mu'}. \tag{1.3.3}$$

For the purpose of including a streaming motion in a magnetized plasma one is concerned with a boost in which the axes in K and K' are parallel, and K' is moving along the 3-axis of K at velocity $-\beta$. In K' the streaming velocity of a plasma at rest in K is then β. The explicit forms for the transformation matrices in this case are, denoting the dependence of β explicitly,

$$L^{\mu'}{}_{\mu}(\beta) = \begin{pmatrix} \gamma & 0 & 0 & -\gamma\beta \\ 0 & 1 & 0 & 0 \\ 0 & 0 & 1 & 0 \\ -\gamma\beta & 0 & 0 & \gamma \end{pmatrix}, \qquad L^{\mu}{}_{\mu'}(\beta) = \begin{pmatrix} \gamma & 0 & 0 & \gamma\beta \\ 0 & 1 & 0 & 0 \\ 0 & 0 & 1 & 0 \\ \gamma\beta & 0 & 0 & \gamma \end{pmatrix}, \tag{1.3.4}$$

with $\gamma = (1-\beta^2)^{-1/2}$.

It is sometimes useful to write the dependence of $\Pi^{\mu\nu}(k)$ on k, in terms of invariants that involve k. In any other inertial frame the components of the response are given by applying a Lorentz transformation to $\Pi^{\mu\nu}(k)$, and by expressing the invariants that involve k in terms of k'.

Streaming Cold Distribution

The contribution to the response 4-tensor from a single cold distribution of particles may be written in the concise form

$$\Pi^{\mu\nu}(k) = -\frac{q^2 n_{\rm pr}}{m}\, G^{\alpha\mu}(k,u)\tau_{\alpha\beta}(ku)G^{\beta\nu}(k,u). \tag{1.3.5}$$

For a streaming distribution, the 4-velocity, u^μ, is interpreted as the streaming 4-velocity of a streaming distribution of cold particles. The proper number density (for a cold plasma) is related to the actual number density, n, in the chosen frame in which the plasma is streaming with Lorentz factor γ by $n_{\rm pr} = n/\gamma$.

An alternative way of writing (1.3.5) is

$$\Pi^{\mu\nu}(k) = -\frac{q^2 n_{\rm pr}}{m}\left\{\tau^{\mu\nu}(ku) - \frac{1}{ku}[u^\mu k_\alpha \tau^{\alpha\nu}(ku) + k_\beta \tau^{\mu\beta}(ku) u^\nu] + k_\alpha k_\beta \tau^{\alpha\beta}(ku)\frac{u^\mu u^\nu}{(ku)^2}\right\}. \tag{1.3.6}$$

A further alternative way of writing (1.3.5) is

$$\Pi^{\mu\nu}(k) = -\frac{q^2 n}{m\gamma}\left\{-\frac{k_D^\mu k_D^\nu}{(ku)^2} + \frac{1}{(ku)^2 - \Omega_0^2}\Big[(ku)^2 g_\perp^{\mu\nu} - ku\,(k_\perp^\mu u^\nu + u^\mu k_\perp^\nu) - k_\perp^2 u^\mu u^\nu - i\epsilon\,\Omega_0\big(ku\, f^{\mu\nu} + k_G^\mu u^\nu - u^\mu k_G^\nu\big)\Big]\right\}, \tag{1.3.7}$$

where the 4-vectors introduced in (1.3.7) are defined by (1.1.18). With the form (1.3.7), the gyrotropic terms are those that depend on the sign, $-\epsilon$, of the charge.

The space components of (1.3.7), for streaming at velocity β, are

$$\Pi^{ij}(k) = -\frac{q^2 n}{m\gamma}\left\{\left[-\frac{\omega^2}{\gamma^2(\omega - k_z\beta)^2} - \frac{k_\perp^2\beta^2}{(\omega - k_z\beta)^2 - \Omega^2}\right] b^i b^j + \frac{(\omega - k_z\beta)^2 g_\perp^{ij} - (\omega - k_z\beta)(k_\perp^i b^j + k_\perp^j b^i)\beta}{(\omega - k_z\beta)^2 - \Omega^2} - \frac{i\epsilon\Omega\big[(\omega - k_z\beta) f^{ij} + (k_G^i b^j - b^j k_G^i)\beta\big]}{(\omega - k_z\beta)^2 - \Omega^2}\right\}, \tag{1.3.8}$$

where $\boldsymbol{b}$ is a unit vector along the 3-axis, $\boldsymbol{k}_\perp$ and $\boldsymbol{k}_G$ are vectors of magnitude $k_\perp$ along the 1- and 2-axes, respectively, with $g_\parallel^{ij} = -b^i b^j$, $g_\perp^{ij}$ diagonal $(-1, -1, 0)$, f^{ij} nonzero only for $f^{12} = -f^{21} = -1$, and with $\Omega = \Omega_0/\gamma$.

Multiple Streaming Cold Components

The generalization of a single cold streaming distribution to a plasma consisting of several cold species in relative motion to each other follows by summing over the relevant contributions to the response 4-tensor. Let α label an arbitrary species, which has charge $q_\alpha = -\epsilon_\alpha |q_\alpha|$, mass m_α, number density n_α, cyclotron frequency Ω_α and 4-velocity $u_\alpha^\mu = [\gamma_\alpha, \gamma_\alpha v_\alpha \boldsymbol{b}]$. With this generalization (1.3.5) implies

$$\Pi^{\mu\nu}(k) = \sum_\alpha -\frac{q_\alpha^2 n_\alpha}{m_\alpha \gamma_\alpha} G^{\eta\mu}(k, u_\alpha)\tau_{\eta\theta}^{(\alpha)}(ku_\alpha) G^{\theta\nu}(k, u_\alpha), \tag{1.3.9}$$

with (1.2.19) translating into

$$\tau^{(\alpha)\mu\nu}(\omega) = g_{\parallel}^{\mu\nu} + \frac{\omega}{\omega^2 - \Omega_\alpha^2}(\omega g_{\perp}^{\mu\nu} - i\epsilon_\alpha \Omega_\alpha f^{\mu\nu}). \tag{1.3.10}$$

The alternative form (1.3.6) gives

$$\begin{aligned} \Pi^{\mu\nu}(k) &= \sum_\alpha -\frac{q_\alpha^2 n_\alpha}{m_\alpha \gamma_\alpha}\Big\{\tau^{(\alpha)\mu\nu}(ku_\alpha) - \frac{1}{ku_\alpha}[u_\alpha^\mu k_\eta \tau^{(\alpha)\eta\nu}(ku_\alpha) \\ &\quad + k_\theta \tau^{(\alpha)\mu\theta}(ku_\alpha)u_\alpha^\nu] + k_\eta k_\theta \tau^{(\alpha)\eta\theta}(ku_\alpha)\frac{u_\alpha^\mu u_\alpha^\nu}{(ku_\alpha)^2}\Big\}. \end{aligned} \tag{1.3.11}$$

The sum over species α may be interpreted as including a sum over different components of a single species with different streaming motions. For example, in the case of an electron gas that consists of two cold counterstreaming electron beams, one can interpret the sum over α as the sum over these two components.

1.3.2 Dielectric Tensor for a Streaming Distribution

In 3-tensor notation the response tensor is related to the dielectric tensor by $K^i{}_j(k) = \delta^i_j + \Pi^i{}_j(k)/\varepsilon_0\omega^2$. In particular, the sum over components in (1.3.7), rewritten in the form (1.3.6), translates into the dielectric tensor

$$\begin{aligned} K^i{}_j(k) = \delta^i_j - \sum_\alpha \frac{\omega_{p\alpha}^2}{\gamma_\alpha \omega^2}\Big\{\tau^{(\alpha)i}{}_j(ku_\alpha) - \frac{v_\alpha}{\omega - k_z v_\alpha}\Big[b^i k_r \tau^{(\alpha)r}{}_j(ku_\alpha) \\ + k^s \tau^{(\alpha)i}{}_s(ku_\alpha)b_j\Big] + \frac{v_\alpha^2[\omega^2 + k_r k^s \tau^{(\alpha)r}{}_s(ku_\alpha)]\, b^i b_j}{(\omega - k_z v_\alpha)^2}\Big\}, \end{aligned} \tag{1.3.12}$$

where $\omega_{p\alpha}^2 = q_\alpha^2 n_\alpha/\varepsilon_0 m_\alpha$ defines the plasma frequency for species α. In matrix form one has

$$\tau^{(\alpha)}(ku_\alpha) = \begin{pmatrix} \dfrac{(\omega - k_z v_\alpha)^2}{(\omega - k_z v_\alpha)^2 - \Omega_\alpha^2/\gamma_\alpha^2} & -\dfrac{i\epsilon_\alpha(\omega - k_z v_\alpha)\Omega_\alpha/\gamma_\alpha}{(\omega - k_z v_\alpha)^2 - \Omega_\alpha^2/\gamma_\alpha^2} & 0 \\ \dfrac{i\epsilon_\alpha(\omega - k_z v_\alpha)\Omega_\alpha/\gamma_\alpha}{(\omega - k_z v_\alpha)^2 - \Omega_\alpha^2/\gamma_\alpha^2} & \dfrac{(\omega - k_z v_\alpha)^2}{(\omega - k_z v_\alpha)^2 - \Omega_\alpha^2/\gamma_\alpha^2} & 0 \\ 0 & 0 & 1 \end{pmatrix}. \tag{1.3.13}$$

The resulting matrix form for the dielectric tensor (1.3.12) can be written

$$K^i{}_j = \begin{pmatrix} S & -iD & Q \\ iD & S & -iR \\ Q & iR & P \end{pmatrix}, \tag{1.3.14}$$

with the components identified as

$$
\begin{aligned}
S &= 1 - \sum_\alpha \frac{\omega_{p\alpha}^2}{\gamma_\alpha \omega^2} \frac{[\gamma_\alpha(\omega - k_z v_\alpha)]^2}{[\gamma_\alpha(\omega - k_z v_\alpha)]^2 - \Omega_\alpha^2}, \\
P &= 1 - \sum_\alpha \frac{\omega_{p\alpha}^2}{\gamma_\alpha \omega^2} \left[\frac{\omega^2}{[\gamma_\alpha(\omega - k_z v_\alpha)]^2} - \frac{\gamma_\alpha^2 k_\perp^2 v_\alpha^2}{[\gamma_\alpha(\omega - k_z v_\alpha)]^2 - \Omega_\alpha^2} \right], \\
D &= \sum_\alpha \epsilon_\alpha \frac{\omega_{p\alpha}^2}{\gamma_\alpha \omega^2} \frac{\gamma_\alpha(\omega - k_z v_\alpha)\Omega_\alpha}{[\gamma_\alpha(\omega - k_z v_\alpha)]^2 - \Omega_\alpha^2}, \\
Q &= - \sum_\alpha \frac{\omega_{p\alpha}^2}{\gamma_\alpha \omega^2} \frac{\gamma_\alpha^2 k_\perp(\omega - k_z v_\alpha)}{[\gamma_\alpha(\omega - k_z v_\alpha)]^2 - \Omega_\alpha^2}, \\
R &= - \sum_\alpha \epsilon_\alpha \frac{\omega_{p\alpha}^2}{\gamma_\alpha \omega^2} \frac{\gamma_\alpha k_\perp v_\alpha \Omega_\alpha}{[\gamma_\alpha(\omega - k_z v_\alpha)]^2 - \Omega_\alpha^2}.
\end{aligned} \tag{1.3.15}
$$

1.3.3 Cold Counterstreaming Electrons and Positrons

In an oscillating model for a pulsar magnetosphere [15], a parallel electric field, $E_z \neq 0$, accelerates the electrons and positrons, causing their streaming velocities to be different. Unlike the case where there is a single streaming velocity, the effect of relative streaming motions cannot be removed by a Lorentz transformation. However, one can transform to a frame in which the electrons and positrons are streaming in opposite directions with the same speed.

Consider the frame in which the streaming velocities, $\pm\beta$, of electrons ($\epsilon = 1$) and positrons ($\epsilon = -1$) are equal and opposite. Let their number densities be (note $+$ for electrons, $-$ for positrons) written $n^\pm = n(1 \pm \bar{\epsilon})$, $n = n^+ + n^-$, $\bar{\epsilon} = (n^+ - n^-)/(n^+ + n^-)$. One can identify two sources of gyrotropy is counterstreaming pair plasma. One is due to a charge imbalance, $\bar{\epsilon} \neq 0$. The other is due to a nonzero current: the current density is $\boldsymbol{J} = -e(n^+\beta^+ - n^-\beta^-)$, with $\beta^\pm = \pm\beta\boldsymbol{b}$ here. The sign of the current-induced gyrotropy is determined by the sign of the velocity, β.

One can write the response tensor in the form [30]

$$
\Pi^{\mu\nu}(k) = \bar{\Pi}^{\mu\nu}(k) - \bar{\epsilon}\hat{\Pi}^{\mu\nu}(k), \tag{1.3.16}
$$

with $K^i{}_j(k) = \delta^i_j + \mu_0 \Pi^i{}_j(k)/\omega^2$. It is convenient to introduce the notation

$$
\omega_p^2 = \frac{e^2 n}{\varepsilon_0 m}, \qquad \omega_0 = \gamma\omega, \qquad \omega_\parallel = \gamma k_z \beta, \qquad \omega_\perp = \gamma k_\perp \beta. \tag{1.3.17}
$$

The components of the response tensor that do not depend on $\bar{\epsilon}$ are

$$\mu_0\bar{\Pi}^1{}_1 = \mu_0\bar{\Pi}^2{}_2 = \omega_p^2 \frac{(\omega_0^2 - \omega_\parallel^2)^2 - \Omega^2(\omega_0^2 + \omega_\parallel^2)}{(\omega_0^2 + \omega_\parallel^2 - \Omega^2)^2 - 4\omega_0^2\omega_\parallel^2},$$

$$\mu_0\bar{\Pi}^3{}_3 = \omega_p^2 \left[\frac{\omega^2(\omega_0^2 + \omega_\parallel^2)}{(\omega_0^2 - \omega_\parallel^2)^2} + \frac{\omega_\perp^2(\omega_0^2 + \omega_\parallel^2 - \Omega^2)}{(\omega_0^2 + \omega_\parallel^2 - \Omega^2)^2 - 4\omega_0^2\omega_\parallel^2}\right],$$

$$\mu_0\bar{\Pi}^1{}_3 = \omega_p^2 \frac{\omega_\parallel\omega_\perp(\omega_0^2 - \omega_\parallel^2 + \Omega^2)}{(\omega_0^2 + \omega_\parallel^2 - \Omega^2)^2 - 4\omega_0^2\omega_\parallel^2},$$

$$\mu_0\bar{\Pi}^1{}_2 = -i\omega_p^2 \frac{\Omega_e\omega_\parallel(\omega_0^2 - \omega_\parallel^2 + \Omega_e^2)}{(\omega_0^2 + \omega_\parallel^2 - \Omega_e^2)^2 - 4\omega_0^2\omega_\parallel^2},$$

$$\mu_0\bar{\Pi}^2{}_3 = i\omega_p^2 \frac{\Omega_e\omega_\perp(\omega_0^2 + \omega_\parallel^2 - \Omega_e^2)}{(\omega_0^2 + \omega_\parallel^2 - \Omega_e^2)^2 - 4\omega_0^2\omega_\parallel^2}. \tag{1.3.18}$$

The components proportional to $-\bar{\epsilon}$ are

$$\mu_0\hat{\Pi}^1{}_1 = \mu_0\hat{\Pi}^2{}_2 = \omega_p^2 \frac{2\omega_0\omega_\parallel\Omega_e^2}{(\omega_0^2 + \omega_\parallel^2 - \Omega_e^2)^2 - 4\omega_0^2\omega_\parallel^2},$$

$$\mu_0\hat{\Pi}^3{}_3 = -\omega_p^2 \left[-\frac{2\omega^2\omega_0\omega_\parallel}{(\omega_0^2 - \omega_\parallel^2)^2} - \frac{2\omega_\perp^2\omega_0\omega_\parallel^2}{(\omega_0^2 + \omega_\parallel^2 - \Omega_e^2)^2 - 4\omega_0^2\omega_\parallel^2}\right],$$

$$\mu_0\hat{\Pi}^1{}_3 = -\omega_p^2 \frac{\omega_0\omega_\perp(\omega_0^2 - \omega_\parallel^2 - \Omega_e^2)}{(\omega_0^2 + \omega_\parallel^2 - \Omega_e^2)^2 - 4\omega_0^2\omega_\parallel^2},$$

$$\mu_0\hat{\Pi}^1{}_2 = -i\omega_p^2 \frac{\Omega_e\omega_0(\omega_0^2 - \omega_\parallel^2 - \Omega_e^2)}{(\omega_0^2 + \omega_\parallel^2 - \Omega_e^2)^2 - 4\omega_0^2\omega_\parallel^2},$$

$$\mu_0\hat{\Pi}^2{}_3 = i\omega_p^2 \frac{2\Omega_e\omega_0\omega_\perp\omega_\parallel}{(\omega_0^2 + \omega_\parallel^2 - \Omega_e^2)^2 - 4\omega_0^2\omega_\parallel^2}. \tag{1.3.19}$$

The response of a counterstreaming pair plasma is gyrotropic even when the number densities are equal ($\bar{\epsilon} = 0$). This is due to the current from the oppositely directed flows of oppositely charged particles; the sign of the current is determined by the sign of β, which is included in $\omega_\parallel$, $\omega_\perp$ in (1.3.18).

1.4 Relativistic Magnetohydrodynamics

The fluid description of a magnetized plasma is referred to as magnetohydrodynamics (MHD). A relativistic generalization of MHD is presented here. The theory is then used to derive the properties of MHD waves in relativistic plasmas. SI units with c included explicitly are used in this section

1.4.1 Covariant Form of the MHD Equations

The relativistic MHD equations can either be postulated [3, 16, 29] or derived from kinetic (Vlasov) theory by taking moments. The zeroth and first moments lead to the continuity equation and the equation of motion for the fluid, respectively. Closing the expansion requires the introduction of another relation, which is identified as a generalized Ohm's law. Apart from Maxwell's equations, the MHD also include an equation of state for the fluid.

Basic Fluid Equations

Let the proper mass density be $\eta(x)$ and the fluid 4-velocity be $u^\mu(x)$. The equation of mass continuity is

$$\partial_\mu[\eta(x)u^\mu(x)] = 0. \tag{1.4.1}$$

The equation of fluid motion depends on the assumed forces on the fluid. Assuming that the only forces are those internal to the system, including electromagnetic forces, the equation of motion can be written in the form of a conservation equation for the energy-momentum tensor:

$$\partial_\mu \left[T_{\text{M}}^{\mu\nu}(x) + T_{\text{EM}}^{\mu\nu}(x)\right] = 0, \tag{1.4.2}$$

where $T_{\text{M}}^{\mu\nu}(x)$ is the energy-momentum tensor for the matter and $T_{\text{EM}}^{\mu\nu}(x)$ is the energy-momentum tensor for the electromagnetic field. The terms in (1.4.2) can be rearranged into the rate of change of the 4-momentum density in the fluid, which arises from the kinetic energy contribution to $T_{\text{M}}^{\mu\nu}(x)$, and thermal and electromagnetic forces that arise from the thermal contribution to $T_{\text{M}}^{\mu\nu}(x)$ and from $T_{\text{EM}}^{\mu\nu}(x)$, respectively.

Energy-Momentum Tensor for the Fluid

The energy-momentum tensor for the matter is identified as

$$T_{\text{M}}^{\mu\nu} = (\eta c^2 + \mathcal{E} + P)u^\mu u^\nu - P g^{\mu\nu}, \tag{1.4.3}$$

where ηc^2 is the proper rest energy density, $\mathcal{E} = U/V$, where V is the volume of the system, is the internal energy density and P is the pressure, and where the dependence on x is implicit. The combined first and second laws of thermodynamics imply the familiar relation $dU = TdS - PdV$, where U is the internal energy, T is the temperature and S is the entropy. Regarding $U(S, V)$ as the state function, with independent variables S, V, implies $T = (\partial U/\partial S)_V$, $P = -(\partial U/\partial V)_S$. In the present case one has $U = V\mathcal{E}$ and $V \propto 1/\eta$. Making the physical assumption that all changes are adiabatic, that is, at constant entropy, the relation $P = -(\partial U/\partial V)_S$

translates into

$$\eta \frac{\partial \mathcal{E}}{\partial \eta} = \mathcal{E} + P, \tag{1.4.4}$$

where constant entropy is implicit. The right hand side is the enthalpy, $U + PV$. Assuming an adiabatic law with an adiabatic index Γ one has

$$\eta \frac{\partial \mathcal{E}}{\partial \eta} = \Gamma \mathcal{E}, \tag{1.4.5}$$

with $\Gamma = 5/3$ for a monatomic nonrelativistic ideal gas, $\Gamma = 4/3$ for a highly relativistic gas.

Electromagnetic Energy-Momentum Tensor

The electromagnetic energy-momentum tensor, which satisfies $\partial_\mu T^{\mu\nu}_{\rm EM} = J_\mu F^{\mu\nu}$, has the canonical form

$$T^{\mu\nu}_{\rm EM} = \frac{1}{\mu_0} \left(F^{\mu}{}_{\alpha} F^{\alpha\nu} + \tfrac{1}{4} g^{\mu\nu} F_{\alpha\beta} F^{\alpha\beta} \right). \tag{1.4.6}$$

The Maxwell tensor, $F^{\mu\nu}$, may be written in terms of the 4-vectors $E^\mu = F^{\mu\nu} u_\nu$, $B^\mu = \mathcal{F}^{\mu\nu} u_\nu$, defined for an arbitrary 4-velocity u, identified here as the fluid 4-velocity. One has

$$F^{\mu\nu} = \frac{E^\mu u^\nu - E^\nu u^\mu}{c} + \epsilon^{\mu\nu\alpha\beta} u_\alpha B_\beta, \qquad \mathcal{F}^{\mu\nu} = B^\mu u^\nu - B^\nu u^\mu - \frac{\epsilon^{\mu\nu\alpha\beta} u_\alpha E_\beta}{c}, \tag{1.4.7}$$

with Maxwell's equations taking the form (1.1.4). The energy-momentum tensor (1.4.6) becomes

$$T^{\mu\nu}_{\rm EM} = -\frac{B^\mu B^\nu}{\mu_0} - \varepsilon_0 E^\mu E^\nu + \left(\tfrac{1}{2} g^{\mu\nu} - u^\mu u^\nu\right) \left(\frac{B^\sigma B_\sigma}{\mu_0} + \varepsilon_0 E^\sigma E_\sigma \right). \tag{1.4.8}$$

Combining (1.4.3) and (1.4.8), the total energy-momentum tensor for the system of fluid and electromagnetic field is

$$\begin{aligned} T^{\mu\nu} = {} & \left(\eta c^2 + \mathcal{E} + P - \frac{B^\sigma B_\sigma}{\mu_0} - \varepsilon_0 E^\sigma E_\sigma \right) u^\mu u^\nu \\ & - \left(P - \frac{B^\sigma B_\sigma}{2\mu_0} - \frac{1}{2} \varepsilon_0 E^\sigma E_\sigma \right) g^{\mu\nu} - \frac{B^\mu B^\nu}{\mu_0} - \varepsilon_0 E^\mu E^\nu. \end{aligned} \tag{1.4.9}$$

Lagrangian Density for Relativistic MHD

Relativistic MHD is amenable to a Lagrangian formulation [1, 11]. The action principle is

$$\delta \int d^4x\, \Lambda(x) = 0, \qquad \Lambda(x) = -\eta c^2 - \mathcal{E} + B^\sigma B_\sigma / 2\mu_0, \tag{1.4.10}$$

where η is the proper mass density and $\mathcal{E}$ is the thermal energy density of the fluid. One may regard the form (1.4.10) as a postulate that defines relativistic MHD. The final term in (1.4.10) arises from the Lagrangian for the electromagnetic field, $\mathcal{L}_{\rm EM} = -F^{\alpha\beta}F_{\alpha\beta}/4\mu_0$, with $F^{\mu\nu} = B^\mu u^\nu - B^\nu u^\mu$ for $E^\mu = 0$.

The derivation of the equation of motion in the form (1.4.2) follows from the fact that the Lagrangian (1.4.10) may be regarded as a functional of η, u^μ and B^μ: the dependence on x is implicit in this functional dependence. The energy-momentum tensor calculated from the Lagrangian (1.4.10) reproduces (1.4.9) for $E^\mu = 0$:

$$\begin{aligned} T^{\mu\nu} &= \left(\eta c^2 + \mathcal{E} + \frac{|\mathbf{B}|^2}{2\mu_0}\right) u^\mu u^\nu - \left(P + \frac{|\mathbf{B}|^2}{2\mu_0}\right) h^{\mu\nu} - \frac{|\mathbf{B}|^2}{\mu_0} b^\mu b^\nu, \\ h^{\mu\nu} &= g^{\mu\nu} - u^\mu u^\nu, \end{aligned} \tag{1.4.11}$$

with $|\mathbf{B}|^2 = -B^\sigma B_\sigma$ and where $h^{\mu\nu}$ projects onto the 3-dimensional hypersurface orthogonal to the fluid 4-velocity. The equation of motion in the form (1.4.2) corresponds to the conservation law $\partial_\mu T^{\mu\nu} = 0$.

1.4.2 Derivation from Kinetic Theory

Fluid equations may be postulated or derived from kinetic theory. The latter approach is outlined here.

Consider a species α, with rest mass m_α, charge q_α, and with distribution function $F_\alpha(x, p)$. Fluid equations are obtained by considering moments of the distribution function. For simplicity in writing, the x dependences of all quantities are suppressed in the following equations.

The zeroth order moment defines the proper number density, $n_{\alpha\rm pr}$, for species α and the corresponding proper mass density is

$$(\eta)_\alpha = m_\alpha n_{\alpha\rm pr}, \qquad n_{\alpha\rm pr} = \int \frac{d^4p}{(2\pi)^4} F_\alpha(p). \tag{1.4.12}$$

The first moment defines the fluid 4-velocity, u^μ_α:

$$n_{\alpha\rm pr} u^\mu_\alpha = \int \frac{d^4p}{(2\pi)^4} F_\alpha(p) \frac{p^\mu}{m_\alpha}. \tag{1.4.13}$$

Each species satisfies a continuity equation of the form (1.2.1), specifically

$$\partial_\mu (n_{\alpha\mathrm{pr}} u^\mu_\alpha) = 0. \tag{1.4.14}$$

The fluid is described by its proper mass density, η, and a fluid 4-velocity, u^μ, given by

$$\eta = \sum_\alpha m_\alpha n_{\alpha\mathrm{pr}}, \qquad u^\mu = \sum_\alpha m_\alpha n_{\alpha\mathrm{pr}} u^\mu_\alpha / \eta, \tag{1.4.15}$$

respectively. The continuity Eq. (1.4.1) for the fluid is then satisfied as a consequence of (1.4.14) with (1.4.15).

The second moment of the distribution defines the energy-momentum tensor for species α. This includes a contribution, $(\eta)_\alpha u^\mu_\alpha u^\nu_\alpha$, that corresponds to the rest mass energy in the rest frame of the fluid. It is convenient to separate the energy-momentum tensor into a part corresponding to its rest energy and a part due to internal motions in the fluid:

$$T^{\mu\nu}_\alpha = T^{\mu\nu}_{\alpha\mathrm{rm}} + T^{\mu\nu}_{\alpha\mathrm{th}}, \qquad T^{\mu\nu}_{\alpha\mathrm{rm}} = (\eta)_\alpha u^\mu_\alpha u^\nu_\alpha, \tag{1.4.16}$$

where 'rm' denotes rest mass and 'th' denotes thermal motions. One has

$$T^{\mu\nu}_{\alpha\mathrm{th}} = \int \frac{d^4p}{(2\pi)^4} F_\alpha(p)\, m_\alpha (u^\mu - u^\mu_\alpha)(u^\nu - u^\nu_\alpha) = (\mathcal{E}_\alpha + P_\alpha) u^\mu_\alpha u^\nu_\alpha - P_\alpha g^{\mu\nu}, \tag{1.4.17}$$

where $\mathcal{E}_\alpha$ is the internal energy density and P_α is the partial pressure for species α. Assuming that the only force is electromagnetic, the equation of fluid motion for species α is

$$\partial_\mu T^{\mu\nu}_\alpha = F^\nu{}_\rho J^\rho_\alpha, \tag{1.4.18}$$

with $J^\mu_\alpha = q_\alpha n_{\alpha\mathrm{pr}} u^\mu_\alpha$. On summing (1.4.18) over all species α, the net 4-force, $F^\nu{}_\rho J^\rho$, on the right hand side may be written as a 4-gradient and included in the left hand side. From Maxwell's equations one has $J^\rho = \partial_\mu F^{\mu\rho}/\mu_0$, and hence this 4-force becomes $F^\nu{}_\rho \partial_\mu F^{\mu\rho}/\mu_0$, which may be rewritten as $-\partial_\mu T^{\mu\nu}_{\mathrm{EM}}$, in terms of the energy-momentum tensor (1.4.6). The equation of motion then reduces to (1.4.2).

1.4.3 Generalized Ohm's Law

A characteristic difference between MHD and kinetic theory is the appeal in MHD to some form of Ohm's law to place a restriction on the electromagnetic field. Two examples of Ohm's law are discussed briefly here: that for a nonrelativistic, collisional, electron-ion plasma, and that for a relativistic, collisionless pair plasma. A two-fluid model is assumed in both cases, with the fluids being electrons and ions, and electrons and positrons, respectively.

In a nonrelativistic electron-ion plasma, the ratio, m_e/m_i, of the mass of the electron to the mass of the ion is a small parameter in which one can expand. The fluid velocity is equal to the velocity of the ions to lowest order in m_e/m_i, and the current density is determined by the flow of the electrons relative to the ions. In the presence of collisions, there is a drag on the electrons that may be represented by a frictional force equal to $-\nu_e$ times the momentum of the electrons, where ν_e is the electron collision frequency. In an isotropic plasma, the effect of the collisions is described by a conductivity $\sigma_0 = \omega_p^2/\varepsilon_0\nu_e$. In a quasi-neutral plasma, when the charge density is assumed negligible, the static response may be written in the 4-tensor form

$$J^\mu = \sigma^\mu{}_\nu E^\nu, \qquad \sigma^{\mu\nu} = \sigma_0(g^{\mu\nu} - u^\mu u^\nu). \tag{1.4.19}$$

The limit of infinite conductivity corresponds to $\sigma_0 \to \infty$, $\nu_e \to 0$, that is, to the collisionless limit. In this limit, for the current to remain finite one requires $E^\mu = 0$.

In the presence of a magnetic field the conductivity is anisotropic. The conductivity tensor may be obtained from the response tensor for a cold, magnetized electron gas by replacing the frequency, ω, in the rest frame by $\omega + i\nu_e$ to take account of the collisions. This corresponds to identifying $\sigma^{\mu\nu}$ as $i\Pi^{\mu\nu}/ku$ with $\Pi^{\mu\nu}$ given by the cold plasma form (1.3.5). In (1.3.5) one makes the replacement $k\tilde{u} \to ku + i\nu_e$, and projects onto the 3-dimensional hyperplane orthogonal to u^μ. This gives

$$\sigma^{\mu\nu} = \frac{i\omega_p^2}{\varepsilon_0(ku + i\nu_e)}\left[\tau^{\mu\nu}(ku + i\nu_e) - u^\mu u^\nu\right], \tag{1.4.20}$$

with $\tau^{\mu\nu}(\omega)$ given by (1.2.19). In the static limit, $ku \to 0$, one has

$$\sigma^{\mu\nu} = \frac{\omega_p^2}{\varepsilon_0\nu_e}\left[-b^\mu b^\nu + \frac{g_\perp^{\mu\nu} + (\Omega_e/\nu_e)f^{\mu\nu}}{1 + \Omega_e^2/\nu_e^2}\right]. \tag{1.4.21}$$

In the limit $\nu_e \to 0$ only the component along $b^\mu b^\nu$ becomes infinite, and this requires the condition $Eb = -E_z = 0$; the Hall term ($\propto f^{\mu\nu}$) remains finite, and the Pedersen terms ($\propto g_\perp^{\mu\nu}$) tends to zero in this limit.

1.4.4 Two-Fluid Model for a Pair Plasma

A two-fluid model for a relativistic electron-positron plasma with no thermal motions enables one to calculate E^μ and to discuss the assumption that it is zero in relativistic MHD. In this case there is no obvious small parameter, such as the mass ratio, that allows one to justify a simple approximation to Ohm's law.

One may rearrange the fluid Eqs. (1.4.14) and (1.4.18) for $\alpha = \pm$, $m_\pm = m_e$, $\mathcal{E}_\pm = 0 = P_\pm$, for electrons and positrons, respectively, into equations for the variables

$$n = n_+ + n_-, \qquad \rho = -e(n_+ - n_-),$$

$$u^\mu = \frac{n_+ u_+^\mu + n_- u_-^\mu}{n}, \qquad J^\mu = -e(n_+ u_+^\mu - n_- u_-^\mu). \tag{1.4.22}$$

The first two of the variables (1.4.22) are the proper number density and the proper charge density, respectively. The equations of continuity (1.4.14) for $\alpha = \pm$ imply continuity equations $\partial_\mu(nu^\mu) = 0$ for mass and $\partial_\mu J^\mu = 0$ for charge. The equations of motion (1.4.18) for $\alpha = \pm$ imply an equation of motion for the fluid of the form (1.4.2) and an equation for the current. The equation of motion for the fluid is

$$\partial_\mu T_{\rm M}^{\mu\nu} = F^\nu{}_\beta J^\beta, \qquad T_{\rm M}^{\mu\nu} = nmu^\mu u^\nu + \frac{m}{ne^2}\frac{(J^\mu - \rho u^\mu)(J^\nu - \rho u^\nu)}{1 - \rho^2/n^2e^2}. \tag{1.4.23}$$

The term $nmu^\mu u^\nu$ is the conventional energy-momentum tensor for a cold fluid, and conventional MHD is justified only if the additional term in (1.4.23) can be neglected.

The generalized Ohm's law is identified by calculating $E^\nu = F^\nu{}_\beta u^\beta$ with u given by (1.4.22). Using the equations of motion (1.4.18) together with (1.4.22) implies

$$E^\nu = \frac{1}{ne}\,\partial_\mu\left[\frac{u^\mu J^\nu + u^\nu J^\mu - \rho(u^\mu u^\nu + J^\mu J^\nu/n^2e^2)}{1 - \rho^2/n^2e^2}\right], \tag{1.4.24}$$

which gives the electric field in the rest frame of the fluid. The assumption $E^\mu = 0$ in conventional MHD applies to a pair plasma only if one can justify neglecting the right hand side of (1.4.24). If the assumption cannot be justified, relativistic MHD is not valid and should not be used. Relativistic MHD can break down for a variety of reasons [19], including that there are too few charges to carry the required current density.

It might be remarked that the ideal MHD assumption that the electric field is zero in the rest frame of the plasma is usually justified by arguing that in the limit of infinite conductivity a nonzero $\mathbf{E}$ would imply an infinite $\mathbf{J}$, which is unphysical. However, in the collisionless limit, $\nu_e \to 0$, only the parallel component of the conductivity becomes infinite in a magnetized plasma. The argument for the perpendicular component of $\mathbf{E}$ is different. A nonzero $\mathbf{E}_\perp$ causes all particles to drift across the field lines with a velocity $-\mathbf{E}\times\mathbf{B}/|\mathbf{B}|^2$. The $\mathbf{E}_\perp$ is removed by Lorentz transforming to the frame in which the plasma is stationary. The plasma is at rest in the frame $\mathbf{E}_\perp = 0$.

1.4.5 MHD Wave Modes

The properties of (small amplitude) waves in relativistic MHD are determined by a wave equation that may be derived from the Lagrangian. Suppose that there is an oscillating part of the fluid displacement, such that one has

$$x^\mu \to x_0^\mu + \delta x^\mu, \qquad \delta x^\mu = \xi^\mu e^{-i\Phi} + \text{c.c.}, \tag{1.4.25}$$

where x_0^μ is the fluid displacement in the absence of the fluctuations. (The subscript 0 is omitted below after making the expansion.) The phase, or eikonal, Φ, satisfies $k_\mu = \partial_\mu \Phi$. The 4-velocity has a small perturbation, given by the derivative of (1.4.25) with respect to proper time. This gives $u^\mu \to u_0^\mu + \delta u^\mu$, with $\delta u^\mu = -i\omega\xi^\mu$ to first order in the perturbation. The normalization $u^2 = 1$ must be preserved, and to lowest order, this requires $\xi u = 0$. In the following it is assumed that ξ^μ, like B^μ, has no component along u^μ.

On averaging the action (1.4.10) over the phase, only terms of even power in ξ remain. The first two terms implied by the phase average are [1]

$$\mathcal{L}^{(0)}(x) = -\eta c^2 - \mathcal{E} + B^\sigma B_\sigma / 2\mu_0, \tag{1.4.26}$$

$$\mathcal{L}^{(2)}(x) = -\left(\eta c^2 + P + \mathcal{E}\right)(ku)^2\, \xi\xi^* - \Gamma P\, k\xi\, k\xi^* - \frac{1}{\mu_0}\{[B\xi\, B\xi^* - B^\sigma B_\sigma\, \xi\xi^*](ku)^2 - \Omega^\sigma \Omega_\sigma\},$$

$$\Omega^\mu = k\xi\, B^\mu - kB\, \xi^\mu, \tag{1.4.27}$$

respectively, with $\tilde{A}^\mu = h^{\mu\nu} A_\nu$ for any 4-vector A^μ.

The wave equation for MHD waves follows from $\partial\mathcal{L}^{(2)}/\partial\xi_\mu^* = 0$ with $\mathcal{L}^{(2)}$ given by (1.4.27). The resulting equation is of the form

$$\tilde{\Gamma}^{\mu\nu}\xi_\nu = 0, \qquad \tilde{\Gamma}^{\mu\nu} = h^{\mu\alpha} h^{\nu\beta} \Gamma_{\alpha\beta}(k, u), \tag{1.4.28}$$

$$\Gamma^{\mu\nu} = \left[-\left(\eta c^2 + P + \mathcal{E} - \frac{B^\sigma B_\sigma}{\mu_0} \right)(ku)^2 + \frac{(kB)^2}{\mu_0} \right] g^{\mu\nu} - \frac{(ku)^2}{\mu_0} B^\mu B^\nu - \left(\Gamma P - \frac{B^\sigma B_\sigma}{\mu_0} \right) k^\mu k^\nu - \frac{kB}{\mu_0}(k^\mu B^\nu + k^\nu B^\mu), \tag{1.4.29}$$

with $B^\sigma B_\sigma = -|\mathbf{B}|^2$.

There are only three independent (orthogonal to u) components of ξ^μ. A set of basis vectors that span the 3-dimensional space, orthogonal to u^μ, consists of the direction of the magnetic field, b^μ, the component of the wave 4-vector orthogonal to both the magnetic field and the fluid velocity, $\kappa_\perp^\mu \propto k^\mu - ku\, u^\mu + kb\, b^\mu$, and the direction orthogonal to these, $a^\mu = \epsilon^\mu{}_{\nu\rho\sigma} u^\nu b^\rho \kappa_\perp^\sigma = f^\mu{}_\nu \kappa_\perp^\nu$. The perpendicular and parallel components of k are introduced by writing $k_\perp = -k\kappa_\perp$, $k_z = -kb$, so that one has $k^\mu = ku\, u^\mu + k_\perp \kappa_\perp^\mu + k_z b^\mu$, and $g^{\mu\nu} = u^\mu u^\nu - \kappa_\perp^\mu \kappa_\perp^\nu - a^\mu a^\nu - b^\mu b^\nu$. Equation (1.4.28) may be written as three simultaneous equations for the components $\xi_\perp = -\xi\kappa_\perp$, $\xi_a = -\xi a$, $\xi_b = -\xi b$. The matrix form of these equations is

$$\begin{pmatrix} A & 0 & B \\ 0 & D & 0 \\ B & 0 & C \end{pmatrix} \begin{pmatrix} \xi_\perp \\ \xi_a \\ \xi_b \end{pmatrix} = 0, \tag{1.4.30}$$

with the matrix components given by

$$D = (\eta c^2 + \mathcal{E} + P + |\boldsymbol{B}|^2/\mu_0)(ku)^2 - k_z^2|\boldsymbol{B}|^2/\mu_0,$$

$$A = (\eta c^2 + \mathcal{E} + P + |\boldsymbol{B}|^2/\mu_0)(ku)^2 - \Gamma P k_\perp^2 - |\boldsymbol{k}|^2|\boldsymbol{B}|^2/\mu_0,$$

$$B = -\Gamma P\, k_\perp k_z, \qquad C = (\eta c^2 + \mathcal{E} + P)(ku)^2 - \Gamma P\, k_z^2, \tag{1.4.31}$$

with $\Gamma = \partial(\ln P)/\partial(\ln \eta)$ for adiabatic changes. The resulting dispersion equation for relativistic MHD is

$$D(AC - B^2) = 0. \tag{1.4.32}$$

The dispersion Eq. (1.4.32) factorizes into $D = 0$, which gives the dispersion relation for the Alfvén mode, and $AC - B^2 = 0$, which has two solutions corresponding to the fast and slow magnetoacoustic modes.

Alfvén and Sound Speeds

It is convenient to define the Alfvén speed and the (adiabatic) sound speed by

$$v_A^2 = \frac{|\boldsymbol{B}|^2}{\mu_0(\eta + \mathcal{E}/c^2 + P/c^2)}, \qquad c_s^2 = \frac{\Gamma P}{\eta + \mathcal{E}/c^2 + P/c^2}. \tag{1.4.33}$$

The dispersion relation for the Alfvén (A) mode becomes

$$\omega_A = \frac{|k_z|v_A}{(1 + v_A^2/c^2)^{1/2}}, \qquad v_\phi^2 = \frac{v_A^2 \cos^2\theta}{1 + v_A^2/c^2}, \tag{1.4.34}$$

where $v_\phi = \omega/|\boldsymbol{k}|$ is the phase speed. The fluid displacement in Alfvén waves is along a^μ. The dispersion equation for the fast and slow modes is

$$\left(1 + \frac{v_A^2}{c^2}\right)\omega^4 - \left[\left(1 + \frac{v_A^2}{c^2}\right)c_s^2 k_z^2 - c_s^2 k_\perp^2 - v_A^2|\boldsymbol{k}|^2\right]\omega^2 + c_s^2 v_A^2 |\boldsymbol{k}|^2 k_z^2 = 0. \tag{1.4.35}$$

Solving for the phase speed, the dispersion relations for the two modes are of the form $v_\phi^2 = v_\pm^2$, with

$$v_\pm^2 = \frac{1}{2(1+v_A^2/c^2)}\left\{ v_A^2 + c_s^2 + \frac{v_A^2 c_s^2}{c^2}\cos^2\theta \right.$$
$$\left. \pm\left[\left(v_A^2 - c_s^2 - \frac{v_A^2 c_s^2}{c^2}\cos^2\theta\right)^2 + 4c_s^2 v_A^2 \sin^2\theta\right]^{1/2}\right\}. \tag{1.4.36}$$

The solution for the fluid displacement in the two modes is

$$\xi_\pm^\mu \propto \sin\psi_\pm \kappa_\perp^\mu + \cos\psi_\pm b^\mu, \tag{1.4.37}$$

$$\tan\psi_\pm = \frac{v_\pm^2 - c_s^2\cos^2\theta}{c_s^2\sin\theta\cos\theta} = \frac{c_s^2\sin\theta\cos\theta}{(1+v_A^2/c^2)v_\pm^2 - v_A^2 - c_s^2\sin^2\theta}. \tag{1.4.38}$$

For either a very low density or a very strong magnetic field, satisfying $|\boldsymbol{B}|^2/\mu_0 \gg \eta c^2 + \mathcal{E} + P$, the conventional Alfvén speed exceeds the speed of light, $v_A \gg c$, and the MHD speed becomes $v_A/(1+v_A^2/c^2)^{1/2}$. At sufficiently high (relativistic) temperature, the adiabatic index is $\Gamma = 4/3$, the pressure satisfies $P = \mathcal{E}/3 \gg \eta c^2$, implying that the sound speed approaches the limit $c_s \to c/\sqrt{3}$.

1.5 Quantum Fluid Theory

A fluid approach to quantum plasmas, referred to here as quantum fluid theory (QFT), has generated an extensive literature since about 2000 [2, 12, 18]. QFT may be derived from moments of a kinetic equation, with the Vlasov equation replaced by an appropriate quantum counterpart. Although a derivation of fluid equations from the Dirac equation had been developed in the 1950s by Takabayasi [28], the later development of QFT started from simpler assumptions. In its simplest form (nonrelativistic, spinless, unmagnetized and longitudinal) the approach is to take moments of the Wigner-Moyal equations that describe such a quantum system. The subsequent extension of QFT involved generalizing it to include other effects in a piecemeal fashion.

1.5.1 Early QFT Theories

It was pointed out by Bohm [6], that a fluid-like description was implicit in alternative interpretations of Schrödinger's theory, discussed by de Broglie and Madelung in 1926. Madelung [17] wrote the wavefunction in the form

$$\psi(\boldsymbol{x},t) = A(\boldsymbol{x},t)\exp[iS(\boldsymbol{x},t)/\hbar], \quad n(\boldsymbol{x},t) = [A(\boldsymbol{x},t)]^2, \quad \boldsymbol{p}(\boldsymbol{x},t) = \nabla S(\boldsymbol{x},t), \tag{1.5.1}$$

where A and S are real functions. The Madelung equations are a continuity equation for $n(\boldsymbol{x},t)$, which may be interpreted as a probability density for the electrons, and a Hamiltonian-Jacobi-like equation for $S(\boldsymbol{x},t)$, which may be reinterpreted as an equation of motion for the fluid momentum $\boldsymbol{p}(\boldsymbol{x},t)$. (Madelung's equations are sometimes regarded as equivalent to Schrödinger's equation, but this seems not to be the case [31].) In QFT, Madelung's equations become the equation of continuity

$$\left[\frac{\partial}{\partial t} + \boldsymbol{v}(\boldsymbol{x},t)\cdot\nabla\right] n(\boldsymbol{x},t) = 0, \tag{1.5.2}$$

with $\boldsymbol{v} = \boldsymbol{p}/m$, and the equation of motion

$$\left[\frac{\partial}{\partial t} + \boldsymbol{v}(\boldsymbol{x},t)\cdot\nabla\right] \boldsymbol{p}(\boldsymbol{x},t) = e\nabla\Phi(\boldsymbol{x},t) + \frac{\hbar^2}{2m_e}\nabla\left(\frac{\nabla^2 A(\boldsymbol{x},t)}{A(\boldsymbol{x},t)}\right), \tag{1.5.3}$$

with $A(\boldsymbol{x},t) = [n(\boldsymbol{x},t)]^{1/2}$. The final term in (1.5.3), referred to as the Bohm term in QFT, is an intrinsically quantum mechanical term that describes the effect of quantum mechanical diffusion and tunneling.

The QFT Eqs. (1.5.2) and (1.5.3) can be derived from moments of kinetic equations that include quantum effects. A quantum counterpart of the classical distribution function is the Wigner function, which satisfies a kinetic equation similar to the Boltzmann equation, re-interpreted as the Vlasov equation in plasma kinetic theory. The Wigner function is defined in terms of the outer product of the Schrödinger wavefunction and its complex conjugate, and it includes neither spin nor relativistic effects. In the generalization to Dirac's theory, the outer product of the wave function and its adjoint is a 4×4 Dirac matrix, referred to as the Wigner matrix in § 8.4.2 of volume 1. This generalization leads to substantial increase in algebraic complexity. Existing versions of QFT that include spin and/or relativistic effects are based on various simplifying approximations.

Wigner-Moyal Equations

The Wigner function is defined in § 8.4 of volume 1. Let $\psi(\boldsymbol{x},t)$ be the one-dimensional wavefunction satisfying the one-dimensional Schrödinger equation. The Wigner function is defined by (ordinary units)

$$f(\boldsymbol{p},\boldsymbol{x},t) = \int d^3\boldsymbol{y}\, e^{i\boldsymbol{p}\cdot\boldsymbol{y}/\hbar}\,\psi\left(\boldsymbol{x}-\tfrac{1}{2}\boldsymbol{y},t\right)\psi^*\left(\boldsymbol{x}+\tfrac{1}{2}\boldsymbol{y},t\right). \tag{1.5.4}$$

The notation used for the Wigner function, $f(\boldsymbol{p},\boldsymbol{x},t)$ in (1.5.4), is the same as for the classical distribution function, but the interpretation is different. One cannot interpret the Wigner function as a probability distribution, as is the case for its classical counterpart, in particular because it can be negative. The wavefunction and its adjoint in (1.5.4) satisfy the Schrödinger equation and its adjoint, respectively.

For the case of an electron in a longitudinal field, described by an electrostatic potential $\Phi(\boldsymbol{x},t)$, the wavefunction satisfies (ordinary units)

$$-i\hbar\frac{\partial}{\partial t}\psi(\boldsymbol{x},t) = -\frac{\hbar^2}{2m_e}\nabla^2\psi(\boldsymbol{x},t) - e\Phi(\boldsymbol{x},t)\psi(\boldsymbol{x},t). \tag{1.5.5}$$

The Wigner function satisfies (ordinary units)

$$\left(\frac{\partial}{\partial t} + \boldsymbol{v}\cdot\nabla\right) f(\boldsymbol{p},\boldsymbol{x},t) = -i\frac{e}{\hbar}\int\frac{d^3\boldsymbol{p}'d^3\boldsymbol{y}}{(2\pi\hbar)^3}\, f(\boldsymbol{p}',\boldsymbol{x},t)e^{(\boldsymbol{p}-\boldsymbol{p}')\cdot\boldsymbol{y}/\hbar} \times\left[\Phi(\boldsymbol{x}-\tfrac{1}{2}\boldsymbol{y},t) - \Phi(\boldsymbol{x}+\tfrac{1}{2}\boldsymbol{y},t)\right], \tag{1.5.6}$$

with Poisson's equation identified as

$$\nabla^2\Phi(\boldsymbol{x},t) = \frac{e}{\varepsilon_0}\left[\int\frac{d^3\boldsymbol{p}}{(2\pi\hbar)^3}\, f(\boldsymbol{p},\boldsymbol{x},t) - n_0\right], \tag{1.5.7}$$

where n_0 is a constant positive background charge. Equations (1.5.6) and (1.5.7) are sometimes referred to a the Wigner-Moyal equations. The QFT equations may be derived by taking moments of (1.5.6).

A simple example of the implications of Eqs. (1.5.6) and (1.5.7) follows by linearizing and Fourier transforming them, and then solving for the dispersion relation for longitudinal waves. For any isotropic distribution this gives (ordinary units)

$$\omega^2 = \omega_p^2 + |\boldsymbol{k}|^2\langle v^2\rangle + \hbar^2|\boldsymbol{k}|^4/2m_e^2, \tag{1.5.8}$$

with $\langle v^2\rangle = 3V_e^2$ for a Maxwellian distribution and $\langle v^2\rangle = 3v_F^2/5$ for a completely degenerate electron distribution. The final term in (1.5.8) arises from the Bohm term. This term has an obvious interpretation in terms of the quantum recoil. This implies a relation between quantum mechanical diffusion in coordinate space and the quantum recoil in momentum space [20].

1.5.2 Generalizations of QFT

The form of QFT outlined above applies only to nonrelativistic plasmas and longitudinal fields. As already mentioned, the generalization to include relativistic effects is discussed briefly in § 8.4 of volume 1, with the Wigner function generalized to a Wigner matrix whose evolution is determined by Dirac's equation in place of Schrödinger's equation. The derivation of quantum fluid equations from relativistic quantum theory [28] has become of renewed interest more recently [7, 9, 25].

1.5.3 Quasi-classical Models for Spin

The effect of the spin of the electron in a magnetic field has been included in QFT using several different quasi-classical approaches. In the simplest approach, the magnetic moment of the electron is identified as $\boldsymbol{m} = g\mu_B \boldsymbol{s}$, with (SI units)

$$\mu_B = \frac{e\hbar}{2m} = 9.274 \times 10^{-24}\,\mathrm{J\,T^{-1}}, \quad g = 2.00232, \tag{1.5.9}$$

where μ_B is the Bohr magneton, and the gyromagnetic ratio, g, differs from 2 due to radiative corrections in QED. For an electron at rest, a classical form for the equation of motion of the spin is

$$\frac{d\boldsymbol{s}}{dt} = \frac{ge}{2m_e}\boldsymbol{s} \times \boldsymbol{B}. \tag{1.5.10}$$

A quasi-classical way of including spin dependence in a kinetic theory [8,23] is based on an earlier theory in a different context [10,14]: generalize phase space from the 6-dimensional **x**–**p** space to a 9-dimensional **x**–**p**–**s** space (the restriction $|\mathbf{s}| = 1$ formally reduces the dimensionality to 8), and introduce a distribution function, $f(\mathbf{x}, \mathbf{p}, \mathbf{s})$, in this space. The Vlasov equation is then generalized to include the evolution of the spin, described by (1.5.10) in the simplest approximation.

Generalization of QFT to include the magnetic field leads to quantum MHD theory. The generalization of (1.5.3) involves including the Lorentz force and the force that results from the gradient of the energy associated with the magnetic moment. This gives

$$\left(\frac{\partial}{\partial t} + \boldsymbol{v} \cdot \nabla\right) \boldsymbol{p} = -e\,(\boldsymbol{E} + \mathbf{v} \times \boldsymbol{B}) + \nabla(\mu_B \boldsymbol{s} \cdot \boldsymbol{B}) + \frac{\hbar^2}{2m}\nabla\left(\frac{\nabla^2 n_e^{1/2}}{n_e^{1/2}}\right), \tag{1.5.11}$$

where arguments $(\boldsymbol{x}, t)$ are omitted. The inclusion of the spin in (1.5.11) is not rigorously justified. The use of fluid theory to describe a (classical or quantum) plasma imposes an intrinsic limitation that cannot be avoided in the magnetized case: the spiraling motion of particles cannot be taken into account. Magnetized fluid theory is reproduced by kinetic theory only when the gyroradii of the particles are assumed negligibly small. The gyroradius is a classical concept, and the quantum counterpart of the small gyroradius limit has not been identified in the context of QFT. (The relevant limit is referred to as the small-x approximation in § 9.4.2.)

Covariant Model for Spin: BMT Equation

A covariant generalization of the equation of motion (1.5.10) for the spin leads to the Bargmann-Michel-Telegdi (BMT) equation [5]. The spin vector, $\boldsymbol{s}$, is interpreted as the space components of a 4-vector in the frame in which the electron is at rest. Writing $s^\mu = [s^0, \boldsymbol{s}]$ in an arbitrary frame, one assumes $s^0 = 0$ in the rest frame, and then $su = 0$ in the rest frame implies $\gamma(s^0 - \boldsymbol{s}\cdot\boldsymbol{v}) = 0$, and hence $s^0 = \boldsymbol{s}\cdot\boldsymbol{v}$ in an arbitrary frame. For an accelerated particle, in its instantaneous rest frame, one has $ds^0/dt = \boldsymbol{s}\cdot d\boldsymbol{v}/dt$, and together with (1.5.10) this determines ds^μ/dt in the instantaneous rest frame. It is straightforward to rewrite the resulting equation in a covariant form. Assuming an equation of motion that omits the final two terms in (1.5.11), this becomes the BMT equation

$$\frac{ds^\mu}{d\tau} = -\frac{e}{m_e}\left[\tfrac{1}{2}gF^{\mu\nu}s_\nu + \left(\tfrac{1}{2}g - 1\right)s_\alpha F^{\alpha\beta}u_\beta u^\mu\right], \qquad \frac{du^\mu}{d\tau} = -\frac{e}{m_e}F^{\mu\nu}u_\nu, \tag{1.5.12}$$

with $d\tau = dt/\gamma$, where τ is the proper time. In this model, the spin does not affect the dynamics, in the sense that there is no term corresponding to the force associated with the gradient of the magnetic energy, $-\frac{1}{2}g\mu_B\boldsymbol{s}\cdot\boldsymbol{B}$.

An alternative covariant form of the magnetic moment is in terms of the second rank 4-tensor

$$m^{\mu\nu} = -\tfrac{1}{2}g\mu_B\epsilon^{\mu\nu\alpha\beta}s_\alpha u_\beta. \tag{1.5.13}$$

In the rest frame, $u^\beta = [1, \boldsymbol{0}]$, in the case where the spin is along the direction of $\boldsymbol{B}$, assumed to be the 3-axis, one has $m^{12} = -m^{21} = \frac{1}{2}g\mu_B s$. Equations (1.5.12) and (1.5.13) imply, $dm^{\mu\nu}/d\tau = 0$, and hence that the magnetic moment in this sense is conserved. This conservation law also applies when the radiative correction $g - 2 \neq 0$ is included.

1.5.4 Spin-Dependent Cold Plasma Response

The classical covariant form of the response of a cold electron gas is calculated in § 1.2, and use of (1.5.12) and (1.5.13) facilitates generalizing that calculation of the response tensor to include the contribution due to the magnetic moments of the electrons in a magnetized electron gas. For simplicity the radiative correction is neglected by setting $g = 2$.

The 4-magnetization of the electron gas is $M^{\mu\nu} = n_e m^{\mu\nu}$. The assumption that the electron gas is magnetized implies that there is a nonzero mean spin, denoted $\bar{s}^\mu$. Let the average magnetization be $M^{\mu\nu} = n_e\bar{m}^{\mu\nu}$, with $\bar{m}^{\mu\nu} = \mu_B\epsilon^{\mu\nu\alpha\beta}\bar{s}_\alpha\bar{u}_\beta$, where an overbar denotes an average value. In the rest frame of the cold electron gas, one has $\bar{u}^\mu = [1, \boldsymbol{0}]$, $\bar{s}^\mu = [0, \boldsymbol{s}]$, implying a 3-magnetization $\boldsymbol{M} = \mu_B n_e\bar{\boldsymbol{s}}$.

The linear response tensor associated with the perturbation of the magnetic moments of the electrons can be evaluated in terms of the linear perturbation, $M^{(1)\mu\nu}(k)$, in the magnetization. The associated 4-current is $J^{(1)\mu}(k) = -ik_\nu M^{(1)\mu\nu}(k)$, and writing this in the form

$$-ik_\rho M^{(1)\mu\rho}(k) = \Pi_{\rm m}^{\mu\nu}(k)A_\nu(k), \tag{1.5.14}$$

defines the relevant contribution $\Pi_{\rm m}^{\mu\nu}(k)$ to the response tensor. In the model used here, the spin does not affect the dynamics, and hence there is assumed to be no perturbation in n_e. The linear perturbation in the magnetization is

$$M^{(1)\mu\nu}(k) = -\mu_B n_e \epsilon^{\mu\nu\alpha\beta}\left[s_\alpha^{(1)}(k)\bar{u}_\beta + \bar{s}_\alpha u_\beta^{(1)}(k)\right]. \tag{1.5.15}$$

The perturbation in the 4-velocity is given by (1.2.15), and for electrons this becomes

$$u^{(1)\mu}(k) = \frac{e}{m_e k\bar{u}}\left[k\bar{u}\,\tau^{\mu\nu}(k\bar{u}) - k_\rho\tau^{\mu\rho}(k\bar{u})\bar{u}^\nu\right]A_\nu(k), \tag{1.5.16}$$

with, from (1.2.19) for electrons,

$$\tau^{\mu\nu}(\omega) = g_\parallel^{\mu\nu} + \frac{\omega}{\omega^2 - \Omega_e^2}\left(\omega g_\perp^{\mu\nu} - i\Omega_e f^{\mu\nu}\right). \tag{1.5.17}$$

The analogous perturbation in the spin 4-vector follows from (1.5.12), with $g = 2$ here:

$$s^{(1)\mu}(k) = \frac{e}{m_e k\bar{u}}\left[k\bar{s}\,\tau^{\mu\nu}(k\bar{u}) - k_\rho\tau^{\mu\rho}(k\bar{u})\bar{s}^\nu\right]A_\nu(k). \tag{1.5.18}$$

Explicit evaluation gives

$$\begin{aligned}\Pi_{\rm m}^{\mu\nu}(k) = -\frac{ien_e}{m_e\,k\bar{u}}k_\rho\epsilon^{\mu\rho}{}_{\alpha\beta}\Big\{&[k\bar{s}\,\tau^{\alpha\nu}(k\bar{u}) - k_\tau\tau^{\alpha\tau}(k\bar{u})\bar{s}^\nu]\bar{u}^\beta \\ &+[k\bar{u}\,\tau^{\beta\nu}(k\bar{u}) - k_\tau\tau^{\beta\tau}(k\bar{u})\bar{u}^\nu]\bar{s}^\alpha\Big\},\end{aligned} \tag{1.5.19}$$

with $\tau^{\mu\nu}$ given by (1.5.17).

The covariant form (1.5.19) applies to a collection of electrons at rest in the frame moving with 4-velocity $\bar{u}^\mu$. As in the case of the cold plasma response, one can reinterpret (1.5.19) in a way that allows one to include an arbitrary distribution of particles in parallel velocity β. One replaces $\bar{u}$ by u, with $u^\mu = \gamma[1,0,0,\beta]$, $\gamma = 1/(1-\beta^2)^{1/2}$, and replaces n_e by the differential proper number density, $d\beta\, g^\epsilon(\beta)/\gamma$, where $g^\epsilon(\beta)$ is the distribution function for electrons, $\epsilon = 1$, or positrons, $\epsilon = -1$. After integrating over β, this generalization of (1.5.19) gives the magnetic moment contribution to the response tensor for the distribution of electrons plus positrons.

This model does not include the spiraling motion of the electrons, and the response tensor (1.5.19) is valid only in the small-gyroradius limit.

Spin-Dependent Response in the Rest Frame

The spin-dependent contribution (1.5.19) to the response tensor simplifies considerably in the rest frame of the (cold) electron gas, when one has $\bar{u}^\mu = [1,\mathbf{0}]$,

$\bar{s}^\mu = [0, \bar{s}\boldsymbol{b}]$, where $\boldsymbol{b} = (0, 0, 1)$ is a unit vector along the magnetic field. One then has $k\bar{u} = \omega$, $k\bar{s} = -k_z\bar{s}$, and the spin-dependent contribution to the response tensor is proportional to the magnetization $M = \mu_B n_e \bar{s}$.

For cold electrons in their rest frame, (1.5.20) reduces to [21]

$$\Pi_{\rm m}^{\mu\nu}(k) = -\frac{eM}{m_e(\omega^2 - \Omega_e^2)} \begin{pmatrix} k_\perp^2 \Omega_e & \omega k_\perp \Omega_e & i\omega^2 k_\perp & 0 \\ \omega k_\perp \Omega_e & (\omega^2 - k_z^2)\Omega_e & i(\omega^2 - k_z^2)\omega & k_\perp k_z \Omega_e \\ -i\omega^2 k_\perp & -i(\omega^2 - k_z^2)\omega & (\omega^2 - k_z^2)\Omega_e & -i\omega k_\perp k_z \\ 0 & k_\perp k_z \Omega_e & i\omega k_\perp k_z & -k_\perp^2 \Omega_e \end{pmatrix}. \tag{1.5.20}$$

The spin-dependent contribution to the dielectric tensor is (ordinary units)

$$[K_{\rm m}]^i{}_j(k) = \frac{\Omega_{\rm m} c^2}{\omega^2(\omega^2 - \Omega_e^2)} \begin{pmatrix} (\omega^2/c^2 - k_z^2)\Omega_e & i(\omega^2/c^2 - k_z^2)\omega & k_\perp k_z \Omega_e \\ -i(\omega^2/c^2 - k_z^2)\omega & (\omega^2/c^2 - k_z^2)\Omega_e & -ik_\perp k_z \omega \\ k_\perp k_z \Omega_e & ik_\perp k_z \omega & -k_\perp^2 \Omega_e \end{pmatrix}, \tag{1.5.21}$$

where the frequency associated with the magnetization is

$$\Omega_{\rm m} = \frac{\mu_0 M}{B}\Omega_e = \frac{\hbar\bar{s}\omega_p^2}{m_e c^2}. \tag{1.5.22}$$

The ratio $\Omega_{\rm m}/\omega_p$ is small except in dense, strongly magnetized plasmas, where the plasmon energy, $\hbar\omega_p$, is a significant fraction of the rest energy, $m_e c^2$, and $\bar{s}$ is of order unity.

The derivation of the response tensor (1.5.20) involves two different assumptions: the BMT equation for the spin evolution and the cold-plasma approximation. An analogous result derived using a nonrelativistic theory [22] differs from (1.5.20) in that $\omega^2/c^2 - k_z^2$ is replaced by $-k_z^2$. A similar result has been derived for a special case using kinetic theory [23]. The validity of the quasi-classical approach for including the spin in the dispersion is questioned in § 9.6, where it is argued that a rigorous theory does not reproduce the result (1.5.20) in any obvious way.

References

1. A. Achterberg, Phys. Rev. A **28**, 2449 (1983)
2. D. Anderson, B. Hall, M. Lisak, M. Marklund, Phys. Rev. E **65**, 046417 (2002)
3. A.M. Anile, *Relativistic Fluids and Magneto-Fluids: With Applications in Astrophysics and Plasma Physics* (Cambridge University Press, Cambridge, 1989)
4. E.V. Appleton, J. Inst. Electr. Eng. **71**, 642 (1932)

5. V. Bargmann, L. Michel, V.L. Telegdi, Phys. Rev. Lett. **2**, 435 (1959)
6. D. Bohm, Phys. Rev. **85**, 166 (1952)
7. G. Brodin, M. Marklund, New J. Phys. **9**, 277 (2008)
8. G. Brodin, M. Marklund, G. Manfredi, Phys. Rev. Lett. **100**, 175001 (2008)
9. G. Brodin, M. Marklund, J. Zamanian, A. Ericsson, P.L. Mana, Phys. Rev. Lett. **101**, 245002 (2008)
10. S.C. Cowley, R.M. Kulsrud, E. Valeo, Phys. Fluid **29**, 430 (1986)
11. R.L. Dewar, Aust. J. Phys. **30**, 533 (1977)
12. F. Haas, G. Manfredi, M.R. Feix, Phys. Rev. E **62**, 2763 (2000)
13. D.R. Hartree, Proc. Camb. Philos. Soc. **27**, 143 (1931)
14. R.M. Kulsrud, E.J. Valeo, S.C. Cowley, Nucl. Fusion **26**, 1443 (1986)
15. A. Levinson, D. Melrose, A. Judge, Q. Luo, Astrophys. J. **31**, 456 (2005)
16. A. Lichnerowicz, *Relativistic Magnetohydrodynamics* (Benjamin, New York, 1967)
17. E. Madelung, Z. Phys. **40**, 32 (1926)
18. M. Marklund, P.K. Shukla, Rev. Mod. Phys. **78**, 591 (2006)
19. A. Melatos, D.B. Melrose, Mon. Not. R. Astron. Soc. **279**, 1168 (1996)
20. D.B. Melrose, A. Mushtaq, Phys. Plasmas **16**, 094508 (2009)
21. D.B. Melrose, A. Mushtaq, Phys. Rev. E **83**, 056404 (2011)
22. A.P. Misra, G. Brodin, M. Marklund, P.K. Shukla, J. Plasma Phys. **76**, 875 (2010)
23. V.N. Oraevsky, V.B. Semikoz, Phys. At. Nucl. **66**, 466 (2003)
24. A.E. Shabad, Ann. Phys. **90**, 166 (1975)
25. P.K. Shukla, B. Eliasson, Physica (Utrecht) **53**, 51 (2010)
26. T.H. Stix, *The Theory of Plasma Waves* (McGraw-Hill, New York, 1962)
27. T.H. Stix, *Waves in Plasmas* (Springer, New York, 1992)
28. T. Takabayasi, Prog. Theor. Phys. Suppl. **4**, 1 (1957)
29. T. Uchida, Phys. Rev. E **56**, 2181 (1997)
30. M.W. Verdon, D.B. Melrose, Phys. Rev. E **77**, 046403 (2008)
31. T.C. Wallstrom, Phys. Rev. A **49**, 1613 (1994)

Chapter 2
Response Tensors for Magnetized Plasmas

The generalization of the covariant classical kinetic theory of plasma responses from an unmagnetized to a magnetized plasma involves a considerable increase in algebraic complexity. As in the unmagnetized case, two different methods of calculation are available and both are useful: the forward-scattering and Vlasov methods. In the forward-scattering method, one includes a perturbing electromagnetic field, $A^{\mu}(k)$, in the equation of motion for a single particle, and expands the single-particle current in powers of $A^{\mu}(k)$. On averaging this current over a distribution of particles, the result is interpreted as the weak turbulence expansion, allowing one to identify the linear and nonlinear response tensors. In the Vlasov method, the perturbation due to $A^{\mu}(k)$ is included in the Vlasov equation, and the distribution function is expanded in $A^{\mu}(k)$, with the nth order term used to calculate the nth order current. The linear ($n = 1$) nonlinear ($n > 1$) terms determine the linear and nonlinear response tensors. A third method is used in the unmagnetized case, based on combining a fluid (cold plasma) approach with a Lorentz transformation; this method does not generalize to the arbitrary magnetized case because it cannot be used to include the effect of the gyration of particles.

Convenient general forms for the response tensors are obtained by expanding in Bessel functions. The response may then be interpreted in terms of dispersive contributions due to gyroresonant interactions. After expanding in Bessel functions, the forward-scattering and Vlasov methods give alternative expressions, related by a partial integration and sum rules for the Bessel functions. An alternative approach for a relativistic thermal (Jüttner) distribution is to evaluate the linear response tensor using a procedure due to Trubnikov that involves no expansion in Bessel functions.

D. Melrose, *Quantum Plasmadynamics: Magnetized Plasmas*, Lecture Notes in Physics 854, DOI 10.1007/978-1-4614-4045-1_2,

2.1 Orbit of a Spiraling Charge

In a uniform magnetostatic field, the unperturbed orbit of a charged particle is a spiraling motion about the magnetic field lines. By solving the covariant form of Newton's equation of motion, one determines the 4-velocity of a particle at proper time τ in terms of its 4-velocity at an arbitrary initial proper time $\tau = 0$. The relation between these two 4-vectors defines a tensor that depends on τ. The Fourier transform of this tensor, denoted $\tau^{\mu\nu}(\omega)$, characterizes the linear response of a spiraling charge. This tensor is used in the calculation of the linear response tensor for a plasma by both the forward-scattering method and the Vlasov method. The 4-current due to the charge is expanded in powers of a perturbing electromagnetic field in § 2.2.

2.1.1 Orbit of a Spiraling Charge

In the absence of the perturbing electromagnetic field, the classical equation of motion for a particle with charge q and mass m in a static electromagnetic field $F_0^{\mu\nu}$ is

$$\frac{du^\mu(\tau)}{d\tau} = \frac{q}{m} F_0^{\mu\nu} u_\nu(\tau), \tag{2.1.1}$$

where τ is the proper time of the particle. For a magnetostatic field, $F_0^{\mu\nu} = Bf^{\mu\nu}$, (2.1.1) becomes

$$\frac{du^\mu(\tau)}{d\tau} = -\epsilon\Omega_0 f^{\mu\nu} u_\nu(\tau), \qquad \epsilon = -\frac{q}{|q|}, \qquad \Omega_0 = \frac{|q|B}{m}, \tag{2.1.2}$$

where ϵ is positive for electrons and negative for positively charged particles. (The choice to write the sign of the charge as $-\epsilon$ is made to be consistent with the notation used in a QED treatment.) The frequency Ω_0 is the nonrelativistic gyrofrequency, also called the cyclotron frequency.

In solving (2.1.2) one is free to choose a frame in which the magnetic field is along the 3-axis and such that the initial 4-velocity, $u^\mu(0) = u_0^\mu$, corresponds to

$$u_0^\mu = (\gamma, \gamma v_\perp \cos\phi_0, \epsilon\gamma v_\perp \sin\phi_0, \gamma v_z), \tag{2.1.3}$$

where ϕ_0 is an initial gyrophase. For a single particle one is free to choose the initial conditions such that $\phi_0 = 0$, and the 3-velocity in the 1–3 plane. The spiraling motion is in a right-hand screw sense, relative to the magnetic field, for electrons ($\epsilon = 1$) and other negative charges, and left hand for positive charges ($\epsilon = -1$). All of γ, $v_\perp$, v_z and $p_\perp = \gamma m v_\perp$, $p_z = \gamma m v_z$ are constants of the motion.

Let the solution for the orbit be written in the form

$$X^\mu(\tau) = x_0^\mu + t^{\mu\nu}(\tau) u_{0\nu}, \tag{2.1.4}$$

where x_0 is a constant 4-vector. It is convenient to consider the motion in space-time projected onto two 2D-subspaces, denoted as the $\|$- and $\perp$-subspaces. These correspond to the 0–3 plane and the 1–2 planes, respectively. The $\|$-motion is constant rectilinear motion, which corresponds to $X^\mu_\|(\tau) = x^\mu_{0\|} + u^\mu_{0\|}\tau$, with $u^\mu_{0\|} = (\gamma, 0, 0, \gamma v_z)$ a constant 4-vector. Hence, on writing $t^{\mu\nu}(\tau) = t^{\mu\nu}_\|(\tau) + t^{\mu\nu}_\perp(\tau)$, one identifies $t^{\mu\nu}_\|(\tau) = g^{\mu\nu}_\|\tau$.

The equation of motion (2.1.1) projected onto the $\perp$-space, that is, the 1–2 plane, leads to a differential equation for $t^{\mu\nu}_\perp(\tau)$:

$$\ddot{t}^{\mu\nu}_\perp(\tau) = -\epsilon\Omega_0 f^\mu{}_\rho \dot{t}^{\rho\nu}_\perp(\tau), \tag{2.1.5}$$

where a dot denotes differentiation with respect to τ. The solution that satisfies the initial conditions is $\dot{t}^{\mu\nu}_\perp(\tau) = g^{\mu\nu}_\perp \cos\Omega_0\tau - \epsilon f^{\mu\nu}\sin\Omega_0\tau$. The derivative of (2.1.4) with respect to τ then has the explicit form

$$u^\mu(\tau) = \dot{t}^{\mu\nu}(\tau)u_{0\nu}, \qquad \dot{t}^{\mu\nu}(\tau) = g^{\mu\nu}_\| + g^{\mu\nu}_\perp\cos\Omega_0\tau - \epsilon f^{\mu\nu}\sin\Omega_0\tau. \tag{2.1.6}$$

Integration gives

$$t^{\mu\nu}(\tau) = g^{\mu\nu}_\|\tau + g^{\mu\nu}_\perp\frac{\sin\Omega_0\tau}{\Omega_0} + \epsilon f^{\mu\nu}\frac{\cos\Omega_0\tau}{\Omega_0}. \tag{2.1.7}$$

The solution for the 4-velocity at proper time τ is

$$u^\mu(\tau) = (\gamma, \gamma v_\perp\cos(\phi_0 + \Omega_0\tau), \epsilon\gamma v_\perp\sin(\phi_0 + \Omega_0\tau), \gamma v_z). \tag{2.1.8}$$

The solution for the orbit is

$$X^\mu(\tau) = x^\mu_0 + (\gamma\tau, R\sin(\phi_0 + \Omega_0\tau), -\epsilon R\cos(\phi_0 + \Omega_0\tau), \gamma v_z\tau), \tag{2.1.9}$$

where $R = \gamma v_\perp/\Omega_0$ is the gyroradius. The handedness of the spiraling motion is specified by the sign ϵ, and is right-hand for $\epsilon = 1$ (a negative charge).

Properties of $t^{\mu\nu}(\tau), \dot{t}^{\mu\nu}(\tau)$

The tensors $t^{\mu\nu}(\tau)$ and $\dot{t}^{\mu\nu}(\tau)$ characterize the spiraling motion of a charge in a magnetic field. The tensor $\dot{t}^{\mu\nu}(\tau)$ satisfies the differential equation (2.1.5). Integrating (2.1.5) gives

$$\dot{t}^{\mu\nu}(\tau) = g^{\mu\nu}_\| - \epsilon\Omega_0 f^\mu{}_\rho t^{\rho\nu}(\tau), \tag{2.1.10}$$

with the initial conditions,

$$\dot{t}^{\mu\nu}(0) = g^{\mu\nu}, \qquad t^{\nu\mu}(0) = \epsilon f^{\mu\nu}/\Omega_0, \tag{2.1.11}$$

implied by (2.1.6) and (2.1.7), respectively. The tensor $\dot{t}^{\mu\nu}(\tau)$ also satisfies

$$\dot{t}^{\mu\nu}(-\tau) = \dot{t}^{\nu\mu}(\tau), \qquad \dot{t}^{\mu}{}_{\nu}(\tau_1)\dot{t}^{\nu\rho}(\tau_2) = \dot{t}^{\mu\rho}(\tau_1+\tau_2). \tag{2.1.12}$$

The contravariant components of tensor $t^{\mu\nu}(\tau)$ are

$$t^{\mu\nu}(\tau) = \frac{1}{\Omega_0}\begin{pmatrix} \Omega_0\tau & 0 & 0 & 0 \\ 0 & -\sin\Omega_0\tau & -\epsilon\cos\Omega_0\tau & 0 \\ 0 & \epsilon\cos\Omega_0\tau & -\sin\Omega_0\tau & 0 \\ 0 & 0 & 0 & -\Omega_0\tau \end{pmatrix}. \tag{2.1.13}$$

The contravariant components of $\dot{t}^{\mu\nu}(\tau)$,

$$\dot{t}^{\mu\nu}(\tau) = \begin{pmatrix} 1 & 0 & 0 & 0 \\ 0 & -\cos\Omega_0\tau & \epsilon\sin\Omega_0\tau & 0 \\ 0 & -\epsilon\sin\Omega_0\tau & -\cos\Omega_0\tau & 0 \\ 0 & 0 & 0 & -1 \end{pmatrix}, \tag{2.1.14}$$

follow by differentiation of (2.1.13).

2.1.2 Characteristic Response Due to a Spiraling Charge

The response of a spiraling charge is characterized by the Fourier transform of $\dot{t}^{\mu\nu}(\tau)$. Specifically, one may regard $\dot{t}^{\mu\nu}(\tau)$ as a causal function, which vanishes for $\tau < 0$, so that its Fourier transform is defined by writing

$$\int_0^\infty d\tau\, e^{i\omega\tau}\, \dot{t}^{\mu\nu}(\tau) = \frac{i}{\omega}\,\tau^{\mu\nu}(\omega). \tag{2.1.15}$$

The integral reproduces the tensor $\tau^{\mu\nu}(\omega)$, given by (1.2.19), viz.

$$\tau^{\mu\nu}(\omega) = g_{\parallel}^{\mu\nu} + \frac{\omega}{\omega^2-\Omega_0^2}\left(\omega g_{\perp}^{\mu\nu} - i\epsilon\Omega_0 f^{\mu\nu}\right). \tag{2.1.16}$$

A matrix representation of $\tau^{\mu\nu}(\omega)$ is given by (1.2.20). Being a causal function, one is to interpret the poles, at $\omega = 0, \pm\Omega_0$, in terms of the Landau prescription: give ω an infinitesimal imaginary part, $\omega \to \omega + i0$, and use the Plemelj formula

$$\frac{1}{\omega-\omega_0+i0} = \wp\,\frac{1}{\omega-\omega_0} - i\pi\,\delta(\omega-\omega_0), \tag{2.1.17}$$

where $\wp$ denotes the Cauchy principal value.

The 4-current density, $J^{\mu}_{\rm sp}(x)$ say, for a single charge q spiraling in a magnetic field is

$$J^{\mu}_{\rm sp}(x) = q\int d\tau\, u^{\mu}(\tau)\,\delta^4(x - X(\tau)). \tag{2.1.18}$$

The Fourier transform of (2.1.18) is

$$J^{\mu}_{\rm sp}(k) = \int d^4x\, J^{\mu}_{\rm sp}(x)\, e^{ikx} = q\int d\tau\, u^{\mu}(\tau)\, e^{ikX(\tau)}. \tag{2.1.19}$$

The explicit forms (2.1.6) for $u^{\mu}(\tau)$ and (2.1.4) for $X^{\mu}(\tau)$ are to be inserted into (2.1.19).

On inserting the expression (2.1.4) for $X^{\mu}(\tau)$, the exponent in (2.1.19) becomes

$$kX(\tau) = kx_0 + (ku_0)_{\parallel}\tau + \left[k_{\mu}t^{\mu\nu}(\tau)u_{0\nu}\right]_{\perp}, \tag{2.1.20}$$

where the notation $(ab)_{\parallel} = g^{\mu\nu}_{\parallel}a_{\mu}b_{\nu}$ is used. Let the wave 3-vector be written in cylindrical polar coordinates, so that the 4-vector has components

$$k^{\mu} = (\omega, k_{\perp}\cos\psi, k_{\perp}\sin\psi, k_z). \tag{2.1.21}$$

Then (2.1.20) contains the term

$$\left[k_{\mu}t^{\mu\nu}(\tau)u_{0\nu}\right]_{\perp} = -k_{\perp}R\sin(\phi_0 + \Omega_0\tau - \epsilon\psi), \tag{2.1.22}$$

with $R = \gamma v_{\perp}/\Omega_0 = p_{\perp}/|q|B$ the gyroradius of the particle. To perform the τ-integral in (2.1.19), one first expands in Bessel functions in such a way that all the dependence on τ is in exponents.

2.1.3 Expansion in Bessel Functions

The expansion in Bessel functions is based on the generating function

$$e^{iz\sin\phi} = \sum_{a=-\infty}^{\infty} e^{ia\phi} J_n(z). \tag{2.1.23}$$

The expansions needed here are (2.1.23) and

$$\begin{pmatrix}\cos\phi\\ i\sin\phi\end{pmatrix} e^{iz\sin\phi} = \sum_{a=-\infty}^{\infty} e^{ia\phi}\begin{pmatrix}(a/z)J_a(z)\\ J'_a(z)\end{pmatrix}, \tag{2.1.24}$$

which follow from (2.1.23) by differentiating with respect to ϕ, z, respectively. One may also obtain (2.1.24) from (2.1.23) by using the recursion relations

$$J_{a-1}(z) + J_{a+1}(z) = 2\frac{a}{z}J_a(z), \tag{2.1.25}$$

$$J_{a-1}(z) - J_{a+1}(z) = 2J_a'(z). \tag{2.1.26}$$

The actual expansion required is for the integrand in (2.1.19), and can be written

$$u^\mu(\tau)e^{ikX(\tau)} = e^{ikx_0}\sum_{a=-\infty}^{\infty} e^{is\epsilon\psi}e^{i[(ku_0)_\parallel - a\Omega_0]\tau}U^\mu(a,k). \tag{2.1.27}$$

The 4-vector $U^\mu(a,k)$ plays an important role in a covariant theory of the response. It has no standard name, and could be referred to as the Fourier-Bessel components of the 4-velocity. Its explicit form is

$$U^\mu(a,k) = \big[\gamma J_a(k_\perp R), \gamma \boldsymbol{V}(a,k)\big], \tag{2.1.28}$$

with the 3-velocity components given by

$$\boldsymbol{V}(a,k) = \bigg(\frac{1}{2}v_\perp\big[e^{-i\epsilon\psi}J_{a-1}(k_\perp R) + e^{i\epsilon\psi}J_{a+1}(k_\perp R)\big], \\ \frac{1}{2}i\epsilon v_\perp\big[e^{-i\epsilon\psi}J_{a-1}(k_\perp R) - e^{i\epsilon\psi}J_{a+1}(k_\perp R)\big], v_z J_a(k_\perp R)\bigg). \tag{2.1.29}$$

The 4-vector (2.1.28) satisfies the identity

$$k_\mu U^\mu(a,k) = \big[(ku)_\parallel - a\Omega_0\big]J_a(k_\perp R). \tag{2.1.30}$$

After expanding in Bessel functions, all the dependence on τ in (2.1.19) is in exponents, such that the τ-integral is trivial, giving a sum of terms involving δ-functions with argument $(ku)_\parallel - a\Omega_0$. The final result for the single particle 4-current (2.1.19) is

$$J^\mu_{\rm sp}(k) = qe^{ikx_0}\sum_{a=-\infty}^{\infty} e^{is\epsilon\psi}U^\mu(a,k)\,2\pi\delta\big[(ku)_\parallel - a\Omega_0\big]. \tag{2.1.31}$$

The identity (2.1.30) ensures that the charge-continuity condition, $k_\mu J^\mu_{\rm sp}(k) = 0$, is satisfied.

2.1.4 Gyroresonance Condition

The gyroresonance condition, implied by the δ-function in (2.1.31), is

$$(ku)_\parallel - a\Omega_0 = 0, \qquad \varepsilon\omega - a|q|B - p_z k_z = 0, \tag{2.1.32}$$

with $p^\mu = mu^\mu$, $p^0 = \varepsilon$, $p^3 = p_z$. The condition (2.1.32) is also sometimes referred to as the Doppler condition.

When viewed as an equation for p_z for given a, ω, k_z, (2.1.32) has the disadvantage that it contains the square root $\varepsilon = (m^2 + p_\perp^2 + p_z^2)^{1/2}$. The gyroresonance condition in (2.1.32) may be rationalized, to remove the square root, by multiplying by $\varepsilon\omega + a|q|B + p_z k_z$. This gives

$$\left(\omega^2 - k_z^2\right) p_z^2 - 2a|q|Bk_z p_z + \left(m^2 + p_\perp^2\right)\omega^2 - a^2 q^2 B^2 = 0. \tag{2.1.33}$$

In $p_\perp$-p_z space, (2.1.33) is a conic section, being an ellipse with minor axis along the $p_\perp$-axis for $\omega^2 - k_z^2 > 0$, and a hyperbola for $\omega^2 - k_z^2 < 0$. Only one arm of the hyperbola is physical for $\omega^2 - k_z^2 > 0$, with the spurious solution being introduced by the rationalization process.

One may also regard (2.1.33) as a quadratic equation for p_z, and solve it for the two solutions $p_z = p_{z\pm}$. These resonant momenta are

$$p_{z\pm} = k_z f_a \pm \omega g_a, \qquad f_a = \frac{a|q|B}{\omega^2 - k_z^2}, \qquad g_a^2 = f_a^2 - \frac{\varepsilon_\perp^2}{\omega^2 - k_z^2}, \tag{2.1.34}$$

with $\varepsilon_\perp^2 = m^2 + p_\perp^2$. The corresponding resonant energies, $\varepsilon_\pm = (\varepsilon_\perp^2 + p_{z\pm}^2)^{1/2}$, are

$$\varepsilon_\pm = \omega f_a \pm k_z g_a, \tag{2.1.35}$$

where the sign of the square root is chosen such that one has $\varepsilon_\pm\omega - a|q|B - p_{z\pm}k_z{=}0$. The physically allowed region for the resonance corresponds to $g_a^2 > 0$, and $\varepsilon_\pm \geq m$.

An alternative way of factoring the gyroresonance condition involves replacing the momentum components $p_\perp, p_z$ by $\varepsilon_\perp, t$, defined by

$$\varepsilon_\perp^2 = m^2 + p_\perp^2, \qquad p_z = \varepsilon_\perp \frac{2t}{1-t^2}, \qquad \varepsilon = \varepsilon_\perp \frac{1+t^2}{1-t^2}, \tag{2.1.36}$$

with $-1 < t < 1$. One has

$$\varepsilon\omega - p_z k_z - a|q|B = (\varepsilon_\perp\omega + a|q|B)\frac{(t-t_+)(t-t_-)}{1-t^2}, \tag{2.1.37}$$

with the resonant values given by

$$t_\pm = \frac{\varepsilon_\perp k_z \pm (\omega^2 - k_z^2) g_a}{\varepsilon_\perp\omega + a|q|B}, \quad p_{z\pm} = \varepsilon_\perp \frac{2t_\pm}{1-t_\pm^2}, \quad \varepsilon_\pm = \varepsilon_\perp \frac{1+t_\pm^2}{1-t_\pm^2}. \tag{2.1.38}$$

The physically allowed solutions are those with $t_\pm$ real and in the range with $-1 < t_\pm < 1$.

2.2 Perturbation Expansions

On including a fluctuating electromagnetic field, described by $A^\mu(k)$, the weak turbulence expansion involves expanding various relevant quantities in powers of $A^\mu(k)$. These expansions are discussed in this section.

2.2.1 Perturbation Expansion of the 4-Current

The equation of motion for a charge can be written

$$\frac{du^\mu(\tau)}{d\tau} = \frac{q}{m}\,F_0^{\mu\nu}u_\nu(\tau) + S^\mu(\tau), \tag{2.2.1}$$

where the perturbing electromagnetic field is included in

$$S^\mu(\tau) = \frac{iq}{m}\int\frac{d^4k_1}{(2\pi)^4}\,e^{-ik_1X(\tau)}k_1u(\tau)\,G^{\mu\nu}\big(k_1,u(\tau)\big)A_\nu(k_1), \tag{2.2.2}$$

with $G^{\mu\nu}(k,u) = g^{\mu\nu} - k^\mu u^\nu/ku$. The tensor $\dot{t}^{\mu\nu}(\tau)$ plays the role of an integrating factor in the sense that it allows one to integrate (2.2.1) to find

$$u^\mu(\tau) = \dot{t}^{\mu\nu}(\tau)u_{0\nu} + \int_0^\tau d\tau'\,\dot{t}^{\mu\nu}(\tau-\tau')S_\nu(\tau'),$$

$$X^\mu(\tau) = x_0^\mu + t^{\mu\nu}(\tau)u_{0\nu} + \int^\tau d\tau''\int^{\tau''}d\tau'\,\dot{t}^{\mu\nu}(\tau''-\tau')S_\nu(\tau'), \tag{2.2.3}$$

which follow by integrating once and twice, respectively.

A perturbation expansion of the orbit in powers of $A(k)$ may be written

$$X^\mu(\tau) = \sum_{n=0} X^{(n)\mu}(\tau), \qquad u^\mu(\tau) = \sum_{n=0} u^{(n)\mu}(\tau), \tag{2.2.4}$$

where $u^{(n)\mu}(\tau) = dX^{(n)\mu}(\tau)/d\tau$ is of nth order in $A(k)$. The expansion may be made by first inserting the expansions (2.2.4) into the expression (2.2.2) for $S^\mu(\tau)$, and expanding it in the same form as (2.2.4):

$$S^\mu(\tau) = \sum_{n=1} S^{(n)\mu}(\tau), \tag{2.2.5}$$

which has no zeroth order term. The nth order terms follow from

$$u^{(n)\mu}(\tau) = \int^\tau d\tau'\,\dot{t}^{\mu\nu}(\tau-\tau')S_\nu^{(n)}(\tau'),$$

$$X^{(n)\mu}(\tau) = \int^\tau d\tau''\int^{\tau''}d\tau'\,\dot{t}^{\mu\nu}(\tau''-\tau')S_\nu^{(n)}(\tau'). \tag{2.2.6}$$

The perturbation expansion involves an iteration. The first order term $S^{(1)\mu}(\tau)$ follows by inserting the zeroth order solutions (2.1.8) and (2.1.9) into the integrand of (2.2.2). The first order terms $u^{(1)\mu}(\tau)$ and $X^{(1)\mu}(\tau)$ follow from $n=1$ in (2.2.6). These terms are then used to find $S^{(2)\mu}(\tau)$ by expanding the integrand of (2.2.2), with the second order terms $u^{(2)\mu}(\tau)$ and $X^{(2)\mu}(\tau)$ given by $n=2$ in (2.2.6). In this way the solution at order n follows from the solutions of order $< n$.

Expansion of the 4-Current

The expansion of the 4-current leads to an nth order term of the form

$$J_{\rm sp}^{(n)\mu}(k) = q\int d\tau\, j^{(n)\mu}(\tau)\, e^{ikX^{(0)}(\tau)}, \tag{2.2.7}$$

with $X^{(0)\mu}(\tau)$ identified with the zeroth order orbit (2.1.4). The zeroth order current is given by (2.1.19), and corresponds to

$$j^{(0)\mu}(\tau) = \dot{t}^{\mu\nu}(\tau)u_{0\nu}, \tag{2.2.8}$$

where $\dot{t}^{\mu\nu}(\tau)$ is given by (2.1.6). The first term is

$$j^{(1)\mu}(\tau) = u^{(1)\mu}(\tau) + ikX^{(1)}(\tau)\,u^{(0)\mu}(\tau), \tag{2.2.9}$$

the second order term is

$$j^{(2)\mu}(\tau) = u^{(2)\mu}(\tau) + ikX^{(1)}(\tau)\,u^{(1)\mu}(\tau) + \{ikX^{(1)}(\tau) - \tfrac{1}{2}[kX^{(1)}(\tau)]^2\}u^{(0)\mu}(\tau), \tag{2.2.10}$$

and so on for the higher order currents.

First-Order Current

The first order current follows from (2.2.7) with (2.2.9). Writing $u^{(1)\mu}(\tau) = dX^{(1)\mu}(\tau)/d\tau$ and partially integrating, the result reduces to

$$J_{\rm sp}^{(1)\mu}(k) = iq\int d\tau\, ku^{(0)}(\tau)\, G^{\alpha\mu}\big(k, u^{(0)}(\tau)\big)\, X_\alpha^{(1)}(\tau)\, e^{ikX^{(0)}(\tau)}, \tag{2.2.11}$$

with

$$ku(\tau)\, G^{\mu\nu}\big(k, u(\tau)\big) = ku(\tau)\, g^{\mu\nu} - k^\mu u^\nu(\tau). \tag{2.2.12}$$

On inserting the first order perturbation in the expression (2.2.11) for the orbit, one finds

$$J_{\rm sp}^{(1)\mu}(k) = -\frac{q^2}{m}\int d\tau \int^{\tau} d\tau'' \int^{\tau''} d\tau' \int \frac{d^4k_1}{(2\pi)^4} e^{i[kX^{(0)}(\tau)-k_1X^{(0)}(\tau')]}$$
$$\times\, i_{\alpha\beta}(\tau''-\tau')\, ku^{(0)}(\tau)\, G^{\alpha\mu}(k,u^{(0)}(\tau))\, k_1u^{(0)}(\tau')\, G^{\beta\nu}(k_1,u^{(0)}(\tau'))\, A_\nu(k_1). \tag{2.2.13}$$

To carry out the integrals over τ, τ' and τ'' in (2.2.13) one first expands the integrand in Bessel functions, using (2.1.27), so that all the dependence on proper time is in exponents. The resulting expression is

$$J_{\rm sp}^{(1)\mu}(k) = \sum_{a,a_1=-\infty}^{\infty} \int \frac{d^4k_1}{(2\pi)^4}\, e^{-i(k-k_1)x_0}\, e^{i\epsilon(a\psi-a_1\psi_1)}$$
$$\times\, G^{\alpha\mu}(a,k,u)\tau_{\alpha\beta}\big((ku)_\parallel - a\Omega_0\big)G^{*\beta\nu}(a_1,k_1,u)$$
$$\times\, 2\pi\delta\big[(ku)_\parallel-(k_1u)_\parallel-(a-a_1)\Omega_0\big], \tag{2.2.14}$$

where $\tau^{\mu\nu}(\omega)$ is given by (2.1.16), and where the azimuthal angles ψ, ψ_1 are defined by (2.1.21). The tensor $G^{\mu\nu}(a,k,u)$ is given by

$$G^{\mu\nu}(a,k,u) = g^{\mu\nu}J_a(k_\perp R) - \frac{k^\mu U^\nu(a,k)}{(ku)_\parallel - a\Omega_0}. \tag{2.2.15}$$

2.2.2 *Small-Gyroradius Approximation*

An important limiting case in which the foregoing formulae simplify considerably is the limit of small gyroradii. The argument of the Bessel functions is $k_\perp R$, and in the limit $R\to 0$ one has $J_a(0)=1$ for $a=0$ and $J_a(0)=0$ for $a\neq 0$. Applying this approximation to $U^\mu(a,k)$, as given by (2.1.28), one finds that $V(a,k)$ has nonzero contributions only for $a=0,\pm 1$:

$$V(0,k) = (0,0,v_z), \qquad V(\pm 1,k) = \tfrac{1}{2}v_\perp e^{\mp i\epsilon\psi}(1,\pm i\epsilon,0). \tag{2.2.16}$$

In this approximation (2.1.28) gives

$$U^\mu(0,k) = u_\parallel^\mu, \qquad U^\mu(\pm 1,k) = \tfrac{1}{2}u_\perp e^{\mp i\epsilon\psi}(0,1,\pm i\epsilon,0), \tag{2.2.17}$$

with $u_\parallel^\mu = \gamma(1,0,0,v_z)$, $u_\perp = \gamma v_\perp$. In the small gyroradius limit, the current (2.1.31) reduces to

$$J_{\rm sp}^\mu(k) = qe^{ikx_0}u_\parallel^\mu\, 2\pi\delta[(ku)_\parallel], \tag{2.2.18}$$

which is equivalent to the current for a charge in constant rectilinear motion. In the small-gyroradius approximation, the function $G^{\mu\nu}(a,k,u)$, defined by (2.2.15), is nonzero only for $a = 0$, when it simplifies to $G^{\mu\nu}(0,k,u) = G^{\mu\nu}(k,u_\parallel) = g^{\mu\nu} - k^\mu u_\parallel^\nu / ku_\parallel$. The first-order current (2.2.14) then simplifies to

$$J_{\rm sp}^{(1)\mu}(k) = \int \frac{d^4k_1}{(2\pi)^4}\, e^{-i(k-k_1)x_0} \times G^{\alpha\mu}(k,u_\parallel)\tau_{\alpha\beta}(ku_\parallel)G^{\beta\nu}(k_1,u_\parallel)\, 2\pi\delta\big[(k-k_1)u_\parallel\big]\, A_{\nu_1}(k_1). \tag{2.2.19}$$

One has

$$G^{\alpha\mu}(k,u_\parallel)\tau_{\alpha\beta}(ku_\parallel)G^{\beta\nu}(k',u_\parallel) = \tau^{\mu\nu}(ku_\parallel) - \frac{u_\parallel^\mu k_\alpha \tau^{\alpha\nu}(ku_\parallel) + \tau^{\mu\beta}(ku_\parallel)k'_\beta u_\parallel^\nu}{ku_\parallel} + \frac{k_\alpha k'_\beta \tau^{\alpha\beta}(ku_\parallel)u_\parallel^\mu u_\parallel^\nu}{(ku_\parallel)^2}, \tag{2.2.20}$$

with $k'u_\parallel = ku_\parallel$ implied by the δ-function in (2.2.19).

2.3 General Forms for the Linear Response 4-Tensor

The forward-scattering and Vlasov methods are used in this section to derive general expressions for the response tensor for a magnetized plasma with arbitrary distribution functions for the particles.

2.3.1 *Forward-Scattering Method for a Magnetized Plasma*

The derivation of the linear response tensor using the forward-scattering method involves averaging the first-order, single-particle current over the distribution of particles. For a magnetized particle the first order, single-particle current is given by (2.2.13), with the unperturbed orbit given by (2.1.4), viz. $X^\mu(\tau) = x_0^\mu + t^{\mu\nu}(\tau)u_{0\nu}$, where x_0 describes the initial conditions and u_0 is the initial 4-velocity. The average over a distribution of particles follows by noting that the distribution $F(x_0, p_0)$ represents the number of world lines (one per particle) threading the 7-dimensional surface $d^4x_0\, d^4p_0/(2\pi)^4 d\tau$ [1]. Hence, the appropriate average follows by replacing the integral over $d\tau$ in (2.2.13) by the integral over $d^4x_0\, d^4p_0/(2\pi)^4$ times $F(x_0, p_0)$. Assuming a uniform distribution in space and time implies that $F(p_0)$ does not depend on x_0. Then x_0 appears only in an exponential factor $\exp[i(k-k_1)x_0]$, and the x_0-integral gives $(2\pi)^4\,\delta^4(k-k_1)$. The k_1-integral in (2.2.13) is performed over the resulting δ-function, with the implied identity $k_1^\mu = k^\mu$ being the forward-scattering condition. This leads to an expression of the form $J^{(1)\mu}(k) = \Pi^{\mu\nu}(k)A_\nu(k)$.

Using this method one identifies the linear response tensor as

$$\Pi^{\mu\nu}(k) = -\frac{q^2}{m}\int \frac{d^4p(\tau)}{(2\pi)^4} F(p) \int^{\tau} d\tau'' \int^{\tau''} d\tau' \exp\left[ik\left(X(\tau) - X(\tau')\right)\right]$$
$$\times \dot{t}_{\alpha\beta}(\tau'' - \tau')\, ku(\tau)\, G^{\alpha\mu}\big(k, u(\tau)\big)\, ku(\tau')\, G^{\beta\nu}\big(k, u(\tau')\big), \qquad (2.3.1)$$

where the superscript (0) on $u(\tau)$ is now omitted, with $u(\tau)$ given by (2.1.6), viz. $u^\mu(\tau) = \dot{t}^{\mu\nu}(\tau)u_{0\nu}$, with $t^{\mu\nu}(\xi)$ given by (2.1.7) and where the dot denotes the derivative with respect to τ. Also in (2.3.1) it is noted that the initial value of τ is related to the initial gyrophase, ϕ_0 say, and one is free to choose $\phi_0 = \Omega_0\tau$ and to write the integral over d^4p_0 as an integral over $d^4p(\tau)$. The distribution $F(p)$ is assumed independent of gyrophase, and hence of τ.

The normalization of $F(p)$ is to the proper number density, $n_{\rm pr}$,

$$n_{\rm pr} = \int \frac{d^4p}{(2\pi)^4} F(p). \qquad (2.3.2)$$

In terms of the conventional distribution function $f(\boldsymbol{p})$ the proper number density and the number density are given by (ordinary units)

$$n_{\rm pr} = \int \frac{d^3\boldsymbol{p}}{(2\pi\hbar)^3\gamma} f(\boldsymbol{p}), \qquad n = \int \frac{d^3\boldsymbol{p}}{(2\pi\hbar)^3} f(\boldsymbol{p}), \qquad (2.3.3)$$

respectively. The factor $(2\pi\hbar)^3$ is included in (2.3.3) to be consistent with the conventional normalization in quantum statistical mechanics. One has $F(p) \propto \delta(p^2 - m^2)\, f(\boldsymbol{p})$, with $p^2 = \varepsilon^2 - \boldsymbol{p}^2$ and $\varepsilon = p^0$.

The dependence on proper times in (2.3.1) simplifies after a partial integration:

$$\Pi^{\mu\nu}(k) = \frac{q^2}{m}\int \frac{d^4p(\tau)}{(2\pi)^4} F(p) \int_0^{\infty} d\xi \exp\left[ik\left(X(\tau) - X(\tau - \xi)\right)\right]$$
$$\times T_{\alpha\beta}(\xi)\, ku(\tau)\, G^{\alpha\mu}\big(k, u(\tau)\big)\, ku(\tau - \xi)\, G^{\beta\nu}\big(k, u(\tau - \xi)\big), \qquad (2.3.4)$$

where it is convenient to introduce the tensor

$$T^{\mu\nu}(\tau) = t^{\mu\nu}(\tau) - t^{\mu\nu}(0). \qquad (2.3.5)$$

The components of $T^{\mu\nu}(\tau)$ follow from (2.1.13):

$$T^{\mu\nu}(\tau) = \frac{1}{\Omega_0}\begin{pmatrix} \Omega_0\tau & 0 & 0 & 0 \\ 0 & -\sin\Omega_0\tau & \epsilon(1 - \cos\Omega_0\tau) & 0 \\ 0 & -\epsilon(1 - \cos\Omega_0\tau) & -\sin\Omega_0\tau & 0 \\ 0 & 0 & 0 & -\Omega_0\tau \end{pmatrix}. \qquad (2.3.6)$$

The property $T^{\mu\nu}(\tau) = -T^{\nu\mu}(-\tau)$ allows one to separate the integrand in (2.3.4) into parts that are even and odd functions of ξ. These correspond to the hermitian and antihermitian parts, respectively, of the response tensor. Thus the antihermitian part of (2.3.4) can be found by replacing the lower limit of the ξ-integral by $-\infty$, and dividing by 2:

$$\Pi^{A\mu\nu}(k) = \frac{q^2}{2m} \int \frac{d^4 p(\tau)}{(2\pi)^4} F(p) \int_{-\infty}^{\infty} d\xi \, \exp\big[ik\big(X(\tau) - X(\tau - \xi)\big)\big] \, T_{\alpha\beta}(\xi)$$
$$\times \, ku(\tau) \, G^{\alpha\mu}\big(k, u(\tau)\big) \, ku(\tau - \xi) \, G^{\beta\nu}\big(k, u(\tau - \xi)\big). \qquad (2.3.7)$$

The hermitian part is implicit in (2.3.4) and it is not particularly helpful to identify it explicitly.

2.3.2 Forward-Scattering Form Summed over Gyroharmonics

Further evaluation of the response tensor in the forward-scattering form (2.3.1) involves expanding in Bessel functions. This enables one to perform the integrals over ξ and τ (equivalent to gyrophase here) explicitly, and interpret the result in terms of a sum over gyroresonant contributions. To perform the τ' and τ'' integrals in (2.3.1) and the ξ-integral in (2.3.4) one first expands in Bessel functions using (2.1.27).

The resulting expression for the linear response tensor is

$$\Pi^{\mu\nu}(k) = -\frac{q^2}{m} \int \frac{d^4 p}{(2\pi)^4} F(p)$$
$$\times \sum_{a=-\infty}^{\infty} G^{\alpha\mu}(a, k, u) \tau_{\alpha\beta}((ku)_\parallel - a\Omega_0) G^{*\beta\nu}(a, k, u), \qquad (2.3.8)$$

where the subscript 0 on u and $p = mu$ is now redundant, and with $G^{\mu\nu}(a, k, u)$ defined by (2.2.15). The 4-vector $U^\mu(a, k)$ is given by (2.1.28) with (2.1.29), and $\tau^{\mu\nu}(\omega)$ is given by (2.1.15).

The form (2.3.8) for the linear response tensor is in a concise notation, and one needs to write it in a more explicit form in order to evaluate it for specific distributions. One step is to substitute the explicit form (2.2.15) for $G^{\mu\nu}$, giving

$$\Pi^{\mu\nu}(k) = -\frac{q^2}{m} \int \frac{d^4 p}{(2\pi)^4} F(p) \sum_{a=-\infty}^{\infty} \Bigg[\tau^{\mu\nu} J_a^2$$
$$- \frac{U^\mu k_\alpha \tau^{\alpha\nu} + \tau^{\mu\beta} k_\beta U^{*\nu}}{(ku)_\parallel - a\Omega_0} J_a + \frac{k_\alpha \tau^{\alpha\beta} k_\beta U^\mu U^{*\nu}}{[(ku)_\parallel - a\Omega_0]^2} \Bigg], \qquad (2.3.9)$$

where the argument $((ku)_\parallel - a\Omega_0)$ and (a,k) of $\tau^{\mu\nu}$ and U^μ, respectively, are omitted for simplicity in writing. Substituting the explicit expression (2.1.15) for $\tau^{\mu\nu}$ gives

$$
\begin{aligned}
\Pi^{\mu\nu}(k) = -\frac{q^2}{m}\int \frac{d^4p}{(2\pi)^4}\, F(p) \sum_{a=-\infty}^{\infty} \Bigg\{ g_\parallel^{\mu\nu}\, J_a^2 \\
-\frac{k_\parallel^\mu U^{*\nu} + k_\parallel^\nu U^\mu}{(ku)_\parallel - a\Omega_0}\, J_a + \frac{(k^2)_\parallel U^\mu U^{*\nu}}{[(ku)_\parallel - a\Omega_0]^2} \\
+\frac{[(ku)_\parallel - a\Omega_0]^2}{[(ku)_\parallel - a\Omega_0]^2 - \Omega_0^2}\left[g_\perp^{\mu\nu}\, J_a^2 - \frac{k_\perp^\mu U^{*\nu} + k_\perp^\nu U^\mu}{(ku)_\parallel - a\Omega_0}\, J_a + \frac{(k^2)_\perp U^\mu U^{*\nu}}{[(ku)_\parallel - a\Omega_0]^2}\right] \\
-\frac{i\epsilon\Omega_0[(ku)_\parallel - a\Omega_0]}{[(ku)_\parallel - a\Omega_0]^2 - \Omega_0^2}\left[f^{\mu\nu}\, J_a^2 + \frac{k_G^\mu U^{*\nu} - k_G^\nu U^\mu}{(ku)_\parallel - a\Omega_0}\, J_a\right]\Bigg\}. \qquad (2.3.10)
\end{aligned}
$$

Further separation into components in the $\parallel$- and $\perp$-subspaces follows by writing

$$
\begin{aligned}
U^\mu(a,k) &= U_\parallel^\mu(a,k) + U_\perp^\mu(a,k), \\
U_\parallel^\mu(a,k) &= \frac{J_a(k_\perp R)}{(k^2)_\parallel}[(ku)_\parallel k_\parallel^\mu - (k_D u)_\parallel k_D^\mu], \\
U_\perp^\mu(a,k) &= \frac{\gamma v_\perp}{k_\perp}\left[\frac{a}{k_\perp R}\, J_a(k_\perp R)\, k_\perp^\mu + i\epsilon\, J_a'(k_\perp R)\, k_G^\mu\right], \qquad (2.3.11)
\end{aligned}
$$

where the four basis 4-vectors $k_\parallel^\mu$, $k_\perp^\mu$, k_G^μ, k_D^μ, defined by (1.1.18), span the 4-dimensional space. Using (2.3.11) and the identities

$$
(k^2)_\parallel g_\parallel^{\mu\nu} = k_\parallel^\mu k_\parallel^\nu - k_D^\mu k_D^\nu, \qquad (k^2)_\perp g_\perp^{\mu\nu} = k_\perp^\mu k_\perp^\nu + k_G^\mu k_G^\nu, \qquad (2.3.12)
$$

with $(k^2)_\perp = -k_\perp^2$, allows one to (2.3.10) explicitly in terms of this choice of basis 4-vectors.

Explicit Forms for the Components of $\Pi^{\mu\nu}(k)$

In practice it is convenient to have explicit expressions for the components of the response tensor.

The components of the response 4-tensor in the forward-scattering form (2.3.10) may be written

$$
\Pi^{\mu\nu}(k) = -\frac{q^2}{m}\int \frac{d^3\mathbf{p}}{(2\pi)^3\gamma}\, f(\mathbf{p}) \sum_{a=-\infty}^{\infty} A^{\mu\nu}(a,k,u), \qquad (2.3.13)
$$

with $A^{\mu\nu}$ given explicitly in Table 2.1, and where the relation $\int d^4p\, F(p)/(2\pi)^4 = \int d^3\mathbf{p}\, f(\mathbf{p})/(2\pi)^3\gamma$ is used.

Table 2.1 The components of $A^{\mu\nu} = A^{\mu\nu}(a, k, u)$ in (2.3.13), with the argument of the Bessel functions, $J_a = J_a(k_\perp R)$, omitted

$$A^{00} = \left[\frac{(\gamma k_z v_z + a\Omega_0)^2 - \gamma^2 k_z^2}{[(ku)_\parallel - a\Omega_0]^2} - \frac{\gamma^2 k_\perp^2}{[(ku)_\parallel - a\Omega_0]^2 - \Omega_0^2}\right] J_a^2$$

$$A^{01} = \left[\frac{\omega(\gamma k_z v_z + a\Omega_0) - \gamma k_z^2}{[(ku)_\parallel - a\Omega_0]^2}\frac{a\Omega_0}{k_\perp} - \frac{\gamma^2 k_\perp(\omega - k_z v_z)}{[(ku)_\parallel - a\Omega_0]^2 - \Omega_0^2}\right] J_a^2$$

$$A^{02} = -i\epsilon\left[\frac{\gamma k_z(\omega v_z - k_z) + \omega a\Omega_0}{[(ku)_\parallel - a\Omega_0]^2}\gamma v_\perp J_a' J_a - \frac{\gamma k_\perp}{[(ku)_\parallel - a\Omega_0]^2 - \Omega_0^2}(\Omega_0 J_a + \gamma k_\perp v_\perp J_a')J_a\right]$$

$$A^{03} = \left[\frac{\gamma(\omega v_z + k_z)a\Omega_0 - \gamma^2\omega k_z(1 - v_z^2)}{[(ku)_\parallel - a\Omega_0]^2} - \frac{\gamma^2 v_z k_\perp^2}{[(ku)_\parallel - a\Omega_0]^2 - \Omega_0^2}\right] J_a^2$$

$$A^{11} = \left[\frac{(k^2)_\parallel}{[(ku)_\parallel - a\Omega_0]^2}\frac{a^2\Omega_0^2}{k_\perp^2} - \frac{(ku)_\parallel^2}{[(ku)_\parallel - a\Omega_0]^2 - \Omega_0^2}\right] J_a^2$$

$$A^{22} = -J_a^2 + \frac{(k^2)_\parallel}{[(ku)_\parallel - a\Omega_0]^2}(\gamma v_\perp J_a')^2 - \frac{(\Omega_0 J_a + \gamma k_\perp v_\perp J_a')^2}{[(ku)_\parallel - a\Omega_0]^2 - \Omega_0^2}$$

$$A^{33} = \left[-\frac{(\gamma\omega - a\Omega_0)^2 - (\omega\gamma v_z)^2}{[(ku)_\parallel - a\Omega_0]^2} - \frac{(k_\perp \gamma v_z)^2}{[(ku)_\parallel - a\Omega_0]^2 - \Omega_0^2}\right] J_a^2$$

$$A^{12} = -i\epsilon\left[\frac{\omega^2 - k_z^2}{[(ku)_\parallel - a\Omega_0]^2}\frac{a\Omega_0}{k_\perp} J_a \gamma v_\perp J_a' - \frac{(ku)_\parallel}{[(ku)_\parallel - a\Omega_0]^2 - \Omega_0^2}(\Omega_0 J_a + \gamma k_\perp v_\perp J_a')J_a\right]$$

$$A^{13} = \left[\frac{\gamma\omega(\omega v_z - k_z) + k_z a\Omega_0}{[(ku)_\parallel - a\Omega_0]^2}\frac{a\Omega_0}{k_\perp} - \frac{\gamma k_\perp v_z\,(ku)_\parallel}{[(ku)_\parallel - a\Omega_0]^2 - \Omega_0^2}\right] J_a^2$$

$$A^{23} = i\epsilon\left[\frac{\gamma\omega(\omega v_z - k_z) + k_z a\Omega_0}{[(ku)_\parallel - a\Omega_0]^2}\gamma v_\perp J_a' - \frac{k_\perp \gamma v_z}{[(ku)_\parallel - a\Omega_0]^2 - \Omega_0^2}(\Omega_0 J_a + \gamma k_\perp v_\perp J_a')\right] J_a$$

$$A^{10} = A^{01} \qquad A^{20} = -A^{02} \qquad A^{30} = A^{03}$$
$$A^{21} = -A^{12} \qquad A^{23} = -A^{32} \qquad A^{31} = A^{13}$$

2.3.3 *Vlasov Method for a Magnetized Plasma*

The Vlasov method leads to a form for the response tensor that is equivalent to but superficially different from that obtained using the forward-scattering method.

Covariant Vlasov Equation

The method is based on the covariant Vlasov equation, which is

$$\left(u^\mu \frac{\partial}{\partial x^\mu} + m \frac{du^\mu}{d\tau} \frac{\partial}{\partial p^\mu}\right) F(x, p) = 0, \tag{2.3.14}$$

with $du^\mu/d\tau$ determined by the equation of motion (2.2.1). On Fourier transforming (2.3.14) with (2.2.1) and using (2.2.2), one obtains

$$\left[-iku + q\, F_0^{\mu\nu}\, u_\nu \frac{\partial}{\partial p^\mu}\right] F(k, p)$$
$$= -iq \int d\lambda^{(2)}\, k_1 u\, G^{\mu\nu}(k_1, u) A_\nu(k_1) \frac{\partial F(k_2, p)}{\partial p^\mu}, \tag{2.3.15}$$

where the right hand side is the convolution of the Lorentz force due to the perturbing electromagnetic field and the perturbed distribution function. One may interpret the operator $q\, F_0^{\mu\nu} u_\nu \partial/\partial p^\mu$ on the left hand side as $d/d\tau$, that is, as the derivative with respect to proper time along the orbit.

One may integrate (2.3.15) once before making an expansion in powers of the amplitude of the fluctuating field. With u interpreted as the unperturbed orbit $u(\tau)$, given by (2.1.6) and (2.1.7), integrating (2.3.15) once gives

$$F\big(k, p(\tau)\big) = -iq\, e^{ikX(\tau)} \int^\tau d\tau'\, e^{-ikX(\tau')} \int d\lambda^{(2)}\, k_1 u(\tau')$$
$$\times\, G^{\alpha\nu}\big(k_1, u(\tau')\big) A_\nu(k_1) \frac{\partial F(k_2, p(\tau'))}{\partial p^\alpha(\tau')}. \tag{2.3.16}$$

It is convenient to omit the subscript 0 on the initial values in (2.1.6) and (2.1.7), so that one has

$$u^\mu(\tau) = \dot{t}^{\mu\nu}(\tau) u_\nu, \quad u^\mu = u^\mu(0) = (\gamma, u_\perp, 0, u_z), \tag{2.3.17}$$

with $u_\perp = \gamma v_\perp$, $u_z = \gamma v_z$ and with $p^\mu = m u^\mu = (\varepsilon, p_\perp, 0, p_z)$. The explicit form for the tensor $\dot{t}^{\mu\nu}(\tau)$ is given by (2.1.6). One also has

$$kX(\tau) = kx_0 + (ku)_\parallel \tau + k_\perp R \sin(\epsilon\psi - \Omega_0 \tau), \tag{2.3.18}$$

with $R = u_\perp/\Omega_0$ the gyroradius.

Linearized Vlasov Equation

In the Vlasov method, the weak turbulence expansion involves an expansion of $F(k, p)$ in powers of $A(k)$:

$$F(k, p) = F(p)\,(2\pi)^4 \delta^4(k) + \sum_{n=1}^{\infty} F^{(n)}(k, p). \tag{2.3.19}$$

On inserting (2.3.19) into (2.3.15) one obtains a hierarchy of equations, starting with the linearized Vlasov equation,

$$\left[-iku + q\,F_0^{\mu\nu} u_\nu \frac{\partial}{\partial p^\mu}\right] F^{(1)}(k,p) = -iq\,ku\,G^{\alpha\nu}(k,u)A_\nu(k)\frac{\partial F(p)}{\partial p^\alpha}, \tag{2.3.20}$$

with the nth term in the expansion determined by

$$\left[-iku + q\,F_0^{\mu\nu} u_\nu \frac{\partial}{\partial p^\mu}\right] F^{(n)}(k,p)$$
$$= -iq \int d\lambda^{(2)}\, k_1 u\, G^{\alpha\nu}(k_1,u)A_\nu(k_1)\frac{\partial F^{(n-1)}(k_2,p)}{\partial p^\alpha}. \tag{2.3.21}$$

The solution of this hierarchy of equations is obtained by substituting (2.3.19) into (2.3.16), with the linearized Vlasov equation the first order term.

The explicit form for the first order term is

$$F^{(1)}\big(k,p(\tau)\big) = -iq\,A_\nu(k)\int_0^\infty d\xi\, e^{ik[X(\tau)-X(\tau-\xi)]}$$
$$\times\, ku(\tau-\xi)\,G^{\alpha\nu}\big(k,u(\tau-\xi)\big)\,\frac{\partial F(p)}{\partial p^\alpha(\tau-\xi)}, \tag{2.3.22}$$

where the variable τ' in (2.3.16) is replaced by $\xi = \tau - \tau'$ in (2.3.22), and where $t^{\mu\nu}(\tau) = t^{\nu\mu}(-\tau)$ is used, cf.(2.1.7).

The linear response tensor is found by writing the induced current in the form

$$J^{(1)\mu}(k) = q\int \frac{d^4p(\tau)}{(2\pi)^4}\, u^\mu(\tau)\, F^{(1)}\big(k,p(\tau)\big), \tag{2.3.23}$$

and equating the right hand side to $\Pi^{\mu\nu}(k)A_\nu(k)$ to identify $\Pi^{\mu\nu}(k)$.

2.3.4 Vlasov Form for the Linear Response Tensor

The resulting expression for the Vlasov form for the linear response tensor, for a distribution of one (unlabeled) species of particles, is

$$\Pi^{\mu\nu}(k) = -iq^2\int \frac{d^4p(\tau)}{(2\pi)^4}\int_0^\infty d\xi\, u^\mu(\tau)\, e^{ik[X(\tau)-X(\tau-\xi)]}$$
$$\times\, ku(\tau-\xi)\,G^{\alpha\nu}\big(k,u(\tau-\xi)\big)\,\frac{\partial F(p)}{\partial p^\alpha(\tau-\xi)}. \tag{2.3.24}$$

The Vlasov form (2.3.24) is equivalent to the forward-scattering form (2.3.4), as may be shown by partially integrating. The relatively lengthy calculation is facilitated somewhat by noting the identity

$$\frac{\partial}{\partial p^\alpha(\tau)}\big[ku(\tau)\, G^{\alpha\nu}(k,u(\tau))\big] = 0. \tag{2.3.25}$$

In evaluating the derivative $\partial F(p)/\partial p^\alpha(\tau-\xi)$ in (2.3.27), one needs to take account of the fact that $F(p)$ is independent of the gyrophase, and hence of τ or ξ. The derivative in (2.3.27) may be written in the form

$$\frac{\partial F(p)}{\partial p^\alpha(\tau-\xi)} = \left[\frac{u_{\perp\alpha}(\tau-\xi)}{u_\perp}\frac{\partial}{\partial p_\perp} + \frac{\partial}{\partial p_\parallel^\alpha}\right] F(p). \tag{2.3.26}$$

Terms proportional to $ku(\tau-\xi)$ in the integrand in (2.3.27) are perfect differentials, $ku(\tau-\xi)\, e^{-ikX(\tau-\xi)} = i d[e^{-ikX(\tau-\xi)}]/d\xi$, and may be integrated trivially. The integrand of the integrated term depends on τ only through $u^\mu(\tau)$, which reduces to $u_\parallel^\mu$ after integrating over gyrophase. With these changes (2.3.24) becomes

$$\Pi^{\mu\nu}(k) = q^2 \int \frac{d^4p}{(2\pi)^4}\, u_\parallel^\mu \left[\frac{u_\parallel^\nu}{u_\perp}\frac{\partial}{\partial p_\perp} - \frac{\partial}{\partial p_{\parallel\nu}}\right] F(p) - iq^2 \int \frac{d^4p(\tau)}{(2\pi)^4}\, u^\mu(\tau)$$
$$\times \int_0^\infty d\xi\, e^{ik[X(\tau)-X(\tau-\xi)]}\, u^\nu(\tau-\xi) \left[\frac{(ku)_\parallel}{u_\perp}\frac{\partial}{\partial p_\perp} + k^\alpha \frac{\partial}{\partial p_\parallel^\alpha}\right] F(p). \tag{2.3.27}$$

The term in (2.3.27) that does not involve the ξ-integral is symmetric in μ,ν, as may be seen by partially integrating the derivative with respect to $p_{\parallel\nu}$. Hence, in (2.3.27) one may replace $u_\parallel^\mu \partial/\partial p_{\parallel\nu}$ by $u_\parallel^\nu \partial/\partial p_{\parallel\mu}$, or by half the sum of the two to give an obviously symmetric form.

The form (2.3.24) includes both hermitian and antihermitian parts, with the latter given by

$$\Pi^{A\mu\nu}(k) = -\frac{iq^2}{2}\int \frac{d^4p(\tau)}{(2\pi)^4}\int_{-\infty}^\infty d\xi\, u^\mu(\tau)\, u^\nu(\tau-\xi)\, e^{ik[X(\tau)-X(\tau-\xi)]}$$
$$\times \left[\frac{(ku)_\parallel}{u_\perp}\frac{\partial}{\partial p_\perp} + k^\alpha \frac{\partial}{\partial p_\parallel^\alpha}\right] F(p). \tag{2.3.28}$$

2.3.5 Vlasov Form Summed over Gyroharmonics

Further reduction of (2.3.27) involves expanding in Bessel functions, using (2.1.27), and performing the ξ-integral. The expression for the linear response tensor obtained from (2.3.27) is

$$\Pi^{\mu\nu}(k) = q^2 \int \frac{d^4 p}{(2\pi)^4} \left[\left(\frac{u_\parallel^\mu u_\parallel^\nu}{u_\perp} \frac{\partial}{\partial p_\perp} - u_\parallel^\mu \frac{\partial}{\partial p_{\parallel\nu}} \right) \right.$$
$$\left. - \sum_{a=-\infty}^{\infty} \frac{U^\mu(a,k) U^{*\nu}(a,k)}{(ku)_\parallel - a\Omega_0} \left(\frac{(ku)_\parallel}{u_\perp} \frac{\partial}{\partial p_\perp} + k_\parallel^\alpha \frac{\partial}{\partial p_\parallel^\alpha} \right) \right] F(p), \tag{2.3.29}$$

with $U^\mu(a,k)$ given by (2.1.28) with (2.1.29). The antihermitian part follows either by applying the Landau prescription to (2.3.29) or by making the expansion in Bessel functions and performing the ξ-integral in (2.3.28):

$$\Pi^{A\mu\nu}(k) = i\pi q^2 \int \frac{d^4 p}{(2\pi)^4} \sum_{a=-\infty}^{\infty} U^\mu(a,k) U^{*\nu}(a,k)$$
$$\times\, \delta\big((ku)_\parallel - a\Omega_0\big) \left(\frac{(ku)_\parallel}{u_\perp} \frac{\partial}{\partial p_\perp} + k_\parallel^\alpha \frac{\partial}{\partial p_\parallel^\alpha} \right) F(p). \tag{2.3.30}$$

The equivalence of the Vlasov form (2.3.29) and the forward-scattering form (2.3.8) follows from a partial integration.

Response 3-Tensor

In non-covariant notation the response may be described by the dielectric tensor, $K^i{}_j(\omega, \boldsymbol{k})$. The relation between the notations implies

$$K^i{}_j(\omega, \boldsymbol{k}) = \delta^i{}_j + \frac{\Pi^i{}_j(k)}{\varepsilon_0 \omega^2}, \tag{2.3.31}$$

with the mixed components $\Pi^i{}_j(k)$, with $k^\mu = [\omega, \boldsymbol{k}]$, numerically equal to minus the contravariant components $\Pi^{ij}(k)$.

The 3-tensor components of (2.3.29) and (2.3.30) follow from the $\mu = i, \nu = j$ components of $U^\mu(a,k)U^{*\nu}(a,k)$. The explicit expression, (2.1.28) with (2.1.29) for $U^\mu(a,k)$ implies

$$
U^i U^{*j} = \begin{pmatrix}
\left(\dfrac{a\Omega_0}{k_\perp}\right)^2 J_a^2 & -i\epsilon\gamma v_\perp \dfrac{a\Omega_0}{k_\perp} J_a J_a' & \gamma v_z \dfrac{a\Omega_0}{k_\perp} J_a^2 \\
i\epsilon\gamma v_\perp \dfrac{a\Omega_0}{k_\perp} J_a J_a' & (\gamma v_\perp)^2 J_s'^2 & i\epsilon\gamma^2 v_\perp v_z J_a J_a' \\
\gamma v_z \dfrac{a\Omega_0}{k_\perp} J_a^2 & -i\epsilon\gamma^2 v_\perp v_z J_a^2 & (\gamma v_z)^2 J_a^2
\end{pmatrix}, \tag{2.3.32}
$$

where the arguments of $U^\mu(a,k)$, $J_a(k_\perp R)$ are omitted. On expressing $F(p)$ in 8-dimensional phase space in terms of the distribution function $f(\boldsymbol{p})$ in 6-dimensional phase space, using $d^4p\,F(p)/(2\pi)^4 = d^3\boldsymbol{p}\,f(\boldsymbol{p})/(2\pi)^3\gamma$, the derivative $\partial F(p)/\partial p^0$ in (2.3.27) is omitted such that the term proportional to $k_\parallel^\alpha \partial F(p)/\partial p_\parallel^\alpha$ is replaced by a term proportional to $\boldsymbol{k}\cdot\partial f(\boldsymbol{p})/\partial \boldsymbol{p}$ for example. The space-components of (2.3.29) become

$$
\Pi^{ij}(k) = q^2 \int \frac{d^3\boldsymbol{p}}{(2\pi)^3\gamma} \left[b^i b^j \gamma \frac{v_z}{v_\perp} \left(v_\perp \frac{\partial}{\partial p_z} - v_z \frac{\partial}{\partial p_\perp} \right) \right.
$$
$$
\left. + \sum_{a=-\infty}^{\infty} \frac{U^i(a,k) U^{*j}(a,k)}{\gamma(\omega - k_z v_z) - a\Omega_0} \left(\frac{\omega - k_z v_z}{v_\perp} \frac{\partial}{\partial p_\perp} + k_z \frac{\partial}{\partial p_z} \right) \right] f(\boldsymbol{p}). \tag{2.3.33}
$$

2.4 Response of a Relativistic Thermal Plasma

The linear response tensor for a relativistic, thermal, magnetized plasma was first calculated by Trubnikov [2]. A covariant generalization of Trubnikov's calculation is presented in this section, starting with the Vlasov form for the response tensor. The evaluation of the forward-scattering form leads to a superficially different result; the two forms are related by identities satisfied by the Trubnikov functions that appear.

2.4.1 *Trubnikov's Response Tensor for a Magnetized Plasma*

A relativistic thermal distribution is the Jüttner distribution [3,4]

$$
F(p) = \frac{(2\pi)^3 n\rho}{m^2 K_2(\rho)}\,\delta(p^2 - m^2)\,\exp[-\rho(p\tilde{u}/m)], \qquad f(\boldsymbol{p}) = \frac{2\pi^2 n\rho\, e^{-\rho\gamma}}{m^3 K_2(\rho)}, \tag{2.4.1}
$$

where $K_2(\rho)$ is a Macdonald function, which is a modified Bessel function, with argument $\rho = m/T$, which is the inverse temperature in units of the rest energy of the particle. (The factors 2π result from the normalization convention that every integral over a component of momentum has a factor $2\pi\hbar$ in the denominator, with $\hbar = 1$ here.) The normalization corresponds to

$$n_{\mathrm{pr}} = \int \frac{d^4p}{(2\pi)^4}\, F(p) = \int \frac{d^3\boldsymbol{p}}{(2\pi)^3}\, \frac{f(\boldsymbol{p})}{\gamma}, \qquad n = \int \frac{d^3\boldsymbol{p}}{(2\pi)^3}\, f(\boldsymbol{p}), \tag{2.4.2}$$

where n_{pr} is the proper number density, and n is the number density in the rest frame.

On inserting the Jüttner distribution (2.4.1) into (2.3.24), one needs to evaluate the derivative of $F(p) \propto \delta(p^2 - m^2)\exp(-\rho\, u\tilde{u})$. The relevant derivative gives

$$ku(\tau')\, G^{\alpha\nu}\big(k, u(\tau')\big)\, \dot{t}_{\alpha}{}^{\beta}(\tau')\, \frac{\partial F(p)}{\partial p^{\beta}} = -\frac{\rho}{m}\big[ku(\tau')\, \tilde{u}^{\nu} - k\tilde{u}\, u^{\nu}(\tau')\big] F(p), \tag{2.4.3}$$

with $\tau' = \tau - \xi$. The first term on the right hand side of (2.4.3) leads to a trivial ξ-integral, due to the integrand being a perfect differential:

$$\int_0^{\infty} d\xi\, ku(\tau - \xi)\, e^{ik[X(\tau) - X(\tau - \xi)]} = i. \tag{2.4.4}$$

In this way (2.3.27) reduces to

$$\Pi^{\mu\nu}(k) = -\frac{q^2\rho}{m} \int \frac{d^4p}{(2\pi)^4}\, F(p) \left[\tilde{u}^{\mu}\tilde{u}^{\nu} + i\, k\tilde{u} \int_0^{\infty} d\xi\, u^{\mu} \dot{t}^{\nu\sigma}(-\xi) u_{\sigma}\, e^{iR(\xi)u}\right], \tag{2.4.5}$$

where the 4-vector

$$R^{\mu}(\xi) = k_{\alpha}\big[t^{\alpha\mu}(-\xi) - t^{\alpha\mu}(0)\big] \tag{2.4.6}$$

is introduced. Using the definition (2.3.5) of $T^{\mu\nu}(\xi)$, an alternative form for (2.4.6) is

$$R^{\mu}(\xi) = k_{\alpha} T^{\mu\alpha}(\xi), \qquad \tilde{R}^{\nu}(\xi) = k_{\beta} T^{\beta\nu}(\xi), \tag{2.4.7}$$

where $\tilde{R}^{\nu}(\xi)$ is defined for a later purpose.

It is convenient to define the function

$$\begin{aligned} I(\rho, \xi, s + s') &= \int \frac{d^4p}{(2\pi)^4}\, F(p)\, e^{iR(\xi)u + (s+s')u} \\ &= \frac{n\rho}{2\pi m^2 K_2(\rho)} \int \frac{d^4p}{(2\pi)^4}\, \delta(p^2 - m^2)\, e^{-[\rho\tilde{u} - iR(\xi) - (s+s')]u}. \end{aligned} \tag{2.4.8}$$

The integral over d^4p reduces to a standard integral for the MacDonald function K_1, with a complex argument:

$$I(\rho, \xi, s+s') = \frac{n\rho}{K_2(\rho)} \frac{K_1\big(r(\xi)\big)}{r(\xi)}, \tag{2.4.9}$$

with the complex function $r(\xi)$ defined by

$$r(\xi) = \big[\big(\rho\tilde{u} - iR(\xi) - (s+s')\big)^2\big]^{1/2}. \tag{2.4.10}$$

In the rest frame of the plasma, and for $s^\mu = 0$, $s'^\mu = 0$, (2.4.10) gives

$$r(\xi) \to \left[(\rho - i\omega\xi)^2 + k_z^2\xi^2 + \frac{2k_\perp^2}{\Omega_0^2}(1 - \cos\Omega_0\xi)\right]^{1/2}.$$

The response tensor (2.4.5) may be re-expressed in terms of the function $I(\rho, \xi)$:

$$\Pi^{\mu\nu}(k) = -\frac{q^2\rho}{m}\left[n\tilde{u}^\mu\tilde{u}^\nu + i\,k\tilde{u}\int_0^\infty d\xi\, \dot{t}^\nu{}_\sigma(-\xi)\hat{u}^\mu\hat{u}^\sigma I(\rho,\xi)\right], \tag{2.4.11}$$

where $\hat{u}^\mu$ denotes differentiating with respect to s_μ and setting $s_\mu = 0$. Evaluating the derivatives of the Macdonald functions using the identity (A.1.13), one has

$$\begin{aligned}
\hat{u}^\mu \frac{K_1\big(r(\xi)\big)}{r(\xi)} &= a^\mu(\xi)\frac{K_2\big(r(\xi)\big)}{r^2(\xi)},\\
\hat{u}^\mu\hat{u}^\nu \frac{K_1\big(r(\xi)\big)}{r(\xi)} &= -g^{\mu\nu}\frac{K_2\big(r(\xi)\big)}{r^2(\xi)} + a^\mu(\xi)a^\nu(\xi)\frac{K_3\big(r(\xi)\big)}{r^3(\xi)},\\
a^\mu(\xi) = \rho\tilde{u}^\mu - iR^\mu(\xi), &\qquad \tilde{a}^\mu(\xi) = \rho\tilde{u}^\mu - i\tilde{R}^\mu(\xi).
\end{aligned} \tag{2.4.12}$$

The resulting expression for the response tensor is

$$\begin{aligned}
\Pi^{\mu\nu}(k) = -\frac{q^2 n\rho}{m}\Bigg[\tilde{u}^\mu\tilde{u}^\nu - i\frac{k\tilde{u}\,\rho}{K_2(\rho)}\int_0^\infty d\xi\\
\times\left\{t^{(1)\mu\nu}(\xi)\frac{K_2\big(r(\xi)\big)}{r^2(\xi)} - t^{(2)\mu\nu}(\xi)\frac{K_3\big(r(\xi)\big)}{r^3(\xi)}\right\}\Bigg],
\end{aligned} \tag{2.4.13}$$

with $\dot{t}^{\mu\nu}(\xi)$ given by (2.1.14) and with

$$t^{(1)\mu\nu}(\xi) = \dot{t}^{\nu\mu}(-\xi) = \dot{T}^{\mu\nu}(\xi), \qquad t^{(2)\mu\nu}(\xi) = a^\mu(\xi)\tilde{a}^\nu(\xi), \tag{2.4.14}$$

with $a^\mu(\xi)$, $\tilde{a}^\mu(\xi)$ defined by (2.4.12). The form (2.4.13) with (2.4.14) is a covariant version of the response tensor originally derived by Trubnikov [2].

The antihermitian part of (2.4.13) is

$$\Pi^{A\mu\nu}(k) = i\frac{q^2 n \rho^2 \omega}{2mK_2(\rho)} \int_{-\infty}^{\infty} d\xi \left[t^{(1)\mu\nu}(\xi) \frac{K_2\big(r(\xi)\big)}{r^2(\xi)} - R^\mu(\xi)\tilde{R}^\nu(\xi) \frac{K_3\big(r(\xi)\big)}{r^3(\xi)} \right]. \tag{2.4.15}$$

For some purposes it is convenient to introduce a matrix representation of the two tensorial quantities inside the integrand in (2.4.13). The following matrix representations apply in the rest frame of the plasma, $\tilde{u}^\mu = [1, \mathbf{0}]$, with the axes oriented to give $\boldsymbol{k} = (k_\perp, 0, k_z)$, $\boldsymbol{b} = (0, 0, 1)$. One has

$$\dot{T}^{\mu\nu}(\xi) = \begin{pmatrix} 1 & 0 & 0 & 0 \\ 0 & -\cos\Omega_0\xi & \epsilon\sin\Omega_0\xi & 0 \\ 0 & -\epsilon\sin\Omega_0\xi & -\cos\Omega_0\xi & 0 \\ 0 & 0 & 0 & -1 \end{pmatrix}, \tag{2.4.16}$$

$$T^{\mu\nu}(\xi) = \begin{pmatrix} \xi & 0 & 0 & 0 \\ 0 & \dfrac{\sin\Omega_0\xi}{\Omega_0} & \epsilon\dfrac{1-\cos\Omega_0\xi}{\Omega_0} & 0 \\ 0 & -\epsilon\dfrac{1-\cos\Omega_0\xi}{\Omega_0} & \dfrac{\sin\Omega_0\xi}{\Omega_0} & 0 \\ 0 & 0 & 0 & -\xi \end{pmatrix}. \tag{2.4.17}$$

Explicit forms for the 4-vectors (2.4.7) are

$$R^\mu(\xi) = \left(\omega\xi, k_\perp\frac{\sin\Omega_0\xi}{\Omega_0}, \epsilon k_\perp\frac{1-\cos\Omega_0\xi}{\Omega_0}, k_z\xi\right),$$
$$\tilde{R}^\nu(\xi) = \left(\omega\xi, k_\perp\frac{\sin\Omega_0\xi}{\Omega_0}, -\epsilon k_\perp\frac{1-\cos\Omega_0\xi}{\Omega_0}, k_z\xi\right). \tag{2.4.18}$$

Manifestly Gauge-Invariant Form

The response tensor in the form (2.4.13) must satisfy the charge-continuity and gauge-invariance conditions, $k_\mu\Pi^{\mu\nu}(k) = 0$, $k_\nu\Pi^{\mu\nu}(k) = 0$. On writing down the relevant conditions one finds that they are not trivially satisfied. These relations imply that certain identities must be satisfied. These identities are of the form

$$f(0)\frac{K_\nu(\rho)}{\rho^\nu} + \int_0^\infty d\xi \left\{ \frac{df(\xi)}{d\xi}\frac{K_\nu\big(r(\xi)\big)}{r^\nu(\xi)} + i f(\xi)\, ka(\xi)\frac{K_{\nu+1}\big(r(\xi)\big)}{r^{\nu+1}(\xi)} \right\} = 0, \tag{2.4.19}$$

with arbitrary $f(\xi)$ and ν. The identity is confirmed by a partial integration, noting that $r^2(\xi) = a^\mu(\xi)a_\mu(\xi)$ with (2.4.12) and (2.4.7) implies $dr(\xi)/d\xi =$

$-ika(\xi)/r(\xi)$. The specific identities required to confirm that (2.4.13) satisfies the charge-continuity and gauge-invariance conditions correspond to $\nu = 2$ and $f(\xi) = 1$, $f(\xi) = \xi$, $f(\xi) = \sin\Omega_0\xi$ and $f(\xi) = 1 - \cos\Omega_0\xi$ in (2.4.19).

To rewrite (2.4.13) in a form that manifestly satisfies the charge-continuity and gauge-invariance conditions, the integral is re-expressed in a form that involves only $K_3\big(r(\xi)\big)/r^3(\xi)$. This involves using $dr(\xi)/d\xi = -ika(\xi)/r(\xi)$ with $\nu = 2$ twice, with $f(\xi) = T^{\mu\nu}(\xi)$, $df(\xi)/d\xi = t^{(1)\mu\nu}(\xi)$, and then with $f(\xi) = \tilde{u}^\mu\tilde{u}^\nu$. The result is

$$\Pi^{\mu\nu}(k) = i\frac{q^2 n\rho^2\, k\tilde{u}}{mK_2(\rho)}\int_0^\infty d\xi\,\big[k_\alpha T^{\mu\alpha}(\xi)\,k_\beta T^{\beta\nu}(\xi)$$
$$- k_\alpha k_\beta T^{\alpha\beta}(\xi)\,T^{\mu\nu}(\xi) - i\rho\, k\tilde{u}\,\tilde{T}_{\mu\nu}(\xi)\big]\frac{K_3\big(r(\xi)\big)}{r^3(\xi)}, \tag{2.4.20}$$

which involves the tensor

$$\tilde{T}^{\mu\nu}(\xi) = T_{\alpha\beta}(\xi)\,G^{\alpha\mu}(k,\tilde{u})G^{\beta\nu}(k,\tilde{u}). \tag{2.4.21}$$

Contracting (2.4.20) with either k_μ or k_ν gives zero as required. In the rest frame, explicit expressions for the tensor quantities in (2.4.21) follow from (2.4.18).

2.4.2 Forward-Scattering Form of Trubnikov's Tensor

Trubnikov's method may also be applied to the forward-scattering form of the response tensor (2.3.4). The steps involved are as follows: insert the form (2.4.1) for the Jüttner distribution into (2.3.4), write $u^\mu(\tau - \xi) = \dot{t}^{\mu\sigma}(-\xi)u_\sigma(\tau)$, include the dependences on $u(\tau)$ in exponential form using $u^\mu = \partial e^{su}/\partial s_\mu$ with $s_\mu \to 0$, evaluate the integral over $d^4p(\tau)$ using (2.4.8), and carry out the differentiations using (2.4.12). After writing $ku\,G^{\alpha\mu}(k,u) = (k^\sigma g^{\alpha\mu} - k^\alpha g^{\sigma\mu})u_\sigma$, $ku\,G^{\beta\nu}(k,u) = (k^\tau g^{\beta\nu} - k^\beta g^{\tau\nu})u_\tau$, this procedure gives

$$\Pi^{\mu\nu}(k) = \frac{q^2 n\rho}{K_2(\rho)m}\int_0^\infty d\xi\, T_{\alpha\beta}(\xi)\,(k^\sigma g^{\alpha\mu} - k^\alpha g^{\sigma\mu})(k^\tau g^{\beta\nu} - k^\beta g^{\tau\nu})$$
$$\times\, \dot{t}_\tau{}^\eta(-\xi)\left[-g_{\sigma\eta}\frac{K_2\big(r(\xi)\big)}{r^2(\xi)} + a_\sigma(\xi)a_\eta(\xi)\frac{K_3\big(r(\xi)\big)}{r^3(\xi)}\right]. \tag{2.4.22}$$

The coefficients of the $K_2\big(r(\xi)\big)/r^2(\xi)$ and $K_3\big(r(\xi)\big)/r^3(\xi)$ terms are

$$\begin{aligned}
&T_{\alpha\beta}(\xi)\,(k^\sigma g^{\alpha\mu} - k^\alpha g^{\sigma\mu})(k^\tau g^{\beta\nu} - k^\beta g^{\tau\nu})\,\dot{t}_\tau{}^\eta(-\xi)g_{\sigma\eta}\\
&\quad= (d/d\xi)[T^{\mu\nu}(\xi)k_\alpha k_\beta T^{\alpha\beta}(\xi) - k_\beta T^{\mu\beta}(\xi)k_\alpha T^{\alpha\nu}(\xi)],\\
&(k^\sigma g^{\alpha\mu} - k^\alpha g^{\sigma\mu})(k^\tau g^{\beta\nu} - k^\beta g^{\tau\nu})\,\dot{t}_\tau{}^\eta(-\xi)a_\sigma(\xi)a_\eta(\xi) = \rho^2(k\tilde{u})^2\tilde{T}^{\mu\nu}(\xi)\\
&\quad-[2i\rho k\tilde{u} + k_\sigma k_\tau T^{\sigma\tau}(\xi)][T^{\mu\nu}(\xi)k_\alpha k_\beta T^{\alpha\beta}(\xi) - R^\mu(\xi)\tilde{R}^\nu(\xi)],
\end{aligned} \tag{2.4.23}$$

respectively, where (2.4.18) and (2.4.21) are used. The fact that the coefficient of $K_2\big(r(\xi)\big)/r^2(\xi)$ in (2.4.22) is a perfect derivative allows one to use the identity (2.4.19) to rewrite (2.4.22) in the form (2.4.20), in which only $K_\nu\big(r(\xi)\big)/r^\nu(\xi)$ with $\nu = 3$ appears. One finds that (2.4.22) reproduces (2.4.20), which is equivalent to (2.4.13). This establishes that (2.3.4) and (2.3.27) are equivalent starting points for the calculation based on Trubnikov's method.

2.4.3 Other Forms of $\Pi^{\mu\nu}(k)$ for a Jüttner Distribution

There are further forms for $\Pi^{\mu\nu}(k)$ that are suggested by methods used to evaluate $\Pi^{\mu\nu}(k)$ in the quantum case. An important practical difference between the non-quantum and quantum cases is that the continuous variable $p_\perp$ is replaced by a discrete variable, $p_n = (2neB)^{1/2}$, with $n = 0, 1, \ldots$ the Landau quantum number. In some ways this makes the quantum case simpler than its non-quantum counterpart: the only continuous variable over which one must integrate is p_z. It is then convenient to write the resonant denominator in terms of solutions for p_z. In the non-quantum limit these solutions correspond to $p_z = p_{z\pm}$, given by (2.1.34), and it is of interest to consider forms for $\Pi^{\mu\nu}(k)$ that are written in this manner.

The resonant denominator, $(ku)_\parallel - a\Omega_0 = (\varepsilon\omega - a|q|B - p_z k_z)/m$, is rationalized by multiplying by $\varepsilon\omega + a|q|B + p_z k_z$. This leads to the resonant denominator being replaced by the resonance condition in the form (2.1.33), that is, by a quadratic function p_z. The integrand in (2.3.29) can then be rewritten using

$$\frac{U^\mu(a,k)U^{*\nu}(a,k)}{(ku)_\parallel - a\Omega_0} = \sum_\pm \pm m \frac{U^\mu(a,k)U^{*\nu}(a,k)}{2\omega g_a(\omega^2 - k_z^2)} \frac{\varepsilon\omega + a|q|B + p_z k_z}{p_z - p_{z\pm}}, \tag{2.4.24}$$

with g_a given by (2.1.34). It is convenient to separate $U^\mu(a,k)$ into $U_\parallel^\mu(a,k) + U_\perp^\mu(a,k)$ using (2.3.11), with p_z appearing only in $U_\parallel^\mu(a,k) = u_\parallel^\mu J_a(k_\perp R)$, with $u_\parallel^\mu = p_\parallel^\mu/m = (\varepsilon, 0, 0, p_z)/m$. After some manipulations, the sum of (2.4.24) over a reduces to

$$\sum_{a=-\infty}^{\infty} \frac{U^\mu(a,k)U^{*\nu}(a,k)}{(ku)_\parallel - a\Omega_0} = \frac{1}{(k^2)_\parallel}\left[u_\parallel^\mu k_\parallel^\nu + u_\parallel^\nu k_\parallel^\mu - (ku)_\parallel g_\parallel^{\mu\nu}\right]$$
$$+ m \sum_\pm \pm \sum_{a=-\infty}^{\infty} \frac{U_\pm^\mu(a,k)U_\pm^{*\nu}(a,k)}{2g_a(k^2)_\parallel} \frac{\varepsilon + \varepsilon_\pm}{p_z - p_{z\pm}}, \tag{2.4.25}$$

with $(k^2)_\parallel = \omega^2 - k_z^2$, with $\varepsilon_\pm$ given by (2.1.35), and with

$$U_{\parallel\pm}^\mu(a,k) = u_{\parallel\pm}^\mu J_a(k_\perp R), \qquad u_{\parallel\pm}^\mu = (\varepsilon_\pm, 0, 0, p_{z\pm})/m, \tag{2.4.26}$$

and $U^{\mu}_{\perp\pm}(a,k) = U^{\mu}_{\perp}(a,k)$ independent of p_z. The resulting expression is

$$\Pi^{\mu\nu}(k) = \frac{q^2 n\rho\, k\tilde{u}}{m(k^2)_\parallel}\,[u^\mu_\parallel k^\nu_\parallel + u^\nu_\parallel k^\mu_\parallel - (ku)_\parallel g^{\mu\nu}_\parallel] + \sum_\pm \pm \sum_{a=-\infty}^{\infty} \frac{q^2 n\rho^2 k\tilde{u}}{2m^3 K_2(\rho)}$$
$$\times \int_0^\infty dp_\perp p_\perp \int_{-\infty}^{\infty} dp_z\, \frac{U^{\mu}_{\pm}(a,k) U^{*\nu}_{\pm}(a,k)}{2g_a(k^2)_\parallel}\, \frac{\varepsilon + \varepsilon_\pm}{p_z - p_{z\pm}}\, e^{-\rho u\tilde{u}}. \tag{2.4.27}$$

2.4.4 Relativistic Plasma Dispersion Functions (RPDFs)

The p_z-integral in (2.4.27) can be expressed in terms of

$$\int dp_z\, \frac{e^{-\rho\gamma}}{p_z - p_\pm}, \qquad \int \frac{dp_z}{\gamma}\, \frac{e^{-\rho\gamma}}{p_z - p_\pm}. \tag{2.4.28}$$

These integrals may be evaluated in terms of the relativistic plasma dispersion function (RPDF) used to evaluate the response tensor for the Jüttner distribution in the unmagnetized case. The definition of this relativistic dispersion function is [5]

$$T(v_0, \rho) = \int_{-1}^{1} dv\, \frac{e^{-\rho\gamma}}{v - v_0}, \tag{2.4.29}$$

with $v_0 = \omega/|\boldsymbol{k}|$ in the unmagnetized case, such that the Cerenkov resonance is at $v = v_0$. In the magnetized case, the resonances occur at $v = v_{z\pm} = p_{z\pm}/\varepsilon_\pm$, determined by (2.1.34) and (2.1.35). An intermediate step in the evaluation is to write p_z in terms of the parameter t introduced in (2.1.36): this involves writing $p_z/\varepsilon_\perp = 2t/(1-t^2)$, so that one has $\gamma = \varepsilon/m = (\varepsilon_\perp/m)(1+t^2)/(1-t^2)$. It is convenient to define a plasma dispersion function

$$J(t_0, \rho_\perp) = \int_{-1}^{1} dt\, \frac{e^{-\rho_\perp(1+t^2)/(1-t^2)}}{t - t_0}, \tag{2.4.30}$$

with $t_0 = t_\pm$ and $\rho_\perp = \rho\varepsilon_\perp/m$. The dispersion integral (2.4.30) can be expressed in terms of (2.4.29):

$$J(t_0, \rho_\perp) = \frac{1}{2}\left[-\frac{(1-v_0^2)^{1/2}}{v_0}\left(2K_1(\rho_\perp) + \frac{(1-v_0^2)}{\rho_\perp} T'(v_0, \rho_\perp)\right) + T(v_0, \rho_\perp)\right], \tag{2.4.31}$$

with $v_0 = 2t_0/(1+t_0^2)$.

The form of the response tensor that results from the use of this procedure is a nonquantum counterpart of a relativistic quantum form, but in this case that nonquantum case is intrinsically more complicated than its quantum counterpart. The reason is that the resulting expression for the response tensor includes an integral over $p_\perp$ with the argument $\rho_\perp$ of $T(v_0, \rho_\perp)$ a function of $p_\perp$. In the quantum case the integral over $p_\perp$ is replaced by a sum over Landau levels, and this complication does not arise.

2.4.5 Strictly-Perpendicular Jüttner Distribution

Trubnikov's method may be used to evaluate the response tensor for strictly-parallel and strictly-perpendicular Jüttner distributions. A strictly-parallel distribution is relevant to pulsars and is discussed in Sect. 2.6.3. Trubnikov's method is used here to evaluate the response tensor for the strictly-perpendicular relativistic thermal distribution

$$F(p) = \frac{n\rho^{1/2}}{(2\pi)^{1/2} mc K_{3/2}(\rho)}\,\delta(p^2 - m^2)\,\delta(p_z)\,\exp[-\rho(p\bar{u})/m]. \tag{2.4.32}$$

When (2.4.32) is inserted into the Vlasov form (2.3.24) for the response tensor, some terms involve the derivative of $\delta(p_z)$. Such terms are to be evaluated only after a partial integration is performed. An alternative procedure, adopted here, is to start from the forward-scattering form (2.3.4) rather than (2.3.24).

On inserting (2.4.32) into (2.3.4), the calculation is closely analogous to that leading to (2.4.22) for the isotropic thermal distribution. There are three notable changes: (1) the orders of all the functions $K_\nu(z)/z^\nu$ are reduced by 1/2; (2) the argument, $r(\xi)$, of these functions is replaced by $r_{\rm P}(\xi)$, which is equal to $r(\xi)$ evaluated at $k_z = 0$; and (3) in the counterpart of (2.4.22) the expression inside the square brackets is replaced by the corresponding expression with the components along the magnetic field set to zero.

In place of (2.4.9) one has

$$I(\rho, \xi) = \frac{n\rho^{1/2}}{K_{3/2}(\rho)}\,\frac{K_{1/2}\big(r_{\rm P}(\xi)\big)}{r_{\rm P}^{1/2}(\xi)}. \tag{2.4.33}$$

In place of (2.4.10) and (2.4.12) one has

$$r_{\rm P}(\xi) = \left[a_{\rm P}^{\mu}(\xi) a_{{\rm P}\mu}(\xi)\right]^{1/2} = \left[(\rho - i\omega\xi)^2 + \frac{2k_\perp^2}{\Omega_0^2}(1 - \cos\Omega_0\xi)\right]^{1/2},$$

$$\hat{u}^\mu\,\frac{K_{1/2}\big(r_{\rm P}(\xi)\big)}{r_{\rm P}^{1/2}(\xi)} = a_{\rm P}^{\mu}(\xi)\,\frac{K_{3/2}\big(r_{\rm P}(\xi)\big)}{r_{\rm P}^{3/2}(\xi)},$$

$$\hat{u}^\mu\hat{u}^\nu\,\frac{K_{1/2}\big(r_{\rm P}(\xi)\big)}{r_{\rm P}^{1/2}(\xi)} = -\tilde{g}^{\mu\nu}\,\frac{K_{3/2}\big(r_{\rm P}(\xi)\big)}{r_{\rm P}^{3/2}(\xi)} + a_{\rm P}^{\mu}(\xi)a_{\rm P}^{\nu}(\xi)\,\frac{K_{5/2}\big(r_{\rm P}(\xi)\big)}{r_{\rm P}^{5/2}(\xi)}, \tag{2.4.34}$$

respectively, and with $a_{\mathrm{P}}^{\mu}(\xi) = a^{\mu}(\xi) - b^{\mu} b^{\mu} a_{\mu}(\xi)$. The subscript P denotes a projection onto the 3D subspace orthogonal to the magnetic field, in particular with $g_{\mathrm{P}}^{\mu\nu} = g^{\mu\nu} + b^{\mu}b^{\nu}$.

The resulting expression for the response tensor that replaces (2.4.22) for the strictly-perpendicular distribution (2.4.32) is

$$
\begin{aligned}
\Pi^{\mu\nu}(k) = \frac{q^2 n\rho^{1/2}}{mK_{3/2}(\rho)} \int_0^{\infty} d\xi \bigg\{ & -\big[T^{\mu\nu}(\xi)k_{\sigma}k_{\tau}\dot{T}_{\mathrm{P}}^{\sigma\tau}(\xi) - k_{\beta}T^{\mu\beta}(\xi)k_{\sigma}\dot{T}_{\mathrm{P}}^{\sigma\nu}(\xi) \\
& - k_{\alpha}T^{\alpha\nu}(\xi)k_{\tau}\dot{T}_{\mathrm{P}}^{\mu\tau}(\xi) + k_{\alpha}k_{\beta}T^{\alpha\beta}(\xi)\dot{T}_{\mathrm{P}}^{\mu\nu}\big]\frac{K_{3/2}\big(r_{\perp}(\xi)\big)}{r_{\perp}^{3/2}(\xi)} \\
& + \big[\big(ka_{\mathrm{P}}(\xi)\big)^2 T^{\mu\nu}(\xi) - ka_{\mathrm{P}}(\xi)a_{\mathrm{P}}^{\mu}(\xi)k_{\alpha}T^{\alpha\nu}(\xi) \\
& - ka_{\mathrm{P}}(\xi)k_{\beta}T^{\mu\beta}(\xi)\tilde{a}_{\mathrm{P}}^{\nu}(\xi) + k_{\alpha}k_{\beta}T^{\alpha\beta}(\xi)a_{\mathrm{P}}^{\mu}(\xi)\tilde{a}_{\mathrm{P}}^{\nu}(\xi)\big]\frac{K_{5/2}\big(r_{\perp}(\xi)\big)}{r_{\perp}^{5/2}(\xi)}\bigg\},
\end{aligned}
\tag{2.4.35}
$$

with $a_{\mathrm{P}}^{\mu}(\xi) = \rho\tilde{u}_{\mathrm{P}}^{\mu} - ik_{\beta}T_{\mathrm{P}}^{\mu\beta}(\xi)$, $\tilde{a}_{\mathrm{P}}^{\nu}(\xi) = \rho\tilde{u}_{\mathrm{P}}^{\nu} - ik_{\alpha}k_{\alpha}T_{\mathrm{P}}^{\alpha\nu}(\xi)$, and $T_{\mathrm{P}}^{\mu\nu}(\xi) = T^{\mu\nu}(\xi) - \xi b^{\mu}b^{\nu}$. Various alternative forms are obtained by using the identity (2.4.19), with $ka(\xi)$ reinterpreted as $ka_{\mathrm{P}}(\xi)$.

In particular, the $\mu = \nu = 3$ term in (2.4.35) simplifies considerably, reducing to

$$
\begin{aligned}
\Pi^{33}(k) = \frac{q^2 n\rho^{1/2}}{mK_{3/2}(\rho)} \int_0^{\infty} d\xi\, \xi \bigg\{ & k_{\sigma}k_{\tau}\dot{T}_{\mathrm{P}}^{\sigma\tau}(\xi)\frac{K_{3/2}\big(r_{\perp}(\xi)\big)}{r_{\perp}^{3/2}(\xi)} \\
& -\big(ka_{\mathrm{P}}(\xi)\big)^2 \frac{K_{5/2}\big(r_{\perp}(\xi)\big)}{r_{\perp}^{5/2}(\xi)}\bigg\} = \frac{q^2 n_{\mathrm{pr}}}{m},
\end{aligned}
\tag{2.4.36}
$$

where (2.4.19) is used. One has $n_{\mathrm{pr}} = nK_{1/2}(\rho)/K_{3/2}(\rho)$, for the distribution (2.4.32).

The form (2.4.35) is a covariant generalization of the result derived by Trubnikov and Yakubov [6]. The Macdonald functions of half-integer order in (2.4.35) can be expressed in terms of a rational function times an exponential function,

$$
\begin{aligned}
K_{1/2}(z) &= \left(\frac{\pi}{2}\right)^{1/2} \frac{e^{-z}}{z^{1/2}}, \qquad K_{3/2}(z) = \left(\frac{\pi}{2}\right)^{1/2} \frac{e^{-z}}{z^{3/2}}(1+z), \\
K_{5/2}(z) &= \left(\frac{\pi}{2}\right)^{1/2} \frac{e^{-z}}{z^{5/2}}(3+3z+z^2).
\end{aligned}
\tag{2.4.37}
$$

2.5 Weakly Relativistic Thermal Plasma

Trubnikov's response tensor contains all relativistic effects for a magnetized thermal plasma, but it is too cumbersome for most practical applications, and approximations to it need to be made. An important application of the general result is in deriving relativistic correction to the response tensor for a nonrelativistic magnetized thermal plasma, referred to here as the weakly relativistic approximation. This involves assuming $\rho \gg 1$ ($T \ll 5 \times 10^9\,$K for electrons or positrons) and expanding in powers of $1/\rho$. When this expansion is made, weakly relativistic effects are described in terms of appropriate RPDFs, referred to as Shkarofsky functions, or Dnestrovskii functions in the case of perpendicular propagation. In this section the response tensor is first evaluated directly for a nonrelativistic thermal distribution, and this nonrelativistic approximation is rederived starting from Trubnikov's response tensor. The corrections and other changes associated with weakly relativistic effects are then discussed.

2.5.1 Nonrelativistic Plasma Dispersion Function

In the nonrelativistic limit, the Jüttner distribution reduces to a Maxwellian distribution. Dispersion in a magnetized thermal plasma involves the (nonrelativistic) plasma dispersion function familiar in the context of an unmagnetized thermal plasma, denoted here by $Z(y)$.

The nonrelativistic approximation corresponds formally to $c \to \infty$, and it is helpful to use ordinary units, rather than natural units, by including c explicitly in discussing this limit. The nonrelativistic limit corresponds to $\gamma \to 1$, $p_\perp \to m v_\perp$, $p_z \to m v_z$. In the Jüttner distribution (2.4.1) one has $\rho = mc^2/T \to \infty$, when the asymptotic approximation to the Macdonald functions apply. For $z \gg 1$ one has

$$K_\nu(z) \approx (\pi/2z)^{1/2} e^{-z}. \tag{2.5.1}$$

An expansion in $1/c^2$ gives $\gamma = 1 + (v_\perp^2 + v_z^2)/2c^2 + \cdots$, so that the Jüttner distribution (2.4.1) gives

$$f(\boldsymbol{p}) = \frac{2\pi^2\hbar^3 n\rho\, e^{-\rho\gamma}}{(mc)^3 K_2(\rho)} \approx \left(\frac{\sqrt{2\pi}\,\hbar}{mV}\right)^3 n e^{-(v_\perp^2+v_z^2)/2V^2}, \tag{2.5.2}$$

which is a Maxwellian distribution with temperature $T = mV^2$. Note that the normalization is the conventional one used in quantum statistical mechanics,

$$n = \int \frac{d^3\boldsymbol{p}}{(2\pi\hbar)^3} f(\boldsymbol{p}) = 2\pi \int_0^\infty dp_\perp\, p_\perp \int_{-\infty}^{\infty} dp_z \frac{f(\boldsymbol{p})}{(2\pi\hbar)^3}, \tag{2.5.3}$$

rather than the one used conventionally in classical kinetic theory, where the factor $(2\pi\hbar)^3$ is omitted in (2.5.3).

Modified Bessel Functions, $I_n(z)$

A major simplification occurs in the evaluation of the response tensor in the nonrelativistic limit for a Maxwellian distribution: the integrals over $p_\perp$ and p_z factorize into two independent integrals. The resonant denominator in the response tensor (2.3.29) reduces to $\omega - a\Omega_0 - k_z v_z$, which does not involve $v_\perp$. The integral over $v_\perp$ involves only the Bessel functions, whose argument is proportional to $v_\perp$. The $v_\perp$-integral can be evaluated in terms of modified Bessel function, $I_a(\lambda)$, using

$$\int_0^\infty \frac{dv_\perp v_\perp}{V^2} \begin{pmatrix} J_a^2(z) \\ v_\perp J_a'(z) J_a(z) \\ v_\perp^2 J_s'^2(z) \end{pmatrix} e^{-v_\perp^2/2V^2}$$
$$= e^{-\lambda} \begin{pmatrix} I_a(\lambda) \\ (\Omega_0/k_\perp)[I_a(\lambda) - I_a'(\lambda)] \\ (\Omega_0/k_\perp)^2\{a^2 I_a(\lambda) - 2\lambda^2[I_a(\lambda) - I_a'(\lambda)]\} \end{pmatrix}, \tag{2.5.4}$$

with $z = k_\perp v_\perp/\Omega_0$, $\lambda = k_\perp^2 V^2/\Omega_0^2$, and where the latter two identities follow from the first using the properties (A.1.3)–(A.1.4) of J_a and (A.1.7)–(A.1.9) of I_a.

Plasma Dispersion Function $Z(y)$

The v_z-integral involves the factor $\exp(-v_z^2/2V^2)$ in the Maxwellian distribution and the resonant denominator which depends on v_z. This integral can be evaluated in terms of the nonrelativistic plasma dispersion function. A conventional definition of this function is the function $Z(y)$, which is often written in terms of a related function $\phi(y) = -yZ(y)$. The definition is [7]

$$Z(y) = -\frac{\phi(y)}{y} = \pi^{-1/2} \int_{-\infty}^{\infty} dt\, \frac{e^{-t^2}}{t-y}, \tag{2.5.5}$$

which is a complex error function. $Z(y)$ satisfies the differential equation

$$\frac{dZ(y)}{dy} = -2[1 + yZ(y)]. \tag{2.5.6}$$

For real y, $Z(y)$ has real and imaginary parts

$$\mathrm{Re}\, Z(y) = -2\, e^{-y^2} \int_0^y dt\, e^{t^2}, \qquad \mathrm{Im}\, Z(y) = i\sqrt{\pi} e^{-y^2}, \tag{2.5.7}$$

respectively. Expansions for large and small y give

$$\mathrm{Re}\, Z(y) = -\frac{1}{y}\begin{cases} 1 + \dfrac{1}{2y^2} + \dfrac{3}{4y^4} + \cdots & \text{for } y \gg 1, \\ 2y^2 + \dfrac{4}{3}y^4 + \cdots & \text{for } y \ll 1. \end{cases} \tag{2.5.8}$$

The specific integral that appears in the response function corresponds to $t \to v_z/\sqrt{2}V$, $y \to y_a = (\omega - a\Omega_0)/\sqrt{2}k_zV$. The argument y_a may be interpreted in terms of the line profile for gyromagnetic absorption at the ath harmonic. Absorption is described by the imaginary part of $Z(y_a)$, and according to (2.5.7) this corresponds to a gaussian profile, centered on $\omega = a\Omega_0$, with a line width $\sqrt{2}\,|k_z|V$ determined by the Doppler effect due to the thermal spread in v_z. The real part of $Z(y_a)$ describes the dispersion corresponding to this absorption. The real and imaginary parts of any causal function, $f(\omega)$, satisfy the Kramers-Kronig relation

$$\mathrm{Re}\, f(\omega) = -\frac{i}{\pi}\wp\int_{-\infty}^{\infty}\frac{d\omega'}{\omega' - \omega}\,\mathrm{Im}\, f(\omega'), \tag{2.5.9}$$

where $\wp$ denotes the Cauchy principal value, with $f(\omega) = Z(y_a)$ here.

2.5.2 *Response Tensor for a Maxwellian Distribution*

The resulting expression for the dielectric tensor (2.3.31), with the contribution of only one (unlabeled) species retained explicitly, corresponds to the 3-tensor components of $\Pi^{\mu\nu}(k)$ in the rest frame of the plasma. These components are

$$\Pi^{ij}(\omega, \boldsymbol{k}) = \frac{q^2 n\omega}{m}\left[\frac{\omega}{k_z^2V^2}b^ib^j + \sum_a \frac{y_a Z(y_a)}{\omega - a\Omega_0}N^{ij}(a, \omega, \boldsymbol{k})\right], \tag{2.5.10}$$

with $y_a = (\omega - a\Omega_0)/\sqrt{2}k_zV$, and with

$$N^{ij}(a, \omega, \boldsymbol{k}) = -e^{-\lambda}\begin{pmatrix} \dfrac{a^2}{\lambda}I_a & -i\epsilon a(I_a' - I_a) & \dfrac{k_\perp}{k_z}\Delta_a\dfrac{a}{\lambda}I_a \\ i\epsilon a(I_a' - I_a) & \dfrac{a^2}{\lambda}I_a - 2\lambda(I_a' - I_a) & i\epsilon\dfrac{k_\perp}{k_z}\Delta_a(I_a' - I_a) \\ \dfrac{k_\perp}{k_z}\Delta_a\dfrac{a}{\lambda}I_a & -i\epsilon\dfrac{k_\perp}{k_z}\Delta_a(I_a' - I_a) & \dfrac{k_\perp^2}{k_z^2}\dfrac{\Delta_a^2}{\lambda}I_a \end{pmatrix}, \tag{2.5.11}$$

with $\Delta_a = (\omega - a\Omega_0)/\Omega_0$, where $-\epsilon$ is the sign of the charge, and where the argument λ of the modified Bessel functions is omitted.

The antihermitian part of the response tensor (2.5.10) follows from the imaginary part of the dispersion function (2.5.7). This gives

$$\Pi^{Aij}(\omega, \mathbf{k}) = i\left(\frac{\pi}{2}\right)^{1/2} \frac{q^2 n\omega}{m|k_z|V} \sum_a e^{-y_a^2} N^{ij}(a, \omega, \mathbf{k}). \tag{2.5.12}$$

The particular case of Landau damping corresponds to $a = 0$. In this case, (2.5.11) simplifies to

$$N^{ij}(0, \omega, \mathbf{k}) = -e^{-\lambda} \begin{pmatrix} 0 & 0 & 0 \\ 0 & -2\lambda(I_0' - I_0) & i\epsilon \dfrac{k_\perp}{k_z} \dfrac{\omega}{\Omega_0}(I_0' - I_0) \\ 0 & -i\epsilon \dfrac{k_\perp}{k_z} \dfrac{\omega}{\Omega_0}(I_0' - I_0) & \dfrac{\omega^2}{k_z^2 V^2} I_0 \end{pmatrix}. \tag{2.5.13}$$

Further simplification occurs in the small gyroradius limit, $\lambda \to 0$, $I_0 \to 1$.

Cold Plasma Limit

The equivalent dielectric tensor may be defined by

$$K^i{}_j(k) = \delta^i_j + \frac{1}{\varepsilon_0 \omega^2} \Pi^i{}_j(k), \tag{2.5.14}$$

with $\mu_0 = 1/\varepsilon_0 c^2$ in SI units. The dielectric tensor implied by (2.5.10) with (2.5.11) in (2.5.14) includes the cold-plasma response tensor in the limit $V \to 0$. The cold plasma response is described by (1.2.27) with (1.2.28). The cold plasma limit of (2.5.10) with (2.5.11) follows from $V \to 0$, which implies $\lambda \to 0$, $y_a \to \infty$, with $y_a Z(y_a) \to -1$ for $y_a \to \infty$. The power series expansion of the modified Bessel functions,

$$I_a(\lambda) = \sum_{k=0}^{\infty} \frac{1}{k!(a+k)!} \left(\frac{\lambda}{2}\right)^{a+2k}, \tag{2.5.15}$$

implies that in the limit $\lambda \to 0$, only $a = 0, \pm 1$. The leading terms suffice to reproduce the cold plasma limit for the component with $i, j = 1, 2$. For the 13- and 31-components, in this limit the contributions from $a = \pm 1$ sum to zero for each species. For the 23- and 32-components, one obtains zero in the cold plasma limit only after summing over species and assuming that the plasma is charge-neutral. Separating the sum over species, α, into sums over the positively charges, $\alpha+$, and negatively charged, $\alpha-$, charge neutrality requires

$$\sum_{\alpha+} q_{\alpha+} n_{\alpha+} = \sum_{\alpha-} |q_{\alpha-}| n_{\alpha-}, \qquad \sum_{\alpha+} \frac{\omega_{p\alpha+}^2}{\Omega_{\alpha+}} = \sum_{\alpha-} \frac{\omega_{p\alpha-}^2}{\Omega_{\alpha-}}, \tag{2.5.16}$$

where $\omega_{p\alpha\pm}$ and $\Omega_{\alpha\pm}$ are the plasma and cyclotron frequencies, respectively. For the 33-component, for $a = 0$ one needs to retain the next order term in the expansion (2.5.8) of $Z(y_0)$ for $y_0 \gg 1$ to reproduce the cold plasma result.

Thermal corrections to the cold plasma limit follow from the result (2.5.11) by expanding in $\lambda \propto V^2$ and in $1/y_a \propto V$. When only the terms of zeroth and first order in V^2 are retained, there are contributions from $a = 0, \pm 1, \pm 2$. The general form of the response tensor to this order is rather cumbersome, and it is usually appropriate to make other simplifying assumptions before making this expansion.

Thermal Corrections to Longitudinal Response Tensor

A particular simplifying assumption is to assume that only the longitudinal part of the response need be retained. The contribution of a single species to the longitudinal part of the dielectric tensor (2.3.31), $K^L(\omega, \boldsymbol{k}) = 1 + \Pi^L(\omega, \boldsymbol{k})/\varepsilon_0\omega^2$, follows from the longitudinal part of (2.5.10) with (2.5.11), which gives

$$\Pi^L(\omega, \boldsymbol{k}) = \frac{q^2 n}{m} \frac{\omega^2}{|\boldsymbol{k}|^2 V^2} \left[1 + \sum_a y_a Z(y_a) \frac{\omega}{\omega - a\Omega_0} e^{-\lambda} I_a(\lambda) \right]. \tag{2.5.17}$$

Expanding in powers of V^2 gives

$$\begin{aligned}\Pi^L(\omega, \boldsymbol{k}) = -\frac{q^2 n}{m} \Bigg\{ &\frac{\omega^2 - \Omega_0^2 \cos^2\theta}{\omega^2 - \Omega_0^2} + \frac{|\boldsymbol{k}|^2 V^2}{\omega^2} \Bigg[3\cos^4\theta - \frac{\omega^2 \sin^2\theta \cos^2\theta}{\Omega_0^2} \\ &- \frac{\omega^4 \sin^4\theta}{\Omega_0^2(\omega^2 - \Omega_0^2)} + \frac{\omega^6(\omega^2 + 3\Omega_0^2)\sin^2\theta\cos^2\theta}{\Omega_0^2(\omega^2 - \Omega_0^2)^3} + \frac{\omega^4 \sin^4\theta}{\Omega_0^2(\omega^2 - 4\Omega_0^2)} \Bigg] \Bigg\}.\end{aligned} \tag{2.5.18}$$

The dispersion relation for longitudinal waves in a magnetized thermal electron gas follows from $\omega^2 + \mu_0 \Pi^L(\omega, \boldsymbol{k}) = 0$ with (2.5.18) evaluated for electrons.

MHD-Like Limit

A different simplifying assumption applies to MHD-like waves in an electron-ion plasma. Adding a label α for species $\alpha = e, i$, with $y_a \to y_{\alpha a}$, this limit corresponds to $y_{e0} \ll 1$, with $y_{ea} \gg 1$ for $a \neq 0$ and $y_{ia} \gg 1$ for all a. For the 23- and 32-components, the cancelation between electron and ion contributions, due to the charge-neutrality condition (2.5.16), no longer occurs, so that these components are nonzero. The resulting approximate expression for the dielectric tensor is

$$K^1{}_1 = 1 + \sum_i \frac{\omega_{pi}^2}{\Omega_i^2} \left[1 + \frac{\omega^2}{\Omega_i^2 - \omega^2} \right], \qquad K^1{}_2 = i \sum_i \frac{\omega_{pi}^2}{\Omega_i^2} \frac{\omega}{\Omega_i}, \qquad K^1{}_3 = 0,$$

$$K^2{}_3 = -i \sum_i \frac{\omega_{pi}^2}{\Omega_i^2} \frac{\Omega_i}{\omega} \frac{k_\perp}{k_z}, \qquad K^3{}_3 = 1 - \sum_i \frac{\omega_{pi}^2}{\omega^2} + \frac{\omega_p^2}{k_z^2 V_e^2}, \tag{2.5.19}$$

with $K^1{}_1 = K^2{}_2$, $K^1{}_2 = -K^2{}_1$, $K^1{}_3 = K^3{}_1$, $K^2{}_3 = -K^3{}_2$, and with $\sum_i \omega_{pi}^2/\Omega_i^2 = c^2/v_A^2$ in ordinary units.

The most important contribution to damping in this limit is Landau damping by electrons. The antihermitian part of the response tensor is given by retaining only the resonance at $a = 0$ is (2.5.12). In the small gyroradius approximation, $\lambda \to 0$, $I_0 \to 1$ (2.5.13) gives

$$K^{Ai}{}_j(\omega, \boldsymbol{k}) = i\left(\frac{\pi}{2}\right)^{1/2} \frac{q^2 n\omega}{m|k_z|V} e^{-\omega^2/2k_z^2V^2} N^i{}_j(0, \omega, \boldsymbol{k}), \tag{2.5.20}$$

with the only nonzero components of $N^i{}_j(0, \omega, \boldsymbol{k})$ being

$$N^2{}_3 = -N^3{}_2 = -i\frac{k_\perp}{k_z}\frac{\omega}{\Omega_0}, \qquad N^3{}_3 = \frac{\omega^2}{k_z^2 V^2}. \tag{2.5.21}$$

2.5.3 *Mildly Relativistic Limit of Trubnikov's Tensor*

It is of interest to rederive the foregoing results starting from Trubnikov's response tensor. The nonrelativistic limit of Trubnikov's response tensor is found by taking the limit $\rho = mc^2/T \gg 1$ in the response tensor in either the forms (2.4.13) or (2.4.22) with (2.4.23). On expanding $r(\xi)$, defined by (2.4.10), in powers of $1/\rho$, one finds

$$r(\xi) = \rho - i\omega\xi + \frac{k_z^2\xi^2}{2\rho} + \frac{k_\perp^2}{\rho\Omega_0^2}(1 - \cos\Omega_0\xi) + O(1/\rho). \tag{2.5.22}$$

With this approximation, $\rho \gg 1$ implies $|r(\xi)| \gg 1$, and the asymptotic limit (2.5.1) of the Macdonald functions is then justified. This gives

$$\frac{K_\nu(r(\xi))}{r^\nu(\xi)} \approx \left(\frac{\pi}{2}\right)^{1/2} \frac{e^{-\rho}}{\rho^{\nu+1/2}} \exp\left(\frac{i\omega\xi}{\rho} - \frac{k_z^2\xi^2}{2\rho} - \frac{k_\perp^2}{\rho\Omega_0^2}(1 - \cos\Omega_0\xi)\right). \tag{2.5.23}$$

The final term involving $\cos\Omega_0\xi$ can be expanded, effectively in gyroharmonics, using the generating function for modified Bessel functions, (A.1.10). The modified Bessel functions, $I_a(\lambda)$, satisfy the differential equation (A.1.7) and the recursion relations (A.1.9). Using these relations, one finds

$$\begin{bmatrix} 1 \\ \cos\Omega_0\xi \\ \sin\Omega_0\xi \\ \cos^2\Omega_0\xi \\ \sin^2\Omega_0\xi \\ \sin\Omega_0\xi\cos\Omega_0\xi \end{bmatrix} e^{\lambda\cos\Omega_0\xi} = \sum_{a=-\infty}^{\infty} \begin{bmatrix} I_a(\lambda) \\ I_a'(\lambda) \\ i(a/\lambda)I_a(\lambda) \\ I_a''(\lambda) \\ -(1/\lambda^2)[a^2 I_a(\lambda) - \lambda I_a'(\lambda)] \\ -i(a/\lambda^2)[I_a(\lambda) - \lambda I_a'(\lambda)] \end{bmatrix} e^{-ia\Omega_0\xi}. \tag{2.5.24}$$

After using (2.5.24) in the form (2.4.13) of Trubnikov's response tensor to write all the ξ-dependence in the integrand in exponential form, the ξ-integral can be performed trivially.

The resulting integral over proper time ξ in either of the forms (2.4.13) or (2.4.22) of Trubnikov's response tensor can be evaluated in terms of the plasma dispersion function (2.5.7). Three integrals appear:

$$\int_0^\infty d\xi \begin{bmatrix} 1 \\ \xi \\ \xi^2 \end{bmatrix} \exp\left[i(\omega - a\Omega_0)\xi - \frac{k_z^2\xi^2}{2\rho}\right] = -i\,\frac{y_a Z(y_a)}{\omega - a\Omega_0} \begin{bmatrix} 1 \\ \Delta_a\Omega_0/k_z^2V^2 \\ \Delta_a^2\Omega_0^2/k_z^4V^4 \end{bmatrix}, \tag{2.5.25}$$

with $\Delta_a = (\omega - a\Omega_0)/\Omega_0 = y_a\sqrt{2}k_zV/\Omega_0$. For the $\mu, \nu = 1,2$ components the integral with unity in the square brackets appears, for the components with either $\mu = 1,2$, $\nu = 0,3$ or $\mu = 0,3$, $\nu = 1,2$ the additional factor ξ appears, and for the $\mu, \nu = 0,3$ components the factor ξ^2 appears. A derivation of the result (2.5.25), with unity in the square brackets, involves the following steps: write $\rho = c^2/V^2$; replace the variable of integration by ξ by $t = k_zV\xi/\sqrt{2}$; complete the square in the exponent, which becomes $-(t - iy_a)^2 - y_a^2$; regard the integral over $0 \le t < \infty$ as a contour integral in complex-t space; and, deform the contour such that it is along the imaginary-t axis from the origin to $\mathrm{Im}\,t = y_a$, and then parallel to the real-t axis to infinity. The portions of the integral along the imaginary axis and parallel to the real axis give ($-i$ times) the real and imaginary parts of the function $Z(y_a)$, respectively, in (2.5.7). To evaluate the integrals with ξ or ξ^2, one includes the additional factors of t or t^2 in the integral representation (2.5.5) for $Z(y)$ to derive the results given in (2.5.25). In this way, one can show that the space components of the form (2.4.13) of Trubnikov's response tensor reproduce the nonrelativistic result (2.5.10) with (2.5.11).

Shkarofsky's Response 3-Tensor

To include weakly relativistic effects one needs to modify the approximations made to reduce the exact expression for the response tensor to its nonrelativistic counterpart. Specifically, in using the generating function for modified Bessel functions, as in (2.5.24), one writes

$$r(\xi) \approx r_0(\xi) + \Lambda(1 - \cos\Omega_0\xi), \qquad r_0(\xi) = [(\rho - i\omega\xi)^2 + k_z^2\xi^2]^{1/2},$$
$$\Lambda = \frac{k_\perp^2}{\Omega_0^2 r_0(\xi)}, \tag{2.5.26}$$

where Λ is implicitly a function of ξ. The results (2.5.24) apply, with $\lambda \to \Lambda$.

With this modification, Trubnikov's response leads to Shkarofsky's [8] approximation to the response 3-tensor. Specifically, the 3-tensor components of (2.4.13) become

$$\Pi^{ij}(k) = -i\frac{q^2 n\omega}{m}\int_0^\infty d\xi\,\frac{e^{\rho-\Lambda}}{[r_0(\xi)/\rho]^{5/2}}\sum_{a=-\infty}^{\infty}\hat{H}_a^{ij}(k)\,e^{-r_0(\xi)-ia\Omega_0\xi},$$

$$\hat{H}_a^{ij}(k) = \begin{pmatrix} \dfrac{a^2 I_a}{\Lambda} & -i\epsilon a(I_a' - I_a) & \dfrac{k_z}{k_\perp}\,a I_a\,\dfrac{\partial}{\partial a} \\ i\epsilon a(I_a' - I_a) & \dfrac{a^2 I_a}{\Lambda} - 2\Lambda(I_a' - I_a) & i\epsilon\dfrac{k_z}{k_\perp}\,\Lambda(I_a' - I_a)\,\dfrac{\partial}{\partial a} \\ \dfrac{k_z}{k_\perp}\,a I_a\,\dfrac{\partial}{\partial a} & -i\epsilon\dfrac{k_z}{k_\perp}\,\Lambda(I_a' - I_a)\,\dfrac{\partial}{\partial a} & I_a\left(1 + k_z\dfrac{\partial}{\partial k_z}\right) \end{pmatrix}, \tag{2.5.27}$$

with $r_0(\xi)$ given by (2.5.26), and where the argument Λ of the modified Bessel functions is omitted for simplicity in writing. The derivatives $\partial/\partial a$ and $\partial/\partial k_z$ are understood to operate on the exponential functions $\exp(-ia\Omega_0\xi)$ and $\exp[-r_0(\xi)]$, respectively.

2.5.4 *Shkarofsky and Dnestrovskii Functions*

Another approximation is needed to allow the integral over ξ in (2.5.27) to be evaluated in terms of a relativistic plasma dispersion function (RPDF). Two classes of such functions are Shkarofsky functions and Dnestrovskii functions. Another form for the RPDFs involves hypergeometric functions § 2.5.5.

Shkarofsky Functions

The approximation made to allow the integral over ξ in (2.5.27) is $\Lambda \ll 1$. The modified Bessel functions are approximated by the leading term in their expansion, $I_a(\Lambda) \approx \Lambda^a/2^a a!$. The dependence of the integrand in (2.5.27) on ξ is made explicit by writing $\Lambda = \lambda\rho/r_0(\xi)$, $\lambda = k_\perp^2/\rho\Omega_0^2$. The integral over ξ can be written in terms of the generalized Shkarofsky functions [9]

$$\mathcal{F}_{q,r}(z,\zeta) = -i\int_0^\infty dt\,\frac{(it)^r}{(1-it)^q}\exp\left[izt - \frac{\zeta t^2}{(1-it)}\right], \tag{2.5.28}$$

where the dimensionless arguments, $z = \rho(\omega - a\Omega_0)/\omega$, $\zeta = k_z/\omega$, are introduced by writing $\xi = t/\omega$, with t dimensionless. The Shkarofsky functions [8] correspond to the special case $r = 0$:

$$\mathcal{F}_q(z,\zeta) = \mathcal{F}_{q,0}(z,\zeta). \tag{2.5.29}$$

The resulting approximation to the space components of the response 4-tensor, expanded to low orders in λ, was written down by Shkarofsky [8], and generalized by Robinson [10]. The latter form includes contributions from harmonics $a = 0, \pm 1, \pm 2$ and is of the form

$$\Pi^{\mu}{}_{\nu}(k) = -\frac{\omega_p^2 \rho}{\varepsilon_0} X^{\mu}{}_{\nu}, \tag{2.5.30}$$

with the components given by

$$\begin{aligned}
X^1{}_1 &= \mathcal{F}^{1+}_{5/2} - \lambda(\mathcal{F}^{1+}_{7/2} - \mathcal{F}^{2+}_{7/2}), \qquad X^2{}_2 = X^1{}_1 + 2\lambda(\mathcal{F}^{0+}_{7/2} - \mathcal{F}^{1+}_{7/2}),\\
X^1{}_2 &= -i\epsilon\big[\mathcal{F}^{1-}_{5/2} - \lambda(2\mathcal{F}^{1-}_{7/2} - \mathcal{F}^{2-}_{7/2})\big],\\
X^3{}_3 &= \mathcal{F}^{0+}_{5/2} - \lambda(\mathcal{F}^{0+}_{7/2} - \mathcal{F}^{1+}_{7/2})\\
&\quad + \frac{k_z^2 c^2 \rho}{\omega^2}\Big[\mathcal{F}^{0+}_{7/2,2} - \lambda(\mathcal{F}^{0+}_{9/2,2} - \mathcal{F}^{1+}_{9/2,2}) + \tfrac{1}{4}\lambda^2(3\mathcal{F}^{0+}_{11/2,2} - 4\mathcal{F}^{1+}_{11/2,2} + \mathcal{F}^{2+}_{11/2,2})\Big],\\
X^1{}_3 &= \frac{k_\perp k_z}{\omega \Omega_0}\big[\mathcal{F}^{1-}_{7/2,1} - \tfrac{1}{2}\lambda(2\mathcal{F}^{1-}_{9/2,1} - \mathcal{F}^{2-}_{9/2,1})\big],\\
X^2{}_3 &= i\epsilon\frac{k_\perp k_z}{\omega \Omega_0}\big[\mathcal{F}^{0+}_{7/2,1} - \mathcal{F}^{1+}_{7/2,1} - \tfrac{1}{4}(3\mathcal{F}^{0+}_{9/2,1} - 4\mathcal{F}^{1+}_{9/2,1} + \mathcal{F}^{2+}_{9/2,1})\big],
\end{aligned} \tag{2.5.31}$$

where the following notation is introduced:

$$\mathcal{F}^{(a\pm)}_{q,r} = \tfrac{1}{2}\big[\mathcal{F}_{q,r}(z_a, k_z^2 c^2 \rho/2\omega^2) \pm \mathcal{F}_{q,r}(z_{-a}, k_z^2 c^2 \rho/2\omega^2)\big], \tag{2.5.32}$$

with $z_{\pm a} = \rho(\omega - a\Omega_0)/\omega$.

Further physical approximations need to be made to the plasma dispersion functions to derive results that are useful for applications. The special case of perpendicular propagation is of particular relevance in identifying the effects that distinguish the weakly relativistic approximation from the nonrelativistic approximation.

Dnestrovskii Functions

For perpendicular propagation, $k_z = 0$, the function (2.5.26) simplifies to $r_0(\xi) = \rho - i\omega\xi$, and the response tensor (2.5.27) simplifies considerably. The relativistic plasma dispersion functions that appear can be written in terms of [11, 12]

$$R_l(z, \lambda, a) = -i \int_0^\infty dt \, \frac{e^{izt - \Lambda}}{(1 - it)^l} I_a(\Lambda), \tag{2.5.33}$$

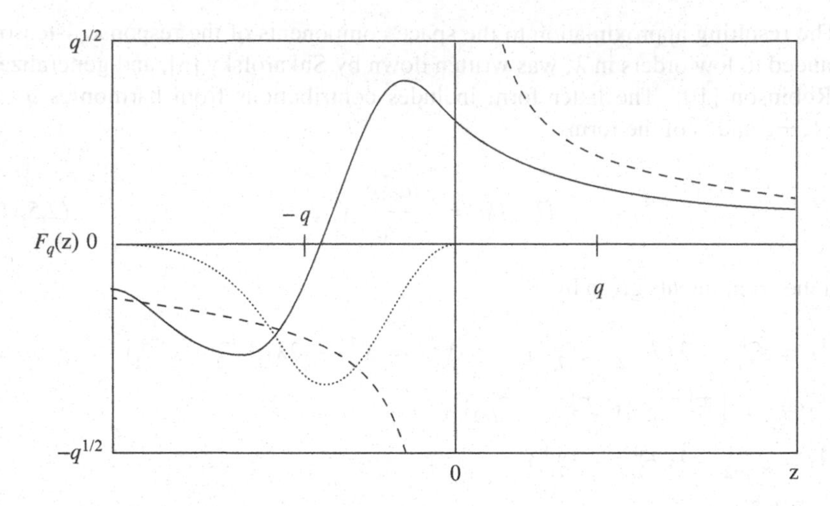

Fig. 2.1 A plot of the Dnestrovskii function $F_q(z)$ for real z, *solid line* for the real part and *dotted line* for the imaginary part; the function $1/z$ is indicated by the *dashed curve* (From [12, 16], reprinted with permission Cambridge University Press)

with $\Lambda = \lambda/(1 - it)$ and $\lambda = \rho k_\perp^2/\Omega_0^2$. An approximate form is obtained by expanding in $\lambda \lesssim 1$, and then each term in the expansion can be written in terms of a Dnestrovskii function [13].

A generalized Dnestrovskii function is defined as the counterpart for perpendicular propagation of the generalized Shkarofsky function (2.5.28) [12]:

$$F_{q,r}(z) = -i \int_0^\infty dt\, \frac{(it)^r e^{izt}}{(1 - it)^q}, \qquad F_q(z) = F_{q,0}(z). \tag{2.5.34}$$

The special case $r = 0$ corresponds to the Dnestrovskii function [13]:

$$F_q(z) = -i \int_0^\infty dt\, \frac{e^{izt}}{(1 - it)^q}. \tag{2.5.35}$$

An alternative definition is [14, 15]

$$F_q(z) = \frac{1}{\Gamma(q)} \int_0^\infty dx\, \frac{x^{q-1} e^{-x}}{x + z}. \tag{2.5.36}$$

Properties of these functions are summarized in Appendix A.2.3.

The Dnestrovskii function $F_q(z)$ is plotted in Fig. 2.1. For comparison, the nonrelativistic counterpart, $1/z$, for $k_z \to 0$ is shown. Features to note are (a) the resonance at $z = 0$ is removed, (b) there is a skewing so that the maximum value occurs below $z = 0$, and (c) damping (which is absent in the nonrelativistic case) is nonzero in a region below $z = 0$. An approximation to the Dnestrovskii function that is useful in understanding dispersion in the weakly relativistic case can be derived from an expansion of $F_q(z)$ in derivatives of the nonrelativistic plasma dispersion function:

$$F_q(z) = -\frac{Z(\psi)}{(2q)^{1/2}} - \frac{Z'''(\psi)}{12q} - \cdots, \tag{2.5.37}$$

with $\psi = (z+q)/(2q)^{1/2}$. Retaining only the leading term in (2.5.37) would imply $F_q(z) = 0$ at $z = -q$, which is close to the actual zero in Fig. 2.1.

Weakly Relativistic Dispersion for $k_z \neq 0$

For $k_z \neq 0$, the Shkarofsky functions and the Dnestrovskii functions are related by an expansion in modified Bessel functions:

$$\mathcal{F}_q(z, a) = \sum_{j=-\infty}^{\infty} e^{-2a}\, I_j(2a)\, F_{q-j}(z), \tag{2.5.38}$$

with $q = a + 5/2$, $z = \rho\Omega_0\Delta_a/\omega$, $a = \rho k_z^2/2\omega^2$. An approximation analogous to (2.5.37) is [17]

$$\mathcal{F}_q(z, a) = -\frac{Z(\psi)}{(4a+2q)^{1/2}} - \frac{(3a+q)Z'''(\psi)}{3(4a+2q)^2} + \cdots, \tag{2.5.39}$$

with $\psi = (z+q)/(4a+2q)^{1/2}$. Comparison of (2.5.37) and (2.5.39) suggests an approximation $\mathcal{F}_q(z, a) \approx F_{q+2a}(z)$ [17].

The Shkarofsky functions include both the nonrelativistic dispersion that exists for $k_z \neq 0$ and the weakly relativistic dispersion described by the Dnestrovskii functions for $k_z = 0$. The center of the line, which is at $\omega = a\Omega_0$ in the nonrelativistic case, is downshifted to $z + q = 0$ for $k_z = 0$, and this is unchanged for $k_z \neq 0$, so that the center of the line is determined approximately by (in ordinary units with $\rho \to c^2/V^2$)

$$\frac{\omega - a\Omega_0}{\omega} \approx -(a + 5/2)\frac{V^2}{c^2}, \tag{2.5.40}$$

for $q = a + 5/2$. The line width, which is $\delta\omega_a = \sqrt{2}\,|k_z|V$ in the nonrelativistic case, is broadened to (ordinary units)

$$\frac{\delta\omega_s}{\omega} = \left[\frac{k_z^2 V^2}{\omega^2} + (s + 5/2)\frac{\omega^2}{\Omega_0^2}\frac{V^2}{c^2}\right]^{1/2}, \tag{2.5.41}$$

with $\omega \approx a\Omega_0$.

A notable feature of the inclusion of relativistic effect occurs at sufficiently high harmonics, $a^2 \gtrsim \rho$. According to (2.5.41), for sufficiently large a the line width is dominated by the relativistic contribution, which increases $\propto a^{3/2}$. In comparison, the separation between neighboring harmonics implied by (2.5.41) decreases more slowly, $\propto a$. Hence, for sufficiently large a, the relativistic broadening exceeds the separation between harmonics. The resulting smoothing out of the harmonics into a continuum is a characteristic relativistic effect.

2.5.5 RPDFs Involving Hypergeometric Functions

An alternative approach to the evaluation of the response tensor for a relativistic thermal distribution was taken by Swanson [18] by considering functions that can be evaluated relatively simply by numerical methods, rather than by approximate analytic methods. The approach in its simplest form applies only for perpendicular propagation. The method involves performing the integral over pitch angle, α, in terms of hypergeometric function, and the assumption of perpendicular propagation is needed in order to use the relevant integral identities. The advantage of the method is that the numerical evaluation of the integral of the resulting hypergeometric functions over a Jüttner distribution is relatively straightforward. The two hypergeometric functions that appear are defined by

$$ {}_1F_2(a;b_1,b_2;x) = \frac{\Gamma(b_1)\Gamma(b_2)}{\Gamma(a)} \sum_{k=0}^{\infty} \frac{\Gamma(a+k)}{\Gamma(b_1+k)\Gamma(b_2+k)} \frac{x^k}{k!}, \tag{2.5.42} $$

$$ \begin{aligned} {}_2F_3(a_1,a_2;b_1,b_2,b_3;x) = {} & \frac{\Gamma(b_1)\Gamma(b_2)\Gamma(b_3)}{\Gamma(a_1)\Gamma(a_2)} \\ & \times \sum_{k=0}^{\infty} \frac{\Gamma(a_1+k)\Gamma(a_1+k)}{\Gamma(b_1+k)\Gamma(b_2+k)\Gamma(b_3+k)} \frac{x^k}{k!}. \end{aligned} \tag{2.5.43} $$

The components of the response tensor in the Vlasov form (2.3.29) involve a sum over harmonics, a, with a product of Bessel function in the numerator and a resonant denominator that is proportional to $a - a_0$, where $a_0 = (ku)_\parallel/\Omega_0$ does not depend on a. The argument of the Bessel functions is proportional to $\sin\alpha$, and for an isotropic distribution, the integral over α can be expressed in terms of a hypergeometric function. The relevant integrals include

$$ \int_{-1}^{1} d\cos\alpha\, [J_a(b\sin\alpha)]^2 = \frac{2b^{2a}}{(2a+1)!}\, {}_1F_2(a+\tfrac{1}{2}; a+\tfrac{3}{2}, 2a+1; -b^2). \tag{2.5.44} $$

Analogous integrals for all the other combinations of Bessel functions that appear in the components of the response tensor can be evaluated similarly in terms of ${}_1F_2$ with different arguments.

An alternative procedure [18] is to perform the infinite sum over a first, using the Newberger sum rule [19]

$$\sum_{a=-\infty}^{\infty} \frac{J_a(z)J_{a-n}(z)}{a-a_0} = \frac{(-1)^{n+1}\pi}{\sin \pi a_0} J_{n-a_0}(z)J_{a_0}(z), \tag{2.5.45}$$

which applies for $n \geq 0$ and arbitrary a_0. Perpendicular propagation, $k_z = 0$ implies that $a_0 = \gamma\omega/\Omega_0$ does not depend on α. The integral over α can then be performed using another set of integrals, including

$$\int_{-1}^{1} d\cos\alpha \, J_{a_0}(b\sin\alpha)J_{-a_0}(b\sin\alpha)$$
$$= \frac{2\sin\pi a_0}{\pi a_0} \, {}_2F_3\left(\tfrac{1}{2}, 1; \tfrac{3}{2}, 1-a_0, 1+a_0; -b^2\right). \tag{2.5.46}$$

Analogous integrals for all the other combinations of Bessel functions that appear in the components of the response tensor can be evaluated in terms of ${}_2F_3$ with different arguments.

The response tensor has six independent components in general, and two of these are zero for perpendicular propagation. The four nonzero components can be expressed in terms of five RPDFs that are integrals of hypergeometric functions over a Jüttner distribution. These RPDFs are

$$\begin{bmatrix} I_1(z,\lambda,\rho) \\ I_2(z,\lambda,\rho) \\ I_3(z,\lambda,\rho) \\ I_4(z,\lambda,\rho) \\ I_5(z,\lambda,\rho) \end{bmatrix} = \frac{\rho}{K_2(\rho)} \int_0^\infty dp p^2 e^{-\rho\gamma} \begin{bmatrix} {}_2F_3(\frac{1}{2}, 1; \frac{3}{2}, 1-a_0, 1+a_0; -b^2) \\ {}_2F_3(\frac{1}{2}, 1; \frac{3}{2}, 1-a_0, a_0; -b^2) \\ {}_2F_3(\frac{1}{2}, 1; \frac{3}{2}, -a_0, 1+a_0; -b^2) \\ a_4(p) {}_2F_3(\frac{1}{2}, 1; \frac{5}{2}, 1-a_0, 1+a_0; -b^2) \\ a_5(p) {}_2F_3(\frac{1}{2}, 1; \frac{5}{2}, 2-a_0, a_0; -b^2) \end{bmatrix} \tag{2.5.47}$$

with $a_4(p) = p^2/3(m^2+p^2)$, $a_5(p) = p^2/3m^2a_0(a_0+1))$, $b^2 = \lambda\rho p^2/m^2$, $\lambda = k_\perp^2/\rho\Omega_0^2$, $z = \rho(1-a_0\Omega_0/\omega)$. The nonzero components of the dielectric tensor for $k_z = 0$ are

$$K^1{}_1 = 1 - \frac{\rho\omega_p^2}{k_\perp^2}[I_1(z,\lambda,\rho) - 1],$$

$$K^2{}_2 = K^1{}_1 - \frac{2\rho\omega_p^2}{k_\perp^2}[1 - I_2(z,\lambda,\rho) + I_5(z,\lambda,\rho)],$$

$$K^3{}_3 = 1 - \frac{\rho\omega_p^2}{\omega^2}I_4(z,\lambda,\rho),$$

$$K^1{}_2 = -i\frac{2\rho\omega_p^2}{k_\perp^2}[I_2(z,\lambda,\rho) - I_3(z,\lambda,\rho)], \tag{2.5.48}$$

with $K^1{}_3 = K^2{}_3 = 0$.

Swanson [18] presented detailed results comparing the exact form for the response tensor, evaluated using (2.5.48), and some of the approximate forms discussed above.

2.6 Pulsar Plasma

A pulsar plasma is an exotic example of an intrinsically relativistic, strongly magnetized electron gas. Despite an extensive literature and many novel ideas, only some of which are discussed in this section, the radio emission mechanism for pulsars is not understood.

2.6.1 Pulsars

Pulsars are strongly magnetized, rotating neutron stars that emit beamed radiation, seen as a pulse as the beam sweeps across the observer's line of sight. About 2000 radio pulsars are known, and these provide a large body of statistical data on pulsar properties. A small fraction of these pulsars are observed at high energies, which can include both pulsed nonthermal radiation and thermal radiation for the very hot ($>10^6$ K) polar-cap regions of the neutron star surface. Although the basic features of a pulsar model are widely accepted, the details of the electrodynamics remain uncertain and controversial. The pulsed emission is attributed to radiation beamed along magnetic field lines in a polar cap region, with the beaming associated with a relativistically outflowing pair (electron–positron) plasma. In the discussion here, the emphasis is on the properties of this so-called pulsar plasma.

The interpretation of pulsars is based on two models for the electrodynamics: the vacuum-dipole model, in which the fields are those of a rotating magnetized star in vacuo, and the corotating-magnetosphere model, in which the only electric field in the magnetosphere is the corotation field, $\boldsymbol{E}_{\rm cor}$, required for rigid corotation.

These two models are both based on unjustifiable neglects – of the plasma in the magnetosphere, and of the inductive electric field, respectively [20]. Nevertheless the two models are used selectively as the basis for pulsar electrodynamics: the vacuum-dipole model is used to treat the slowing down of the rotation and to estimate the surface magnetic field and the age of the pulsar, and the corotating-magnetosphere as the basis for detailed models for the distribution of plasma in the magnetosphere.

Vacuum-Dipole Model

The basic observable parameters for a pulsar are its period, P, and the period derivative, $\dot{P}$. Based on a magnetic-dipole model, the surface magnetic field is estimated to be $B_* = 3.2 \times 10^{15} (P\dot{P})^{1/2}$ T, and the characteristic age as $\propto P/2\dot{P}$. Pulsars are separated into three classes:. Most are normal pulsars with $P \sim 10^{-2}$–10 s, $B \sim 10^7$–10^9 T with characteristic ages ranging from $\sim 10^3$–10^5 year for young pulsars and up to $> 10^7$ year for old pulsars. Recycled or millisecond pulsars, which are very old pulsars spun up in binary systems, have $P \lesssim 100$ ms and $B_* \lesssim 10^6$ T. There is a small class of magnetars with $P \approx 10$ s and $B_* \gtrsim 10^{10}$ T.

The polar cap is defined by the field lines that extend beyond the light cylinder, $r = R_L = Pc/2\pi$. Assuming that the field lines are dipolar out to the light cylinder, the last closed field line is the line $r = r_L \sin^2\theta$. Where this field line intersects the stellar surfaces, at $r = R_*$, defines the polar-cap angle, $\theta = \theta_{\rm PC}$,

$$\theta_{\rm PC} = \arcsin(R_*/r_L)^{1/2} = \arcsin(2\pi R_*/Pc) \tag{2.6.1}$$

on the surface of the star. The field lines at $\theta < \theta_{\rm PC}$ define the polar cap. The maximum potential available can be estimated from the potential between the field line at the magnetic pole and a field line at the edge of the polar cap. This is

$$\Phi_{\max} = R_*^2 \Omega_* B_* \sin^2\theta_{\rm PC} = R_*^3 \Omega_*^2 B_*/c. \tag{2.6.2}$$

For most pulsars, the corresponding maximum energy, $e\Phi_{\max}$, is very large compared with the energy of particles needed to trigger a pair cascade, and only a small fraction of $\Phi_{\max}$ needs to relocate along field lines to trigger pair creation in a localized "gap". For example, for parameters for the Crab pulsar one finds $\Phi_{\max} \approx 10^{16}$ V, with somewhat smaller values for other pulsars. Normal pulsars are thought to cease radiation when $\Phi_{\max}$ drops below the threshold (the "death line") for effective pair creation.

Corotating-Magnetosphere Model

In a corotating model, it is assumed that the corotation electric field,

$$\mathbf{E}_{\rm cor} = -(\boldsymbol{\Omega}_* \times \mathbf{x}) \times \mathbf{B}, \tag{2.6.3}$$

is present in the magnetosphere. The divergence of $\mathbf{E}_{\rm cor}$ requires that the Goldreich-Julian charge density

$$\rho_{\rm GJ}(t,\boldsymbol{x}) = -2\varepsilon_0\boldsymbol{\omega}\cdot\boldsymbol{B}(t,\boldsymbol{x}) + \varepsilon_0(\boldsymbol{\omega}\times\boldsymbol{x})\cdot{\rm curl}\,\boldsymbol{B}(t,\boldsymbol{x}), \tag{2.6.4}$$

be present in the magnetosphere. In a conventional model, just above the stellar surface, $\rho_{\rm GJ}$ is provided by charges of a single sign drawn from the stellar surface. With the surface the only source of charge, it is not possible to satisfy (2.6.4) at greater heights. An additional source of charge is needed, and this is provided by a pair cascade. The details of the cascade are model-dependent, with the common feature in all models being the appearance of $E_\parallel \neq 0$, which accelerates charges to sufficiently high energies for them to emit gamma rays that decay into electron-positron pairs. The charge density required is set up through electrons and positrons being separated by $E_\parallel \neq 0$. In older models, the pair creation is assumed quasi-stationary and confined to local regions called gaps. When the quasi-stationary assumption is relaxed, the inclusion of the displacement current leads to an instability [21–24], resulting in large amplitude electric oscillations (LAEWs) [25]. In a model involving LAEWs the pair creation is confined in time, rather than space, to near the phase of the LAEW where the particles have their maximum energy.

Properties of Pulsar Plasma

The pulsar plasma is identified as consisting of the pairs produced in this cascade. These pairs can escape along the open field lines, ultimately forming a pulsar wind, and need to be continuously replaced. The properties of the pulsar plasma are poorly determined. These properties include the number density and mean energy of the pairs. The number density can be expressed as M times the Goldreich-Julian number density. This corresponds to a plasma frequency $M^{1/2}$ times the characteristic plasma frequency defined by $\rho_{\rm GJ}$,

$$\omega_{\rm GJ} = (\Omega_*\Omega_e)^{1/2}, \tag{2.6.5}$$

where factors of order unity are ignored. The surface value of $\omega_{\rm GJ} \propto (\dot{P}/P)^{1/4}$ is relatively insensitive to the properties of pulsar, and it falls off $\propto (R_*/r)^{3/2}$ above the surface. The values of the multiplicity, M, and of the mean Lorentz factor, $\bar{\gamma}$, of the secondary pairs are plausibly in the ranges $M \sim 10^3$–10^5 and $\bar{\gamma} \sim 10$–10^3 [26–28].

Due to the superstrong magnetic field in a pulsar, the lifetime for an electron to radiate away its perpendicular energy through gyromagnetic emission is very short. All the electrons and positrons in the pulsar plasma are assumed to be in their lowest Landau state, corresponding to a one-dimensional (1D) distribution with $p_\perp = 0$.

2.6.2 *Response Tensor for a 1D Pair Plasma*

The response tensor for an arbitrary 1D distribution of electrons and positrons follows from (1.3.7). The distribution of electrons (+) and positrons (−) are described by 1D distribution functions, $g_\pm(u)$, with $u = p_z/m = \gamma v$, $\gamma = (1-v^2)^{-1/2}$. The normalization is chosen to be such that one has

$$n_{\mathrm{pr}\pm} = \int du\,\frac{g_\pm(u)}{\gamma} = \int \frac{d^3p}{(2\pi)^3}\frac{f(p_\perp,p_z)}{\gamma} = \int \frac{d^4p}{(2\pi)^4}F_\pm(p),$$
$$n_\pm = \int du\,g_\pm(u) = \int \frac{d^3p}{(2\pi)^3}\,f(p_\perp,p_z) = \int \frac{d^4p}{(2\pi)^4}\,\gamma\,F_\pm(p), \tag{2.6.6}$$

which are the proper number densities and the actual number densities, respectively.
In the 1D case, the response tensor (2.3.9) can be written as

$$\Pi^{\mu\nu}(k) = \sum_\pm -\frac{e^2 n_\pm}{m}\bigg\langle\frac{1}{\gamma}\bigg\{-\frac{k_D^\mu k_D^\nu}{(ku)^2} + \frac{1}{(ku)^2-\Omega_e^2}\big[(ku)^2\,g_\perp^{\mu\nu}$$
$$- ku\,(k_\perp^\mu u^\nu + u^\mu k_\perp^\nu) - k_\perp^2 u^\mu u^\nu - i\,\epsilon\Omega_e\big(ku\,f^{\mu\nu} + k_G^\mu u^\nu - u^\mu k_G^\nu\big)\big]\bigg\}\bigg\rangle_\pm, \tag{2.6.7}$$

with

$$n_\pm\langle K\rangle_\pm = \int du\,g_\pm(u)\,K, \tag{2.6.8}$$

for any function K, with $\epsilon = \pm$ for electrons and positrons, respectively.

RPDFs for a 1D Distribution

With $ku = \gamma(\omega - k_z v)$, the denominators that appear in (2.6.7) can be rewritten using

$$(ku)^2 = k_z^2\gamma^2(z-v)^2, \quad (ku)^2 - \Omega_e^2 = k_z^2\gamma^2(1+y^2)(v-z_+)(v-z_-), \tag{2.6.9}$$

with $z = \omega/k_z$, $y = \Omega_e/k_z$,

$$z_\pm = \frac{z \pm y(1+y^2-z^2)^{1/2}}{1+y^2}. \tag{2.6.10}$$

The result may be expressed in terms of three relativistic plasma dispersion functions (RPDFs), defined by

$$W(z) = \left\langle \frac{1}{\gamma^3(z-v)^2} \right\rangle, \quad R(z) = \left\langle \frac{1}{\gamma(z-v)} \right\rangle, \quad S(z) = \left\langle \frac{1}{\gamma^2(z-v)} \right\rangle. \quad (2.6.11)$$

Dielectric Tensor in Terms of RPDFs

Consider the special case where the distributions of electrons and positrons is the same except for their number densities, specifically, $g_+(u)/n_+ = g_-(u)/n_-$. The difference between the number densities of electrons and positrons can be included by re-interpreting the sign ϵ as its average over the two distributions, $\epsilon = (n_+ - n_-)/(n_+ + n_-)$. The resulting expression for the dielectric tensor is [29]

$$K^1{}_1 = K^2{}_2 = 1 - \frac{\omega_p^2}{\omega^2}\frac{1}{1+y^2}\left[\left\langle\frac{1}{\gamma}\right\rangle + \frac{(z-z_+)^2R(z_+) - (z-z_-)^2R(z_-)}{z_+ - z_-}\right],$$

$$K^3{}_3 = 1 - \frac{\omega_p^2}{\omega^2}\left\{z^2W(z) + \frac{\tan^2\theta}{1+y^2}\left[\left\langle\frac{1}{\gamma}\right\rangle + \frac{z_+^2R(z_+) - z_-^2R(z_-)}{z_+ - z_-}\right]\right\},$$

$$K^1{}_2 = i\epsilon\frac{\omega_p^2}{\omega^2}\frac{y}{1+y^2}\left[\frac{(z-z_+)S(z_+) - (z-z_-)S(z_-)}{z_+ - z_-}\right],$$

$$K^1{}_3 = -\frac{\omega_p^2}{\omega^2}\frac{1}{1+y^2}\left[\frac{(z-z_+)z_+R(z_+) - (z-z_-)z_-R(z_-)}{z_+ - z_-}\right],$$

$$K^2{}_3 = -i\epsilon\frac{\omega_p^2}{\omega^2}\frac{y\tan\theta}{1+y^2}\left[\frac{z_+S(z_+) - z_-S(z_-)}{z_+ - z_-}\right], \quad (2.6.12)$$

with $K^2{}_1 = -K^1{}_2$, $K^3{}_1 = K^1{}_3$, $K^2{}_3 = -K^3{}_2$.

Low-Frequency Limit

In the radio emission region of the pulsar magnetosphere, although this region is poorly determined, the cyclotron frequency is very large compared with radio frequencies, and the low-frequency approximation to the dielectric tensor is appropriate.

At frequencies well below the cyclotron frequency, the approximations $y \to \infty$, $z_\pm \to \pm 1$ lead to simplification to the general form of the response tensor (2.6.12). With $\langle v \rangle = 0$, the RPDFs associated with the cyclotron resonance simplify according to

$$R(\pm 1) = \mp\langle\gamma\rangle, \qquad S(\pm 1) = \mp 1. \quad (2.6.13)$$

In this limit the dielectric tensor (2.6.12) may be approximated by

$$K^1{}_1 = 1 + \frac{\omega_p^2}{\omega^2}\frac{k_z^2}{\Omega_e^2}\big(z^2\langle\gamma\rangle - 2z\langle\gamma v\rangle + \langle\gamma v^2\rangle\big),$$

$$K^3{}_3 = 1 - \frac{\omega_p^2}{\omega^2}\left[z^2 W(z) - \frac{k_\perp^2}{\Omega_e^2}\langle\gamma v^2\rangle\right],$$

$$K^1{}_3 = -\frac{\omega_p^2}{\omega^2}\frac{k_\perp k_z}{\Omega_e^2}\big(z\langle\gamma v\rangle - \langle\gamma v^2\rangle\big),$$

$$K^1{}_2 = -i\epsilon\,\frac{\omega_p^2}{\omega^2}\frac{k_z c}{\Omega_e}\big(z - \langle v\rangle\big), \qquad K^2{}_3 = i\epsilon\,\frac{\omega_p^2}{\omega^2}\frac{k_\perp c}{\Omega_e}\langle v\rangle. \tag{2.6.14}$$

In the absence of a streaming motion, $\langle v\rangle = 0$, the averages involve the two parameters

$$v_A = \langle\gamma\rangle^{-1/2}\frac{\Omega_e}{\omega_p}, \qquad \langle\gamma v^2\rangle = \langle\gamma\rangle\,\delta v^2. \tag{2.6.15}$$

The parameter v_A is interpreted as the Alfvén speed; it generalizes the definition $v_A = \Omega_e/\omega_p$ for a cold plasma to a relativistic plasma with $\langle\gamma\rangle > 1$. The parameter δv^2 describes the intrinsic spread in velocities; for $\delta v^2 \ll 1$ the spread is nonrelativistic, and it is highly relativistic for $\delta v^2 \approx 1$.

2.6.3 Specific Distributions for a 1D Electron Gas

Wave dispersion in a pulsar plasma depends on the form of the 1D distribution of electrons and positrons. However, the wave properties are not particularly sensitive to the actual form of the distribution function, provided that the electrons are highly relativistic. The dispersion is sensitive to relative streaming motions between different distributions, as discussed in § 1.3. The RPDFs are compared here for several different non-streaming distributions.

1D Jüttner Distribution

A 1D Jüttner distribution is

$$g(u) = \frac{e^{-\rho\gamma}}{2K_1(\rho)}, \qquad \left\langle\frac{1}{\gamma}\right\rangle = \frac{K_0(\rho)}{K_1(\rho)}, \qquad \langle\gamma\rangle = \frac{K_0(\rho) + K_2(\rho)}{2K_1(\rho)}, \tag{2.6.16}$$

where two relevant examples of averages are given. The RPDF associated with dispersion in the parallel direction is

$$W(z) = \frac{T'(z,\rho)}{2K_1(\rho)}, \qquad T(z,\rho) = \int_{-1}^{1} dv\,\frac{e^{-\rho\gamma}}{v - z}. \tag{2.6.17}$$

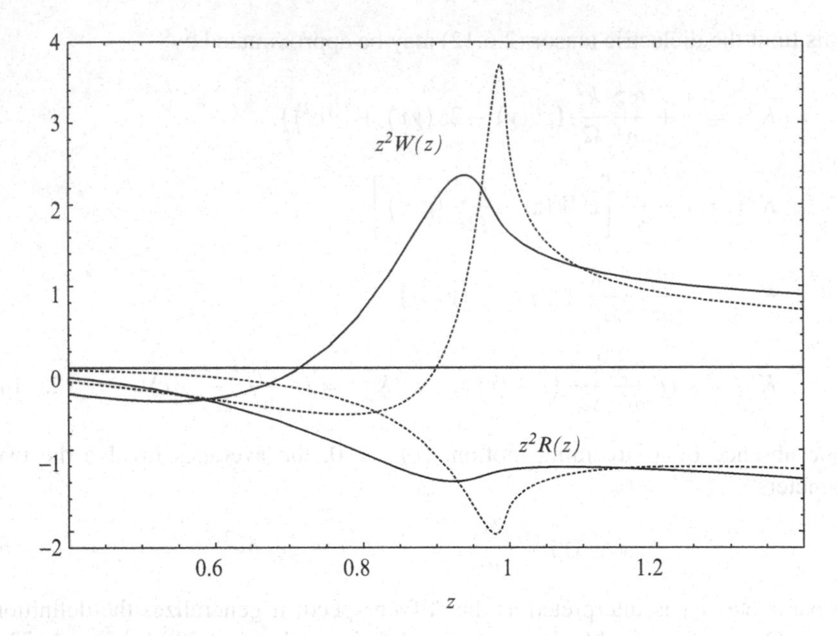

Fig. 2.2 Plots of $z^2W(z)$ and $z^2R(z)$ for a Jüttner distribution with $\rho = 1$ (*dotted*) and $\rho = 2.5$ (*solid*) (From [29], reprinted with permission Cambridge University Press)

The two RPDFs associated with the cyclotron resonance are

$$R(z) = \frac{2K_0(\rho)T(z,\rho)}{(1-z^2)2K_1(\rho)}, \qquad S(z) = -\frac{1}{z}\left[\frac{T'(z,\rho)}{2K_1(\rho)}\right]. \tag{2.6.18}$$

The properties of the RPDF $T(z,\rho)$ are discussed in § 4.4 of volume 1.

The RPDFs $z^2W(z)$ and $z^2R(z)$ are illustrated in Fig. 2.2 for two mildly relativistic Jüttner distributions.

1D Form of Trubnikov's Response Tensor

An alternative treatment of a 1D Jüttner distribution involves applying Trubnikov's method to evaluate the response tensor for a strictly-parallel distribution. The distribution function, $F(p)$ or $f(\boldsymbol{p})$, is proportional to $\delta(p_\perp^2)$, implying that all particles have $p_\perp = 0$. The steps involved in the derivation of the response tensor for such a distribution with a relativistic thermal distribution for the remaining parallel motion closely follow the foregoing derivation for an isotropic distribution.

A strictly-parallel Jüttner distribution corresponds to

$$g(p_z) = \frac{n\, e^{-\rho\gamma}}{2mK_1(\rho)} = \frac{n_{\rm pr}\, e^{-\rho\gamma}}{2mK_0(\rho)}. \tag{2.6.19}$$

The evaluation of the counterpart of the result (2.4.22) for the response tensor differs from the isotropic case in that the orders of the Macdonald functions are reduced by unity. Specifically, (2.4.22) is replaced by

$$\Pi^{\mu\nu}(k) = \frac{q^2 n\rho}{K_1(\rho)m} \int_0^\infty d\xi \left\{ -\left[(k^2)_\| T^{\mu\nu}(\xi) - k_\beta T^{\mu\beta}(\xi)k^\nu - k^\nu k_\alpha T^{\alpha\nu}(\xi) \right.\right.$$
$$\left. + k_\alpha k_\beta T^{\alpha\beta}(\xi) g_\|^{\mu\nu}\right] \frac{K_1(r_\|(\xi))}{r_\|(\xi)} + \left[(ka_\|(\xi))^2 T^{\mu\nu}(\xi) - ka_\|(\xi)a_\|^\mu(\xi)k_\alpha T^{\alpha\nu}(\xi)\right.$$
$$\left.\left. - ka_\|(\xi)k_\beta T^{\mu\beta}(\xi)a_\|^\nu(\xi) + k_\alpha k_\beta T^{\alpha\beta}(\xi)a_\|^\mu(\xi)a_\|^\nu(\xi)\right] \frac{K_2(r_\|(\xi))}{r_\|^2(\xi)} \right\}, \tag{2.6.20}$$

with $a_\|^\mu(\xi) = g_\|^{\mu\nu} a_\nu(\xi)$, $r_\|^2(\xi) = a_\|^\mu(\xi)a_{\|\mu}(\xi)$. If one separates into $\|$- and $\perp$-subspaces by writing

$$T^{\alpha\beta}(\xi) = g_\|^{\alpha\beta}\xi + T_\perp^{\alpha\beta}(\xi), \tag{2.6.21}$$

then $T_\perp^{\alpha\beta}(\xi)$, which corresponds to the central four components in (2.3.6), is the only part that involves trignometric functions.

Various alternative forms of (2.6.20) follow by appealing to the identity (2.4.19), in this case with $ka(\xi)$ reinterpreted as $ka_\|(\xi)$.

Water-Bag and Bell Distributions

A water-bag distribution is one in which the distribution function is a constant between two limits and zero otherwise, so that the distribution function is discontinuous at these limits. In a bell distribution the step functions at the two limits are replaced by a power-law variation. These distribution functions are of the form

$$g(u) = (u_m^2 - u^2)^p \, H(u_m^2 - u^2), \tag{2.6.22}$$

where $H(x)$ is the step function, and where the two limits are at $u = \pm u_m$. The water-bag distribution corresponds to $p = 0$, and $p = 1, 2, 3$ have been called hard bell, soft bell and squishy bell, respectively. These distributions have the advantage that the RPDFs can be evaluated in terms of powers and logarithmic functions [29, 30]. The argument of the logarithms is $(z - v_m)/(z + v_m)$, with $u_m = \gamma_m v_m$, and this has an imaginary part for $-v_m < z < v_m$, where resonance is possible.

The RPDFs $z^2 W(z)$ and $z^2 R(z)$ are illustrated in Fig. 2.3 for two soft bell distributions with the same values of $\langle\gamma\rangle$ as the two Jüttner distributions in Fig. 2.2.

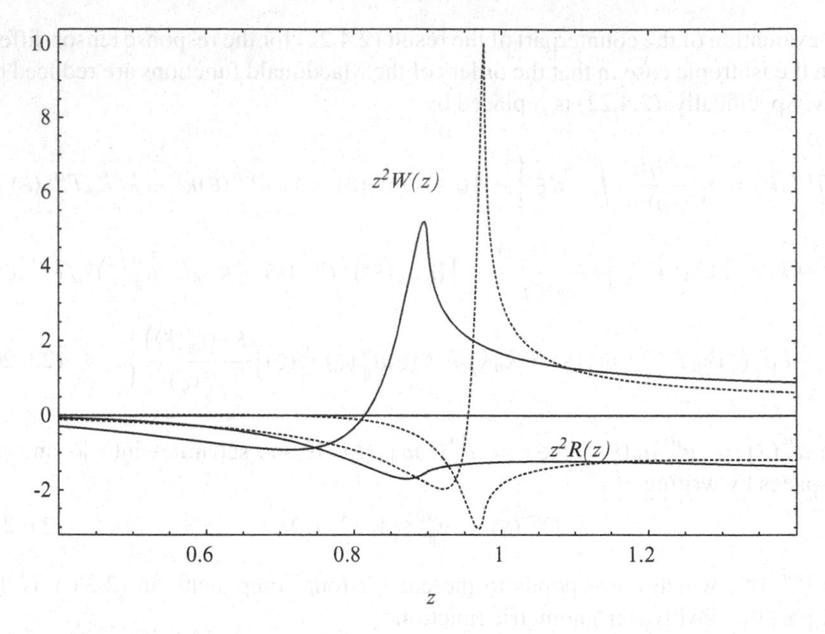

Fig. 2.3 As for Fig. 2.2 but for a soft bell distribution with $u_m = 0.977098$ (*dotted*) and $u_m = 0.9$ (*solid*) (From [29], reprinted with permission Cambridge University Press)

2.7 Response Tensor for a Synchrotron-Emitting Gas

Highly relativistic electrons in a magnetic field emit and absorb synchrotron radiation. Dispersion is related to dissipation, and the dispersion due to highly relativistic electrons is related to synchrotron absorption. For a thermal distribution, such dispersion can be treated using Trubnikov's response tensor in the limit $\rho \ll 1$ [31]. However, synchrotron-emitting particles typically have power-law energy spectra, and not thermal spectra. The response tensor for an arbitrary nonthermal distribution of synchrotron-emitting particles can be evaluated by making the highly relativistic approximation [32, 33], which involves Airy functions.

2.7.1 Synchrotron Approximation

The synchrotron approximation involves expanding in inverse powers of the Lorentz factor of the radiating electron.

Method of Stationary Phase

In the synchrotron approximation, most of the emission is strongly beamed in the forward direction, implying that the difference between the angle of emission and

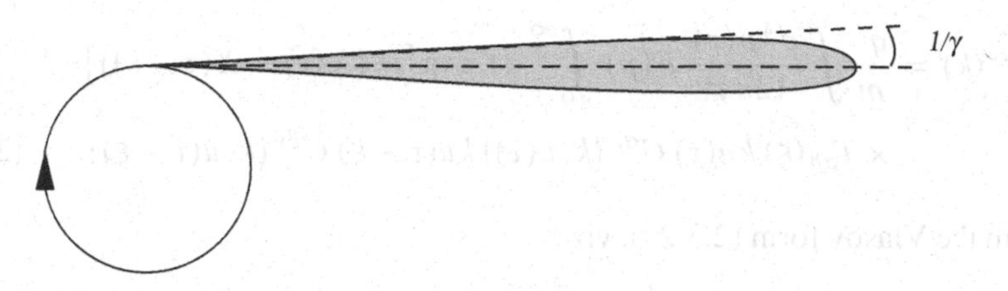

Fig. 2.4 Radiation from a highly relativistic particle moving in a circle is seen as a sequence of pulses, one per gyroperiod, by observers whose line of sight is within an angle $\approx 1/\gamma$ of the plane of the circle

the pitch angle, $|\theta - \alpha|$, is of first order in $1/\gamma$, cf. Fig. 2.4. An observer at $\theta \approx \alpha$ receives a pulse of radiation each time the particle's motion is directly towards the observer, who can see the particle only for a fraction $\approx 1/\gamma$ of its orbit each gyroperiod. Consider the integral over proper time, ξ, in the expression (2.3.4) or (2.3.24) for the response tensor. The proper time parameterizes the gyrophase, and the foregoing argument implies that the dominant contribution to the ξ-integral is from a small range of ξ corresponding to emission in the direction of the observer. It is appropriate to approximate such an integral using the method of stationary phase.

The method of stationary phase applies to an integral, over z say, in which the integrand contains a phase factor, $\exp[if(z)]$ say, such that the integral is dominated by points of stationary phase, where $f'(z) = df(z)/dz$ is zero. Assuming only one stationary phase point, at $z = z_0$, the stationary phase approximation corresponds to

$$\int dz\, G(z)\, e^{if(z)} \approx G(z_0)\left(\frac{i\pi}{f''(z_0)}\right)^{1/2} e^{if(z_0)}, \tag{2.7.1}$$

where $G(z)$ is a slowly varying function. The derivation of (2.7.1) involves expanding in a Taylor series about $z = z_0$, with $f'(z_0) = 0$. Only terms up to second order are retained in the phase $f(z) = f(z_0) + \frac{1}{2}(z - z_0)^2 f''(z_0)$, and $G(z) \approx G(z_0)$ is regarded as a constant. The resulting integral gives (2.7.1). In the application considered here, integrals appear in which $G(z)$ has a zero near $z = z_0$. To evaluate such integrals one appeals either to Hermite integration or, more simply, to a partial integration to find

$$\int dz \begin{bmatrix} (z - z_0)^2 \\ (z - z_0)^4 \end{bmatrix} e^{if(z)} \approx \left(\frac{i\pi}{f''(z_0)}\right)^{1/2} \begin{bmatrix} i/f''(z_0) \\ -3/[f''(z_0)]^2 \end{bmatrix} e^{if(z_0)}, \tag{2.7.2}$$

and so on, with integrals of odd powers of $z - z_0$ vanishing.

Synchrotron Approximation to the Response Tensor

The response tensor in the synchrotron limit is obtained by starting either from (2.7.10), viz.

$$\Pi^{\mu\nu}(k) = \frac{q^2}{m}\int \frac{d^4p(\tau)}{(2\pi)^4}\,F(p)\int_0^\infty d\xi\,\exp\left[ik\big(X(\tau)-X(\tau-\xi)\big)\right]$$
$$\times\, T_{\alpha\beta}(\xi)\,ku(\tau)\,G^{\alpha\mu}\big(k,u(\tau)\big)\,ku(\tau-\xi)\,G^{\beta\nu}\big(k,u(\tau-\xi)\big), \qquad (2.7.3)$$

or from the Vlasov form (2.3.24), viz.

$$\Pi^{\mu\nu}(k) = -iq^2\int \frac{d^4p(\tau)}{(2\pi)^4}\int_0^\infty d\xi\,u^\mu(\tau)\,e^{ik[X(\tau)-X(\tau-\xi)]}$$
$$\times\, ku(\tau-\xi)\,G^{\alpha\nu}\big(k,u(\tau-\xi)\big)\,t_\alpha{}^\beta(\tau-\xi)\,\frac{\partial F(p)}{\partial p^\beta}. \qquad (2.7.4)$$

The phase factor in the integrand is $k[X(\tau)-X(\tau-\xi)]$, with the orbit given by $X^\mu(\tau) = x_0^\mu + t^{\mu\nu}(\tau)u_{0\nu}$. The initial 4-velocity, u_0^μ includes an arbitrary initial gyrophase, ϕ, and $X^\mu(\tau)$ involves trignometric functions of both $\phi-\Omega_0\tau$ and $\phi-\Omega_0(\tau-\xi)$. The integral over $d^4p(\tau)$ contains an integral over gyrophase, which can be written either as an integral over ϕ or as an integral over $\Omega_0\tau$.

Using (2.1.21) and (2.1.22) one finds

$$k[X(\tau)-X(\tau-\xi)] = (ku_0)_\parallel\xi - k_\perp R[\sin(\phi-\epsilon\psi)-\sin(\phi-\epsilon\psi-\Omega_0\xi)], \qquad (2.7.5)$$

with $\phi=\phi_0+\Omega_0\tau$. The points of stationary phase with respect to ϕ occur at

$$\cos(\phi-\epsilon\psi) = \cos(\phi-\epsilon\psi-\Omega_0\xi). \qquad (2.7.6)$$

One solution is at $\phi-\epsilon\psi = \frac{1}{2}\Omega_0\xi$. There are two solutions each period, and the two solutions contribute equally. Thus the integral over ϕ is approximated by making the stationary phase approximation for one of these solutions, and multiplying the result by 2 to take account of the other. After applying the method of stationary phase to the ϕ-integral, the phase factor (2.7.5) is approximated by

$$k[X(\tau)-X(\tau-\xi)] = (ku_0)_\parallel\xi - 2k_\perp R\sin(\tfrac{1}{2}\Omega_0\xi). \qquad (2.7.7)$$

Writing $k_\perp = |\boldsymbol{k}|\sin\theta$, $v_\perp = v\sin\alpha$, (2.7.7) gives

$$k[X(\tau)-X(\tau-\xi)] = \gamma\xi[(\omega-|\boldsymbol{k}|v\cos(\alpha-\theta)]$$
$$-\gamma|\boldsymbol{k}|v\sin\alpha\sin\theta[\xi-2\sin(\tfrac{1}{2}\Omega_0\xi)]. \qquad (2.7.8)$$

The integral over pitch angle can also be approximated using the method of stationary phase. The points of stationary phase in the α-integral occur at

$$\sin(\alpha-\theta) = \cos\alpha\sin\theta\left[1-\frac{2\sin(\frac{1}{2}\Omega_0\xi)}{\Omega_0\xi}\right]. \qquad (2.7.9)$$

Thus the condition for stationary phase for the α-integral can be approximated by

$$\sin(\alpha - \theta) = \pi_1, \qquad \pi_1 = \frac{(\Omega_0\xi)^2}{24} \sin\theta\cos\theta. \tag{2.7.10}$$

As for the ϕ-integral, there are two stationary phase points that contribute equally, and only one need be retained with the result multiplied by 2.

In evaluating the integrals by the method of stationary phase, it is convenient to write

$$y = \Omega_0\xi, \qquad a = \frac{\gamma(\omega - |\boldsymbol{k}|v)}{\Omega_0}, \qquad b = \frac{\gamma|\boldsymbol{k}|v\sin^2\theta}{8\Omega_0}, \tag{2.7.11}$$

and to change variables to $\delta\phi = \phi - \xi/2$, $\delta\alpha = \alpha - \theta - \pi_1$. The phase factor in (2.7.4) reduces to

$$k[X(\tau) - X(\tau - \xi)] = ay + \frac{by^3}{3} + 4by\,\frac{(\delta\alpha)^2}{\sin^2\theta} + 4by\,\frac{\sin\alpha}{\sin\theta}\,(\delta\phi)^2, \tag{2.7.12}$$

and in the integrand one makes the replacement $\Omega_0\tau = \delta\phi + \frac{1}{2}y$, $\alpha = \delta\alpha + \theta + \pi_1$. The variables $y = \Omega_0\xi$, $\delta\alpha$ and $\delta\phi$ are all small quantities of $O(\gamma^{-1})$. The correction π_1 is $O(\gamma^{-2})$.

2.7.2 Expansion About a Point of Stationary Phase

On making the foregoing approximations, one needs to evaluate the slowly varying functions in the integrand in (2.7.3) or (2.7.4) before performing the integral over ξ or $y = \Omega_0\xi$.

It is conventional to describe a distribution of highly relativistic particles in terms of the energy spectrum, which can be written in terms of Lorentz factors. Let $N(\gamma)d\gamma$ be the number density of particles in the range γ to $\gamma + d\gamma$, with $\phi(\alpha)$ the pitch-angle distribution. The normalization conditions are

$$\int d\gamma\, N(\gamma) = 1, \qquad \int d\cos\alpha\,\phi(\alpha) = 2. \tag{2.7.13}$$

The integral over pitch angle is evaluated by the method of stationary phase, and $\phi(\alpha)$ is evaluated at $\alpha = \theta + \delta\alpha + \pi_1$, giving

$$\phi(\alpha) = \phi(\theta)\left[1 + \delta\alpha\, g(\theta)\frac{\cos\theta}{\sin\theta}\right], \qquad \sin\alpha = \sin\theta\left[1 + \delta\alpha\,\frac{\cos\theta}{\sin\theta}\right], \tag{2.7.14}$$

with $g(\theta) = \tan\theta\,\phi'(\theta)/\phi(\theta)$.

The contribution to dispersion from the highly relativistic particles is usually of interest only for relatively high frequency waves, where the waves are transverse to an excellent approximation. It is convenient to project the response tensor onto the two transverse 4-vectors

$$t^\mu = [0, \boldsymbol{t}], \quad \boldsymbol{t} = (\cos\theta, 0, -\sin\theta), \qquad a^\mu = [0, \boldsymbol{a}], \quad \boldsymbol{a} = (0, 1, 0). \quad (2.7.15)$$

The components along these 4-vectors are denoted as the t- and a-components, respectively. For the response tensor in the forward-scattering form (2.3.4) the following expansions are required:

$$T^{\mu\nu}(\xi) = \frac{1}{\Omega_0}\begin{pmatrix} -y & \frac{1}{2}\epsilon\, y^2 \\ -\frac{1}{2}\epsilon\, y^2 & -y \end{pmatrix}, \qquad k_\alpha k_\beta T^{\alpha\beta}(\xi) = \frac{|\boldsymbol{k}|^2 \sin^2\theta}{\Omega_0}\frac{y^3}{6},$$

$$k_\beta T^{\mu\beta}(\xi) = \frac{|\boldsymbol{k}|\sin\theta}{\Omega_0}(-\cos\theta\, \tfrac{1}{6}y^3, \tfrac{1}{2}\epsilon\, y^2),$$

$$k_\alpha T^{\alpha\nu}(\xi) = \frac{|\boldsymbol{k}|\sin\theta}{\Omega_0}(-\cos\theta\, \tfrac{1}{6}y^3, -\tfrac{1}{2}\epsilon\, y^2), \quad (2.7.16)$$

with $y = \Omega_0\xi$. The matrix representation in (2.7.16) is restricted to the 2D transverse subspace, corresponding to $\mu, \nu = t, a$, with t^μ, a^μ, defined by (2.7.15). The Vlasov involves a derivative of the distribution function; this gives

$$ku(\tau)\, G^{\alpha\nu}(k, u(\tau))\, \dot{t}_\alpha{}^\beta(\tau)\frac{\partial}{\partial p^\beta} = [ku(\tau)\, \tilde{u}^\nu - k\tilde{u}\, u^\nu(\tau)]\frac{1}{m\gamma}\frac{\partial}{\partial\gamma}$$

$$-[ku(\tau)\, a^\nu(\tau) - ka(\tau)\, u^\nu(\tau)]\frac{1}{m\gamma^2\beta^2}\frac{\partial}{\partial\alpha},$$

$$a^\mu(\tau) = \frac{d}{d\alpha}u^\mu(\tau), \qquad ka(\tau) = \gamma|\boldsymbol{k}|\beta\,\delta\alpha, \quad (2.7.17)$$

where only the leading term in the approximate expression for $ka(\tau)$ is retained.

2.7.3 Transverse Components of the Response Tensor

There are only three independent transverse components for the response tensor, and it is convenient to write them in terms of $t^{\mu\nu} = \mu_0\Pi^{\mu\nu}$, with the 2D transverse subspace spanned by the 4-vectors t^μ, a^μ, defined by (2.7.15). The three independent components contain a class of integrals that need to be evaluated by the method of stationary phase. This class is defined by

$$I^{(N)}(a, b) = \int_0^\infty dy\, y^N\, \exp[iay + i\tfrac{1}{3}by^3], \quad (2.7.18)$$

with N an integer.

The forward-scattering form for the response tensor gives

$$t^{\mu\nu}(k) = -\frac{\omega_{p0}^2\Omega_0}{\omega}\phi(\theta)\int d\gamma\,\frac{N(\gamma)}{\gamma^2\beta}\,J^{\mu\nu}(a,b), \tag{2.7.19}$$

with $\beta \to 1$ here. There are three independent components:

$$\begin{aligned}
J^{tt}(a,b) &= \tfrac{4}{3}bI^{(1)}(a,b) - a, \quad J^{aa}(a,b) = 4bI^{(1)}(a,b) + \tfrac{8}{3}ia^2I^{(0)}(a,b) + \tfrac{5}{3}a,\\
J^{ta}(a,b) &= \tfrac{1}{2}\epsilon\cos\theta\,\{[\tfrac{16}{9}ia^2I^{(1)}(a,b) - \tfrac{20}{9}aI^{(0)}(a,b) - 2iI^{(-1)}(a,b) + 2i]\\
&\quad + g(\theta)\,[-\tfrac{4}{3}aI^{(0)}(a,b) - 2iI^{(-1)}(a,b) + i\tfrac{1}{3}]\}.
\end{aligned} \tag{2.7.20}$$

The Vlasov form (2.7.4) leads to

$$\begin{aligned}
t^{\mu\nu}(k) &= t^{\mu\nu}_{(\gamma)}(k) + t^{\mu\nu}_{(\alpha)}(k),\\
t^{\mu\nu}_{\gamma}(k) &= -\frac{\omega_{p0}^2\Omega_0}{\omega}\phi(\theta)\int d\gamma\,\gamma\beta\frac{d}{d\gamma}\left[\frac{N(\gamma)}{\gamma^2\beta}\right]h^{\mu\nu},\\
t^{ta}_{\alpha}(k) &= \frac{\omega_{p0}^2\Omega_0}{\omega}\phi(\theta)\int d\gamma\,\frac{N(\gamma)}{\gamma}\,(-\tfrac{1}{2}\epsilon\cos\theta)\,[-g(\theta)\,iI^{(-1)}(a,b)],
\end{aligned} \tag{2.7.21}$$

with $h^{\mu\nu}$ given by

$$\begin{pmatrix} h^{tt}\\ h^{aa}\\ h^{ta}\end{pmatrix} = \begin{pmatrix} -aI^{(-1)}(a,b) - bI^{(1)}(a,b)\\ -aI^{(-1)}(a,b) - 3bI^{(1)}(a,b)\\ -\tfrac{1}{2}\epsilon\cos\theta[(2+g(\theta))iI^{(-1)}(a,b) + \tfrac{4}{3}aI^{(0)}(a,b) - i\tfrac{4}{3}]\end{pmatrix}. \tag{2.7.22}$$

The forms (2.7.19) and (2.7.21) are related to each other by a partial integration. However, the two expressions (2.7.19) and (2.7.21) are not equivalent because the constant terms in the diagonal components in (2.7.19) are not reproduced by the partial integration of (2.7.21). This inconsistency reflects a more fundamental problem in taking the ultrarelativistic limit.

Ultrarelativistic Limit

In the ultrarelativistic limit the gyroradii of the particles become arbitrarily large and their gyroperiods become arbitrarily long. The particle motions are better approximated by constant rectilinear motion than by gyromagnetic motion in this extreme limit. The inconsistencies that arise in taking the ultrarelativistic limit are related to this breakdown in the assumption that the motions are gyromagnetic. One procedure that avoids the inconsistencies is to note that in the extreme limit the system is effectively unmagnetized, and to use this fact to remove the terms that cause the inconsistencies. Specifically, in taking the ultrarelativistic limit one first subtracts the constant terms in (2.7.19) and (2.7.21) that remain in this limit, ensuring that the remaining terms are well behaved. One then adds the known result for the response tensor for an unmagnetized plasma to replace the subtracted terms. This procedure involves separating the response into two parts: the unmagnetized part and the magnetic correction to it. This separation corresponds to

$$t^{\mu\nu}(k) = t_0^{\mu\nu}(k) + t_{\rm mag}^{\mu\nu}(k), \tag{2.7.23}$$

where $\mu, \nu = 1, 2$ span only the 2D transverse plane. The unmagnetized part corresponds to $t_0^{\mu\nu}(k) = \omega_{p0}^2 \delta^{\mu\nu}$ at high frequencies, where ω_{p0} is the proper plasma frequency. The magnetized part, $t_{\rm mag}^{\mu\nu}(k)$, in (2.7.23) describes the dispersion associated with the synchrotron emitting particles. The method developed here gives $t_{\rm mag}^{\mu\nu}(k)$ when terms independent of the magnetic field are discarded.

2.7.4 Airy Integral Approximation

The integrals $I^{(n)}(a,b)$ that appear in (2.7.19) and (2.7.21) may be evaluated in terms of the Airy functions ${\rm Ai}\,(z)$ and ${\rm Gi}\,(z)$, cf.̃(A.1.21). Identifying $z = a/b^{1/3}$, one finds

$$\begin{aligned}
I^{(0)}(a,b) &= \pi b^{-1/3}\big[{\rm Ai}\,(z) + i{\rm Gi}\,(z)\big],\\
I^{(1)}(a,b) &= -i\pi b^{-1/3}\big[{\rm Ai}'\,(z) + i{\rm Gi}'\,(z)\big],\\
I^{(-1)}(a,b) &= i\pi \int_0^z dz'\,\big[{\rm Ai}\,(z') + i{\rm Gi}\,(z')\big].
\end{aligned} \tag{2.7.24}$$

The Airy function ${\rm Ai}\,(z)$ can be represented as a Bessel function of order $1/3$, and in the case of relevance to synchrotron emission these are Macdonald functions. The relevant representations are

$$\begin{aligned}
{\rm Re}\,I^{(0)}(a,b) &= \frac{1}{\sqrt{3}}\left(\frac{a}{b}\right)^{1/2} K_{1/3}(R), \qquad {\rm Im}\,I^{(1)}(a,b) = \frac{1}{\sqrt{3}}\frac{a}{b}\,K_{2/3}(R),\\
{\rm Im}\,I^{(-1)}(a,b) &= -\frac{1}{\sqrt{3}}\int_R^\infty dt\,K_{1/3}(t), \qquad R = \frac{2a^{3/2}}{3b^{1/2}}.
\end{aligned} \tag{2.7.25}$$

These functions appear in the discussion of synchrotron emission and absorption in §4.4.

Response at High Frequencies

The hermitian part of the response involves the function Gi (z). Approximate forms of Gi (z) are available for large and small z. The asymptotic expansion for $z \gg 1$ is given by (A.1.23), and the expansion for $z \ll 1$ by (A.1.24). The case of most interest is high frequencies, corresponding to $a \gg 1$. In this case the asymptotic expansion gives

$$I^{(0)}(a,b) = \frac{i}{a}\left(1 + \frac{2b}{a^3}\right), \qquad I^{(1)}(a,b) = -\frac{1}{a^2}, \qquad I^{(-1)}(a,b) = -\ln a, \tag{2.7.26}$$

with $a = \omega/2\Omega_0\gamma$.

The leading terms in the hermitian part of the response tensor in the form (2.7.19) become

$$J^{tt}(a,b) = -\frac{2}{3}\frac{\Omega_0}{\omega}\gamma^3\sin^2\theta, \qquad J^{aa}(a,b) = 7J^{tt}(a,b),$$
$$J^{ta}(a,b) = -\tfrac{1}{2}\,i\epsilon\cos\theta[1 + g(\theta)]\,\ln\left(\frac{\omega}{2\Omega_0\gamma}\right), \tag{2.7.27}$$

where the logarithmic term is assumed to dominate in $t^{ta}(k)$. A result similar to (2.7.27) with (2.7.19) was derived in a different way by Sazonov [32].

Dispersion in a synchrotron-emitting electron gas is determined primarily by the tt- and aa-components. The natural modes are approximately linearly polarized, with electric vectors along the t- and a-directions. The circularly polarized component can be regarded as a correction of order $1/\gamma$.

2.7.5 *Power-Law and Jüttner Distributions*

Dispersion in a synchrotron-emitting gas is of most interest for a power-law distribution. It is of interest to compare this with the case of a highly relativistic Jüttner distribution.

Power-Law Distribution

For present purposes a power-law distribution is defined by

$$N(\gamma) = \begin{cases} K\gamma^{-a} & \text{for } \gamma_1 < \gamma < \gamma_2, \\ 0 & \text{otherwise,} \end{cases}$$

$$K = \begin{cases} (a-1)/(\gamma_1^{1-a} - \gamma_2^{1-a})^{-1} & \text{for } a \neq 1, \\ \ln(\gamma_2/\gamma_1) & \text{for } a = 1, \end{cases} \tag{2.7.28}$$

Inserting (2.7.28) into (2.7.19) with (2.7.27) gives

$$t^{\mu\nu}(k) = \omega_{p0}^2 \delta^{\mu\nu} + \Delta t^{\mu\nu}(k), \tag{2.7.29}$$

with $\delta^{\mu\nu}$ the unit tensor in the 2D subspace, and with

$$\begin{pmatrix} \Delta t^{tt}(k) \\ \Delta t^{aa}(k) \end{pmatrix} = \frac{\omega_{p0}^2 \Omega_0^2}{\omega^2} \frac{2\zeta\phi(\theta)\sin^2\theta}{3} \begin{pmatrix} 1 \\ 7 \end{pmatrix},$$

$$\Delta t^{ta}(k) = \frac{i\epsilon}{2} \frac{\omega_{p0}^2 \Omega_0}{\omega} \cos\theta\phi(\theta)[1+g(\theta)] \frac{a-1}{a-2} \frac{1}{\gamma_1} \ln\left(\frac{\omega}{\Omega_0 \gamma_1 \sin\theta}\right), \tag{2.7.30}$$

$$\zeta = \begin{cases} \dfrac{a-1}{a-2}\gamma_1 & \text{for } a > 2, \\ \dfrac{a-1}{2-a}(\gamma_1^{1-a} - \gamma_2^{1-a})^{-1} \left(\dfrac{\omega}{\Omega_0 \sin\theta}\right)^{(2-a)/2} & \text{for } a < 2, \end{cases} \tag{2.7.31}$$

with the proviso that the latter approximation applies only for frequencies $(\omega/\Omega_0 \sin\theta)^{1/2} \lesssim \gamma_2$.

Ultrarelativistic Thermal Distribution

Application of the foregoing method to a Jüttner distribution with a highly relativistic temperature, $\rho \ll 1$, is of formal interest in comparing the method with Trubnikov's method.

For a highly relativistic Jüttner distribution one has

$$N(\gamma) \approx \frac{1}{2}\gamma^2 \rho^3 e^{-\rho\gamma}. \tag{2.7.32}$$

The result (2.7.32) follows from the Jüttner distribution (2.7.28) for $\gamma \gg 1$ and $\rho \ll 1$, with $K_2(\rho) \approx 2/\rho^2$ for $\rho \ll 1$.

The resulting expression for the transverse components are of the form (2.7.23), with the magnetized part, $t_{\text{mag}}^{\mu\nu}(k)$, having diagonal components

$$t_{\text{mag}}^{11}(k) = \frac{2\omega_{p0}^2 \Omega_0^2 \sin^2\theta}{\rho^2 \omega^2}, \qquad t_{\text{mag}}^{22}(k) = 7t_{\text{mag}}^{11}(k), \tag{2.7.33}$$

and off-diagonal components

$$t^{12}_{\rm mag}(k) = -t^{21}_{\rm mag}(k) = i\epsilon\cos\theta\,\frac{\omega_{p0}^2\Omega_0}{\omega}\,\ln\left(\frac{\omega\rho}{2\Omega_0}\right). \tag{2.7.34}$$

Alternatively, the result (2.7.33) and (2.7.34) may be derived by making the ultra-relativistic approximation in Trubnikov's response tensor. A relevant approximation to the function $r(\xi)$, defined by (2.4.10), is

$$r^2(\xi) = r_0^2(\xi) + \delta r^2(\xi), \qquad r_0^2(\xi) = -2i\omega\rho\xi,$$
$$\delta r^2(\xi) = -\frac{\omega^2\sin^2\theta}{\Omega_0^2}\frac{(\Omega_0\xi)^4}{12}, \tag{2.7.35}$$

where a term ρ^2 is neglected, and with $k_z^2 \to \omega^2\cos^2\theta$, $k_\perp^2 \to \omega^2\sin^2\theta$. In evaluating $t^{\mu\nu}_{\rm mag}(k)$ one is to subtract the contribution that is nonzero for $B \to 0$. Inspection of the response tensor (2.4.13) shows that the term proportional to $K_2\big(r(\xi)\big)/r^2(\xi)$ has a nonzero contribution for $B \to 0$, and subtracting it using (2.7.35) gives

$$\frac{K_2\big(r(\xi)\big)}{r^2(\xi)} - \frac{K_2\big(r_0(\xi)\big)}{r_0^2(\xi)} = -\frac{\delta r^2(\xi)}{2}\frac{K_3\big(r_0(\xi)\big)}{r_0^3(\xi)}. \tag{2.7.36}$$

The integral over ξ in (2.4.13) can then be evaluated, either using the identity

$$\int_0^\infty dx\, x^\mu\, K_\nu(ax) = 2^{\mu-1}a^{-\mu-1}\,\Gamma\left(\frac{1+\mu+\nu}{2}\right)\Gamma\left(\frac{1+\mu-\nu}{2}\right), \tag{2.7.37}$$

or by first making the approximation $K_n(r_0) \approx 2^{n-1}(n-1)!/r_0^n$ for $|r_0| \ll 1$. The result (2.7.33) for the diagonal terms is reproduced. For the off-diagonal term, and integral of the form $\int d\xi\,\xi\,K_2(r_0)/r_0^2$ appears, and with $r_0^2 \propto \xi$ and $K_2(r_0) \approx 2/r_0^2$, this integral is logarithmically divergent. The result (2.7.34) is reproduced, except for the argument of the logarithm. This argument is not well determined: as in (2.7.30) it effectively depends on a lower energy cutoff, which is ill-defined for a relativistic thermal distribution.

2.8 Nonlinear Response Tensors

Covariant forms for the quadratic and cubic response tensors may be derived either for a cold plasma or for an arbitrary distribution of particle by extending the methods used in § 1.2 and § 2.1, respectively.

2.8.1 Quadratic Response Tensor for a Cold Plasma

The quadratic response tensor for a cold plasma can be derived from the second order terms in the expansion (1.2.11) of the induced current, that is, from

$$J^{(2)\mu}(k) = q\left[n\, u^{(2)\mu}(k) + n^{(2)}(k)\, \tilde{u}^{\mu} + \int d\lambda^{(2)}\, n^{(1)}(k_1) u^{(1)\mu}(k_2)\right]. \tag{2.8.1}$$

The first order terms $n^{(1)}(k)$, $u^{(1)\mu}(k)$ are given by (1.2.13) and (1.2.15). The second order term $n^{(2)}(k)$ follows from the quadratic terms in (1.2.4),

$$k\tilde{u}\, n^{(2)}(k) = -\int d\lambda^{(2)}\, n^{(1)}(k_1)\, ku^{(1)}(k_2). \tag{2.8.2}$$

The second order term $u^{(2)\mu}(k)$ follows from the quadratic terms in (1.2.5), which give

$$\begin{aligned} u^{(2)\mu}(k) = \tau^{\mu}{}_{\rho}(k\tilde{u}) \Bigg[& -\int d\lambda^{(2)}\, k_2 u^{(1)}(k_1)\, u^{(1)\rho}(k_2) \\ & -\frac{q}{m}\int d\lambda^{(2)}\, k_1 u^{(1)}(k_2)\, G^{\rho\nu}\big(k_1, u^{(1)}(k_2)\big) A_{\nu}(k_1)\Bigg]. \end{aligned} \tag{2.8.3}$$

The quadratic response tensor follows by writing (2.8.1) in the form defined by the second order term in the weak turbulence expansion (1.1.31). One needs to symmetrize the result over $\nu, k_1 \leftrightarrow \rho, k_2$, to avoid the result depending on the details of the calculation. It is convenient to write $k_0 = -k$, so that one has $k_0^{\mu} + k_1^{\mu} + k_2^{\mu} = 0$. The result is then symmetric under $\mu, k_0 \leftrightarrow \nu, k_1 \leftrightarrow \rho, k_2$. The resulting form is

$$\Pi^{(2)\mu\nu\rho}(k_0, k_1, k_2) = -\frac{q^3 n}{2m^2} G^{\alpha\mu}(k_0, \tilde{u}) G^{\beta\nu}(k_1, \tilde{u}) G^{\gamma\rho}(k_2, \tilde{u})\, f_{\alpha\beta\gamma}(k_0, k_1, k_2, \tilde{u}), \tag{2.8.4}$$

with

$$\begin{aligned} f^{\alpha\beta\gamma}(k_0, k_1, k_2, u) = -\Bigg[& \frac{k_{1\sigma}}{k_0 u}\tau^{\sigma\alpha}(k_0 u)\tau^{\beta\gamma}(k_2 u) + \frac{k_{2\sigma}}{k_0 u}\tau^{\sigma\alpha}(k_0 u)\tau^{\gamma\beta}(k_1 u) \\ & + \frac{k_{0\sigma}}{k_1 u}\tau^{\sigma\beta}(k_1 u)\tau^{\alpha\gamma}(k_2 u) + \frac{k_{0\sigma}}{k_2 u}\tau^{\sigma\gamma}(k_2 u)\tau^{\beta\alpha}(k_1 u) \\ & + \frac{k_{1\sigma}}{k_2 u}\tau^{\sigma\gamma}(k_2 u)\tau^{\alpha\beta}(k_0 u) + \frac{k_{2\sigma}}{k_1 u}\tau^{\sigma\beta}(k_1 u)\tau^{\gamma\alpha}(k_0 u)\Bigg], \end{aligned} \tag{2.8.5}$$

where $u^{\mu} = u_{\parallel}^{\mu}$ is confined to the 0–3 plane.

It is straightforward to use the cold plasma approach to derive an expression for the cubic response tensor. The result (2.8.4) with (2.8.5) for the quadratic response tensor and the corresponding cubic response tensor may be derived from the expressions derived below using the forward-scattering method, by assuming a δ-function distribution of particles.

2.8.2 Higher Order Currents

Using the forward-scattering method (§ 2.3.1), the nonlinear response tensors can be calculated for an arbitrary distribution of particles from the nonlinear terms in the expansion of the single-particle current.

A general form for the nth order current, which reduces to the linear current (2.2.14) for $n = 1$, is

$$J_{\rm sp}^{(n)\mu}(k) = \frac{q^{n+1}}{m^n}\int\frac{d^4k_1}{(2\pi)^4}\cdots\int\frac{d^4k_n}{(2\pi)^4}\,e^{-i(k-k_1-\cdots-k_n)x_0}$$
$$\times\sum_{a,a_1,\ldots,a_n} e^{i\epsilon(a\psi-a_1\psi_1-\cdots-a_n\psi_n)}\,\beta^{(n)\mu\nu_1\ldots\nu_n}(a,k;a_1,k_1;\ldots;a_n,k_n)$$
$$\times 2\pi\delta\big[(ku)_\parallel-(k_1u)_\parallel-\cdots(k_nu)_\parallel-(a-a_1-\cdots-a_n)\Omega_0\big]. \tag{2.8.6}$$

The nth order induced current in the plasma due to a single species of particle is found by multiplying $J_{\rm sp}^{(n)\mu}(k)$ by $d^4x_0d^4p\,F(p)$, with $p = mu$, and integrating. The integral over x_0 is trivial, giving $(2\pi)^4\delta^4(k-k_1-\cdots-k_n)$. The integrals over $k_1,\ldots,k_n$ combined with this δ-function correspond to the n-fold convolution integral (1.1.32).

The explicit expression for the integrand in (2.8.6) for $n = 2$ is

$$\beta^{(2)\mu\nu_1\nu_2}(a,k;a_1,k_1;a_2,k_2)$$
$$= G^{\alpha\mu}(a,k,u)G^{*\beta\nu}(a_1,k_1,u)G^{*\gamma\rho}(a_2,k_2,u)\,f_{\alpha\beta\gamma}(a,k;a_1,k_1;a_2,k_2;u),$$
$$f^{\alpha\beta\gamma}(a,k;a_1,k_1;a_2,k_2;u) = \frac{1}{2}\bigg[\frac{k_{1\sigma}\tau^{\alpha\sigma}(\tilde\omega)\tau^{\beta\gamma}(\tilde\omega_2)}{\tilde\omega}+\frac{k_\sigma\tau^{\sigma\beta}(\tilde\omega_1)\tau^{\alpha\gamma}(\tilde\omega_2)}{\tilde\omega_1}$$
$$+\frac{k_{1\sigma}\tau^{\sigma\gamma}(\tilde\omega_2)\tau^{\alpha\beta}(\tilde\omega)}{\tilde\omega_2}+(\nu,a_1,k_1)\leftrightarrow(\rho,a_2,k_2)\bigg], \tag{2.8.7}$$

with $\tilde\omega = (ku)_\parallel - a\Omega_0$, $\tilde\omega_n = (k_nu)_\parallel - a_n\Omega_0$, and where the final entry in (2.8.7) implies three additional terms obtained from those written by making the interchanges indicated.

The explicit expressions for $n = 3$ is

$$\beta^{(3)\mu\nu\rho\sigma}(k_0, k_1, k_2, k_3) = \sum_{a_0+a_1+a_2+a_3=0} e^{-i\epsilon(a_0\psi_0+a_1\psi_1+a_2\psi_2+a_3\psi_3)}$$

$$\times G^{*\alpha\mu}(a_0, k_0, u) G^{*\beta\nu}(a_1, k_1, u) G^{*\gamma\rho}(a_2, k_2, u) G^{*\delta\sigma}(a_3, k_3, u)$$

$$\times f_{\alpha\beta\gamma\delta}(a_0, k_0; a_1, k_1; a_2, k_2; a_3, k_3; u), \tag{2.8.8}$$

$$f^{\alpha\beta\gamma\delta}(a_0, k_0; a_1, k_1; a_2, k_2; a_3, k_3; u)$$

$$= \left(\frac{k_{1\eta}\tau^{\eta\alpha}(\tilde{\omega}_0) g^{\beta}{}_{\phi}}{\tilde{\omega}_0} - \frac{k_{1\phi}\tau^{\beta\alpha}(\tilde{\omega}_0) + k_{0\phi}\tau^{\alpha\beta}(\tilde{\omega}_1)}{\tilde{\omega}_0 + \tilde{\omega}_1} + \frac{k_{0\eta}\tau^{\eta\beta}(\tilde{\omega}_1) g^{\alpha}{}_{\phi}}{\tilde{\omega}_1} \right) \tau^{\theta\phi}(\tilde{\omega}_0 + \tilde{\omega}_1)$$

$$\times \left(\frac{k_{3\eta}\tau^{\eta\gamma}(\tilde{\omega}_2) g^{\delta}{}_{\theta}}{\tilde{\omega}_2} - \frac{k_{2\theta}\tau^{\gamma\delta}(\tilde{\omega}_3) + k_{3\theta}\tau^{\delta\gamma}(\tilde{\omega}_2)}{\tilde{\omega}_2 + \tilde{\omega}_3} + \frac{k_{2\eta}\tau^{\eta\delta}(\tilde{\omega}_3) g^{\gamma}{}_{\theta}}{\tilde{\omega}_3} \right)$$

$$+ (1, \nu) \leftrightarrow (2, \rho) + (1, \nu) \leftrightarrow (3, \sigma)$$

$$+ \frac{k_{0\theta} k_{0\eta} \tau^{\alpha\beta}(\tilde{\omega}_1) \tau^{\theta\gamma}(\tilde{\omega}_2) \tau^{\eta\delta}(\tilde{\omega}_3)}{\tilde{\omega}_2 \tilde{\omega}_3} + 11 \text{ other terms}, \tag{2.8.9}$$

where $+(1, \nu) \leftrightarrow (2, \rho) + (1, \nu) \leftrightarrow (3, \sigma)$ refers to two other terms obtained from that written by making the interchanges indicated, and where "11 other terms" refers to those obtained by completely symmetrizing over $(0, \mu)$, $(1, \nu)$, $(2, \rho)$, $(3, \sigma)$.

2.8.3 Quadratic Response Tensor for Arbitrary Distribution

The general expression for the quadratic response tensor for an arbitrary distribution, $F(p)$, of magnetized particles may be written in a form that is closely analogous to the cold plasma form (2.8.4) with (2.8.5). With $k_0 = -k$, this form is

$$\Pi^{(2)\mu\nu\rho}(k_0, k_1, k_2) = -\frac{q^3}{2m^2} \int \frac{d^4p}{(2\pi)^4} F(p) \sum_{a_0+a_1+a_2=0} e^{-i\epsilon(a_0\psi_0+a_1\psi_1+a_2\psi_2)}$$

$$\times G^{*\alpha\mu}(a_0, k_0, u) G^{*\beta\nu}(a_1, k_1, u) G^{*\gamma\rho}(a_2, k_2, u) f_{\alpha\beta\gamma}(a_0, k_0; a_1, k_1; a_2, k_2; u), \tag{2.8.10}$$

$$f^{\alpha\beta\gamma}(a_0,k_0;a_1,k_1;a_2,k_2;u) = -\left[\frac{k_{1\sigma}\tau^{\sigma\alpha}(\tilde{\omega}_0)\tau^{\beta\gamma}(\tilde{\omega}_2)}{\tilde{\omega}_0}\right.$$
$$+\frac{k_{2\sigma}\tau^{\sigma\alpha}(\tilde{\omega}_0)\tau^{\gamma\beta}(\tilde{\omega}_1)}{\tilde{\omega}_0}+\frac{k_{0\sigma}\tau^{\sigma\beta}(\tilde{\omega}_1)\tau^{\alpha\gamma}(\tilde{\omega}_2)}{\tilde{\omega}_1}+\frac{k_{0\sigma}\tau^{\sigma\gamma}(\tilde{\omega}_2)\tau^{\beta\alpha}(\tilde{\omega}_1)}{\tilde{\omega}_2}$$
$$\left.+\frac{k_{1\sigma}\tau^{\sigma\gamma}(\tilde{\omega}_2)\tau^{\alpha\beta}(\tilde{\omega}_0)}{\tilde{\omega}_2}+\frac{k_{2\sigma}\tau^{\sigma\beta}(\tilde{\omega}_1)\tau^{\gamma\alpha}(\tilde{\omega}_0)}{\tilde{\omega}_1}\right], \tag{2.8.11}$$

with $\tilde{\omega}_n = (k_n u)_\parallel - a_n \Omega_0$.

Only the contribution from one species is retained in (2.8.10). Due to the result being proportional to $1/m^2$, the contribution of ions to the nonlinear response is intrinsically small compared with the contribution of electrons or positrons. The dependence on q^3 suggests that electrons and positrons contribute with opposite sign, but this is only partially the case. Unlike the unmagnetized case, where the quadratic response tensor for a pure pair plasma (identical distributions of electrons and positrons) is zero, in the magnetized case there is an additional dependence on ϵ in the gyrotropic term in $\tau^{\mu\nu}(\omega)$, specifically, in the term $-i\epsilon\Omega_0 f^{\mu\nu}$ in (2.1.16). Even for a pure pair plasma, the gyrotropic terms imply a nonzero quadratic response.

References

1. R.L. Dewar, Aust. J. Phys. **30**, 533 (1977)
2. B.A. Trubnikov, Dissertation, Moscow Institute of Engineering and Physics (1958); AEC-tr-4073, US Atomic Energy Commission, Oak Ridge, Tennessee (1960)
3. F. Jüttner, Ann. der Phys. **34**, 856 (1911)
4. J.L Synge, *The Relativistic Gas* (North-Holland, Amsterdam, 1957)
5. B.B. Godfrey, B.S. Newberger, K.A. Taggart, IEEE Trans. Plasma Phys. **PS-3**, 60, 68 (1975)
6. B.A.Trubnikov, V.B. Yakubov, Plasma Phys. **5**, 7 (1963)
7. D.B. Fried, S.D. Conte, *The Plasma Dispersion Function* (Academic, New York, 1961)
8. I.P. Shkarofsky, Phys. Fluid **9**, 561 (1966)
9. V. Krivenski, A. Orefice, J. Plasma Phys. **30**, 125 (1983)
10. P.A. Robinson, J. Plasma Phys. **35**, 187 (1986)
11. P.A. Robinson, J. Math. Phys. **27**, 1206 (1986)
12. P.A. Robinson, J. Math. Phys. **28**, 1203 (1987)
13. V.N. Dnestrovskii, D.P. Kostomorov, N.V. Skrydlov, Sov. Phys. Tech. Phys. **8**, 691 (1964)
14. M. Bornatici, U. Raffina, Plasma Phys. Control. Fusion **30**, 115 (1988)
15. M.J. Bruggen-Kerkhof, L.P.J. Kamp, F.W. Sluijter, J. Phys. A **26**, 5505 (1993)
16. P.A. Robinson, J. Plasma Phys. **37**, 435, 449 (1987)
17. P.A. Robinson, J. Math. Phys. **30**, 2484 (1989)
18. D.G. Swanson, Plasma Phys. Control. Fusion **44**, 1329 (2002)
19. B.S. Newberger, J. Math. Phys. **23**, 1278 (1982)
20. D.B. Melrose, R. Yuen, Astrophys. J. **745**, 169 (2012)
21. P.A. Sturrock, Astrophys. J. **164**, 529 (1971)
22. A. Levinson, D. Melrose, A. Judge, Q. Luo, Astrophys. J. **631**, 456 (2005)
23. A.M. Beloborodov, C. Thompson, Astrophys. J. **657**, 967 (2007)

24. A.N. Timokhin, Mon. Not. R. Astron. Soc. **408**, 2092 (2010)
25. Q. Luo, D.B. Melrose, Mon. Not. R. Astron. Soc. **387**, 1291 (2008)
26. B. Zhang, A.K. Harding, Astrophys. J. **532**, 1159 (2000)
27. J.A. Hibschman, J. Arons, Astrophys. J. **560**, 871 (2001)
28. P.N. Arendt Jr., J.A. Eilek, Astrophys. J. **581**, 451 (2002)
29. M.P. Kennett, D.B. Melrose, Q. Luo, J. Plasma Phys. **64**, 333 (2000)
30. M. Gedalin, D.B. Melrose, E. Gruman, Phys. Rev. E **57**, 3399 (1998)
31. D.B. Melrose, J. Plasma Phys. **58**, 735 (1997)
32. V.N. Sazonov, Sov. Phys. JETP **20**(9), 587 (1969)
33. D.B. Melrose, Phys. Rev. E **56**, 3527 (1997)

Chapter 3
Waves in Magnetized Plasmas

The linear response tensor contains all information on the linear response of a medium. In particular, it determines the properties of the natural wave modes of the medium. Magnetized plasmas can support a large variety of different wave modes. There is no systematic classification of wave modes, leading to a confusing variety of names. Some modes are given historical names (e.g., Alfvén, Bernstein and Langmuir waves), some are given names associated with the theory used to derive them (e.g., cold-plasma, magnetoionic and MHD waves), and many are given names descriptive of the wave itself (e.g., longitudinal, lower-hybrid and electron-cyclotron waves). Moreover, there is arbitrariness in the definition of a wave mode: a single dispersion curve can be interpreted as one mode in one limit and as another mode in another limit. Even the concept of a wave mode is ill-defined in the presence of damping (or growth); for example, there are many natural peaks in the spectrum of fluctuations in a thermal plasma and when a particular peak is to be interpreted as a natural wave mode is ill-defined. Let the properties of an arbitrary wave mode, labeled as mode M, be regarded as a function of the independent variable $\boldsymbol{k}$. A weakly-damped wave mode is characterized by its dispersion relation, $\omega = \omega_M(\boldsymbol{k})$, its polarization vector, $\boldsymbol{e}_M(\boldsymbol{k})$, the ratio $R_M(\boldsymbol{k})$ of electric to total energy, and its absorption coefficient $\gamma_M(\boldsymbol{k})$. These quantities are frame-dependent, and a choice of frame (usually the rest frame of the plasma) needs to be made to derive them.

General properties of wave dispersion in a magnetized plasma are discussed in § 3.1. The theory is applied to a cold plasma in § 3.2 and to an electronic plasma in § 3.3. Weakly relativistic effects are discussed in § 3.4. Some aspects of wave dispersion in a pulsar plasma are treated in § 3.5. The weak-anisotropy limit is discussed in § 3.6.

D. Melrose, *Quantum Plasmadynamics: Magnetized Plasmas*, Lecture Notes in Physics 854, DOI 10.1007/978-1-4614-4045-1_3,

3.1 Wave Dispersion

In this section, the formal theory of wave dispersion is summarized. The wave properties are derived by reducing the covariant formalism to a conventional 3-tensor approach in the rest frame of the plasma.

3.1.1 *Invariant Dispersion Equation*

The covariant theory of wave dispersion is based on the wave equation, derived from the Fourier transform of the covariant form of Maxwell's equation. Following the notation used in volume 1, the wave equation is

$$\Lambda^{\mu\nu}(k)A_\nu(k) = -\mu_0 J^\mu_{\text{ext}}(k), \tag{3.1.1}$$

where the current is separated into an induced part and an extraneous part; the linear induced current is included on the left hand side of (3.1.1), with

$$\Lambda^\mu{}_\nu(k) = k^2\delta^\mu_\nu - k^\mu k_\nu + t^\mu{}_\nu(k), \qquad t^{\mu\nu}(k) = \mu_0 \Pi^{\mu\nu}(k), \tag{3.1.2}$$

and the extraneous part remains as a source term in (3.1.1). The source term is set to zero to obtain the homogeneous wave equation, which is to be solved for the wave properties.

Without the source term, (3.1.1) becomes the homogeneous wave equation, $\Lambda^\mu{}_\nu(k) = 0$, which may be interpreted as four simultaneous equations for the four components of $A^\mu(k)$. These may be written in matrix form, and then the condition for a solution to exist is that the determinant of the matrix of coefficients vanish. However, the gauge-invariance and charge-continuity conditions imply that $A^\mu(k) \propto k^\mu$ is a trivial solution, and that this determinant is identically zero. Non-trivial solutions are determined by the requirement that the matrix of cofactors vanish. Let $\lambda^{\mu\nu}(k)$ be the matrix of cofactors of $\Lambda^{\mu\nu}(k)$; it is necessarily of the form

$$\lambda^{\mu\nu}(k) = \lambda(k)\, k^\mu k^\nu, \tag{3.1.3}$$

where $\lambda(k)$ is an invariant. The invariant form of the dispersion equation is $\lambda(k) = 0$.

An explicit form for $\lambda(k)$ may be obtained in terms of the traces of powers of $t^{\mu\nu}(k)$, defined by (3.1.2). Let the trace of the nth power be denoted $t^{(n)}(k)$, so that one has $t^{(1)}(k) = t^\mu{}_\mu(k)$, $t^{(2)}(k) = t^\mu{}_\nu(k)t^\nu{}_\mu(k)$, and so on. The explicit form is

$$\lambda(k) = k^4 + k^2 t^{(1)}(k) + \frac{1}{2}\left\{\left[t^{(1)}(k)\right]^2 - t^{(2)}(k)\right\}$$
$$+\frac{1}{6k^2}\left\{\left[t^{(1)}(k)\right]^3 - 3t^{(1)}(k)t^{(2)}(k) + 2t^{(3)}(k)\right\}. \tag{3.1.4}$$

A non-trivial solution for $A^\mu(k)$ requires that one construct the second order matrix of cofactors. Denoting this by $\lambda^{\mu\nu\alpha\beta}(k)$, it satisfies

$$\Lambda^\mu{}_\rho(k)\lambda^{\rho\nu\alpha\beta}(k) = \lambda(k)\left[g^{\mu\alpha}k^\nu k^\beta - g^{\mu\beta}k^\nu k^\alpha\right]. \tag{3.1.5}$$

A solution satisfying a specific gauge condition $G_\mu A^\mu(k) = 0$ is found by projecting onto $G_\mu G_\alpha$, and identifying the solution in this gauge with any column of the matrix representation of $G_\mu G_\alpha \lambda^{\mu\nu\alpha\beta}(k)$. The temporal gauge corresponds to $G^\mu = [0, \mathbf{1}]$.

Cold Unmagnetized Plasma

The covariant theory is inconvenient for detailed calculations in general cases. To illustrate this point, consider the simple case of a cold unmagnetized plasma, which corresponds to $\tau_{\alpha\beta} \to g_{\alpha\beta}$ in (1.2.16). The response tensor for a cold unmagnetized plasma is

$$t^{\mu\nu}(k) = \mu_0 \Pi^{\mu\nu}(k) = -\omega_p^2\left(g^{\mu\nu} - \frac{k^\mu \bar{u}^\nu + k^\nu \bar{u}^\mu}{k\bar{u}} + \frac{k^2 \bar{u}^\mu \bar{u}^\nu}{(k\bar{u})^2}\right), \tag{3.1.6}$$

with $\omega_p^2 = q^2 n/\varepsilon_0 m$. The traces in (3.1.4) can be evaluated relatively simply in this case. They are

$$t^{(n)}(k) = (-\omega_p^2)^n\left[\left(\frac{k^2}{(k\bar{u})^2}\right)^n + 2\right], \tag{3.1.7}$$

with $n = 1, 2, 3$. Then (3.1.4) gives

$$\lambda(k) = \left(1 - \frac{\omega_p^2}{(k\bar{u})^2}\right)(k^2 - \omega_p^2)^2. \tag{3.1.8}$$

The solutions of $\lambda(k) = 0$ for an unmagnetized cold plasma are $(k\bar{u})^2 = \omega_p^2$ and two degenerate solutions $k^2 = \omega_p^2$. One can construct the second order matrix of cofactors, e.g. using (2.2.26) of volume 1. On inserting the solution $(k\bar{u})^2 = \omega_p^2$ into the resulting expression for $\lambda^{\mu\nu\rho\sigma}(k)$, one can show that this solution corresponds to a longitudinal wave in the rest frame. Inserting the solution $k^2 = \omega_p^2$, $\lambda^{\mu\nu\rho\sigma}(k)$ vanishes, due to this solution being degenerate, corresponding to two degenerate modes which can be shown to be transverse in the rest frame of the plasma. Thus the covariant theory reproduces results that are well known, but even in this simple case the covariant theory is unnecessarily cumbersome. In more general cases it is usually simpler to reduce the covariant theory to a 3-tensor formalism, and use this for detailed calculations.

A conventional treatment of wave dispersion involves choosing the temporal gauge, $A^0(k) = 0$, thereby reducing the wave equation to a set of three simultaneous equations. The determinant of the matrix of coefficients gives the dispersion equation, and a particular solution of it defines a wave mode. The polarization 3-vector is found by constructing the matrix of cofactors, choosing any column of it, evaluating the entries using the relevant dispersion relation, and normalizing the resulting 3-vector to unity.

3-Tensor Form of the Wave Equation

A 3-tensor form for the wave equation is obtained by choosing a specific gauge and frame. A convenient choice is the temporal gauge, $A^0(k) = 0$. Only the space components of the wave equation (3.1.1) are then relevant:

$$\Lambda^i{}_j(k)A^j(k) = -\mu_0 J^i_{\rm ext}(k). \tag{3.1.9}$$

In the rest frame of the plasma, it is convenient to divide both sides of (3.1.9) by ω^2, so that it becomes

$$\left[-\frac{|\boldsymbol{k}|^2}{\omega^2}(\delta^i_j + \kappa^i\kappa_j) + K^i{}_j(k)\right] A^j(k) = -\frac{\mu_0}{\omega^2}\, J^i_{\rm ext}(k), \tag{3.1.10}$$

with $\boldsymbol{\kappa} = \boldsymbol{k}/|\boldsymbol{k}|$, and where the equivalent dielectric tensor is defined by (1.2.27). The use of the mixed tensor components on the left hand side of (3.1.10) is convenient in that K^i_j corresponds to the conventional 3-tensor components of the dielectric tensor. For $\boldsymbol{\kappa}$ in the 1–3 plane, at angle θ to $\boldsymbol{B}$, one has

$$\boldsymbol{\kappa} = (\sin\theta, 0, \cos\theta), \qquad \kappa^i\kappa_j = -\begin{pmatrix} \sin^2\theta & 0 & \sin\theta\cos\theta \\ 0 & 0 & 0 \\ \sin\theta\cos\theta & 0 & \cos^2\theta \end{pmatrix}. \tag{3.1.11}$$

The homogeneous wave equation follows from (3.1.10) by setting the source term to zero. The condition for a solution to exist is that the determinant of the 3×3 matrix be set to zero. This determinant is equal to $\lambda(k)/\omega^6$, where $\lambda(k) = 0$ is the dispersion equation in the covariant theory. Introducing the refractive index, which is $n = |\boldsymbol{k}|c/\omega$ in ordinary units, and choosing $\boldsymbol{\kappa}$ as in (3.1.11), one has

$$\lambda(k) = \omega^6 \begin{vmatrix} -n^2\cos^2\theta + K^1{}_1(k) & K^1{}_2(k) & n^2\sin\theta\cos\theta + K^1{}_3(k) \\ K^2{}_1(k) & -n^2 + K^2{}_2(k) & K^2{}_3(k) \\ n^2\sin\theta\cos\theta + K^3{}_1(k) & K^3{}_2(k) & -n^2\sin^2\theta + K^3{}_3(k) \end{vmatrix}. \tag{3.1.12}$$

As shown in § 2.2.4 of volume 1, (3.1.12) is equivalent to

$$\lambda(k) = \omega^6 \left\{n^4 K^L(k) - n^2[K^L(k)K_1(k) - K_2^L(k)] + \det[K^i{}_j(k)]\right\}, \tag{3.1.13}$$

with $K_1(k) = K^s{}_s(k)$ the trace of the dielectric tensor, $K^L(k) = -\kappa_i\kappa^j K^i{}_j(k)$ its longitudinal part, $K_2^L(k) = -\kappa_i\kappa^j K^i{}_s(k) K^s{}_j(k)$ the longitudinal part of its square, and $\det[K^i{}_j(k)]$ its determinant.

3.1.2 Polarization 3-Vector

In the covariant theory of wave dispersion (§ 2.3 of volume 1), the condition for a solution to exist is that the dispersion equation, $\lambda(k) = 0$, be satisfied. Let $k^\mu = k^\mu_M$ be the solution for the mode M, with $k^\mu_M = [\omega_M(\boldsymbol{k}), \boldsymbol{k}]$ when one chooses $\boldsymbol{k}$ as the independent variable and ω as the dependent variable. Then one can solve the wave equation for $A^\mu(k_M)$. The amplitude, phase and gauge are all arbitrary. The polarization vector is defined by this solution with appropriately chosen amplitude, phase and gauge.

For $\lambda(k) = 0$, the second order matrix of cofactors is of rank 2 implying that it can be expressed in terms of two 4-vectors. These 4-vectors are k^μ_M and the polarization 4-vector $e^\mu_M((k)$. The symmetry properties $\lambda^{\mu\alpha\nu\beta} = -\lambda^{\alpha\mu\nu\beta} = -\lambda^{\mu\alpha\beta\nu}$ imply

$$\lambda^{\mu\alpha\nu\beta}(k_M) \propto \left[e^\mu_M(k)k^\nu_M - e^\nu_M(k)k^\mu_M\right]\left[e^\alpha_M(k)k^\beta_M - e^\beta_M(k)k^\alpha_M\right]^*. \tag{3.1.14}$$

One can identify the polarization vector in the arbitrary G-gauge, with $e^\mu_M G_\mu = 0$ by definition, by contracting (3.1.14) with $G_\alpha G_\beta$. This gives

$$e^\mu_M(k)e^{*\nu}_M(k) \propto G_\alpha G_\beta \lambda^{\mu\alpha\nu\beta}(k_M). \tag{3.1.15}$$

The temporal gauge, which corresponds to $G^\mu = [1, \boldsymbol{0}]$, is the only choice that allows a simple normalization in general. It implies $e^\mu_M(k) = [0, \boldsymbol{e}_M(\boldsymbol{k})]$ allowing the normalization $e^\mu_M(k)e_{M\mu}(k) = -|\boldsymbol{e}_M(\boldsymbol{k})|^2 = -1$. With this choice, one may use (3.1.15) to identify

$$\lambda^{i0}{}_{j0}(k_M) = -\lambda^{0s}{}_{0s}(k_M)\, e^i_M(k)e^*_{Mj}. \tag{3.1.16}$$

The tensor (3.1.14) can then be written as

$$\lambda^{\mu\nu\alpha\beta}(k_M) = -\frac{\lambda^{0\sigma}{}_{0\sigma}(k_M)}{\omega^2_M(k)}\left[e^\mu_M(k)k^\nu_M - e^\nu_M(k)k^\mu_M\right]\left[e^\alpha_M(k)k^\beta_M - e^\beta_M(k)k^\alpha_M\right]^*. \tag{3.1.17}$$

A specific form for $\lambda^{0i}{}_{0j}$ is given by (2.2.35) of volume 1:

$$\begin{aligned}\lambda^{0i}{}_{0j} &= n^4\kappa^i\kappa_j - n^2(\kappa^i\kappa_j K_1 + \delta^i_j K^L - \kappa^i\kappa_r K^r{}_j - \kappa_j\kappa^s K^i{}_s) \\ &\quad + \frac{1}{2}\delta^i_j\,[(K_1)^2 - K_2] + K^i{}_s K^s{}_j - K_1 K^i{}_j.\end{aligned} \tag{3.1.18}$$

On inserting the dispersion relation for mode M into (3.1.18), the result is proportional to the outer product $e^i_M e^*_{Mj}$, and hence may be used to construct $\boldsymbol{e}_M$. This is equivalent to using a 3-tensor formalism to construct $\boldsymbol{e}_M$.

With the choice of the temporal gauge, and with $|\boldsymbol{e}_M(\boldsymbol{k})|^2 = 1$, the gauge and normalization of the solution for the 4-potential, $A^\mu(k_M) \propto e^\mu_M(k)$, are specified, but the phase remains arbitrary. A unique choice for the phase is possible due to the Onsager relations. For $\boldsymbol{k}$, $\boldsymbol{b}$ in the 1–3 plane, the Onsager relations imply that the 2-component is out of phase with the 1- and 3-components, allowing one to choose the phase by specifying that the 1- and 3-components are real, and that the 2-component is imaginary. The 3-polarization vector, chosen in this way, can be written in the form

$$\boldsymbol{e}_M = \frac{L_M \boldsymbol{\kappa} + T_M \boldsymbol{t} + i\boldsymbol{a}}{(L_M^2 + T_M^2 + 1)^{1/2}}, \tag{3.1.19}$$

with $\boldsymbol{t}, \boldsymbol{a}$ defined by (1.1.25). For the specific choice of coordinate axes indicated, one has

$$\boldsymbol{\kappa} = (\sin\theta, 0, \cos\theta), \qquad \boldsymbol{t} = (\cos\theta, 0 - \sin\theta), \qquad \boldsymbol{a} = (0, 1, 0). \tag{3.1.20}$$

The parameter T_M is the axial ratio of the polarization ellipse for mode M; the sign of T_M determines the handedness of the polarization, in a screw sense relative to $\boldsymbol{\kappa}$, with $T_M > 0$ and $T_M < 0$ corresponding to right and left hand, respectively.

3.1.3 Ratio of Electric to Total Energy

The ratio of electric to total energy, $R_M(\boldsymbol{k})$, appears naturally in the theory of emission and absorption processes in plasmas. Physically, this may be attributed to the total energy in waves in the mode M consisting of three parts, electric, magnetic and induced-particle kinetic energies. Only the electric part is involved in the work done by a current in an emission or absorption process. $R_M(\boldsymbol{k})$ is related to $\lambda^{0s}{}_{0s}(k_M)$ which appears naturally in the normalization of the polarization vector through (3.1.16). The specific relation is

$$R_M(k) = \left. \frac{\lambda^{0s}{}_{0s}(k)}{\omega \partial\lambda(k)/\partial\omega} \right|_{k=k_M}. \tag{3.1.21}$$

Several alternative forms for $R_M(k)$ are derived in § 2.3 of volume 1. One of these is

$$[R_M(k)]^{-1} = \left[2 - \frac{1}{\varepsilon_0\omega}\frac{\partial}{\partial\omega}\Pi_M(k)\right]\bigg|_{\omega=\omega_M}, \qquad \Pi_M(k) = e^*_{M\mu} e_{M\nu} \Pi^{\mu\nu}(k_M), \tag{3.1.22}$$

where only the hermitian part of $\Pi^{\mu\nu}(k)$ is to be retained. In terms of the dielectric tensor, (3.1.22) becomes

$$[R_M(k)]^{-1} = \frac{1}{\omega}\frac{\partial}{\partial\omega}[\omega^2 K_M(k)]\bigg|_{\omega=\omega_M}, \qquad K_M(k) = -e^*_{Mi}e^j_M K^{Hi}{}_j(k), \tag{3.1.23}$$

where H denotes the hermitian part. For a medium that is not spatially dispersive, a further alternative form is

$$[R_M(k)]^{-1} = 2[1 - |\boldsymbol{\kappa}\cdot\boldsymbol{e}_M|^2] n_M(\omega,\boldsymbol{\kappa}) \frac{\partial}{\partial\omega}[\omega n_M(\omega,\boldsymbol{\kappa})], \tag{3.1.24}$$

where $n_M(\omega,\boldsymbol{\kappa})$ is the refractive index for the mode M, found by re-expressing $|\boldsymbol{k}|/\omega_M(\boldsymbol{k})$ as a function of independent variable ω and $\boldsymbol{\kappa} = \boldsymbol{k}/|\boldsymbol{k}|$.

3.1.4 Absorption Coefficient

The foregoing theory of wave dispersion neglects dissipation. In effect, $\Pi^{\mu\nu}(k)$ is interpreted as the hermitian part of the tensor. The inclusion of the antihermitian part, $\Pi^{A\mu\nu}(k)$, leads to damping of the waves. In this context, damping also includes growth, interpreted as negative damping. Implicit in the approach is that the waves are weakly damped, implying that the damping can be included as a perturbation. The damping can then be described by a single parameter, called the absorption coefficient, $\gamma_M(\boldsymbol{k})$.

As discussed in § 2.4 of volume 1, the inclusion of dissipation allows one to identify the total energy in waves, and this is implicit in the identification of $R_M(k)$. The procedure is to calculate the rate work is done by the current associated with the dissipation, and to identify this with the rate at which the total energy in the waves changes, allowing the total energy to be identified. The calculation gives the rate, $Q^\mu_M(k)$, that 4-momentum is transferred to the waves as

$$Q^\mu_M(k) = -2i\,\frac{R_M(k)N_M(k)}{\varepsilon_0\omega_M(k)}\,k^\mu_M\,\Pi^A_M(k_M), \tag{3.1.25}$$

$$\Pi^A_M(k_M) = e^*_{M\alpha}(k)e_{M\beta}(k)\Pi^{A\alpha\beta}(k_M), \tag{3.1.26}$$

where $\Pi^A_M(k_M)$ is an imaginary quantity. The power $Q^0_M(k)$ causes the energy $W_M(k)$ to vary exponentially. If this variation is purely temporal, the variation is described by a factor $\exp[-\gamma_M(k)t]$. The absorption coefficient is

$$\gamma_M(k) = -\frac{Q^0_M(k)}{W_M(k)} = 2i\,\frac{R_M(k)}{\varepsilon_0\omega_M(k)}\,\Pi^A_M(k_M). \tag{3.1.27}$$

When the response is described in terms of the dielectric tensor, the result (3.1.27) translates into

$$\gamma_M(k) = 2i\,R_M(k)\,\omega_M(k)\,K^A_M(k_M), \quad K^A_M(k) = -e^*_{Mi}(k)e^j_M(k)\,K^{Ai}{}_j(k_M). \tag{3.1.28}$$

Damping can occur in both time and space, depending on the boundary condition. Damping is purely temporal (in a given frame) if the waves are uniformly excited everywhere initially, and the damping is purely spatial if there is a time-independent point source for the waves. More generally, a mixture of temporal and spatial damping occurs, and this can be described in terms of an imaginary part, $\mathrm{Im}\, k_M^\mu$, of the wave 4-vector, such that the wave amplitude varies as $\exp[\mathrm{Im}\, k_M^\mu x_\mu]$. The wave energy, which is proportional to the square of the amplitude, varies as $\exp[2\, \mathrm{Im}\, k_M^\mu x_\mu]$. Temporal and spatial damping are related by

$$\mathrm{Im}\, \omega_M - \mathrm{Im}\, \{\boldsymbol{k}\} \cdot \boldsymbol{v}_{gM}(\boldsymbol{k}) = -\tfrac{1}{2}\gamma_M(k), \qquad \boldsymbol{v}_{gM}(\boldsymbol{k}) = \frac{\partial \omega_M(\boldsymbol{k})}{\partial \boldsymbol{k}}, \tag{3.1.29}$$

where $\boldsymbol{v}_{gM}(\boldsymbol{k})$ is the group velocity for waves in the mode M. The weak-damping condition requires that the damping be weak in time, $|\gamma_M(\boldsymbol{k})| \ll \omega_M(\boldsymbol{k})$, and in space. The validity of these requirements can usually be checked only a posteriori.

3.2 Waves in Cool Electron-Ion Plasmas

The simplest model for wave dispersion in a magnetized plasma is one consisting of electrons and positively charged ions in which the thermal motions of all particles are neglected, called a cold plasma. The cold-plasma wave modes are of interest both in themselves, and as the basis for classification of waves in a magnetized plasma more generally. The properties of the cold plasma wave modes are summarized in this section. Thermal modifications are included in the low-frequency limit, leading to three MHD-like modes. A more detailed discussion of waves in a cold electronic plasma is given in § 3.3.

3.2.1 *Cold Plasma Dispersion Equation*

The dispersion equation, $\lambda(k) = 0$ with $\lambda(k)$ given by (3.1.13), reproduces a standard form [1, 2] for the cold plasma dispersion equation. This form can be derived directly from the homogeneous wave equation in the form $\Lambda^i_{\ j}(k)A^j(k) = 0$, with

$$\Lambda^i_{\ j} = \begin{pmatrix} S - n^2\cos^2\theta & -iD & n^2\sin\theta\cos\theta \\ iD & S - n^2 & 0 \\ n^2\sin\theta\cos\theta & 0 & P - n^2\sin^2\theta \end{pmatrix}, \tag{3.2.1}$$

where arguments are omitted. Setting the determinant of $\Lambda^i_{\ j}$ to zero gives the cold plasma dispersion equation

$$An^4 - Bn^2 + C = 0, \tag{3.2.2}$$

with the coefficients given by

$$A = S\sin^2\theta + P\cos^2\theta, \qquad B = (S^2 - D^2)\sin^2\theta + PS(1+\cos^2\theta),$$
$$C = P(S^2 - D^2). \tag{3.2.3}$$

With the dependent variable chosen to be n^2, the dispersion equation (3.2.2) is a quadratic equation for n^2, and the two solutions are

$$n^2 = n_\pm^2 = \frac{B \pm F}{2A}, \qquad F = \left(B^2 - 4AC\right)^{1/2}, \tag{3.2.4}$$

where the $\pm$ labeling is arbitrary.

The solutions (3.2.4) correspond to propagating waves only for $n^2 > 0$. In principle, $n^2 < 0$ can be due either to imaginary ω for real $\boldsymbol{k}$, or imaginary $\boldsymbol{k}$ for real ω, and only the latter occurs in a cold plasma. Regions with $n^2 < 0$ corresponds to evanescence: an evanescent wave oscillates in time and decays exponentially in space.

Polarization Vector

The polarization vector written in the form (3.1.19) involves the parameters L_M, T_M. Explicit expressions for these are found using the matrix of cofactors, $\lambda^j{}_i$, of $\Lambda^i{}_j$. One sets $n^2 = n_M^2$, with $M = \pm$, in any of the columns of $\lambda^i{}_j$ and normalizes the result appropriately. Choosing the middle column, $\lambda^i{}_2$, gives

$$L_M = \frac{(P - n_M^2)D\sin\theta}{An_M^2 - PS}, \qquad T_M = \frac{DP\cos\theta}{An_M^2 - PS}. \tag{3.2.5}$$

The parameter T_M is the axial ratio of the polarization ellipse, and the two modes are orthogonal in the sense that their polarization ellipses are orthogonal, $T_+T_- = -1$. Except for special cases, cold plasma waves have a nonzero longitudinal component, described by L_M. The polarization vectors themselves are not orthogonal, with $\boldsymbol{e}_\pm^* \cdot \boldsymbol{e}_\mp \neq 0$ due to the longitudinal parts.

The expression (3.2.5) for the axial ratio implies that $1/T$ is a linear function of n^2. It follows that because n^2 satisfies a quadratic equation, $1/T$ and hence T also satisfy quadratic equations. For some purposes it is convenient to choose T as the independent variable in place of n^2. One then solves the quadratic equation

$$T^2 - \frac{(PS - S^2 + D^2)\sin^2\theta}{PD\cos\theta}\,T - 1 = 0, \tag{3.2.6}$$

for $T = T_\pm$. The solutions are

$$T_\pm = \frac{(PS - S^2 + D^2)\sin^2\theta \pm F}{2PD\cos\theta} = \frac{-2PD\cos\theta}{(PS - S^2 + D^2)\sin^2\theta \mp F},$$
$$F^2 = (PS - S^2 + D^2)^2\sin^4\theta + 4P^2D^2\cos^2\theta. \tag{3.2.7}$$

The solutions for n^2 are then found by inverting (3.2.5) to find n_M^2 and L_M in terms of T_M. Three superficially different but equivalent relations between n^2 and T can be found by choosing each of the three columns of the matrix of cofactors, λ^i_j. Choosing $j = 1, 2, 3$, constructing T as a function of n^2 and inverting to find n^2 as a function of T gives

$$n_M^2 = P\frac{S\cos\theta + DT_M}{P\cos\theta + DT_M\sin^2\theta} = \frac{P}{A}\left(S + \frac{D\cos\theta}{T_M}\right) = \frac{S^2 - D^2}{S - DT_M\cos\theta}, \tag{3.2.8}$$

respectively. Two forms of the resulting relation between L and T are

$$L_M = \frac{\sin\theta}{A}[(P - S)T_M\cos\theta - D] = \frac{\sin\theta}{P}\frac{(PS - S^2 + D^2)T_M\cos\theta - PD}{S - DT_M\cos\theta}. \tag{3.2.9}$$

3.2.2 Parallel and Perpendicular Propagation

In the limit of parallel propagation, $\sin\theta = 0$, the wave properties simplify. The dispersion equation (3.2.2) with (3.2.3) reduces to $P[(n^2 - S)^2 - D^2] = 0$ for $\sin\theta = 0$. The solution $P = 0$ corresponds to longitudinal oscillations at $\omega = \omega_p$, and the other two solutions correspond to oppositely circularly polarized modes. This is a general feature of parallel propagation in an arbitrary magnetized plasma: the polarization of the modes for parallel propagation is either longitudinal or transverse and circular. The solutions of $(n^2 - S)^2 - D^2 = 0$ may be written $n^2 = R_\pm$, with $R_\pm$ given by (1.2.31).

The handedness of the circular polarization of the wave modes is described by the sign of the axial ratio, with $T > 0$ for right-hand polarization. The wave modes of a cold plasma are affected by the sense in which particles gyrate, and this is determined by a screw sense relative to the direction of the magnetic field; electrons spiral in a right-hand sense with this convention. The cold plasma mode that has a resonance at $\omega = \Omega_\alpha$, for species α, has a handedness in this sense determined by the sign of the charge of species α, being right-hand for electrons and left-hand for ions. The two conventions for handedness coincide when the angle, θ, between $\boldsymbol{\kappa}$ and $\boldsymbol{b}$ is acute and are opposite when θ is obtuse, that is, for $\cos\theta > 0$ and $\cos\theta < 0$, respectively.

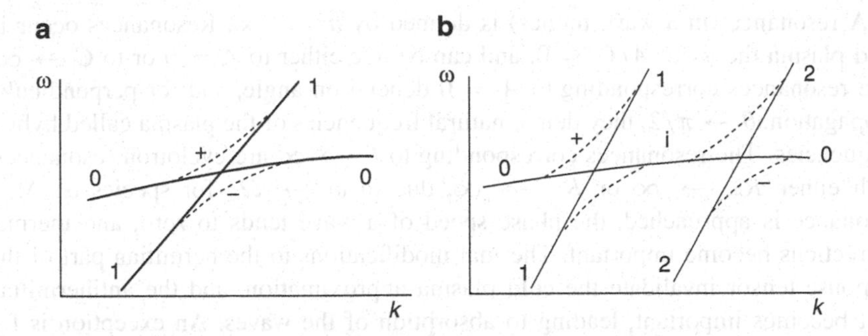

Fig. 3.1 Schematic illustration of the dispersion curves for nearly parallel propagating modes. **(a)** For two modes labeled 0 and 1, near the cross-over point: the *solid curves* are for $\theta = 0$ and *light dashed curves* show how the curves reconnect for $\theta \neq 0$ into two modes labeled + and −. **(b)** For three modes, labeled 0, 1 and 2, with the *dashed curves* defining three reconnected modes labeled +, i and −

A subtle point is that the solutions for the wave properties in the limit of parallel propagation are not the same as the limit for parallel propagation of the solutions for oblique propagation. That is, if one first sets $\sin\theta = 0$ and solves for the wave modes, one gets a different labeling of the modes than is obtained by solving for $\sin\theta \neq 0$ and taking the limit $\sin\theta \to 0$. The solutions for $\sin\theta = 0$ cross each other, at a number of points, depending on the number of ionic species, but the dispersion curves for $\sin\theta \neq 0$ never cross. One such crossing occurs at the plasma frequency, where P changes sign. In a cold plasma with multiple ion species, R_- has resonances at each of the ion cyclotron frequencies, where it changes sign by passing through infinity; there are solutions of $R_+ = R_-$, implying $D = 0$ in between the cyclotron frequencies. The actual solutions for $\sin\theta \to 0$, deviate away from the crossing point, as illustrated schematically in Fig. 3.1.

For perpendicular propagation, $\cos\theta = 0$, the solutions of the dispersion equation (3.2.2) reduce to $n^2 = P$, $n^2 = (S^2 - D^2)/S$. The fact that the solution $n^2 = P$ does not depend on the magnetic field led to the mode being called 'ordinary', with the other mode being called 'extraordinary' in the context of a cold electron gas. These names continue to be used in connection with the magnetoionic modes.

3.2.3 *Cutoffs and Resonances*

A cutoff is defined as a zero of refractive index. The dispersion equation in the form (3.2.2) implies that cutoff frequencies are determined by $C/A = 0$, which is equivalent to $C = 0$. (One can have $A \to \infty$ due to $S \to \infty$, but this also implies $C \to \infty$ and $C/A \to \infty$.) With $C = P(S^2 - D^2)$, cutoffs occur at $P = 0$ and $R_\pm = 0$.

A resonance (in a wave modes) is defined by $n^2 \to \infty$. Resonances occur in cold plasma theory at $A/C = 0$, and can be due either to $A = 0$ or to $C \to \infty$. The resonances corresponding to $A = 0$ depend on angle, and for perpendicular propagation, $\theta \to \pi/2$, they define natural frequencies of the plasma called hybrid frequencies. The resonances corresponding to $C \to \infty$ are cyclotron resonances, with either $R_+ \to \infty$ or $R_- \to \infty$, due to $\omega \to \Omega_\alpha$ for species α. As a resonance is approached, the phase speed of a wave tends to zero, and thermal corrections become important. Thermal modifications to the hermitian part of the response tensor invalidate the cold plasma approximation, and the antihermitian part becomes important, leading to absorption of the waves. An exception is for perpendicular propagation; this is due to thermal corrections associated with gyroresonances appearing in terms of the form $k_z^2 V^2/\omega^2$, where V is the thermal speed, so that there is no gyromagnetic absorption at perpendicular propagation (provided relativistic effects are neglected). Thermally modified waves near the parallel cyclotron resonances are called parallel cyclotron resonant waves. There is one such resonant wave mode for each species.

Cutoffs and resonance separate the cold plasma modes into distinct branches. Each branch is bounded from below by a cutoff and from above by a resonance. Exceptions are the two lowest frequency branches, which extend down to $\omega = 0$, and the two highest frequency branches that extend to arbitrarily large ω.

In the presence of inhomogeneity, waves are refracted away from the direction of decreasing refractive index and towards the direction of increasing refractive index. Hence they refract away from a cutoff, and towards a resonance, where resonant absorption becomes important.

3.2.4 Hybrid Waves

For perpendicular propagation, $\cos\theta = 0$, the condition $A = 0$ for a resonance reduces to $S = 0$. For an electron-ion plasma, $S = 0$ implies

$$1 - \frac{\omega_p^2}{\omega^2 - \Omega_e^2} - \sum_i \frac{\omega_{pi}^2}{\omega^2 - \Omega_i^2} = 0, \tag{3.2.10}$$

where the sum is over all ionic species, with the ith species having plasma frequency ω_{pi} and cyclotron frequency Ω_i. The upper-hybrid (UH) frequency is the solution of (3.2.10) found by neglecting the ions:

$$\omega_{\text{UH}}^2 = \omega_p^2 + \Omega_e^2. \tag{3.2.11}$$

The lower-hybrid (LH) frequency is found by assuming $\omega^2 \ll \Omega_e^2$. When there is only one ionic species present, $S = 0$ gives

$$\omega_{\rm LH}^2 = \Omega_i^2 + \frac{\omega_{pi}^2 \Omega_e^2}{\omega_{\rm UH}^2}. \tag{3.2.12}$$

In a multi-ion plasma there are further hybrid frequencies between the ion-cyclotron resonances.

Upper-Hybrid Waves

For oblique propagation, the condition $A = 0$ for a resonance can be solved for the resonant frequency, ω, as a function of θ. When the contribution of the ions is neglected there are two such solutions. Writing these two resonant frequencies as $\omega_\pm(\theta)$, one finds

$$\omega_\pm^2(\theta) = \tfrac{1}{2}(\omega_p^2 + \Omega_e^2) \pm \tfrac{1}{2}\left[(\omega_p^2 + \Omega_e^2)^2 - 4\omega_p^2 \Omega_e^2 \cos^2\theta\right]^{1/2}. \tag{3.2.13}$$

The solution $\omega_+(\theta)$ is equal to $\omega_{\rm UH}$ at $\theta = \pi/2$; the frequency decreases with increasing $\cos^2\theta$, and approaches the maximum of ω_p and Ω_e for parallel propagation. The solution $\omega_-(\theta)$ gives zero for $\theta = \pi/2$, and this is replaced by $\omega_{\rm LH}$ when the contribution of the ions is included. The frequency $\omega_-(\theta)$ increases with increasing $\cos^2\theta$, and approaches the minimum of ω_p and Ω_e for parallel propagation. These solutions are referred to as upper- and lower-hybrid waves, respectively. In the cold plasma approximation, both modes are longitudinal. However, thermal corrections need to be considered for both, and electromagnetic corrections become significant away from the resonance.

Lower-Hybrid Waves

Lower-hybrid waves are of particular interest in connection with instabilities that can couple electrons and ions together. The generic feature that allows this coupling is that lower-hybrid waves can be driven unstable by a nonthermal distribution of ions, with an anisotropic or ring-type distribution for example, with the waves being (Landau) damped by nonthermal electrons, leading to acceleration of the electrons.

A variety of approximate dispersion relations for lower-hybrid waves are given in the literature. A recent example is $\omega = \omega_{\rm LH}(\boldsymbol{k})$ with [3]

$$\omega_{\rm LH}(\boldsymbol{k}) = 1 + \frac{m_i}{2m_e}\cos^2\theta - \frac{\omega_p^2}{2|\boldsymbol{k}|^2 c^2} + \left(\frac{3}{2}\frac{T_i}{T_e} + \frac{3}{8}\right)\frac{|\boldsymbol{k}|^2 V_e^2}{\Omega_e^2}. \tag{3.2.14}$$

The terms on the right hand side of (3.2.14) are corrections due to non-perpendicular, non-electrostatic and ion and electron thermal effects, respectively.

3.2.5 *Low-Frequency Cold-Plasma Waves*

The contribution of the ions to wave dispersion is negligible at high frequencies but it becomes important at low frequencies, comparable with or below the cyclotron frequency of ions. In the low-frequency limit one has $\omega \ll \omega_p$, corresponding to P large and negative. The appropriate limit of cold plasma theory is found by expanding in powers of $1/P$. To lowest order (3.2.7) gives

$$T_{\pm} = \frac{S\sin^2\theta \pm [S^2\sin^4\theta + 4D^2\cos^2\theta]^{1/2}}{2D\cos\theta}, \tag{3.2.15}$$

where the $\pm$ labeling is arbitrary. To the same order (3.2.8) gives

$$n^2 = \frac{1}{\cos^2\theta}\left(S + \frac{D\cos\theta}{T}\right) = \frac{S^2 - D^2}{S - DT\cos\theta}, \tag{3.2.16}$$

and $L = T\tan\theta$.

Relevant sums over the ionic species give

$$\sum_i \frac{\omega_{pi}^2}{\Omega_i^2} = \frac{c^2}{v_A^2}, \qquad \sum_i \frac{\omega_{pi}^2}{\Omega_i} = \frac{\omega_p^2}{\Omega_e}, \tag{3.2.17}$$

where the inertia of the electrons is neglected in the Alfvén speed, v_A, and where the latter condition follows from charge neutrality. One has (in ordinary units)

$$S = 1 + \frac{c^2}{v_A^2} + \sum_i \frac{\omega_{pi}^2\omega^2}{\Omega_i^2(\Omega_i^2 - \omega^2)}, \qquad D = -\sum_i \frac{\omega_{pi}^2\omega}{\Omega_i(\Omega_i^2 - \omega^2)}. \tag{3.2.18}$$

At frequencies $\omega \ll \Omega_i$ one has $D \ll S$, and a further approximation involves expanding in powers of D/S.

The two low-frequency branches of the cold plasma modes are counterparts of two MHD wave modes, with the $+$ mode corresponding to the Alfvén mode and the $-$ mode to the fast magnetoacoustic mode. (There is no counterpart of a sound wave in a cold plasma.) The solution for the Alfvén (A) mode corresponds to

$$n_A^2 = \frac{S}{\cos^2\theta} + \frac{D^2\cos^2\theta}{S\sin^2\theta}, \qquad T_A = \frac{S\sin^2\theta}{D\cos\theta}, \qquad L_A = T_A\tan\theta. \tag{3.2.19}$$

The solution for the magnetoacoustic (m) mode corresponds to

$$n_{\rm m}^2 = S - \frac{D^2}{S\sin^2\theta}, \qquad T_{\rm m} = -\frac{D\cos\theta}{S\sin^2\theta}, \qquad L_{\rm m} = T_{\rm m}\tan\theta. \tag{3.2.20}$$

These reduce to the MHD modes (with the sound speed neglected, $c_s/v_A \to 0$) in the limit $\omega \to 0$, corresponding to $D = 0$. The modification to the MHD dispersion relations for $\omega/\Omega_i \neq 0$ becomes substantial as the ion cyclotron frequency is approached. This modifies the Alfvén mode, which becomes a parallel ion-cyclotron mode. The fast mode does not encounter an ion-cyclotron resonance, and evolves into the whistler mode as the frequency increases to above the ion-cyclotron range.

3.2.6 Inertial and Kinetic Alfvén Waves

The approximation $P \to \infty$ made in deriving (3.2.19) and (3.2.20) leads to both modes having zero parallel electric field, $\boldsymbol{b} \cdot \boldsymbol{e}_M = 0$. Zero parallel electric field is consistent with ideal MHD, which requires $\boldsymbol{E} \cdot \boldsymbol{b} = 0$. A nonzero parallel electric field arises when the assumption $P \to \infty$ is relaxed. For Alfvén waves, $\boldsymbol{b} \cdot \boldsymbol{e}_A \neq 0$ can arise from either inertial effects or kinetic effects.

The properties of inertial Alfvén waves can be found by expanding cold plasma theory in powers of $1/P$ and retaining terms of first order in $1/P$. The simplest relevant approximation corresponds to setting $D = 0$, and then the dispersion relation for the Aflvén mode becomes $n^2 = PS/A$. With (ordinary units) $S = c^2/v_A^2$ and $P = -\omega_p^2/\omega^2$, $1/n^2 = A/PS$ corresponds to (ordinary units) $\omega^2/|\boldsymbol{k}|^2c^2 = (v_A^2/c^2)\cos^2\theta - (\omega^2/\omega_p^2)\sin^2\theta$. This gives the dispersion relation for inertial Aflvén waves (IAWs)

$$\omega_{\mathrm{IAW}}^2 \approx \frac{k_z^2 v_A^2}{1 + k_\perp^2 \lambda_e^2}, \tag{3.2.21}$$

where $\lambda_e = c/\omega_p$ (ordinary units) is the skin depth. The non-zero parallel component of the polarization vector can be found by noting that the generic form (3.1.19) for the polarization implies that the ratio of the z to x components of the polarization vector is $(L_M \cos\theta - T_M \sin\theta)/(L_M \sin\theta + T_M \cos\theta)$. In the approximation $D \to 0$, the relation (3.2.9) between L_M and T_M implies

$$\frac{e_z}{e_x} = -\frac{S}{P}\tan\theta = \frac{k_z^2\lambda_e^2}{1 + k_\perp^2\lambda_e^2}\tan\theta, \tag{3.2.22}$$

where (3.2.21) is used.

Thermal motions are neglected in the cold plasma approximation, and the approximation is valid only if the phase speed of the waves is much greater than the thermal speed. For Alfvén waves this requires that the Alfvén speed be much greater than the thermal speed of electrons. When this inequality is reversed, the longitudinal response of the unmagnetized electrons changes from $K^L = 1 - \omega_p^2/\omega^2$ to $K^L = 1 + 1/|\boldsymbol{k}|^2\lambda_{De}^2$. For magnetized electrons the electronic contribution to $K^3{}_3$ changes from $-\omega_p^2/\omega^2$ at high phase speed to $1/k_z^2\lambda_{De}^2$ at low phase speed, with $\lambda_{De} = V_e/\omega_p$. When this modification is included in deriving the properties of the waves they are called kinetic Alfvén waves (KAWs). The form of these modifications can be identified by repeating the derivation for IAWs with P replaced by $1/k_z^2\lambda_{De}^2 \gg 1$. This gives the dispersion relation

$$\omega_{\rm KAW}^2 \approx k_z^2 v_A^2 (1 + k_\perp^2 R_g^2), \tag{3.2.23}$$

with $R_g^2 = \lambda_{De}^2 c^2/v_A^2$. The corresponding ratio of the z to x components of the polarization vector for KAWs is

$$\frac{e_z}{e_x} = -k_z^2 R_g^2 \tan\theta. \tag{3.2.24}$$

The approximation made in (3.2.23) and (3.2.24) involves including only one thermal effect: a modification of $K^3{}_3$ for phase speed less than V_e. Two other thermal effects are of comparable order and need to be considered in a more detailed treatment. One is a nonzero contribution to $K^2{}_3 = -K^3{}_2$, which is included in the treatment of MHD-like waves below. The other is a nonzero thermal gyroradius, V_i/Ω_i for the ions. This is relevant because R_g^2 in (3.2.23) is equal to T_e/T_i times V_i^2/Ω_i^2, and V_i^2/Ω_i^2 is clearly of the same order. The thermal motion of the ions was included in an early treatment of KAWs [4] which gave $R_g^2 = (V_i^2/\Omega_i^2)(T_e/T_i + 3/4)$.

3.2.7 MHD-Like Waves

The three MHD waves, discussed in § 1.4, include the effect of thermal motions through a nonzero pressure, giving the (adiabatic) sound speed (1.4.33) which corresponds to $c_s^2 = \Gamma P/\eta$ in the nonrelativistic case. In an unmagnetized collisionless plasma, sound-like waves exist for phase speeds between the ion and electron thermal speeds. The longitudinal response can then be approximated by $K^L = 1 - \omega_{pi}^2/\omega^2 + 1/|\boldsymbol{k}|^2\lambda_{De}^2$, with $\lambda_{De} = V_e/\omega_p$. The solution $\omega_s(\boldsymbol{k}) = |\boldsymbol{k}|^2 v_s^2/(1 + |\boldsymbol{k}|^2\lambda_{De}^2)$, with $v_s^2 = \omega_{pi}^2\lambda_{De}^2$, becomes sound like for $|\boldsymbol{k}|^2\lambda_{De}^2 \ll 1$. These waves are called ion acoustic or ion sound waves. Analogous waves exist in a magnetized plasma, leading to three MHD-like modes with the ion sound speed, v_s, playing the role of the sound speed.

An approximate form of the dielectric tensor that applies in the MHD-like limit is written down in (2.5.19). For $\omega^2 \ll \Omega_i^2$, further approximation to (2.5.19) gives (ordinary units)

$$K^i{}_j(\boldsymbol{k}) = \frac{c^2}{v_A^2}\begin{pmatrix} 1 & i\dfrac{\omega}{\Omega_i} & 0 \\ -i\dfrac{\omega}{\Omega_i} & 1 & -i\dfrac{\Omega_i}{\omega}\tan\theta \\ 0 & i\dfrac{\Omega_i}{\omega}\tan\theta & -\dfrac{\Omega_i^2}{\omega^2}\left(1 - \dfrac{c^2}{n^2 v_s^2\cos^2\theta}\right) \end{pmatrix}. \tag{3.2.25}$$

The approximations made in deriving (3.2.25) include assuming that the ions are cold and unmagnetized. The final entry $c^2/n^2v_s^2\cos^2\theta$ arises from approximating the response of the electrons $1/k_z^2\lambda_{De}^2$, rather than $-\omega_p^2/\omega^2$ in the cold plasma limit.

In solving for the wave properties, it is convenient to introduce $N^2 = |\boldsymbol{k}|^2v_A^2/\omega^2$ and to write the wave equation as $\Lambda^i{}_j e^j = 0$ with

$$\Lambda^i{}_j = \frac{c^2}{v_A^2}\begin{pmatrix} 1 - N^2\cos^2\theta & i\dfrac{\omega}{\Omega_i} & N^2\sin\theta\cos\theta \\ -i\dfrac{\omega}{\Omega_i} & 1 - N^2 & -i\dfrac{\Omega_i}{\omega}\tan\theta \\ N^2\sin\theta\cos\theta & i\dfrac{\Omega_i}{\omega}\tan\theta & -\dfrac{\Omega_i^2}{\omega^2}\left(1 - \dfrac{v_A^2}{v_s^2N^2\cos^2\theta}\right) - N^2\sin^2\theta \end{pmatrix}. \tag{3.2.26}$$

To lowest order in ω^2/Ω_i^2, setting the determinant of $\Lambda^i{}_j$ to zero gives the dispersion equation

$$(1 - N^2\cos^2\theta)\left[\left(1 - \frac{v_A^2}{v_s^2N^2\cos^2\theta}\right)(1 - N^2) + \tan^2\theta\right] = 0. \tag{3.2.27}$$

The solution $N^2\cos^2\theta = 1$ corresponds to $\omega^2 = |\boldsymbol{k}|^2v_A^2$ and is the Alfvén mode. The other factor in (3.2.27) has solutions

$$\omega^2 = |\boldsymbol{k}|^2v_\pm^2, \quad v_\pm^2 = \tfrac{1}{2}(v_A^2 + v_s^2) \pm \tfrac{1}{2}[(v_A^2 + v_s^2)^2 - 4v_A^2v_s^2\cos^2\theta]^{1/2}. \tag{3.2.28}$$

The dispersion relations (3.2.28) are analogous to those for the fast and slow MHD waves, (1.4.36), with the sound speed replaced by the ion sound speed.

A perturbation expansion in ω^2/Ω_i^2 gives a correction to the dispersion relation for Alfvén waves

$$1 - N^2\cos^2\theta = -\frac{\omega^2}{\Omega_i^2}\cot^2\theta\left(1 - \frac{v_s^2}{v_A^2}\sec^4\theta\right). \tag{3.2.29}$$

This dispersion relation is similar to (3.2.23), but the angular dependence is not the same as in (3.2.23). This is due to the inclusion of the $K^2{}_3 = -K^3{}_2$ components in the MHD-like theory.

The polarization vectors for the MHD-like modes can be found by constructing the matrix of cofactors, $\Lambda^i{}_j$, using (3.2.26), inserting the dispersion relation in any of the columns, and normalizing to unity. For the Alfvén mode one finds, for $\omega^2 \ll \Omega_i^2(v_A^2/v_s^2)\tan^2\theta$,

$$\boldsymbol{e}_A = \left(1, -i\frac{\omega}{\Omega_i}\cot^2\theta, -\frac{\omega^2}{\Omega_i^2}\frac{v_s^2}{v_A^2}\frac{1}{\cos\theta\sin\theta}\right), \tag{3.2.30}$$

where the y- and z-components are small corrections to $\boldsymbol{e}_A = (1, 0, 0)$. The ratio of e_z/e_x is similar in form to (3.2.24) but is not the same because of the inclusion of the $K^2{}_3 = -K^3{}_2$ components. The polarization vectors for the fast and slow modes for $v_A^2 \gg v_s^2$ and $\omega \ll \Omega_i$ are $\boldsymbol{e}_m = (0, i, 0)$ and $\boldsymbol{e}_s = \boldsymbol{\kappa} = (\sin\theta, 0, \cos\theta)$, and the lowest order corrections give

$$\boldsymbol{e}_m = \left(\frac{\omega}{\Omega_i}\frac{1}{\sin^2\theta}, i, \frac{\omega}{\Omega_i}\frac{v_s^2}{v_A^2}\sin\theta\cos\theta\right),$$

$$\boldsymbol{e}_s = \left(\sin\theta, -i\frac{\Omega_i}{\omega}\frac{v_s^2}{v_A^2}\sin\theta\cos^2\theta, \cos\theta\right). \tag{3.2.31}$$

In this approximation the ratio of electric to total energy can be approximated by

$$R_A \approx v_A^2/2c^2 \approx R_m,$$
$$R_s \approx (v_A^2/2c^2)(\omega^2/\Omega_i^2\cos^2\theta)[1 + (\Omega_i^2/\omega^2)(v_s^4/v_A^4)\sin^2\theta\cos^2\theta].$$

Landau damping by electrons is the most important damping mechanism for the MHD-like waves. The antihermitian part of the response tensor is given by (2.5.20) with the only nonzero components given by (2.5.21). This leads to absorption coefficients

$$\gamma_A = \omega\left(\frac{\pi}{2}Z_i\frac{m_e}{m_i}\right)^{1/2}\frac{v_s}{v_A}\frac{\omega^2}{\Omega_i^2}(\tan^2\theta + \cot^2\theta),$$

$$\gamma_m = \omega\left(\frac{\pi}{2}Z_i\frac{m_e}{m_i}\right)^{1/2}\frac{v_s}{v_A}\frac{\sin^2\theta}{|\cos\theta|}, \qquad \gamma_s = \omega\left(\frac{\pi}{2}Z_i\frac{m_e}{m_i}\right)^{1/2}, \tag{3.2.32}$$

where $Z_i e$ is the ionic charge. At low frequencies, $\omega \ll \Omega_i$, the Alfvén mode is much more weakly damped than the other two modes.

3.3 Waves in Cold Electronic Plasmas

At frequencies well above the ion plasma and ion cyclotron frequencies, the contribution of ions to dispersion in a plasma can be neglected in comparison with the contribution of the electrons. The magnetoionic theory describes dispersion in a cold electron gas, and the waves are often called magnetoionic waves. The properties of magnetoionic waves are discussed in this section.

3.3.1 Magnetoionic Waves

The name "magnetoionic" is an anachronism: magnetoionic theory was developed before the present-day meaning of "ion" became accepted. Ions, in the modern-day sense, play no role in the magnetoionic theory. Magnetoionic theory corresponds to cold plasma theory with only the contribution of the electrons retained.

In treating the magnetoionic theory here, an admixture of (cold) positrons is included. This is achieved by re-interpreting the sign of the charge, $-\epsilon$, as the average charge per particle, that is, as $\epsilon = (n^+ - n^-)/(n^+ + n^-)$, where $n^\pm$ are the number densities of electrons and positrons, respectively. (The $\pm$ labeling corresponds to the sign ϵ used in QED, with $\epsilon = +1$ for the particle and $\epsilon = -1$ for the antiparticle, which are electron and positron, respectively, here.) An electron gas corresponds to $\epsilon = 1$, a pure pair plasma (equal numbers of electrons and positrons) corresponds to $\epsilon = 0$, and a positron gas corresponds to $\epsilon = -1$.

The dispersion equation (3.2.2) for a cold plasma becomes the dispersion equation for the magnetoionic waves when (1.2.38) is used to express S, D, P in terms of the magnetoionic parameters, $X = \omega_p^2/\omega^2$, $Y = \Omega_e/\omega$, specifically $S = 1 - X/(1-Y^2)$, $D = -\epsilon XY/(1-Y^2)$, $P = 1 - X$. The magnetoionic dispersion equation becomes

$$
\begin{aligned}
&An^4 - Bn^2 + C = 0, A = 1 - \frac{X}{1-Y^2}(1 - Y^2\cos^2\theta),\\
&B = \left[1 - \frac{2X}{1-Y^2} + \frac{X^2}{(1-Y^2)^2}(1 - \epsilon^2 Y^2)\right]\sin^2\theta\\
&\qquad + (1-X)\left(1 - \frac{X}{1-Y^2}\right)(1 + \cos^2\theta),\\
&C = (1-X)\left[1 - \frac{2X}{1-Y^2} + \frac{X^2}{(1-Y^2)^2}(1 - \epsilon^2 Y^2)\right].
\end{aligned}
\tag{3.3.1}
$$

The solutions in the form (3.2.4), $n^2 = n_\pm^2 = (B \pm F)/2A$, $F = (B^2 - 4AC)^{1/2}$ define two magnetoionic modes.

An alternative derivation of the wave properties for a cold plasma involves first solving the quadratic equation (3.2.6) for the axial ratio, T. In the magnetoionic theory (3.2.6) becomes

$$
T^2 + RT - 1 = 0, \qquad R = \frac{(1-E)Y\sin^2\theta}{\epsilon(1-X)\cos\theta}, \qquad E = \frac{(1-\epsilon^2)X}{1-Y^2}. \tag{3.3.2}
$$

The solutions of (3.3.2) are $T = \frac{1}{2}[\pm(4+R^2)^{1/2} - R] = 1/\{\frac{1}{2}[\pm(4+R^2)^{1/2} + R]\}$. These are

$$
\begin{aligned}
&T = T_\sigma = \frac{\epsilon Y(1-X)\cos\theta}{\frac{1}{2}(1-E)Y^2\sin^2\theta - \sigma\Delta} = -\frac{\frac{1}{2}(1-E)Y^2\sin^2\theta + \sigma\Delta}{\epsilon Y(1-X)\cos\theta},\\
&\Delta^2 = \tfrac{1}{4}(1-E)^2Y^4\sin^4\theta + \epsilon^2(1-X)^2Y^2\cos^2\theta,
\end{aligned}
\tag{3.3.3}
$$

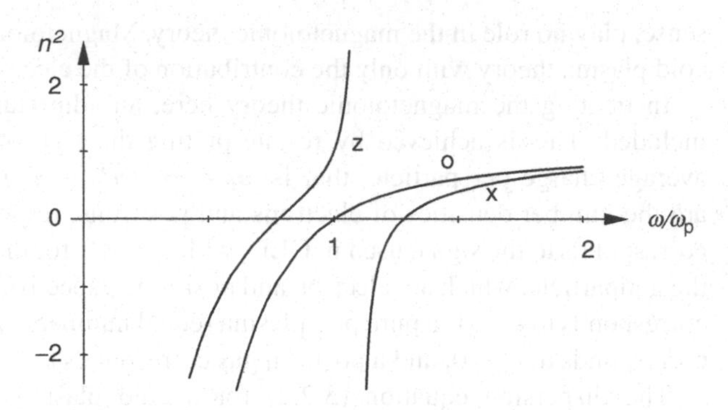

Fig. 3.2 The dispersion curves for the o, x and z branches of the magnetoionic modes are indicated schematically for $\omega_p < \Omega_e$. The whistler branch (called helicon waves in a laboratory plasma) is off the diagram in the *upper left hand corner*

with $\sigma = \pm 1$. The three relations (3.2.8) give alternative expressions for n_σ^2 by setting $M \to \sigma = \pm$ and substituting the expressions (1.2.38) for P, S, D.

3.3.2 Four Branches of Magnetoionic Modes

The magnetoionic waves in an electron gas ($\epsilon = 1$, $E = 0$) have four branches: two high frequency branches, referred to here as the o mode and the x mode, and two lower frequency branches, referred to as the z mode and the whistler mode. The o mode and the x mode exist above respective cutoffs, the z mode exists between a cutoff and a resonance, and the whistler mode exists below a resonance. These branches are illustrated in Fig. 3.2.

The cutoff frequencies satisfy $C = 0$, with C given by (3.3.1). For $\epsilon = 1$ there are three (positive frequency) solutions. The solution corresponding to $1 - X = 0$ is $\omega = \omega_\mathrm{p}$, which is the cutoff for the o mode. Two positive frequency solutions arise from the factor in square brackets in (3.3.1). Writing these as $\omega = \omega_\mathrm{x}$ and $\omega = \omega_\mathrm{z}$, one finds

$$\omega_\mathrm{x} = \tfrac{1}{2}\Omega_\mathrm{e} + \tfrac{1}{2}(4\omega_\mathrm{p}^2 + \Omega_\mathrm{e}^2)^{1/2}, \quad \omega_\mathrm{z} = -\tfrac{1}{2}\Omega_\mathrm{e} + \tfrac{1}{2}(4\omega_\mathrm{p}^2 + \Omega_\mathrm{e}^2)^{1/2}. \tag{3.3.4}$$

These are the cutoff frequencies for the x and z modes, respectively.

The resonant frequencies satisfy $A/C = 0$. The solutions of $A = 0$, which are independent of ϵ, are the resonant frequencies $\omega_\pm(\theta)$, defined by (3.2.13). The higher-frequency resonance is between the larger of ω_p, Ω_e and ω_UH, and is the resonance in the z mode. The lower-frequency resonance is below the smaller of ω_p, Ω_e, and is the resonance in the whistler mode. For the lower resonant frequency, the neglect of the ions is not justified when $\omega_-(\theta)$ is comparable with or less than the lower-hybrid frequency.

The solutions of the dispersion equation in the limit $\sin\theta = 0$ are different from the limit $\sin\theta \to 0$ of the solutions for $\sin\theta \neq 0$. The dispersion relations for $\sin\theta = 0$ are $n^2 = 1 - X/(1 \pm Y)$ for two transverse modes, and $X = 1$ for a

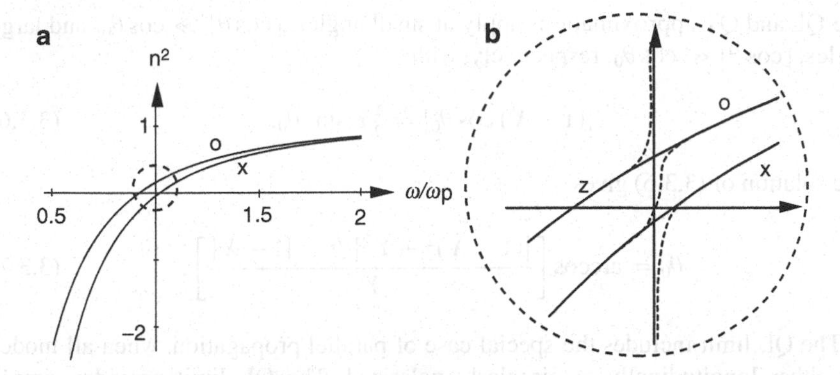

Fig. 3.3 (**a**) For $\theta = 0$ there are two curves plus a vertical line (not shown) at $\omega = \omega_p$. (**b**) Circled portion of (**a**) magnified; the *dashed lines* is for $\theta \neq 0$; for $\theta = 0$ the o mode and the z mode join at $\omega = \omega_p$

longitudinal mode. The dispersion curve $n^2 = 1 - X/(1+Y)$ crosses the solution $X = 1$ at $n^2 = Y/(1+Y)$. The two solutions for nonzero angle θ, reduce to these solutions for arbitrarily small θ, except that a reconnection occurs, as illustrated schematically in Fig. 3.1a. In the limit $\sin\theta \to 0$, the solution $n^2 = 1 - X/(1+Y)$ corresponds to the o mode for $\omega > \omega_p$ and to the z mode for $\omega < \omega_p$; the solution $X = 1$ corresponds to the o mode for $n^2 < Y/(1+Y)$, and to the z mode for $n^2 > Y/(1+Y)$. The change in the dispersion curves from $\sin\theta = 0$ to $\sin\theta \neq 0$ is indicated schematically in Fig. 3.3.

3.3.3 *QL and QT Limits*

A useful approximation is based on whether the modes are nearly circularly polarized, $|T_\pm| = 1$, or nearly linearly polarized, $T_\pm = 0, \infty$. In the now outdated conventions of the magnetoionic theory these were called the quasi-longitudinal (QL) and quasi-transverse (QT) limits. Although the labels QL and QT are used here, they should be interpreted as the nearly-circular and nearly-linear approximations, respectively.

For an electron gas ($\epsilon = 1$, $E = 0$), it follows from (3.3.2) that for $|Y \sin^2\theta/(1-X)\cos\theta| \ll 1$ the two solutions for the axial ratio are $T \approx \pm 1$, and that for $|Y \sin^2\theta/(1-X)\cos\theta| \gg 1$, the two solutions approach $T = 0, \infty$. It follows from (3.3.3) that one has

$$\Delta \approx \begin{cases} |(1-X)Y\cos\theta| & \text{QL limit,} \\ \frac{1}{2}Y^2\sin^2\theta & \text{QT limit.} \end{cases} \tag{3.3.5}$$

The QL and QT approximations apply at small angles, $|\cos\theta| \gg \cos\theta_0$, and large angles, $|\cos\theta| \ll \cos\theta_0$, respectively, with

$$|(1-X)\cos\theta_0| = \tfrac{1}{2}Y\sin^2\theta_0. \tag{3.3.6}$$

The solution of (3.3.6) gives

$$\theta_0 = \arccos\left[\frac{[(1-X)^2+Y^2]^{1/2}-|1-X|}{Y}\right]. \tag{3.3.7}$$

The QL limit includes the special case of parallel propagation, when all modes are either longitudinally or circularly polarized. The QL limit provides simple approximations to the dispersion relations for the (nearly) circularly polarized modes for oblique propagation, $\theta \ll \theta_0$. For example, in the high-frequency limit $X, Y \ll 1$ the modes with $T_\sigma = \sigma$, with $\sigma = \pm 1$, have

$$n_\sigma^2 = 1 - \frac{X}{1-\sigma Y\cos\theta}, \qquad L_\sigma = \frac{XY\sin\theta}{1-X}\,\frac{1}{1-\sigma Y\cos\theta}. \tag{3.3.8}$$

The QT limit includes perpendicular propagation. The ordinary mode in magnetoionic theory, is defined as the mode with dispersion relation $n^2 = P$ for $\theta = \pi/2$. Solving for the two modes for $\theta = \pi/2$ gives $n^2 = P$, $n^2 = (S^2 - D^2)/S$, with $T \to \infty$ for the ordinary mode, and $T \to 0$, $L \to D/S$ for the extraordinary mode.

Transition Angle

The axial ratio, T, which completely describes the transverse part of the polarization of a natural wave mode, depends only on the parameter R in the quadratic equation (3.3.2). The degree of linear polarization is $r_l = (T^2-1)/(T^2+1)$ and the degree of circular polarization is $r_c = 2T/(T^2+1)$. The variations of $|r_l|$ and $|r_c|$ for either of the two solutions of (3.3.2) as a function of R are plotted in Fig. 3.4. One has $|r_c/r_l| = 2/|R|$, implying that the degrees of polarization are equal for $|R| = 2$. It is convenient to define a transition angle, θ_c, corresponding to $|R| = 2$. Then a given mode is approximately circularly polarized in one sense for $\theta \lesssim \theta_c$, approximately linearly polarized for $\theta_c \lesssim \theta \lesssim \pi - \theta_c$, and approximately circularly polarized in the opposite sense for $\pi - \theta_c \lesssim \theta \leq \pi$.

Writing $|R| = r\sin^2\theta/\cos\theta$, with $r = (1-E)Y/\epsilon(1-X) \approx Y/\epsilon$, where the approximation applies for $X \ll 1$, the transition angle satisfies $2 = |r|\sin^2\theta_c/\cos\theta_c$, which gives

$$\theta_c = \arccos\left(\frac{(1+r^2)^{1/2}-1}{|r|}\right) \approx \begin{cases} \pi/2 - |r| & \text{for } r \ll 1, \\ (2/|r|)^{1/2} & \text{for } r \gg 1. \end{cases} \tag{3.3.9}$$

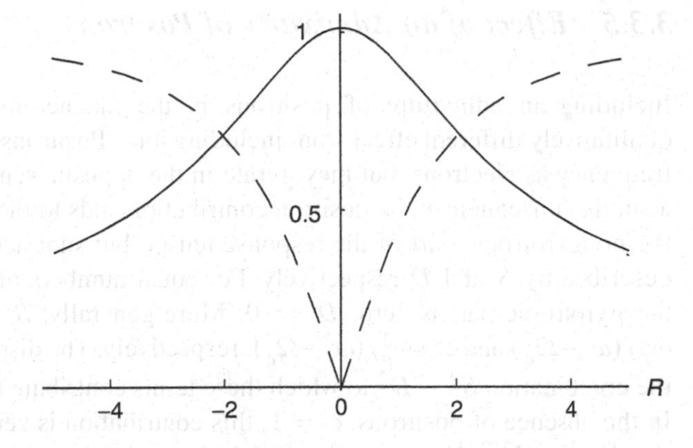

Fig. 3.4 The magnitudes of the degrees of circular (*solid line*) and linear (*dashed line*) are plotted for either mode as a function of R [23]

For $|r| \ll 1$ the polarization is nearly circular except for a small range of angles $\approx \Omega_e/\omega$ around $\pi/2$, and for $|r| \gg 1$, the polarization is nearly linear except in small cones about parallel and anti-parallel propagation.

3.3.4 High-Frequency Limit

At high frequency the magnetoionic waves may be approximated by expanding in powers of $X \ll 1$, $Y \ll 1$. At sufficiently high frequency $X \propto 1/\omega^2$ becomes much smaller than $Y \propto 1/\omega$, and the wave properties may be approximated by taking the limit $X \to 0$ in (3.3.2). The solutions (3.3.3) for $X = 0$ become

$$T_\sigma = \frac{\epsilon \cos\theta}{\frac{1}{2}Y\sin^2\theta - \sigma\delta} = -\frac{\frac{1}{2}Y\sin^2\theta + \sigma\delta}{\epsilon\cos\theta}, \qquad \delta = \left(\tfrac{1}{4}Y^2\sin^4\theta + \epsilon^2\cos^2\theta\right)^{1/2}. \tag{3.3.10}$$

An expansion in Y is valid except for a small range of angles about perpendicular propagation, specifically, except for $|\cos\theta| \lesssim \frac{1}{2}Y$. The leading term in the expansion of (3.3.10) gives $T_\sigma = -\epsilon\sigma\cos\theta/|\epsilon\cos\theta|$, corresponding to circular polarization. In this approximation, the refractive indices become

$$n_\sigma^2 = 1 - X(1 - \sigma Y). \tag{3.3.11}$$

The longitudinal part of the polarization, L_σ, is very small, being proportional to XY. The neglect of the longitudinal part of the polarization is the basis for the weak-anisotropy approximation, discussed in § 3.6.

3.3.5 *Effect of an Admixture of Positrons*

Including an admixture of positrons in the magnetoionic theory [5–7] has a qualitatively different effect from including ions. Positrons have the same cyclotron frequency as electrons, but they gyrate in the opposite sense to electrons. This has a subtle consequence: the positron contribution adds to the electron contribution in the nongyrotropic part of the response tensor, but subtracts in the gyrotropic part, described by S and D respectively. For equal numbers of electrons and positrons the gyrotropic part is zero, $D = 0$. More generally, S and D include terms $\propto \omega_p^2/(\omega^2 - \Omega_e^2)$ and $\propto \epsilon\omega_p^2/(\omega^2 - \Omega_e^2)$, respectively. The dispersion equation involves the combination $S^2 - D^2$ to which these terms contribute $(1 - \epsilon^2)\omega_p^4/(\omega^2 - \Omega_e^2)^2$. In the absence of positrons, $\epsilon = 1$, this contribution is zero, and there is no terms $\propto 1/(\omega^2 - \Omega_e^2)^2$. However, for $\epsilon^2 \neq 1$, the dispersion equation contains a term that diverges quadratically at $\omega = \Omega_e$, and this introduces a resonance at the cyclotron frequency that is not present for an electron gas. There is also an additional cutoff, leading to an additional branch of the magnetoionic waves that is a unique feature of a cold electron/positron gas.

The cutoffs occur at $C = 0$, corresponding to either $P = 0$ or $S^2 - D^2 = 0$. The cutoff at $P = 0$, $X = 1$ or $\omega = \omega_p$, is unaffected by the presence of positrons. The two cutoffs (3.3.4) in an electron gas correspond to $S^2 - D^2 = 0$, and for $\epsilon^2 \neq 1$ there are three cutoffs in general. These satisfy

$$(1 - Y^2 - X)^2 - \epsilon^2 X^2 Y^2 = 0, \tag{3.3.12}$$

which factorizes into two cubic equation for ω:

$$\omega^3 - \omega(\omega_p^2 + \Omega_e^2) \pm \epsilon\omega_p^2\Omega_e = 0. \tag{3.3.13}$$

The cutoffs are given by the real solutions for $\omega > 0$ of one or other of these equations. A graphical solution is indicated in Fig. 3.5, where (3.3.12) is plotted as a function of $x = \omega/\Omega_e$. There are three solutions for $\epsilon \neq 0$. For $\epsilon = 1$ two of these are the cutoffs ω_x, ω_z given by (3.3.4), and the third is at Ω_e, but this is spurious, arising by multiplying by $(1 - Y^2)^2$ in deriving (3.3.12). For $\epsilon \neq \pm 1, 0$ the three cutoffs are all different and none is equal to Ω_e. For $\epsilon \to 0$ the three cutoffs approach each other, and coincide in the limit $\epsilon = 0$ at the upper-hybrid frequency $\omega = \omega_{\text{UH}}$.

The modification of the magnetoionic modes due to an admixture of positrons is indicated by the dashed curves in Fig. 3.6. The additional mode introduced by the resonance at $\omega = \Omega_e$ may be interpreted as a cyclotron resonant mode, and it is denoted by c in Fig. 3.6.

The wave properties simplify considerably for a pure pair plasma, $\epsilon = 0$. For $\epsilon = 0$ one has $D = 0$ in (1.2.38), and the equivalent dielectric tensor (1.2.29) reduces to the same form as for a uniaxial crystal. The dispersion equation becomes

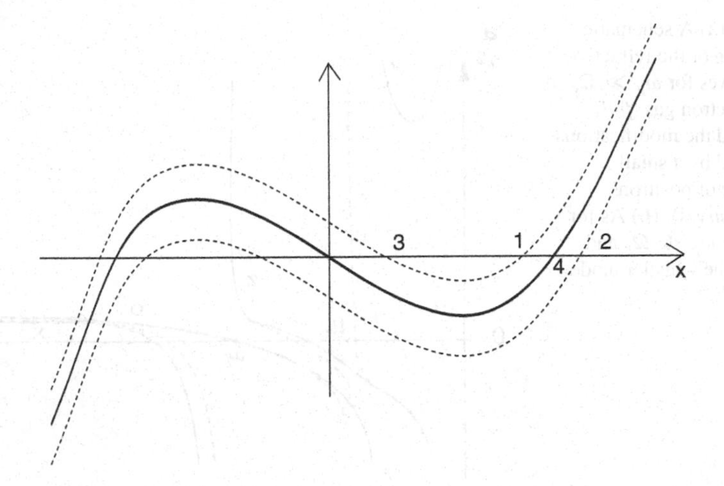

Fig. 3.5 The cubic equation (3.3.13) is plotted schematically as a function of $x = \omega/\Omega_e$. The *two dashed curves* are for $\epsilon = \pm 1$ and they give positive frequency cutoffs at points 1, 2 and 3. As $|\epsilon|$ is decreased, the *dashed curves* approach each other, and coincide as the *solid curve* for $\epsilon = 0$, when there is only one cutoff at 4

$(n^2 - S)(An^2 - PS) = 0$. For $\theta \neq 0$, the ordinary mode (as defined in magnetoionic theory) has dispersion relation and polarization vector

$$n_{\rm o}^2 = \frac{(\omega^2 - \omega_p^2)(\omega^2 - \omega_p^2 - \Omega_e^2)}{[\omega^2 - \omega_-^2(\theta)]\,[\omega^2 - \omega_+^2(\theta)]}, \qquad \boldsymbol{e}_{\rm o} = \frac{(P\cos\theta, 0, -S\sin\theta)}{(P^2\cos^2\theta + S^2\sin^2\theta)^{1/2}}, \tag{3.3.14}$$

and the extraordinary mode (as defined in magnetoionic theory) has

$$n_{\rm x}^2 = \frac{\omega^2 - \omega_p^2 - \Omega_e^2}{\omega^2 - \Omega_e^2}, \qquad \boldsymbol{e}_{\rm x} = (0, i, 0), \tag{3.3.15}$$

where the coordinates axes have $\boldsymbol{B}$ along the 3-axis and $\boldsymbol{k}$ in the 1–3 plane, and where $\omega_\pm^2(\theta)$ are given by (3.2.13).

There is an inconsistency in the labeling of the modes: the conventional labeling of the modes from magnetoionic theory is opposite to that for the labeling implied by the analogy with a uniaxial crystal. For a uniaxial crystal the 'ordinary' mode is the one that does not depend on angle, and the 'extraordinary' mode is the one that does depend on angle. With this convention, the labels o and x in (3.3.14) and (3.3.15) would be reversed.

For $\epsilon = 0$ there are two cutoff frequencies, at $\omega = \omega_p$ in the o mode, and at $\omega = (\omega_p^2 + \Omega_e^2)^{1/2}$. The latter is a double solution, and corresponds to a cutoff that is common to both modes.

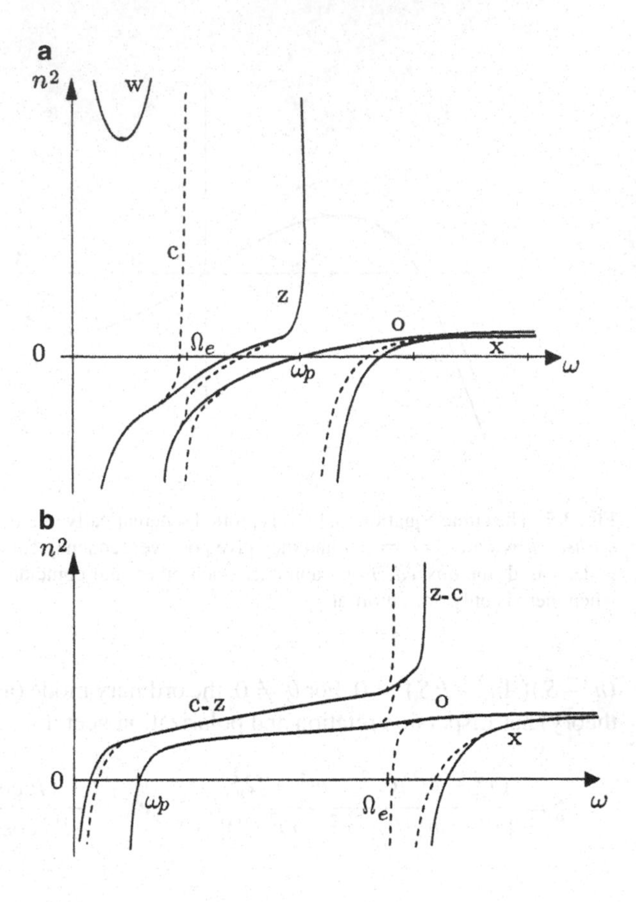

Fig. 3.6 (**a**) A schematic illustration of the refractive index curves for $\omega_p \gg \Omega_e$ in a cold electron gas (*full curve*) and the modifications introduced by a small admixture of positrons (*dashed curves*). (**b**) As for (**a**) but for $\omega_p \ll \Omega_e$ and omitting the whistler mode

3.3.6 Lorentz Transformation of Magnetoionic Waves

In the presence of streaming, one may treat wave dispersion in two alternative ways. The more general way is to construct the dispersion equation with the streaming motions included in the response tensor, and solve for the wave modes directly. This is the only method that applies when different species are streaming at different velocities. The other method applies when all species are streaming with the same velocity. Then one may solve for the wave properties in the rest frame, and Lorentz transform these properties to the frame in which the plasma is streaming.

Consider the magnetoionic waves in a frame in which the (cold) electron gas is streaming; the ions play no role and can be ignored. Let the rest frame of the plasma be the unprimed frame and letthe primed frame be the laboratory frame in which the plasma is streaming. A Lorentz transformation between the wave 4-vectors in the two frames gives

$$\begin{pmatrix} \omega' \\ k'_\perp \\ k'_z \end{pmatrix} = \begin{pmatrix} \gamma(\omega + k_z\beta) \\ k_\perp \\ \gamma(k_z + \omega\beta) \end{pmatrix}, \qquad \begin{pmatrix} \omega \\ k_\perp \\ k_z \end{pmatrix} = \begin{pmatrix} \gamma(\omega' - k'_z\beta) \\ k'_\perp \\ \gamma(k'_z - \omega'\beta) \end{pmatrix}. \tag{3.3.16}$$

Given the refractive index $n = n_\sigma(\omega, \theta)$ in the rest frame, one uses it to write $k_z = n\cos\theta$, $k_\perp = n\sin\theta$ as functions of ω, θ. One then finds $k'_z = n'\cos\theta'$, $k'_\perp = n'\sin\theta'$ from (3.3.16) and uses these to find n' and θ' as functions of ω, θ. It is straightforward to write down the solutions implicitly, but finding explicit solutions usually requires making simplifying approximations.

Rather than Lorentz transform directly, it is sometimes helpful to write the solutions in the rest frame in terms of invariants, and then to evaluate these invariants in the moving frame. The quantities n^2, $\sin^2\theta$, ω in the rest frame, whose 4-velocity is $\tilde{u}^\mu$, can be written in invariant forms by using the identifications

$$n^2 \to \frac{(k\tilde{u})^2 - k^2}{(k\tilde{u})^2}, \qquad \sin^2\theta \to \frac{k_\perp^2}{(k\tilde{u})^2 - k^2}, \qquad \omega \to k\tilde{u}. \tag{3.3.17}$$

In the rest frame one has $\tilde{u}^\mu = [1, \mathbf{0}]$. This allows one to rewrite the dispersion equation in an arbitrary frame by making the replacements $k \to k'$, $\tilde{u} \to u$, with $u^\mu = [\gamma, \gamma\boldsymbol{\beta}]$, and with $k'_\perp = k_\perp$. However, the dispersion equation $An^4 - Bn^2 + C = 0$ is no longer a quadratic equation in the new variables, e.g., in n'^2, θ', ω', and solving it for the dispersion relations is not straightforward.

Provided one is interested only in waves with high frequency, such that the refractive index is of order unity, it is sometimes convenient to rewrite the dispersion equation (3.3.1) for $n^2 - 1 + X$, rather than for n^2. The advantage is that one has (in ordinary units) $n^2 - 1 + X = -(k^2c^2 - \omega_p^2)/\omega^2$, where $k^2c^2 - \omega_p^2$ is an invariant. Then (3.3.1) becomes

$$a(n^2 - 1 + X)^2 - b(n^2 - 1 + X) + c = 0, \tag{3.3.18}$$

with $a = A$, $b = B - 2A(1 - X)$, $c = C - B(1 - X) + A(1 - X)^2$. The solutions of the quadratic equation (3.3.18) give the dispersion relations in invariant form (ordinary units)

$$k^2 = k_\pm^2 = \frac{\omega_p^2}{c^2}\left[1 - \frac{b \pm (b^2 - 4ac)^{1/2}}{2aX}\right]. \tag{3.3.19}$$

The dispersion relations in the primed frame are then formally given by (3.3.19) with $k^2 = \omega'^2/c^2 - |\boldsymbol{k}'|^2$ and with a, b, c expressed in terms of the primed variables. Although the solutions are implicit, provided that n^2 is not too different from $1 - X$, one can solve (3.3.19) iteratively to find explicit solutions.

3.3.7 *Transformation of the Polarization Vector*

The transformation of a polarization vector from one inertial frame to another requires that the Lorentz transformation be complemented by a gauge transformation. This arises because a polarization 3-vector is defined in the temporal gauge, and the gauge condition is not preserved by the Lorentz transformation. Consider the transformation of the polarization 3-vector, in the general form

$$\boldsymbol{e}_{\pm} = \frac{(L_{\pm}\boldsymbol{\kappa} + T_{\pm}\boldsymbol{t} + i\boldsymbol{a})}{(L_{\pm}^2 + T_{\pm}^2 + 1)^{1/2}}, \tag{3.3.20}$$

from the rest (unprimed) frame of the plasma to a (primed) frame in which the plasma is streaming along the magnetic field lines.

It is straightforward to apply a Lorentz transformation to the vectors $\boldsymbol{\kappa}, \boldsymbol{t}, \boldsymbol{a}$, giving, say, $\boldsymbol{\kappa}_{\rm t}, \boldsymbol{t}_{\rm t}, \boldsymbol{a}_{\rm t}$, respectively. One needs to make a gauge transformation to the temporal gauge in the primed frame (§ 2.6.4 of volume 1). The transformed vectors are

$$\boldsymbol{\kappa}_{\rm t} = \frac{\gamma\omega}{\omega'|\boldsymbol{k}|}\,(|\boldsymbol{k}|'\sin\theta', 0, |\boldsymbol{k}|'\cos\theta' - \omega'\beta), \tag{3.3.21}$$

$$\boldsymbol{t}_{\rm t} = \frac{1}{n}\left(n'\cos\theta' - \frac{\beta(1-n'^2)}{1-n'\beta\cos\theta'}, 0, -n'\sin\theta'\right), \qquad \boldsymbol{a}_{\rm t} = \boldsymbol{a}. \tag{3.3.22}$$

One may rewrite (3.3.22) as

$$\boldsymbol{t}_{\rm t} = \frac{1}{n}\left(n' - \frac{(1-n'^2)\beta\cos\theta'}{1-n'\beta\cos\theta'}\right)\boldsymbol{t}' - \frac{1}{n}\frac{(1-n'^2)\beta\sin\theta'}{1-n'\beta\cos\theta'}\boldsymbol{\kappa}', \qquad \boldsymbol{a}_{\rm t} = \boldsymbol{a}', \tag{3.3.23}$$

with $\boldsymbol{\kappa}' = (\sin\theta', 0\cos\theta')$, $\boldsymbol{t}' = (\cos\theta', 0, -\sin\theta')$, $\boldsymbol{a}' = (0, 1, 0)$. In the primed frame the polarization vector is given by

$$\boldsymbol{e}'_{\pm} = \frac{(L_{\pm}\boldsymbol{\kappa}_{\rm t} + T_{\pm}\boldsymbol{t}_{\rm t} + i\boldsymbol{a}_{\rm t})}{(L_{\pm}^2 + T_{\pm}^2 + 1)^{1/2}}, \tag{3.3.24}$$

with $L_{\pm}, T_{\pm}$ rewritten as functions of the primed variables. Using (3.3.21) and (3.3.23) one may rewrite (3.3.24) in the form $\boldsymbol{e}'_{\pm} = (L'_{\pm}\boldsymbol{\kappa}' + T'_{\pm}\boldsymbol{t}' + i\boldsymbol{a}')/(L'^2_{\pm} + T'^2_{\pm} + 1)^{1/2}$, with $L'_{\pm}, T'_{\pm}$ identified in terms of $L_{\pm}, T_{\pm}$ by relatively cumbersome expressions.

3.4 Waves in Weakly Relativistic Thermal Plasmas

There is a rich variety of natural wave modes in a magnetized thermal plasma. These modes include modified forms of the modes of a cold magnetized plasma, and additional modes that depend intrinsically on thermal effects. In this section the properties of waves in a weakly relativistic, magnetized, thermal, electron gas are discussed.

3.4.1 Cyclotron-Harmonic Modes

Emphasis is placed here on cyclotron-harmonic modes.

Wave Modes for Perpendicular Propagation

Cyclotron harmonic modes near perpendicular propagation, often called Bernstein modes, were first discussed by Gross [8] and Bernstein [9], who considered longitudinal modes. Two further classes of cyclotron harmonic modes, related to ordinary and extraordinary modes, were identified by Dnestrovskii and Kostomarov [10, 11]. The properties of all three classes of cyclotron harmonic modes in the nonrelativistic case were described by Puri et al. [12, 13]. Here we follow [14–16] in referring to these as the Gross-Bernstein (GB) modes, and the Dnestrovskii-Kostomarov (DK) modes.

Cyclotron harmonic modes were first identified for strictly perpendicular propagation in a nonrelativistic plasma, and it is appropriate to start by considering this case. Setting $k_z = 0$ in (2.5.27), the 13-, 23-, 31- and 32-components of the response 3-tensor are zero. The dispersion equation then becomes

$$\begin{vmatrix} K^1{}_1 & K^1{}_2 & 0 \\ K^2{}_1 & K^2{}_2 - n^2 & 0 \\ 0 & 0 & K^3{}_3 - n^2 \end{vmatrix} = (K^3{}_3 - n^2)[K^1{}_1(K^2{}_2 - n^2) - K^1{}_2\, K^2{}_1] = 0, \quad (3.4.1)$$

where $K^1{}_2 = -K^2{}_1$ is imaginary. There are two relevant solutions

$$n^2 = K^3{}_3, \qquad K^1{}_1\,(K^2{}_2 - n^2) - |K^1{}_2|^2 = 0, \quad (3.4.2)$$

corresponding to the ordinary and extraordinary modes, respectively. The solution of (3.4.2) for the extraordinary mode, satisfying

$$n^2 = K^2{}_2 - \frac{|K^1{}_2|^2}{K^1{}_1}, \quad (3.4.3)$$

are the DK modes, which are transverse. The GB modes are longitudinal.

Cyclotron-Harmonic Modes in a Nonrelativistic Plasma

The cyclotron-harmonic modes have their simplest form for perpendicular propagation in a nonrelativistic thermal electron gas. The relevant dielectric tensor is Shkarofsky's approximation (2.5.27) to Trubnikov's response tensor. In (2.5.27) one sets $k_z = 0$ for perpendicular propagation, and $r_0(\xi) = \rho - i\omega\xi$ in the exponential function and $r_0(\xi) = \rho$ elsewhere, corresponding to the nonrelativistic limit. In the nonrelativistic limit, the quantity Λ, introduced in (2.5.26), reduces to $\lambda = k_\perp^2 V^2/\Omega_0^2$, with $V^2 = 1/\rho$. Then (2.5.27) gives

$$K^i_{\ j} = \delta^i_j - \frac{\omega_p^2}{\omega^2}\sum_{a=-\infty}^{\infty}\frac{h^i_{\ j}(\lambda)e^{-\lambda}}{\Delta_a}, \qquad \lambda = \frac{k_\perp^2}{\rho\Omega_0^2} = \frac{n^2\omega^2}{\rho\Omega_0^2}, \qquad \Delta_a = \frac{\omega - a\Omega_0}{\omega},$$

$$h^1{}_1(\lambda) = a^2 I_a(\lambda)/\lambda, \qquad h^1{}_2 = -i\epsilon s[I_a(\lambda) - I_a'(\lambda)],$$
$$h^2{}_2 = h^1{}_1 - 2\lambda[I_a(\lambda) - I_a'(\lambda)], \qquad h^3{}_3 = I_a(\lambda), \tag{3.4.4}$$

with $\epsilon = 1$ for an electron gas.

Ordinary Modes

On inserting (3.4.4) in the dispersion relation (3.4.2) for ordinary mode waves, one obtains

$$n^2 = 1 - \frac{\omega_p^2}{\omega^2}\sum_{a=-\infty}^{\infty}\frac{I_a(\lambda)e^{-\lambda}}{\Delta_a}. \tag{3.4.5}$$

The factor Δ_a in (3.4.5) becomes arbitrarily small in the limit $\omega \to a\Omega_e$, and no matter how small the numerator, which contains the factor

$$I_a(\lambda)e^{-\lambda} \approx \begin{cases} \lambda^a/2^a a! & \text{for } \lambda \to 0, \\ 1/(2\pi\lambda)^{1/2} & \text{for } \lambda \to \infty, \end{cases} \tag{3.4.6}$$

there is always a solution of (3.4.5). Hence, there is one ordinary mode per harmonic $a \geq 1$.

The qualitative form of the dispersion relations is illustrated in Fig. 3.7. Each mode has both a cutoff ($n^2 \to 0$, $\lambda \to 0$) and a resonance ($n^2 \to \infty$, $\lambda \to \infty$). The properties of these modes can be summarized as follows:

1. As the cutoff at the ath harmonic is approached, the dispersion relation (3.4.5) is approximated by

$$\lambda^a = -2^a a!\Delta_a(\omega_p^2 - a^2\Omega_e^2)/\omega_p^2. \tag{3.4.7}$$

This implies that for $\lambda \to 0$ the dispersion curve approaches the harmonic from below for $a\Omega_e < \omega_p$ and from above for $a\Omega_e > \omega_p$, as illustrated in Fig. 3.7.

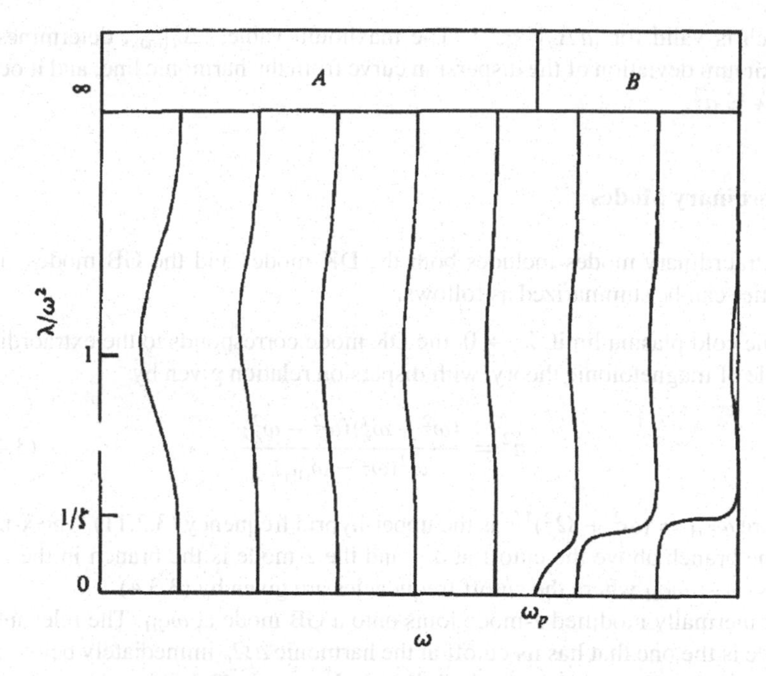

Fig. 3.7 Schematic, classical dispersion of ordinary modes. The maximum deviation of each individual mode from the nearest harmonic for $\lambda \approx a^2$ has been exaggerated for clarity. Regions A and B correspond to $\omega < \omega_p$ and $\omega > \omega_p$, respectively. The cold plasma o mode has $\lambda \approx a^2/\rho$, and is approximated by the segments of dispersion curve lying outside the immediate neighbourhood of the harmonics in region B (From [15], reprinted with permission Cambridge University Press)

2. As the resonance at the ath harmonic is approached, the dispersion relation (3.4.5) is approximated by

$$\lambda^{3/2} = -\omega_p^2/\Omega_e^2(2\pi)^{1/2}\Delta_a\rho. \tag{3.4.8}$$

This implies that for $\lambda \to \infty$ the dispersion curve approaches the harmonic from below for $a\Omega_e < \omega_p$ and from above for $a\Omega_e > \omega_p$, as illustrated in Fig. 3.7.

3. For $\omega \geq \omega_p$ each dispersion curve has nearly horizontal portions in which ω varies slowly with λ at small and large λ, plus a portion in which the dispersion curve goes from just above one harmonic to just below the next harmonic. The latter portions of the curves for each harmonic form an envelope approximately along $n^2 = 1 - \omega_p^2/\omega^2$.
4. Close to the ath harmonic the sum in (3.4.5) is approximated by retaining only the terms labeled 0 and a. This gives

$$\Delta_a = \frac{\omega_p^2 I_a(\lambda)e^{-\lambda}}{a^2\Omega_e^2 - \omega_p^2 I_0(\lambda)e^{-\lambda} - \lambda\rho\Omega_e^2}, \qquad |\Delta_a|_{\max} \approx \frac{\omega_p^2}{\Omega_e^2(2\pi e)^{1/2}\rho a^3}, \tag{3.4.9}$$

which is valid for $|a\Delta_a| \ll 1$. The maximum value, $|\Delta_a|_{\max}$, determines the maximum deviation of the dispersion curve from the harmonic line, and it occurs for $\lambda \approx a^2$.

Extraordinary Modes

The extraordinary modes includes both the DK modes and the GB modes. Their properties can be summarized as follows:

1. In the cold plasma limit, $\lambda \to 0$, the DK mode corresponds to the extraordinary mode of magnetoionic theory, with dispersion relation given by

$$n^2 = \frac{(\omega^2 - \omega_{\rm x}^2)(\omega^2 - \omega_{\rm z}^2)}{\omega^2(\omega^2 - \omega_{\rm UH}^2)}, \tag{3.4.10}$$

where $\omega_{\rm UH} = (\omega_p^2 + \Omega_e^2)^{1/2}$ is the upper-hybrid frequency (3.2.11). The x-mode is the branch above the cutoff at $\omega_{\rm x}$, and the z-mode is the branch in the range $\omega_{\rm z} < \omega < \omega_{\rm UH}$ where the cutoff frequencies are given by (3.3.4).
2. The thermally modified z-mode joins onto a GB mode at $\omega_{\rm UH}$. The relevant GB mode is the one that has its cutoff at the harmonic $a\Omega_e$ immediately below $\omega_{\rm UH}$.
3. Close to the ath harmonic, for $|a\Delta_a| \ll 1$, the sum in (3.4.5) is approximated by retaining only the terms labeled ± 1 and a. This gives

$$(A - y)^2 - \xi^2(B + y)^2 + Ey^2 = 0, \qquad y = \frac{\omega_p^2}{\Omega_e^2}\frac{\lambda^{a-1}}{2^a a!}\frac{e^{-\lambda}}{\Delta_a},$$

$$A = 1 - \frac{\omega_p^2 I_1(\lambda)e^{-\lambda}}{\omega\lambda}\left(\frac{1}{\omega - \Omega_e} + \frac{1}{\omega + \Omega_e}\right), \qquad E = \frac{(a+2)^2}{a^2(a+1)},$$

$$B = \frac{\omega_p^2[I_1'(\lambda) - I_1(\lambda)]e^{-\lambda}}{\omega}\left(\frac{1}{\omega - \Omega_e} - \frac{1}{\omega + \Omega_e}\right). \tag{3.4.11}$$

4. Near the cutoff at the ath harmonic, the dispersion curves for the DK and GB modes approach each other ($K^2{}_2 \to K^1{}_1$, $|K^1{}_2|^2/K^1{}_1 \to 0$), and the two solutions of (3.4.11) give the cutoffs for the DK and GB modes. The two modes that emerge from cutoffs at the ath harmonic have dispersion relations, for $\lambda \to 0$,

$$\lambda^{s-1} = 2^a a!\Delta_a \frac{\Omega_e^2}{\omega_p^2}\frac{A + \xi^2 B \pm [(A + \xi^2 B)^2 - (1 - \xi^2 + E)(A^2 - \xi^2 B^2)]^{1/2}}{1 - \xi^2 + E}, \tag{3.4.12}$$

where A, B are evaluated at $\lambda = 0$. For $a\Omega_e < \omega_{\rm z}$ both modes emerge below the harmonic, for $\omega_{\rm z} < a\Omega_e < \omega_{\rm x}$ one emerges below the harmonic and the other above the harmonic, and for $a\Omega_e > \omega_{\rm x}$ both emerge above the harmonic.

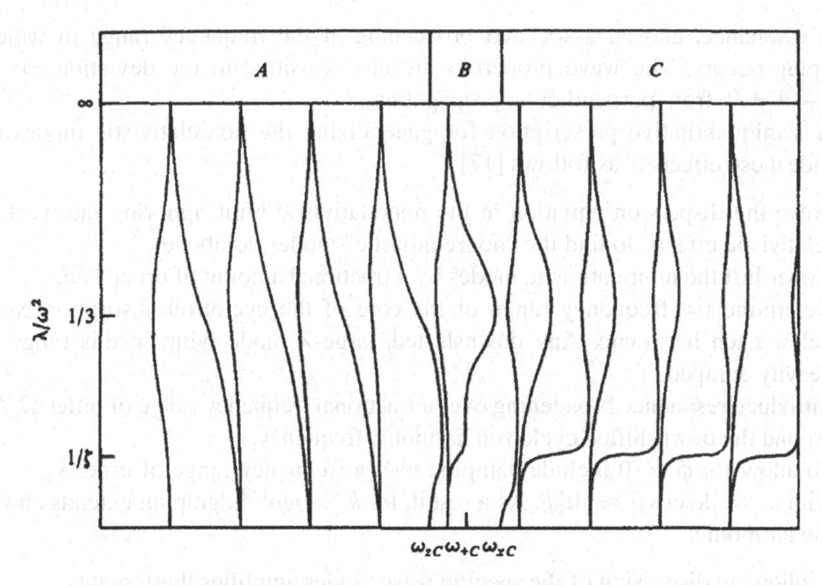

Fig. 3.8 Schematic classical dispersion of extraordinary modes. For modes localized near a particular harmonic, the maximum deviation from that harmonic has been exaggerated for clarity. Regions A, B and C correspond to $\omega < \omega_{zC}$, $\omega_{zC} < \omega < \omega_{xC}$ and $\omega_{xC} < \omega$, respectively. For $\lambda \gtrsim a^2$, Gross-Bernstein and Dnestrovskii-Kostomarov modes approach the harmonics from above and below respectively. The cold plasma x and z modes are approximated by the segments of dispersion curve having $\lambda \approx a^2/\rho$ and lying outside the immediate neighbourhood of the harmonics in regions B and C respectively (From [16] reprinted with permission Cambridge University Press)

5. For $\lambda \to \infty$ the resonances correspond to

$$\lambda^{3/2} = \begin{cases} \omega_p^2/\Omega_e^2(2\pi)^{1/2}\Delta_a\rho & \text{for GB modes,} \\ -\omega_p^2/\Omega_e^2(2\pi)^{1/2}\Delta_a\rho & \text{for DK modes.} \end{cases} \tag{3.4.13}$$

   Thus the GB mode approaches the harmonic from above and the DK mode approaches the harmonic from below.

The qualitative form of the dispersion curves for the extraordinary mode are illustrated in Fig. 3.8.

3.4.2 Inclusion of Weakly Relativistic Effects

The foregoing properties of cyclotron wave modes in a nonrelativistic plasma are modified when either weakly relativistic effects [14–16] or non-perpendicular propagation [17] are included. Weakly relativistic effects, which are included through the Shkarofsky functions, include a frequency downshift, a broadening of each resonance, and an associated broadening of the frequency range in which damping occurs. The wave properties are also sensitive to the deviation, $\phi = \pi/2 - \theta \neq 0$, from perpendicular propagation.

A semiquantitative prescription for generalizing the nonrelativistic modes to include these effects is as follows [17]:

1. Solve the dispersion equation in the nonrelativistic limit, ignoring the weakly relativistic effects, to find the "nonrelativistic" modes, as above.
2. Downshift the nonrelativistic modes by a fractional amount of order $1/\rho$.
3. Determine the frequency range of the core of the cyclotron absorption band below each harmonic. Any downshifted, large-λ mode lying in this range is heavily damped.
4. Introduce resonance broadening over a fractional frequency range of order Ω_e/ρ around the downshifted cyclotron harmonic frequency.
5. To allow for $\phi \neq 0$ include damping over a frequency range of order $k_z/\rho^{1/2}$, with $k_z = |\mathbf{k}| \cos\theta \approx |\mathbf{k}|\phi$. As a result, for $k_z > \omega\rho^{1/2}$ damping extends above the harmonic.

The following discussion of the specific wave modes amplifies these points.

Modified Ordinary Modes

Weakly relativistic effects are included in the dispersion relation (3.4.2) for the ordinary mode by evaluating $K^3{}_3$ using the approximate form (2.5.31). A somewhat improved approximation [17] gives

$$n^2 = 1 - \frac{\omega_p^2 I_a(\lambda) e^{-\lambda}\rho}{\omega^2}\left(1 + 2a\frac{\partial}{\partial a}\right)\mathcal{F}_q(z_a, a), \tag{3.4.14}$$

$$q = 5/2 + (a^2 + \lambda^2)^{1/2} - \lambda - \frac{\lambda}{2(a^2 + \lambda^2)}, \tag{3.4.15}$$

with $a = k_z^2\rho/2\omega^2$ and with the Shkarofsky function approximated by the form (2.5.39).

The properties of $F_q(z)$ illustrated in Fig. 2.1 underlie the points 1–5 above in describing weak relativistic effects on the dispersion. Using (3.4.14), these weakly relativistic effects are confined to the range

$$|(z_a + q)/(4a + 2q)^{1/2}| \lesssim 3. \tag{3.4.16}$$

As a result the large-λ ordinary modes cannot exist outside the range $a \lesssim a_{\max}$, $\phi \lesssim \phi_{\max}$, with

$$a_{\max}^3 \approx \frac{\omega_p^2}{\Omega_e^2}\frac{1}{7 + 6(7 + \frac{2}{3}\rho^2\tan^2\phi)}, \qquad \phi_{\max}^2 = \frac{3}{2\rho^2}\left[\frac{(2\rho|\Delta_a|_{\max} - 7)^2}{36} - 7\right]. \tag{3.4.17}$$

It follows that weakly damped ordinary mode waves exist only for $\omega_p^2 \gtrsim 23\Omega_e^2$, and then only for $\omega \ll \omega_p$.

Dnestrovskii-Kostomarov Modes

The DK modes are modified in a similar way to ordinary mode waves by weakly relativistic effects. Weakly damped DK modes exist only for [16]

$$a \lesssim \frac{1}{8}\left(\frac{\omega_p}{\Omega_e}\right)^{3/2}. \tag{3.4.18}$$

In the range where they exist, the DK mode approximate their nonrelativistic counterparts closely.

Inclusion of an admixture of positrons affects only $K^1{}_2 = -K^2{}_1$. It follows that the presence of positrons affects the DK modes, but not the GB modes.

Gross-Bernstein Modes

The weakly relativistic dispersion relation for the GB modes is

$$0 = 1 - \frac{\omega_p^2 \rho I_a(\lambda) e^{-\lambda}}{\omega^2 \lambda} \sum_a a^2 \mathcal{F}_{q-1}(z_a, a), \tag{3.4.19}$$

with q given by (3.4.15), and with the Shkarofsky function approximated by (2.5.38). The modifications from the nonrelativistic GB modes are similar to those for the ordinary modes. The maximum harmonic for which GB modes exist for perpendicular propagation, and the maximum value, $\phi_{\max}$, for which off-angle propagation is possible for given a are [17]:

$$a^3 \lesssim \frac{3\omega_p^2}{\rho \Omega_e^2} \frac{1}{15 + \rho^2 \tan^2 \phi}, \qquad \phi^2 \lesssim \phi_{\max}^2 = \frac{3}{2\rho^2}\left(\frac{\omega_p^2 \rho}{\Omega_e^2 a^3} - 5\right). \tag{3.4.20}$$

In summary, weakly relativistic effects are severely limiting on the range of existence of the cyclotron harmonic wave modes. This is due primarily to damping just below the cyclotron harmonic associated with the relativistic downshift in the cyclotron frequency, Ω_e/γ, compared with the nonrelativistic case, Ω_e. The nonrelativistic limit for perpendicular propagation implies no damping except exactly at $\omega = a\Omega_e$, and weakly relativistic effects imply damping in a range below $a\Omega_e$, as illustrated in Fig. 2.1. The inclusion of a small angular deviation, $\phi = \pi/2 - \theta$, from perpendicular propagation allows damping at $\omega \neq a\Omega_e$, which is qualitatively similar to the weakly relativistic effect. The damping allowed for $\phi \neq 0$ (that is, $k_z \neq 0$) is symmetric in frequency about the cyclotron line, whereas the damping due to the relativistic effect, $\Omega_e \to \Omega_e/\gamma$, is always below the nonrelativistic resonance.

3.5 Waves in Pulsar Plasma

The plasma in a pulsar magnetosphere has properties that are qualitatively different from plasmas in other contexts, leading to the development of a subfield of plasma dispersion theory specifically directed towards interpreting pulsar radio emission.

3.5.1 Cold-Plasma Model

In most models for the radio emission, the condition $\omega \ll \Omega_e$ is assumed to apply in the source region of radio emission. The simplest model for the wave dispersion is to assume a cold plasma in the rest frame of the pairs, use the cold-plasma model to determine the wave properties in this frame, and apply a Lorentz transformation to determine the properties in the pulsar frame.

The properties of the cold-plasma wave modes are derived in § 3.3.1. In the pulsar context, these properties apply with $\beta_A = \Omega_e/\omega_p \gg 1$ and $|\epsilon| = 1/M \ll 1$ where M is the multiplicity. The modes are referred to as the O and X modes [18]. They have relatively simple properties in three ranges: low frequencies, $\omega \ll \omega_p$, intermediate frequencies, $\omega_p \ll \omega \ll \Omega_e$, and high frequencies, $\omega \gg \Omega_e$.

Low-Frequency Modes

At low and intermediate frequencies, one has $S \approx 1/\beta_0^2$, $D \approx -\epsilon Y/\beta_A^2$, $P = 1 - X$, where $\beta_0 = \beta_A/(1+\beta_A^2)^{1/2}$ is the MHD speed, with $\beta_0 \approx 1$ for $\beta_A \gg 1$. Except for a small range of angles about $\sin\theta = 0$, one has $R^2 \gg 4$ in (3.3.2), implying that the two modes are approximately linearly polarized. The identification of the $\pm$ modes depends on the sign of R, which is determined by the sign $\sigma = -(1-X)\epsilon\cos\theta/|(1-X)\epsilon\cos\theta|$. For $\sigma = 1$, $T_+ \approx R$ corresponds to the O mode, and $T_- \approx -1/R$ corresponds to the X mode; for $\sigma = -1$, $T_- \approx -R = |R|$ corresponds to the O mode, and $T_+ \approx 1/R = -1/|R|$ corresponds to the X mode.

At low frequencies, $\omega \ll \omega_p$, the two modes are MHD-like. In the limit $Y \to \infty$, using the first and last of (3.2.8) with $T_+ \to \infty$ and $T_- \to 0$, respectively, one obtains the approximations

$$n_{\rm O}^2 \approx \frac{PS}{A} \approx \frac{1-X}{1-X\cos^2\theta}, \qquad n_{\rm X}^2 \approx S \approx 1. \tag{3.5.1}$$

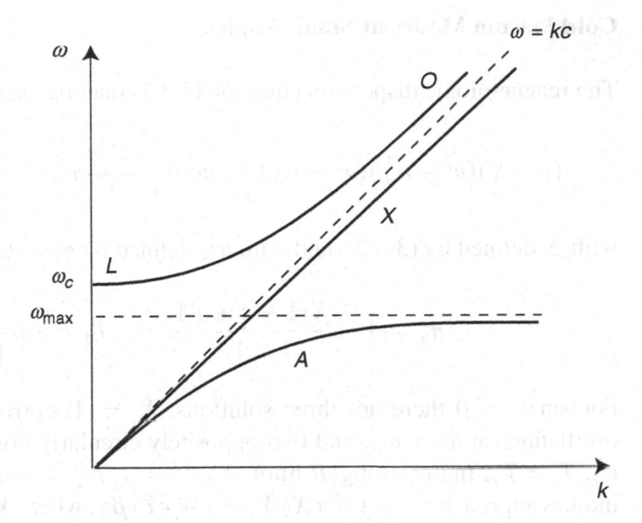

Fig. 3.9 The dispersion curves for the X, Alfvén and O modes are shown schematically for a pulsar plasma for a small value of $\langle\gamma\rangle$ and a relatively small θ. The resonance in the Alfvén mode is denoted by $\omega_{\rm max}$ and the cutoff in the longitudinal (L) portion of the O mode by $\omega_{\rm c}$. The O mode crosses the light line (*dashed diagonal line*) for very small θ, but not otherwise. For $\langle\gamma\rangle \gg 1$ the qualitative shape of the dispersion curves remains the same, but drawn out along the light line

The polarization vectors in the same approximation are

$$e_{\rm O} = \frac{L_{\rm O}\boldsymbol{\kappa} + T_{\rm O}\boldsymbol{t}}{(L_{\rm O}^2 + T_{\rm O}^2)^{1/2}}, \qquad \frac{L_{\rm O}}{T_{\rm O}} = -\frac{X\sin\theta\cos\theta}{1 - X\cos^2\theta}, \qquad e_{\rm X} = i\boldsymbol{a}. \tag{3.5.2}$$

The dispersion curves for the O and X modes are illustrated in Fig. 3.9. The O mode dispersion relation (3.5.1) separates into a lower-frequency range, $\omega^2 < \omega_p^2\cos^2\theta$, and a higher-frequency range $\omega^2 > \omega_p^2$. At low frequencies, $\omega \ll \omega_p$, the dispersion relation becomes $n^2 = 1/\cos^2\theta$, polarized along the 1-axis. These properties correspond to the Alfvén mode in the limit $\beta_A \to \infty$. The upper limit of the low-frequency branch is at the resonance at $\omega^2 = \omega_p^2\cos^2\theta$, where the polarization becomes longitudinal. There is a stop band, $\omega_p^2\cos^2\theta < \omega^2 < \omega_p^2$, between the lower and upper branches. The upper branch starts at a cutoff at $\omega^2 = \omega_p^2$. At higher frequencies, still satisfying $\omega \ll \Omega_e$, the dispersion relation approaches $n_{\rm O}^2 \approx 1 - (\omega_p^2/\omega^2)\sin^2\theta$, which is equivalent to $\omega^2 - |\boldsymbol{k}|^2 \approx \omega_p^2\sin^2\theta$. The X mode has vacuum-like properties, $n_{\rm X}^2 \approx 1/\beta_0^2 \approx 1$, and its polarization is strictly transverse in this approximation.

3.5.2 *Effect of the Cyclotron Resonance*

The foregoing properties of the two modes are derived under the assumption that the modes are nearly linearly polarized, corresponding to $R \gg 2$ in (3.3.2). The opposite condition, $R \ll 2$, is satisfied at sufficiently small angles, $\sin\theta \ll 1$, where the two modes are oppositely circularly polarized. As the ratio ω/Ω_e increases along the escape path, the range of angles where the modes are nearly circularly polarized increases, and near the cyclotron resonance it extend to nearly all angles.

Cold Plasma Modes at Small Angles

The magnetoionic dispersion equation (3.3.1) may be written in the form

$$(1-X)(n^2-n_+^2)(n^2-n_-^2)-\sin^2\theta\frac{XY^2}{1-Y^2}n^2(n^2-1+E)=0, \tag{3.5.3}$$

with E defined by (3.3.2), and with $n_\pm^2$ defined by $\sigma=\pm 1$ in

$$n_\sigma^2=1-\frac{X(1+\sigma|\epsilon|Y)}{1-Y^2}, \qquad T_\sigma=-\sigma\frac{\epsilon\cos\theta}{|\epsilon\cos\theta|}. \tag{3.5.4}$$

For $\sin\theta = 0$ there are three solutions, $X = 1$, corresponding to longitudinal oscillations at $\omega=\omega_p$, and two oppositely circularly polarized modes with $n^2 = n_\sigma^2$, $T=T_\sigma$. In the strong-B limit, $X/Y^2 = 1/\beta_A^2 \ll 1$ and $Y \gg 1$, the refractive indices approach $n^2 = 1 \mp \epsilon X/Y = 1 \mp \epsilon Y/\beta_A^2$, where $Y \ll \beta_A^2$ ($\omega \gg \omega_p/\beta_A$) is assumed. The handedness of the circular polarization is determined by the sign $\sigma = -(1-X)\epsilon\cos\theta/|(1-X)\epsilon\cos\theta|$, which reverses at $\omega=\omega_p$. As $\sin\theta$ increases, $|R|$ increases, and for $|R|\gtrsim 4$ these modes become the O and X modes.

The effect of a Lorentz transformation to the pulsar frame for an outward streaming motion is to modify the small range of angles where the modes are nearly circularly polarized. The small forward cone shrinks due to the Lorentz transform from the plasma rest frame, and the backward cone widens. As the cyclotron resonance is approached, the backward cone can widen to extend into forward angles in the pulsar frame.

Cold-Plasma Modes Near the Cyclotron Resonance

Escaping pulsar radiation necessarily passes through a region where the wave frequency is equal to the cyclotron frequency. The cyclotron resonance is smoothed out when a spread in Lorentz factors is included. Nevertheless the cold plasma assumption provides a useful guide to identifying how the polarization varies as the ratio of the wave frequency to the cyclotron frequency varies from well below to well above unity.

As $Y = \Omega_e/\omega$ varies from $Y \gg 1$ to $Y \ll 1$, the parameter R, defined by (3.3.2) with $E = 0$, varies from $|R| \gg 1$ to $|R| \ll 1$. The axial ratios vary from $T_\pm \approx R, -1/R$ for $|R| \gg 1$ to $T_\pm = \pm 1 + \frac{1}{2}R$, corresponding to a change from nearly linear to nearly circular polarization. It follows that the nearly linear polarization of the wave modes at intermediate frequencies changes to the nearly circular polarization characteristic of the wave modes of any magnetized plasma at high frequencies, $\omega \gg \omega_p, \Omega_e$. Due to the rapid change in the shape of the polarization ellipse with Ω_e, an inhomogeneity involving a gradient in B can be effective in causing a wave in one mode to couple with the other mode, such that it becomes a mixture of the two modes. Such mode coupling is said to be weak when the effect of the inhomogeneity is unimportant, and waves in a given mode remain in that mode; this implies that the polarization follows the change in the polarization of the mode as the cyclotron resonance is crossed. In the opposite limit, when mode coupling is strong, the initial linear polarization is preserved, implying that the waves change from one mode to the other as the cyclotron resonance is crossed. The interpretation of the observed pulsar polarization, involving jumps between orthogonal polarizations that can be significantly elliptical seems to require that mode coupling be relatively strong at the cyclotron resonance allowing a partial conversion of linear into circular polarization [19].

3.5.3 *Effect of a Spread in Lorentz Factors*

At frequencies well below the cyclotron resonance, the wave properties in a pulsar plasma are modified from those of a cold plasma through the RDPF $z^2W(z)$. As illustrated in Figs. 2.2 and 2.3, this function is dominated by a peak just below $z = 1$. As in the cold plasma case, at low and intermediate frequencies, the effect of the gyrotropic terms is small, and a useful first approximation is to neglect them.

Wave Dispersion in Non-gyrotropic Approximation

In the non-gyrotropic case, the dispersion equation reduces to

$$\begin{vmatrix} K^1{}_1 - n^2\cos^2\theta & 0 & K^1{}_3 + n^2\sin\theta\cos\theta \\ 0 & K^2{}_2 - n^2 & 0 \\ K^3{}_1 + n^2\sin\theta\cos\theta & 0 & K^3{}_3 - n^2\sin^2\theta \end{vmatrix} = 0, \tag{3.5.5}$$

with the components of the dielectric tensor given by (1.3.15), with $\epsilon = 0$, $\langle\beta\rangle = 0$, $\langle\gamma\beta\rangle = 0$ here. The dispersion equation (3.5.5) factors into two equations

$$K^2{}_2 - n^2 = 0, \tag{3.5.6}$$

which describes the X mode, and

$$K^1{}_1\,K^3{}_3 - K^1{}_3\,K^3{}_1 - n^2\big[K^1{}_1\sin^2\theta + K^3{}_3\cos^2\theta\big] = 0, \tag{3.5.7}$$

which has a high-frequency branch identified as the O mode, and a low-frequency branch identified as the Alfvén mode.

X Mode

On inserting the expression (1.3.15) for $K^i{}_j$, the dispersion relation for the X mode becomes

$$n_X^2 = \frac{1 + 1/\beta_A^2}{1 - (\delta\beta^2/\beta_A^2)\cos^2\theta}, \tag{3.5.8}$$

where $\delta\beta^2$ characterizes the spread in velocities, with $\delta\beta^2 \to 1$ when the spread is highly relativistic. The X mode is strictly transverse, polarized orthogonal to $\boldsymbol{B}$. For $\beta_A^2 \gg \delta\beta^2 \lesssim 1$, (3.5.8) may be approximated by $n_X^2 = 1/\beta_0^2$, $\beta_0 = \beta_A/(1+\beta_A^2)^{1/2}$, which is the dispersion relation for magnetoacoustic waves in a cold plasma.

Equation (3.5.7) describes the Alfvén and O modes. For parallel propagation, $\sin\theta \to 0$, (3.5.7) factors into $K^3{}_3 = 0$ and $n^2 = K^1{}_1$. The mode defined by $K^3{}_3 = 0$ is longitudinal. The mode defined by $n^2 = K^1{}_1$ is the Alfvén mode, which is degenerate with the X mode for $\sin\theta = 0$; the degeneracy is broken and the modes become oppositely circularly polarized when gyrotropic effects are included. The dispersion curves for the parallel longitudinal mode intersects both the other modes for $\sin\theta = 0$, and for small $\sin\theta \neq 0$ they reconnect to form the O mode and the Alfvén mode on the high and low frequency sides of a stop band. The X mode passes continuously through this band, reversing its handedness across it.

Longitudinal Mode

The dispersion relation for the parallel longitudinal (L) mode is [20]

$$\omega = \omega_L(z), \qquad \omega_L^2(z) = \omega_p^2 W(z). \tag{3.5.9}$$

The L mode has a cutoff at $\omega = \omega_c$ given by

$$\omega_c^2 = \omega_L^2(\infty) = \omega_p^2 \langle \gamma^{-3} \rangle. \tag{3.5.10}$$

The dispersion curve crosses the light line at $\omega = \omega_1$, given by

$$\omega_1^2 = \omega_L^2(1) = \omega_p^2 \langle \gamma \rangle (1 + \delta\beta^2). \tag{3.5.11}$$

As for Langmuir waves in a nonrelativistic plasma, the parallel L mode has a maximum frequency at a phase speed of order the mean speed of the particles ($\approx (\delta\beta^2)^{1/2}$), which is very close to the speed of light in a highly relativistic plasma. Landau damping results from resonance at $\omega = k_z v$, and is strong for phase speeds near and below the mean speed of the particles. As a consequence, the L mode effectively ceases to exist for phases speeds near and below this maximum.

Parallel Alfvén Mode

The dispersion relation for the parallel Alfvén mode may be written in terms of the refractive index as $n^2 = n_A^2$ with

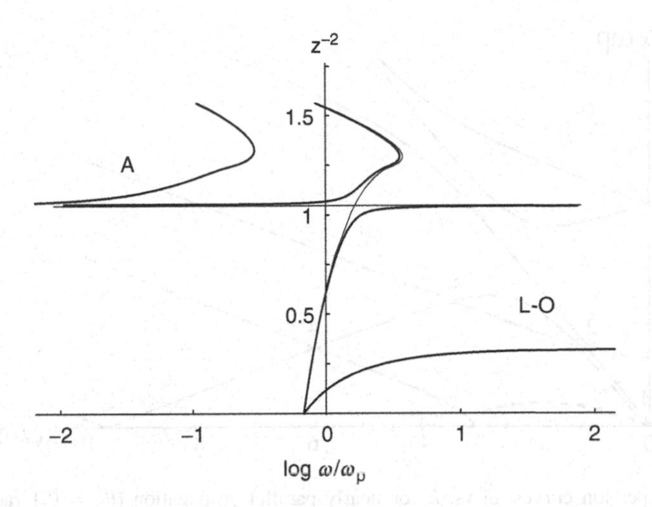

Fig. 3.10 The dispersion relations for the Alfvén (A) mode and the O modes are shown for three values of θ: the faint curve corresponds to $\theta = 0$, where the resonance (turnover at high z^{-2}) is in the L mode and the Alfvén mode is a horizontal line that crosses the faint curve at $\omega = \omega_1$; for a very small but non-zero value of θ these two dispersion curves reconnect and separate, as shown by the inner pair of *solid curves*, with the O mode extending slightly into the region $z^{-2} > 1$; the outer pair of *solid curves* are for a much larger value of θ

$$n_A^2 = \frac{1 + \beta_A^2 + \delta\beta^2}{\beta_A^2}. \tag{3.5.12}$$

The parallel L mode and Alfvén modes intersect at the cross-over frequency $\omega = \omega_{\rm co} = \omega_L(1/n_A)$, which may be expressed in terms of the frequencies, $\omega_{\rm c}$ given by (3.5.10), at which the parallel L mode crosses the light line, and ω_1 given by (3.5.11), at which the parallel L mode has its cutoff. One has

$$\omega_{\rm co} = \omega_p^2 W(1/n_A) \approx \begin{cases} \omega_1^2 & \text{for } n_A - 1 \ll \delta n, \\ \omega_{\rm c}^2 (n_A - 1)^{-2} & \text{for } n_A - 1 \gg \delta n, \end{cases} \tag{3.5.13}$$

with n_A given by (3.5.12) and with $\delta n = \omega_{\rm c}/\omega_1 \approx 1/\langle\gamma\rangle^{1/2}$. For $n_A - 1 \gg 1/\langle\gamma\rangle^{1/2}$, Landau damping by the bulk pair plasma is strong, and the waves are strongly damped except at very small angles. For $n_A - 1 \ll 1/\langle\gamma\rangle^{1/2}$ one has $\omega_{\rm co} \approx \omega_1$.

Obliquely Propagating Modes

For slightly oblique propagation, as illustrated in Fig. 3.10, the two parallel modes reconnect, and may be regarded as a higher-z mode and a lower-z mode, where $z = \omega/k_z$ is the phase speed. The higher phase-speed mode is longitudinal near the

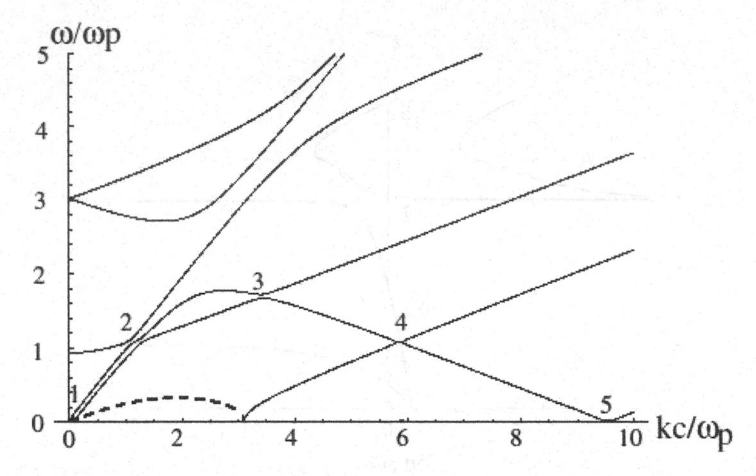

Fig. 3.11 Dispersion curves, ω vs. k for nearly parallel propagation ($\theta = 0.1$ rad) in a cold counter-streaming plasma with $\beta = 0.3$, where $\omega_p = 10$ and $\Omega_e = 30$ (arbitrary units). *Dashed lines* are imaginary parts and *solid lines* are real parts. *Numbers* indicate regions expanded in Fig. 3.12 (From [21], copyright American Physical Society)

cutoff frequency and becomes nearly transverse at frequencies $\omega \gtrsim \omega_{co}$. This branch is labeled the O mode in Fig. 3.10. The lower phase-speed mode corresponds to the oblique Alfvén mode, with

$$z^2 = \frac{\omega^2}{k_z^2} = \frac{1}{n_A^2}\left[1 - \frac{\omega^2}{\omega_{\rm co}^2}\left(1 - \frac{\delta\beta^2}{\beta_A^2}\tan^2\theta\right)\right], \tag{3.5.14}$$

at low frequencies $\omega \ll \omega_{\rm co}$. At higher frequencies the mode is limited by the maximum frequency for the oblique Alfvén mode, as illustrated in Fig. 3.10.

The minimum frequency for the O mode is the cutoff frequency, as illustrated in Fig. 3.10. At higher frequencies the phase speed of the O mode increases, with the mode being superluminal ($z > 1$) except for a small range of angles $\theta \approx 0$. Analytic approximations to the dispersion relation of the O mode at $\omega \gtrsim \omega_{\rm co}$ are derived in the weak-anisotropy limit, cf. § 3.6.

3.5.4 Wave Modes of a Counter-Streaming Pair Plasma

The inclusion of counter-streaming leads to a rich variety of dispersive effects, even in the cold approximation [21], as illustrated in Fig. 3.11. In order to illustrate these effects, the parameters chosen in Fig. 3.11 correspond to $\Omega_e/\omega_p = 3$, $\beta = 0.3$; the topological structure of the dispersion curves is insensitive to the choice of these parameters.

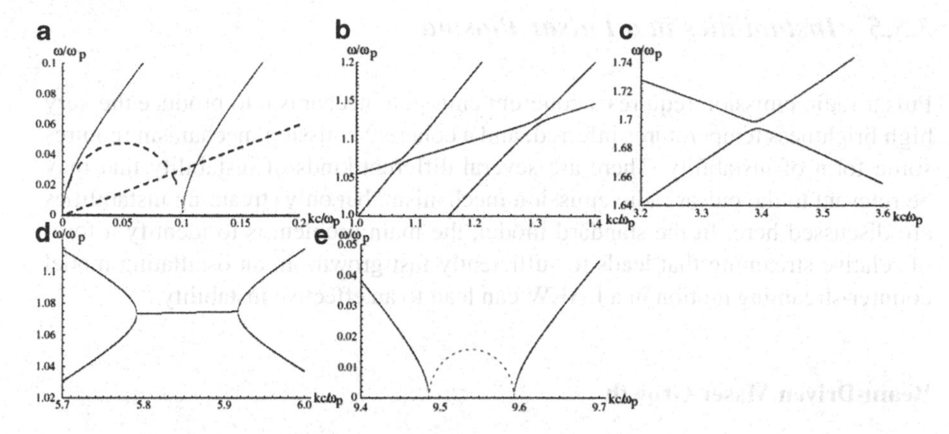

Fig. 3.12 Dispersion curves, ω versus k for propagation at a small angle ($\theta = 0.1$ rad) in a cold counter-streaming plasma with $\beta = 0.3$, where $\omega_p = 10$ and $\Omega = 30$. These subfigures show close-up views of the numbered regions in Fig. 3.11, and display the behavior in the interaction regions (From [21], copyright American Physical Society)

The cutoffs, $|\boldsymbol{k}| = 0$, in Fig. 3.11 are near $\omega = 0, \omega_p, \Omega_e$. As β is reduced to zero, the modes with cutoffs at $\omega = 0, \omega_p, \Omega_e$ reduce to the cutoffs in the Alfvén, O and X modes, respectively. The number of different modes reduces in this limit due to modes becoming coincident, which occurs with the two that cut off near Ω_e, or disappearing, which occurs for the beam modes that intersect $\omega = 0$ in Fig. 3.11.

An important feature of dispersion in a cold plasma with counter-streaming is that there can be intrinsically growing modes: a complex conjugate pair of solutions arises when two real modes merge (as some parameter is varied) to become a pair of modes with one growing and the other decaying. Complex modes appear in Fig. 3.11 at the points labeled 1–5. New modes appear in qualitatively different ways at nonzero frequency, as illustrated in cases 2–5 in Fig. 3.12. In some cases two real modes reconnect to create two different real modes, as in cases 2 and 3. One class of unstable mode is purely growing, with zero real frequency, as illustrated by the dashed curve in Fig. 3.11 and by cases 1 and 5 in Fig. 3.12. In other cases two real modes become a complex conjugate pair of modes, one of which is necessarily a growing mode, as in case 4 in Fig. 3.12.

Consider, for example, the specific dispersion curve in Fig. 3.12 that begins at $\omega = 0$ for $k = 0$. This branch is initially imaginary, becoming real and propagating at $k \simeq 1$. As the intersection labeled 2 is approached, it deflects away from the other modes and remains a single real mode, and similarly at intersection 3. At intersection 4, it joins another mode and forms a pair of complex conjugate solutions, which then become real again for higher k.

3.5.5 *Instabilities in a Pulsar Plasma*

Pulsar radio emission requires a coherent emission mechanism to produce the very high brightness temperatures inferred, and a coherent emission mechanism requires some form of instability. There are several different kinds of instability that may be relevant to the pulsar radio emission mechanism, but only streaming instabilities are discussed here. In the standard model, the main problem is to identify a form of relative streaming that leads to sufficiently fast growth. In an oscillating model counter-streaming motion in a LAEW can lead to an effective instability.

Beam-Driven Maser Growth

Consider a beam with number density $n_b \ll n_e$, mean Lorentz factor γ_b and spread $\Delta\gamma_b$ about the mean propagating through the secondary pair plasma with number density n_e and mean Lorentz factor $\langle\gamma\rangle \ll \gamma_b$ in its rest frame. A distribution function for an idealized beam with $\gamma_b \gg 1$ is

$$g(\gamma) = \frac{n_b\, e^{-(\gamma-\gamma_b)^2/2(\Delta\gamma_b)^2}}{(2\pi)^{1/2}\Delta\gamma_b}, \tag{3.5.15}$$

where the normalization is $n_b = \int d\gamma\, g(\gamma)$. The absorption coefficient for waves in a mode M for the distribution (3.5.15) is

$$\gamma_M(k) = 2i\, \frac{R_M(k)}{\varepsilon_0\omega_M(k)}\, \Pi^A_M(k_M), \tag{3.5.16}$$

where $\Pi^A_M(k_M)$ is the antihermitian part of the response tensor for the distribution (3.5.15) projected onto the polarization vector for the mode M and evaluated at the dispersion relation $k^\mu = k^\mu_M$ for the mode M. This gives

$$\gamma_M(k) = \frac{2R_M(k)|e_{M\parallel}|^2}{\varepsilon_0 m\omega_M(k)}\, n_b \int \frac{d\gamma}{\gamma}\, \delta(k_M u)\, \frac{dg(\gamma)}{d\gamma}, \tag{3.5.17}$$

with $e_{M\parallel}(k) = \boldsymbol{b}\cdot\boldsymbol{e}_M(k)$. On inserting (3.5.15) into (3.5.17), and carrying out the integral over the δ-function, it is convenient to change notation to ω, θ as the independent variables, omitting the label M for the mode and making the approximation $R_M(k) \to \frac{1}{2}$. One finds

$$\gamma(\omega,\theta) = \frac{\omega_{pb}^2}{\omega}\, \frac{|e_\parallel|^2(\gamma_0-\gamma_b)z\gamma_0}{2(2\pi)^{1/2}(\Delta\gamma_b)^3}\, e^{-(\gamma_0-\gamma_b)^2/2(\Delta\gamma_b)^2}, \tag{3.5.18}$$

with $\gamma_0 = (1-z^2)^{-1/2}$, $z = \omega/k_z$, $\omega_{pb}^2 = q^2 n_b/\varepsilon_0 m$. Wave growth occurs for $\gamma_0 < \gamma_b$, which corresponds to $z < \beta_b$.

The maximum growth rate for this maser instability is near $\gamma_0 = \gamma_b - \Delta\gamma_b$, and is given approximately by

$$\gamma_{\max}(\omega, \theta) = -\frac{\omega_{pb}^2}{\omega} \frac{|e_\parallel|^2 \gamma_b}{2(2\pi e)^{1/2}(\Delta\gamma_b)^2}, \tag{3.5.19}$$

where $\Delta\gamma_b \ll \gamma_b$ is assumed. The bandwidth of the growing waves is $\Delta\omega \approx \omega\Delta\gamma_b/\gamma_b^3$. The ratio of the maximum growth rate to the bandwidth of the growing waves is

$$\left|\frac{\gamma_{\max}(\omega, \theta)}{\Delta\omega}\right| \approx \frac{n_b}{n_e} \frac{\gamma_b^4}{\langle\gamma\rangle(\Delta\gamma_b)^3} \frac{|e_\parallel|^2}{2(2\pi e)^{1/2}}, \tag{3.5.20}$$

where $\omega \approx \omega_p \gamma_b^{1/2}$ is assumed. The maser treatment is valid only if the ratio (3.5.20) of the growth rate greater to the bandwidth of the growing waves is small. A growth rate greater than the bandwidth of the growing waves requires a reactive instability.

Reactive Beam Instability

When the velocity spread of the beam is neglected, the beam is cold, and the only possible instability is reactive. For a cold beam with number density n_b, velocity β_b and Lorentz factor γ_b propagating through a cold background plasma, the dispersion equation for parallel, longitudinal waves is

$$1 - \frac{\omega_p^2}{\omega^2} - \frac{n_b}{n_e} \frac{\omega_p^2}{\gamma_b^3(\omega - k_z\beta_b)^2} = 0. \tag{3.5.21}$$

The dispersion equation is a quartic equation for ω. We are interested in the case where the beam is weak, in the sense $n_b/\gamma_b^3 \ll n_e$.

In the limit of arbitrarily large $k_z\beta_b$, the four solutions of (3.5.21) approach $\omega = \pm\omega_p, k_z\beta_b \pm \omega_p(n_b/n_e\gamma_b^3)^{1/2}$. The solution near $\omega = -\omega_p$ is of no interest here, and it is removed by approximating the quartic equation by the cubic equation

$$(\omega - \omega_p)(\omega - k_z\beta_b)^2 - \frac{n_b}{2n_e\gamma_b^3}\omega_p\omega^2 = 0. \tag{3.5.22}$$

The solutions of the cubic equation simplify in two cases: the 'resonant' case $k_z\beta_b \approx \omega_p$, and the 'nonresonant' case $\omega \ll \omega_p$. The approximate solutions for the growth rate in these two cases are

$$\omega \approx \begin{cases} \omega_p + i\dfrac{\omega_p}{\gamma_b}\dfrac{\sqrt{3}}{2}\left(\dfrac{n_b}{2n_e}\right)^{1/3}, & \text{resonant,} \\ k_z\beta_b + i\dfrac{\omega_p}{\gamma_b^{3/2}}\left(\dfrac{n_b}{2n_e}\right)^{1/2}, & \text{nonresonant.} \end{cases} \tag{3.5.23}$$

The nonresonant version of the beam instability applies only at frequencies below the resonant frequency, $\omega = k_z\beta_b$. As the resonant frequency is approached the nonresonant instability transforms into the faster-growing resonant instability.

Both maser and reactive beam instabilities are possible for the O mode in the range where its phase velocity is less than the beam velocity. Difficulties with the mechanism are in identifying a possible beam to drive the instability, and in accounting for fast enough wave growth. Both problems are alleviated in an oscillating model where counter-streaming of electrons and positrons occurs naturally.

3.5.6 Counter-Streaming Instabilities

An oscillating model for the pulsar magnetosphere involves counter-streaming electrons and positrons. This differences from the weak-beam case in that there is no background plasma at rest. Various instabilities arise in this case, as illustrated in Fig. 3.12. A simple analytic model involves equal number densities and oppositely directed velocities, $\pm\beta_b$. In this case one has $W(z) = (1/2\gamma_b^3)[1/(z-\beta_b)^2 + 1/(z+\beta_b)^2]$. The dispersion equation for parallel longitudinal waves, $1-(\omega_p^2/\omega^2)z^2W(z) = 0$, gives

$$(\omega^2 - k_z^2\beta_b^2)^2 - \frac{\omega_p^2}{\gamma_b^3}(\omega^2 + k_z^2\beta_b^2) = 0. \tag{3.5.24}$$

The two solutions, $\omega^2 = \omega_\pm^2$ say, are

$$\omega_\pm^2 = k_z^2\beta_b^2 + \frac{\omega_p^2}{2\gamma_b^3}\left\{1 \pm \left[1 + \frac{8k_z^2\beta_b^2\gamma_b^3}{\omega_p^2}\right]^{1/2}\right\}. \tag{3.5.25}$$

The higher frequency branch has a cutoff ($k_z = 0$) at $\omega = \omega_p/\gamma_b^3$, with the frequency increasing monotonically with increasing k_z, given approximately by

$$\omega \approx \begin{cases} [\omega_p^2/\gamma_b^3 + 3k_z^2\beta_b^2]^{1/2} & \text{for } 8k_z^2\beta_b^2 \ll \omega_p^2/\gamma_b^3, \\ k_z\beta_b & \text{for } 8k_z^2\beta_b^2 \gg \omega_p^2/\gamma_b^3. \end{cases} \tag{3.5.26}$$

There is no reactive instability associated with this branch.

The lower frequency branch of (3.5.25) has a frequency that is real for $k_z^2\beta_b^2 > \omega_p^2/\gamma_b^3$, and imaginary for $k_z^2\beta_b^2 < \omega_p^2/\gamma_b^3$:

$$\omega \approx \begin{cases} ik_z\beta_b & \text{for } 8k_z^2\beta_b^2 \ll \omega_p^2/\gamma_b^3, \\ k_z\beta_b - \omega_p/2^{1/2}\gamma_b^3 & \text{for } 8k_z^2\beta_b^2 \gg \omega_p^2/\gamma_b^3. \end{cases} \tag{3.5.27}$$

Thus the lower frequency branch implies a beam-type instability for $k_z^2\beta_b^2 < \omega_p^2/\gamma_b^3$. The maximum growth rate is for $k_z^2\beta_b^2 = 3\omega_p^2/8\gamma_b^3$, and is

$$\Gamma_{\max} = \frac{1}{2\sqrt{2}}\frac{\omega_p}{\gamma_b^{3/2}}, \qquad k_z = \frac{\sqrt{3}}{2\sqrt{2}}\frac{\omega_p}{\gamma_b^{3/2}\beta_b}. \tag{3.5.28}$$

It follows that the maximum growth rate is smaller the higher the Lorentz factor of the flow, so that maximum growth is when the flow is nonrelativistic or mildly relativistic, $\gamma_b \approx 1$.

Growth of Alfvén Waves

A beam instability can generate Alfvén waves. The resonance at $z = \omega/k_z = \beta_b$ occurs in the O mode for $\beta_b > 1/n_A$, with n_A given by (3.5.12), and in the Alfvén mode for $\beta_b < 1/n_A$. For Alfvén waves near the cross-over frequency one has $\omega \approx \omega_1$, with ω_1 given by (3.5.11).

For Alfvén waves at low frequency the conditions for beam-driven growth are special. The Cerenkov resonance requires $z = \beta$, that is, $\omega = k_z\beta_b$, and for low frequency Alfvén waves this implies $\beta_b = 1/n_A$. In a pulsar magnetosphere, the quantity n_A varies with position in the magnetosphere, and in particular with radial distance, r, along a given magnetic field line. A particular beam satisfies the condition $\beta_b = 1/n_A$ only at one point along a given field line, and if this condition is satisfied, Alfvén waves over a wide frequency range resonate simultaneously with the beam. In the case of a weak beam, this suggests that beam-driven Alfvén turbulence develops only in localized regions where the resonance condition is satisfied. Alfvén waves cannot escape directly from the plasma, and some nonlinear process is required to convert the energy in the Alfvén turbulence into escaping radiation in either the O or X-modes.

3.6 Weak-Anisotropy Approximation

The weak-anisotropy approximation, described in § 2.5.7 of volume 1, is based on the assumption that the waves can be approximated as transverse. The wave equation then reduces to a two-dimensional equation for the two transverse components of $A^\mu(k)$. The polarization of the wave modes is described by a single parameter, chosen to be the axial ratio, $T_\pm = -1/T_\mp$, of the polarization ellipse of one of the modes. The dispersion relations, written in terms of the refractive index, involve an isotropic part, that is the same for both modes, and a part that has opposite sign for the two modes and depends on $T_\pm$.

3.6.1 Projection onto the Transverse Plane

The relevant choice of gauge is the radiation gauge, where there are components only in the transverse plane, with zero time-like and longitudinal components. One may choose the two 4-vectors t^μ, a^μ, defined by (1.1.26), as basis vectors. The wave equation (3.1.2) reduces to the 2-dimensional equation

$$[k^2\delta^\mu_\nu + t^\mu{}_\nu(k)]A^\nu(k) = 0, \tag{3.6.1}$$

where μ, ν run over only the transverse components t, a. The axial ratio, T, is defined in terms of the ratio of the t- and a-components of A^μ, with $A^t : A^a = T : i$. Hence, (3.6.1) implies $(k^2 + t^t{}_t)T + it^t{}_a = 0$, $t^a{}_t T + i(k^2 + t^a{}_a) = 0$. Eliminating T between these two equations gives the dispersion equation

$$k^4 + (t^t{}_t + t^a{}_a)k^2 + t^t{}_t t^a{}_a - t^t{}_a t^a{}_t = 0. \tag{3.6.2}$$

Equation (3.6.2) reproduces the dispersion equation (2.5.24) of volume 1, viz.

$$k^4 + k^2 t^{(1)} + \tfrac{1}{2}\left\{[t^{(1)}]^2 - t^{(2)}\right\} = 0, \tag{3.6.3}$$

with the traces given explicitly by

$$t^{(1)} = t^t{}_t + t^a{}_a, \qquad t^{(2)} = (t^t{}_t)^2 + (t^a{}_a)^2 + 2t^t{}_a t^a{}_t. \tag{3.6.4}$$

Eliminating k^2 between the two equations leads to a quadratic equation for the axial ratio:

$$T^2 - \frac{t^t{}_t - t^a{}_a}{it^t{}_a}T + 1 = 0, \tag{3.6.5}$$

where $it^t{}_a = -it^a{}_t$ is real.

The solutions of (3.6.3) with (3.6.4) are

$$k^2 = k^2_\pm = -\frac{1}{2}(t^t{}_t + t^a{}_a) \pm \tfrac{1}{2}\left[(t^t{}_t - t^a{}_a)^2 + 4t^t{}_a t^a{}_t\right]^{1/2}. \tag{3.6.6}$$

The solutions of (3.6.5) are

$$T_\pm = \frac{-(t^t{}_t - t^a{}_a) \pm \left[(t^t{}_t - t^a{}_a)^2 + 4t^t{}_a t^a{}_t\right]^{1/2}}{-2it^t{}_a}, \tag{3.6.7}$$

with $T_+T_- = -1$. The two components of the equations imply

$$k^2_\pm = -t^a{}_a - it^t{}_a T_\pm = -t^t{}_t + \frac{it^t{}_a}{T_\pm}. \tag{3.6.8}$$

Rest Frame of the Plasma

There is a preferred frame in which to apply the weak-anisotropy approximation: the rest frame of the plasma. To see this, suppose that one tries to apply the weak-anisotropy approximation in two different frames, the rest frame and a primed frame in which that plasma has a bulk velocity. The transformation of the polarization vector, described by (3.3.21)–(3.3.23), implies that if there is no longitudinal part in the rest frame, there is a longitudinal part in the primed frame. Hence, it is inconsistent to apply the weak-anisotropy approximation in both frames. That the rest frame is the preferred frame can be seen by considering the case of a cold unmagnetized plasma. In the rest frame, one can separate the response into longitudinal and transverse parts, and the dispersion relation for transverse waves can be written as either $k^2 = \omega_p^2$ or $n^2 = 1 - \omega_p^2/\omega^2$. In the primed frame the dispersion relation becomes $k'^2 = \omega_p^2$ or $n'^2 = 1 - \omega_p^2/\omega'^2$, but the waves are not transverse, having a longitudinal component. The weak-anisotropy limit involves a perturbation (to break the degeneracy of two transverse modes) about this isotropic case. The theory is valid only if the unperturbed solution corresponds to a transverse wave, and this is only the case in the rest frame of the plasma. The error introduced by applying the weak-anisotropy approximation in the primed frame is small: according to (3.3.23), the longitudinal part of the polarization in the primed frame is proportional to $n'^2 - 1 = -\omega_p^2/\omega'^2$. To avoid inconsistency one needs to add a proviso when applying the weak-anisotropy approximation in a frame other than the rest frame: terms of order ω_p^2/ω'^2 can be retained in the unperturbed refractive index, but must be neglected in determining the axial ratio of the polarization vector.

3.6.2 Stokes Parameters

A transverse wave propagating through a weakly anisotropic plasma has its polarization modified by generalized Faraday rotation. One can write an arbitrary transverse polarization vector as a sum of components in the two modes, together with the phase difference between them (the overall phase is arbitrary). The different refractive indices of the two modes cause the relative phase to change systematically as a function of distance along the ray path, and this leads to generalized Faraday rotation. This may be described either in terms of the two amplitudes and the relative phase difference (called the Jones calculus in optics) or in terms of the Stokes parameters (the Mueller calculus in optics). The relation between these two descriptions is summarized in § 2.5.7 of volume 1.

The transfer equation for the Stokes vector $S_A = (I, Q, U, V)$ is

$$\frac{dS_A}{d\ell} = r_{AB}S_B - \mu_{AB}S_B, \tag{3.6.9}$$

where ℓ denotes distance along the ray path, and with

$$r_{AB} = \begin{pmatrix} 0 & 0 & 0 & 0 \\ 0 & 0 & -\rho_V & \rho_U \\ 0 & \rho_V & 0 & -\rho_Q \\ 0 & -\rho_U & \rho_Q & 0 \end{pmatrix}, \qquad \mu_{AB} = \begin{pmatrix} \mu_I & \mu_Q & \mu_U & \mu_V \\ \mu_Q & \mu_I & 0 & 0 \\ \mu_U & 0 & \mu_I & 0 \\ \mu_V & 0 & 0 & \mu_I \end{pmatrix}, \tag{3.6.10}$$

which describe generalized Faraday rotation and absorption, respectively. The parameters in (3.6.10) follow by writing $t^{\mu\nu} = t_0^{\mu\nu} + \Delta t^{\mu\nu}$, where $t_0^{\mu\nu}$ describes the isotropic part, implying a refractive index n_0, separating the anisotropic part into hermitian and antihermitian parts, writing

$$\rho^{\mu\nu} = \frac{1}{n\omega c} \Delta t^{H\mu\nu}, \qquad \mu^{\mu\nu} = \frac{i}{n\omega c} \Delta t^{A\mu\nu}, \tag{3.6.11}$$

and making the transformation from SU2 to O4 using the Pauli matrices,

$$I^{\mu\nu} = \tfrac{1}{2} \sum_A S_A \sigma_A^{\mu\nu}, \qquad S_A = \sigma_A^{\mu\nu} I_{\mu\nu}, \tag{3.6.12}$$

with

$$\sigma_I^{\mu\nu} = \begin{pmatrix} 1 & 0 \\ 0 & 1 \end{pmatrix}, \qquad \sigma_Q^{\mu\nu} = \begin{pmatrix} 1 & 0 \\ 0 & 1 \end{pmatrix},$$

$$\sigma_U^{\mu\nu} = \begin{pmatrix} 0 & 1 \\ 1 & 0 \end{pmatrix}, \qquad \sigma_V^{\mu\nu} = \begin{pmatrix} 0 & -i \\ i & 0 \end{pmatrix}. \tag{3.6.13}$$

The polarization of the radiation may be described by a point on the Poincaré sphere. Writing $q = Q/I, u = U/I, v = V/I$, the mapping of a polarization onto a point on the unit sphere follows from

$$(q, u, v) = (\sin 2\chi \, \cos 2\psi, \sin 2\chi \, \sin \psi, \cos 2\chi), \tag{3.6.14}$$

with the angle 2ψ identified as the longitude and the angle 2χ as the colatitude. The natural modes correspond to antipodal points on the sphere.

Wave Modes as Stokes Eigenvectors

The properties of natural wave modes in the absence of damping are implicit in the square matrix r_{AB} in (3.6.10). Two of the four eigenvalues of r_{AB} are zero. The eigenvectors corresponding to eigenvalue zero are conserved quantities: one of these is I = constant, which is obviously the case in the absence of damping, and the other is $\rho_Q Q + \rho_U U + \rho_V V$, which is discussed below. The two non-trivial eigenvalues are $\pm\Delta k$, where $\Delta k = (\rho_Q^2 + \rho_U^2 + \rho_V^2)^{1/2}$ is the difference in wave number between the two modes. The corresponding eigenvectors describe the polarization of the two natural modes in terms of the Stokes parameters. It is convenient to write

$$(\rho_Q, \rho_U, \rho_V) = -\Delta k\,(\cos 2\chi_B\,\cos 2\psi_B, \cos 2\chi_B\,\sin 2\psi_B, \sin 2\chi_B), \qquad (3.6.15)$$

with

$$\Delta k = (\rho_Q^2 + \rho_U^2 + \rho_V^2)^{1/2}, \qquad \cos 2\chi_B = \frac{T^2 - 1}{T^2 + 1}, \qquad \sin 2\chi_B = \frac{2T}{T^2 + 1}. \qquad (3.6.16)$$

The orthogonal polarizations of the two modes define a "mode axis" through the sphere, with end points at $2\chi_B, 2\psi_B$ and $\pi - 2\chi_B, 2\psi_B + \pi$. In the absence of damping, the evolution described by (3.6.9) corresponds to generalized Faraday rotation: the polarization point rotates about the mode axis at constant colatitude relative to it.

As already noted, in the absence of damping, $\rho_Q Q + \rho_U U + \rho_V V$ is conserved. A physical interpretation of this conserved quantity is in terms of the mixture of the two natural modes. With I = constant, the other conserved quantity is

$$\cos 2\zeta = \cos 2\psi_B\,\cos 2\chi\,\cos 2(\psi - \psi_B) + \sin 2\chi_B\,\sin 2\chi. \qquad (3.6.17)$$

The extrema $\cos 2\zeta = \pm 1$ correspond to radiation purely in one or other mode, and $\cos 2\zeta = 0$ corresponds to an equal mixture of the two modes.

In the presence of polarization-dependent damping, the polarization of the radiation changes, for $\cos 2\zeta \neq \pm 1$, due to the mixture of the two modes evolving as one mode is more strongly damped than the other. When the damping is negative, the combination of generalized Faraday rotation and polarization-dependent maser growth has unexpected implications on the resulting polarization [22].

3.6.3 High-Frequency Waves

The weak-anisotropy limit applies at sufficiently high frequencies. Specific applications are to high-frequency waves in a cold plasma, in a synchrotron-emitting gas (discussed in § 2.7), and in a pulsar plasma.

High-Frequency Waves in Cold Plasma

In order to discuss the range of validity of the weak-anisotropy approximation, it is useful to apply it to the magnetoionic waves. The exact solutions can then be compared with the weak-anisotropy approximation to them.

In the rest frame of a cold plasma, one has

$$\frac{t^t{}_t}{\omega^2} = S\cos^2\theta + P\sin^2\theta - 1, \quad \frac{t^a{}_a}{\omega^2} = S - 1, \quad \frac{t^t{}_a}{\omega^2} = -\frac{t^a{}_t}{\omega^2} = -iD\cos\theta. \tag{3.6.18}$$

In the quadratic equation (3.6.5) for T, the coefficient of the term proportional to T becomes, in the weak-anisotropy approximation,

$$-\frac{t^t{}_t - t^a{}_a}{it^t{}_a} = -\frac{(P-S)\sin^2\theta}{D\cos\theta} = \frac{Y\sin^2\theta}{\epsilon\cos\theta}, \tag{3.6.19}$$

where the final expression follows from (1.2.38). Comparison with the exact form (3.2.6) shows that $(PS - S^2 + D^2)/PD$ is approximated by $(P - S)/D$. Making this comparison for the magnetoionic case, the exact coefficient $(1 - E)$ $Y/(1-X)$ in (3.3.2) is approximated by Y. The latter approximation is well justified at high frequency where X and E are small. The refractive index is given by writing $k_\sigma^2 = \omega^2(1 - n_\sigma^2)$ with $\sigma = \pm$, implying

$$n_\sigma^2 = S\cos^2\theta + P\sin^2\theta - \frac{D\cos\theta}{T_\sigma}. \tag{3.6.20}$$

The expressions for T_σ and n_σ^2 reproduce the results (3.3.10) and (3.3.11) in the high-frequency limit.

High-Frequency Waves in Pulsar Plasma

Pulsar plasma and a synchrotron-emitting gas are two examples of highly relativistic plasmas. For the latter case, the projection onto the transverse (t-a) plane leads to (2.7.19) with (2.7.20), or to (2.7.21) with (2.7.22). The application of the weak-anisotropy approximation to a pulsar plasma is discussed here.

Before discussing details of the application to relativistic plasmas, it is relevant to note some qualitative differences between the relativistic and nonrelativistic cases. A mathematical difference is that the 13- and 23-components of t^i_j are zero in the nonrelativistic case and nonzero in the relativistic case. When they are nonzero, one has

$$t^t{}_t = \cos^2\theta\, t^1{}_1 + \sin^2\theta\, t^3{}_3 - 2\sin\theta\cos\theta\, t^1{}_3, \qquad t^t{}_a = \cos\theta\, t^1{}_2 + \sin\theta\, t^2{}_3, \tag{3.6.21}$$

with $a^t{}_a = t^2{}_2$. A qualitative difference arises from the relative magnitudes of $t^t{}_t - t^a{}_a$ and $t^t{}_a$, whose ratio depends on angle, and for most angles it is small in the nonrelativistic case and large in the relativistic case. As a consequence, the natural modes are nearly circularly polarized at most angles in the nonrelativistic case, the exception being a small range of angles about $\theta = \pi/2$, whereas the natural modes are nearly linearly polarized at most angles in the relativistic case, the exception being a small range of angles about parallel propagation, $\sin\theta \approx 0$. Thus, qualitatively, the relativistic case is effectively the obverse of the cold-plasma case, with the relative range of angles over which the QL and QT approximations reversed. An implication is that for most angles of propagation in a relativistic plasma, generalized Faraday rotation is primarily in χ, as discussed above in connection with the QT limit in a cold plasmas.

For a pulsar plasma, the relevant response tensor is that for a 1D pair plasma, given by (2.6.7). Projecting (2.6.7) onto the transverse plane gives

$$\begin{aligned} t^{\mu\nu}(k) = \sum_{\pm} \frac{\mu_0 e^2 n_\pm}{m} \bigg\langle \frac{1}{\gamma} \bigg\{ \frac{\omega^2 \sin^2\theta}{(ku)^2} t^\mu t^\nu \\ + \frac{1}{(ku)^2 - \Omega_e^2} \big[(\cos\theta\, ku - \sin\theta\, k_\perp \gamma v)^2 t^\mu t^\nu + (ku)^2\, a^\mu a^\nu \\ - i\,\epsilon\Omega_e (\cos\theta\, ku - \sin\theta\, k_\perp \gamma v)(t^\mu a^\nu - a^\mu t^\nu) \big] \bigg\} \bigg\rangle_{\pm}, \end{aligned} \tag{3.6.22}$$

with $\cos\theta\, ku - \sin\theta\, k_\perp \gamma v = \gamma(\omega\cos\theta - |\boldsymbol{k}|v)$. The combinations $t^t{}_t - t^a{}_a, t^t{}_a$ implied by (3.6.22) can be rewritten in terms of the variables $z = \omega/k_z$, $y = \Omega_e/k_z$ using (2.6.9) with (2.6.10) and the definitions (2.6.11) of the RPDFs $W(z)$, $R(z)$, $S(z)$. One finds

$$\begin{aligned} t^t{}_t - t^a{}_a = -\omega_p^2 \bigg\{ z^2 W(z) \sin^2\theta \\ + \frac{\tan^2\theta}{(1+y^2)} \bigg[\bigg\langle \frac{1}{\gamma} \bigg\rangle + \frac{1}{z_+ - z_-} \sum_{\alpha=\pm} \alpha(z_\alpha^2 - z^2) R(z_\alpha) \bigg] \bigg\}, \\ t^t{}_a = -i\,\epsilon\,\omega_p^2 \frac{y}{(1+y^2)\cos\theta} \frac{1}{z_+ - z_-} \sum_{\alpha=\pm} \alpha(z_\alpha - z\cos^2\theta) S(z_\alpha), \end{aligned} \tag{3.6.23}$$

with $z = \omega/k_z$, $y = \Omega_e/k_z$, and $z_\pm$ defined by (2.6.10).

At low frequencies the O and X modes in a pulsar plasma are nearly linearly polarized. This case may be treated by neglecting the gyrotropic term $t^t{}_a$. The inclusion of gyrotropy $t^t{}_a \neq 0$, causes the modes to be elliptically polarized, with $T_\pm = \frac{1}{2}R \pm \frac{1}{2}(R^2+4)^{1/2}$, and R given by

$$R = \frac{t^t{}_t - t^a{}_a}{i t^t{}_a} = -\frac{\sin^2\theta}{\epsilon\cos\theta} \frac{\Omega_e}{\omega} \bigg[z^2 W(z) - \frac{\omega^2}{\omega_p^2} \frac{1-\delta\beta^2}{\beta_A^2} \bigg]. \tag{3.6.24}$$

The modes are nearly circularly polarized for $R \ll 1$ and nearly linearly polarized for $R \gg 1$. Let θ_c be the angle around which this change from circular to linear occurs. Setting $R = 1$ in (3.6.24) and assuming $v_A \gg 1, \delta v^2 \approx 1$ and $\omega \ll \Omega_e$, one finds that $\theta_c \approx (\epsilon\, \omega/\Omega_e \langle\gamma\rangle)^{1/2}$ is small, as expected. The result (3.6.24) applies in the rest frame of the plasma. On Lorentz transforming to the pulsar frame in which the plasma is streaming with a Lorentz factor γ_p, the angle at which the transition from circular to linear polarization occurs is strongly modified, with the angle shrinking in the forward direction and broadening in the backward direction [23]. The angle corresponding to $R = 1$ also broadens at higher frequencies, becoming large near the cyclotron resonance.

3.6.4 Mode Coupling

Mode coupling results from gradients in the plasma parameters causing gradients in the properties of the natural modes. In the presence of weak gradients, one identifies the wave modes at each point as those of a locally homogeneous plasma. The effect of the gradients in the plasma parameters is taken into account by introducing mode coupling. The concept of a mode coupling is somewhat counter-intuitive, in that it suggests that a physical process called 'mode coupling' exists. This is not the case. The relevant physical process in an anisotropic medium is the independent propagation of two orthogonal modes in a birefringent medium. This actively changes the polarization of radiation passing through the medium, due to generalized Faraday rotation. Gradients in the wave properties reduce the effectiveness of the medium in causing the two modes to propagate independently. The concept of mode coupling is particularly confusing when it is 'strong' and negates the effect of the anisotropy, so that the initial polarization is preserved as in an isotropic medium. Mode coupling is not a physical effect, but rather the partial negation of the physical effects associated with the independent propagation of the two modes in a homogeneous, birefringent medium.

The important gradient in a weakly anisotropic plasma is in the polarization of the natural modes. The polarization vectors for the two modes may be written

$$\boldsymbol{e}_\pm = \sin\chi_\pm \boldsymbol{t} + i\cos\chi_\pm \boldsymbol{a}, \quad \sin\chi_\pm = \frac{T_\pm}{(1+T_\pm^2)^{1/2}}, \quad \cos\chi_\pm = \frac{1}{(1+T_\pm^2)^{1/2}}. \tag{3.6.25}$$

Without loss of generality one can choose $T_+ > 0$, with $T_- = -1/T_+ < 0$. Then one has $\chi_+ - \chi_- = \pi/2$. The derivative $\boldsymbol{e}'_\pm$, of $\boldsymbol{e}_\pm$ with respect to distance along the direction of the gradient, depends on the derivative $\chi'_\pm = \pm\chi'_+$, and on the derivatives θ', ϕ' of the polar angles of $\boldsymbol{\kappa}$ relative to the magnetic field direction $\boldsymbol{b}$. One has

$$\frac{\partial \boldsymbol{t}}{\partial \theta} = -\boldsymbol{\kappa}, \qquad \frac{\partial \boldsymbol{t}}{\partial \phi} = \cos\theta\, \boldsymbol{a}, \qquad \frac{\partial \boldsymbol{a}}{\partial \phi} = -\cos\theta\, \boldsymbol{t} - \sin\theta\, \boldsymbol{\kappa}. \tag{3.6.26}$$

The derivative of (3.6.25) gives

$$\begin{aligned} \boldsymbol{e}'_{\pm} = \chi'_{\pm}(\cos\chi_{\pm}\boldsymbol{t} - i\sin\chi_{\pm}\boldsymbol{a}) + \phi'\cos\theta(\sin\chi_{\pm}\boldsymbol{a} - i\cos\chi_{\pm}\boldsymbol{t}) \\ -(\theta'\sin\chi_{\pm} + i\phi'\cos\chi_{\pm}\sin\theta)\boldsymbol{\kappa}. \end{aligned} \tag{3.6.27}$$

Ignoring the component along $\boldsymbol{\kappa}$, (3.6.27) gives

$$\begin{pmatrix} e'_+ \\ e'_- \end{pmatrix} = \begin{pmatrix} -i\phi'\cos\theta\sin 2\chi_+ & \chi'_+ - i\phi'\cos\theta\cos 2\chi_+ \\ -\chi'_+ + i\phi'\cos\theta\cos 2\chi_+ & -i\phi'\cos\theta\sin 2\chi_+ \end{pmatrix} \begin{pmatrix} e_+ \\ e_- \end{pmatrix}. \tag{3.6.28}$$

It follows from the off-diagonal terms in (3.6.28) that the coupling rate per unit length from one mode to the other due to the inhomogeneity is proportional to the rate of change of the shape of the polarization ellipse, described by χ'_+, or to the rate of twisting of the magnetic field, described by ϕ'. These changes are opposed by the components in the two modes getting out of phase as the rate per unit length $\Delta k = \omega(n_+ - n_-)$. Mode coupling is strong or weak depending on which of these two rates, respectively, is the larger.

Mode coupling in the weak-anisotropy approximation may be described in terms of the Stokes parameters by allowing the matrices r_{AB}, μ_{AB} in (3.6.9) to be functions of distance along the ray path. Given a model for the medium, one can integrate (3.6.9) with (3.6.10) between two points, ℓ_1, ℓ_2 along the ray path, to find $S_A(\ell_2)$ in terms of $S_A(\ell_1)$. The relation can be written $S_A(\ell_2) = M_{AB}(\ell_1, \ell_2) S_B(\ell_1)$, where $M_{AB}(\ell_1, \ell_2)$ is a 4×4 matrix, called the Mueller matrix. Mode coupling is identified by writing $S_A(\ell_1)$ and $S_A(\ell_2)$ as mixtures of the natural modes defined by a locally homogeneous medium at ℓ_1 and ℓ_2, respectively, and comparing these ratios at the two points. In weak mode coupling, radiation with unit amplitude purely in one mode at ℓ_1 develops a small admixture of the other mode at ℓ_2. In strong mode coupling, radiation with unit amplitude purely in one mode at ℓ_1 becomes a mixture of the two modes with their amplitudes varying rapidly between 0 and ≈ 1 as a function of $\ell_2 - \ell_1$.

References

1. T.H. Stix, *The Theory of Plasma Waves* (McGraw-Hill, New York, 1962)
2. D.B. Melrose, R.C. McPhedran, *Electromagnetic Processes in Dispersive Media* (Cambridge University Press, Cambridge, 1991)
3. A.L. Verdon, I.H. Cairns, D.B. Melrose, P.A. Robinson, Phys. Plasma **16**, 052105 (2009)
4. A. Hasegawa, J. Geophys. Res. **81**, 5083 (1976)
5. G.A. Stewart, E.W. Liang, J. Plasma Phys. **47**, 295 (1992)
6. G.P. Zank, R.G. Greaves, Phys. Rev. E **51**, 6079 (1995)
7. D.B. Melrose, Plasma Phys. Cont. Fusion **39**, A93 (1997)

8. E.P. Gross, Phys. Rev. **82**, 232 (1951)
9. B. Bernstein, Phys. Rev. **109**, 10 (1958)
10. V.N. Dnestrovskii, D.P. Kostomorov, Sov. Phys. JETP **13**, 98 (1961)
11. V.N. Dnestrovskii, D.P. Kostomorov, Sov. Phys. JETP **14**, 1089 (1962)
12. S. Puri, F. Leuterer, M. Tutter, J. Plasma Phys. **9**, 89 (1973)
13. S. Puri, F. Leuterer, M. Tutter, J. Plasma Phys. **14**, 169 (1975)
14. P.A. Robinson, J. Math. Phys. **28**, 1203 (1987)
15. P.A. Robinson, J. Plasma Phys. **37**, 435 (1987)
16. P.A. Robinson, J. Plasma Phys. **37**, 449 (1987)
17. P.A. Robinson, Phys. Fluid **31**, 107 (1988)
18. J.J. Barnard, J. Arons, Astrophys. J. **302**, 138 (1986)
19. C. Wang, D. Lai, J. Han, Mon. Not. Roy. Astron. Soc. **403**, 569 (2010)
20. D.B. Melrose, M.E. Gedalin, Astrophys. J. **521**, 351 (1999)
21. M.W. Verdon, D.B. Melrose, Phys. Rev. E **77**, 046403 (2008)
22. D.B. Melrose, A.C. Judge, Phys. Rev. E **70**, 056408 (2004)
23. D.B. Melrose, Q. Luo, Mon. Not. Roy. Astron. Soc. **352**, 519 (2004)

Chapter 4
Gyromagnetic Processes

Gyromagnetic emission is the generic name for emission due to the spiraling motion of a particle in a magnetic field. Gyromagnetic emission by nonrelativistic particles is referred to as cyclotron emission, which is dominated by the fundamental and first few harmonics of the cyclotron frequency. Gyromagnetic emission by highly relativistic particles is referred to as synchrotron emission, which is dominated by very high harmonics which overlap and form a continuum. These emission processes are treated in this chapter. The generalization of Thomson scattering to the scattering of waves in a magnetized plasma is also discussed.

General formulae for gyromagnetic emission are written down in § 4.1. The special case of emission in vacuo is treated in § 4.2. Cyclotron emission is discussed in § 4.3 with emphasis on a relativistic effect that is important for electron cyclotron maser emission. Synchrotron emission is treated in § 4.4. Scattering of waves by electrons in a magnetic field is discussed in § 4.5.

4.1 Gyromagnetic Emission

Gyromagnetic emission is described here in terms of a semi-classical probability of emission. The semi-classical approach facilitates the derivation of kinetic equations that describe the effect of gyromagnetic emission and absorption on the distributions of waves and particles.

4.1.1 Probability of Emission for Periodic Motion

A particle spiraling in a magnetic field is one example of a general class of periodic motions. It is convenient to start with the general case of a charge in periodic motion emitting radiation and then to apply the general result to the particular case of gyromagnetic emission.

D. Melrose, *Quantum Plasmadynamics: Magnetized Plasmas*, Lecture Notes in Physics 854, DOI 10.1007/978-1-4614-4045-1_4,

A general formula is derived in § 5.1 of volume 1 for emission of radiation in an arbitrary wave mode M due to an arbitrary extraneous current $J_{\rm ext}$. The emission formula may be written in terms of the probability per unit time that a wave quantum is emitted in the mode M in the range $d^3\boldsymbol{k}/(2\pi)^3$. An explicit expression for this quantity is

$$w_M(k) = \frac{\mu_0 R_M(k)}{T|\omega_M(\boldsymbol{k})|} |e^*_{M\mu}(k) J^\mu_{\rm ext}(k_M)|^2, \tag{4.1.1}$$

where T is an arbitrarily long normalization time. For a particle whose orbit is $x = X(\tau)$, implying the 4-velocity $u^\mu(\tau) = dX^\mu(\tau)/d\tau$, the current is

$$J^\mu(k) = q \int d\tau\, u^\mu(\tau)\, e^{ikX(\tau)}. \tag{4.1.2}$$

For periodic motion at a proper frequency ω_0, the orbit is of the form

$$X^\mu(\tau) = x_0^\mu + \bar{u}^\mu \tau + \tilde{X}^\mu(\chi), \quad \tilde{X}^\mu(\chi + 2\pi) = \tilde{X}^\mu(\chi), \quad \chi = \omega_0 \tau, \tag{4.1.3}$$

with the 4-velocity given by

$$u^\mu(\tau) = \bar{u} + \tilde{u}^\mu(\chi), \qquad \tilde{u}^\mu(\chi) = \omega_0\, d\tilde{X}^\mu(\chi)/d\chi. \tag{4.1.4}$$

One may expand the periodic motion in Fourier series by writing

$$u^\mu(\tau)\, e^{ik\tilde{X}(\omega_0\tau)} = \sum_{a=-\infty}^{\infty} U^\mu(a,k)\, e^{ia\omega_0\chi},$$

$$U^\mu(a,k) = \frac{\omega_0}{2\pi} \int_0^{2\pi/\omega_0} d\tau\, u^\mu(\tau)\, e^{-ia\omega_0\tau}\, e^{ik\tilde{X}(\omega_0\tau)}. \tag{4.1.5}$$

The 4-current becomes

$$J^\mu(k) = q e^{ikx_0} \sum_{a=-\infty}^{\infty} U^\mu(a,k)\, 2\pi\delta(k\bar{u} - a\omega_0), \tag{4.1.6}$$

which is identified with $J^\mu_{\rm ext}(k)$ in (4.1.2).

Square of δ-Function

On inserting (4.1.6) into (4.1.1), the square of the δ-function appears, and this is rewritten using $[2\pi\delta(k\bar{u} - a\omega_0)]^2 = (T/\bar{\gamma})2\pi\delta(k\bar{u} - a\omega_0)$, where $\bar{u}^\mu = [\bar{\gamma}, \bar{\gamma}\bar{\boldsymbol{v}}]$ implies $k\bar{u} = \bar{\gamma}(\omega - \boldsymbol{k}\cdot\bar{\boldsymbol{v}})$ in the δ-function. The probability of emission becomes

$$w_M(k) = \frac{\mu_0 q^2 R_M(k)}{\bar{\gamma}|\omega_M(k)|} \sum_a |e^*_{M\mu}(k) U^\mu(a, k_M)|^2 2\pi\delta(k_M\bar{u} - a\omega_0). \tag{4.1.7}$$

The probability (4.1.7) describes emission by a particle executing an arbitrary periodic motion, at proper frequency ω_0 with an average (over the periodic motion) 4-velocity $\bar{u}$. The emission separates into contributions from harmonics of the oscillation frequency, with the harmonic number, a, taking on all integer values. However, depending on the wave properties, emission is possible at a specific harmonic only if the resonance condition, described by the δ-function in (4.1.7), can be satisfied. For example, for waves with phase speed greater than the speed of light, $\omega/|\boldsymbol{k}| > 1$, $k\bar{u}$ is positive for $|\boldsymbol{v}| < 1$, and emission at $s \le 0$ is forbidden.

Probability of Gyromagnetic Emission

Gyromagnetic motion is an example of such periodic motion, consisting of circular motion perpendicular to $\boldsymbol{B}$ plus constant rectilinear motion parallel to $\boldsymbol{B}$. The circular motion about the magnetic field is periodic in the gyrophase, ϕ, which evolves a $\phi = \Omega_0\tau = \Omega t$, with $\Omega = \Omega_0/\gamma$, where $\Omega_0 = |q|B/m$ is the cyclotron frequency. The average motion corresponds to $\bar{u}^\mu \to u^\mu_\parallel$. The current (2.1.31) due to a spiraling charge is of the form (4.1.6), specifically,

$$J^\mu(k) = q e^{ikx_0} \sum_{a=-\infty}^{\infty} e^{ia\epsilon\psi} U^\mu(a,k)\, 2\pi\delta\big[(ku)_\parallel - a\Omega_0\big], \tag{4.1.8}$$

with the 4-vector $U^\mu(a,k)$ having the explicit form given by (2.1.28). The probability of gyromagnetic emission in the mode M at the ath gyroharmonic follows from (4.1.7):

$$w_M(a,k,p) = \frac{q^2 R_M(k)}{\hbar\varepsilon_0\gamma\omega_M(k)} \big|e^*_{M\mu}(k) U^\mu(a,k_M)\big|^2 2\pi\delta\big[(k_M u)_\parallel - a\Omega_0\big]. \tag{4.1.9}$$

4.1.2 Gyroresonance Condition

The gyroresonance condition, also called the Doppler condition, implied by the δ-function in (4.1.9) is $(ku)_\parallel - a\Omega_0 = 0$ with $k = k_M$ for waves in the mode M. Classically, the resonance condition is interpreted by noting that $(ku)_\parallel$ is the frequency of the wave in the frame in which the gyrocenter of the particle is at rest. Hence the resonance corresponds to the wave frequency being an integral multiple, $a = 0, \pm1, \pm2, \ldots$, of the gyrofrequency of the particle in the inertial frame in which the motion of the gyrocenter is at rest. Resonance at $a > 0$ is referred to as the normal Doppler effect, and resonance at $a < 0$ as the anomalous Doppler effect.

(The resonance at $a = 0$ is sometimes referred to as the Cerenkov condition, but this can lead to confusion with the resonance condition for an unmagnetized particle.)

In a semi-classical theory, in which the particles are treated quantum mechanically and the waves are treated classically, the resonance condition follows from conservation of energy and momentum during the emission of a wave quantum. A relativistic quantum treatment of a particle in a magnetic field leads to energy eigenvalues (ordinary units are used to discuss the quantum recoil)

$$\varepsilon_n(p_z) = (m^2c^4 + p_z^2c^2 + 2n\hbar\Omega_0 mc^2)^{1/2}, \tag{4.1.10}$$

where $n = 0, 1, 2, \ldots$ is the Landau quantum number. For spin-$\frac{1}{2}$ particles one has $2n = 2l + 1 + s$, where $l = 0, 1, 2, \ldots$ describes the orbital motion and $s = \pm 1$ is the spin quantum number. The classical limit corresponds to $\hbar \to 0$, $n \to \infty$, $\hbar n \to p_\perp^2/\Omega_0 m$.

As a result of emission, p_z changes to $p_z' = p_z - \hbar k_z$ and the energy of the particle decreases by $\hbar\omega$. Suppose that the initial energy is given by (4.1.10) and that the final Landau quantum number is $n' = n - a$, where a is an integer. The final energy satisfies (in ordinary units)

$$\varepsilon_{n'}(p_z') = \varepsilon_{n-a}(p_z - \hbar k_z) = \varepsilon_n(p_z) - \hbar\omega. \tag{4.1.11}$$

On squaring (4.1.11) and using (4.1.10) one obtains (in ordinary units)

$$\varepsilon_n(p_z)\omega - \hbar p_z k_z - a\Omega_0 mc^2 - \hbar(\omega^2 - k_z^2c^2)/2 = 0. \tag{4.1.12}$$

An expansion in $\hbar$ gives (in ordinary units)

$$(ku)_\parallel - a\Omega_0 - \hbar(\omega^2 - k_z^2c^2)/2mc^2 + \cdots = 0, \tag{4.1.13}$$

where the neglected terms, denoted $+\cdots$, are of higher order in $\hbar$. The lowest order term, $(ku)_\parallel - a\Omega_0 = 0$, in (4.1.13) reproduces the classical Doppler condition.

Quantum mechanically, the normal and anomalous Doppler effects correspond to the particle jumping to lower and higher n, respectively, on emission of a wave quantum. That is, with $n \to n - a$ in (4.1.10), the contribution $2n\hbar\Omega_0 mc^2$ of the motion perpendicular to the field lines decreases for $a > 0$ and increases for $a < 0$. The total energy of the particle must decrease, and in the anomalous Doppler effect this occurs due to the decrease in p_z^2 exceeding the increase in $p_\perp^2 = 2n\hbar\Omega_0 m$.

4.1.3 Quantum Recoil

The quantum recoil corresponds to the term of order $\hbar$ in (4.1.13). This is interpreted as the recoil due to the emission. The resonance conditions for absorption differs from that for emission in that the quantum recoil has the opposite sign; this corresponds to reversing the signs of ω, k_z in (4.1.11)–(4.1.13). In a relativistic quantum treatment of the response of an electron gas, there are contributions corresponding to virtual emission and virtual absorption, and these contribute through resonant denominators with both signs of the recoil term.

An important difference between a relativistic and a nonrelativistic treatment relates to the form of the quantum recoil. Suppose that one assumes the nonrelativistic form for the energy of a particle (ordinary units): $\varepsilon_n(p_z) = mc^2 + p_n^2/2m + p_z^2/2m$. With $p_n = (2n|q|B\hbar)^{1/2}$, one has $p_n^2/2m = n\hbar\Omega_0$, and the Landau quantum number, $n = l + \frac{1}{2}(s+1)$, can be interpreted as $l + \frac{1}{2}$ from the quantization of the circular motion, which is simple harmonic with frequency $\Omega_0 = eB/m$ and $l = 0, 1, 2, \ldots$, and $\frac{1}{2}s$ from the spin. When one considers an emission process, described by (4.1.11), repeating the calculation using the nonrelativistic form for the energy leads to, in place of (4.1.13),

$$\omega - k_z v_z - a\Omega_0 + \hbar k_z^2/2m = 0, \tag{4.1.14}$$

with $v_z = p_z/m$. This does not agree with the limit of (4.1.13) in which the Lorentz factor is set to unity in writing $(ku)_\parallel = \gamma(\omega - k_z v_z) \to \omega - k_z v_z$; the relativistically correct form has $-(\omega^2/c^2 - k_z^2)$ in place of k_z^2 in the quantum recoil.

There is a paradox: the nonrelativistic limit of the relativistic quantum treatment is inconsistent with the nonrelativistic quantum treatment. The resolution is that the nonrelativistic limit is formally $c^2 \to \infty$, and in this limit the additional term ω^2/c^2 should indeed be neglected. However, the implication is that a strictly nonrelativistic quantum treatment of emission, absorption and dispersion is incorrect when applied to waves with $\omega^2/c^2 \gtrsim k_z^2$. Put another way, nonrelativistic theory requires $c^2 \to \infty$, and it is formally inconsistent to combine nonrelativistic theory with Maxwell's equation, in which finite c plays an essential role.

4.1.4 Resonance Ellipses

The gyroresonance condition is amenable to a graphical interpretation. If one plots the resonance curve in $v_\perp$–v_z space for given ω and k_z, the resonance condition for each harmonic defines a resonance ellipse. The resonance ellipse corresponds to all the values of $v_\perp$ and v_z for which resonance with a wave at given ω, k_z and a is possible. That is, a given wave resonates with all particles that lie on the resonance ellipse that it defines. Similarly, a given particle resonates with all waves that define resonance ellipses that pass through the representative point of the particle in $v_\perp$–v_z space.

The resonance ellipse is centered on the v_z axis at $v_z = v_c$, with semi-major axis A perpendicular to the v_z axis, and with eccentricity $e = (A^2 - B^2)^{1/2}/A$. This ellipse is described by

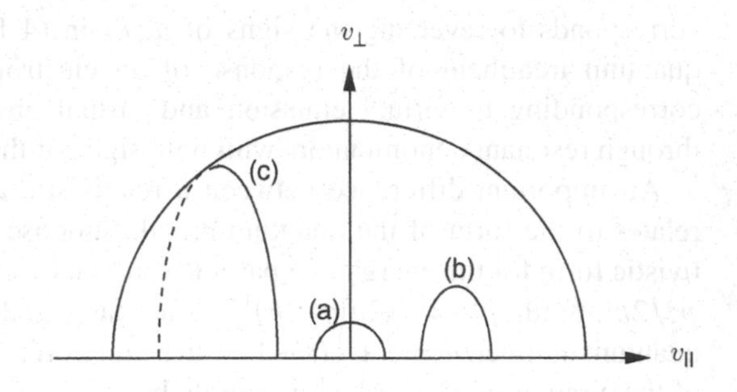

Fig. 4.1 Examples of resonance ellipses: (**a**) a semicircle centered on the origin, (**b**) an ellipse inside $v = 1$, (**c**) an ellipse touching $v = 1$

$$\frac{(v_z - v_c)^2}{B^2} + \frac{v_\perp^2}{A^2} = 1, \qquad v_c = \frac{\omega k_z}{a^2\Omega_0^2 + k_z^2},$$

$$A^2 = \frac{a^2\Omega_0^2 + k_z^2 - \omega^2}{a^2\Omega_0^2 + k_z^2}, \qquad e^2 = \frac{k_z^2}{a^2\Omega_0^2 + k_z^2}. \tag{4.1.15}$$

Some examples of resonance ellipses in $v_\perp$–v_z space are illustrated schematically in Fig. 4.1. For $\omega^2 < k_z^2$ the resonance ellipse touches the circle, $v = 1$, and the outer segment of the curve is nonphysical, as shown by the dashed segment in Fig. 4.1. The resonance ellipse in v_z-$v_\perp$ space is convenient when considering nonrelativistic and mildly relativistic particles, but not for relativistic particles, which are near the unit circle, $v \approx 1$.

Alternatively, one may plot the resonance condition in p_z-$p_\perp$ space. In this case the resonance condition becomes

$$\frac{(p_z - p_c)^2}{m^2\tilde{A}^2} + \frac{p_\perp^2}{m^2\tilde{A}^2} = 1, \qquad p_c = \frac{a\Omega_0 m k_z}{\omega^2 - k_z^2},$$

$$\tilde{A}^2 = \frac{\omega^2(a^2\Omega_0^2 - \omega^2 + k_z^2)}{(\omega^2 - k_z^2)^2}, \qquad \tilde{B}^2 = \frac{a^2\Omega_0^2 - \omega^2 + k_z^2}{\omega^2 - k_z^2}. \tag{4.1.16}$$

For $\omega^2 > k_z^2$, (4.1.16) defines an ellipse with major axis along the p_z-axis. For $\omega^2 < k_z^2$, $\tilde{B}^2$ in (4.1.16) is negative and the curve is an hyperbola. The dashed segment of curve (c) in Fig. 4.1 correspond to a nonphysical arm of the hyperbola in $p_\perp$-p_z space. For emission in vacuo, with $k_z = \omega \cos\theta$, (4.1.16) reduces to

$$\frac{(|\boldsymbol{p}|\cos\alpha - p_c)^2}{m^2\tilde{A}^2} + \frac{|\boldsymbol{p}|^2\sin^2\alpha}{m^2\tilde{b}^2} = 1, \qquad p_c = \frac{a\Omega_0 m\cos\theta}{\omega\sin^2\theta},$$

$$\tilde{A}^2 = \frac{a^2\Omega_0^2 - \omega^2\sin^2\theta}{\omega^2\sin^4\theta}, \qquad \tilde{B}^2 = \tilde{A}^2\sin^2\theta, \tag{4.1.17}$$

where the cylindrical coordinates $p_\perp$, p_z are rewritten in terms of polar coordinates $|\mathbf{p}|$, α. It follows that emission at given ω and θ is possible in vacuo only at harmonics, a, which satisfy $\omega < a\Omega_0/\sin\theta$.

4.1.5 Differential Changes

In semi-classical theory, changes due to emission and absorption may be treated in terms of a differential operator. In the unmagnetized case, the change due to $p \to p - k$ is treated using the differential operator $k^\alpha \partial/\partial p^\alpha$. In the magnetized case the changes for emission are $p_\parallel^\mu \to p_\parallel^\mu - k_\parallel^\mu$ and $n \to n - a$. The differential operator corresponds to

$$k^\alpha \frac{\partial}{\partial p^\alpha} \to k_\parallel^\alpha \frac{\partial}{\partial p_\parallel^\alpha} + a \frac{\partial}{\partial n}.$$

In the classical limit, $2n\Omega_0 m$ is interpreted as $p_\perp^2 = m^2\gamma^2 v_\perp^2$. Hence, the n-derivative reduces to its classical counterpart

$$a \frac{\partial}{\partial n} = \frac{a\Omega}{v_\perp} \frac{\partial}{\partial p_\perp} = \frac{(kp)_\parallel}{p_\perp} \frac{\partial}{\partial p_\perp}, \tag{4.1.18}$$

where in the final form the Doppler condition is used.

The transfer equation for the waves due to gyromagnetic emission and absorption by a distribution of particles can be written in the form

$$\partial_\mu T_M^{\mu\nu}(k) = S_M^\nu(k) - \gamma_M(k)\, P_M^\nu(k), \tag{4.1.19}$$

where $T_M^{\mu\nu}(k) = v_{gM}^\mu(k) k_M^\nu N_M(k)$ is the energy-momentum tensor for the waves, and $P_M^\nu(k) = k_M^\nu N_M(k)$ is the 4-momentum in the waves. In (4.1.19) spontaneous emission is described by the term

$$S_M^\nu(k) = \int \frac{d^4p}{(2\pi)^4}\, k_M^\nu\, \gamma w_M(k,p)\, F(p). \tag{4.1.20}$$

The final term in (4.1.19) describes absorption, with the absorption coefficient given by

$$\gamma_M(k) = -\sum_{a=-\infty}^{\infty} \int \frac{d^4p}{(2\pi)^4}\, \gamma\, w_M(a,k,p)\, \hat{D} F(p),$$

$$\hat{D} = \frac{a\Omega}{v_\perp} \frac{\partial}{\partial p_\perp} + k_\parallel^\alpha \frac{\partial}{\partial p_\parallel^\alpha}. \tag{4.1.21}$$

There is a factor k_M^ν that is common to all three terms in (4.1.19), and when this factor is omitted the transfer equation describes the evolution of the occupation number, $N_M(k)$, for waves in the mode M. The transfer equation becomes

$$\frac{\mathrm{D}N_M(k)}{\mathrm{D}t} = \dot{N}_M(k) - \gamma_M(k)\, N_M(k), \tag{4.1.22}$$

with $\mathrm{D}/\mathrm{D}t = v^{\mu}_{gM}(k)\partial_\mu$ and with the spontaneous emission described by

$$\dot{N}_M(k) = \int \frac{d^4p}{(2\pi)^4}\,\gamma w_M(k,p)\,F(p). \tag{4.1.23}$$

4.1.6 *Quasilinear Equations*

The covariant version of the quasilinear equation that describes the effect of gyromagnetic emission and absorption on a distribution of magnetized particles is derived in an analogous manner to the quasilinear equation for the unmagnetized case, cf. § 5.2 of volume 1. The quasilinear equation for the waves is given by either (4.1.20) or (4.1.23).

The quasilinear equation for the particles is obtained by considering transitions $q \leftrightarrow q'$, and $q'' \leftrightarrow q$, with $q; q'; q''$ denoting $p^{\mu}_{\parallel}, n;\ p^{\mu}_{\parallel} - k^{\mu}_{\parallel}, n-a;\ p^{\mu}_{\parallel} + k^{\mu}_{\parallel}, n+a$, respectively. The probability of emission for these transitions is $w_M(a,k,p)$ and $w_M(a,k,p'')$, respectively. Let n_q denote the occupation number of the particles in the state $q = p^{\mu}_{\parallel}, n$. Transitions $q \to q'$ and $q \to q''$ decrease n_q at rates $\propto w_M(a,k,p)n_q[1 + N_M(k)]$ and $\propto w_M(a,k,p'')n_q N_M(k)$, respectively, and transitions $q' \to q$ and $q'' \to q$ increase n_q at rates $\propto w_M(a,k,p)n_{q'} N_M(k)$ and $\propto w_M(a,k,p'')n_{q''}[1 + N_M(k)]$, respectively. On expanding the net rate of change of n_q in powers of $\hbar$ and a/n, the leading terms give the quasilinear equation

$$\begin{aligned}\frac{dF(p)}{d\tau} = {} & \frac{1}{p_\perp}\frac{\partial}{\partial p_\perp}\left\{p_\perp\left[-A_\perp(p)\,F(p) + D_{\perp\perp}(p)\frac{\partial F(p)}{\partial p_\perp} + D^{\nu}_{\perp\parallel}(p)\frac{\partial F(p)}{\partial p^{\nu}_{\parallel}}\right]\right\}\\ & + \frac{\partial}{\partial p^{\mu}_{\parallel}}\left[-A^{\mu}_{\parallel}(p)\,F(p) + D^{\mu}_{\parallel\perp}(p)\frac{\partial F(p)}{\partial p_\perp} + D^{\mu\nu}_{\parallel\parallel}(p)\frac{\partial F(p)}{\partial p^{\nu}}\right],\end{aligned} \tag{4.1.24}$$

where the subscript M denoting the wave mode is omitted, and n_q is re-interpreted in terms of the distribution function $F(p)$. The coefficients

$$\begin{pmatrix} A_\perp(p) \\ A^{\mu}_{\parallel}(p) \end{pmatrix} = -\sum_{a=-\infty}^{\infty}\int \frac{d^3\boldsymbol{k}}{(2\pi)^3}\,\gamma w_M(a,k,p)\begin{pmatrix} a\Omega/v_\perp \\ k^{\mu}_{\parallel} \end{pmatrix} \tag{4.1.25}$$

describe the effect of spontaneous emission. The quasilinear diffusion coefficients in (4.1.24) are

$$\begin{pmatrix} D_{\perp\perp}(p) \\ D^{\mu}_{\parallel\perp}(p) \\ D^{\mu\nu}_{\parallel\parallel}(p) \end{pmatrix} = \sum_{a=-\infty}^{\infty}\int \frac{d^3\boldsymbol{k}}{(2\pi)^3}\,\gamma w_M(a,k,p)\,N_M(k)\begin{pmatrix} (a\Omega/v_\perp)^2 \\ (a\Omega/v_\perp)k^{\mu}_{\parallel} \\ k^{\mu}_{\parallel}k^{\nu}_{\parallel} \end{pmatrix}, \tag{4.1.26}$$

with $D^{\mu}_{\perp\parallel}(p) = D^{\mu}_{\parallel\perp}(p)$.

The kinetic equations (4.1.22) and (4.1.25) imply that energy and parallel momentum are conserved for the sum of the particle and wave systems. The conservation laws apply separately for spontaneous emission and the induced processes. Momentum perpendicular to the magnetic field is not conserved.

4.2 Gyromagnetic Emission in Vacuo

The classical theory of gyromagnetic emission can be treated exactly, in the sense that the power radiated and some other important quantities can be calculated in closed form, starting from the probability (4.1.9). Besides being of interest in themselves, exact results are useful as a basis for various approximations to gyroemission, including the synchrotron limit.

4.2.1 Gyromagnetic Emission of Transverse Waves

For transverse waves in vacuo there are two degenerate states of polarization. All relevant information on the polarization is retained by modifying the probability (4.1.9) so that it becomes the polarization tensor

$$w^{\alpha\beta}(a,k,p) = \frac{q^2\,\tilde{U}_a^{*\alpha}\tilde{U}_a^{\beta}}{2\varepsilon_0\gamma\omega}\,2\pi\delta\big[(ku)_\parallel - a\Omega_0\big], \tag{4.2.1}$$

where $\tilde{U}_a^{\mu}$ is $U^{\mu}(a,k)$, as given by (2.1.27) with (2.1.28), projected onto the transverse plane. The projection onto the transverse plane is carried out by (a) setting both the time-component and the longitudinal component of $U^{\mu}(a,k)$ to zero, and (b) choosing two basis vectors to span the remaining two-dimensional transverse plane. Let the two transverse directions be those introduced in (1.1.25), specifically,

$$\boldsymbol{t} = (\cos\theta, 0, -\sin\theta), \qquad \boldsymbol{a} = (0,1,0), \tag{4.2.2}$$

with $\boldsymbol{k} = \omega(\sin\theta, 0, \cos\theta)$. This corresponds to $\psi = 0$ in (2.1.28), which gives

$$\tilde{U}_a^{\alpha} = \left(\frac{\cos\theta - v_z}{\sin\theta}\,J_a(ax), i\epsilon v_\perp J_a'(ax)\right), \qquad x = \frac{v_\perp\sin\theta}{1 - v_z\cos\theta}, \tag{4.2.3}$$

with $\epsilon = 1$ for electrons, and where the resonance condition, $(ku)_\parallel - a\Omega_0 = 0$, is used.

Emission at the ath harmonic is completely polarized. The polarization vector corresponds to an elliptical polarization with axial ratio

$$T = \epsilon \frac{\cos\theta - v_z}{v_\perp \sin\theta} \frac{J_a(ax)}{J'_a(ax)}, \tag{4.2.4}$$

relative to the t-direction. For nonrelativistic particles, the Bessel functions are approximated by the leading term in their power series expansion, and (4.2.4) gives $T \approx \epsilon(\cos\theta - v_z)/(1 - v_z \cos\theta) \approx \epsilon \cos\theta$. Emission by an electron ($\epsilon = 1$) has a right hand ($T > 0$) circular component for $\cos\theta > 0$. Emission by a highly relativistic particle is dominated by a broad range of high harmonics, and although the emission at each harmonic is completely polarized, the resulting sum over harmonics leads to partial polarization. The polarization in this case needs to be determined in other ways, rather than using (4.2.4).

Power in Gyromagnetic Emission in Vacuo

Starting from the probability (4.2.1) for emission in vacuo it is possible to evaluated several quantities explicitly by performing the sum over a and integral over $\boldsymbol{k}$-space. Consider the power radiated, written as a polarization tensor:

$$P^{\alpha\beta} = \sum_{a=1}^{\infty} \int \frac{d^3\boldsymbol{k}}{(2\pi)^3} \, \omega \, w^{\alpha\beta}(a, \boldsymbol{k}, \boldsymbol{p}). \tag{4.2.5}$$

The total power radiated is $P = P^{tt} + P^{aa}$, and the power radiated in either linear polarization is P^{tt}, P^{aa}. One may evaluate the integral in (4.2.5) by introducing polar coordinates: the integral over azimuthal angle, ϕ, is trivial, and the integral over $|\boldsymbol{k}| = \omega$ is performed over the δ-function. One finds

$$P^{\alpha\beta} = \sum_{a=1}^{\infty} \frac{q^2 a^2 \Omega_0^2}{4\pi\varepsilon_0 \gamma^2} \int_{-1}^{1} d\cos\theta \, \frac{\tilde{U}_a^{*\alpha} \tilde{U}_a^{\beta}}{(1 - v_z \cos\theta)^3}. \tag{4.2.6}$$

The P^{tt}, P^{aa} terms can be evaluated exactly, and only these components are considered in the remainder of the calculation. It is possible to perform the calculation either by integrating over $\cos\theta$ and then performing the sum over a, or by carrying out these steps in the opposite order. Both procedures are summarized below.

The $\cos\theta$-integral in (4.2.6) is performed after making a Lorentz transformation to the frame in which the gyrocenter is at rest. The power radiated is an invariant, and so is unchanged by this transformation. Let the particle have speed v in the laboratory frame and speed v' in this rest frame. Let the emitted wave be at an angle θ' in this rest frame. One has

$$\cos\theta' = \frac{\cos\theta - v_z}{1 - v_z \cos\theta}, \qquad v' \sin\theta' = \frac{v_\perp \sin\theta}{1 - v_z \cos\theta}, \qquad v' = \frac{v_\perp}{(1 - v_z^2)^{1/2}}. \tag{4.2.7}$$

One has $d\cos\theta/(1-v_z\cos\theta)^2 = d\cos\theta'/(1-v_z^2)$, and $1/(1-v_z\cos\theta) = (1+v_z\cos\theta')/(1-v_z^2)$. The remainder of the integrands for P^{tt}, P^{aa} are even functions of $\cos\theta'$, so that the term $v_z\cos\theta'$ does not contribute to them. One requires the following integral identities:

$$\int_{-1}^{1} d\cos\theta' \begin{pmatrix} J_t^2(av'\sin\theta') \\ J_t^2(av'\sin\theta')/\sin^2\theta' \end{pmatrix} = 2\int_0^{v'} dy \begin{pmatrix} J_{2t}(2ay)/v' \\ J_{2t}(2ay)/y \end{pmatrix}. \tag{4.2.8}$$

Using the recursion relations (2.1.25) and (2.1.26) for the Bessel functions, one obtains

$$P^{tt,aa} = \sum_{a=1} \frac{2q^2a^2\Omega_0^2}{4\pi\varepsilon_0} v'^2(1-v'^2)\, G_a^{tt,aa},$$

$$G_a^{tt} = \frac{1}{v'^2}\int_0^{v'} dy\, \frac{J_{2a}(2ay)}{y} - \frac{1}{v'^3}\int_0^{v'} dy\, J_{2a}(2ay),$$

$$G_a^{aa} = \frac{1}{av'} J_{2a}'(2av') - \frac{1}{v'^2}\int_0^{v'} \frac{dy}{y} J_{2a}(2ay) + \frac{1}{v'}\int_0^{v'} dy\, J_{2a}(2ay). \tag{4.2.9}$$

The next step is to perform the sum over a. The relevant sums are Kapteyn series, and the following series were first evaluated explicitly for the present purpose by Schott [9]:

$$\sum_{a=1} a^2 J_{2a}(2ay) = \frac{y^2(1+y^2)}{2(1-y^2)^4}, \qquad \sum_{a=1} a J_{2a}'(2ay) = \frac{y}{2(1-y^2)^2}, \tag{4.2.10}$$

which apply for $0 \le y < 1$. The remaining integral is elementary, and the final result for the power radiated is

$$P = P^{tt} + P^{aa} = \frac{q^2\Omega_0^2 p_\perp^2}{6\pi\varepsilon_0 m^2} = \frac{q^2\Omega_0^2\gamma^2 v_\perp^2}{6\pi\varepsilon_0}, \tag{4.2.11}$$

$$\frac{P^{tt} - P^{aa}}{P^{tt} + P^{aa}} = -\frac{2 + v_\perp^2 - 2v_z^2}{4(1-v_z^2)} = -\frac{2m^2 + 3p_\perp^2}{4(m^2 + p_\perp^2)}. \tag{4.2.12}$$

The total power P is well known, and can be derived from a generalization of the Larmor formula; it depends only on the perpendicular component, $p_\perp$, of momentum, and is independent of p_z.

Formula (4.2.12) shows that the power in gyromagnetic emission is preferentially polarized along $\boldsymbol{a}$ rather than along $\boldsymbol{t}$. In the nonrelativistic limit (4.2.12) implies a degree of polarization $\approx -1/2$, but this is not particularly meaningful; it refers to an average over all angles and according to (4.2.4) the polarization depends relatively strongly on angle in the nonrelativistic limit. In the highly relativistic limit, the degree of polarization implied by (4.2.12) is $\approx -3/4$. This is a characteristic value for the polarization of synchrotron emission; the actual polarization of synchrotron emission is a function of frequency.

Angular Distribution of Gyromagnetic Emission

An alternative procedure involves performing the sum over a in (4.2.6) before carrying out the integral over θ. The required summation formulae were also evaluated by Schott [9] for this purpose:

$$\sum_{a=1} a^2 J_a^2(ax) = \frac{x^2(4+x^2)}{16(1-x^2)^{7/2}}, \qquad \sum_{a=1} a^2 J_a'^2(ax) = \frac{4+3x^2}{16(1-x^2)^{5/2}}, \tag{4.2.13}$$

for $0 \le x < 1$. One finds

$$P = \frac{q^2\Omega_0^2 v_\perp^2}{64\pi\varepsilon_0\gamma^2} \int_{-1}^{1} d\cos\theta \left\{ \frac{(\cos\theta - v_z)^2[4(1-v_z\cos\theta)^2 + v_\perp^2\sin^2\theta]}{[(1-v_z\cos\theta)^2 - v_\perp^2\sin^2\theta]^{7/2}} + \frac{4(1-v_z\cos\theta)^2 + 3v_\perp^2\sin^2\theta}{[(1-v_z\cos\theta)^2 - v_\perp^2\sin^2\theta]^{5/2}} \right\}. \tag{4.2.14}$$

The integral over $\cos\theta$ is performed using standard integral identities. The result (4.2.11) and (4.2.12) is reproduced.

4.2.2 Radiation Reaction to Gyromagnetic Emission

Gyromagnetic emission in vacuo is one specific case where quasilinear theory can be compared with the radiation reaction force. The quasilinear coefficients may be used to determine the mean rates of change of $p_\perp$, p_z, ε, α for a particle due to its gyromagnetic emission, and these rates can be compared with analogous results derived from the radiation reaction force.

Quasilinear Coefficients

Using the techniques outlined above, it is possible to evaluate the quasilinear coefficients $A_\perp(p)$, $A_\parallel^\mu(p)$, as given by (4.1.25), for gyromagnetic emission in vacuo. The coefficients (4.5.7) are

$$\begin{pmatrix} A_\perp(p) \\ A_\parallel^0(p) \\ A_\parallel^3(p) \end{pmatrix} = -\sum_{a=1}^{\infty} \int \frac{d^3\boldsymbol{k}}{(2\pi)^3} \begin{pmatrix} a\Omega_0/\gamma v_\perp \\ \omega \\ \omega\cos\theta \end{pmatrix} \gamma w^{\alpha\beta}(a, \boldsymbol{k}, \boldsymbol{p}). \tag{4.2.15}$$

The evaluation of these coefficients closely parallels the evaluation of P. One finds

$$A_\perp(p) = -\frac{1 - v_z^2}{v_\perp^2}\gamma P, \qquad A_\parallel^0(p) = -\gamma P, \qquad A_\parallel^3(p) = -\gamma v_z P, \tag{4.2.16}$$

with P given by (4.2.11).

The mean rate of change of the 4-momentum of the particle per unit time follows from (4.2.16), which implies

$$\left\langle \frac{dp_\perp}{dt} \right\rangle = -\frac{1 - v_z^2}{v_\perp} P, \qquad \left\langle \frac{d\varepsilon}{dt} \right\rangle = -P, \qquad \left\langle \frac{dp_z}{dt} \right\rangle = -\gamma v_z P. \tag{4.2.17}$$

For a nonrelativistic particle, $v_\perp, |v_z| \ll 1$, (4.2.17) implies that the rate of change of p_z is small in comparison with the rate of change of $p_\perp$, with $v_\perp \langle dp_\perp/dt \rangle \approx \langle d\varepsilon/dt \rangle$. However, for a highly relativistic particle, $v \to 1$, (4.2.17) implies $\langle dp_\perp/dt \rangle \approx -\sin\alpha\, P$, $\langle dp_z/dt \rangle \approx -\cos\alpha\, P$, where α is the pitch angle. In this case the rate of change in the pitch angle is small, such that the particle loses energy essentially at constant α. This is seen directly by comparing the rates of change of p and α:

$$\left\langle \frac{dp}{dt} \right\rangle = -\frac{P}{v}, \qquad \left\langle \frac{d\alpha}{dt} \right\rangle = -\frac{P}{m\gamma^3 v^2} \cot\alpha, \tag{4.2.18}$$

with $P \propto \sin^2\alpha$ given by (4.2.14). Comparing the rates of change of p and α, (4.2.18) implies that they are in the ratio $p^{-1}\langle dp/dt \rangle : \langle d\alpha/dt \rangle = \tan\alpha : 1/\gamma^2 v$, so that the change in α is small for $\gamma^2 \gg 1$.

A distribution of particles that is initially isotropic becomes anisotropic as a result of gyromagnetic losses. For highly relativistic particles, each particle moves along a nearly radial line in $\boldsymbol{p}$-space, that is, with $|\boldsymbol{p}|$ decreasing at nearly constant α. The anisotropy arises because $P \propto \sin^2\alpha$ implies that particles with larger α move to smaller $|\boldsymbol{p}|$ faster than those with smaller α. Particles that lie initially on a circle in $p_\perp$-p_z space, with $p_\perp^2 + p_z^2 = p_0^2$ initially say, define an ellipse with semi-major axis along the p_z-axis fixed at p_0, and with the axial ratio of the ellipse decreasing with time. For nonrelativistic particles the trajectory is along a line that is radial in the cylindrical sense: decreasing $p_\perp$ at constant p_z.

The diffusion coefficients in the quasilinear equation (4.1.24) cannot be evaluated in the same way even for emission in vacuo. The reason is that these terms depend on the distribution of waves, e.g., described by the occupation number $N(\omega, \theta)$. After performing the integral over ω using the δ-function, as in the derivation of (4.2.6), one has $\omega \to a\Omega_0/\gamma(1 - v_z\cos\theta)$, so that the occupation number becomes an implicit function of both a and θ, precluding the sum over a being performed exactly and precluding the integral over $\cos\theta$ being performed exactly even for an isotropic distribution of radiation.

Generalized Larmor Formula

The power radiated in vacuo by an accelerated charge can be treated by combining the Larmor formula with a Lorentz transformation. This approach leads to a general formula for the power radiated in vacuo, but it does not give any information on how the radiation is distributed in frequency and angle, nor on its polarization.

The Larmor formula gives the power radiated in vacuo by an accelerated charge in its instantaneous rest frame:

$$P(t) = \frac{q^2|\mathbf{a}(t)|^2}{6\pi\varepsilon_0}, \tag{4.2.19}$$

where $\boldsymbol{a}(t)$ is the instantaneous acceleration. The acceleration in (4.2.19) is related to the force acting on the particle by $\boldsymbol{a}(t) = \boldsymbol{F}(t)/m$. In the instantaneous rest frame one has $\boldsymbol{v} = 0$, $\gamma = 1$, and $\boldsymbol{a}(t) = \dot{\boldsymbol{v}}(t)$, where the dot implies differentiation with respect to t. In terms of 4-vectors, in the rest frame one has $u^\mu = [1, \mathbf{0}]$, $a^\mu = du^\mu/d\tau = [0, \dot{\boldsymbol{v}}]$, with $d\tau = dt$ in this frame. Using $a^\mu a_\mu = -\dot{v}^2$, (4.2.19) becomes

$$P = -\frac{q^2 a^\mu a_\mu}{6\pi\varepsilon_0}, \tag{4.2.20}$$

in the rest frame. The left hand side of (4.2.20) is the ratio of the time components of two 4-vectors, and hence is an invariant, and the right hand side is already in invariant form. Hence the special theory of relativity implies that (4.2.20) applies in all inertial frames. It is therefore the desired generalization of the Larmor formula.

In the laboratory frame, in which the particle has instantaneous velocity $\boldsymbol{v}$, (4.2.20) is rewritten in 3-vector notation by noting the relations

$$a^\mu = \gamma^4[\boldsymbol{v}\cdot\dot{\boldsymbol{v}}, \boldsymbol{v}\cdot\dot{\boldsymbol{v}}\,\boldsymbol{v} + (1-v^2)\dot{\boldsymbol{v}}], \qquad a^\mu a_\mu = -\gamma^6\big[\dot{\boldsymbol{v}}^2 - |\boldsymbol{v}\times\dot{\boldsymbol{v}}|^2\big]. \tag{4.2.21}$$

The generalization of the Larmor formula is

$$P = \frac{q^2}{6\pi\varepsilon_0}\,\frac{\dot{\boldsymbol{v}}^2 - |\boldsymbol{v}\times\dot{\boldsymbol{v}}|^2}{(1-v^2)^3}. \tag{4.2.22}$$

The acceleration is a somewhat artificial quantity for a relativistic particle and it is more relevant to write (4.2.22) in terms of the 4-force, $dp^\mu/d\tau$, acting on the particle. One has $m^2|\boldsymbol{a}|^2 = -(dp^\mu/d\tau)(dp_\mu/d\tau)$, and hence the Larmor formula has the covariant generalization

$$P = -\frac{q^2}{6\pi\varepsilon_0 m^2}\,\frac{dp^\mu}{d\tau}\,\frac{dp_\mu}{d\tau}, \tag{4.2.23}$$

which applies in an arbitrary frame.

The power in gyromagnetic radiation follows by substituting the equation of motion, $dp^\mu/d\tau = qF_0^{\mu\nu}u_\nu$, into (4.2.23). Writing $F_0^{\mu\nu} = Bf^{\mu\nu}$, the right hand side of (4.2.23) contains a factor

$$f^{\mu\alpha}u_\alpha f_\mu{}^\beta u_\beta = g_\perp^{\alpha\beta}u_\alpha u_\beta = -u_\perp^2 = -p_\perp^2/m^2.$$

Then (4.2.23) reproduces the result (4.2.11) for the power radiated, P.

Radiation Reaction 4-Force

The radiation reaction force in conventional classical electrodynamics is

$$\boldsymbol{F}_{\rm react}(t) = \frac{q^2\ddot{\boldsymbol{v}}(t)}{6\pi\varepsilon_0}, \tag{4.2.24}$$

where a dot denotes differentiation with respect to t. The covariant generalization to the radiation reaction 4-force was written down by Dirac [2]:

$$\mathcal{F}^\mu_{\rm react} = \frac{q^2}{6\pi\varepsilon_0 m}\left[\frac{d^2p^\mu}{d\tau^2} + \frac{dp^\nu}{d\tau}\frac{dp_\nu}{d\tau}\frac{p^\mu}{m^2}\right]. \tag{4.2.25}$$

On inserting the equation of motion $dp^\mu/d\tau = qF_0^{\mu\nu}u_\nu$, and proceeding as in the derivation of (4.2.23), one finds

$$\mathcal{F}^\mu_{\rm react} = -\frac{q^4B^2}{6\pi\varepsilon_0 m^3}\frac{m^2p_\perp^\mu + p_\perp^2 p^\mu}{m^2} = -P\,\frac{m^2p_\perp^\mu + p_\perp^2 p^\mu}{mp_\perp^2}, \tag{4.2.26}$$

which is the radiation reaction 4-force for gyromagnetic emission in vacuo. Comparison of (4.2.26) and (4.2.16) shows that the parallel components of $\mathcal{F}^\mu_{\rm react}$ and $A_\parallel^\mu$ are equal. The perpendicular components $\mathcal{F}^\mu_{\rm react}$ reproduce $A_\perp^\mu$ provided that one averages (4.2.26) over gyrophase, and notes the identity $(m^2 + p_\perp^2)/p_\perp^2 = (1-v_z^2)/v_\perp^2$. This confirms that in the case of gyromagnetic emission the terms in the quasilinear equation (4.1.24) that describe the effects of spontaneous emission are equivalent to radiation reaction 4-force in conventional classical electrodynamics.

4.3 Cyclotron Emission

Cyclotron emission is gyromagnetic emission from nonrelativistic particles. In this section, cyclotron emission and absorption in a nonrelativistic plasma are discussed.

4.3.1 Emissivity in a Magnetoionic Mode

The only radiation that can escape from a nonrelativistic magnetized plasma is in the magnetoionic modes, specifically, either in the x mode or the o mode. The relevant wave properties are summarized in § 3.3.1. The polarization vector, (3.1.19), and the ratio of electric energy are

$$e_\sigma^\mu = \frac{L_\sigma \kappa^\mu + T_\sigma t^\mu + i a^\mu}{(L_\sigma^2 + T_\sigma^2 + 1)^{1/2}}, \qquad R_\sigma = \frac{1 + L_\sigma^2 + T_\sigma^2}{2(1 + T_\sigma^2) n_\sigma \, \partial(\omega n_\sigma)/\partial\omega}, \tag{4.3.1}$$

with σ labeling the magnetoionic waves, the high frequency branches of which are labeled o, x. It is conventional to introduce the emissivity, $\eta_\sigma(a, \omega, \theta)$, to describe the emission by a single particle at the ath harmonic in the mode σ. The emissivity is the power radiated per unit frequency and per unit solid angle and hence it is related to the probability for gyromagnetic emission by

$$2\pi \int_0^\infty d\omega \int_{-1}^1 d\cos\theta \, \eta_\sigma(a, \omega, \theta) = \int \frac{d^3\boldsymbol{k}}{(2\pi)^3} \, \omega \, w_\sigma(a, \boldsymbol{k}, \boldsymbol{p}), \tag{4.3.2}$$

with $w_\sigma(a, \boldsymbol{k}, \boldsymbol{p})$ given by (4.1.9). A factor $\partial[\omega n_\sigma]/\partial\omega$ from the change of variable of integration from $|\boldsymbol{k}| = n_\sigma \omega$ to ω in (4.3.2) cancels with this factor in (4.3.1). The emissivity at the ath harmonic reduces to

$$\begin{aligned} \eta_\sigma(a, \omega, \theta) = \frac{q^2 n_\sigma \omega^2 v^2 \sin^2\alpha}{8\pi^2 \varepsilon_0 (1 + T_\sigma^2)} & \left| \frac{L_\sigma \sin\theta + T_\sigma(\cos\theta - n_\sigma v \cos\alpha)}{n_\sigma v \sin\alpha \sin\theta} J_a + \epsilon J_a' \right|^2 \\ & \times \delta[\omega(1 - n_\sigma v \cos\theta \cos\alpha) - a\Omega_0], \end{aligned} \tag{4.3.3}$$

where the arguments (ω, θ) of n_σ, T_σ and L_σ are omitted, and similarly the argument $(\omega/\Omega_0) n_\sigma v \sin\alpha \sin\theta$ of J_a and J_a' is omitted. The sign ϵ is equal to $+1$ for electrons, $\Omega_0 \to \Omega_e$.

A form of the transfer equation in terms of these variables is

$$v_{g\sigma} \frac{dI_\sigma(\omega, \theta)}{d\ell} = J_\sigma(\omega, \theta) - \gamma_\sigma(\omega, \theta) I_\sigma(\omega, \theta), \tag{4.3.4}$$

where I_σ is the specific intensity (power per unit solid angle per unit frequency per unit area) in the mode, and where ℓ denotes distance along the ray path. The coefficient $J_\sigma(\omega, \theta)$ is the volume emissivity, and both it and the absorption coefficient (4.1.21) can be expressed in terms of the emissivity. In ordinary units the volume emissivity becomes

$$J_\sigma(\omega, \theta) = \sum_a J_\sigma(a, \omega, \theta),$$

$$J_\sigma(a, \omega, \theta) = 2\pi \int_0^\infty d|\boldsymbol{p}| \, |\boldsymbol{p}|^2 \int_{-1}^1 d\cos\alpha \, \eta_\sigma(a, \omega, \theta) \frac{f(|\boldsymbol{p}|, \cos\alpha)}{(2\pi\hbar)^3}, \tag{4.3.5}$$

and absorption coefficient becomes

$$\gamma_\sigma(\omega,\theta) = \sum_a \gamma_\sigma(a,\omega,\theta),$$

$$\gamma_\sigma(a,\omega,\theta) = -2\pi \int_0^\infty d|\boldsymbol{p}|\,|\boldsymbol{p}|^2 \int_{-1}^1 d\cos\alpha \sum_a \frac{(2\pi c)^3 \eta_\sigma(a,\omega,\theta)}{v n_\sigma^2 \omega^2 \partial(\omega n_\sigma)/\partial\omega}$$
$$\times \left(\frac{\partial}{\partial|\boldsymbol{p}|} + \frac{\cos\alpha - n_\sigma v\cos\theta}{|\boldsymbol{p}|\sin\alpha}\frac{\partial}{\partial\alpha}\right)\frac{f(|\boldsymbol{p}|,\cos\alpha)}{(2\pi\hbar)^3}. \tag{4.3.6}$$

The factor $(2\pi\hbar)^3$ is included in (4.3.5) and (4.3.6) to be consistent with the normalization used in quantum statistical mechanics, where the integral over the 6-dimensional phase space involves $d^3\boldsymbol{x}d^3\boldsymbol{p}/(2\pi\hbar)^3$. (For electrons $f(\boldsymbol{p})$ is equal to $2n(\boldsymbol{p})$, where $n(\boldsymbol{p})$ is the electron occupation number and the factor 2 is from the sum over spin states.) The factor $(2\pi\hbar)^3$ is omitted in classical statistical mechanics, where the integral over phase space involves $d^3\boldsymbol{x}d^3\boldsymbol{p}$.

4.3.2 Gyromagnetic Emission by Thermal Particles

In the strictly nonrelativistic limit, cyclotron emission by nonthermal electrons occurs at lines centered on the cyclotron frequency and its harmonics, with the intensity of the lines decreasingly rapidly with increasing harmonic number a. These lines have a finite width as a result of Doppler broadening due to the random thermal motions of the electrons along $\boldsymbol{B}$.

A thermal distribution function of particles in the strictly nonrelativistic limit ($\gamma = 1$, $\boldsymbol{p} = m\boldsymbol{v}$) is the Maxwellian distribution (ordinary units with $\rho = mc^2/T$, and T the temperature in energy units)

$$\frac{f(\boldsymbol{p})}{(2\pi\hbar)^3} = \frac{n\rho^{3/2}}{(2\pi)^{3/2}(mc)^3}\, e^{-\rho v^2/2c^2}, \tag{4.3.7}$$

where the factor $(2\pi\hbar)^3$ is included for the same reason as in (4.3.5) and (4.3.6). For a thermal distribution the volume emissivity and the absorption coefficient are proportional to each other (Kirchhoff's law), with the specific relation being (ordinary units)

$$\gamma_\sigma(\omega,\theta) = \frac{(2\pi)^3 \rho c}{m\omega^2 n_\sigma^2 \partial(\omega n_\sigma)/\partial\omega}\, J_\sigma(\omega,\theta). \tag{4.3.8}$$

In the following the volume emissivity at the ath harmonic, $J_\sigma(a,\omega,\theta)$, is calculated, and the absorption coefficient follows from (4.3.8).

The integrals in (4.3.5) over velocity space for the distribution (4.3.7) separates in cylindrical coordinates. This follows by writing (4.3.7) in the form

$\exp(-\rho v^2/2) = \exp(-\rho v_\perp^2/2)\exp(-\rho v_z^2/2)$ and performing the $v_\perp$-integral over the ordinary Bessel functions using

$$\int_0^\infty dv_\perp v_\perp e^{-\frac{1}{2}\rho v_\perp^2}\begin{bmatrix} J_a^2(z) \\ zJ_a(z)J_a'(z) \\ z^2 J_a'^2(z) \end{bmatrix} = e^{-\lambda}\begin{bmatrix} I_a(\lambda) \\ \lambda[I_s'(\lambda) - I_a(\lambda)] \\ a^2 I_a(\lambda) - 2\lambda^2[I_a'(\lambda) - I_a(\lambda)] \end{bmatrix}, \quad (4.3.9)$$

with $z = k_\perp v_\perp/\Omega_0$ and where the argument of the modified Bessel functions, I_a, is $\lambda_\sigma = (\omega n_\sigma \sin\theta V/\Omega_0)^2$, with $V = 1/\rho^{1/2}$ the thermal speed. The resulting expression for the emission coefficient is

$$J_\sigma(a,\omega,\theta) = \left(\frac{2}{\pi}\right)^{1/2}\frac{\omega_p^2 m}{\omega c}\frac{n_\sigma A_\sigma(a,\omega,\theta)}{\rho^{1/2}|\cos\theta|}e^{-\rho(\omega-a\Omega_0)^2/2\omega^2 n_\sigma^2\cos^2\theta}, \quad (4.3.10)$$

$$\begin{aligned} A_\sigma(a,\omega,\theta) = \frac{e^{-\lambda_\sigma}}{1+T_\sigma^2}\Bigg\{&\left[\frac{\omega}{\Omega_0}(L_\sigma\cos\theta - T_\sigma\sin\theta)\tan\theta + aT_\sigma\sec\theta\right]^2\frac{I_a}{\lambda_\sigma} \\ &+ 2\left[\frac{\omega}{\Omega_0}(L_\sigma\cos\theta - T_\sigma\sin\theta)\tan\theta + aT_\sigma\sec\theta\right](I_a' - I_a) \\ &+ \epsilon\left[\left(\frac{a^2}{\lambda_\sigma} + 2\lambda_\sigma\right)I_a - 2\lambda_M I_s'\right]\Bigg\}. \end{aligned} \quad (4.3.11)$$

The general form (4.3.11) is unnecessarily cumbersome for most purposes and approximations need to be made.

One simplifying assumption is the small gyroradius limit, $\lambda_\sigma \ll 1$. This corresponds to the argument of the modified Bessel functions being small, when only the leading terms in the power series expansion of the modified Bessel functions are retained. In the small-gyroradius limit, (4.3.11) reduces to

$$A_\sigma(a,\omega,\theta) \approx \frac{(\frac{1}{2}\lambda_\sigma)^{a-1}}{2a!(1+T_\sigma^2)}\left[\frac{\omega}{\Omega_e}(L_\sigma\cos\theta - T_\sigma\sin\theta)\tan\theta + aT_\sigma\sec\theta + a\epsilon\right]^2. \quad (4.3.12)$$

Further simplification occurs when (1) the longitudinal component can be neglected ($L_\sigma \to 0$), and (2) by setting $\omega = a\Omega_0$, in which case the quantity in square brackets reduces to $a(T_\sigma\cos\theta - \epsilon)$, with $\epsilon = 1$ for electrons.

The exponential function in (4.3.10) determines the line profile. The line emission at the ath harmonic is centered on $\omega = a\Omega_0$ and it is Doppler broadened with a characteristic width

$$(\Delta\omega)_a = \frac{n_\sigma|\cos\theta|}{\rho^{1/2}}a\Omega_0. \quad (4.3.13)$$

The Doppler broadening is due to the component of the random thermal motions along the magnetic field lines. In the strictly nonrelativistic limit the perpendicular motions do not contribute to the line width, whereas relativistic effects imply a nonzero broadening due to the so-called transverse Doppler effect.

4.3.3 *Semirelativistic Approximation*

The strictly nonrelativistic limit breaks down near perpendicular propagation even for $v \ll 1$. This may be seen by noting that in the nonrelativistic approximation, $\gamma = 1$, the resonance condition (4.1.11) reduces to a vertical line in velocity space, and for perpendicular propagation the resonance condition reduces to a semicircle centered on the origin, cf. Fig. 4.1.

The simplest useful approximation to the resonance condition that allows one to treat perpendicular propagation is the semirelativistic approximation $\gamma = 1 + \frac{1}{2}v^2$. This leads to the Doppler condition being approximated by

$$\omega(1 - n_\sigma v_z \cos\theta) - a\Omega_e\left(1 - \tfrac{1}{2}v_\perp^2 - \tfrac{1}{2}v_z^2\right) = 0. \qquad (4.3.14)$$

It follows from (4.3.14) that in the semirelativistic approximation, a resonance ellipse is approximated by a circle with its center at $v_z = v_c$, $v_\perp = 0$, and radius v_0, given by

$$v_c = n_\sigma \cos\theta, \qquad v_0^2 = n_\sigma^2 \cos^2\theta + 2(a\Omega_e - \omega)/a\Omega_e. \qquad (4.3.15)$$

Relativistic Frequency Downshift and Line Broadening

Two intrinsically relativistic effects on cyclotron emission are (1) a frequency downshift and (2) a line broadening due to the transverse Doppler effect. Both of these are treated here assuming perpendicular propagation.

The line width may be characterized by the spread in frequency around the mean frequency. This involves moments of the frequency distribution, with the Nth moment defined by

$$\langle\omega^N(a,\theta)\rangle_\sigma = \int_0^\infty d\omega\,\omega^{N-1} J_\sigma(a,\omega,\theta) \Big/ \int_0^\infty d\omega\,\omega^{-1} J_\sigma(a,\omega,\theta). \quad (4.3.16)$$

The mean frequency is given by the moment $N = 1$, and it determines the line center. The second moment, $N = 2$, determines the variance in the frequency: the bandwidth may be characterized by the square root of the variance, $\Delta\omega = (\langle\omega^2\rangle - \langle\omega\rangle^2)^{1/2}$. The moments (4.3.16) may be evaluated approximately for emission at $\theta = \pi/2$ using the method of steepest descent.

The method of steepest descent applies to an integral of the form

$$I = \int dy\, G(y) \exp[-F(y)], \tag{4.3.17}$$

where $G(y)$ is slowly varying and $\exp[-F(y)]$ is sharply peaked about some value $y = y_0$. The range of integration is assumed to cover the important contribution from this peak. The value of y_0 is determined by the solution of $F'(y_0) = 0$, where the prime denotes differentiation. The integral is approximated by

$$I = \left(\frac{2\pi}{F''(y_0)}\right)^{1/2} G(y_0) \exp[-F(y_0)]. \tag{4.3.18}$$

The specific integrals of interest here are written as integrals over $\alpha^2 = 2(a\Omega_e - \omega)/a\Omega_e$, implying $F(\alpha^2) = \rho\alpha^2/2 - a\ln(\alpha^2)$ for the x mode and $F(\alpha^2) = \rho\alpha^2/2 - (a+1)\ln(a^2)$ for the o mode. One finds

$$\langle\omega(a,\pi/2)\rangle = a\Omega_e \left[1 - \frac{(a + \frac{1}{2} \pm \frac{1}{2})}{\rho}\right], \tag{4.3.19}$$

for these two modes, respectively. The frequency downshift of the center of the cyclotron line, by fraction $\approx a/\rho$, may be attributed to the relativistic decrease in the gyrofrequency, Ω_e/γ.

There is a relativistic broadening, described by

$$[\Delta\omega(a,\pi/2)]^2 = (a\Omega_e)^2 \frac{a + 1 \pm \frac{1}{2}}{\rho^2}, \tag{4.3.20}$$

that may be attributed to the transverse Doppler effect. The total line broadening for $\theta \neq \pi/2$ is

$$[\Delta\omega(a,\theta)]^2 = (a\Omega_e)^2 \left[\frac{\cos^2\theta}{\rho} + \frac{a + \frac{3}{2} \pm \frac{1}{2}}{\rho^2}\right]. \tag{4.3.21}$$

In a nonrelativistic plasma, $\rho \gg 1$, the relativistic broadening is significant only for angles $|\theta - \pi/2| \lesssim a^{1/2}/\rho$.

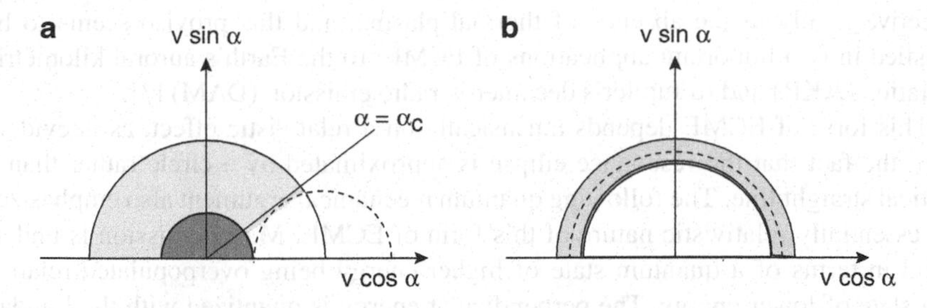

Fig. 4.2 (a) A nonthermal loss-cone distribution, with loss cone angle α_0, is assumed to be confined to the *lightly shaded* region, with the lower-energy thermal distribution filling the *darkly shaded* region. The *dotted semicircle* indicates a resonance ellipse that samples only regions where $\partial f/\partial p_\perp$ is positive. (b) A shell distribution with a resonance ellipse inside the shell in the region where f is an increasing function of v

4.3.4 Electron Cyclotron Maser Emission

Electron cyclotron absorption can be negative under relatively mild conditions in a nonrelativistic plasma. This results in electron cyclotron maser emission (ECME) near the fundamental or low harmonics of the cyclotron frequency. An important type of ECME is due to the intrinsically relativistic effect [6, 12, 14] and this is treated using the semirelativistic approximation (4.3.14).

Consider the absorption coefficient (4.1.21) in the semirelativistic approximation. The integral over momentum space in (4.1.21) reduces to an integral around the resonance ellipse. The sign of the absorption coefficient is determined by a weighted average of the value of $-\hat{D}f$ around the resonance ellipse, with $\hat{D}$ given by (4.1.21). Specifically, one has

$$-\hat{D}f(p_\perp, p_z) = -\left(\frac{a\Omega_e}{v_\perp}\frac{\partial}{\partial p_\perp} + k_z\frac{\partial}{\partial p_z}\right)f(p_\perp, p_z). \tag{4.3.22}$$

For $|k_z| \ll \omega$ the contribution from the p_z-derivative is small compared with the contribution from the $p_\perp$-derivative in (4.3.22). It follows that if the resonance ellipse for a wave with given k_z, ω, a is such that $\partial f/\partial p_\perp$ is positive everywhere around the ellipse, then the absorption coefficient is necessarily negative. Two examples of such an ellipse are illustrated schematically in Fig. 4.2, (a) for a loss-cone distribution, that is, a distribution in which the number of particles decreases inside a loss cone $\alpha < \alpha_c$, and (b) for a shell distribution, where the particles are confined to a shell in velocity space. The resonance ellipse in Fig. 4.2a is chosen not to intersect the region where the thermal electrons are located; waves corresponding to resonance ellipses that intersect this region suffer strong thermal gyromagnetic absorption. The resonance ellipse in Fig. 4.2b is for perpendicular propagation; in a magnetoionic medium such waves are below the cutoff frequency for the x mode and so cannot escape. ECME due to a shell distribution leads to escaping radiation effectively only in the absence of thermal plasma, and this proviso seems to be satisfied in two important applications of ECME: to the Earth's auroral kilometric radiation (AKR) and to Jupiter's decametric radio emission (DAM) [3].

This form of ECME depends intrinsically on a relativistic effect, as is evident from the fact that the resonance ellipse is approximated by a circle rather than a vertical straight line. The following quantum mechanical argument also emphasizes the essentially relativistic nature of this form of ECME. Maser emission is understood in terms of a quantum state of higher energy being overpopulated relative to a state of lower energy. The perpendicular energy is quantized with the Landau quantum number n introduced in (4.1.10). Let the occupation number be $N(n)$. Stimulated emission $n \to n - a$ is proportional to $N(n)$ and true absorption $n - a \to n$ is proportional to $N(n - a)$, so that the net absorption is proportion to $N(n) - N(n-a)$. (In the classical limit one has $N(n)-N(n-a) = a\partial N(n)/\partial n$ with $\partial/\partial n \to (\Omega_0/v_\perp)\partial/\partial p_\perp$.) This suggests that for $N(n) > N(n - a)$ the absorption is negative. This conclusion is correct in a relativistically correct treatment, but not in a strictly nonrelativistic theory. This follows from the nonrelativistic counterpart of (4.1.10), $\varepsilon_n = m + p_z^2/2m + n\Omega_0$, which implies that the harmonics are equally spaced in the strictly nonrelativistic case. The emission at $\omega = a\Omega_0$ comes from all transitions $n \to n - a$, that is, from all values of n. Hence, one needs to sum over n, to include the effect of transitions $n \leftrightarrow n - a$ for all n. This sum gives

$$\sum_{n=a}^{\infty} N(n) - N(n - a) = N(\infty) - N(0). \tag{4.3.23}$$

One necessarily has $N(\infty) = 0$ and $N(0) \geq 0$, so that the absorption is nonnegative and there can be no maser action.

When the correct relativistic formula (4.1.10) for the energy ε_n is used, it implies that each transition $n \leftrightarrow n - a$ occurs at a slightly different frequency, which is referred to an anharmonicity. Hence maser action between two specific quantum states can be considered in isolation from transitions for all other n. Then $N(n) > N(n-a)$ suffices for the absorption to be negative for the relevant frequency of transition. Thus, quantum mechanically, this form of ECME is attributed to the anharmonicity in cyclotron transitions implied by the relativistically correct form of the energy (4.1.10).

4.4 Synchrotron Emission

Synchrotron emission is gyromagnetic emission by highly relativistic particles. In this section synchrotron emission and absorption are treated as the highly relativistic limit of the treatment of gyromagnetic emission and absorption. The case of a power-law energy distribution of particles is of most interest in practice. The case of a relativistic thermal distribution is also discussed. Ordinary units are used in this section.

4.4.1 Synchrotron Emissivity

The emissivity in the synchrotron case is given by making the Airy-integral approximation to the Bessel functions. In this limit the harmonic number, a, is large and is regarded as a continuous variable, $a = \gamma\omega\sin^2\theta/\Omega_0$. The resulting expression for the average emissivity for synchrotron emission is

$$\bar{\eta}^{\alpha\beta}(\omega,\theta,\gamma) = \frac{\sqrt{3}q^2\Omega_0\xi\sin\theta\phi(\theta)}{64\pi^3\varepsilon_0\gamma c}F^{\alpha\beta}(\omega,\theta,\gamma), \tag{4.4.1}$$

with ξ a modified Lorentz factor. In (4.4.1), α, β are the two transverse components, which can be denoted as t, a or $\parallel, \perp$, repectively, where $\parallel, \perp$ refer to directions in the transverse plane relative to the projection of $\boldsymbol{B}$ onto the plane. The components of $F^{\alpha\beta}$ are

$$\begin{aligned}
F^{tt,aa}(\omega,\theta,\gamma) &= R\int_R^\infty dt\, K_{5/3}(t) \mp R\, K_{2/3}(R),\\
F^{ta}(\omega,\theta,\gamma) &= \frac{2i\epsilon\cot\theta}{3n\xi}\left[(2+g(\theta))\int_R^\infty dt\, K_{1/3}(t) + 2R\, K_{1/3}(R)\right],\\
R &= \frac{2a}{3\xi^3\sin^3\theta} = \frac{\omega}{\omega_c}, \qquad g(\theta) = \frac{\tan\theta\phi'(\theta)}{\phi(\theta)},\\
\omega_c &= \frac{3}{2}\Omega_0\xi^2\sin\theta, \qquad \xi = (1-n^2v^2/c^2)^{-1/2},
\end{aligned} \tag{4.4.2}$$

with $F^{at} = -F^{ta}$, and with $\boldsymbol{t} = (\cos\theta, 0, -\sin\theta)$, $\boldsymbol{a} = (0,1,0)$. The refractive index, n, is assumed very close to unity, but $n \neq 1$ needs to be retained to include the Razin effect; the modified Lorentz factor, ξ reduces to γ for $n \to 1$.

The functional dependence on the parameter R in the total emissivity (summed over the two states of polarization) is described by a function $F(R)$,

$$F(R) = \frac{1}{2}[F^{tt}(R) + F^{aa}(R)] = R\int_R^\infty dt\, K_{5/3}(t). \tag{4.4.3}$$

Limiting cases for $F(R)$ are

$$F(R) = \begin{cases} \dfrac{4\pi}{\sqrt{3}\Gamma(1/3)}\left(\dfrac{R}{2}\right)^{1/3}\left[1 - \dfrac{1}{2}\Gamma(1/3)\left(\dfrac{R}{2}\right)^{2/3} + \cdots\right] & R \ll 1,\\[2ex] \left(\dfrac{\pi R}{2}\right)^{1/2} e^{-R}\left[1 + \dfrac{55}{72R} + \cdots\right] & R \gg 1. \end{cases} \tag{4.4.4}$$

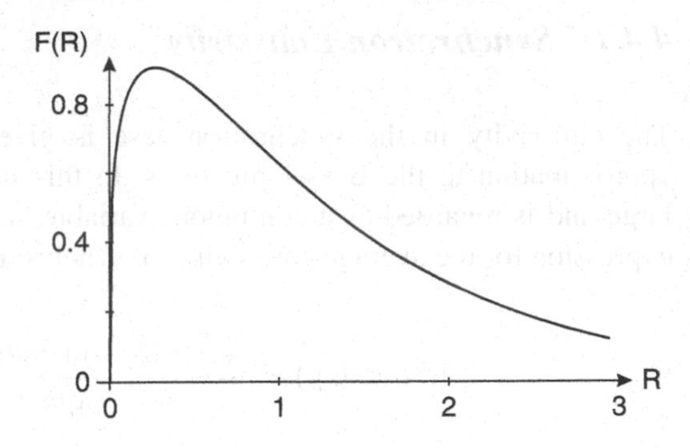

Fig. 4.3 The function $F(R)$ defined by (4.4.3)

In between these limiting cases, there is a maximum at $F(0.29) = 0.92$. The function $F(R)$ is plotted in Fig. 4.3. A simple analytic approximation to it is $F(R) \approx 1.8R^{0.3}e^{-R}$.

Power-Law Distribution

For highly relativistic particles one has $\gamma \gg 1$, with $\varepsilon = \gamma mc^2$, $|\boldsymbol{p}| = \gamma mc$. An isotropic distribution of particles can be described by the energy spectrum, $N(\gamma)$, written in terms of the Lorentz factor, γ. The number density of particles is given by

$$\int \frac{d^3\boldsymbol{p}}{(2\pi\hbar)^3} f(\boldsymbol{p}) = \int d\gamma\, N(\gamma), \tag{4.4.5}$$

which relates the two notations. For an anisotropic distribution, a pitch-angle dependence, $\phi(\alpha)$, is introduced, and the relation between the notations becomes

$$\frac{f(\boldsymbol{p})}{(2\pi\hbar)^3} = \frac{N(\gamma)\phi(\alpha)}{2\pi(mc)^3\gamma^2}. \tag{4.4.6}$$

A power-law energy spectrum for highly relativistic particles is of the form (2.7.28), viz. $N(\gamma) = K\gamma^{-a}$ for $\gamma_1 \leq \gamma \leq \gamma_2$. The energy-integrals in the expression (2.7.1) with (4.4.1) for the synchrotron emission coefficient may be evaluated in closed form for $\gamma_1 = 0$, $\gamma_2 = \infty$ using the standard integral

$$\int_0^\infty dx\, x^\mu K_\nu(ax) = 2^{\mu-1}a^{-\mu-1}\,\Gamma\left(\frac{1+\mu+\nu}{2}\right)\Gamma\left(\frac{1+\mu-\nu}{2}\right), \tag{4.4.7}$$

and identities satisfied by the Γ-function:

$$\Gamma(1+x) = x\Gamma(x), \qquad \Gamma(1-x)\Gamma(x) = \frac{\pi}{\sin(\pi x)}. \tag{4.4.8}$$

One obtains

$$J^{\alpha\beta}(\omega,\theta) = A(\theta)\, j^{\alpha\beta}(a)\left(\frac{2\omega}{3\Omega_0 \sin\theta}\right)^{-(a-1)/2}, \tag{4.4.9}$$

with

$$\begin{aligned}
A(\theta) &= K\,\frac{\sqrt{3}q^2\Omega_0 \sin\theta\,\phi(\theta)}{64\pi^3\varepsilon_0 c},\\
j^{tt}(a) &= \frac{2/3}{a+1}2^{(a-3)/2}\,\Gamma\left(\frac{3a+7}{12}\right)\Gamma\left(\frac{3a-1}{12}\right),\quad j^{aa}(a) = \frac{a+5/3}{2/3}\,j^{tt}(a),\\
j^{ta}(a) &= -2i\epsilon\cot\theta\left(\frac{2\omega}{3\Omega_0\sin\theta}\right)^{-1/2}\frac{2^{(a-2)/2}(a+2+g(\theta))}{3a}\\
&\quad\times\Gamma\left(\frac{3a+8}{12}\right)\Gamma\left(\frac{3a+4}{12}\right),
\end{aligned} \tag{4.4.10}$$

with $j^{at}(a) = -j^{ta}(a)$. The fact that the frequency dependence of the ta-term differs from the tt- and aa-terms by a factor $\approx (\omega/\Omega_0 \sin\theta)^{-1/2}$ may be understood by noting that this factor results from the first order term in an expansion in $1/\gamma$ with $\omega \approx \Omega_0\gamma^2 \sin\theta$.

4.4.2 Synchrotron Absorption

The relation between the synchrotron absorption coefficient and the synchrotron emissivity is of the same form as the relation (4.3.6) for an arbitrary wave mode σ. Rewriting (4.3.6) as a polarization tensor gives an absorption coefficient, $\gamma^{\alpha\beta}(\omega,\theta)$, which can be written as the absorption coefficient per unit length, $\mu^{\alpha\beta}(\omega,\theta) = \gamma^{\alpha\beta}(\omega,\theta)/c$. There are contributions from derivatives of the distribution function with respect to $|\boldsymbol{p}| \to \gamma mc$ and α, with the latter contributing only for an anisotropic distribution.

For an isotropic distribution the synchrotron absorption coefficient, in the form of a polarization tensor, is

$$\mu^{\alpha\beta}(\omega,\theta) = -\frac{(2\pi)^4\phi(\theta)}{\omega^2 m}\int d\gamma\, N(\gamma)\,\eta^{\alpha\beta}(\omega,\theta,\gamma)\,\frac{d\ln[N(\gamma)/\gamma^2]}{d\gamma}, \tag{4.4.11}$$

with the synchrotron emissivity, $\bar{\eta}^{\alpha\beta}(\omega,\theta,\gamma)$, given by (4.4.1). Thus, (4.4.11) becomes

$$\mu^{\alpha\beta}(\omega,\theta) = -\frac{(2\pi)^4}{\omega^2 mc}\int d\gamma\,\gamma^2\,\frac{\sqrt{3}q^2\Omega_0\xi\sin\theta\,\phi(\theta)}{64\pi^3\varepsilon_0\gamma}\,F^{\alpha\beta}(\omega,\theta,\gamma)\,\frac{d[N(\gamma)/\gamma^2]}{d\gamma}, \tag{4.4.12}$$

with $F^{\alpha\beta}(\omega,\theta,\gamma)$ given by (4.4.2). For an anisotropic distribution, the derivative with respect to α gives the derivative of the pitch angle distribution at $\alpha = \theta$, $\phi'(\theta) = d\phi(\theta)/d\theta$. For the tt and aa components, the anisotropy gives a correction of order $1/\gamma$, which can be neglected. For the ta-component it is of the same order in $1/\gamma$ as the ta-terms in (4.4.12), and it needs to be retained. This additional contribution is

$$\mu_{\alpha}^{ta}(\omega,\theta) = \frac{2i\epsilon(2\pi)^4 c\phi'(\theta)}{3\omega^2 mc}\int\frac{d\gamma N(\gamma)}{\gamma}\frac{\sqrt{3}q^2\Omega_0\sin\theta\phi(\theta)}{64\pi^3\varepsilon_0 c}\int_R^{\infty} dt\, K_{1/3}(t), \tag{4.4.13}$$

with $R = \omega/\omega_c$, with ω_c given by (4.4.2).

The absorption coefficient may be evaluated explicitly for the power-law distribution (2.7.28), giving

$$\mu^{\alpha\beta}(\omega,\theta) = \frac{(2\pi)^3}{\omega^2}\frac{K\sqrt{3}q^2\Omega_0\sin\theta\phi(\theta)}{64\pi^3\varepsilon_0 c} j_{\rm ab}^{\alpha\beta}(a)\left(\frac{2\omega}{3\Omega_0\sin\theta}\right)^{-a/2}, \tag{4.4.14}$$

with $j_{\rm ab}^{\alpha\beta}(a) = (a+2)j^{\alpha\beta}(a+1)$ for the tt- and aa-components and with $j_{\rm ab}^{12}(a) = (a+3)\big(a+2+g(\theta)\big)j^{12}(a+1)/\big(a+3+g(\theta)\big)$.

In the case of strong Faraday rotation, as discussed below, the absorption coefficients for the two natural modes are relevant, and these may be constructed from the components of $\mu^{\alpha\beta}$. Let the two modes be labeled $\pm$ with polarization vectors $\boldsymbol{e}_{\pm} = (T_{\pm}\boldsymbol{t} + i\boldsymbol{a})/(T_{\pm}^2+1)^{1/2}$ with $T_+T_- = -1$. The absorption coefficients, $\mu_{\pm}$ for the two modes are related to $\mu^{\alpha\beta}$ by

$$\mu_{\pm}(\omega,\theta) = \frac{T_{\pm}^2\mu^{tt}(\omega,\theta) + 2iT_{\pm}\mu^{ta}(\omega,\theta) + \mu^{aa}(\omega,\theta)}{T_{\pm}^2+1}. \tag{4.4.15}$$

In the approximation in which the natural modes are the magnetoionic modes in the circularly polarized limit, one has $T_{\pm} = \mp\epsilon\cos\theta/|\epsilon\cos\theta|$, with the $+$-mode being the o mode for an electron gas, $\epsilon = 1$.

Polarization-Dependent Absorption

Synchrotron emission and absorption are both polarization dependent, and include an unpolarized part and linearly and circularly polarized parts. A convenient way of treating polarized emission is in terms of the Stokes parameters, I, Q, U, V, cf. §3.6, with the polarization described by the three parameters $q = Q/I, u = U/I, v = V/I$. Absorption can be described by four absorption coefficients, $\mu_I, \mu_Q, \mu_U, \mu_V$, with μ_I describing polarization-independent absorption, and with the other three parameters describing the polarization-dependence of the absorption. Polarization-dependent absorption is included in the transfer equation (3.6.9) for the Stokes vector, which is generalized here to

$$dS_A/d\ell = J_A + \rho_{AB}S_B - \mu_{AB}S_B, \tag{4.4.16}$$

with $A, B = I, Q, U, V$, and with the sum over B implied; J_A is a column matrix that describes spontaneous emission and ρ_{AB} and μ_{AB} are square matrices written down in (3.6.10). The emission coefficients, J_A, and absorption coefficients, μ_{AB}, may be constructed by writing the synchrotron emissivity and the absorption coefficient as polarization tensors, and translating into the notation used in (4.4.16). The translation from the form involving polarization tensors to the form (4.4.16) involving the Stokes parameters follows by using the Pauli matrices (3.6.13). In the synchrotron case, the two transverse polarizations are chosen perpendicular and parallel to the projection of the magnetic field onto the transverse plane, and the components J_U and μ_U are then zero.

There is a qualitative difference between two limiting cases depending on which of the terms $\rho_{AB} S_B$ and $\mu_{AB} S_B$ in (4.4.16) is the more dominant. These terms describe generalized Faraday rotation and synchrotron absorption, respectively. The two limiting cases lead to different expressions for the polarization of a self-absorbed synchrotron source.

If generalized Faraday rotation dominates, then to a first approximation the absorption may be neglected. The components in the two natural modes of the plasma then propagate independently of each other, and each is described by its own absorption coefficient. For the magnetoionic o and x modes, and the transfer equations are

$$\frac{dI_{\mathrm{o,x}}}{d\ell} = J_{\mathrm{o,x}} - \mu_{\mathrm{o,x}} I_{\mathrm{o,x}}. \tag{4.4.17}$$

In the self-absorbed limit, emission and absorption are in balance, and (4.4.17) implies that the specific intensities in the two modes are $I_{\mathrm{o,x}} = J_{\mathrm{o,x}}/\mu_{\mathrm{o,x}}$.

In the opposite limiting case, when generalized Faraday rotation is negligible compared with absorption, the transfer equation may be approximated by (4.4.16) with the term involving ρ_{AB} omitted. In this case, in the self-absorbed limit, the solution of the transfer equation (4.4.16) is

$$S_A = \mu_{AB}^{-1} J_B, \tag{4.4.18}$$

where μ_{AB}^{-1} is the inverse of $\mu_{AB} = \gamma_{AB}/c$. The general result is relatively cumbersome. It simplifies if one chooses coordinate axes such that $J_U = 0$, $\mu_U = 0$, and if one assumes that J_V and μ_V are of order $1/\gamma$ compared with J_I, J_Q and μ_I, μ_Q, respectively, and ignores terms of order $1/\gamma^2$. In the more general case, where both synchrotron absorption and Faraday rotation are included, (4.4.18) is replaced by $S_A = [\mu - \rho]_{AB}^{-1} J_B$, where $[\mu - \rho]_{AB}^{-1}$ is the inverse of $\mu_{AB} - \rho_{AB}$. In practice, the direction of the magnetic field varies through the source, and one needs to take this into account in integrating (4.4.16) along the line of sight through the source.

4.4.3 Synchrotron Absorption: Thermal

Synchrotron absorption by a (highly relativistic) thermal distribution of particles may be treated both directly and using the antihermitian part of Trubnikov's response tensor.

The volume emissivity and the absorption coefficient for a thermal distribution are related by Kirchhoff's law (2.7.19), and only one of them need be calculated. The absorption coefficient is

$$\gamma^{\alpha\beta}(\omega,\theta)=\frac{(2\pi)^3 n\rho^2}{\omega^2 K_2(\rho)}\int_1^\infty d\gamma\,\gamma(\gamma^2-1^2)^{1/2}\,e^{-\rho\gamma}\,\bar{\eta}^{\alpha\beta}(\omega,\theta,\gamma), \tag{4.4.19}$$

where the thermal distribution (2.4.1) is assumed, and where the synchrotron emissivity, $\bar{\eta}^{\alpha\beta}$, is given by (4.4.2). The highly relativistic limit corresponds to $\rho \ll 1$ and one may assume $\gamma \gg 1$ in the integrand in (4.4.19) with the lower limit of the integration approximated by zero. For simplicity, only the two linearly polarized components for $\theta=\pi/2$ are considered here, and the refractive index is assumed to be unity.

On inserting the expression (4.4.2) for the emissivity, and noting $R \propto 1/\gamma^2$, one may partially integrate and introduce the variable $x=1/\rho^2\gamma^2$ to write

$$\begin{aligned}\int d\gamma\gamma^2 e^{-\rho\gamma}\,R\left[\int_R^\infty dt\,K_{5/3}(t)\pm K_{2/3}(R)\right]\\ =\int_0^\infty dx\,e^{-1/\sqrt{x}}\left[K_{5/3}(\nu x)\pm K'_{2/3}(\nu x)\right],\end{aligned} \tag{4.4.20}$$

with $\nu=2\omega\rho^2/3\Omega_0$. The resulting integral simplifies in two limits:

$$K_{5/3}(\nu x)\pm K'_{2/3}(\nu x)=\begin{cases}\dfrac{\Gamma(5/3)}{2^{1/3}(\nu x)^{5/3}}\,(1,3) & \text{for } \nu\ll 1,\\[2ex] \left(\dfrac{2\pi}{\nu x}\right)^{1/2}\dfrac{e^{-\nu x}}{3\nu x}\,(1,3\nu x) & \text{for } \nu\gg 1,\end{cases} \tag{4.4.21}$$

where the recursion relation (A.1.12) are used. The remaining integral is then performed using the method of steepest descents.

The limit $\nu \ll 1$ in (4.4.21) corresponds to frequencies below the characteristic frequency $\omega \approx \Omega_0/\rho^2$ of emission by a thermal particle with $\gamma \approx 1/\rho$. In this limit (4.4.19) reduces to

$$\gamma^{tt,aa}(\omega,\pi/2)=\frac{3\pi\omega_p^2}{4\Omega_0\left(4P^5\right)^{-1/3}}\,(1,3), \tag{4.4.22}$$

with $P=9\omega/2\rho\Omega_0$. In the high frequency limit $\nu \gg 1$, (4.4.22) is replaced by

$$\gamma^{tt,aa}(\omega,\pi/2) = \frac{3\pi\omega_p^2\rho^2}{4\sqrt{2}\Omega_0}\,\frac{e^{-\rho P^{1/3}}}{\rho P^{4/3}}\left(1,\rho P^{1/3}\right). \tag{4.4.23}$$

In (4.4.22) and (4.4.23) the factor $K_2(\rho)$ in the normalization of the Jüttner distribution is approximated by $2/\rho^2$, which applies for $\rho \ll 1$.

Synchrotron Absorption: Trubnikov's Form

The foregoing results may be derived in a different manner by starting from the antihermitian part of Trubnikov's response tensor, given by (2.4.15). Besides using this to treat gyromagnetic absorption in the synchrotron limit, Trubnikov also applied it to the nonrelativistic and mildly relativistic cases.

Starting from (2.4.15), one has

$$\gamma^{tt}(\omega,\theta) + \gamma^{aa}(\omega,\theta) = \frac{\sqrt{\pi}\omega_p^2\rho^{5/2}}{\Omega_0(r_0 r_0'')^{1/2}}\,e^{\rho-\sqrt{r_0}},$$

$$r_0 = \left[\frac{\omega}{\Omega_0}\sin\theta(1-\cos\Omega_0\xi_0)\right]^2 - \rho^2\cot^2\theta,$$

$$r_0'' = -2\frac{\omega^2}{\Omega_0^2}\sin^2\theta(1-\cos\Omega_0\xi_0). \tag{4.4.24}$$

For $\omega/\rho\Omega_0 \gg 1$, $r_0' = 0$ has the approximate solution

$$\Omega_0\xi_0 = i\frac{3}{\sin\theta}\left(\frac{\sin\theta}{P}\right)^{1/3}\left[1-\frac{3}{20\sin^2\theta}\left(\frac{\sin\theta}{P}\right)^{2/3}+\cdots\right], \tag{4.4.25}$$

with $P = 9\omega/2\rho\Omega_0$. On substituting (4.4.25) into (4.4.24), and separating into the two modes for $\theta = \pi/2$, one finds

$$\gamma^{tt,aa}(\omega,\pi/2) = \frac{\sqrt{\pi}\omega_p^2\rho^{1/2}}{\Omega_0}\,\frac{3}{2P}\,e^{-\rho\left(P^{1/3}-1+9/20P^{1/3}\right)}\left(\frac{1}{\rho P^{1/3}},1\right), \tag{4.4.26}$$

which is the result derived by Trubnikov [11].

The synchrotron case is the highly relativistic limit $\rho \ll 1$, when the asymptotic approximation is made to the Macdonald functions in (2.4.15):

$$K_2(r(\xi)) \approx \frac{2}{r^2(\xi)} - \frac{1}{2} + \cdots, \qquad K_3(r(\xi)) \approx \frac{8}{r^3(\xi)} - \frac{1}{r(\xi)} + \cdots. \tag{4.4.27}$$

The resulting approximation to the synchrotron absorption coefficient for a thermal distribution is

$$\gamma^{tt,aa}(\omega,\pi/2) = \frac{\omega_p^2\rho^2}{K_2(\rho)}\int_{-\infty}^{\infty} d\xi \left[\frac{1}{r^4(\xi)}, \frac{1}{r^4(\xi)} + \frac{\xi^4\omega^2\Omega_0^2}{r^6(\xi)}\right],$$

$$r(\xi) \approx \left[\rho^2 - 2i\rho\omega\xi - \frac{1}{12}\omega^2\,\Omega_0^2\xi^4\right]^{1/2}, \tag{4.4.28}$$

with $\theta = \pi/2$ assumed here. Trubnikov [11] evaluated the integrals in terms of the residue at the pole $r(\xi) = 0$, which for $\rho^2 \ll 1$ occurs at $\xi_1 = 2i(3\rho/\Omega_0\omega\sin^2\theta)^{1/3}$. The specific integrals are

$$\int_{-\infty}^{\infty} d\xi \begin{pmatrix} 1/r^4(\xi) \\ \xi^4/r^6(\xi) \end{pmatrix} = \begin{pmatrix} (3\pi/2\rho^4\Omega_0)\,(\sin^2\theta/4P^5)^{1/3} \\ (3^5\pi/4\rho^6\Omega_0^5)(2^3\sin^4\theta\, P^{11})^{1/3} \end{pmatrix}, \tag{4.4.29}$$

with $P = (9\omega/2\rho\Omega_0)\sin^2\theta$. Thus, (4.4.28) with (4.4.29) reproduces (4.4.22) for $\theta = \pi/2$ and $K_2(\rho) = 2/\rho^2$.

4.4.4 Razin Suppression

The presence of a cold plasma causes a suppression of synchrotron emission, referred to as the Razin effect [8]. In the context of synchrotron radiation, the suppression effect may be seen by noting that (4.4.4) implies that synchrotron radiation falls off exponentially for $R \gtrsim 1$ with $R = \omega/\omega_c$, $\omega_c = (3/2)\Omega_0\xi^2\sin\theta$ and with

$$\xi = (1 - n^2v^2/c^2)^{-1/2} \approx \gamma\left(1 + \frac{\gamma^2\omega_p^2}{\omega^2}\right)^{-1/2}. \tag{4.4.30}$$

The medium is unimportant for $\omega \gg \gamma\omega_p$, when one has $\omega_c = (3/2)\Omega_0\gamma^2\sin\theta$ and $R \propto \omega$, implying that synchrotron emission falls off exponentially at high frequency. However, for $\omega \lesssim \gamma\omega_p$ one has $\xi \approx \omega/\omega_p$ and $R \propto 1/\omega$, so that the emission also falls off exponentially at low frequency. The characteristic frequency below which this suppression effect occurs is determined by setting $\omega = \gamma\omega_p = \omega_c$, and is called the Razin-Tsytovich frequency,

$$\omega_{RT} = 2\omega_p^2/3\Omega_e\sin\theta. \tag{4.4.31}$$

More generally, any form of emission by a highly relativistic particle in a plasma is suppressed at $\omega \lesssim \gamma\omega_p$.

The Razin effect also applies to emission by mildly relativistic particles. Consider gyromagnetic emission at the ath harmonic in a medium with refractive index n. The emission coefficient depends on n through a multiplicative factor of n^{2a-1}, which corresponds to 2^{2a}-multipole emission. For large a, and for $n = (1 - \omega_p^2/\omega^2)^{1/2}$ with $\omega^2 \gg \omega_p^2$, one has

$$(1 - \omega_p^2/\omega^2)^{a-1/2} \approx e^{-a\omega_p^2/\omega^2}, \tag{4.4.32}$$

which implies suppression for $a\omega_p^2/\omega^2 \gtrsim 1$. With $\omega \approx a\Omega_e$ for cyclotron emission at the ath harmonic, this also implies suppression for $\omega \lesssim \omega_p^2/\Omega_e$.

Maser Synchrotron Emission

Synchrotron absorption cannot be negative under realistic conditions. To see this, consider the absorption coefficient in the form

$$\begin{aligned}\gamma^{\alpha\beta}(\omega,\theta) &= -\frac{(2\pi)^4\phi(\theta)}{\omega^2 m}\int d\gamma\, \gamma^2 \bar{\eta}^{\alpha\beta}(\omega,\theta,\gamma)\,\frac{d}{d\gamma}\frac{N(\gamma)}{\gamma^2}\\ &= \frac{(2\pi)^4\phi(\theta)}{\omega^2 m}\int d\gamma\,\frac{N(\gamma)}{\gamma^2}\frac{d}{d\gamma}\big[\gamma^2\,\bar{\eta}^{\alpha\beta}(\omega,\theta,\gamma)\big],\end{aligned} \tag{4.4.33}$$

where the second form follows from the first by a partial integration. Ignoring the circular polarization, absorption can be negative only if at least one of the two eigenvalues, $\frac{1}{2}(\gamma^{tt} \pm \gamma^{aa})$, of $\gamma^{\alpha\beta}$ is negative. When the Razin effect is ignored, $\xi \to \gamma$, it is straightforward to carry out the derivative in the second form of (4.4.33), using (4.4.1) with (4.4.2), and one finds that the eigenvalues cannot be negative [13].

This proof breaks down when the Razin effect is important, and it is then possible in principle for absorption to be negative for $\omega \lesssim \gamma\omega_p$ [5, 15]. One also requires $d[N(\gamma)/\gamma^2]/d\gamma > 0$. There is no known case where this effect occurs. It seems implausible that both the necessary conditions for it to occur could be satisfied simultaneously.

4.5 Thomson Scattering in a Magnetic Field

Inclusion of the magnetic field in the theory of scattering of waves by particles affects both the waves and the particles. The waves are the natural modes of the magnetized plasma, and waves in one mode may be scattered into waves in the same mode or in another mode. The spiraling motion of a scattering particle needs to be taken into account, and this has two notable effects: it introduces an additional length, the gyroradius of the scattering particle, and it allows transitions at frequencies that differ by a harmonic of the cyclotron frequency. Two limiting cases are for magnetized particles, when the gyroradius is smaller than other relevant lengths, and unmagnetized particles, when the gyroradius is larger than other relevant lengths. Scattering by unmagnetized particles may be approximated by ignoring the spiraling motion and treating the unperturbed motion of the scattering particles as rectilinear. For strongly magnetized scattering particles, a simplifying approximation is to assume that the gyroradius is zero.

4.5.1 Probability for Thomson Scattering

Scattering of waves may be treated as emission of the scattered waves, due to a current that is of first order in the amplitude of the unscattered waves. The theory for an unmagnetized particle is presented in § 5.5 of volume 1, and the generalization to a magnetized particle involves identifying the current with the first order current for a magnetized particle. Thomson scattering by highly relativistic particles is called inverse Compton scattering in the astrophysical literature.

Current Associated with Thomson Scattering

The current associated with Thomson scattering by a charged particle in a magnetic field, denoted $J^{(1)\mu}_{\rm sp}(k)$, is calculated in § 2.2 and is given by (2.2.14). There is an additional current that needs to be included, associated with the screening of the scattering particle in the plasma. This is referred to as nonlinear scattering. The total current, in the calculation of the scattering probability, is the sum of the contributions from Thomson scattering and nonlinear scattering.

Nonlinear Scattering

In simple cases, nonlinear scattering can be interpreted as scattering due to a quasi-particle associated with the (Debye) screening field of the scattering charge. For an electron scattering long-wavelength waves, the quasi-particle looks like a positively charged electron co-moving with the scattering electron, and the two currents interfere strongly, leaving the dominant scattering due to the electron-like quasi-particle associated with a scattering ion. At short wavelengths nonlinear scattering is unimportant, and the scattering can be approximated by Thomson scattering by unscreened electrons.

The current associated with nonlinear scattering is due to the quadratic response of the plasma to the field of the unscattered wave and the self-consistent field of the scattering particle. This current is

$$J^{(\rm nl)\mu}(k) = 2\int d\lambda^{(2)}\, \Pi^{(2)\mu}{}_{\nu\rho}(-k,k_1,k_2)A^{(q)\nu}(k_1)A^{\rho}_{M'}(k_2). \qquad (4.5.1)$$

The field $A^{(q)\mu}(k)$ is found by solving the inhomogeneous wave equation with the unperturbed current due to the motion of the charge q as the source term. This gives

$$A^{(q)\mu}(k) = -q\, e^{ikx_0} \sum_{a=-\infty}^{\infty} e^{ia\epsilon\psi} D^{\mu\nu}(k)\, U^{\nu}(a,k)\, 2\pi\delta\big[(ku)_{\parallel} - a\Omega_0\big], \qquad (4.5.2)$$

where (2.1.31) is used. The scattering current is the sum of (2.2.14) and (4.5.2). The calculation of the probability for scattering then proceeds as in the unmagnetized case.

Scattering Probability

The resulting expression for the scattering probability is

$$w_{MP}(a,p,k,k') = \frac{q^4}{\varepsilon_0^2 m^2 \gamma} \frac{R_M(k) R_P(k')}{\omega_M(k)\omega_P(k')} |\tilde{a}_{MP}(a,k,k',p)|^2$$
$$\times\, 2\pi\delta\big[(ku)_{\parallel} - (k'u)_{\parallel} - a\Omega_0\big], \qquad (4.5.3)$$

where M, P label the unscattered and scattered wave modes, whose dispersion relations are implicit in $k^{\mu} = k_M^{\mu}$, $k'^{\mu} = k_P'^{\mu}$, with

$$\tilde{a}_{MP}(a,k,k',p) = e^*_{M\mu}(k) e_{P\nu}(k') \Bigg[e^{-i\epsilon a\psi} \sum_{t=-\infty}^{\infty} e^{-i\epsilon t(\psi+\psi')}$$
$$\times\, G^{\alpha\mu}(a-t,k,u)\tau_{\alpha\beta}\big((ku)_{\parallel} - (a-t)\Omega_0\big) G^{*\beta\nu}(t,k',u)$$
$$- \frac{2m}{q} \Pi^{\mu\nu\rho}(k,k',k-k') U_{\rho}(a,k-k') \Bigg], \qquad (4.5.4)$$

and with $G^{\mu\nu}(a,k,u)$ given by (2.2.15), $\tau^{\mu\nu}(\omega)$ given by (2.1.15) and $U^{\mu}(a,k,u)$ by (2.1.28) with (2.1.29). The term involving $\Pi^{\mu\nu\rho}$ is due to nonlinear scattering.

The resonance condition in (4.5.4) may be interpreted either from a purely classical viewpoint or from a semi-classical viewpoint. The classical interpretation is that the difference between the frequencies of the scattered and unscattered waves in the rest frame of the gyrocenter of the scattering particle is an integral multiple of the gyrofrequency. The semi-classical interpretation is in terms of conservation of energy and momentum, and follows from (4.1.10)–(4.1.13) by replacing the emission of a wave quantum, k^{μ}, by emission of the beat disturbance, $k - k'$.

The scattering probability (4.5.3) is symmetric under the interchange of the roles of the scattered and unscattered waves, $M \leftrightarrow P$ (and $a \leftrightarrow -a$). It also describes double emission, that is, simultaneous emission of waves with wave 4-vectors k, $-k'$.

4.5.2 Quasilinear Equations for Scattering

Quasilinear equations for scattering by a magnetized particle may be derived using semi-classical theory, in a way that is closely analogous to the unmagnetized case discussed in volume 1. These equations include the kinetic equations for both the scattered and unscattered waves:

$$\frac{\mathrm{D}N_M(\boldsymbol{k})}{\mathrm{D}t} = \sum_{a=-\infty}^{\infty}\int\frac{d^4p}{(2\pi)^4}\int\frac{d^3\boldsymbol{k}'}{(2\pi)^3}\,\gamma w_{MP}(a)\Big\{\big[N_P(\boldsymbol{k}')-N_M(\boldsymbol{k})\big]F(p)$$

$$+\,N_M(\boldsymbol{k})N_P(\boldsymbol{k}')\left[\frac{(k-k')u_\parallel}{u_\perp}\frac{\partial}{\partial p_\perp}+(k-k')^\alpha_\parallel\frac{\partial}{\partial p^\alpha_\parallel}\right]F(p)\Big\}, \tag{4.5.5}$$

$$\frac{\mathrm{D}N_P(\boldsymbol{k}')}{\mathrm{D}t} = -\sum_{a=-\infty}^{\infty}\int\frac{d^4p}{(2\pi)^4}\int\frac{d^3\boldsymbol{k}}{(2\pi)^3}\,\gamma w_{MP}(a)\Big\{\big[N_P(\boldsymbol{k}')-N_M(\boldsymbol{k})\big]F(p)$$

$$+\,N_M(\boldsymbol{k})N_P(\boldsymbol{k}')\left[\frac{((k-k')u)_\parallel}{u_\perp}\frac{\partial}{\partial p_\perp}+(k-k')^\alpha_\parallel\frac{\partial}{\partial p^\alpha_\parallel}\right]F(p)\Big\}. \tag{4.5.6}$$

The kinetic equation for the particles due to wave-particle scattering is described by the quasilinear equation with the coefficients (4.1.25) and (4.1.26) replaced by

$$\begin{pmatrix}A_\perp(p)\\ A^\mu_\parallel(p)\end{pmatrix} = -\sum_a\int\frac{d^3\boldsymbol{k}}{(2\pi)^3}\,\gamma w_{MP}(a)\big[N_M(\boldsymbol{k})-N_P(\boldsymbol{k}')\big]\begin{pmatrix}a\Omega/v_\perp\\ (k-k')^\mu_\parallel\end{pmatrix}, \tag{4.5.7}$$

for the terms that describe the effect of spontaneous emission alone, and

$$\begin{pmatrix}D_{\perp\perp}(p)\\ D^\mu_{\parallel\perp}(p)\\ D^{\mu\nu}_{\parallel\parallel}(p)\end{pmatrix} = \sum_a\int\frac{d^3\boldsymbol{k}}{(2\pi)^3}\,\gamma w_{MP}(a)\,N_M(\boldsymbol{k})N_P(\boldsymbol{k}')\begin{pmatrix}(a\Omega/v_\perp)^2\\ \left(\dfrac{a\Omega}{v_\perp}\right)(k-k')^\mu_\parallel\\ (k-k')^\mu_\parallel(k-k')^\nu_\parallel\end{pmatrix}, \tag{4.5.8}$$

for the diffusion coefficients, with $w_{MP}(a)$ denoting $w_{MP}(a,\boldsymbol{k},\boldsymbol{k}',p)$.

The quasilinear equations conserve the energy, the parallel momentum and the number of photons in a system consisting of waves in the modes M and P and the scattering particles.

Magnetized and Unmagnetized Particles

The general form of the probability (4.5.3) with (4.5.4) is too cumbersome to be of practical use and one needs to make various simplifying assumptions to reduce it to a directly useful form. One complicating feature is the sum over Bessel functions.

This may be simplified in two limiting cases, referred to as magnetized and unmagnetized particles, respectively. For unmagnetized particles, the gyroradius of the scattering particle is effectively assumed infinite, in which case the motion of the particle is approximated by constant rectilinear motion, and Thomson scattering in a magnetic field is replaced by its unmagnetized counterpart. Formally, this limit corresponds to large argument of the Bessel functions, $(\boldsymbol{k}-\boldsymbol{k}')_\perp R \gg 1$, when the sum is dominated by high harmonics, of the same order as this argument, $a \approx (\boldsymbol{k}-\boldsymbol{k}')_\perp R$.

For magnetized particles the small gyroradius approximation is made, $(\boldsymbol{k}-\boldsymbol{k}')_\perp R \ll 1$, such that only the leading terms in the power series expansion of the Bessel functions need be retained. The first order current is approximated by (2.2.19). The corresponding approximation to the probability (4.5.3) is

$$w_{MP}(p,k,k') = \frac{q^4}{\varepsilon_0^2 m^2 \gamma} \frac{R_M(k) R_P(k')}{\omega_M(k)\omega_P(k')} |\tilde{a}_{MP}(k,k',p)|^2 2\pi\delta\big[(k_M - k'_P)u_\parallel\big], \quad (4.5.9)$$

where the nonlinear scattering term is neglected, and with

$$\tilde{a}_{MP}(k,k',p) = e^*_{M\mu}(k) e_{P\nu}(k') G^{\alpha\mu}(k_M, u_\parallel)\tau_{\alpha\beta}(k_M u_\parallel) G^{\beta\nu}(k'_P, u_\parallel). \quad (4.5.10)$$

The right hand side of (4.5.10) may be rewritten using (2.2.20).

Scattering by Nonrelativistic Particles

Simplification occurs when the scattering particle is nonrelativistic. Provided that the refractive index of neither wave is large, one has $k_M u_\parallel \approx \omega_M(k)$, $k'_P u_\parallel \approx \omega_P(k')$, so that the δ-function in (4.5.9) implies that the change in frequency is small. For nonrelativistic particles, the form (4.5.10) with (2.2.20) may be approximated by

$$\tilde{a}_{MP}(k,k',p) = e^*_{M\mu}(k) e_{P\nu}(k')\tau_{\mu\nu}(k_M u_\parallel), \quad (4.5.11)$$

with $k_M u_\parallel = k'_P u_\parallel$. Further simplifications involve approximations to the wave properties. Specific cases discussed here are the scattering of magnetoionic waves by nonrelativistic electrons, and the scattering of the perpendicular and parallel modes of the birefringent vacuum in strongly magnetized, low density plasmas.

4.5.3 Scattering Cross Section

In describing Thomson scattering, and inverse Compton scattering, it is conventional to introduce the differential scattering cross section [1, 7]. Such a cross section is written down here for scattering in a magnetized plasma.

The differential scattering cross section relates the energy flux in the scattered radiation, per unit solid angle about its ray direction, to the energy flux in the incident radiation, per unit solid angle about its ray direction. The energy flux is along the group velocity, which is at an angle, θ_r, to the magnetic field that is different from the wave-normal angle, θ, in general. One has

$$\boldsymbol{v}_{gM}(\boldsymbol{k}) = \frac{\partial \omega_M(\boldsymbol{k})}{\partial \boldsymbol{k}} = (\sin\theta_r \cos\psi, \sin\theta_r \sin\psi, \cos\theta_r) v_{gM}. \tag{4.5.12}$$

For simplicity in writing, the argument of the group speed, v_{gM}, and a label on the ray angle, θ_r, denoting the mode, M, are omitted. For the mode P, the group speed is written v'_{gP}, where the prime denotes that the argument depend on the primed variables. The cross section depends on the Jacobian $\partial \cos\theta_r / \partial \cos\theta$ of the transformation from the wave-normal to the ray angle. The differential scattering cross section is identified as

$$\Sigma_{MP} = \sum_a \int_0^\infty \frac{d\omega\, \omega^2 n_M^2}{(2\pi)^3 \omega' v'_{gP}} \frac{\partial(\omega n_M)}{\partial \omega} \frac{\partial \cos\theta}{\partial \cos\theta_r} \frac{\partial \cos\theta'}{\partial \cos\theta'_r} w_{MP}(a), \tag{4.5.13}$$

where arguments are omitted for simplicity in writing. The integral over ω is performed over the δ-function in the probability (4.5.3). The cross section is too cumbersome to be useful in most cases where the plasma dispersion is important.

A subtle point concerns the cancelation of two factors $n_M \partial(\omega n_M)/\partial\omega$ in the cross section, one from the group velocity, v_{gM}, and one from the ratio of electric to total energy, R_M, in the probability $w_{MP}(a)$. (A similar cancelation occurs for $n'_P \partial(\omega' n'_P)/\partial\omega'$.) At sufficiently high frequencies, $n_M \partial(\omega n_M)/\partial\omega$ approaches unity, v_{gM} approaches the speed of light and R_M approaches one half. In the opposite limit, near a resonance, $n_M \partial(\omega n_M)/\partial\omega$ becomes very large, and v_{gM}, R_M become small. The cancelation of these two factors in the cross section may be interpreted as the strength of the coupling becoming weak, due to small R_M, being offset by the energy flux in the waves becoming small, due to small v_{gM}, allowing a long time for the interaction. The cross section is rarely useful in such cases where the plasma dispersion has a large effect.

4.5.4 Scattering of Magnetoionic Waves

In many applications, high frequency waves in magnetized plasmas are well approximated by treating the plasma as a cold electron gas, so that the waves are described by the magnetoinic theory § 3.3.1. For this case, it is convenient to change the labeling of the modes $M, P \to \sigma, \sigma'$. The wave properties in (4.3.1) include the polarization parameters T_σ, L_σ, and the refractive index n_σ, and are given by (3.3.3)–(3.3.4). On inserting the wave properties (4.3.1) into the probability (4.5.9) with $M, P \to \sigma, \sigma'$, to avoid loss of generality one needs to allow the two waves to be in different azimuthal planes, and this is achieved by assuming the wave normal direction to be $\boldsymbol{\kappa} = (\sin\theta\cos\psi, \sin\theta\sin\psi, \cos\theta)$, so that one has

$$\boldsymbol{t} = (\cos\theta\cos\psi, \cos\theta\sin\psi, -\sin\theta), \qquad \boldsymbol{a} = (-\sin\psi, \cos\psi, 0), \tag{4.5.14}$$

and similarly for $\boldsymbol{\kappa}', \boldsymbol{t}', \boldsymbol{a}'$ in terms of θ', ψ'. The probability (4.5.9) depends on the azimuthal angle, $\psi - \psi'$, between the scattered and unscattered wave, and assuming azimuthal symmetry in the particle and wave distributions, this dependence is of no interest. Using the wave properties (4.3.1), assuming the scattering particle to be at rest and averaging over azimuthal angles, the probability (4.5.9) reduces to [7]

$$\langle w_{\sigma\sigma'}(\boldsymbol{k}, \boldsymbol{k}')\rangle = \frac{q^4}{4\varepsilon_0^2 m^2} f_{\sigma\sigma'}(\omega, \theta, \theta')\, 2\pi\delta(\omega - \omega'), \tag{4.5.15}$$

where the angular brackets denote the average over azimuthal angle. The dependence on the wave properties is included in

$$\begin{aligned} f_{\sigma\sigma'}(\omega, \theta, \theta') = {} & \left[(1 + T_\sigma^2)\, n_\sigma \frac{\partial}{\partial\omega}(\omega n_\sigma)\, (1 + T_{\sigma'}^2)\, n_{\sigma'} \frac{\partial}{\partial\omega}(\omega n_{\sigma'}) \right]^{-1} \\ & \times \left\{ \frac{1 + Y^2}{2(1 - Y^2)^2} [(a_\sigma a_{\sigma'} + 1)^2 + (a_\sigma + a_{\sigma'})^2] \right. \\ & \left. + \frac{2Y}{(1 - Y^2)^2} (a_\sigma a_{\sigma'} + 1)(a_\sigma + a_{\sigma'}) + (b_\sigma b_{\sigma'})^2 \right\}, \end{aligned} \tag{4.5.16}$$

with $Y = \Omega_e/\omega$ and where the parameters describing the longitudinal and transverse components of the polarization appear in

$$a_\sigma = L_\sigma \sin\theta + T_\sigma \cos\theta, \qquad b_\sigma = L_\sigma \cos\theta - T_\sigma \sin\theta, \tag{4.5.17}$$

and similarly for the primed quantities.

Scattering of High-Frequency Waves

At sufficiently high frequencies, the magnetoionic waves become nearly circularly polarized, except for a small range of angles about perpendicular propagation, with refractive indices close to unity. A generalization that is important in very low density plasmas with $\omega_p \ll \Omega_e$ is when the frequency is much greater than the plasma frequency, ω_p, but not necessarily small in comparison with the cyclotron frequency, Ω_e. This limiting case is described by expanding in $X \ll 1$ in the formulae in § 3.3.1 that describe the magnetoionic theory. This gives

$$T_\sigma \approx -\frac{\frac{1}{2}Y\sin^2\theta + \sigma(\frac{1}{4}Y^2\sin^4\theta + \cos^2\theta)^{1/2}}{\cos\theta}, \qquad L_\sigma \approx 0, \quad n_\sigma^2 \approx 1. \quad (4.5.18)$$

Further simplification occurs for $Y \ll 1$, when one has $T_\sigma = -\sigma\cos\theta/|\cos\theta|$, corresponding to circular polarization. In this limit, (4.5.16) simplifies to

$$f_{\sigma\sigma'}(\omega,\theta,\theta') \approx \tfrac{1}{8}\{(1+\cos^2\theta)(1+\cos^2\theta')+2\sin^2\theta\sin^2\theta'+4\sigma\sigma'|\cos\theta\cos\theta'|\}. \quad (4.5.19)$$

The sign $\sigma\sigma'$ is equal to $+1$ if the scattered wave is in the same mode as the unscattered wave, and equal to -1 if the mode changes. Scattering in which there is no change in mode is preferred.

In the high-frequency limit, the scattering is equivalent to Thomson scattering in the absence of a magnetic field. This may be seen by considering isotropic, unpolarized initial radiation, and averaging over the angular distribution of the scattered radiation. The average over polarizations implies the term involving $\sigma\sigma'$ in (4.5.18) gives zero, and the averages over $\cos\theta$ and $\cos\theta'$ imply $\langle f_{\sigma\sigma'}(\omega,\theta,\theta')\rangle = 4/3$. The scattering cross section then reduces to the Thomson cross section.

4.5.5 Resonant Thomson Scattering

A specific case of interest in astrophysics is the scattering of waves with frequencies of order the cyclotron frequency by nonrelativistic electrons in a plasma with $\omega_p \ll \Omega_e$. The factor $(1-Y^2)^2$ in the denominator in (4.5.16) suggests that the scattering cross section diverges $\propto 1/(1-Y)^2 \propto 1/(\omega-\Omega_e)^2$ for $\omega \to \Omega_e$. The enhanced scattering when this factor becomes large is referred to as resonant scattering.

In a cold plasma, as the resonance is approached, the axial ratios, T_σ, for the two modes approach $T_+ = -1/\cos\theta$, $T_- = \cos\theta$, corresponding to the o and x modes, respectively. One has $a_\mathrm{o} = 0$, $a_\mathrm{x} = \cos^2\theta$. In this case, (4.5.16) implies $f_{\mathrm{oo}'} = f_{\mathrm{ox}'} = f_{\mathrm{xo}'} = 0$, and

$$f_{\mathrm{xx}'}(\omega,\theta,\theta') \approx \frac{\Omega_e^2}{8(\omega-\Omega_e)^2}\,\frac{(1+\cos^2\theta\cos^2\theta')^2+(\cos^2\theta+\cos^2\theta')^2}{(1+\cos^2\theta)(1+\cos^2\theta')} \quad (4.5.20)$$

near the resonance at $\omega = \Omega_e$. In this case, only the x mode is involved in resonant scattering. The enhancement associated with resonant scattering is limited by the dispersion itself: sufficiently near the resonance, the approximations (4.5.18) in which the refractive index is set to unity and the longitudinal part of the polarization is neglected is not justified. Near the resonance one has

$$n_\mathrm{x}^2 \approx 1 - \frac{X}{1-Y}, \qquad L_\mathrm{x} \approx \frac{X\sin\theta}{1-Y}, \quad (4.5.21)$$

and both diverge at the resonance. From (4.5.21) it follows that the neglect of the terms proportional to X is invalid for $1 - Y \lesssim X$. This suggests that the maximum enhancement is by a factor $\approx 1/4(1-Y)^2 \approx 1/4X^2 = (\Omega_e/\omega_p)^4/4$. However, this estimate neglects the derivatives of the refractive index in (4.5.16)

$$n_{\rm x}\frac{\partial}{\partial\omega}(\omega n_x) \approx 1 + \frac{X}{2(1-Y)^2}, \tag{4.5.22}$$

suggesting that the neglect of the terms involving X is valid only for $2(1-Y^2) \ll X$, and that the enhancement is limited to a much smaller factor $\approx 1/4(1-Y)^2 \approx 1/2X = (\Omega_e/\omega_p)^2/2$. Note that the factor (4.5.22) does not appear in the cross section (4.5.13), due to the cancelation of two effects, and the implied large enhancement of the cross section for $X < 1 - Y < (X/2)^{1/2}$ needs to be interpreted with care.

Resonant scattering can give a large enhancement only if thermal effects are neglected. Thermal effects modify the dispersion near the cyclotron resonance, limiting the enhancement. To be consistent, when thermal effects are included in the wave dispersion, they also need to be included in the scattering itself. Specifically, one needs to average the probability for scattering over the thermal distribution of particles assumed to determine the wave dispersion. Averaging the enhancement factor for resonant scattering over a thermal distribution leads to

$$\frac{\omega_e^2}{(2\pi)^{1/1}V_e}\int_{-\infty}^{\infty}\frac{dv_z\, e^{-v_z^2/2V_e^2}}{(\omega-\Omega_e-k_zv_z)^2} = -\frac{\Omega_e^2}{k_z^2V_e^2}[1+y_e\phi(y_e)], \tag{4.5.23}$$

with $y_e = (\omega - \Omega_e)/\sqrt{2}\,|k_z|V_e$, and where $Z(y)$ is a form of the nonrelativistic plasma dispersion function. The maximum value of the integral (4.5.23) is for y of order unity. This implies that the maximum enhancement factor in resonant scattering is limited by thermal effects to $\approx c^2/V_e^2$ in ordinary units.

Another case of interest is where the density of the scattering electrons is so low that their contribution to the wave dispersion can be neglected in comparison with the contributions of the birefringent vacuum, as discussed in § 8.4. The two modes of the birefringent vacuum may be labeled $\perp$, $\parallel$, with $T_\perp = 0$, $T_\parallel = \infty$. In this case, (4.5.16) gives

$$\begin{pmatrix} f_{\perp\perp'} \\ f_{\perp\parallel'} \\ f_{\parallel\perp'} \\ f_{\parallel\parallel'} \end{pmatrix} = \frac{1+Y^2}{2(1-Y^2)^2}\begin{pmatrix} 1 \\ \cos^2\theta \\ \cos^2\theta' \\ \cos^2\theta\cos^2\theta' \end{pmatrix} + \begin{pmatrix} 0 \\ 0 \\ 0 \\ \sin^2\theta\sin^2\theta' \end{pmatrix}. \tag{4.5.24}$$

It follows that there is a resonance at the cyclotron frequency in all four scattering channels.

Such resonant scattering is thought to play a role in pair production in pulsars [4, 10]. The scattering particles are highly relativistic and they scatter thermal photons from the neutron-star surface into high energy photons. In the rest frame of the scattering particle the initial and final frequencies satisfy $\omega \approx \omega' \approx \Omega_e$, and the boost in frequency occurs for the scattered photons in a forward cone on transforming a to frame in which the scattering particle is highly relativistic.

References

1. V. Canuto, J. Lodenquai, M. Ruderman, Phys. Rev. D **3**, 2303 (1971)
2. P.A.M. Dirac, Proc. R. Soc. A **167**, 148 (1938)
3. R.E. Ergun, C.W. Carlson, J.P. McFadden, G.T. Delory, R.J. Strangeway, P.L. Pritchett, Astrophys. J. **538**, 456 (2000)
4. Q. Luo, Astrophys. J. **468**, 338 (1996)
5. R. McCray, Science **154**, 1320 (1966)
6. D.B. Melrose, *Instabilities in Space and Laboratory Plasmas* (Cambridge University Press, Cambridge/New York, 1986)
7. D.B. Melrose, W.N. Sy, Astrophys. Space Sci. **17**, 343 (1972)
8. V.A. Razin, Izv. Vuzov. Radiofiz. **3**, 584 (1960)
9. G.A. Schott, *Electromagnetic Radiation* (Cambridge University Press, Cambridge, 1912)
10. S.J. Sturner, Astrophys. J. **446**, 292 (1995)
11. B.A. Trubnikov, Dissertation, Moscow Institute of Engineering and Physics (1958); AEC-tr-4073, US Atomic Energy Commission, Oak Ridge, Tennessee (1960)
12. R.Q. Twiss, Aust. J. Phys. **11**, 564 (1958)
13. J.P. Wild, S.F. Smerd, A.A. Weiss, Annu. Rev. Astron. Astrophys. **1**, 291 (1963)
14. C.S. Wu, L.C. Lee, Astrophys. J. **230**, 621 (1979)
15. V.V. Zheleznyakov, Sov. Phys. JETP **24**, 381 (1967)

Chapter 5
Magnetized Dirac Electron

In this chapter the Dirac equation is solved for an electron in the presence of a magnetostatic field. To solve the Dirac equation requires a specific choice of gauge for the magnetostatic field, and a specific choice of the spin operator. It is then possible to separate the wavefunction into a gauge- and spin-dependent factor and a reduced wavefunction that satisfies a reduced form of the Dirac equation that is independent of the choice of gauge and spin operator. In generalizing QED to include the magnetic field exactly, the conventional momentum representation for the Feynman amplitudes is not available, because momentum perpendicular to the magnetic field is not conserved. However, the separation of the wavefunction allows an analogous separation of the electron propagator and the vertex functions, with the gauge-independent part closely analogous to the momentum-space representation in the unmagnetized case. The gauge-dependent part (partially) describes the location of the center of gyration of the electron, and how it changes in a QED interaction, and such information is rarely of interest, and is simply ignored when using the reduced theory.

The Dirac equation in a magnetostatic field is written down in § 5.1, and solutions are found for a convenient but implicit choice of spin operator. Eigenfunctions for well-defined spin operators are derived in § 5.2. The electron propagator in a magnetostatic field is written down in § 5.3, and evaluated explicitly for the magnetized vacuum. In § 5.4, the vertex function is written down and factorized into a gauge- and spin-dependent part; the gauge-independent part is evaluated explicitly for the different spin eigenfunctions. The reduced, gauge-independent formalism is developed in § 5.5. Feynman rules for QPD processes in a magnetic field are summarized in § 5.6.

Natural units ($\hbar = 1$, $c = 1$) are used here, except where stated otherwise.

D. Melrose, *Quantum Plasmadynamics: Magnetized Plasmas*, Lecture Notes in Physics 854, DOI 10.1007/978-1-4614-4045-1_5,

5.1 Dirac Wavefunctions in a Magnetostatic Field

Explicit solutions of the Dirac equation in the presence of a magnetostatic field, $\boldsymbol{B}$, depend on the choice of gauge for the vector potential, $\boldsymbol{A}(\boldsymbol{x})$, for $\boldsymbol{B}$, and on the choice of spin operator. However, the energy eigenvalues are independent of both choices. In this section solutions of the Dirac equation are derived in the Landau gauge, and for an implicit choice of spin operator.

5.1.1 *Review of the Dirac Equation for B = 0*

The Dirac wavefunction, $\Psi(x)$, has four complex components, which are written as a column matrix. Observable quantities are represented by operators which are 4×4 matrices. The Dirac matrices, γ^μ, are four such matrices that are assumed to transform as a 4-vector under a Lorentz transformation, and which satisfy

$$\gamma^\mu \gamma^\nu + \gamma^\nu \gamma^\mu = 2g^{\mu\nu}, \tag{5.1.1}$$

where it is implicit that the unit 4×4 matrix multiplies $2g^{\mu\nu}$ on the right hand side. The covariant form of the Dirac equation is

$$(i\not{\partial} - m)\,\Psi(x) = (\hat{\not{p}} - m)\,\Psi(x) = 0, \tag{5.1.2}$$

where the slash notation is defined by

$$\not{A} = \gamma^\mu A_\mu, \qquad \not{\partial} = \gamma^\mu \partial_\mu, \tag{5.1.3}$$

for any 4-vector A^μ. The Dirac Hamiltonian is identified as

$$\hat{H} = \boldsymbol{\alpha} \cdot \hat{\boldsymbol{p}} + \beta m, \qquad \boldsymbol{\alpha} = \gamma^0 \boldsymbol{\gamma}, \quad \beta = \gamma^0, \tag{5.1.4}$$

with $\hat{\boldsymbol{p}} = -i\,\partial/\partial\boldsymbol{x}$. The Dirac adjoint of the wavefunction is defined by

$$\overline{\Psi}(x) = \Psi^\dagger(x)\gamma^0, \tag{5.1.5}$$

and the adjoint of the Dirac equation in the form (5.1.2) becomes

$$\overline{\Psi}(x)\,(\hat{\not{p}} - m) = 0, \tag{5.1.6}$$

where the operators operate to the left.

A specific choice for the Dirac matrices needs to be made for the purposes of detailed calculations, and here the standard representation is chosen. It corresponds to

$$\gamma^0 = \begin{pmatrix} 1 & 0 & 0 & 0 \\ 0 & 1 & 0 & 0 \\ 0 & 0 & -1 & 0 \\ 0 & 0 & 0 & -1 \end{pmatrix}, \qquad \gamma^1 = \begin{pmatrix} 0 & 0 & 0 & 1 \\ 0 & 0 & 1 & 0 \\ 0 & -1 & 0 & 0 \\ -1 & 0 & 0 & 0 \end{pmatrix},$$

$$\gamma^2 = \begin{pmatrix} 0 & 0 & 0 & -i \\ 0 & 0 & i & 0 \\ 0 & i & 0 & 0 \\ -i & 0 & 0 & 0 \end{pmatrix}, \quad \gamma^3 = \begin{pmatrix} 0 & 0 & 1 & 0 \\ 0 & 0 & 0 & -1 \\ -1 & 0 & 0 & 0 \\ 0 & 1 & 0 & 0 \end{pmatrix}. \tag{5.1.7}$$

A convenient way of writing these and other 4×4 matrices is in terms of block matrices. Let $\mathbf{0}$ and $\mathbf{1}$ be the null and unit 2×2 matrices. One writes

$$\boldsymbol{\Sigma} = \begin{pmatrix} \boldsymbol{\sigma} & \mathbf{0} \\ \mathbf{0} & \boldsymbol{\sigma} \end{pmatrix}, \qquad \rho_x = \begin{pmatrix} \mathbf{0} & \mathbf{1} \\ \mathbf{1} & \mathbf{0} \end{pmatrix},$$

$$\rho_y = \begin{pmatrix} \mathbf{0} & -i\mathbf{1} \\ i\mathbf{1} & \mathbf{0} \end{pmatrix}, \qquad \rho_z = \begin{pmatrix} \mathbf{1} & \mathbf{0} \\ \mathbf{0} & -\mathbf{1} \end{pmatrix}, \tag{5.1.8}$$

where the 2×2 matrices

$$\sigma_x = \begin{pmatrix} 0 & 1 \\ 1 & 0 \end{pmatrix}, \quad \sigma_y = \begin{pmatrix} 0 & -i \\ i & 0 \end{pmatrix}, \quad \sigma_z = \begin{pmatrix} 1 & 0 \\ 0 & -1 \end{pmatrix}, \tag{5.1.9}$$

are the usual Pauli matrices. In this representation one has

$$\gamma^\mu = [\rho_z, i\rho_y \boldsymbol{\Sigma}], \qquad \boldsymbol{\alpha} = \rho_x \boldsymbol{\Sigma}, \qquad \beta = \rho_z. \tag{5.1.10}$$

5.1.2 The Dirac Equation in a Magnetostatic Field

The minimal coupling procedure for including an electromagnetic field with 4-potential $\boldsymbol{A}(x)$ in the Dirac equation is to replace $\hat{p}^\mu$ by $\hat{p}^\mu + e\hat{A}^\mu(x)$, where $-e$ is the charge on an electron. The Dirac Hamiltonian (5.1.4) becomes

$$\hat{H} = \boldsymbol{\alpha} \cdot [\boldsymbol{p} + e\boldsymbol{A}(x)] + \beta m - e\phi(x). \tag{5.1.11}$$

A variety of choices is possible for $A^\mu = [\phi, \boldsymbol{A}]$ for a uniform, magnetostatic field, $\boldsymbol{B}$, and all involve a dependence on at least one component of the 4-vector x. All conventional choices satisfy the Coulomb gauge, $\operatorname{div} \boldsymbol{A} = 0$, but this does not uniquely determined the gauge. With the magnetostatic field along the z-axis, one choice of gauge is

$$\boldsymbol{A} = (0, Bx, 0), \tag{5.1.12}$$

which is called the Landau gauge here. Other choices of gauge are related to (5.1.12) by adding the gradient of a scalar to $\boldsymbol{A}$. One alternative choice of gauge is

$$\boldsymbol{A} = (-By, 0, 0), \tag{5.1.13}$$

and another is the cylindrical gauge

$$\boldsymbol{A} = \tfrac{1}{2}(-By, Bx, 0) = \tfrac{1}{2}B\varpi(-\sin\phi, \cos\phi, 0), \tag{5.1.14}$$

with $\varpi = (x^2 + y^2)^{1/2}$ and $x = \varpi\cos\phi$, $y = \varpi\sin\phi$.

The introduction of a magnetostatic field leads to the Hamiltonian depending on a spatial coordinate, with the specific coordinate depending on the choice of gauge. This leads to a conceptual complication: our description of the system depends in a nontrivial way on the choice of gauge. In the absence of the field, there are plane wave solutions that depend on the components of x in the form $\exp(-iPx)$, where the components of P^μ are constants of the motion. In the presence of a magnetostatic field, the component of P^μ conjugate to the component of x that appears in the Hamiltonian is not conserved. With the choice of the Landau gauge (5.1.12), one is free to seek solutions of the form $\exp(-iEt + iP_y y + iP_z z)$, where E, P_y, P_z are constants of the motion, but the momentum, P_x say, conjugate to x is not a constant of the motion. Alternatively, with the choice (5.1.14), one is free to seek solutions of the form $\exp(-iEt + iP_z z + iP_\phi\phi)$, where E, P_z and the momentum P_ϕ conjugate to ϕ are constants of the motion, but the momentum conjugate to the coordinate ϖ is not conserved. (It is possible to choose the temporal gauge, with $\boldsymbol{A} = -\boldsymbol{B}t$ dependent on time, and then P_x, P_y, P_z are constants of the motion, but the energy, $P^0 = E$, is not conserved.) From these remarks it is clear that the interpretation of the momentum components perpendicular to $\boldsymbol{B}$ requires care. For the Landau gauge, P_y is interpreted as specifying the x-component of the center of gyration, and then the uncertainty principle implies that we have no information on the value of the conjugate momentum, P_x. With the choice (5.1.14), it is the radial distance of the center of gyration from a particular field line that is specified, and we then have no information on the conjugate (radial) momentum.

The detailed discussion below is for the Landau gauge (5.1.12), and some specific results for the cylindrical gauge (5.1.14) are noted.

5.1.3 Construction of the Wavefunctions

One is free to assume a wavefunction of the form

$$\Psi(t, \boldsymbol{x}) = f(x)\, e^{-iEt + iP_y y + iP_z z} = f(x)\, e^{-i\epsilon(\varepsilon t - p_y y - p_z z)}, \tag{5.1.15}$$

where $\epsilon = \pm 1$ is the sign of the energy, whose magnitude is ε. The function $f(x)$ is a column matrix whose components are denoted by f_1, f_2, f_3, f_4. On inserting the trial solution (5.1.15) into the Dirac equation in the form

$$\left(i\frac{\partial}{\partial t} - \hat{H}\right)\Psi(t, \boldsymbol{x}) = 0, \tag{5.1.16}$$

with $\hat{H}$ given by (5.1.11) and (5.1.12) in the Coulomb gauge, one requires

$$\begin{pmatrix} -\epsilon\varepsilon + m & 0 & \epsilon p_z & \hat{\mathcal{O}}_1 \\ 0 & -\epsilon\varepsilon + m & \hat{\mathcal{O}}_2 & -\epsilon p_z \\ \epsilon p_z & \hat{\mathcal{O}}_1 & -\epsilon\varepsilon - m & 0 \\ \hat{\mathcal{O}}_2 & -\epsilon p_z & 0 & -\epsilon\varepsilon - m \end{pmatrix} \begin{pmatrix} f_1(x) \\ f_2(x) \\ f_3(x) \\ f_4(x) \end{pmatrix} = 0, \tag{5.1.17}$$

$$\hat{\mathcal{O}}_1 = -i\left(\frac{\partial}{\partial x} + \epsilon p_y + eBx\right), \qquad \hat{\mathcal{O}}_2 = -i\left(\frac{\partial}{\partial x} - \epsilon p_y - eBx\right). \tag{5.1.18}$$

It is convenient to write

$$\xi = (eB)^{1/2}\left(x + \frac{\epsilon p_y}{eB}\right), \tag{5.1.19}$$

so that (5.1.17) reduces to

$$\begin{aligned} &(-\epsilon\varepsilon + m) f_1 + \epsilon p_z f_3 - i(eB)^{1/2}(\xi + d/d\xi) f_4 = 0, \\ &(-\epsilon\varepsilon + m) f_2 - \epsilon p_z f_4 + i(eB)^{1/2}(\xi - d/d\xi) f_3 = 0, \\ &(-\epsilon\varepsilon - m) f_3 + \epsilon p_z f_1 - i(eB)^{1/2}(\xi + d/d\xi) f_2 = 0, \\ &(-\epsilon\varepsilon - m) f_4 - \epsilon p_z f_2 + i(eB)^{1/2}(\xi - d/d\xi) f_1 = 0. \end{aligned} \tag{5.1.20}$$

Operating on the first and third of these equations with $(\xi - d/d\xi)$ and on the second and fourth with $(\xi + d/d\xi)$, the four first order equations are replaced by two second order equations:

$$\begin{aligned} &\left[\frac{d^2}{d\xi^2} + \frac{\varepsilon^2 - m^2 - p_z^2}{eB} - (\xi^2 + 1)\right] f_{1,3} = 0, \\ &\left[\frac{d^2}{d\xi^2} + \frac{\varepsilon^2 - m^2 - p_z^2}{eB} - (\xi^2 - 1)\right] f_{2,4} = 0. \end{aligned} \tag{5.1.21}$$

Equations (5.1.21) are of the same form as Schrödinger's equation for a simple harmonic oscillator. The solutions are simple harmonic oscillator wavefunctions. As is familiar for a simple harmonic oscillator, normalizable solutions exist only for discrete energy eigenvalues, specifically, $(n + \frac{1}{2})\omega$ for an oscillator with frequency ω. The differential equations (5.1.21) have normalizable solutions only if the constant n, defined by

$$n = \frac{\varepsilon^2 - m^2 - p_z^2}{2eB}, \tag{5.1.22}$$

has non-negative integral values. It is convenient to introduce another non-negative integer, l, by writing

$$2n \mp 1 = 2l + 1. \tag{5.1.23}$$

The interpretation of n and l is simplest in the nonrelativistic limit: n is the Landau quantum number that determines the perpendicular energy of the particle, and it is composed of a gyromotion, which is simple harmonic motion with energy $(l + \frac{1}{2})\Omega$, and a spin part, $\frac{1}{2}s\Omega$, with $s = \pm 1$. In the relativistic case the corresponding contributions are to $p_\perp^2$ and they are $(2l + 1)eB$ and seB, respectively.

The normalized solutions are the harmonic oscillator wavefunctions

$$v_n(\xi) = \frac{1}{(\sqrt{\pi}2^n n!)^{1/2}} H_n(\xi)\, e^{-\xi^2/2}, \tag{5.1.24}$$

where H_n is a hermite polynomial. The differential operators in (5.1.20) become the raising and lowering operators that satisfy

$$\left(\xi + \frac{d}{d\xi}\right) v_n(\xi) = \sqrt{2n}\, v_{n-1}(\xi), \quad \left(\xi - \frac{d}{d\xi}\right) v_n(\xi) = \sqrt{2(n+1)}\, v_{n+1}(\xi). \tag{5.1.25}$$

A general solution of (5.1.20) may be written in the form

$$f(x) = \begin{pmatrix} C_1 v_{n-1}(\xi) \\ C_2 v_n(\xi) \\ C_3 v_{n-1}(\xi) \\ C_4 v_n(\xi) \end{pmatrix}, \tag{5.1.26}$$

where $C_1, \ldots, C_4$ are constants. For convenience, so that (5.1.26) includes the ground state $n = 0$, it is assumed that $v_{-1}(\xi)$ is identically zero.

All the states except the ground state are doubly degenerate, as may be seen by writing (5.1.23) in the form $n = l + \frac{1}{2}(1 + s)$, with $s = \pm 1$ as the spin eigenvalue. The ground state, $n = 0$, has $l = 0$, $s = -1$, and states with $n > 1$ are doubly degenerate with $s = \pm 1$, $l = n - \frac{1}{2}(1 + s)$. The particle energy eigenvalues are $\varepsilon = \varepsilon_n(p_z)$, with

$$\varepsilon_n(p_z) = (m^2 + p_z^2 + 2neB)^{1/2}, \qquad n = l + \tfrac{1}{2}(1 + s). \tag{5.1.27}$$

With the sign ϵ included in $P^0 = \epsilon\varepsilon$, the energy eigenvalues for positrons are the same as for electrons: $\varepsilon = \varepsilon_n(p_z)$. Note that this fixes an ambiguity in the choice of sign of the spin of the positron relative to the electron: the ground state is

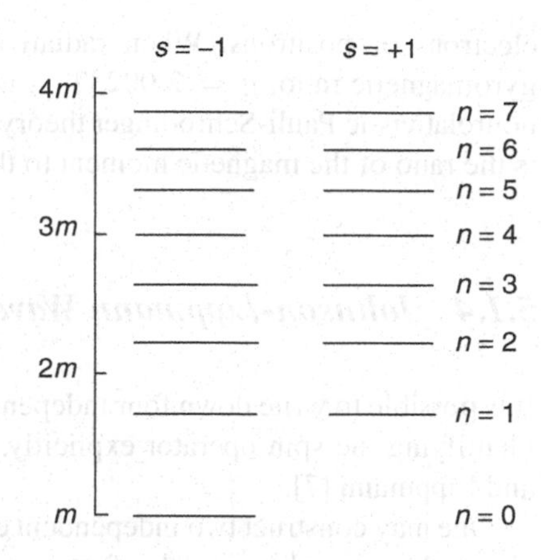

Fig. 5.1 The energy eigenvalues for $B/B_c = 1$, $p_z = 0$ for $n = 0, 1, \ldots, 7$. The uneven spacing between the levels is referred to as the anharmonicity. The levels for spin states $s = +1, s = -1$ are degenerate except for the ground state, $n = 0$, which has $s = -1$

$l = 0$, $s = -1$ for both an electron and a positron. For convenience in notation, the abbreviation $\varepsilon_n(p_z) \to \varepsilon_n$ is used when no confusion should result.

The critical magnetic field, B_c, may be defined by writing (5.1.27) in the form $\varepsilon_n(p_z) = m(1 + p_z^2/m^2 + 2nB/B_c)^{1/2}$, with $B_c = m^2/e$. In SI units this field is

$$B_c = \frac{m^2c^2}{\hbar e} = 4.4 \times 10^9\,\mathrm{T}. \tag{5.1.28}$$

The energy eigenvalues are illustrated in Fig. 5.1, showing the two branches with $s = -1$ and $s = +1$. The spacing between the energy eigenvalues decreases as n increases. This is a relativistic effect. In contrast, in the nonrelativistic case, the difference between neighboring eigenvalues is Ω independent of n, as for a simple harmonic oscillator. As a consequence the relativistic dependence of the energy spacing on n is sometimes referred to as the anharmonicity.

Magnetic Moment of the Electron

The energy eigenvalues (5.1.27) may be written $\varepsilon_n = (m^2 + p_z^2 + p_n^2)^{1/2}$, with $p_n = (2neB)^{1/2}$, and the two contributions to $n = l + \frac{1}{2}(1 + s)$ may be interpreted as an orbital part, described by $l = 0, 1, 2, \ldots$, and a spin part, described by $s = \pm 1$. The orbital part describes the perpendicular motion, which is simple harmonic motion. The remaining part is interpreted as the magnetic energy, $-\boldsymbol{\mu} \cdot \boldsymbol{B}$, due to a magnetic dipole $\boldsymbol{\mu}$ in the magnetic field. The Dirac theory predicts a magnetic moment (in SI units) $\frac{1}{2}g\mu_B s$, where g is the gyromagnetic ratio, and where (SI units)

$$\mu_B = e\hbar/2m_e = 2.74 \times 10^{-24}\,\mathrm{J\,T^{-1}} \tag{5.1.29}$$

is the Bohr magneton. The magnetic moment for a positron with spin s is $-\frac{1}{2}g\mu_B s$, implying that the magnetic-moment eigenvalue can be written $\frac{1}{2}g\mu_B \epsilon s$ for either electrons or positrons. When radiative corrections are taken into account, the gyromagnetic ratio, $g = 2.00232\ldots$ is slightly greater than 2. In contrast, in the nonrelativistic Pauli-Schrödinger theory the gyromagnetic ratio of the electron, that is the ratio of the magnetic moment to the spin, is undetermined.

5.1.4 Johnson-Lippmann Wavefunctions

It is possible to write down four independent solutions of the Dirac equation without identifying the spin operator explicitly. The procedure used here follows Johnson and Lippmann [7].

One may construct two independent eigenstates for the doubly degenerate energy eigenvalues by choosing the first two columns of the matrix of coefficients in (5.5.11). This gives

$$\begin{pmatrix} C_1 \\ C_2 \\ C_3 \\ C_4 \end{pmatrix} = c_n \left[\frac{1+s}{2} \begin{pmatrix} \epsilon\varepsilon_n + m \\ 0 \\ \epsilon p_z \\ i p_n \end{pmatrix} + \frac{1-s}{2} \begin{pmatrix} 0 \\ \epsilon\varepsilon_n + m \\ -i p_n \\ -\epsilon p_z \end{pmatrix} \right], \tag{5.1.30}$$

with the normalization coefficient identified as $c_n = 1/[2\epsilon\varepsilon_n(\epsilon\varepsilon_n + m)V]^{1/2}$, where the normalization is to an energy ε in the volume V. The four solutions, written $\Psi_q^\epsilon(t, \boldsymbol{x})$ with q denoting the quantum numbers p_z, n, s collectively, are

$$\Psi_q^\epsilon(t, \boldsymbol{x}) = \frac{e^{-i\epsilon\varepsilon t + i\epsilon p_y y + i\epsilon p_z z}}{[2\epsilon\varepsilon_n(\epsilon\varepsilon_n + m)V]^{1/2}}$$
$$\times \left[\frac{1+s}{2} \begin{pmatrix} (\epsilon\varepsilon_n + m)v_{n-1}(\xi) \\ 0 \\ \epsilon p_z v_{n-1}(\xi) \\ i p_n v_n(\xi) \end{pmatrix} + \frac{1-s}{2} \begin{pmatrix} 0 \\ (\epsilon\varepsilon_n + m)v_n(\xi) \\ -i p_n v_{n-1}(\xi) \\ -\epsilon p_z v_n(\xi) \end{pmatrix} \right]. \tag{5.1.31}$$

The solutions in the form (5.1.31) are referred to here as the Johnson-Lippmann [7] wavefunctions.

Although the four solutions do not correspond to any sensibly defined spin operator, in the nonrelativistic limit they may be interpreted loosely in terms of spin up and down ($s = \pm 1$) for an electron. However, this implicit spin operator loses physical significance for an electron that is not at rest, and it does not describe the spin of a positron in any meaningful way. When one is interested in spin-dependent effects, the choice (5.1.31) is not appropriate, and a specific choice of spin operator needs to be made (§ 5.2). The choice (5.1.31) may be used when one is not interested in the spin and where either a sum or an average over the spin states is performed.

5.1.5 Orthogonality and Completeness Relations

The orthogonality relation between the wavefunctions for different eigenstates is of the general form

$$\int d^3x\, [\Psi_q^{\epsilon}(t,\boldsymbol{x})]^{\dagger} \Psi_{q'}^{\epsilon'}(t,\boldsymbol{x}) = \delta^{\epsilon\epsilon'}\delta_{qq'}, \tag{5.1.32}$$

where q and q' denote two sets of eigenvalues collectively. The completeness relation for the wavefunctions is

$$\sum_{\epsilon,q} \Psi_q^{\epsilon}(t,\boldsymbol{x})[\Psi_q^{\epsilon}(t,\boldsymbol{x}')]^{\dagger} = \delta^3(\boldsymbol{x}-\boldsymbol{x}'). \tag{5.1.33}$$

In the present case some of the quantum numbers are discrete (ϵ, l, s) and some are continuous (p_y, p_z). The sum and the Kronecker δ are appropriate for discrete quantum numbers, and for continuous quantum numbers, these are replaced by integrals and Dirac δ-functions, respectively. To rewrite the sums as integrals, and Kronecker δs as Dirac δs, one needs to take the normalization conditions into account and ensure that the resulting integrals and Dirac δs are dimensionless.

Consider the case where the particle is confined to a large but finite box, in which case all the quantum numbers are discrete. Let the sides of the box be of length L_x, L_y, L_z in the x, y, z directions, respectively. The eigenvalues p_y, p_z are discrete, with values $p_y = n_y 2\pi/L_y$, $p_z = n_z 2\pi/L_z$ with $n_y, n_z = 0, \pm1, \pm2, \ldots$. The sum over all states includes sums over n_y, n_z. To identify the corresponding integrals and Dirac δ-functions in the continuous limit, the box is allowed to extend to infinity. The basic identification is that the difference between n_y, n_y+1 corresponds to a difference $\delta p_y = 2\pi/L_y$ in p_y, and the difference between n_z, n_z+1 corresponds to a difference $\delta p_z = 2\pi/L_z$ in p_z. The sum over states becomes

$$\sum_q = \sum_{\epsilon,s=\pm} \sum_{n=0}^{\infty} L_y L_z \int \frac{dp_y}{2\pi} \int \frac{dp_z}{2\pi}. \tag{5.1.34}$$

The Kronecker δ that expresses the orthogonality of the states becomes

$$\delta_{qq'} = \delta_{ss'}\delta_{nn'}\frac{2\pi}{L_y}\delta(p_y - p_y')\frac{2\pi}{L_z}\delta(p_z - p_z'). \tag{5.1.35}$$

These results apply for the Landau gauge. There are analogous results for other choices of gauge. In particular, for the cylindrical gauge, $L_y \int dp_y/2\pi$ in (5.1.34) and $(2\pi/L_y)\delta(p_y - p_y')$ in (5.1.35) are replaced by $\sum_r$ and $\delta_{rr'}$, where r is the radial quantum number introduced in (5.2.26) below.

The normalization is implicit in (5.1.31), and evaluating the integral in (5.1.32) the factor on the right hand side of (5.1.32) is $(1/eB)^{1/2}L_y L_z/V$. In the cylindrical gauge, the natural normalization of the wavefunctions corresponds to $1/AeB$

charges in the volume $V = AL_z$ with $A = \pi R^2$ where R is the radius of the normalization cylinder. The normalization condition leads to the identifications

$$L_x = \left(\frac{1}{eB}\right)^{1/2}, \qquad A = \pi R^2 = \frac{1}{eB}, \tag{5.1.36}$$

in these two cases, respectively.

5.2 Spin Operators and Eigenfunctions

In this section solutions are constructed for specific spin operators. Two operators are chosen: the helicity operator and the parallel (to $\boldsymbol{B}$) component of the magnetic-moment operator. The solutions are derived in the Landau gauge (5.1.12); a solution is also written down for the choice of the cylindrical gauge (5.1.14). An important point, first made by Sokolov and Ternov [14], is that there is a preferred choice of spin operator for an electron in a magnetic field, and this is the component of the magnetic-moment operator along the magnetic field. The eigenstates of other spin operators precess about the magnetic field. Other arguments for this preference have also been given [4–6]. The helicity states are written down here for comparison, and to emphasize that the solutions depend on the choice of spin operator.

5.2.1 Helicity Eigenstates in a Magnetic Field

The helicity operator, $\boldsymbol{\Sigma}\cdot\hat{\boldsymbol{p}}$, in the absence of a magnetic field is the time-component of a 4-vector. This operator commutes with the Hamiltonian and hence is a constant of the motion. As a consequence there are well-defined simultaneous eigenstates of both the Hamiltonian and the helicity operator. When a magnetostatic field is included, using the minimal coupling assumption, the helicity operator becomes

$$\hat{h} = \boldsymbol{\Sigma}\cdot(\hat{\boldsymbol{p}} + e\boldsymbol{A}). \tag{5.2.1}$$

The helicity is a constant of the motion in the presence of a magnetic field, provided that there is no electric field. Hence we can construct simultaneous eigenstates of this operator and of the Hamiltonian.

Evaluating the helicity operator in the Landau gauge gives

$$\boldsymbol{\Sigma}\cdot[\hat{\boldsymbol{p}} + e\boldsymbol{A}] = \begin{pmatrix} \epsilon p_z & \hat{\mathcal{X}}_+ & 0 & 0 \\ \hat{\mathcal{X}}_- & -\epsilon p_z & 0 & 0 \\ 0 & 0 & \epsilon p_z & \hat{\mathcal{X}}_+ \\ 0 & 0 & \hat{\mathcal{X}}_- & -\epsilon p_z \end{pmatrix}, \quad \hat{\mathcal{X}}_\pm = -i\sqrt{eB}\left(\frac{\partial}{\partial\xi} \pm \xi\right). \tag{5.2.2}$$

Let the eigenvalues of the helicity operator be $\pm h$, with the magnitude, h, to be determined. The sign of the spin eigenvalue depends on the spin quantum number, $\sigma = \pm 1$. The sign $\sigma = -1$ is required for the ground state, and one finds this condition is satisfied only if the helicity eigenvalue is written as $\sigma\epsilon Ph$ with $P = p_z/|p_z|$. Hence, the eigenvalue equation is

$$\boldsymbol{\Sigma} \cdot [\hat{\boldsymbol{p}} + e\boldsymbol{A}]\Psi_q^\epsilon(t, \boldsymbol{x}) = \sigma\epsilon Ph\, \Psi_q^\epsilon(t, \boldsymbol{x}). \tag{5.2.3}$$

The only change from the Johnson-Lippmann wavefunctions is in the coefficients C_i in (5.1.26). For the helicity states, in place of (5.1.30), the coefficients C_i are determined by (5.2.3), which is regarded as an eigenvalue equation. The explicit form of this equation is

$$\begin{pmatrix} \epsilon p_z - \sigma\epsilon Ph & -ip_n & 0 & 0 \\ ip_n & -\epsilon p_z - \sigma\epsilon Ph & 0 & 0 \\ 0 & 0 & \epsilon p_z - \sigma\epsilon Ph & -ip_n \\ 0 & 0 & ip_n & -\epsilon p_z - \sigma\epsilon Ph \end{pmatrix} \begin{pmatrix} C_1 \\ C_2 \\ C_3 \\ C_4 \end{pmatrix} = 0. \tag{5.2.4}$$

The determinant of the matrix of coefficients in (5.2.4) is $(h^2 - p_n^2 - p_z^2)^2$. Setting this to zero, it follows that the eigenvalues of the helicity operator are doubly degenerate with eigenvalues of magnitude

$$h = \left(p_n^2 + p_z^2\right)^{1/2} = \left(\varepsilon_n^2 - m^2\right)^{1/2}. \tag{5.2.5}$$

If the eigenvalues were not degenerate, the eigenfunctions could be constructed from the inverse of the square matrix in (5.2.4), but the degeneracy precludes this procedure because the inverse of the matrix of coefficients (5.2.4) is singular.

Simultaneous eigenfunctions of the helicity and energy may be constructed by starting with an arbitrary linear combinations of the doubly degenerate Johnson-Lippmann wavefunctions (5.1.31). The ratio of the coefficients in the combination is found by applying the helicity operator and requiring that the eigenvalues be σPh. The solution is determined only to within an arbitrary phase factor for each eigenfunction. For any choice of spin operator, the ground state ($n = 0$) wavefunction has the same form as for the Johnson-Lippmann wavefunctions, and this requires the presence of the sign ϵP in (5.2.3). The helicity eigenfunctions may be written in a variety of equivalent forms, by making particular choices of the overall and relative phases of the different eigenfunction, and by using the identity

$$p_n = [(h + \sigma|p_z|)(h - \sigma|p_z|)]^{1/2}. \tag{5.2.6}$$

Specific simultaneous eigenfunctions of the helicity operator (eigenvalues $\sigma\epsilon Ph$) and of the Hamiltonian (eigenvalues $\epsilon\varepsilon_n$) correspond to

$$\begin{pmatrix} C_1 \\ C_2 \\ C_3 \\ C_4 \end{pmatrix} = \frac{1}{[2h2\varepsilon_n V]^{1/2}} \begin{pmatrix} [\varepsilon_n + \epsilon m]^{1/2} (h + \sigma |p_z|)^{1/2} \\ i\sigma\epsilon P [\varepsilon_n + \epsilon m]^{1/2} (h - \sigma |p_z|)^{1/2} \\ \sigma P [\varepsilon_n - \epsilon m]^{1/2} (h + \sigma |p_z|)^{1/2} \\ i\epsilon [\varepsilon_n - \epsilon m]^{1/2} (h - \sigma |p_z|)^{1/2} \end{pmatrix}, \tag{5.2.7}$$

with ε_n given by (5.1.27), with h given by (5.2.5) and with $P = p_z/|p_z|$.

5.2.2 Magnetic-Moment Eigenstates

The magnetic-moment operator in the absence of a magnetic field is

$$\hat{\mu} = m\boldsymbol{\Sigma} - i\gamma \times \hat{\boldsymbol{p}}. \tag{5.2.8}$$

In the presence of a magnetic field the minimal coupling assumption implies that the magnetic-moment operator and its z-component in the Landau gauge are

$$\hat{\mu} = m\boldsymbol{\Sigma} - i\gamma \times (\hat{\boldsymbol{p}} + e\boldsymbol{A}), \qquad \hat{\mu}_z = \begin{pmatrix} m & 0 & 0 & \hat{\mathcal{X}}_+ \\ 0 & -m & -\hat{\mathcal{X}}_- & 0 \\ 0 & -\hat{\mathcal{X}}_+ & m & 0 \\ \hat{\mathcal{X}}_- & 0 & 0 & -m \end{pmatrix}, \tag{5.2.9}$$

respectively, with $\hat{\mathcal{X}}_\pm$ defined by (5.2.2). The simultaneous eigenvalues of the operator (5.2.9) and the Hamiltonian are found in the same way as for the helicity operator. Suppose that eigenvalues of $\hat{\mu}_z$ are $s\lambda$, with λ yet to be determined. The operator $\hat{\mathcal{X}}_\pm$ has eigenvalues $\mp i p_n$, and in place of (5.2.4), one finds

$$\begin{pmatrix} m - s\lambda & 0 & 0 & -ip_n \\ 0 & -m - s\lambda & -ip_n & 0 \\ 0 & ip_n & m - s\lambda & 0 \\ ip_n & 0 & 0 & -m - s\lambda \end{pmatrix} \begin{pmatrix} C_1 \\ C_2 \\ C_3 \\ C_4 \end{pmatrix} = 0. \tag{5.2.10}$$

The determinant of the matrix of coefficients gives $(\lambda^2 - m^2 - p_n^2)^2$. Hence, there are degenerate eigenfunctions with eigenvalues $s\lambda$, $s = \pm 1$, with $\lambda = \varepsilon_n^0$,

$$\varepsilon_n^0 = \left(m^2 + p_n^2\right)^{1/2} = \left(\varepsilon_n^2 - p_z^2\right)^{1/2}. \tag{5.2.11}$$

The eigenfunctions are linear combinations of the Johnson-Lippmann wavefunctions. The ground state ($n = 0$, $s = -1$) must be the same (to within an arbitrary phase) for all choices of spin operator. One finds that simultaneous eigenfunctions of the Hamiltonian and magnetic-moment operators correspond to

$$\begin{pmatrix} C_1 \\ C_2 \\ C_3 \\ C_4 \end{pmatrix} = \frac{1}{V^{1/2}} \begin{pmatrix} a_{\epsilon s}\, b_s \\ -is\, a_{-\epsilon s}\, b_{-s} \\ a_{-\epsilon s}\, b_s \\ is\, a_{\epsilon s}\, b_{-s} \end{pmatrix},$$

$$a_\pm = P_\pm \left(\frac{\varepsilon_n \pm \varepsilon_n^0}{2\varepsilon_n} \right)^{1/2}, \qquad b_s = \left(\frac{\varepsilon_n^0 + sm}{2\varepsilon_n^0} \right)^{1/2}. \tag{5.2.12}$$

where the identities $p_z = 2\varepsilon_n a_{\epsilon s} a_{-\epsilon s}$, $p_n = 2\varepsilon_n^0 b_s b_{-s}$ are used. The sign

$$P_\pm = \tfrac{1}{2}(1 + P) \pm \tfrac{1}{2}(1 - P), \qquad P = p_z/|p_z|, \tag{5.2.13}$$

is equal to unity for a_+ and to P for a_-. The overall phase of either eigenfunction is arbitrary, and so are the relative phase of the four eigenfunctions; these phases are chosen for convenience in writing down the form (5.2.12).

An alternative form for the solutions (5.2.12), written down by Sokolov and Ternov [14], involves the sum and difference of $(1 \pm p_z/\varepsilon_n)^{1/2}$ in place of the $a_\pm$. The identities

$$\varepsilon_n^0 = (\varepsilon_n \pm p_z)^{1/2}(\varepsilon_n \mp p_z)^{1/2}, \qquad \varepsilon_n \pm \varepsilon_n^0 = \tfrac{1}{2}\left[(\varepsilon_n + p_z)^{1/2} \pm (\varepsilon_n - p_z)^{1/2}\right]^2, \tag{5.2.14}$$

relate the two notations. The choice of eigenfunctions made by Sokolov and Ternov [14] is

$$\begin{pmatrix} C_1 \\ C_2 \\ C_3 \\ C_4 \end{pmatrix} = \frac{1}{2\sqrt{2V}} \begin{pmatrix} B_1(A_1 + A_2) \\ -iB_2(A_1 - A_2) \\ B_1(A_1 - A_2) \\ iB_2(A_1 + A_2) \end{pmatrix}, \tag{5.2.15}$$

$$A_1 = \left(1 + \frac{\epsilon p_z}{\varepsilon_n}\right)^{1/2}, \qquad A_2 = s\epsilon \left(1 - \frac{\epsilon p_z}{\varepsilon_n}\right)^{1/2},$$

$$B_1 = \left(1 + \frac{sm}{\varepsilon_n^0}\right)^{1/2}, \qquad B_2 = s\left(1 - \frac{sm}{\varepsilon_n^0}\right)^{1/2}. \tag{5.2.16}$$

The relative phases of the four eigensolutions are again chosen for convenience in writing.

The magnetic-moment eigenfunctions and the Johnson-Lippmann wavefunctions are equivalent for nonrelativistic electrons. This may be seen by setting $\epsilon = 1$ in (5.2.12), and making the nonrelativistic approximation in the form

$$\varepsilon_n = m + \frac{p_z^2 + p_n^2}{2m}, \qquad \varepsilon_n^0 = m + \frac{p_n^2}{2m}. \tag{5.2.17}$$

When only first order terms in p_z/m, p_n/m are retained, the wavefunctions are equivalent. This justifies the use of the Johnson-Lippmann wavefunctions for nonrelativistic electrons. For positrons the Johnson-Lippmann wavefunctions do not correspond to any physically relevant spin eigenfunctions, and they should not be used even for nonrelativistic positrons.

5.2.3 Eigenstates in the Cylindrical Gauge

Suppose that in place of the Landau gauge (5.1.12) one chooses the cylindrical gauge (5.1.14), viz.

$$A = \tfrac{1}{2}(-By, Bx, 0) = \tfrac{1}{2}B\varpi(-\sin\phi, \cos\phi, 0), \qquad \varpi = (x^2 + y^2)^{1/2}, \tag{5.2.18}$$

with $x = \varpi\cos\phi$, $y = \varpi\sin\phi$. In this case, z and ϕ are ignorable coordinates. In place of (5.1.15) an appropriate trial wavefunction is

$$\Psi(t, \boldsymbol{x}) = g(\varpi, \phi)\exp(-i\epsilon\varepsilon t + i\epsilon p_z z). \tag{5.2.19}$$

In place of (5.1.17), the Dirac equation gives

$$\begin{pmatrix} -\epsilon\varepsilon + m & 0 & \epsilon p_z & \hat{\mathcal{D}}_1 \\ 0 & -\epsilon\varepsilon + m & \hat{\mathcal{D}}_2 & -\epsilon p_z \\ \epsilon p_z & \hat{\mathcal{D}}_1 & -\epsilon\varepsilon - m & 0 \\ \hat{\mathcal{D}}_2 & -\epsilon p_z & 0 & -\epsilon\varepsilon - m \end{pmatrix} \begin{pmatrix} g_1(\varpi,\phi) \\ g_2(\varpi,\phi) \\ g_3(\varpi,\phi) \\ g_4(\varpi,\phi) \end{pmatrix} = 0,$$

$$\hat{\mathcal{D}}_1 = -ie^{-i\phi}\left(\frac{\partial}{\partial\varpi} - \frac{i}{\varpi}\frac{\partial}{\partial\phi} + \tfrac{1}{2}eB\varpi\right),$$

$$\hat{\mathcal{D}}_2 = -ie^{i\phi}\left(\frac{\partial}{\partial\varpi} + \frac{i}{\varpi}\frac{\partial}{\partial\phi} - \tfrac{1}{2}eB\varpi\right). \tag{5.2.20}$$

The dependence on ϕ is satisfied by the choice

$$g_1(\varpi,\phi) = g_1(\varpi)\exp[i(a-1)\phi], \qquad g_2(\varpi,\phi) = g_2(\varpi)\exp[ia\phi],$$
$$g_3(\varpi,\phi) = g_3(\varpi)\exp[i(a-1)\phi], \qquad g_4(\varpi,\phi) = g_4(\varpi)\exp[ia\phi], \tag{5.2.21}$$

with $a = 0, \pm1, \pm2, \ldots$. In place of (5.1.20) one finds

$$(-\epsilon\varepsilon + m)g_1 + \epsilon p_z g_3 - i\left(\frac{d}{d\varpi} + \frac{a}{\varpi} + \tfrac{1}{2}eB\varpi\right)g_4 = 0,$$

$$(-\epsilon\varepsilon + m)g_2 - \epsilon p_z g_4 - i\left(\frac{d}{d\varpi} - \frac{a-1}{\varpi} - \tfrac{1}{2}eB\varpi\right)g_3 = 0,$$

$$(-\epsilon\varepsilon - m)g_3 + \epsilon p_z g_1 - i\left(\frac{d}{d\varpi} + \frac{a}{\varpi} + \tfrac{1}{2}eB\varpi\right)g_2 = 0,$$

$$(-\epsilon\varepsilon - m)g_4 - \epsilon p_z g_2 - i\left(\frac{d}{d\varpi} - \frac{a-1}{\varpi} - \tfrac{1}{2}eB\varpi\right)g_1 = 0. \qquad (5.2.22)$$

In place of (5.1.21) one finds

$$\left[\frac{d^2}{d\varpi^2} + \frac{1}{\varpi}\frac{d}{d\varpi} - \frac{(a-1)^2}{\varpi^2} + eB(2n-a) - \tfrac{1}{4}e^2B^2\varpi^2\right]g_{1,3} = 0,$$

$$\left[\frac{d^2}{d\varpi^2} + \frac{1}{\varpi}\frac{d}{d\varpi} - \frac{a^2}{\varpi^2} + eB(2n-a+1) - \tfrac{1}{4}e^2B^2\varpi^2\right]g_{2,4} = 0. \qquad (5.2.23)$$

The normalizable solutions of (5.2.23) are generalized Laguerre polynomials $L_n^\nu(x)$, specifically the functions

$$J_\nu^n(x) = [n!/(n+\nu)!]^{1/2}\exp\left(-\tfrac{1}{2}x\right)x^{\frac{1}{2}\nu}L_n^\nu(x). \qquad (5.2.24)$$

In place of (5.1.26) one obtains the solutions

$$g(\varpi,\phi) = \begin{pmatrix} C_1 J_{n-r-1}^r\left(\tfrac{1}{2}eB\varpi^2\right)e^{i(n-r-1)\phi} \\ C_2 J_{n-r}^r\left(\tfrac{1}{2}eB\varpi^2\right)e^{i(n-r)\phi} \\ C_3 J_{n-r-1}^r\left(\tfrac{1}{2}eB\varpi^2\right)e^{i(n-r-1)\phi} \\ C_4 J_{n-r}^r\left(\tfrac{1}{2}eB\varpi^2\right)e^{i(n-r)\phi} \end{pmatrix}, \qquad (5.2.25)$$

with the so-called radial quantum number identified as

$$r = n - a. \qquad (5.2.26)$$

The construction of specific spin eigenstates involves only the determination of the ratios of the C_i, and these are the same for all choices of gauge, including the Landau and cylindrical gauges. Hence, the values (5.1.30), (5.2.7) and (5.2.12) for the C_i, for the Johnson-Lippmann, helicity and magnetic-moment eigenfunctions, respectively, also apply for the choice of the cylindrical gauge.

5.2.4 Average over the Position of the Gyrocenter

The gauge-dependent quantum number that (partially) described the position of the gyrocenter. One can establish the relation between the gauge-dependent quantum number and the position of the gyrocenter by performing appropriate averages.

For the Landau gauge, on inserting the wavefunctions (5.1.31) into

$$\langle x \rangle = \int d^3\boldsymbol{x}\, x\, [\Psi_q^\epsilon(t,\boldsymbol{x})]^\dagger \Psi_q^\epsilon(t,\boldsymbol{x}), \tag{5.2.27}$$

the y- and z-integrals are trivial. The integral over x is rewritten as an integral over $\xi = (eB)^{1/2}(x - \epsilon p_y/eB)$, which reduces to

$$\langle x \rangle = -\frac{\epsilon p_y}{eB}. \tag{5.2.28}$$

Thus the mean x-position of the gyrocenter, $\langle x \rangle$, is determined by p_y. By calculating $\langle x^2 \rangle$ one can estimate the fluctuations in the position of the gyrocenter about this mean. In the cylindrical gauge the mean value of the square of the radial distance ($\varpi^2 = x^2 + y^2$) is [14]

$$\langle \varpi^2 \rangle = \frac{1}{eB}(2n + 2r + 1). \tag{5.2.29}$$

In the non-quantum limit $\langle \varpi^2 \rangle$ is equal to $R^2 + a^2$, where R is the radius of gyration and a is the radial distance of the center of gyration from the z-axis. The radius of gyration in the non-quantum limit is $R = (2n/eB)^{1/2}$, and hence the radial distance of the center of gyration from the z-axis is related to $(2r/eB)^{1/2}$. This justifies the interpretation of r as specifying the mean radial position of the center of gyration.

For most purposes the position of the center of gyration is irrelevant and one either ignores it or averages over it. In the Landau gauge this average involves the operation

$$\hat{O}_{\rm av} = \frac{1}{L_x}\int_{-L_x/2}^{L_x/2} dx \to \frac{2\pi}{eBL_x}\int \frac{dp_y}{2\pi}, \tag{5.2.30}$$

where (5.2.28) is used. The coefficient on the right hand side of (5.2.30) reduces to $2\pi/(eB)^{1/2}$ with the normalization condition, $L_x = 1/(eB)^{1/2}$. Similarly, in the cylindrical gauge the average involves the operation

$$\hat{O}_{\rm av} = \frac{1}{A}\int_0^{2\pi} d\phi \int_0^\infty d\varpi\, \varpi \to \frac{2\pi}{AeB}\sum_r, \tag{5.2.31}$$

where the integrand is assumed to be independent of ϕ, and where $A = 1/eB$ is the normalization area, so that the coefficient on the right hand side of (5.2.31) reduces to 2π.

5.3 Electron Propagator in a Magnetostatic Field

The electron propagator is derived in this section both for an arbitrary magnetized electron gas, and for the magnetized vacuum. The explicit form of the propagator depends on the choices of gauge and of spin operator. The gauge-dependence can be expressed in terms of a phase factor. Explicit evaluation is possible for the propagator in the magnetized vacuum, which can be written in several superficially different forms, one of which involves a single integral over elementary functions.

5.3.1 *Statistically Averaged Electron Propagator*

The statistically averaged electron propagator in coordinate space is

$$\bar{G}(x,x') = \sum_{\epsilon,q} \Psi_q^\epsilon(x)\overline{\Psi}_q^\epsilon(x') \int \frac{dE}{2\pi}\, e^{-iE(t-t')} \times \left[\frac{1-n_q^\epsilon}{E-\epsilon(\varepsilon_q - i0)} + \frac{n_q^\epsilon}{E-\epsilon(\varepsilon_q + i0)}\right]. \tag{5.3.1}$$

In (5.3.1) the electrons ($\epsilon = 1$) and positrons ($\epsilon = -1$) have energy eigenvalues ε_q and occupation numbers n_q^ϵ, where q describes any appropriate set of quantum numbers for an electron in a magnetostatic field. On inserting the wavefunctions given in § 5.2 for a particular choice of gauge and of spin operator into (5.3.1), one obtains an explicit form for the propagator. Although the resulting expression for the propagator is gauge dependent, the gauge-dependent part may be separated out into a single multiplicative function, $\phi(x,x')$. To see this, it is helpful to consider the explicit form in both the Landau and cylindrical gauges.

In the Landau gauge, the sum over states includes an integral over p_y, which is evaluated using

$$\frac{1}{(eB)^{1/2}} \int dp_y\, v_n(\xi)v_n(\xi')\, e^{i\epsilon p_y(y-y')} = \phi(x,x')\, e^{-R^2/4}\, L_n(R^2/2),$$
$$R^2 = eB[(x-x')^2 + (y-y')^2], \tag{5.3.2}$$

where L_n is the Laguerre polynomial of order n. All the gauge dependence appears in the function

$$\phi(x,x') = \exp\left[-ieB\tfrac{1}{2}(x+x')(y-y')\right]. \tag{5.3.3}$$

The phase factor (5.3.3) applies in the Landau gauge $A^\mu = (0, 0, Bx, 0)$, and it may be written

$$\phi(x,x') = \exp\left[-ie\int_x^{x'} dx''_\mu A^\mu(x'')\right], \tag{5.3.4}$$

where the integral is along the straight line between the two end points. (For example, write $x''^{\mu} = x^{\mu} + \alpha(x' - x)^{\mu}$, $x^{\mu} = (t, x, y, z)$, etc., and integrate over $0 \le \alpha \le 1$.) A generalization, introduced by Schwinger [13], is

$$\phi(x, x') = \exp\left\{-ie \int_{x}^{x'} dx''_{\mu}\left[A^{\mu}(x'') + \tfrac{1}{2}F^{\mu\nu}x''_{\nu}\right] - \tfrac{1}{2}iex_{\mu}x'_{\nu}F^{\mu\nu}\right\}, \quad (5.3.5)$$

which applies along an arbitrary path between the end points.

In the analogous derivation in the cylindrical gauge, (5.1.14), the sum is over the radial quantum number r, and is performed using

$$\sum_{r=0}^{\infty} J^{r}_{n-r}\left(\frac{eB\varpi^2}{2}\right) J^{r}_{n-r}\left(\frac{eB\varpi'^2}{2}\right) e^{i(n-r)(\phi-\phi')} = \phi(x, x')e^{-\frac{1}{4}R^2} L_n\left(\frac{R^2}{2}\right), \quad (5.3.6)$$

with $\varpi^2 = x^2 + y^2$, $\varpi'^2 = x'^2 + y'^2$. The gauge-dependence in (5.3.6) is written in terms of the factor $\phi(x, x')$, again given by (5.3.4) but now for the cylindrical gauge.

Gauge-Independent Form for the Propagator

The sum over states in the expression (5.3.1) for the propagator gives

$$\sum_{q} \Psi^{\epsilon}_{q}(x)\overline{\Psi}^{\epsilon}_{q}(x') = (i\partial\!\!\!/ + eA\!\!\!/(x) + m)\,\phi(x, x') \sum_{n=0}^{\infty} \frac{eB}{2\pi} \int \frac{dp_z}{2\pi} \frac{e^{i\epsilon p_z(z-z')}}{2\epsilon\varepsilon_n}$$
$$\times\, e^{-R^2/4}\left[\mathcal{P}_{+}L_{n-1}(R^2/2) + \mathcal{P}_{-}L_n(R^2/2)\right], \qquad \mathcal{P}_{\pm} = \frac{1}{2}(1 \pm \Sigma_z), \quad (5.3.7)$$

with $L_n(R^2/2)$ assumed to be identically zero for $n < 0$. The differential operator $(i\partial\!\!\!/ + eA\!\!\!/ + m)$ operates on all quantities to its right when (5.3.7) is inserted into (5.3.1). This differential operator is gauge-dependent, and it may be commuted with the gauge-dependent factor, $\phi(x, x')$,

$$(i\partial\!\!\!/ + eA\!\!\!/(x) + m)\,\phi(x, x') = \phi(x, x')(i\partial\!\!\!/ + eb\!\!\!/(x - x') + m),$$
$$b^{\mu}(x) = \left(0, \tfrac{1}{2}\mathbf{B} \times \mathbf{x}\right). \quad (5.3.8)$$

The remaining operator, $(i\partial\!\!\!/ + eb\!\!\!/(x) + m)$, which operates on all quantities to its right when (5.3.7) is inserted into (5.3.1), is independent of the choice of gauge. Hence, all the gauge dependence remains in the phase factor $\phi(x, x')$. In general, the occupation numbers in (5.3.1) depend on n, p_z and are different for electrons ($\epsilon = 1$) and positrons ($\epsilon = -1$), so that no further evaluation of (5.3.7) is possible in general.

Géhéniau Form for the Electron Propagator

In vacuo the occupation numbers in (5.3.1) are zero, and the sum over n and the integral over p_z in (5.3.7) may be evaluated explicitly. The first step is to perform the integral over E in (5.3.1), which gives a step function

$$\int \frac{dE}{2\pi} \frac{e^{-iE(t-t')}}{E - \epsilon(\varepsilon_q - i0)} = -i\epsilon\, H\big(\epsilon(t-t')\big)\, e^{-i\epsilon\varepsilon_n(t-t')}. \tag{5.3.9}$$

The multiplicative factor of ϵ cancels with a corresponding factor in (5.3.7). The integral over p_z and the sum over n can now be performed explicitly.

The integral over p_z is rewritten using the identity

$$\int \frac{dp_z}{\varepsilon_n} e^{-i\epsilon[\varepsilon_n(t-t')-p_z(z-z')]} = \int_0^\infty \frac{d\lambda}{\lambda} \exp\left[-i\epsilon\frac{t-t'}{|t-t'|}\left(\frac{\rho^2\lambda}{2} + \frac{(\varepsilon_n^0)^2}{2\lambda}\right)\right],$$

$$\rho^2 = (t-t')^2 - (z-z')^2, \qquad \left(\varepsilon_n^0\right)^2 = m^2 + 2neB. \tag{5.3.10}$$

In view of the step function in (5.3.9), the integral is nonzero only for $\epsilon(t-t')/|t-t'| = 1$, and the exponent in (5.3.10) simplifies accordingly. The sum over n is performed using a generating function for the Laguerre polynomials,

$$\sum_{n=0}^{\infty} e^{i\alpha n} L_n(R^2/2) = i\frac{e^{\frac{1}{2}i\alpha}}{2\sin\frac{1}{2}\alpha}\, e^{R^2/4}\, e^{-\frac{1}{2}i\cot\frac{1}{2}\alpha}, \tag{5.3.11}$$

with $\alpha = -eB/\lambda$.

The resulting expression for the propagator for the magnetized vacuum is

$$\begin{aligned} G(x,x') &= -\phi(x,x')(i\partial\!\!\!/ + eA\!\!\!/(x-x') + m) \int_0^\infty \frac{d\lambda}{8\pi^2} \frac{1 - i\Sigma_z \tan(eB/2\lambda)}{(2\lambda/eB)\tan(eB/2\lambda)} \\ &\quad \times \exp\left\{\frac{ieB[(x-x')^2+(y-y')^2]}{4\tan(eB/2\lambda)} + \frac{i\lambda}{2}[(z-z')^2-(t-t')^2] - \frac{im^2}{2\lambda}\right\}. \end{aligned} \tag{5.3.12}$$

Then (5.3.12) gives the Géhéniau form propagator [1,2,8,9]

$$G(x,x') = \phi(x,x')\Delta(x-x'), \tag{5.3.13}$$

with the gauge-independent part, $\Delta(x)$, given by

$$\Delta(x) = -\frac{eB}{16\pi^2}\int_0^\infty \frac{d\lambda}{\lambda} \mathcal{B}(\lambda,x) \exp\left\{\frac{-ieB(x^2)_\perp}{4\tan(eB/2\lambda)} - \frac{i\lambda(x^2)_\parallel}{2} - \frac{im^2}{2\lambda}\right\}, \tag{5.3.14}$$

with $(x^2)_\perp = -x^2 - y^2$, $(x^2)_\parallel = t^2 - z^2$, and with

$$\mathcal{B}(\lambda, x) = \left[(\gamma x)_\perp \tfrac{1}{2} eB \cot(eB/2\lambda) + \lambda(\gamma x)_\parallel + e\not{\phi}(x)\right]\left[\cot(eB/2\lambda) - i\Sigma_z\right]$$
$$= \begin{pmatrix} (\lambda t + m)C_- & 0 & -\lambda z C_- & -R_- C_+ \\ 0 & (\lambda t + m)C_+ & -R_+ C_- & \lambda z C_+ \\ \lambda z C_- & R_- C_+ & (-\lambda t + m)C_- & 0 \\ R_+ C_- & -\lambda z C_+ & 0 & (-\lambda t + m)C_+ \end{pmatrix},$$
$$R_\pm = \tfrac{1}{2} eB(x + iy), \qquad C_\pm = \cot(eB/2\lambda) \pm i. \tag{5.3.15}$$

To ensure convergence of the integral, λ is to be interpreted as $(1 + i0)\lambda$.

5.3.2 *Electron Propagator as Green Function*

The propagator in a magnetized vacuum can also be constructed as a Green function[1] by solving the inhomogeneous Dirac equation with the external electromagnetic field included, that is, by solving

$$(i\not{\partial} + e\not{A}(x) - m)G(x, x') = \delta^4(x - x'). \tag{5.3.16}$$

It is convenient to introduce a new function $S(x, x')$ by writing

$$G(x, x') = (i\not{\partial} + e\not{A}(x) + m)S(x, x'), \tag{5.3.17}$$

such that (5.3.16) is replaced by

$$\left[D^\mu D_\mu + m^2 - eS^{\mu\nu} F_{\mu\nu}(x)\right] S(x, x') = \delta^4(x - x'),$$
$$D^\mu = \partial^\mu - ieA^\mu(x), \qquad S^{\mu\nu} F_{\mu\nu} = i\boldsymbol{\alpha} \cdot \boldsymbol{E} - \boldsymbol{\Sigma} \cdot \boldsymbol{B}. \tag{5.3.18}$$

The form (5.3.18) applies for an arbitrary static electromagnetic field. On specializing to a static magnetic field and choosing the Landau gauge (5.1.12) and (5.3.18) reduces to

$$\left[\frac{\partial^2}{\partial t^2} - \frac{\partial^2}{\partial x^2} - \frac{\partial^2}{\partial z^2} - \left(\frac{\partial}{\partial y} + ieBx\right)^2 + m^2 + e\boldsymbol{\Sigma} \cdot \boldsymbol{B}\right] S(x, x') = \delta^4(x - x'). \tag{5.3.19}$$

[1]"Green function", rather than "Green's function", was proposed by J.D. Jackson, *Classical Electrodynamics*, Third Edition, John Wiley & Sons Inc. (1999).

A solution of (5.3.19) is found by firstly solving for the corresponding Klein-Gordon equation, which is similar to (5.3.19) but with the term $-e\boldsymbol{\Sigma}\cdot\boldsymbol{B}$ omitted. Let the solution of the Klein-Gordon equation be $G_0(x,x')$. One has

$$\left[\frac{\partial^2}{\partial t^2}-\frac{\partial^2}{\partial x^2}-\frac{\partial^2}{\partial z^2}-\left(\frac{\partial}{\partial y}+ieBx\right)^2+m^2\right]G_0(x,x')=\delta^4(x-x'). \quad (5.3.20)$$

The solution, $S(x,x')$, of (5.3.19) follows from the solution of (5.3.20) for $G_0(x,x')$ by formally replacing m^2 by $m^2+e\boldsymbol{\Sigma}\cdot\boldsymbol{B}$.

The solution of the inhomogeneous equation (5.3.20) for $G_0(x,x')$ is constructed by first considering the solutions of the homogeneous equation. With a trial solution of the form $f(x)\exp[-i(Et-P_yy-P_zz)]$, one finds that solutions exist only for $E=\epsilon\varepsilon_n$, $\varepsilon_n=[m^2+p_z^2+(2n+1)eB]^{1/2}$ for $n\geq 0$, with $P_y=\epsilon p_y$, $P_z=\epsilon p_z$ and with the solution for a given n having $f(x)=v_n(\xi)$, with $v_n(\xi)$ the simple harmonic oscillator wavefunction (5.1.26), and with ξ given by (5.1.19). The identity

$$\sum_{n=0}^{\infty}v_n(\xi)v_n(\xi')=\delta(\xi-\xi') \quad (5.3.21)$$

allows one to write

$$\delta^3(\boldsymbol{x}-\boldsymbol{x}')=(eB)^{1/2}\sum_{n=0}^{\infty}v_n(\xi)v_n(\xi')\int\frac{dp_ydp_z}{(2\pi)^2}\,e^{i\epsilon[p_y(y-y')+p_z(z-z')]}, \quad (5.3.22)$$

with $\xi=(eB)^{1/2}(x-\epsilon p_y/eB)$, $\xi'=(eB)^{1/2}(x'-\epsilon p_y/eB)$. In order that the time dependence satisfy (5.3.20) one requires that $G_0(x,x')$ be continuous at $t=t'$ with a discontinuous first derivative,

$$\left.\frac{\partial G_0(x,x')}{\partial t}\right|_{t=t'}=\delta^3(\boldsymbol{x}-\boldsymbol{x}'), \quad (5.3.23)$$

as required by the integral of (5.3.20) over t. On integrating (5.3.23) over time the choice of the sign of $(t-t')/|t-t'|$ determines whether the propagator is in its retarded ($(t-t')/|t-t'|>0$), advanced ($(t-t')/|t-t'|<0$) or Feynman ($\epsilon(t-t')/|t-t'|>0$) forms. Here we require the Feynman form, which is

$$G_0(x,x')=i\epsilon H\big(\epsilon(t-t')\big)(eB)^{1/2}\sum_{n=0}^{\infty}\int\frac{dp_ydp_z}{(2\pi)^2}\frac{v_n(\xi)v_n(\xi')}{\epsilon\varepsilon_n}$$
$$\times\, e^{-i\epsilon[\varepsilon_n(t-t')-p_y(y-y')-p_z(z-z')]}. \quad (5.3.24)$$

The p_y-integral is the same as in (5.3.2), and the p_z-integral is closely analogous to (5.3.10). The sum over n is performed using (5.3.11). The resulting expression is

$$
\begin{aligned}
G_0(x,x') &= -\frac{eB}{2\pi}\,\phi(x,x')\int_0^\infty \frac{d\lambda}{8\pi^2}\,\frac{e^{-im^2/2\lambda}}{(2\lambda/eB)\sin(eB/2\lambda)} \\
&\quad\times \exp\left\{\frac{ieB[(x-x')^2+(y-y')^2]}{4\tan(eB/2\lambda)} + \frac{i\lambda}{2}\left[(z-z')^2-(t-t')^2\right]\right\}.
\end{aligned}
\tag{5.3.25}
$$

The propagator has a physical intepretation as the propagator for a charged ($q = -e$) spinless particle that satisfies the Klein-Gordon equation.

The electron propagator is obtained from (5.3.25) by noting that the replacement $m^2 \to m^2 + e\boldsymbol{\Sigma}\cdot\boldsymbol{B}$, with $\boldsymbol{\Sigma}\cdot\boldsymbol{B} = \Sigma_z B$, converts the scalar function into a 4×4 matrix by introducing an additional factor

$$
e^{-ie\Sigma_z B/2\lambda} = \cos(eB/2\lambda) - i\Sigma_z \sin(eB/2\lambda) \tag{5.3.26}
$$

into the integrand. The additional factor (5.3.26) converts $G_0(x,x')$, as given by (5.3.25), into $S(x,x')$, and then the propagator (5.3.12) follows from (5.3.17) and (5.3.8).

5.3.3 *Spin Projection Operators*

The Dirac matrices $\mathcal{P}_\pm$, introduced in (5.3.7), play the role of projection operators onto the eigenstates of Σ_z. With $\mathcal{P}_\pm = \frac{1}{2}(1\pm\Sigma_z)$ and $\Sigma_z^2 = 1$, one has

$$
(\mathcal{P}_\pm)^2 = \mathcal{P}_\pm, \quad \mathcal{P}_+\mathcal{P}_- = 0, \quad \mathcal{P}_+ + \mathcal{P}_- = 1, \quad \mathcal{P}_+ - \mathcal{P}_- = \Sigma_z. \tag{5.3.27}
$$

The projection operators have the standard matrix representations

$$
\mathcal{P}_+ = \begin{pmatrix} 1 & 0 & 0 & 0 \\ 0 & 0 & 0 & 0 \\ 0 & 0 & 1 & 0 \\ 0 & 0 & 0 & 0 \end{pmatrix}, \qquad \mathcal{P}_- = \begin{pmatrix} 0 & 0 & 0 & 0 \\ 0 & 1 & 0 & 0 \\ 0 & 0 & 0 & 0 \\ 0 & 0 & 0 & 1 \end{pmatrix}. \tag{5.3.28}
$$

These projection operators commute with the Dirac matrices $\gamma_\parallel^\mu$, but not with $\gamma_\perp^\mu$. Specifically, the relations

$$
\gamma_\parallel^\mu \Sigma_z = \Sigma_z \gamma_\parallel^\mu, \qquad \gamma_\perp^\mu \Sigma_z = -\Sigma_z \gamma_\perp^\mu, \tag{5.3.29}
$$

imply

$$
\gamma_\parallel^\mu \mathcal{P}_\pm = \mathcal{P}_\pm \gamma_\parallel^\mu, \qquad \gamma_\perp^\mu \mathcal{P}_\pm = \mathcal{P}_\mp \gamma_\perp^\mu. \tag{5.3.30}
$$

5.4 Vertex Function in a Magnetic Field

In this section, the vertex function for an electron in a magnetic field is evaluated explicitly for the three choices of spin wavefunctions (Johnson-Lippmann, helicity and magnetic moment) made in § 5.2.

5.4.1 *Definition of the Vertex Function*

In developing QED for an electron gas in a magnetostatic field it is convenient to use the vertex formalism. This formalism is based on the fact that a Dirac matrix, γ^μ say, always appears in matrix multiplication along an electron line between a Dirac wavefunction and an adjoint Dirac wavefunction. Consider an electron line with vertices corresponding to γ^μ at x' and γ^ν at x. With the propagator in the form (5.3.1), one may associate the adjoint wavefunction, $\overline{\Psi}{}^\epsilon_q(\boldsymbol{x}')$, with γ^μ and the wavefunction $\Psi^\epsilon_q(\boldsymbol{x})$ with γ^ν. (The time-dependence of the wavefunction, $e^{-i\epsilon\varepsilon_q}$, is omitted in defining the vertex function.) The other electron line joining the vertex at x' corresponds to either an initial electron, a final positron or to another propagator, and in all three cases there is another wavefunction, $\Psi^{\epsilon'}_{q'}(\boldsymbol{x}')$ say, associated with it. In an analogous manner, the electron line joining the vertex at x corresponds to a final electron, an initial positron or to another propagator, and in all three cases there is another wavefunction, $\overline{\Psi}{}^{\epsilon''}_{q''}(\boldsymbol{x})$ say, associated with it. It follows that all wavefunctions and adjoint wavefunctions are paired together with a γ-matrix. Specifically, this leads to a (coordinate-space) vertex function of the form $\overline{\Psi}{}^{\epsilon'}_{q'}(\boldsymbol{x})\gamma^\mu\Psi^\epsilon_q(\boldsymbol{x})$.

A momentum-space representation of the vertex function is introduced by Fourier transforming:

$$[\gamma^{\epsilon'\epsilon}_{q'q}(\boldsymbol{k})]^\mu = \int d^3\boldsymbol{x}\, e^{-i\boldsymbol{k}\cdot\boldsymbol{x}}\, \overline{\Psi}{}^{\epsilon'}_{q'}(\boldsymbol{x})\gamma^\mu\Psi^\epsilon_q(\boldsymbol{x}). \tag{5.4.1}$$

The evaluation of this function depends on the specific choice of gauge and of the spin operator. The vertex function (5.4.1) factorizes into a gauge-dependent factor and a gauge-independent part

$$\left[\gamma^{\epsilon'\epsilon}_{q'q}(\boldsymbol{k})\right]^\mu = d^{\epsilon'\epsilon}_{q'q}(\boldsymbol{k})\left[\Gamma^{\epsilon'\epsilon}_{q'q}(\boldsymbol{k})\right]^\mu, \tag{5.4.2}$$

where the factor $d^{\epsilon'\epsilon}_{q'q}(\boldsymbol{k})$ contains all the gauge-dependent factors. The gauge-independent vertex function, $[\Gamma^{\epsilon'\epsilon}_{q'q}(\boldsymbol{k})]^\mu$, remains dependent on the choice of spin operator. The normalization of the factor $d^{\epsilon'\epsilon}_{q'q}(\boldsymbol{k})$ is determined by requiring that it satisfy the identity

$$d^{\epsilon'\epsilon}_{q'q}(\boldsymbol{k}_1+\boldsymbol{k}_2) = \sum_{\tilde{q}''} d^{\epsilon'\epsilon''}_{q'q''}(\boldsymbol{k}_2)\, d^{\epsilon''\epsilon}_{q''q}(\boldsymbol{k}_1), \tag{5.4.3}$$

where the sum over $\tilde{q}''$ is over the gauge-dependent quantum number (p_y'' or r'') and p_z''.

Symmetry Properties of the Vertex Function

The reality condition for Fourier transforms implies that the vertex function satisfies the identity

$$[\gamma^{\epsilon'\epsilon}_{q'q}(\boldsymbol{k})]^{*\mu} = [\gamma^{\epsilon\epsilon'}_{qq'}(-\boldsymbol{k})]^{\mu}. \tag{5.4.4}$$

Both the gauge-dependent and the gauge-independent parts in (5.4.2) separately satisfy this property:

$$[d^{\epsilon'\epsilon}_{q'q}(\boldsymbol{k})]^{*} = d^{\epsilon\epsilon'}_{qq'}(-\boldsymbol{k}), \qquad [\Gamma^{\epsilon'\epsilon}_{q'q}(\boldsymbol{k})]^{*\mu} = [\Gamma^{\epsilon\epsilon'}_{qq'}(-\boldsymbol{k})]^{\mu}. \tag{5.4.5}$$

A further possible symmetry property follows by changing the signs ϵ, ϵ', but such symmetry depends on the choice of spin eigenfunctions. There is no such symmetry for the Johnson-Lippmann wavefunctions, cf. (5.1.30), and the symmetry properties can be seen by inspection of (5.2.7) for the helicity states, and by inspection of (5.2.12) for the magnetic-moment states.

The vertex function also satisfies the relation

$$k_{\mu}\left[\Gamma^{\epsilon'\epsilon}_{q'q}(\boldsymbol{k})\right]^{\mu} = (\omega - \epsilon\varepsilon_q + \epsilon'\varepsilon'_{q'})\left[\Gamma^{\epsilon'\epsilon}_{q'q}(\boldsymbol{k})\right]^{0}. \tag{5.4.6}$$

The right hand side of (5.4.6) is zero only when the resonance condition, $\omega - \epsilon\varepsilon_q + \epsilon'\varepsilon'_{q'} = 0$, is satisfied.

Landau Gauge

To illustrate the factorization (5.4.2), consider its explicit forms in the Landau gauge and in the cylindrical gauge. The wavefunctions in the Landau gauge are given by (5.1.31). Let the components of $\boldsymbol{k}$ be written

$$\boldsymbol{k} = (k_\perp \cos\psi, k_\perp \sin\psi, k_z) = |\boldsymbol{k}|(\sin\theta\cos\psi, \sin\theta\sin\psi, \cos\theta). \tag{5.4.7}$$

(The azimuthal angle ψ may be set to zero without loss of generality only when a single wave is involved.) Then in (5.4.1) with (5.4.7), the integrals over y and z are trivial. The integral over x reduces to a standard integral [3]

$$\int_{-\infty}^{\infty} dx\, e^{-ik_x x}\, v_{n'}(\xi')v_n(\xi) = (eB)^{-1/2} e^{ik_x(\epsilon p_y + \epsilon' p_y')/2eB} \times \{ie^{i\psi}\}^{n-n'} J^{n}_{n'-n}(k_\perp^2/2eB), \tag{5.4.8}$$

with $k_y = \epsilon p_y - \epsilon' p'_y = k_x \tan\psi$ and where the function $J^n_\nu(x)$ is defined by (5.2.24). The functions $J^n_\nu(x)$ play an important role in the theory, and their properties are summarized in § A.1.5. The gauge-dependent factor for the Landau gauge is identified as

$$d^{\epsilon'\epsilon}_{q'q}(\boldsymbol{k}) = \frac{e^{ik_x(\epsilon p_y + \epsilon' p'_y)/2eB}}{V(eB)^{1/2}} \, 2\pi\delta(\epsilon p_y - \epsilon' p'_y - k_y)\, 2\pi\delta(\epsilon p_z - \epsilon' p'_z - k_z). \tag{5.4.9}$$

Cylindrical Gauge

In the cylindrical gauge, the definition (5.4.1) of the vertex function is unchanged, but it is to be evaluated in cylindrical polar coordinates ϖ, ϕ, z, rather than in cartesian coordinates. The normalization factor changes in accord with (5.1.36). On inserting the wavefunctions (5.2.19) with (5.2.25) into (5.4.1), the integral over z is trivial, and the remaining integrals are of the form

$$\begin{aligned}\int_0^{2\pi} d\phi \int_0^\infty d\varpi\, \varpi\, e^{-i[k_\perp \varpi \cos(\psi-\phi)+(n'-n-r'+r)\phi]} J^{r'}_{n'-r'}\left(\tfrac{1}{2}eB\varpi^2\right) J^{r}_{n-r}\left(\tfrac{1}{2}eB\varpi^2\right) \\ = \frac{2\pi}{eB}\{-ie^{-i\psi}\}^{r-r'} J^{r}_{r'-r}(k_\perp^2/2eB)\{-ie^{-i\psi}\}^{n'-n} J^{n}_{n'-n}\left(k_\perp^2/2eB\right).\end{aligned} \tag{5.4.10}$$

The gauge-dependent factor for the cylindrical gauge is identified as

$$d^{\epsilon'\epsilon}_{q'q}(\boldsymbol{k}) = \frac{2\pi\,\{-ie^{-i\psi}\}^{r-r'} J^{r}_{r'-r}(k_\perp^2/2eB)}{VeB} \, 2\pi\delta(\epsilon p_z - \epsilon' p'_z - k_z). \tag{5.4.11}$$

5.4.2 Gauge-Dependent Factor Along an Electron Line

The multiplicative property (5.4.3) implies that the gauge-dependence involves only the initial and final states. It is straightforward to show this for the explicit forms, (5.4.9) for the Landau gauge, and (5.4.11) for the cylindrical gauge.

Consider two vertices along an electron line. The Dirac matrices are written according to matrix multiplication in the direction opposite to the arrow on the electron line. Suppose the initial, intermediate and final states are labeled with quantum numbers q, q'', q', respectively. The sum over the intermediate state gives

$$\sum_{q''}\left[\gamma^{\epsilon'\epsilon''}_{q'q''}(\boldsymbol{k}_2)\right]^\nu \left[\gamma^{\epsilon''\epsilon}_{q''q}(\boldsymbol{k}_1)\right]^\mu = \sum_{q''} d^{\epsilon'\epsilon''}_{q'q''}(\boldsymbol{k}_2)\, d^{\epsilon''\epsilon}_{q''q}(\boldsymbol{k}_1)\left[\Gamma^{\epsilon'\epsilon''}_{q'q''}(\boldsymbol{k}_2)\right]^\nu \left[\Gamma^{\epsilon''\epsilon}_{q''q}(\boldsymbol{k}_1)\right]^\mu. \tag{5.4.12}$$

The sum separates into a gauge-dependent part and a gauge-independent part. For the Landau gauge, the gauge-dependent part reduces to

$$\sum_{q''} d^{\epsilon'\epsilon''}_{q'q''}(\boldsymbol{k}_2)\, d^{\epsilon''\epsilon}_{q''q}(\boldsymbol{k}_1) = L_y L_z \int \frac{dp''_y}{2\pi} \int \frac{dp''_z}{2\pi}\, d^{\epsilon'\epsilon''}_{q'q''}(\boldsymbol{k}_2)\, d^{\epsilon''\epsilon}_{q''q}(\boldsymbol{k}_1), \tag{5.4.13}$$

where the sum over q'' reduces to the integrals over p''_y, p''_z. On inserting the explicit form (5.4.9), the integrals are trivial. The result may be written in the form

$$\sum_{q''} d^{\epsilon'\epsilon''}_{q'q''}(\boldsymbol{k}_2)\, d^{\epsilon''\epsilon}_{q''q}(\boldsymbol{k}_1) = e^{i(\boldsymbol{k}_1\times\boldsymbol{k}_2)_z/2eB}\, d^{\epsilon'\epsilon}_{q'q}(\boldsymbol{k}_1+\boldsymbol{k}_2), \tag{5.4.14}$$

where $L_x = 1/(eB)^{1/2}$ is used. The result (5.4.14) is independent of the choice of gauge.

The result (5.4.14) generalizes to an arbitrary number of vertices along an electron line. Specifically, for n vertices along a line, with 3-momentum $\boldsymbol{k}_i$ emitted at the ith vertex, one has

$$\begin{aligned}
&\sum_{q_1,\dots,q_n} \left[\gamma^{\epsilon'\epsilon_n}_{q'q_n}(\boldsymbol{k}_n)\right]^{\nu_n} \left[\gamma^{\epsilon_n\epsilon_{n-1}}_{q_nq_{n-1}}(\boldsymbol{k}_{n-1})\right]^{\nu_{n-1}} \dots \left[\gamma^{\epsilon_2\epsilon_1}_{q_2q_1}(\boldsymbol{k}_2)\right]^{\nu_2} \left[\gamma^{\epsilon_1\epsilon}_{q_1q}(\boldsymbol{k}_1)\right]^{\nu_1} \\
&= d^{\epsilon'\epsilon}_{q'q}(\textstyle\sum_i \boldsymbol{k}_i)\, e^{i\sum_{i<j}(\boldsymbol{k}_i\times\boldsymbol{k}_j)_z/2eB} \displaystyle\sum_{q_1,\dots,q_n} \left[\Gamma^{\epsilon'\epsilon_n}_{q'q_n}(\boldsymbol{k}_n)\right]^{\nu_n} \\
&\quad\times \left[\Gamma^{\epsilon_n\epsilon_{n-1}}_{q_nq_{n-1}}(\boldsymbol{k}_{n-1})\right]^{\nu_{n-1}} \dots \left[\Gamma^{\epsilon_2\epsilon_1}_{q_2q_1}(\boldsymbol{k}_2)\right]^{\nu_2} \left[\Gamma^{\epsilon_1\epsilon}_{q_1q}(\boldsymbol{k}_1)\right]^{\nu_1}.
\end{aligned} \tag{5.4.15}$$

It follows that the choice of gauge affects the description only of the initial and final states.

5.4.3 Vertex Function for Arbitrary Spin States

The form for the vertex function $[\Gamma^{\epsilon'\epsilon}_{q'q}(\boldsymbol{k})]^\mu$, defined by (5.4.2) with (5.4.1), follows by assuming the wavefunction to be of the form (5.1.26). Specifically, $\Psi^\epsilon_q(\boldsymbol{x})$ is assumed to be the column matrix $C_1 v_{n-1}(\xi)$, $C_2 v_n(\xi)$, $C_3 v_{n-1}(\xi)$, $C_4 v_n(\xi)$, and $\overline{\Psi}^{\epsilon'}_{q'}(\boldsymbol{x})$ is assumed to be the row matrix $C'^*_1 v_{n'-1}(\xi)$, $C'^*_2 v_{n'}(\xi)$, $-C'^*_3 v_{n'-1}(\xi)$, $-C'^*_4 v_{n'}(\xi)$. The integrals follow from (5.4.8), which implies

$$\begin{aligned}
&\int_{-\infty}^{\infty} dx\, e^{-ik_x x} \begin{bmatrix} v_{n'}(\xi')v_n(\xi) \\ v_{n'-1}(\xi')v_{n-1}(\xi) \\ v_{n'-1}(\xi')v_n(\xi) \\ v_{n'}(\xi')v_{n-1}(\xi) \end{bmatrix} = D\{ie^{i\psi}\}^{n-n'} \begin{bmatrix} J^n_{n'-n} \\ J^{n-1}_{n'-n} \\ ie^{i\psi}J^n_{n'-n-1} \\ -ie^{-i\psi}J^{n-1}_{n'-n+1} \end{bmatrix}, \\
&D = (eB)^{-1/2} e^{ik_x(\epsilon p_y+\epsilon' p'_y)/2eB},
\end{aligned} \tag{5.4.16}$$

where the argument of the J-functions is $x = k_\perp^2/2eB$. The factor D is incorporated into the gauge-dependent factor $d^{\epsilon'\epsilon}_{q'q}(\boldsymbol{k})$ in (5.4.2). Explicit evaluation of the vertex function then gives

$$\begin{aligned}
\Gamma^\mu = c^*_{n'}c_n\Big[&(C_1'^*C_1 + C_3'^*C_3)J^{n-1}_{n'-n} + (C_2'^*C_2 + C_4'^*C_4)J^n_{n'-n},\\
&(C_1'^*C_4 + C_3'^*C_2)ie^{i\psi}J^n_{n'-n-1} + (C_2'^*C_3 + C_4'^*C_1)(-ie^{-i\psi})J^{n-1}_{n'-n+1},\\
&-i(C_1'^*C_4 + C_3'^*C_2)ie^{i\psi}J^n_{n'-n-1} + i(C_2'^*C_3 + C_4'^*C_1)(-ie^{-i\psi})J^{n-1}_{n'-n+1},\\
&(C_1'^*C_3 + C_3'^*C_1)J^{n-1}_{n'-n} - (C_2'^*C_4 + C_4'^*C_2)J^n_{n'-n}\Big],\\
c^*_{n'} = (-ie^{-i\psi})^{n'},\qquad &c_n = (ie^{i\psi})^n.
\end{aligned}\tag{5.4.17}$$

The general form (5.4.17) may be used to write down explicit forms for the vertex function for the three choices of spin eigenfunctions discussed in § 5.2. Only the magnetic-moment eigenfunctions are used explicitly here.

The explicit form for the vertex function for the magnetic-moment eigenstates (5.2.12) is

$$\begin{aligned}
[\Gamma^{\epsilon'\epsilon}_{q'q}(\boldsymbol{k})]^\mu &= (ie^{i\psi})^{n-n'}\Big(A^{\epsilon'\epsilon(+)}_{q'q}\,J^{(0)}_{q'q}(\boldsymbol{k}),\\
&\qquad -A^{\epsilon'\epsilon(-)}_{q'q}\,J^{(+)}_{q'q}(\boldsymbol{k}), -iA^{\epsilon'\epsilon(-)}_{q'q}\,J^{(-)}_{q'q}(\boldsymbol{k}), B^{\epsilon'\epsilon}_{q'q}\,J^{(0)}_{q'q}(\boldsymbol{k})\Big),\\
A^{\epsilon'\epsilon(\pm)}_{q'q} &= a'_{\epsilon's'}a_{\epsilon s} \pm a'_{-\epsilon's'}a_{-\epsilon s},\qquad B^{\epsilon'\epsilon}_{q'q} = a'_{\epsilon's'}a_{-\epsilon s} + a'_{-\epsilon's'}a_{\epsilon s},\\
J^{(0)}_{q'q}(\boldsymbol{k}) &= b'_{s'}b_sJ^{n-1}_{n'-n}(x) + s'sb'_{-s'}b_{-s}J^n_{n'-n}(x),\\
J^{(\pm)}_{q'q}(\boldsymbol{k}) &= s'b'_{-s'}b_se^{-i\psi}J^{n-1}_{n'-n+1}(x) \pm sb'_{s'}b_{-s}e^{i\psi}J^n_{n'-n-1}(x),\\
a_\pm &= P_\pm\left(\frac{\varepsilon_n \pm \varepsilon^0_n}{2\varepsilon_n}\right)^{1/2},\qquad b_s = \left(\frac{\varepsilon^0_n + sm}{2\varepsilon^0_n}\right)^{1/2},
\end{aligned}\tag{5.4.18}$$

with $P_\pm = \frac{1}{2}(1+P) \pm \frac{1}{2}(1-P)$, $P = p_z/|p_z|$ introduced in (5.2.13).

Gauge-Invariance Condition

The vertex function in the form (5.4.18) is used widely below, and to illustrate its properties it is instructive to derive the identity (5.4.6) explicitly, Specifically, the identity implies

$$k_\mu[\Gamma^{\epsilon'\epsilon}_{q'q}(\boldsymbol{k})]^\mu = (\omega - \epsilon\varepsilon_q + \epsilon'\varepsilon'_{q'})\,b'(-ie^{-i\psi})^{n'}ie^{i\psi})^n\,A^{\epsilon'\epsilon(+)}_{q'q}\,J^{(0)}_{q'q}(\boldsymbol{k}).\tag{5.4.19}$$

Noting the implicit relation $k_z = \epsilon p_z - \epsilon' p'_z$, the identity (5.4.19) requires

$$\Big[(\epsilon\varepsilon_q - \epsilon'\varepsilon'_{q'})A^{\epsilon'\epsilon(+)}_{q'q} - (\epsilon p_z - \epsilon' p'_z)B^{\epsilon'\epsilon}_{q'q}\Big]\, J^{(0)}_{q'q}(\boldsymbol{k})$$
$$-k_\perp A^{\epsilon'\epsilon(-)}_{q'q}\Big[\cos\psi\, J^{(+)}_{q'q}(\boldsymbol{k}) + i\,\sin\psi\, J^{(-)}_{q'q}(\boldsymbol{k})\Big] = 0. \tag{5.4.20}$$

The first term in (5.4.20) may be rewritten using $a^2_{\epsilon s} + a^2_{-\epsilon s} = 1$, $p_z = 2\varepsilon_n a_{\epsilon s} a_{-\epsilon s}$, and similarly for the primed variables. One finds the identity

$$(\epsilon\varepsilon_q - \epsilon'\varepsilon'_{q'})A^{\epsilon'\epsilon(+)}_{q'q} - (\epsilon p_z - \epsilon' p'_z)B^{\epsilon'\epsilon}_{q'q} = (s\varepsilon^0_n - s'\varepsilon^0_{n'})A^{\epsilon'\epsilon(-)}_{q'q}. \tag{5.4.21}$$

The second term in (5.4.20) is rewritten using the properties of the J-functions. First note that the quantity inside the square brackets is independent of ψ, and equal to $J^{(+)}_{q'q}(\boldsymbol{k})$ with $\psi = 0$. Next, use the identities (A.1.29) and (A.1.30) to write

$$k_\perp J^{n-1}_{n'-n+1}(x) = p_{n'} J^{n-1}_{n'-n}(x) - p_n J^{n}_{n'-n}(x),$$
$$k_\perp J^{n}_{n'-n-1}(x) = -p_n J^{n-1}_{n'-n}(x) + p_{n'} J^{n}_{n'-n}(x). \tag{5.4.22}$$

Using $p_n = 2\varepsilon^0_n b_s b_{-s}$, and similarly for the primed variables, it is straightforward to show that second term in (5.4.20) is proportional to $J^{(0)}_{q'q}(\boldsymbol{k})$, with the coefficient equal to minus the right hand side of (5.4.21). This completes the proof.

5.4.4 *Sum over Initial and Final Spin States*

In many applications one is not interested in the spin of the particles, and it is appropriate to average over the spin states. The transition rate for any specific process involving an electron involves the outer product of a vertex function and its complex conjugate, or a number of such outer products, each of which may be treated separately. It is useful to write

$$\frac{[C_{n'n}(\epsilon' p'_\parallel, \epsilon p_\parallel, k)]^{\mu\nu}}{2\epsilon'\epsilon\varepsilon'_{n'}\varepsilon_n} = \sum_{s,s'} [\Gamma^{\epsilon'\epsilon}_{q'q}(\boldsymbol{k})]^\mu [\Gamma^{\epsilon'\epsilon}_{q'q}(\boldsymbol{k})]^{*\nu}, \tag{5.4.23}$$

with $p'^\mu_\parallel = [\varepsilon'_{n'}, 0, 0, p'_z]$, $p^\mu_\parallel = [\varepsilon_n, 0, 0, p_z]$. Given a specific choice of spin operator, explicit evaluation of the sums in (5.4.23) is straightforward but tedious. The result is summarized in Table 5.1.

Table 5.1 The components of $\left[C^{\epsilon'\epsilon}_{n'n}(\epsilon' p'_\parallel, \epsilon p_\parallel)\right]^{\mu\nu}$, denoted $C^{\mu\nu} = C^{*\nu\mu}$, are tabulated; J^n_ν denotes $J^n_\nu(k_\perp^2/2eB)$

C^{00}	$(\epsilon'\varepsilon'_{n'}\epsilon\varepsilon_n + \epsilon'\epsilon\, p'_z p_z + m^2)\big[(J^{n-1}_{n'-n})^2 + (J^n_{n'-n})^2\big] + 2p_{n'}p_n J^{n-1}_{n'-n}J^n_{n'-n}$
C^{01}	$-p_{n'}\epsilon\varepsilon_n\big[J^{n-1}_{n'-n}e^{i\psi}J^{n-1}_{n'-n+1} + J^n_{n'-n}e^{-i\psi}J^n_{n'-n-1}\big]$
	$-\epsilon'\varepsilon'_{n'}p_n\big[J^n_{n'-n}e^{i\psi}J^{n-1}_{n'-n+1} + J^{n-1}_{n'-n}e^{-i\psi}J^n_{n'-n-1}\big]$
C^{02}	$ip_{n'}\epsilon\varepsilon_n\big[J^{n-1}_{n'-n}e^{i\psi}J^{n-1}_{n'-n+1} - J^n_{n'-n}e^{-i\psi}J^n_{n'-n-1}\big]$
	$+i\epsilon'\varepsilon'_{n'}p_n\big[J^n_{n'-n}e^{i\psi}J^{n-1}_{n'-n+1} - J^{n-1}_{n'-n}e^{-i\psi}J^n_{n'-n-1}\big]$
C^{03}	$(\epsilon'\varepsilon'_{n'}\epsilon p_z + \epsilon\varepsilon_n\epsilon' p'_z)\big[(J^{n-1}_{n'-n})^2 + (J^n_{n'-n})^2\big]$
C^{11}	$(\epsilon'\varepsilon'_{n'}\epsilon\varepsilon_n - \epsilon' p'_z\epsilon p_z - m^2)\big[(J^{n-1}_{n'-n+1})^2 + (J^n_{n'-n-1})^2\big]$
	$+2p_{n'}p_n\cos(2\psi)\,J^{n-1}_{n'-n+1}J^n_{n'-n-1}$
C^{22}	$(\epsilon'\varepsilon'_{n'}\epsilon\varepsilon_n - \epsilon' p'_z\epsilon p_z - m^2)\big[(J^{n-1}_{n'-n+1})^2 + (J^n_{n'-n-1})^2\big]$
	$-2p_{n'}p_n\cos(2\psi)\,J^{n-1}_{n'-n+1}J^n_{n'-n-1}$
C^{33}	$(\epsilon'\varepsilon'_{n'}\epsilon\varepsilon_n + \epsilon' p'_z\epsilon p_z - m^2)\big[(J^{n-1}_{n'-n})^2 + (J^n_{n'-n})^2\big] - 2p_{n'}p_n J^{n-1}_{n'-n}J^n_{n'-n}$
C^{12}	$-i(\epsilon'\varepsilon'_{n'}\epsilon\varepsilon_n - \epsilon' p'_z\epsilon p_z - m^2)\big[(J^{n-1}_{n'-n+1})^2 - (J^n_{n'-n-1})^2\big]$
	$+2p_{n'}p_n\sin(2\psi)\,J^{n-1}_{n'-n+1}J^n_{n'-n-1}$
C^{13}	$-p_{n'}\epsilon p_z\big[J^{n-1}_{n'-n}e^{-i\psi}J^{n-1}_{n'-n+1} + J^n_{n'-n}e^{i\psi}J^n_{n'-n-1}\big]$
	$-\epsilon' p'_z p_n\big[J^n_{n'-n}e^{-i\psi}J^{n-1}_{n'-n+1} + J^{n-1}_{n'-n}e^{i\psi}J^n_{n'-n-1}\big]$
C^{23}	$-ip_{n'}\epsilon p_z\big[J^{n-1}_{n'-n}e^{-i\psi}J^{n-1}_{n'-n+1} - J^n_{n'-n}e^{i\psi}J^n_{n'-n-1}\big]$
	$-i\epsilon' p'_z p_n\big[J^n_{n'-n}e^{-i\psi}J^{n-1}_{n'-n+1} - J^{n-1}_{n'-n}e^{i\psi}J^n_{n'-n-1}\big]$

It is convenient to write the results summarized in Table 5.1 in a covariant form, involving $P^\mu_\parallel = [\epsilon\varepsilon_n, 0, 0, \epsilon p_\parallel]$, $P'^\mu_\parallel = [\epsilon'\varepsilon'_{n'}, 0, 0, \epsilon' p'_\parallel]$. One has

$$
\begin{aligned}
&\left[C_{n'n}(P'_\parallel, P_\parallel, k)\right]^{\mu\nu} \\
&= \left\{P'^\mu_\parallel P^\nu_\parallel + P^\mu_\parallel P'^\nu_\parallel - [(P'P)_\parallel - m^2]g^{\mu\nu}_\parallel\right\}\left[(J^{n-1}_{n'-n})^2 + (J^n_{n'-n})^2\right] \\
&\quad - g^{\mu\nu}_\perp\left[(P'P)_\parallel - m^2\right]\left[(J^{n-1}_{n'-n+1})^2 + (J^n_{n'-n-1})^2\right] \\
&\quad + if^{\mu\nu}\left[(P'P)_\parallel - m^2\right]\left[(J^{n-1}_{n'-n+1})^2 - (J^n_{n'-n-1})^2\right] \\
&\quad + p_{n'}p_n\left\{g^{\mu\nu}_\parallel 2J^{n-1}_{n'-n}J^n_{n'-n} + (e^\mu_+ e^\nu_+ + e^\mu_- e^\nu_-)J^{n-1}_{n'-n+1}J^n_{n'-n-1}\right\} \\
&\quad - p_n P'^\mu_\parallel\left[J^{n-1}_{n'-n}J^n_{n'-n-1}\,e^\nu_+ + J^n_{n'-n}J^{n-1}_{n'-n+1}\,e^\nu_-\right] \\
&\quad - p_n P'^\nu_\parallel\left[J^{n-1}_{n'-n}J^n_{n'-n-1}\,e^\mu_- + J^n_{n'-n}J^{n-1}_{n'-n+1}\,e^\mu_+\right] \\
&\quad - p_{n'} P^\mu_\parallel\left[J^{n-1}_{n'-n}J^{n-1}_{n'-n+1}\,e^\nu_- + J^n_{n'-n}J^n_{n'-n-1}\,e^\nu_+\right] \\
&\quad - p_{n'} P^\nu_\parallel\left[J^{n-1}_{n'-n}J^{n-1}_{n'-n+1}\,e^\mu_+ + J^n_{n'-n}J^n_{n'-n-1}\,e^\mu_-\right],
\end{aligned}
\tag{5.4.24}
$$

where, for simplicity in writing, the argument $x=k_\perp^2/2eB$ of the J-functions is omitted in (5.4.24). The 4-vectors $e_\pm^\mu$, whose components are $e_\pm^\mu = e^{\mp i\psi}[0,1,\pm i,0]$, may be written in the frame-independent form $e_\pm^\mu = (k_\perp^\mu \pm ik_G^\mu)/k_\perp$, with $k_\perp^\mu = g_\perp^{\mu\nu}k_\nu = k_\perp[0,\cos\psi,\sin\psi,0]$, and $k_G^\mu = f^{\mu\nu}k_\nu = k_\perp[0,-\sin\psi,\cos\psi,0]$.

The tensor $[C_{n'n}(P_\parallel', P_\parallel, k)]^{\mu\nu}$ satisfies a symmetry property that involves interchanging n and n'. Starting from the explicit form (5.4.24) and making the replacement $n' \leftrightarrow n$ involves the following transformations of the J-functions:

$$\begin{pmatrix} J_{n-n'}^{n'-1} \\ J_{n-n'}^{n'} \end{pmatrix} = (-)^{n'-n}\begin{pmatrix} J_{n'-n}^{n-1} \\ J_{n'-n}^{n} \end{pmatrix}, \qquad \begin{pmatrix} J_{n-n'+1}^{n'-1} \\ J_{n-n'-1}^{n'} \end{pmatrix} = (-)^{n'-n+1}\begin{pmatrix} J_{n'-n-1}^{n} \\ J_{n'-n+1}^{n-1} \end{pmatrix}. \tag{5.4.25}$$

It follows by inspection that (5.4.24) satisfies

$$[C_{nn'}(P_\parallel', P_\parallel, k)]^{\mu\nu} = [C_{n'n}(-P_\parallel, -P_\parallel', k)]^{\nu\mu} = [C_{n'n}(-P_\parallel, -P_\parallel', k)]^{*\mu\nu}. \tag{5.4.26}$$

The tensor $[C_{n'n}(P_\parallel', P_\parallel, k)]^{\mu\nu}$ also satisfies the identities

$$\begin{aligned} k_\mu[C_{n'n}(P_\parallel', P_\parallel, k)]^{\mu\nu} &= (\omega - \epsilon\varepsilon_n + \epsilon'\varepsilon_{n'}')[C_{n'n}(P_\parallel', P_\parallel)]^{0\nu}, \\ k_\nu[C_{n'n}(P_\parallel', P_\parallel, k)]^{\mu\nu} &= (\omega - \epsilon\varepsilon_n + \epsilon'\varepsilon_{n'}')[C_{n'n}(P_\parallel', P_\parallel, k)]^{\mu 0}. \end{aligned} \tag{5.4.27}$$

To establish that the identities (5.4.27) are satisfied for the form (5.4.24) one needs to use the relations (5.4.22) and $P_\parallel'^\mu - P_\parallel^\mu + k_\parallel^\mu = [\omega - \epsilon\varepsilon_n + \epsilon'\varepsilon_{n'}', \mathbf{0}]$.

5.5 Ritus Method and the Vertex Formalism

Three complicating factors in QED for a magnetized system, compared with an unmagnetized system, are (1) momentum is not conserved at the microscopic level, (2) the detailed results depend on the choice of the gauge used to describe the magnetic field, and (3) the detailed theory depends on the choice of spin operator. The fact that momentum is not conserved precludes a momentum-space representation, which is an essential ingredient in conventional QED for unmagnetized electrons. In most applications, the choice of gauge is of no physical relevance: specifically, the choice of gauge leads to a quantum number (p_y in the Landau gauge) that partly describes the location of the center of gyration of the particle, and in a homogeneous system this location is of no physical relevance. The choice of spin operator is important for some purposes, but in many applications the spin is of no interest and it is desirable to have a theory that applies directly to unpolarized particles, as is possible in the unmagnetized case.

There are two related formalisms that allow these complications to be partially overcome, specifically allowing a gauge-independent, momentum-space-like representation. These are the vertex formalism, based on the vertex function introduced in § 5.4, and the Ritus method [11, 12]. The idea that underlies the Ritus method is to isolate the gauge dependence and the dependence of the theory on the choice of the spin operator from the other dependences. The remaining 'reduced' wavefunctions lead to a 'reduced' propagator whose Fourier transform does exist. The theory is then analogous in structure to the unmagnetized case. In contrast, in the vertex method all the spatial dependence is associated with a vertex function, rather than the electron propagator, so that the spatial Fourier transform of the vertex function exists, with the propagator reduced to a function that depends on the time difference between two vertices and is independent of the positions of the two vertices.

5.5.1 Factorization of the Dirac Equation

The Dirac wavefunction in a magnetic field may be factorized into a function that depends on the choice of gauge for the vector potential, $\boldsymbol{A}(\boldsymbol{x})$, and a part that is a gauge-independent Dirac spinor. Such a separation was first introduced by Ritus [11, 12], cf. also [10]. The gauge-independent part may be further factorized into a spin-independent part and a Dirac spinor that corresponds to the column matrix (C_1, C_2, C_3, C_4) introduced in (5.1.26) and (5.1.30). The reduced Dirac spinor, (C_1, C_2, C_3, C_4), is evaluated for specific spin operators in § 5.2.

Let the Dirac wavefunction be written as the product

$$\Psi_q^\epsilon(x) = e^{-i\epsilon(\varepsilon_n t - p_z z)} \mathcal{V}_g^\epsilon(\boldsymbol{x}, n, p_z) \varphi_s^\epsilon(n, p_z), \tag{5.5.1}$$

where $\mathcal{V}_g^\epsilon(\boldsymbol{x}, n, p_z)$ is a diagonal matrix, and where g denotes a gauge-dependent quantum number. For an arbitrary choice of gauge, the Dirac equation separates into two equations. One of these is the reduced Dirac equation

$$[\not{P}_n^\epsilon - m]\varphi_s^\epsilon(n, p_z) = 0, \qquad [P_n^\epsilon]^\mu = (\epsilon\varepsilon_n, 0, p_n, \epsilon p_z), \tag{5.5.2}$$

where $\varphi_s^\epsilon(n, p_z)$ is a reduced wavefunction. The other equation determines the diagonal matrix $\mathcal{V}_g^\epsilon(\boldsymbol{x}, n, p_z)$ in (5.5.1). This equation depends explicitly on the choice of gauge for $\boldsymbol{A}$. For an arbitrary gauge this equation is

$$\left[\alpha_x\left(-i\frac{\partial}{\partial x} + eA_x\right) + \alpha_y\left(-i\frac{\partial}{\partial y} + eA_y\right)\right]\mathcal{V}_g^\epsilon(\boldsymbol{x}, n, p_z) = p_n\alpha_y\mathcal{V}_g^\epsilon(\boldsymbol{x}, n, p_z), \tag{5.5.3}$$

with $p_n = (2neB)^{1/2}$. The two Eqs. (5.5.2) and (5.5.3) are discussed separately below, starting with the gauge-dependent part.

Gauge-Dependent Part of the Wavefunction

After multiplying by α_y, (5.5.3) becomes

$$\left[-i\Sigma_z\left(-i\frac{\partial}{\partial x}+eA_x\right)+\left(-i\frac{\partial}{\partial y}+eA_y\right)\right]\mathcal{V}_g^\epsilon(\boldsymbol{x},n,p_z)=p_n\mathcal{V}_g^\epsilon(\boldsymbol{x},n,p_z). \tag{5.5.4}$$

In the Landau gauge, (5.1.12), the gauge-dependent quantum number corresponds to $g\to\epsilon p_y$. One seeks a solution of the form $\mathcal{V}_g^\epsilon(\boldsymbol{x},n,p_z)\propto e^{i\epsilon p_y y}$. Noting that the solution (5.1.26) may be written in the form

$$f(x)=\begin{pmatrix}C_1 v_{n-1}(\xi)\\ C_2 v_n(\xi)\\ C_3 v_{n-1}(\xi)\\ C_4 v_n(\xi)\end{pmatrix}=\begin{pmatrix}v_{n-1}(\xi)&0&0&0\\0&v_n(\xi)&0&0\\0&0&v_{n-1}(\xi)&0\\0&0&0&v_n(\xi)\end{pmatrix}\begin{pmatrix}C_1\\C_2\\C_3\\C_4\end{pmatrix}, \tag{5.5.5}$$

one identifies the column matrix (C_1,C_2,C_3,C_4) with $\varphi_s^\epsilon(n,p_z)$ in (5.5.1). This leads to the identification

$$\mathcal{V}_g^\epsilon(\boldsymbol{x},n,p_z)=e^{i\epsilon p_y y}\begin{pmatrix}v_{n-1}(\xi)&0&0&0\\0&v_n(\xi)&0&0\\0&0&v_{n-1}(\xi)&0\\0&0&0&v_n(\xi)\end{pmatrix}. \tag{5.5.6}$$

For some purposes it is more convenient to write (5.5.6) in the form

$$\mathcal{V}_g^\epsilon(\boldsymbol{x},n,p_z)=e^{i\epsilon p_y y}\big[\mathcal{P}_+v_{n-1}(\xi)+\mathcal{P}_-v_n(\xi)\big],\qquad \mathcal{P}_\pm=\tfrac{1}{2}(1\pm\Sigma_z). \tag{5.5.7}$$

The spin projection operators, $\mathcal{P}_\pm$, are introduced in (5.3.7), and their properties are summarized in (5.3.27), (5.3.28), and (5.3.30).

The result corresponding to (5.5.6) for the cylindrical gauge is

$$\mathcal{V}_r^\epsilon(\boldsymbol{x},n,p_z)=e^{i(n-r)\phi+i\epsilon p_z z}$$
$$\times\begin{pmatrix}J_{n-r-1}^r e^{-i\phi/2}&0&0&0\\0&J_{n-r}^r e^{i\phi/2}&0&0\\0&0&J_{n-r-1}^r e^{-i\phi/2}&0\\0&0&0&J_{n-r}^r e^{i\phi/2}\end{pmatrix}, \tag{5.5.8}$$

where the argument $\frac{1}{2}eB\varpi^2$ of the J-functions is omitted for simplicity in writing. In this case the additional quantum number, $g\to r$, is the radial quantum number (5.2.26), which is discrete.

For an arbitrary choice of gauge, $\mathcal{V}_g^\epsilon(\boldsymbol{x},n,p_z)$ satisfies orthogonality and completeness relations,

$$\int d^3x\, \mathcal{V}_{g'}^{\epsilon\dagger}(\boldsymbol{x}, n', p_z')\, \mathcal{V}_g^{\epsilon}(\boldsymbol{x}, n, p_z) = \frac{2\pi}{L_z}\delta(p_z' - p_z)\,\delta_{n'n}\,\delta_{g'g}, \tag{5.5.9}$$

$$L_z \sum_{g,n} \int \frac{dp_z}{2\pi}\, \mathcal{V}_g^{\epsilon}(\boldsymbol{x}', n, p_z)\, \mathcal{V}_g^{\epsilon\dagger}(\boldsymbol{x}, n, p_z) = \delta^3(\boldsymbol{x}' - \boldsymbol{x}). \tag{5.5.10}$$

The sum over g becomes the integral over $L_y\, dp_y/2\pi$ for the Landau gauge and a sum over the radial quantum number for the cylindrical gauge.

5.5.2 Reduced Wavefunctions

The reduced Dirac equation (5.5.2) follows from the Dirac equation and the identity (5.5.4). The resulting equation for the coefficients $C_1, \ldots, C_4$ is

$$\begin{pmatrix} \epsilon\varepsilon_n - m & 0 & -\epsilon p_z & i p_n \\ 0 & \epsilon\varepsilon_n - m & -i p_n & \epsilon p_z \\ -\epsilon p_z & i p_n & \epsilon\varepsilon_n + m & 0 \\ -i p_n & \epsilon p_z & 0 & \epsilon\varepsilon_n + m \end{pmatrix} \begin{pmatrix} C_1 \\ C_2 \\ C_3 \\ C_4 \end{pmatrix} = 0. \tag{5.5.11}$$

The square matrix in (5.5.11) may be written in the form $\epsilon\varepsilon_n - \epsilon\alpha_z p_z - \alpha_y p_n - \beta m$, and after multiplying by $\beta = \gamma^0$, this may be written in the more concise form $(\not{P}_n^{\epsilon} - m)\varphi_s^{\epsilon}(n, p_z) = 0$, reproducing (5.5.2) with the reduced wavefunction identified as

$$\varphi_s^{\epsilon}(n, p_z) = \begin{pmatrix} C_1 \\ C_2 \\ C_3 \\ C_4 \end{pmatrix}. \tag{5.5.12}$$

The specific form for the reduced wavefunction (5.5.12) depends on the choice of spin operator. For the Johnson-Lippmann wavefunctions it is given by (5.1.30), for the helicity eigenstates it is given by (5.2.7), and for the magnetic-moment eigenstates it is given by (5.2.12). In particular, the eigenvalue equation for the reduced magnetic-moment wavefunction is

$$[m\boldsymbol{\Sigma} - i\epsilon\gamma \times \boldsymbol{\Pi}_n^{\epsilon}(p_z)]_z \varphi_s^{\epsilon}(n, p_z) = s\lambda\varphi_s^{\epsilon}(n, p_z), \tag{5.5.13}$$

with, in the standard representation,

$$[m\boldsymbol{\Sigma} - i\epsilon\gamma \times \boldsymbol{\Pi}_n^{\epsilon}(p_z)]_z = \begin{pmatrix} m & 0 & 0 & -i p_n \\ 0 & -m & -i p_n & 0 \\ 0 & i p_n & m & 0 \\ i p_n & 0 & 0 & -m \end{pmatrix}, \qquad \boldsymbol{\Pi}_n^{\epsilon}(p_z) = (0, \epsilon p_n, p_z). \tag{5.5.14}$$

Solving (5.5.13) with (5.5.14) reproduces (5.2.12).

The normalization for the reduced wavefunctions is closely analogous to the normalization for (unmagnetized) free particles. Specifically, one has

$$\varphi_s^{\epsilon\dagger}(n,p_z)\,\varphi_{s'}^{\epsilon'}(n,p_z) = \frac{\delta^{\epsilon\epsilon'}\delta_{ss'}}{V}, \quad \bar{\varphi}_s^{\epsilon}(n,p_z)\,\varphi_{s'}^{\epsilon'}(n,p_z) = \frac{m\delta^{\epsilon\epsilon'}\delta_{ss'}}{\epsilon\varepsilon_n V}. \tag{5.5.15}$$

When summing over intermediate states, or summing or averaging over initial or final polarization states for an electron (or positron), the sum is over the outer product of a wavefunction and its Dirac conjugate. This sum gives

$$\sum_{s=\pm}\varphi_s^{\epsilon}(n,p_z)\,\bar{\varphi}_s^{\epsilon}(n,p_z) = \frac{\not{P}_n^{\epsilon}+m}{2\epsilon\varepsilon_n V}, \tag{5.5.16}$$

with $[P_n^{\epsilon}]^{\mu} = (\epsilon\varepsilon_n, 0, p_n, \epsilon p_z)$, and with, analogous to (5.5.11),

$$\not{P}_n^{\epsilon}+m = \begin{pmatrix} \epsilon\varepsilon_n(p_z)+m & 0 & -\epsilon p_z & ip_n \\ 0 & \epsilon\varepsilon_n(p_z)+m & -ip_n & \epsilon p_z \\ \epsilon p_z & -ip_n & -\epsilon\varepsilon_n(p_z)+m & 0 \\ ip_n & -\epsilon p_z & 0 & -\epsilon\varepsilon_n(p_z)+m \end{pmatrix}. \tag{5.5.17}$$

In the magnetized case, the operator $\not{P}_n^{\epsilon}+m$ plays a role somewhat analogous to $\epsilon\not{p}+m$ in the unmagnetized case.

5.5.3 *Reduced Propagator in the Ritus Method*

Although the electron propagator has no momentum-space representation in a magnetic field, the Ritus method leads to a reduced propagator that plays a similar role to a momentum-space propagator. The absence of a formal momentum-space propagator is due to the coordinate-space form of the propagator depending separately on the space-time points, x, x', and the Fourier transform can be performed only for functions that depend on $x-x'$, rather than on x, x' separately. In a magnetized vacuum, the form (5.3.8) for the propagator shows that it may be separated into a gauge-dependent function, $\phi(x,x')$, times a reduced propagator that depends only on the difference $x-x'$, which can be Fourier transformed.

Using the Ritus method, a momentum-space representation for the statistically averaged electron propagator is possible provided that the electrons are unpolarized. On factorizing the wavefunctions in (5.3.1) using (5.5.1), one has

$$\Psi_q^{\epsilon}(\boldsymbol{x})\overline{\Psi}_{,q}^{\;\epsilon}(\boldsymbol{x}') = e^{i\epsilon p_z z}V_g^{\epsilon}(\boldsymbol{x},n,p_z)\varphi_s^{\epsilon}(n,p_z)\bar{\varphi}_s^{\epsilon}(n,p_z)V_g^{\epsilon\dagger}(\boldsymbol{x}',n,p_z)\,e^{-i\epsilon p_z z'}. \tag{5.5.18}$$

In the Ritus method, the factors $e^{i\epsilon p_z z}\mathcal{V}_g^\epsilon(\boldsymbol{x}, n, p_z)$ and $\mathcal{V}_g^{\epsilon\dagger}(\boldsymbol{x}', n, p_z)\, e^{-i\epsilon p_z z'}$ are treated separately. The remaining factors, $\varphi_s^\epsilon(n, p_z)\bar{\varphi}_s^\epsilon(n, p_z)$, are incorporated into a reduced propagator, $\bar{G}_q(E)$, which is defined by writing (5.3.1) with (5.5.18) in the form

$$\bar{G}(x, x') = \sum_{\epsilon q} \frac{1}{V} \int \frac{dE}{2\pi}\, e^{-iE(t-t')} \times e^{i\epsilon p_z z}\mathcal{V}_g^\epsilon(\boldsymbol{x}, n, p_z)\, \bar{G}_q(E)\, \mathcal{V}_g^{\epsilon\dagger}(\boldsymbol{x}', n, p_z)\, e^{-i\epsilon p_z z'}. \tag{5.5.19}$$

The reduced propagator is identified as

$$\bar{G}_q^\epsilon(E) = \sum_s \varphi_s^\epsilon(n, p_z)\bar{\varphi}_s^\epsilon(n, p_z)\left[\frac{1-n_q^\epsilon}{E-\epsilon(\varepsilon_q - i0)} + \frac{n_q^\epsilon}{E-\epsilon(\varepsilon_q + i0)}\right]. \tag{5.5.20}$$

The factorization process is useful only if the reduced propagator is independent of the choice of spin operator, and this requires that the electrons be unpolarized. Then the spin appears only in $\varphi_s^\epsilon(n, p_z)\bar{\varphi}_s^\epsilon(n, p_z)$, and the sum over s is performed using (5.5.15). The propagator in the Ritus formalism, rewritten $\bar{G}_q^\epsilon(E) \to \bar{G}_n^\epsilon(E, p_z)$, becomes

$$\bar{G}_n^\epsilon(E, p_z) = \frac{\not{P}_n^\epsilon + m}{2\epsilon\varepsilon_n}\left[\frac{1-n_n^\epsilon(\epsilon p_z)}{E-\epsilon(\varepsilon_n - i0)} + \frac{n_n^\epsilon(\epsilon p_z)}{E-\epsilon(\varepsilon_n + i0)}\right], \tag{5.5.21}$$

with $[P_n^\epsilon]^\mu = (\epsilon\varepsilon_n(p_z), 0, p_n, \epsilon p_z)$. The reduced propagator, $\bar{G}_n^\epsilon(E, p_z)$, depends only on the quantum numbers ϵ, n, p_z and the energy E in the internal particle line.

5.5.4 Propagator in the Vertex Formalism

In contrast with the Ritus method, where the reduced wavefunctions are retained in the reduced electron propagator, in the vertex formalism the propagator is independent of the details of the states, and depends only on the (virtual) energy of the electron (or positron). Specifically, let $\mathcal{G}_q^\epsilon(E)$ be the propagator in the vertex formalism in the absence of any statistical average, and let $\bar{\mathcal{G}}_q^\epsilon(E)$ be the statistical average of this propagator. Starting from the propagator $\bar{G}(x, x')$, given by (5.3.1), the wavefunctions $\Psi_q^\epsilon(\boldsymbol{x})$, $\overline{\Psi}_q^\epsilon(\boldsymbol{x}')$ are deleted (and transferred to the vertex functions) so that the propagator becomes

$$\bar{G}(x, x') = \sum_{\epsilon q} \Psi_q^\epsilon(\boldsymbol{x})\overline{\Psi}_q^\epsilon(\boldsymbol{x}') \int \frac{dE}{2\pi}\, e^{-iE(t-t')}\, \bar{\mathcal{G}}_q^\epsilon(E), \tag{5.5.22}$$

In the absence of any statistical average one identifies

$$\mathcal{G}_q^\epsilon(E) = \frac{1}{E - \epsilon(\varepsilon_q - i0)}, \tag{5.5.23}$$

where q denotes the quantum numbers for the electron in the intermediate state, with only n, p_z relevant to the energy $\varepsilon_q \to \varepsilon_n(p_z)$. The statistical average over unpolarized particles, with q identified as n, p_z, gives

$$\bar{\mathcal{G}}_n^\epsilon(E, p_z) = \frac{1 - n_n^\epsilon(\epsilon p_z)}{E - \epsilon(\varepsilon_n(p_z) - i0)} + \frac{n_n^\epsilon(\epsilon p_z)}{E - \epsilon(\varepsilon_n(p_z) + i0)}. \tag{5.5.24}$$

Comparison of (5.5.20) and (5.5.21) with (5.5.23) and (5.5.24) shows that the propagator in the vertex formalism corresponds to deleting the factor $[\not{P}_n^\epsilon + m]/2\epsilon\varepsilon_n$ associated with the sum over the spins of the outer product to the reduced wavefunctions in the reduced propagator. In the vertex formalism, the reduced wavefunctions are included in the vertex functions, and not in the propagators.

5.5.5 *Vertex Matrix in the Ritus Method*

The vertex function $[\gamma_{q'q}^{\epsilon'\epsilon}(\boldsymbol{k})]^\mu$ factorizes into gauge-dependent and a gauge-independent parts, $[\Gamma_{q'q}^{\epsilon'\epsilon}(\boldsymbol{k})]^\mu$, as shown explicitly in (5.4.2). In the Ritus method, a further factorization of the gauge-independent vertex function facilitates a momentum-space-like representation of the theory [10–12]. This factorization allows one to separate the reduced wavefunctions from $[\Gamma_{q'q}^{\epsilon'\epsilon}(\boldsymbol{k})]^\mu$. The remaining quantity is a Dirac matrix, which is the vertex matrix in the Ritus method.

The factorization of the wavefunction, in the form (5.5.1), allows one to make the factorization (5.4.1) of the vertex function explicit by writing

$$[\Gamma_{q'q}^{\epsilon'\epsilon}(\boldsymbol{k})]^\mu = V \bar{\varphi}_{s'}^{\epsilon'}(n', p_z')\, \mathcal{J}_{n'n}^\mu(\boldsymbol{k}_\perp)\, \varphi_s^\epsilon(n, p_z), \tag{5.5.25}$$

which defines the vertex matrix $\mathcal{J}_{n'n}^\mu(\boldsymbol{k}_\perp)$. For example, the form (5.4.17) for the vertex function for an arbitrary choice of spin operator may be written in the form (5.5.25) by identifying the reduced wavefunction as the column matrix (C_1, C_2, C_3, C_4) and the adjoint wavefunction as $(C_1'^*, C_2'^*, -C_3'^*, -C_4'^*)$.

In writing down an explicit form for $\mathcal{J}_{n'n}^\mu(\boldsymbol{k}_\perp)$ it is useful to project onto the $\|$ and $\perp$-subspaces, by writing $\gamma_\|^\mu = g_\|^{\mu\nu}\gamma_\nu$, $\gamma_\perp^\mu = g_\perp^{\mu\nu}\gamma_\nu$, so that these correspond to $\gamma_\|^\mu = (\gamma^0, 0, 0, \gamma^3)$, $\gamma_\perp^\mu = (0, \gamma^1, \gamma^2, 0)$. One has

$$\mathcal{J}_{n'n}^\mu(\boldsymbol{k}_\perp) = \gamma_\|^\mu \mathcal{J}_{n'n}^\|(\boldsymbol{k}_\perp) + \gamma_\perp^\mu \mathcal{J}_{n'n}^\perp(\boldsymbol{k}_\perp), \tag{5.5.26}$$

where $\mathcal{J}^{\parallel}_{n'n}(\boldsymbol{k}_\perp)$, $\mathcal{J}^{\perp}_{n'n}(\boldsymbol{k}_\perp)$ are diagonal 4×4 matrices:

$$\mathcal{J}^{\parallel}_{n'n}(\boldsymbol{k}_\perp) = (-ie^{-i\psi})^{n'-n}\begin{pmatrix} J^{n-1}_{n'-n} & 0 & 0 & 0 \\ 0 & J^{n}_{n'-n} & 0 & 0 \\ 0 & 0 & J^{n-1}_{n'-n} & 0 \\ 0 & 0 & 0 & J^{n}_{n'-n} \end{pmatrix}, \tag{5.5.27}$$

$$\mathcal{J}^{\perp}_{n'n}(\boldsymbol{k}_\perp) = (-ie^{-i\psi})^{n'-n}$$
$$\times\begin{pmatrix} -ie^{-i\psi}J^{n-1}_{n'-n+1} & 0 & 0 & 0 \\ 0 & ie^{i\psi}J^{n}_{n'-n-1} & 0 & 0 \\ 0 & 0 & -ie^{-i\psi}J^{n-1}_{n'-n+1} & 0 \\ 0 & 0 & 0 & ie^{i\psi}J^{n}_{n'-n-1} \end{pmatrix}, \tag{5.5.28}$$

where the arguments of the J-functions is $k_\perp^2/2eB$.

The matrices $\gamma^\mu_\parallel$, $\mathcal{J}^{\parallel}_{n'n}(\boldsymbol{k}_\perp)$ commute, but the matrices $\gamma^\mu_\perp$, $\mathcal{J}^{\perp}_{n'n}(\boldsymbol{k}_\perp)$ do not commute. If one writes the matrix products in (5.5.26) in the opposite order, one needs to replace $\mathcal{J}^{\perp}_{n'n}(\boldsymbol{k}_\perp)$ by

$$\tilde{\mathcal{J}}^{\perp}_{n'n}(\boldsymbol{k}_\perp) = (-ie^{-i\psi})^{n'-n}$$
$$\times\begin{pmatrix} ie^{i\psi}J^{n}_{n'-n-1} & 0 & 0 & 0 \\ 0 & -ie^{-i\psi}J^{n-1}_{n'-n+1} & 0 & 0 \\ 0 & 0 & ie^{i\psi}J^{n}_{n'-n-1} & 0 \\ 0 & 0 & 0 & -ie^{-i\psi}J^{n-1}_{n'-n+1} \end{pmatrix}. \tag{5.5.29}$$

The matrix (5.5.29) also appears in the symmetry relations

$$\mathcal{J}^{\parallel}_{nn'}(-\boldsymbol{k}_\perp) = \left[\mathcal{J}^{\parallel}_{n'n}(\boldsymbol{k}_\perp)\right]^*, \qquad \mathcal{J}^{\perp}_{nn'}(-\boldsymbol{k}_\perp) = \left[\tilde{\mathcal{J}}^{\perp}_{n'n}(\boldsymbol{k}_\perp)\right]^*. \tag{5.5.30}$$

Using the relations (5.5.30) in (5.5.26), one finds

$$\mathcal{J}^{\mu}_{nn'}(-\boldsymbol{k}_\perp) = \gamma^\mu_\parallel\left[\mathcal{J}^{\parallel}_{n'n}(\boldsymbol{k}_\perp)\right]^* + \gamma^\mu_\perp\left[\tilde{\mathcal{J}}^{\perp}_{n'n}(\boldsymbol{k}_\perp)\right]^*. \tag{5.5.31}$$

An alternative way of writing $\mathcal{J}^{\mu}_{n'n}(\boldsymbol{k}_\perp)$ is in terms of the projection matrices $\mathcal{P}_\pm$ introduced in (5.5.7). One has

$$\mathcal{J}^{\mu}_{n'n}(\boldsymbol{k}_\perp) = (-ie^{-i\psi})^{n'-n}\Big\{\gamma^\mu_\parallel\Big[J^{n-1}_{n'-n}(x)\,\mathcal{P}_+ + J^{n}_{n'-n}(x)\,\mathcal{P}_-\Big]$$
$$+\gamma^\mu_\perp\Big[-ie^{-i\psi}J^{n-1}_{n'-n+1}(x)\,\mathcal{P}_+ + ie^{i\psi}J^{n}_{n'-n-1}(x)\,\mathcal{P}_-\Big]\Big\}. \tag{5.5.32}$$

These matrices satisfy the identities

$$\mathcal{P}_{\pm}\gamma_{\parallel}^{\mu} = \gamma_{\parallel}^{\mu}\mathcal{P}_{\pm}, \qquad \mathcal{P}_{\pm}\gamma_{\perp}^{\mu} = \gamma_{\perp}^{\mu}\mathcal{P}_{\mp}. \tag{5.5.33}$$

Using these relations, an alternative form of (5.5.32), with the matrix products written in the opposite order, is

$$\mathcal{J}_{n'n}^{\mu}(\boldsymbol{k}_{\perp}) = (-ie^{-i\psi})^{n'-n}\{[J_{n'-n}^{n-1}(x)\mathcal{P}_{+} + J_{n'-n}^{n}(x)\mathcal{P}_{-}]\gamma_{\parallel}^{\mu}$$
$$+[-ie^{-i\psi}J_{n'-n+1}^{n-1}(x)\mathcal{P}_{-} + ie^{i\psi}J_{n'-n-1}^{n}(x)\mathcal{P}_{+}]\gamma_{\perp}^{\mu}\}. \tag{5.5.34}$$

5.5.6 Calculation of Traces Using the Ritus Method

An example where the Ritus method may be used is in an alternative derivation in the sum (5.4.23), leading to the quantity $[C_{n'n}^{\epsilon'\epsilon}(p_z,\boldsymbol{k})]^{\mu\nu}$, defined by (5.4.23) as the sum over spins of the outer product of the vertex function with itself. The Ritus method gives

$$[C_{n'n}^{\epsilon'\epsilon}(p_z,\boldsymbol{k})]^{\mu\nu} = \mathrm{Tr}\left[\frac{\not{P}_{n'}^{\epsilon'} + m}{2\epsilon'\varepsilon'_{n'}}\mathcal{J}_{n'n}^{\nu}(\boldsymbol{k}_{\perp})\frac{\not{P}_{n}^{\epsilon} + m}{2\epsilon\varepsilon_{n}}\mathcal{J}_{nn'}^{\mu}(-\boldsymbol{k}_{\perp})\right], \tag{5.5.35}$$

with $[P_{n'}^{\epsilon'}]^{\mu} = [\epsilon'\varepsilon'_{n'}, 0, p_{n'}, \epsilon' p'_z]$, $\varepsilon'_{n'} = \varepsilon_{n'}(p'_z)$, and similarly for the unprimed quantity $[P_{n}^{\epsilon}]^{\mu}$. The evaluation of (5.5.35) is analogous to the corresponding unmagnetized case, where one needs to evaluate the trace of the product $(\epsilon\not{p} + m)\gamma^{\mu}(\epsilon'\not{p}' + m)\gamma^{\nu}$. However, new features appear associated with the projection operators, and in order to calculate (5.5.35), the general problem of evaluating traces that include $\mathcal{P}_{\pm}$ needs to be addressed.

In the absence of projection operators, the standard technique for calculating traces of products of Dirac matrices is based on the identities

$$\mathrm{Tr}[\gamma^{\mu}\gamma^{\nu}] = 4g^{\mu\nu}, \qquad \mathrm{Tr}[\gamma^{\mu}\gamma^{\nu}\gamma^{\rho}\gamma^{\sigma}] = 4[g^{\mu\nu}g^{\rho\sigma} - g^{\mu\rho}g^{\nu\sigma} + g^{\mu\sigma}g^{\rho\nu}], \tag{5.5.36}$$

together with the trace of any product of an odd number of γ-matrices being zero. One way of deriving (5.5.36) involves choosing 16 independent matrices to span the Dirac spin space, for example $1, \gamma^{\mu}, [\gamma^{\mu}, \gamma^{\nu}], \gamma^5\gamma^{\mu}, \gamma^5$, and noting that only the unit matrix has a nonzero trace, equal to 4. If one includes a projection operator anywhere in the sequence of γ-matrices, this projects the unit matrix onto a 2-dimensional subspace, so that the trace of the unit matrix is reduced to 2. This reduces the value of the traces in (5.5.36) by one half. However, the projection introduces a second matrix with a nonzero trace: Σ_z has zero trace, but $\mathcal{P}_{\pm}\Sigma_z =$

$\pm\mathcal{P}_\pm$ has trace ± 2. With $\gamma^1\gamma^2 = -i\Sigma_z = -\gamma^2\gamma^1$, this introduces an additional contribution whenever two of the indices in (5.5.36) take on the values 1 and 2. It is useful to distinguish between $\|$- and $\perp$-components, achieved here by writing $\gamma_\|^\mu = [\gamma^0, 0, 0, \gamma^3]$, $\gamma_\perp^\mu = [0, \gamma^1, \gamma^2, 0]$. Including a projection operator in the first of (5.5.36) gives

$$\mathrm{Tr}\left[\gamma_\|^\mu \gamma_\|^\nu \mathcal{P}_\pm\right] = 2g_\|^{\mu\nu}, \qquad \mathrm{Tr}\left[\gamma_\perp^\mu \gamma_\perp^\nu \mathcal{P}_\pm\right] = 2[g_\perp^{\mu\nu} \pm i f^{\mu\nu}], \tag{5.5.37}$$

with $f^{\mu\nu}$, defined by (1.1.16) in terms of the Maxwell tensor for the background magnetic field, having nonzero components $f^{12} = -f^{21} = -1$. An alternative expression is

$$\mathrm{Tr}\left[\gamma_\perp^\mu \gamma_\perp^\nu \mathcal{P}_\pm\right] = -2e_\mp^\mu e_\pm^\nu, \qquad e_\pm^\mu = e^{\mp i\psi}(0, 1, \pm i, 0). \tag{5.5.38}$$

The second of (5.5.36) becomes

$$\begin{aligned}\mathrm{Tr}[\gamma^\mu\gamma^\nu\gamma^\rho\gamma^\sigma\mathcal{P}_\pm] = 2[g^{\mu\nu}g^{\rho\sigma} - g^{\mu\rho}g^{\nu\sigma} + g^{\mu\sigma}g^{\rho\nu}] \pm 2i[g^{\mu\nu}f^{\rho\sigma} + f^{\mu\nu}g^{\rho\sigma} \\ -g^{\mu\rho}f^{\nu\sigma} - f^{\mu\rho}g^{\nu\sigma} + g^{\mu\sigma}f^{\rho\nu} + f^{\mu\sigma}g^{\rho\nu}].\end{aligned} \tag{5.5.39}$$

The projection operator may be moved to any other location, using the commutation relations $\mathcal{P}_\pm M_\| = M_\|\mathcal{P}_\pm$, with $M_\| = \gamma^0$ or γ^3, or $\mathcal{P}_\pm M_\perp = M_\perp\mathcal{P}_\mp$, with $M_\perp = \gamma^1$ or γ^2.

The evaluation of $\left[C_{n'n}^{\epsilon'\epsilon}(p_z, \boldsymbol{k})\right]^{\mu\nu}$, given by (5.5.35), involves using the Ritus method to evaluate the trace in (5.5.35). The result reproduces (5.4.24).

5.6 Feynman Rules for QPD in a Magnetized Plasma

Rules for treating QED processes in a uniformly magnetized plasma are formulated in this section. First, the rules for the unmagnetized case are summarized.

5.6.1 Rules for an Unmagnetized System

The rules for the unmagnetized case are formulated in § 7.1 of volume 1.

The rules for Feynman diagrams are:

1. The initial state is to the right of the diagram and the final state is to the left. For a given process (specified initial and final states) all diagrams with the specified number and kind of particles and wave quanta in the initial and final states are to be drawn.

2. An electron is represented by a solid line with an arrow pointing from right to left and a positron is represented by a solid line with an arrow pointing from left to right. The direction of the arrow along a solid line is continuous.
3. A photon (any wave quantum) is represented by a dashed line.
4. An electron and a photon line join at an electron-photon vertex, which has a 4-tensor index $(\mu, \nu, \ldots)$ and a space-time point associated with it.
5. The nth order nonlinear response of the medium is represented by an $(n+1)$-photon vertex, which is a circle with $n+1$ photon lines joining onto it.
6. Any photon line begins or terminates at a vertex, either joining an electron-positron line at electron-photon vertex, or a m-photon vertex.
7. An m-photon vertex represents a statistical average over an m-sided closed particle loop, and closed particle loops are omitted in diagrams in QPD so that their effect is not counted twice.
8. The order of a diagram is equal to the number of its vertices in the absence of m-photon vertices. An m-photon vertex contributes $m-2$ to the order.
9. For diagrams in momentum space all lines are labeled with the 4-momentum of the particles, rather than the vertices being labeled with the space-time points. Four-momentum is conserved at a vertex.
10. The integral $d^4P/(2\pi)^4$ over any undetermined 4-momentum, P, in a closed loop or associated with an external field $A^\mu(P)$ is to be performed.
11. An interaction with an external field is described by a vertex with the photon line replaced by a squiggly line joined to an "x" that denotes the source of the external field.

In the generalization to the magnetized case, the plane wavefunction for a free electron in the unmagnetized plasma, is replaced by the wavefunction for an electron in a magnetostatic field found by solving the Dirac equation exactly. For a free particle, the only relevant quantum numbers are the 4-momentum $P^\mu = [\epsilon\varepsilon, \epsilon\mathbf{p}]$ and the spin s, and for a magnetized electrons the set of quantum numbers includes n, p_z, ϵ, s plus a gauge-dependent quantum number (p_y or r here).

The rules for constructing $S_{\rm fi}$ in coordinate space are essentially unchanged from the unmagnetized case, as given in § 7.1 of volume 1.

Rule 1

The contributions from all diagrams with the specified number and kind of initial and final particles are to be added in determining $S_{\rm fi}$.

Rule 2

Each electron-photon vertex corresponds to a factor $ie\,\gamma_\mu$, where μ is the 4-tensor index associated with the vertex. The integrals are to be performed over the space-time coordinates associated with each vertex.

Rule 3

An incoming electron line corresponds to $\Psi_q^+(x)\, e^{-i\varepsilon_q t}$, and incoming positron line to $\overline{\Psi}_q^-(x)\, e^{-i\varepsilon_q t}$, an outgoing electron line to $\Psi_q^+(x)\, e^{i\varepsilon_q t}$, and an outgoing positron line to $\overline{\Psi}_q^+(x)\, e^{i\varepsilon_q t}$, where q denotes the quantum numbers.

Rule 4

An incoming photon line in the mode M, joining at a vertex labeled (x, μ), corresponds to a factor $a_M\, e_M^\mu\, e^{-ik_M x}$, with $a_M = [\mu_0 R_M / V\omega_M]^{1/2}$, and an outgoing photon line to $a_M\, e_M^{*\mu}\, e^{ik_M x}$. In an interaction with an external field $A^\mu(x)$, a factor $A^\mu(x)$ is included in place of these photon factors.

Rule 5

An internal electron-positron line pointing from x_1 to x_2 corresponds to the propagator $i\,G(x_2, x_1)$. An internal photon line between vertices (x_1, μ) and (x_2, ν) corresponds to the propagator $-iD^{\mu\nu}(x_2 - x_1)$.

Rule 6

The Dirac spinors are written according to matrix multiplication along the direction opposite to the arrow along each solid line. An extra minus sign is to be included for each closed electron-positron loop. The overall phase of the amplitude is unimportant, but two diagrams that differ only by the interchange of two external electron-positron lines must have opposite signs.

Rule 7

An m-photon vertex corresponds to a factor

$$-\frac{i}{m}\int d^4x_0 \cdots d^4x_{m-1} \int \frac{d^4k_0}{(2\pi)^4} \cdots \frac{d^4k_{m-1}}{(2\pi)^4}\, e^{i(k_0x_0+\cdots+k_{m-1}x_{m-1})} \times (2\pi)^4\, \delta^4(k_0 + \cdots + k_{m-1})\,\Pi^{(m-1)\mu_0\ldots\mu_{m-1}}(k_0, \ldots, k_{m-1}). \tag{5.6.1}$$

Neglect of Gauge-Dependent Factors

A formal complication in the magnetized case is the gauge-dependent factor $d_{q'q}^{\epsilon'\epsilon}(\boldsymbol{K})$ in $S_{\rm fi}$ associated with each electron line in a Feynman diagram, where

q, q' correspond to the initial and final states of the electron, and where $\boldsymbol{K}$ is the net momentum transfer along the line. This factor appears squared in the transition rate $\propto |S_{\rm fi}|^2$. In the Landau gauge, the y- and z-components of 3-momentum are conserved at each vertex, and hence between the initial and final states. In the factor $d^{\epsilon'\epsilon}_{q'q}(\boldsymbol{K})$, conservation of y-momentum is expressed through the factor $2\pi\delta(\epsilon p_y - \epsilon' p'_y - K_y)$. After squaring, one has

$$|d^{\epsilon'\epsilon}_{q'q}(\boldsymbol{K})|^2 = \frac{L_y L_z}{V^2 eB}\, 2\pi\delta(\epsilon p_y - \epsilon' p'_y - K_y)\, 2\pi\delta(\epsilon p_z - \epsilon' p'_z - K_z). \tag{5.6.2}$$

For an electron line that connects an electron in the initial (unprimed) state to an electron in the final (primed) state, the density of final states factor is $L_y L_z dp'_y dp'_z/(2\pi)^2$. The p'_y- and p'_z-integral may be performed over the δ-functions in (5.6.2). The resulting factor is $(L_y L_z)^2/V^2 eB = 1/L_x^2 eB$, which is equal to unity for the choice $L_x = 1/(eB)^{1/2}$. It follows that the factor $|d^{\epsilon'\epsilon}_{q'q}(\boldsymbol{K})|^2$ may be replaced by unity, provided that it is implicit that the y- and z-components of momentum are conserved. Provided that one is not interested in the change in the position of the gyrocenter, one may simply ignore the gauge-dependent factor by replacing $|d^{\epsilon'\epsilon}_{q'q}(\boldsymbol{K})|^2$ by unity. However, this also discards information on the z-component of momentum, and it is usually appropriate to retain information on the z-component of momentum explicitly. This is achieved by replacing (5.6.2) and the density of final states factor for the line by

$$|\bar{d}^{\epsilon'\epsilon}_{q'q}(\boldsymbol{K})|^2 = \frac{2\pi}{VeB}\, 2\pi\delta(\epsilon p_z - \epsilon' p'_z - K_z), \qquad \bar{D}_{\rm f} = \frac{eBV}{2\pi}\int \frac{dp'_z}{2\pi}, \tag{5.6.3}$$

respectively.

The foregoing argument applies to an electron line connecting an electron in the initial state to an electrons in the final state, and the other possibilities for a single electron line may be treated in a similar manner. For a positron ($\epsilon = \epsilon' = -1$) line, the foregoing argument applies with primed and unprimed quantities interchanged. For an electron line corresponding to a pair in the initial state, the density of final states is $\bar{D}_{\rm f} = 1$. For an electron line corresponding to a pair in the final state, K_z is specified by the initial conditions, and the density of final states factor is $\bar{D}_{\rm f} = (eBV/2\pi)^2 \int dp_z/2\pi \int dp'_z/2\pi$.

When there are two or more electron lines, the result (5.6.3) applies separately to each line. For example, consider Møller scattering, where there are two electrons in the initial state, and two electrons in the final state. In the absence of a magnetic field, the details of the scattering depend on the relative positions of the two particles; for example, the scattering angle increases as the distance of closest approach decreases. Nevertheless, in a momentum-space description, the relative position is not specified: the momentum transfer is a free parameter, and it determines both the scattering angle and the distance of closest approach. Thus, the relative position of the scattering particles is not specified independently, but is determined by the theory itself. Similarly, although the relative position of the gyrocenters affects the scattering, in the momentum-space-like description, this relative position is determined by other parameters and does not need to be specified independently.

5.6.2 Modified Rules for Magnetized Systems

Modified rules for processes in a magnetic field are as follows:

The differential transition rate in the S-matrix formalism is given by

$$w_{\mathrm{i}\to\mathrm{f}} = \lim_{T\to\infty} \frac{|S_{\mathrm{fi}}|^2}{T} \prod_{\mathrm{f}} D_{\mathrm{f}}, \tag{5.6.4}$$

where T is the normalization time. In the case of an unmagnetized system, 4-momentum is conserved, and it is conventional to write the scattering matrix as $S_{\mathrm{fi}} = \delta_{\mathrm{fi}} + i\,(2\pi)^4\,\delta^4(p_{\mathrm{f}} - p_{\mathrm{i}})\,T_{\mathrm{fi}}$. In a magnetic field, perpendicular 3-momentum is not conserved and this procedure cannot be used. Energy (and parallel momentum) is conserved in a magnetostatic field, and one may incorporate this into the theory by writing the S-matrix element in the form

$$S_{\mathrm{fi}} = \delta_{\mathrm{fi}} + i\,2\pi\,\delta(E_{\mathrm{f}} - E_{\mathrm{i}})\,\mathcal{T}_{\mathrm{fi}}, \tag{5.6.5}$$

where E_{i}, E_{f} are the energies of the initial and final states, respectively. On inserting (5.6.5) into (5.6.4), the square of $\delta(E_{\mathrm{f}} - E_{\mathrm{i}})$ gives $T\,2\pi\delta(E_{\mathrm{f}} - E_{\mathrm{i}})$, and the factor of T cancels with that in the transition rate (5.6.4). This leads to the following rule.

Rule 8a

With $\mathcal{T}_{\mathrm{fi}}$ defined by (5.6.5), the transition probability per unit time is given by

$$w_{\mathrm{i}\to\mathrm{f}} = 2\pi\,\delta(E_{\mathrm{f}} - E_{\mathrm{i}})\,|\mathcal{T}_{\mathrm{fi}}|^2 \prod_{\mathrm{f}} D_{\mathrm{f}}. \tag{5.6.6}$$

When one is not interested in the locations of the gyrocenters, the gauge-dependent factors are ignored: $|d^{\epsilon'\epsilon}_{q'q}(\boldsymbol{K})|^2$ and the density of states factor are replaced according to (5.6.3). In the following, this replacement is assumed to be made and the overbar is omitted on D_{f}. It is often convenient to assume that the integral over the δ-function is performed, so that $\epsilon' p'_z = \epsilon p_z - K_z$ is implicit. This leads to the following rule.

Rule 8b

For each electron line, $i\mathcal{T}_{\rm fi}$ contains a product of Dirac matrices, with $[\Gamma^{\epsilon'\epsilon}_{q'q}(\boldsymbol{k})]^\mu$ associated with a vertex with 4-tensor index μ, between states ϵ, q and ϵ', q', and with $\boldsymbol{k}$ the 3-momentum emitted. The set q represents the quantum numbers n, s, p_z, and the sum over these quantum numbers is performed for each internal electron line.

Rule 8c

For an incoming photon line, $i\mathcal{T}_{\rm fi}$ contains a factor $a_M = [\mu_0 R_M / V\omega_M]^{1/2}$ and a polarization vector e^μ_M, for a photon in the mode M at a vertex with 4-tensor index μ. For each outgoing photon line, $\mathcal{T}_{\rm fi}$ contains a factor $a_M\, e^{*\mu}_M$.

Rule 9a

In the vertex formalism the matrix element of the product of Dirac matrices, representing vertices and propagators, along an electron line is replaced by a product of vertex functions, of the form $ie[\Gamma^{\epsilon'\epsilon}_{q'q}(\boldsymbol{k})]^\mu$, and reduced propagators. The vertex functions are matrix elements that are independent of the Dirac algebra. An internal electron line is represented by a factor $i\mathcal{G}^\epsilon_q(E)$, with the reduced propagator given by

$$\mathcal{G}^\epsilon_q(E) = \frac{1}{E - \epsilon(\varepsilon_n - i0)}, \tag{5.6.7}$$

with $\varepsilon_n = (m^2 + p_z^2 + 2neB)^{1/2}$. After statistical averaging, and with a minor change in notation, the propagator (5.6.7) becomes

$$\begin{aligned}\bar{\mathcal{G}}^\epsilon_n(E, p_z) &= \frac{1 - n^\epsilon_n(\epsilon p_z)}{E - \epsilon(\varepsilon_n - i0)} + \frac{n^\epsilon_n(\epsilon p_z)}{E - \epsilon(\varepsilon_n + i0)} \\ &= \wp\frac{1}{E - \epsilon\varepsilon_n} - i\epsilon[1 - 2n^\epsilon_n(\epsilon p_z)]\pi\delta(E - \epsilon\varepsilon_n),\end{aligned} \tag{5.6.8}$$

where $\wp$ denoted the Cauchy principal value. The reduced propagator is a scalar that is also independent of the Dirac algebra. In the vertex formalism, the order in which the functions are written is unimportant, although for bookkeeping purposes it is usually convenient to follow the order in which the functions appear along the electron line.

One integrates, $\int dE/2\pi$, over any undetermined energy, E, in a closed loop.

Rule 9b

The Ritus method differs from the vertex formalism in that the reduced wavefunctions, $\varphi_s^\epsilon(n, p_z)$, are associated with the propagators rather than the vertex functions. A vertex between unprimed and primed electron states is represented by a factor $\mathcal{J}^\mu_{n'n}(\boldsymbol{k}_\perp)$, which is defined such that its matrix element with respect to the reduced wavefunctions is $[\Gamma^{\epsilon'\epsilon}_{q'q}(k)]^\mu$, as written down in (5.5.25). An internal electron line is represented by a factor $i\mathcal{G}^\epsilon_n(E, p_z)$, which involves the sum over spin states of the outer product of the reduced wavefunction and its Dirac conjugate. This sum is performed using (5.5.15), giving

$$\mathcal{G}^\epsilon_n(E, p_z) = \frac{\not{P}^\epsilon_n + m}{2\epsilon\varepsilon_n} \frac{1}{E - \epsilon(\varepsilon_n - i0)}, \tag{5.6.9}$$

with $[P^\epsilon_n]^\mu = (\epsilon\varepsilon_n, 0, p_n, \epsilon p_z)$. When a statistical average is performed, (5.5.23) is replaced by

$$\bar{G}^\epsilon_n(E, p_z) = \frac{\not{P}^\epsilon_n + m}{2\epsilon\varepsilon_n} \left[\frac{1 - n^\epsilon_n(\epsilon p_z)}{E - \epsilon(\varepsilon_n - i0)} + \frac{n^\epsilon_n(\epsilon p_z)}{E - \epsilon(\varepsilon_n + i0)} \right], \tag{5.6.10}$$

where the occupation number, $n^\epsilon_n(\epsilon p_z)$, is assumed independent of the choice of spin operator.

Rule 10

An internal photon line corresponds to a factor $-iD_{\mu\nu}(k)$. Explicit forms for the photon propagator $D_{\mu\nu}(k)$ are gauge-dependent. For a medium with linear response 4-tensor $\Pi^{\mu\nu}(k)$, the propagator in the G-gauge, with gauge condition $G_\mu\, A^\mu(k) = 0$, is

$$D^{\mu\nu}(k) = \mu_0 \frac{G_\alpha G_\beta}{(Gk)^2} \frac{\lambda^{\mu\alpha\nu\beta}(k)}{\lambda(k)}, \tag{5.6.11}$$

where $\lambda(k)k^\mu k^\nu$ and $\lambda^{\mu\nu\rho\sigma}(k)$ are the first and second order, respectively, matrices of cofactors of $\Lambda^{\mu\nu}(k) = k^2 g^{\mu\nu} - k^\mu k^\nu + \mu_0\, \Pi^{\mu\nu}(k)$. The temporal, Coulomb and Lorenz gauges correspond to $G^\mu = [1, \boldsymbol{0}]$, $G^\mu = [0, \boldsymbol{k}]$ and $G^\mu = k^\mu$, respectively.

5.6.3 Probability of Transition

Any specific process may be described in terms of a probability of transition, which is defined such that crossing symmetries are built into it. For an emission process, the probability of transition is the probability of spontaneous emission, which is identical to the probability of stimulated emission and the probability of true absorption, so that detailed balance applies trivially. The probabilities for crossed processes to emission, including one-photon pair creation and decay, are obtained from the probability of emission by changing the sign of the appropriate 4-momenta, and it is convenient to include the signs explicitly in the probability. Two other features of the probability are that it is defined (a) such that it is independent of the normalization volume, V, and (b) such that conservation of 3-momentum is implicit, whereas conservation of energy is explicit in a δ-function. The probability of transition is identified from a QPD calculation by identifying the differential rate per unit time, $w_{\mathrm{i}\to\mathrm{f}}$, for the process with the same quantity expressed in terms of the probability of transition.

Let $w^{\epsilon'\epsilon}_{q'q}(\boldsymbol{k})$, with $\epsilon' = \epsilon = 1$, denote the probability of spontaneous emission of a photon in the mode M in the range $d^3\boldsymbol{k}/(2\pi)^3$ by an electron in an initial state q, with transition to a final state q'. The differential rate of transition is $w^{++}_{q'q}(\boldsymbol{k})\, d^3\boldsymbol{k}/(2\pi)^3$ which is identified with the transition rate, $w_{\mathrm{i}\to\mathrm{f}}$, for the emission process calculated using the QPD rules in order to identify $w^{++}_{q'q}(\boldsymbol{k})$. For emission by an electron, the density of final states factor is the product of $Vd^3\boldsymbol{k}/(2\pi)^3$ for the photon, and $(VeB/2\pi)(dp'_z/2\pi)$ for the final electron. Conservation of (the z-component of) momentum is explicit in the transition rate, $w_{\mathrm{i}\to\mathrm{f}}$, through a δ-function, specifically, $(2\pi/VeB)\,2\pi\delta(p'_z - p_z + k_z)$ here. Conservation of parallel momentum is implicit in the probability of transition, and this is achieved by noting that this factor in the transition rate combines with the sum over the final states, through the factor $(VeB/2\pi)(dp'_z/2\pi)$ for the final electron, to give unity, specifically,

$$\frac{VeB}{2\pi}\int\frac{dp'_z}{2\pi}\,\frac{2\pi\delta(p'_z - p_z + k_z)}{VeB} = 1.$$

Both factors are omitted in the probability. The rules for the transition rate in QPD then imply the probability of transition

$$w^{\epsilon'\epsilon}_{Mqq'}(\boldsymbol{k}) = V|a_M(\boldsymbol{k})\,\mathcal{T}_{\mathrm{fi}}|^2 2\pi\,\delta(\epsilon'\varepsilon_{q'} - \epsilon\varepsilon_q + \omega_M), \tag{5.6.12}$$

with $\epsilon' = \epsilon = 1$. The probability is such that it does not depend on the normalization volume, V, due to the explicit power of V (from the density of final states factor $Vd^3\boldsymbol{k}/(2\pi)^3$) canceling with a factor $1/V$ in $|a_M(\boldsymbol{k})|^2$. This implies that, in ordinary units, the probability of emission has the dimensions L^3T^{-1}, so that $w^{\epsilon'\epsilon}_{Mqq'}(\boldsymbol{k})d^3\boldsymbol{k}/(2\pi)^3$ has the dimensions T^{-1}.

The transition rate calculated in QPD assumes a given initial electron, so that it corresponds to the rate of emission of photons by the electron. For a distribution of initial electrons, the relevant quantity of interest is the rate the distribution of electrons emits photons per unit volume. This rate per unit volume is found by multiplying the transition rate per electron by $(eB/2\pi)(dp_z/2\pi)n_q$, and integrating over p_z. The result may be used to identify the probability averaged over the initial distribution of electrons; this averaged probability is $w^{++}_{Mqq'}(\boldsymbol{k})n^+_q$. In the second quantization formalism (§ 8.2 of volume 1), the occupation number n^+_q arises from a statistical average of the outer product of the creation and annihilation operators, $\hat{a}^{+\dagger}_q \hat{a}^+_q$, and the statistical average of the operators in the opposite order, $\hat{a}^+_q \hat{a}^{+\dagger}_q$, gives $1 - n^+_q$ for fermions. The latter is relevant to the effect of electrons in the final state, which suppress the emission due to the Pauli exclusion principle. Similarly, the presence of photons in the final state gives a statistical contribution $1 + N_M(\mathbf{k})$, where the unit term is attributed to spontaneous emission and the other term is interpreted as stimulated emission. The total probability of emission, after statistically averaging over the electrons and photons, is

$$w^{++}_{Mqq'}(\boldsymbol{k})n^+_q\left(1 - n^+_{q'}\right)[1 + N_M(\mathbf{k})].$$

Detailed balance requires that the probability (5.6.12) also apply to the absorption process, in the sense that the probability of true absorption is equal to the probability of spontaneous emission times the occupation number, $N_M(\boldsymbol{k})$, of the photons. In the absorption process, an electron in the initial state q' absorbs a photon with frequency $\omega_M(\boldsymbol{k})$, with transition to the final state q. The matrix element $\mathcal{T}_{\rm fi}$ for this process is the complex conjugate of that for the emission process, and hence the differential rate per unit time, $w_{\rm i\to f}$, for the absorption process differs from that for the emission process only through the density of final states. For the absorption process there is only one density of final states factor, $(VeB/2\pi)(dp_z/2\pi)$ for the final electron. As for spontaneous emission, the integral over this factor combines with the δ-function that expresses momentum conservation, specifically $(2\pi/VeB)\,2\pi\delta(p'_z - p_z + k_z)$, and neither appears explicitly in the probability. In the transition rate, $w_{\rm i\to f}$, it is implicit that there is a single photon in the initial state. The absorption rate for a distribution of photons is found by multiplying by $[Vd^3\boldsymbol{k}/(2\pi)^3]N_M(\boldsymbol{k})$ and integrating. To identify the probability of true absorption one equates this absorption rate to the probability times $N_M(\boldsymbol{k})d^3\boldsymbol{k}/(2\pi)^3$. The resulting expression for this probability of absorption is identical to the probability of spontaneous emission, (5.6.12) with $\epsilon' = \epsilon = 1$, confirming that detailed balance is built into the theory.

The statistical average over a distribution of electrons and photons gives the averaged probability of true absorption as

$$w^{++}_{Mqq'}(\boldsymbol{k})n^+_{q'}(1 - n^+_q)N_M(\mathbf{k}).$$

Kinetic equations for the photons and electrons are written down in § 6.1.3 using these statistically averaged probabilities.

The probability of spontaneous emission by a positron is found from the expression (5.6.12) by setting $\epsilon' = \epsilon = -1$. In this case the role of the initial and final states is interchanged, such that q' is interpreted as the initial state, and q is interpreted as the final state. The other processes described by the probability (5.6.12) have $\epsilon' = -\epsilon = \pm 1$ and correspond to one photon pair creation and annihilation. For pair annihilation, the differential transition rate, $w_{i\to f}$, is for a given electron and a given positron in the initial state and in order to identify the probability of pair annihilation one includes distributions of electrons and positrons. The differential transition rate per unit volume is found from the transition rate for a single initial pair by multiplying by a factor $(eB/2\pi)(dp_z/2\pi)n_q^+$ for the electrons and a factor $(eB/2\pi)(dp_z'/2\pi)n_{q'}^-$ for the positrons, in addition to the density of final states factor $Vd^3\boldsymbol{k}/(2\pi)^3$, times $1 + N_M(\mathbf{k})$ to take account of induced emission. The probability of pair annihilation so identified is given by (5.6.12) with $\epsilon = +1, \epsilon' = -1$.

The only other first order processes in QPD are photon splitting and the cross process of coalescence of two photons into one. The probability of these processes is the same as in the unmagnetized case, which is also the same as in a semi-classical treatment. The properties of the wave modes and the detailed expression for the quadratic and cubic response tensors are affected by the presence of the magnetic field, but otherwise the QPD theory of processes involving only photons is the same as the semi-classical theory for wave-wave interactions.

5.6.4 Probabilities for Second-Order Processes

Second order processes involve four external lines, and these are either two electron lines and two photon lines, four electron lines or four photon lines. Processes with two electron lines and two photon lines correspond to Compton scattering and related crossed processes, including double emission and two-photon pair creation and annihilation. Processes involving four electron lines, correspond to electron–electron scattering and related crossed processes, including electron positron scattering and the trident process (emission of a pair by and electron). Processes involving four photon lines correspond to photon-photon scattering and related crossed processes, and the QPD theory is equivalent to the semi-classical theory for such processes.

The argument leading to the identification of the probability of Compton scattering is closely analogous to the argument leading to the expression (5.6.12) for the probability of spontaneous emission. Given the matrix element $\mathcal{T}_{\rm fi}$ for Compton scattering, the probability is

$$w_{MM'qq'}^{\epsilon'\epsilon}(\boldsymbol{k},\boldsymbol{k}') = V^2|a_M(\boldsymbol{k})\,a_{M'}(\boldsymbol{k}')\,\mathcal{T}_{\rm fi}|^2 2\pi\,\delta(\epsilon'\varepsilon_{q'} - \epsilon\varepsilon_q + \omega_M - \omega_{M'}), \quad (5.6.13)$$

with $\epsilon' = \epsilon = +1$ and where the labels M', M denote the modes of the unscattered and scattered photons, respectively. The factor V^2 cancels with factors $1/V$ in $|a_M(\boldsymbol{k})|^2$ and $|a_{M'}(\boldsymbol{k}')|^2$, such that the probability is independent of V. Conservation of (the z-component of) 3-momentum is implicit. In ordinary units, the probability of Compton scattering has the dimensions L^6T^{-1}.

For electron-electron scattering, with the initial states labeled q_1, q_2 and final states labeled q_1', q_2', the probability is given by

$$w_{q_1 q_2 q_1' q_2'} = 2\pi\delta(\varepsilon_{q_1} + \varepsilon_{q_2} - \varepsilon_{q_1'} - \varepsilon_{q_2'})\,|\mathcal{T}_{\mathrm{fi}}|^2, \tag{5.6.14}$$

where conservation of the z-component of momentum, $p_{1z}+p_{2z}-p'_{1z}-p'_{2z}=0$, is implicit. In ordinary units, this scattering probability also has the dimensions L^6T^{-1}.

References

1. J. Géhéniau, Physica (Utrecht) **16**, 822 (1950)
2. J. Géhéniau, M. Demeur, Physica (Utrecht) **17**, 71 (1951)
3. I.S. Gradshteyn, I.M. Ryzhik, *Table of Integrals, Series, and Products* (Academic, New York, 1965)
4. C. Graziani, Astrophys. J. **412**, 351 (1993)
5. H. Herold, Phys. Rev. D **19**, 2868 (1979)
6. H. Herold, H. Ruder, G. Wunner, Astron. Astrophys. **115**, 90 (1982)
7. M.H. Johnson, B.A Lippmann, Phys. Rev. **76**, 828 (1949)
8. A.O.G. Källén, in *Handbuch der Physik*, vol. 5, part 1, ed. by A.S. Flügge (Springer, Berlin, 1958)
9. Y. Katayama, Prog. Theor. Phys. **6**, 309 (1951)
10. A.J. Parle, Aust J. Phys. **40**, 1 (1987)
11. V.I. Ritus, JETP Lett. **12**, 289 (1970)
12. V.I. Ritus, Ann. Phys. **69**, 555 (1972)
13. J. Schwinger, Phys. Rev. **82**, 664 (1951)
14. A.A. Sokolov, I.M. Ternov, *Synchrotron Radiation* (Pergamon Press, Oxford, 1968)

Chapter 6
Quantum Theory of Gyromagnetic Processes

First order processes are described by a Feynman diagram with a single electron-photon vertex. In the presence of a static magnetic field, such a diagram describes six allowed processes: emission and absorption of a photon by an electron or a positron, and one-photon pair creation or annihilation. These are referred to here as the first-order gyromagnetic processes.

General results for gyromagnetic processes are written down in § 6.1, including the probability of gyromagnetic processes, and kinetic equations that describe the effects of gyromagnetic emission and absorption and pair creation. The theory is applied to the quantum theory of cyclotron emission in § 6.2 and to the quantum theory of synchrotron emission in § 6.3. One-photon pair creation and annihilation are discussed in § 6.4.

6.1 Gyromagnetic Emission and Pair Creation

In this section the general probability of first-order gyromagnetic processes is identified, and kinetic equations are written down for each of the processes. Some general results relating to the kinematics of gyromagnetic processes are also derived, including explicit solutions of the resonance condition.

6.1.1 Probability of Gyromagnetic Transition

The probability of gyromagnetic emission follows directly from (5.6.12) and the other rules. One finds

$$w^{\epsilon'\epsilon}_{Mqq'}(\boldsymbol{k}) = \frac{e^2 R_M(\boldsymbol{k})}{\varepsilon_0|\omega_M(\boldsymbol{k})|}\,|e^*_{M\mu}(\boldsymbol{k})[\Gamma^{\epsilon'\epsilon}_{q'q}(\boldsymbol{k})]^\mu|^2 2\pi\delta[\epsilon\varepsilon_q - \epsilon'\varepsilon_{q'} - \omega_M(\boldsymbol{k})], \quad (6.1.1)$$

D. Melrose, *Quantum Plasmadynamics: Magnetized Plasmas*, Lecture Notes in Physics 854, DOI 10.1007/978-1-4614-4045-1_6,

where ϵ, ϵ' label the signs of the particle energies, and q, q' denote the other quantum numbers that describe the state of the electron or positron. The label M for the wave mode of the wave quantum is omitted below where no confusion should result. The set q denotes the parallel momenta, p_z, the principal quantum number, n, and a spin quantum number, s, and q' denotes p'_z, n', s'. Conservation of parallel momentum is implicit in (6.1.1) and sometimes it is relevant to make this explicit by including a factor

$$\int \frac{dp'_z}{2\pi} 2\pi\delta(\epsilon p_z - \epsilon' p'_z - k_z)$$

in the probability. The gauge-dependent quantum numbers do not appear in (6.1.1) in the vertex function, $[\Gamma^{\epsilon'\epsilon}_{q'q}(\boldsymbol{k})]^\mu$, which is, however, dependent on the choice of the spin operator. For the choice of the magnetic-moment operator, the vertex function is given by (5.4.18). The spin-dependent part may be isolated by writing the vertex function in the form (5.5.25).

The dimensions of the probability are L^3T^{-1}. To exhibit these dimensions explicitly, one includes relevant powers of c and $\hbar$. Omitting all arguments $\boldsymbol{k}$, and writing (6.1.1) in SI units gives

$$w^{\epsilon'\epsilon}_{Mqq'} = \frac{e^2c^2R_M}{\varepsilon_0\hbar|\omega_M|}\left|e^*_{M\mu}(\boldsymbol{k})[\Gamma^{\epsilon'\epsilon}_{q'q}]^\mu\right|^2 2\pi\delta\big[\omega_M - (\epsilon\varepsilon_q - \epsilon'\varepsilon_{q'})/\hbar\big], \qquad (6.1.2)$$

where $[\Gamma^{\epsilon'\epsilon}_{q'q}]^\mu$ is dimensionless, and with $\epsilon' p'_z = \epsilon p_z - \hbar k_z$ implicit.

Gyromagnetic Emission by Electrons and Positrons

Gyromagnetic emission by an electron corresponds to $\epsilon = \epsilon' = +1$ in (6.1.1). The unprimed state is the initial state and the primed state is the final state. Gyromagnetic emission by a positron corresponds to $\epsilon = \epsilon' = -1$. The roles of the initial and final states are reversed, with the unprimed state being the final state and the primed state being the initial state. One may rewrite the probability for a positron in the same form as for an electron, so that the unprimed state is the initial state and the primed state is the final state, by setting $\epsilon = \epsilon' = -1$ in (6.1.1), interchanging primed and unprimed quantities, and changing the sign of $\boldsymbol{k}$, using

$$\omega_M(-\boldsymbol{k}) = -\omega_M(\boldsymbol{k}), \qquad e_{M\mu}(-\boldsymbol{k}) = e^*_{M\mu}(\boldsymbol{k}),$$

$$R_M(-\boldsymbol{k}) = R_M(\boldsymbol{k}), \qquad [\Gamma^{\epsilon'\epsilon}_{q'q}(-\boldsymbol{k})]^\mu = [\Gamma^{\epsilon'\epsilon}_{q'q}(\boldsymbol{k})]^{*\mu}. \qquad (6.1.3)$$

The only net change in (6.1.1) is the change $\Gamma^{++}_{q'q}(\boldsymbol{k})$ for an electron to $\Gamma^{--}_{q'q}(\boldsymbol{k})$ for a positron, as one expects. If the particles are unpolarized, the emission by a positron is the same as the emission by an electron provided one makes the replacement $e^*_{M\mu}(\boldsymbol{k}) \to e_{M\mu}(\boldsymbol{k})$ in (5.4.2), which corresponds to reversing the handedness of the waves. This is as one expects on the basis of a classical picture in which the positrons spiral about the magnetic field lines in the opposite sense to the electrons, so that the handedness of the radiation that a positron emits is opposite to the handedness of the emission by an electron.

Unpolarized Particles

When one is not interested in the spins of the particles, it is appropriate to average over initial spins and sum over final spins. Sums over spins are performed for the outer product of a vertex function and its complex conjugate in (5.4.23). For an arbitrary wave mode, this gives

$$\sum_{s,s'} |e^*_{M\mu}(k)[\Gamma^{\epsilon'\epsilon}_{q'q}(k)]^\mu|^2 = 2\varepsilon'_{n'}\varepsilon_n\big[C^{\epsilon'\epsilon}_{n'n}(p_z,k)\big]_M,$$
$$\big[C^{\epsilon'\epsilon}_{n'n}(p_z,k)\big]_M = e^*_{M\mu}(k)e_{M\mu}(k)\big[C^{\epsilon'\epsilon}_{n'n}(p_z,k)\big]^{\mu\nu}, \tag{6.1.4}$$

with $\big[C^{\epsilon'\epsilon}_{n'n}(p_z,k)\big]^{\mu\nu}$ given explicitly by (5.4.24).

For unpolarized particles, (6.1.1) is replaced by

$$\bar{w}^{\epsilon'\epsilon}_{Mqq'}(k) = \frac{\mu_0 e^2 R_M(k)}{2|\omega_M(k)|}\big[C^{\epsilon'\epsilon}_{n'n}(p_z,k)\big]_M 2\varepsilon'_{n'}\varepsilon_n\, 2\pi\delta\big(\epsilon\varepsilon_n - \epsilon'\varepsilon'_{n'} - \omega_M(k)\big), \tag{6.1.5}$$

with $\varepsilon_n = \varepsilon_n(p_z) = (m^2 + p_z^2 + p_n^2)^{1/2}$, $p_n^2 = 2neB$, and with $\varepsilon'_{n'} = \varepsilon_{n'}(p'_z)$. The factor 2 in the denominator in (6.1.5) applies to emission or absorption, and arises from averaging over the initial spins and summing over the final spins of the particle. For one-photon pair production one sums over the spins of both particles, so that the factor $1/2$ is omitted in (6.1.5), and for pair decay into one photon one averages over the spins of both particles, so that the factor $1/2$ is replaced by $1/4$ in (6.1.5).

Reduction to the Nonquantum Limit

It is of interest to derive the nonquantum limit from the relativistic quantum case. Specifically, consider the reduction of the probability (6.1.5) to its semi-classical counterpart (4.1.9). The semi-classical result for gyromagnetic emission for an electron at harmonic a may be written

$$w_M(a,k,p) = \frac{\mu_0 e^2 R_M(k)}{\gamma\omega_M(k)}\big|e^*_{M\mu}(k)U^\mu(s,k_M)\big|^2 2\pi\delta\big[(k_M u)_\parallel - a\Omega_e\big], \tag{6.1.6}$$

and the objective is to derive (6.1.6) from (6.1.1) with $\epsilon = \epsilon' = 1$. One step is the replacement of the resonance condition, expressed through the δ-function in (6.1.5), by its classical counterpart, expressed through the δ-function in (6.1.6). The resonance condition

$$\omega - \epsilon\varepsilon_n + \epsilon'\varepsilon'_{n'} = 0, \tag{6.1.7}$$

needs to be expanded in powers of $\hbar$. Including $\hbar$ explicitly gives $\omega \rightarrow \hbar\omega$, $\varepsilon_n = \varepsilon_n(p_z) = (m^2c^4 + p_z^2c^2 + 2neB\hbar c^2)^{1/2}$, $\varepsilon'_{n'} = \varepsilon_{n'}(p'_z)$, $\epsilon' p'_z = \epsilon p_z - \hbar k_z$. One assumes $\hbar \rightarrow 0$ and $n, n' \rightarrow \infty$ with $n\hbar \rightarrow p_\perp^2/2eB$ and $n - n' = a$ remaining finite. The expansion in powers of $\hbar$ involves the Taylor series

$$\varepsilon'_{n'} = \left(1 - \hat{D} + \frac{1}{2}\hat{D}^2 + \cdots\right)\varepsilon_n, \qquad \hat{D} = a\frac{\partial}{\partial n} + \hbar k_z \frac{\partial}{\partial p_z}, \tag{6.1.8}$$

with $\partial/\partial n \rightarrow (\hbar eB/p_\perp)(\partial/\partial p_\perp)$. For gyromagnetic emission by an electron, to lowest order in $\hbar$ this gives $\hbar\omega - \varepsilon_n + \varepsilon'_{n'} \rightarrow \hbar[(ku)_\parallel - a\Omega_0]$ with $p_\parallel^\mu = [\varepsilon_n, 0, 0, p_z] = mu_\parallel^\mu$, $\Omega_0 = eB/m$. The other step is to relate the 4-vector $U^\mu(a,k)$ in (6.1.6) to the vertex function in (6.1.1). This involves the large-n approximation to the J-functions: writing the argument as $x = \hbar k_\perp^2/2eB = z^2/4n$, where $z = k_\perp p_\perp/eB$ is the argument of the Bessel functions in $U^\mu(a,k)$. Then the J-functions are expanded in Bessel functions using (A.1.53), which is an expansion in z/n. The expansion gives

$$J_\nu^n\left(\frac{z^2}{4n}\right) = \left[\frac{(n+\nu)!}{n!n^\nu}\right]^{1/2} \sum_{j=0}^{\infty} b_j \left(\frac{z}{2n}\right)^j J_{\nu+j}(z), \tag{6.1.9}$$

with $b_0 = 1$, $b_1 = -\frac{1}{2}(\nu+1)$, and higher values given by (A.1.53), and where Stirling's approximation to the factor $(n+\nu)!/n!n^\nu$ gives unity for $n \rightarrow \infty$. The function J_ν^{n-1} differs from J_ν^n only by a quantum correction:

$$J_\nu^{n-1}\left(\frac{z^2}{4n}\right) - J_\nu^n\left(\frac{z^2}{4n}\right) = -\frac{k_\perp}{p_\perp} J'_\nu(z). \tag{6.1.10}$$

To lowest order only the terms in the vertex function that involve no spin flip contribute, and these terms are independent of the choice of spin operator. For the magnetic-moment operator, the vertex function (5.4.18) for an electron, $\epsilon = \epsilon' = 1$, and for no spin flip, $s' = s = \pm 1$, gives, to lowest order in $\hbar$,

$$[\Gamma_{q'q}^{++}]^\mu = \frac{(ie^{i\psi})^{-a}}{\gamma} U^\mu(a,k), \tag{6.1.11}$$

in the large-n limit. Alternatively, the form in (6.1.4) in which the average over spins is performed may be evaluated by making the large-n approximation to (5.4.24); this gives

$$\frac{[C_{n'n}^{++}(p_z, \mathbf{k})]^{\mu\nu}}{2\varepsilon'_{n'}\varepsilon_n} = \frac{1}{\gamma^2} U^\mu(a,k)U^{*\nu}(a,k). \tag{6.1.12}$$

6.1.2 Resonant Momenta and Energies

The conditions under which gyromagnetic emission is possible in general are determined by the resonance condition. The resonance condition (6.1.7) determines the properties of the initial and final particle for given ω, k_z and given n, n'. To find the values of p_z, ε_n or $p'_z, \varepsilon'_{n'}$ for which resonance at given ω, k_z, n, n' is possible, one needs first to eliminate the square roots in (6.1.7). One way of achieving this is to note that all resonance conditions are included in $D(\omega, \varepsilon_n, \varepsilon'_{n'}) = 0$, with

$$\begin{aligned} D(\omega, \varepsilon_n, \varepsilon'_{n'}) &= (\omega - \epsilon\varepsilon_n + \epsilon'\varepsilon'_{n'})(\omega + \epsilon\varepsilon_n - \epsilon'\varepsilon'_{n'}) \\ &\quad \times (\omega - \epsilon\varepsilon_n - \epsilon'\varepsilon'_{n'})(\omega + \epsilon\varepsilon_n + \epsilon'\varepsilon'_{n'}) \\ &= \omega^4 - 2\omega^2(\varepsilon_n^2 + \varepsilon_{n'}'^2) + (\varepsilon_n^2 - \varepsilon_{n'}'^2)^2. \end{aligned} \tag{6.1.13}$$

Using $\epsilon' p'_z = \epsilon p_z - k_z$, $D(\omega, \varepsilon_n, \varepsilon'_{n'}) = 0$ gives a quadratic equation for either p_z or p'_z. (The kinematic restrictions implied by the resonance conditions were discussed in detail by Péres Rojas and Shabad [15], whose solutions differ from the following only in choice of notation.)

The quadratic equation that determines p_z is

$$p_z^2 - 2\epsilon k_z p_z f_{nn'} - (\omega^2 - k_z^2) f_{nn'}^2 + \frac{\omega^2}{\omega^2 - k_z^2}(\varepsilon_n^0)^2 = 0, \tag{6.1.14}$$

and the quadratic equation that determines p'_z is

$$p_z'^2 + 2\epsilon' k_z p_z f_{n'n} - (\omega^2 - k_z^2) f_{n'n}^2 + \frac{\omega^2}{\omega^2 - k_z^2}(\varepsilon_{n'}^0)^2 = 0, \tag{6.1.15}$$

with $(\varepsilon_n^0)^2 = m^2 + 2neB$ and

$$f_{nn'} = \frac{(\varepsilon_n^0)^2 - (\varepsilon_{n'}^0)^2 + \omega^2 - k_z^2}{2(\omega^2 - k_z^2)} = \frac{(n - n')eB}{(k^2)_\parallel} + \frac{1}{2}, \qquad f_{n'n} = 1 - f_{nn'}. \tag{6.1.16}$$

It is convenient to write

$$g_{nn'}^2 = g_{n'n}^2 = \frac{[\omega^2 - k_z^2 - (\varepsilon_n^0 - \varepsilon_{n'}^0)^2][\omega^2 - k_z^2 - (\varepsilon_n^0 + \varepsilon_{n'}^0)^2]}{4(\omega^2 - k_z^2)^2}. \tag{6.1.17}$$

The solutions of (6.1.14) may be written $\epsilon p_z = p_{nn'}^{\pm}$, $\epsilon' p'_z = p_{n'n}^{\pm}$. The relevant solutions are

$$p_{nn'}^{\pm} = k_z f_{nn'} \pm \omega g_{nn'}, \qquad p_{n'n}'^{\pm} = -k_z f_{n'n} \pm \omega g_{n'n}. \tag{6.1.18}$$

The energies $\varepsilon^{\pm}_{nn'} = \varepsilon_n(p^{\pm}_{nn'})$, $\varepsilon'^{\pm}_{n'n} = \varepsilon'_{n'}(p^{\pm}_{n'n})$ are determined by

$$[\varepsilon^{\pm}_{nn'}]^2 = (\omega f_{nn'} \pm k_z g_{nn'})^2, \qquad [\varepsilon'^{\pm}_{n'n}]^2 = (\omega f_{n'n} \mp k_z g_{n'n})^2. \tag{6.1.19}$$

The signs of the energies must be chosen such that the resonance condition $\omega - \epsilon\varepsilon_n + \epsilon'\varepsilon'_{n'} = 0$ is satisfied. The identity $f_{nn'} + f_{n'n} = 1$ implies the solutions $\epsilon\varepsilon_n = \varepsilon^{\pm}_{nn'}$, $\epsilon'\varepsilon'_{n'} = \varepsilon'^{\pm}_{n'n}$, with

$$\varepsilon^{\pm}_{nn'} = \varepsilon_n(p^{\pm}_{nn'}) = \omega f_{nn'} \pm k_z g_{nn'}, \qquad \varepsilon'^{\pm}_{n'n} = \varepsilon'_{n'}(p^{\pm}_{n'n}) = -\omega f_{n'n} \pm k_z g_{n'n}. \tag{6.1.20}$$

The quantum recoil is neglected in the nonquantum limit. The resonance condition $\varepsilon\omega - aeB - p_z k_z = 0$ may be solved for the resonant values of p_z and ε for fixed $p_\perp$. The solutions correspond to $p^{\pm}_{nn'}$, given by (6.1.18), and $\varepsilon^{\pm}_{nn'}$, given by (6.1.19), and with $n' = n - a$ and with

$$f_{n\,n-a} \to \frac{aeB}{(k^2)_\parallel}, \qquad g^2_{n\,n-a} \to f^2_{n\,n-a} - \frac{\varepsilon^2_\perp}{(k^2)_\parallel}, \tag{6.1.21}$$

with $\varepsilon^2_\perp = m^2 + p^2_\perp$. It follows that the term $+\frac{1}{2}$ in (6.1.16) is associated with the quantum recoil, and is intrinsically quantum mechanical.

Allowed Resonances

The quadratic equations (6.1.14) and (6.1.15) may be written in the forms

$$(p_z - \epsilon k_z f_{nn'})^2 = (p'_z + \epsilon' k_z f_{n'n})^2 = \omega^2 g^2_{nn'}. \tag{6.1.22}$$

A necessary condition for resonance to be possible is that $g^2_{nn'}$ be non-negative. From (6.1.17), this condition requires either

$$\omega^2 - k_z^2 \le (\varepsilon^0_n - \varepsilon^0_{n'})^2, \tag{6.1.23}$$

which is the condition for gyromagnetic emission to be allowed, or

$$\omega^2 - k_z^2 \ge (\varepsilon^0_n + \varepsilon^0_{n'})^2. \tag{6.1.24}$$

which is the condition for one-photon pair creation to be allowed. Resonance is forbidden for $(\varepsilon^0_n - \varepsilon^0_{n'})^2 < \omega^2 - k_z^2 < (\varepsilon^0_n + \varepsilon^0_{n'})^2$, referred to as the dissipation-free region, where waves are undamped due to either gyromagnetic absorption or pair creation.

For gyromagnetic emission in vacuo by an electron or a positron, $\epsilon = \epsilon'$, the condition $\omega^2 - k_z^2 > 0$ excludes resonances with $a = n - n' \le 0$. For $\omega^2 - k_z^2 > 0$ one may make a Lorentz transformation to the frame in which k_z is zero, and in

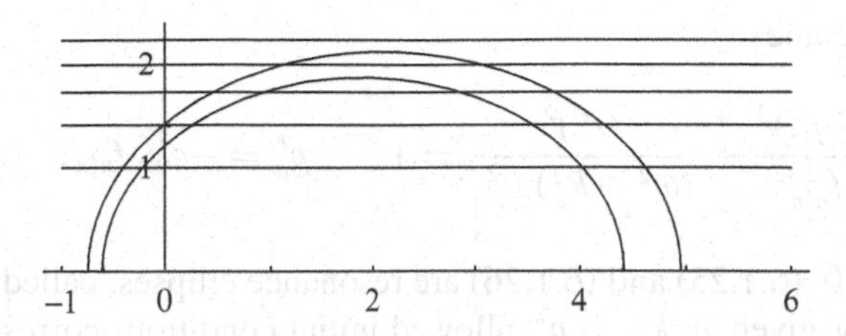

Fig. 6.1 Initial and final resonance ellipse in $\varepsilon^0 = (m^2 + p_\perp^2)^{1/2}$–$p_z$ space. The *horizontal lines* represent the physically allowed values $p_\perp^2 = 2neB$ with $n = 0$–4. The interpretation is discussed in the text

this frame the condition that the energies given by (6.1.20) be non-negative requires $f_{nn'} > 0$, $f_{n'n} < 0$, and the identity $f_{n'n} = 1 - f_{nn'}$ then implies that gyromagnetic emission is possible only for $0 < f_{nn'} < 1$, and this requires $2(n-n')eB > \omega^2 - k_z^2$. Thus for $\omega^2 - k_z^2 > 0$ gyromagnetic emission is possible only for $n > n'$. This corresponds to the normal Doppler effect, with the anomalous Doppler effect being forbidden in vacuo. Gyromagnetic emission is allowed for $\omega^2 - k_z^2 \leq 0$, requiring $n = n'$ for $\omega^2 - k_z^2 = 0$ and emission by the anomalous Doppler effect, $n < n'$, requires $\omega^2 - k_z^2 < 0$.

For one-photon pair creation, $\epsilon = -\epsilon' = 1$, (6.1.24) requires $\omega^2 - k_z^2 > 0$ and the same argument leads to the requirements $f_{nn'} > 0$, $f_{n'n} > 0$ for the energies to be positive. The identity $f_{n'n} = 1 - f_{nn'}$ then requires $0 < f_{nn'} < 1$. For pair creation, there is no restriction on the sign of $n - n'$, but the magnitude is restricted by $2|n - n'|eB < \omega^2 - k_z^2$.

Graphical Solutions of Resonance Condition

A graphical solution of the resonance condition can be useful and informative. Consider momentum space, specifically $p_\perp$–p_z space. The physically allowed states correspond to the discrete values $p_\perp^2 = p_n^2 = 2neB$, $n = 0, 1, \ldots$, which correspond to horizontal lines in $p_\perp$–p_z space. It is found more convenient to choose the vertical axis to be $\varepsilon^0 = (m^2 + p_\perp^2)^{1/2}$, so that the horizontal lines start with $\varepsilon^0 = m$ for $n = 0$ in ε^0–p_z space, as illustrated in Fig. 6.1. The resonance condition may be plotted on the same diagram in two different ways, one corresponding to the initial conditions, and the other to the final conditions.

Suppose that the forms (6.1.14) and (6.1.15) for the resonance condition are expressed in terms of ε^0 by writing $\varepsilon_n^2(p_z) = (\varepsilon^0)^2 + p_z^2$ and $\varepsilon_{n'}^2(p_z') = (\varepsilon^0)^2 + p_z^2 - 2\epsilon' p_z k_z + k_z^2 - 2(n-n')eB$. Then (6.1.14) becomes

$$\frac{(p_z - p_c)^2}{\omega^2 f_{nn'}^2} + \frac{(\varepsilon^0)^2}{(\omega^2 - k_z^2) f_{nn'}^2} = 1, \qquad p_c = \epsilon' k_z f_{nn'}, \tag{6.1.25}$$

and (6.1.15) becomes

$$\frac{(p'_z - p'_c)^2}{\omega^2 f^2_{n'n}} + \frac{(\varepsilon^0)^2}{(\omega^2 - k_z^2) f^2_{n'n}} = 1 \qquad p'_c = -\epsilon k_z f_{n'n}. \tag{6.1.26}$$

For $\omega^2 - k_z^2 > 0$, (6.1.25) and (6.1.26) are resonance ellipses, called the initial and final ellipse. For given ω, k_z, n, n', allowed initial conditions correspond to points in ε^0–p_z space where the initial ellipse intersects the line corresponding to $\varepsilon^0 = \varepsilon^0_n$. Similarly, allowed final conditions correspond to points in ε^0–p_z space where the final ellipse intersects the line corresponding to $\varepsilon^0 = \varepsilon^0_{n'}$.

An example of such a graphical plot is shown in Fig. 6.1 for gyromagnetic emission by an electron, $\epsilon = \epsilon' = 1$. For given $n - n'$, the initial and final ellipses depend on ω and k_z. The initial ellipse intersects the line $\varepsilon^0 = \varepsilon^0_n$ only for $(\omega^2 - k_z^2) f^2_{n'n} > (\varepsilon^0_n)^2$, and two resonances are possible when this condition is satisfied. The intersections of the initial ellipse with the line $\varepsilon^0 = \varepsilon^0_n$ correspond to the solutions $p^{\pm}_{nn'}$, given by (6.1.18), and the intersections of the final ellipse with line $\varepsilon^0 = \varepsilon^0_{n'}$ correspond to the solutions $p^{\pm}_{n'n}$. These two solutions coincide at threshold, where the initial ellipse is tangent to the line $\varepsilon^0 = \varepsilon^0_n$ and the final ellipse is tangent to the line $\varepsilon^0 = \varepsilon^0_{n'}$. These threshold conditions correspond to $g_{nn'} = 0$.

6.1.3 *Kinetic Equations for Gyromagnetic Processes*

For a distribution of electrons emitting and absorbing photons through gyromagnetic transitions, the evolution of the system of photons and electrons is described by a set of kinetic equations. The kinetic equations for gyromagnetic emission in the nonquantum limit are written down in § 4.1, notably the quasilinear equation (4.1.24) for the particles. The fully quantum counterparts of these equation are derived as follows.

Kinetic Equations for Gyromagnetic Emission

Consider the effect of transitions $q \to q'$ and the inverse transition $q' \to q$ on the occupation number, $N_M(\boldsymbol{k})$, of the wave quanta emitted in the former transition and absorbed in the latter transition. The probability $w^{++}_{Mqq'}(\boldsymbol{k})$ is the same for both transitions. Let the occupation number for electrons in the state q be n^+_q. The rate of increase of $N_M(\boldsymbol{k})$ is determined by the probability per unit time of emission,

$$w^{++}_{Mqq'}(\boldsymbol{k})[1 + N_M(\boldsymbol{k})]n^+_q(1 - n^+_{q'}),$$

minus the probability per unit time of absorption

$$w^{++}_{Mqq'}(\boldsymbol{k})N_M(\boldsymbol{k})n^+_{q'}(1 - n^+_q).$$

If more than one transition with given q and different q' gives emission of the same wave quanta (same $\boldsymbol{k}$ and same wave mode M), then one is to sum over all such states q'. The resulting kinetic equation for the waves is

$$\frac{DN_M(\boldsymbol{k})}{Dt} = \sum_{qq'} w^{++}_{Mqq'}(\boldsymbol{k})\{n^+_q(1-n^+_{q'}) + N_M(\boldsymbol{k})[n^+_q - n^+_{q'}]\}, \tag{6.1.27}$$

which may be written in the form

$$\frac{DN_M(\boldsymbol{k})}{Dt} = \beta_M(\boldsymbol{k}) - \gamma_M(\boldsymbol{k})N_M(\boldsymbol{k}),$$

$$\beta_M(\boldsymbol{k}) = \sum_{qq'} w^{++}_{Mqq'}(\boldsymbol{k})\,n^+_q, \qquad \gamma_M(\boldsymbol{k}) = -\sum_{qq'} w^{++}_{Mqq'}(\boldsymbol{k})\,[n^+_q - n^+_{q'}], \tag{6.1.28}$$

where $\beta_M(\boldsymbol{k})$ is an emission coefficient (the rate per unit time that the photon occupation number increases due to spontaneous emission), and $\gamma_M(\boldsymbol{k})$ is the absorption coefficient.

The sums over the states q, q' in (6.1.28) warrant comment. The sum over q is over the initial electrons, and this includes the integral over $(eB/2\pi)dp_z/2\pi$, as well as the sums over n, s. In applications one is often interested in specific transitions, for example, the cyclotron transition $n = 1, s = -1$ to $n' = 0, s' = -1$, and then the sum over q reduces to the integral over $(eB/2\pi)dp_z/2\pi$ alone. The sum over q' is over all possible final states that are allowed. This sum is usually redundant because the final state is uniquely determined. In the example of cyclotron emission $n = 1, s = -1$ to $n' = 0, s' = -1$, the final state is the ground state, and p'_z is implicitly determined by conservation of momentum, so that the sum over q' can be omitted.

The rate of change of the occupation number, n^+_q, of the electrons is given by the difference between the probabilities per unit time of transitions to the state q, due to emission, $q'' \to q$, from higher energy states, and absorption, $q \to q'$, from lower energy states, minus the probabilities per unit time of the inverse transitions. This gives

$$\frac{dn^+_q}{dt} = -\int \frac{d^3\boldsymbol{k}}{(2\pi)^3}\Bigg[\sum_{q'} w^{++}_{Mqq'}(\boldsymbol{k})\{n^+_q(1-n^+_{q'}) + N_M(\boldsymbol{k})(n^+_q - n^+_{q'})\}$$
$$-\sum_{q''} w^{++}_{Mq''q}(\boldsymbol{k})\{n^+_{q''}(1-n^+_q) + N_M(\boldsymbol{k})(n^+_{q''} - n^+_q)\}\Bigg], \tag{6.1.29}$$

where q' denotes the quantum numbers $n', s', p'_z = p_z - k_z$ and q'' denotes $n'', s'', p''_z = p_z + k_z$. The kinetic equations (6.1.28) and (6.1.29) reproduce their unmagnetized counterparts, when the states are identified according to $q \to \boldsymbol{p}$, $q' \to \boldsymbol{p} - \boldsymbol{k}$, $q'' \to \boldsymbol{p} + \boldsymbol{k}$.

The kinetic equations for positrons follow from those for electrons by making the replacements:

$$w^{++}_{Mq'q}(\boldsymbol{k}) \to w^{--}_{Mqq'}(\boldsymbol{k}), \qquad n^+_q \to n^-_{q'}, \qquad n^+_{q'} \to n^-_q.$$

Kinetic Equations for One-Photon Pair Creation

The kinetic equations for one-photon pair creation and decay follow from arguments similar to those for gyromagnetic emission. On including the occupation numbers, the probability of pair annihilation transitions is $w^{+-}_{Mqq'}(\boldsymbol{k})\, n^+_q n^-_{q'}[1 + N_M(\boldsymbol{k})]$, and the probability of pair-creation transitions is $w^{+-}_{Mqq'}(\boldsymbol{k})\, N_M(\boldsymbol{k})(1 - n^+_q)(1 - n^-_{q'})$. A pair-annihilation transition increases the occupation number, $N_M(\boldsymbol{k})$, and decreases the occupation numbers of both electrons, n^+_q, and positrons, n^-_q, and a pair-creation transition has the opposite effect. Adding up the rates of change, the resulting kinetic equation for the photons is

$$\frac{\mathrm{D}N_M(\boldsymbol{k})}{\mathrm{D}t} = \sum_{qq'} w^{+-}_{Mqq'}(\boldsymbol{k})[n^+_q n^-_{q'} - N_M(\boldsymbol{k})(1 - n^+_q - n^-_{q'})]. \tag{6.1.30}$$

The term involving $n^+_q n^-_{q'}$ describes the effect of pair annihilation, and the term involving $1 - n^+_q - n^-_{q'}$ describes the effect of pair creation.

The corresponding kinetic equations for the electrons and positrons are

$$\frac{dn^+_q}{dt} = \int \frac{d^3\boldsymbol{k}}{(2\pi)^3} \sum_{q'} w^{+-}_{Mqq'}(\boldsymbol{k})\big[N_M(\boldsymbol{k})(1 - n^+_q - n^-_{q'}) - n^+_q n^-_{q'}\big],$$

$$\frac{dn^-_{q'}}{dt} = \int \frac{d^3\boldsymbol{k}}{(2\pi)^3} \sum_{q} w^{+-}_{Mqq'}(\boldsymbol{k})\big[N_M(\boldsymbol{k})(1 - n^+_q - n^-_{q'}) - n^+_q n^-_{q'}\big]. \tag{6.1.31}$$

As in (6.1.30), the term involving $n^+_q n^-_{q'}$ describes the effect of pair annihilation, and the term involving $1 - n^+_q - n^-_{q'}$ describes the effect of pair creation. The Pauli exclusion principle requires $n^+_q, n^-_{q'} \le 1$. Nevertheless it is possible for $1 - n^+_q - n^-_{q'}$ to be negative, which is the condition for a maser-like process to occur. This results in an exponential growth of photons and an exponential decay of the number of electrons and positrons, so that $1 - n^+_q - n^-_{q'}$ becomes less negative, with the maser-like process ceasing for $1 - n^+_q - n^-_{q'} = 0$.

6.1.4 Anharmonicity and Quantum Oscillations

Several different quantum effects can be important in cyclotron emission in strong magnetic fields, and these became of direct interest in connection with observed emission from X-ray pulsars, notably Her X-1 [14].

Anharmonicity

A feature of the relativistic theory of gyromagnetic emission and absorption, compared with the nonrelativistic theory, is that each transition, $n \to n'$ for emission by an electron, occurs at a different frequency. Consider the nonrelativistic quantum limit: the transitions between electron states with quantum numbers n and n' involves emission (or absorption) of a wave quantum that satisfies the resonance condition $\omega - k_z v_z = (n - n')\Omega_e$. For given k_z, the resonant frequency then depends on only the difference $n - n'$, and not on n, n' separately. This is due to the energy of the electron being a sum of the kinetic energy, $p_z^2/2m$, associated with the motion along the field line, and the energy $(n + \frac{1}{2})\hbar\Omega_e$ (in ordinary units) associated with the gyromotion, plus an energy $\pm\frac{1}{2}\hbar\Omega_e$ associated with the spin. The energy eigenvalues are harmonically related, such that the difference between neighboring levels has a fixed value, $\hbar\Omega_e$, and emission occurs only at (Doppler shifted) harmonics of this difference. When relativistic quantum effects are taken into account, this degeneracy between levels is broken. Specifically, for given p_z and k_z, the frequency determined by the resonance condition, (6.1.7), gives a different value of ω for every different choice of n and n'. The resonant frequency depends on n or n', as well as on $n - n'$. This effect is sometimes called anharmonicity.

Quantum Oscillations

A characteristic property of the quantum theory of gyromagnetic emission and absorption, and of one-photon pair creation and annihilation in a magnetic field, is that the transition rates have square-root singularities at the thresholds where each new value of n, n' becomes allowed. These thresholds are determined by $g_{nn'} = 0$, with $g_{nn'}^2$ given by (6.1.17).

The gyromagnetic emission and absorption coefficients (6.1.28) involve sums over the initial and final states, which include integrals over p_z and p_z'. These integrals are performed over two δ-functions in the probability (6.1.1). Suppose the integral over p_z', is performed over $\delta(\epsilon' p_z' - \epsilon p_z + k_z)$, as is implicit in (6.1.1). Then the integral over p_z is of the generic form

$$I_{nn'}^{\epsilon\epsilon'} = \int dp_z\, F_{nn'}^{\epsilon\epsilon'}(p_z)\, \delta[\epsilon\varepsilon_n(p_z) - \epsilon'\varepsilon_{n'}(p_z') - \omega], \tag{6.1.32}$$

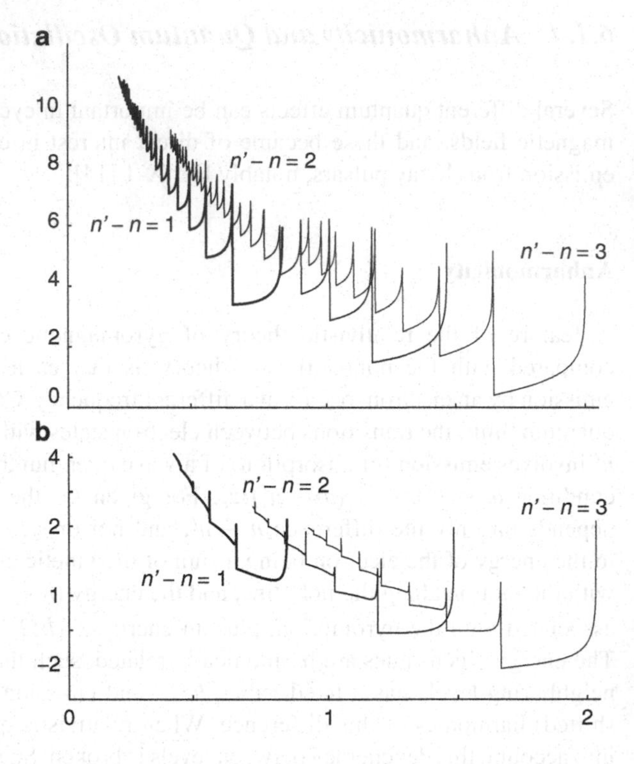

Fig. 6.2 The dependence of the logarithm of the emission coefficient on ω/Ω_e is shown for $B/B_c = 0.5$ for the first three harmonics. The calculations were performed for $\theta = \pi/2$ and radiation polarized (**a**) perpendicular to $\boldsymbol{B}$, and (**b**) along $\boldsymbol{B}$. The thresholds for the lowest transitions correspond to $\omega/m = \sqrt{2}-1$, $\sqrt{3}-1$, $\sqrt{4}-1$ for $n'-n = 1, 2, 3$, respectively, with $\Omega_e/m = B/B_c$

where $F^{\epsilon\epsilon'}_{nn'}(p_z)$ is arbitrary. Both the emission and absorption coefficients (6.1.28) are of the form (6.1.32), as is any other integral that involves a sum of the probability (6.1.1) over the initial and final states. Performing the integral over the δ-function in (6.1.32) gives

$$I^{\epsilon\epsilon'}_{nn'} = \sum_{\pm} F^{\epsilon\epsilon'}_{nn'}(p^{\pm}_{nn'}) \left| \frac{\epsilon p_z}{\varepsilon_n(p_z)} - \frac{\epsilon' p'_z}{\varepsilon_{n'}(p'_z)} \right|^{-1}_{\pm}, \tag{6.1.33}$$

where $\epsilon p_z - \epsilon' p'_z - k_z = 0$ and the resonance condition expressed by the δ-function in (6.1.32) are used. The solutions (6.1.18) and (6.1.20) imply

$$\left| \frac{\epsilon p_z}{\varepsilon_n(p_z)} - \frac{\epsilon' p'_z}{\varepsilon_{n'}(p'_z)} \right|^{-1}_{\pm} = \left| \frac{\varepsilon_n(p_z)\varepsilon_{n'}(p'_z)}{k_z \epsilon \varepsilon_n(p_z) - \omega p_z} \right|_{\pm} = \frac{\varepsilon^{\pm}_{nn'}\varepsilon'^{\pm}_{n'n}}{(\omega^2 - k_z^2)\, g_{nn'}}. \tag{6.1.34}$$

The function $g^2_{nn'}$, given by (6.1.17) has zeros for each value of n', n, implying that (6.1.34) have a square-root singularity at each threshold. This singular behavior is sometimes called quantum oscillations in the case of gyromagnetic emission. An example is illustrated in Fig. 6.2 for $B/B_c = 0.5$.

For the emission and absorption coefficients (6.1.28) evaluation using (6.1.34) gives

$$\beta_M(\boldsymbol{k}) = \sum_{\pm} \frac{\mu_0 e^3 B}{(2\pi)^2} \left[\frac{R|\boldsymbol{e}^* \cdot \boldsymbol{\Gamma}^{++}_{q'q}(\boldsymbol{k})|^2}{\omega(\omega - k_z^2)\, g_{nn'}} \right]_M \varepsilon^{\pm}_{nn'} \varepsilon'^{\pm}_{n'n} n_{n\pm}, \tag{6.1.35}$$

$$\gamma_M(\boldsymbol{k}) = -\sum_{\pm} \frac{\mu_0 e^3 B}{(2\pi)^2} \left[\frac{R|\boldsymbol{e}^* \cdot \boldsymbol{\Gamma}^{++}_{q'q}(\boldsymbol{k})|^2}{\omega(\omega - k_z^2)\, g_{nn'}} \right]_M \varepsilon^{\pm}_{nn'} \varepsilon'^{\pm}_{n'n} (n_{n\pm} - n'_{n'\pm}), \tag{6.1.36}$$

respectively, with $\varepsilon^{\pm}_{nn'}, \varepsilon'^{\pm}_{n'n}$ defined by (6.1.20), and where subscript M indicates that R, $\mathbf{e}$ and ω are to be evaluated for the mode M, and where only the contributions for a specific transition $n, s \to n', s'$ is retained explicitly.

On physical grounds one expects that the quantum oscillations, being an intrinsically quantum phenomenon, to become unobservable for sufficiently nonrelativistic particles. As the classical limit is approached the square-root singularities, that lead to the spiky structure in Fig. 6.2, become densely packed, with a square-root singularity at the threshold for each new n, n'. The quantum oscillations become increasingly unobservable when the peaks are washed out by line-broadening effects. In a nonrelativistic thermal plasma the most important line-broadening effect is Doppler broadening, and the quantum oscillations become unobservable when the Doppler width exceeds the separation between neighboring peaks in the quantum oscillations. The classical Doppler broadening is zero for perpendicular propagation, and then the transverse Doppler effect (due to the spread in Lorentz factors) needs to be taken into account. In the absence of Doppler broadening, there is a natural linewidth for emission by an individual particle: the finite half-life of the upper state introduces an uncertainty in the energy of the transition, and hence of its frequency. When this linewidth exceeds the separation between the peaks associated with the quantum oscillations, the quantum oscillations become intrinsically unobservable [2].

6.2 Quantum Theory of Cyclotron Emission

Gyromagnetic emission from nonrelativistic electrons is called cyclotron emission. In a relativistic quantum treatment there are subtleties in what constitutes the 'nonrelativistic' approximation, and to avoid confusion, the term 'cyclotron' approximation is used here. In practice, the most important approximation is that the J-functions $J^n_\nu(x)$ defined by (5.2.24), are approximated by the leading term in their expansion in powers of x. In this section, after a general introduction to the quantum theory of cyclotron emission, radiative transitions between low Landau levels for an electron in vacuo are discussed in detail, and some expressions for cyclotron emission in a cold plasma are written down.

6.2.1 Cyclotron Approximation

A gyromagnetic transition by an electron may be defined as a transition between an initial state with quantum numbers $q = p_z, n, s$, and a final state with quantum numbers $q' = p'_z, n', s'$, with $p'_z = p_z - k_z$. The parallel motion is unimportant in classifying gyromagnetic transitions: one may make a Lorentz transformation to the frame $p_z = 0$, and provided one has $|k_z| \ll m$ in this frame, the parallel recoil motion is nonrelativistic. The perpendicular momentum is quantized, $p_\perp = p_n = (2neB)^{1/2}$, and the nonrelativistic condition $p_\perp \ll m$ requires $2nB/B_c \ll 1$. In superstrong fields, $B/B_c \gtrsim 1$, even the first Landau level, $n = 1$, has a relativistic energy, and the nonrelativistic approximation is not valid. For weaker fields the nonrelativistic approximation places a limit on $n \ll B_c/2B$.

The transitions $n, s \to n', s'$ may be described in terms of the ladders in Fig. 5.1. Each step in a ladder represents a state n. The left hand ladder corresponds to $s{=}-1$, and includes the ground state $n = 0$, and the right ladder corresponds to $s = +1$ and starts from $n = 1$. The normal Doppler effect is defined such that emission leads to a transition in which n decreases to $n' < n$, and the anomalous Doppler effect to a transition in which n increases to $n' > n$. One can classify transitions as non-spin-flip for $s' = s$, spin-flip for $s = 1, s' = -1$, in which the jump is from the right ladder to the left ladder, and reverse spin-flip for $s = -1, s' = 1$, in which the jump is from the left ladder to the right ladder.

The value of $j = n - n'$ determines the number of steps down which the electron jumps in a transition. The normal Doppler effect corresponds to $j > 0$, and is the only case allowed for $\omega^2 - k_z^2 > 0$. The anomalous Doppler effect, in which the particle moves from a lower to a higher Landau level ($n' > n$), is possible for $k_z^2 > \omega^2$. The case $j = 0$, in which $n = n'$ does not change, is sometimes referred to as the Cerenkov effect, although this can cause confusion with the unmagnetized case. Emission at $j \le 0$ is allowed only for waves with refractive index greater than unity, and is not considered here.

A major simplification in the cyclotron approximation follows from the assumption that the argument of the J-functions in the vertex function (5.4.18) is small, $x = k_\perp^2/2eB \ll 1$. One has

$$x = \frac{k_\perp^2}{2eB} = \frac{k^2 \sin^2\theta}{2m\Omega_e} = \frac{1}{2}\left(\frac{k_\perp}{m}\right)^2 \left(\frac{B}{B_c}\right)^{-1} = \left(\frac{k_\perp}{\Omega_e}\right)^2 \frac{B}{2B_c}. \tag{6.2.1}$$

In the cyclotron approximation, $k_\perp, k_z, \omega$ are assumed of order the cyclotron frequency, Ω_e, and then (6.2.1) implies that x is of order B/B_c. The formal expansion parameter may be regarded as B/B_c. This approximation obviously breaks down in supercritical magnetic fields, $B/B_c \gtrsim 1$.

The power series expansions of the J-functions converge rapidly for $x \ll 1$, and each function may be approximated by its leading term. For example, for $x \ll 1$, (A.1.51) implies that $J^n_{n'-n}(x)$ with $n' = n - j$ is approximated by

$$J^{n}_{-j}(x) = (-)^{j} J^{n-j}_{j}(x) \approx \frac{(-)^{j}}{j!} \left(\frac{n!}{(n-j)!} \right)^{1/2} x^{j/2}. \qquad (6.2.2)$$

The approximation (6.2.2) applies for any integer values of n, j subject to the restrictions $n \geq 0$, $j \leq n$, with $J^{n}_{\nu}(x)$ identically zero for $n < 0$ or $\nu < -n$.

Cyclotron Approximation to the Vertex Function

The foregoing approximations are applied to the vertex function in the form (5.4.18). This involves assuming that p_z/m, p_n/m are small and expanding in powers of them [13]. The momentum, k_z, of the wave quantum is assumed to be of the same order as the momentum of the particle, so that $k_\perp/m$, k_z/m, p'_z/m are of the same order as p_z/m. For the quantity $p_n/m = (2nB/B_c)^{1/2}$ to be small for modest values of n, one also needs to make the weak-field approximation $B \ll B_c$.

The expression (5.4.18) for the vertex function involve factors $a'_\pm, a_\pm$ and $b'_\pm, b_\pm$, as well as J-functions. For an electron, $\epsilon = \epsilon' = 1$, the as and bs are approximated by

$$a'_+ \approx a_+ \approx 1, \qquad b'_+ \approx b_+ \approx 1,$$

$$a'_- \approx \frac{p'_z}{2m}, \quad a_- \approx \frac{p_z}{2m}, \qquad b'_- \approx \frac{p_{n'}}{2m}, \quad b_- \approx \frac{p_n}{2m}, \qquad (6.2.3)$$

where only terms up to first order in small quantities are retained. These approximations allow one to order the coefficients of the J-functions in (5.4.18) in powers of the small quantities.

For a transition that involves a jump of j rungs down the ladder in Fig. 5.1, one has $n' = n - j$. The leading J-functions are those with lower index $n' - n + 1 = -(j-1)$, and these appear only in the $\mu = 1, 2$ components of $[\Gamma^{++}_{q'q}(k)]^{\mu}$. For $s = s'$ these are the only terms that need be retained. However, for $s \neq s'$, the coefficients of these terms are one order higher in the small quantities, and the $\mu = 3$ component is then of the same order as the $\mu = 1, 2$ components. (The $\mu = 0$ component is redundant when one chooses the temporal gauge for the polarization 4-vector of the waves.)

The relevant approximation to the vertex function (5.4.18) is different for transitions involving no spin flip ($s = s'$), those involving a spin flip ($s = 1$, $s' = -1$) and those involving a reverse spin flip ($s = -1$, $s' = 1$). There are three factors in the vertex function (5.4.18), and each of these needs to be evaluated for the four different combinations of spins for specified $j = n - n'$. The first is the phase factor $(ie^{i\psi})^{j}$. The second factor involves the combination of a-factors, which reduce to, for $\epsilon = \epsilon' = 1$,

$$a'_{s'}a_s - a'_{-s'}a_{-s} = s\left[\tfrac{1}{2}(1 + s's) + \tfrac{1}{2}(1 - s's)\,(p'_z - p_z)/2m,\right],$$
$$a'_{s'}a_{-s} + a'_{-s'}a_s = \tfrac{1}{2}(1 + s's)\,(p'_z + p_z)/2m + \tfrac{1}{2}(1 - s's). \qquad (6.2.4)$$

The other factor involves the J-functions. Retaining only the lowest order terms in $x^{1/2}, (B/B_c)^{1/2}$ for each value of s, s', one finds

$$b'_{s'}b_s J^{n-1}_{n'-n} + s'sb'_{-s'}b_{-s}J^{n}_{n'-n} = \frac{(-)^j}{j!}\left(\frac{l!}{(l-j)!}\right)^{1/2} x^{j/2}$$

$$\times \begin{cases} 1 & \text{for } s = s', \\ -j(B/B_c)^{1/2}/(l+1-j)^{1/2} & \text{for } s = -s' = 1, \\ \mathrm{O}(B/B_c, x) & \text{for } s = -s' = -1, \end{cases}$$

$$b'_{-s'}b_s e^{-i\psi} J^{n-1}_{n'-n+1} \mp ss'b'_{s'}b_{-s}e^{i\psi}J^{n}_{n'-n-1}$$

$$= \frac{(-)^{j-1}}{(j-1)!}\left(\frac{l!}{(l-j)!}\right)^{1/2} x^{(j-1)/2}$$

$$\times e^{-i\psi} \begin{cases} (B/B_c)^{1/2} & \text{for } s = s', \\ 1/(l+1-j)^{1/2} & \text{for } s = -s' = 1, \\ \mathrm{O}(B/B_c, x) & \text{for } s = -s' = -1. \end{cases} \tag{6.2.5}$$

In the entries for the reverse spin flip ($s = -1, s' = 1$) in (6.2.5) the leading terms cancel, and the remaining terms are of higher order than those retained in the following discussion.

Multipole Expansion of the Vertex Function

Let $\mathbf{\Gamma}^{++}_{n',s';n,s}(k)$ denote the space components of the vertex function. For non-spin-flip transitions the leading terms give

$$\mathbf{\Gamma}^{++}_{n-j,s;n,s}(k) = (-i)^j e^{i(j-1)\psi}\left(\frac{B}{2B_c}\right)^{1/2}\left(\frac{l!}{(l-j)!}\right)^{1/2}\frac{x^{(j-1)/2}}{(j-1)!}(1,i,0), \tag{6.2.6}$$

with $n = l + (s+1)/2$. With the single-particle current for the emitting electron proportional to the vertex function (6.2.6), and $x = k_\perp^2/2eB$, it follows that the current is proportional to $k_\perp^{j-1}$. This allows a multipole interpretation: on expanding a current in powers of $|\mathbf{k}|$, terms that depend on $|\mathbf{k}|^{j-1}$ correspond to either 2^j-electric multipole or 2^{j-1}-magnetic multipole. The non-spin-flip term (6.2.6) is the 2^j-electric multipole term, with $j = n - n' = l - l'$, so that j corresponds to the change in the orbital quantum number. Specifically, the transition with $s' = s$ and $j = l - l' = 1$ corresponds to electric dipole radiation, the transition with $s' = s$ and $j = l - l' = 2$ corresponds to electric quadrupole radiation, and so on.

For the spin-flip transition, $s = 1 \to s' = -1$, (6.2.6) is replaced by

$$\boldsymbol{\Gamma}^{++}_{n-j,-1;n,1}(\boldsymbol{k}) = (-i)^j e^{i(j-1)\psi}\left(\frac{l!}{(l-j+1)!}\right)^{1/2} \frac{x^{(j-1)/2}}{(j-1)!}\frac{(k_z, ik_z, -k_\perp)}{2m}, \tag{6.2.7}$$

which is the 2^{j-1}-magnetic multipole term. In particular, the transition $s = 1, s' = -1$ for $j = n - n' = 1$ involves no change in the orbital quantum number, $l' = l$, and this corresponds to magnetic dipole radiation. The reverse spin-flip transition, $s = -1 \to s' = 1$, is of higher order and is effectively forbidden in the cyclotron approximation.

Probability of Cyclotron Emission

The probability of gyromagnetic emission, (6.1.1), becomes the probability of cyclotron emission when the cyclotron approximation is made, in particular, when the approximation (6.2.6), (6.2.7) is made to the vertex function. The probability depends on the properties of the emitted waves, and it is convenient to assume a general form that applies when spatial dispersion is unimportant. For a wave in a mode M, this corresponds to $k_\perp = n_M\omega \sin\theta$, $k_z = n_M\omega\cos\theta$ and

$$\boldsymbol{e}_M = \frac{L_M\boldsymbol{\kappa} + T_M\boldsymbol{t} + i\boldsymbol{a}}{1 + L_M^2 + T_M^2}, \qquad R_M = \frac{1 + L_M^2 + T_M^2}{2(1 + T_M^2)n_M\partial(\omega n_M)/\partial\omega}, \tag{6.2.8}$$

with $\boldsymbol{\kappa} = (\sin\theta\cos\phi, \sin\theta\sin\psi, \cos\theta)$, $\boldsymbol{t} = (\cos\theta\cos\phi, \cos\theta\sin\phi, -\sin\theta)$, $\boldsymbol{a} = (-\sin\psi, \cos\phi, 0)$ here.

For an electron, the probability of a non-spin-flip transition, $n \to n' = n - j$, is

$$w^{\rm nsf}_{Mn(n-j)}(\boldsymbol{k}) = \frac{\mu_0 e^2(1 + L_M\sin\theta + T_M\cos\theta)^2}{(1 + T_M^2)2\omega n_M\partial(\omega n_M)/\partial\omega} n_M^{2(j-1)}\sin^{2(j-1)}\theta$$
$$\times \frac{l!}{(l-j)!\,[(j-1)!]^2}\left(\frac{B}{2B_c}\right)^{2-j}\left(\frac{\omega}{m}\right)^{2(j-1)} 2\pi\delta(\varepsilon_n - \varepsilon'_{n-j} - \omega), \tag{6.2.9}$$

where the nonrelativistic approximation is not made to the δ-function. The probability of a transition $n \to n' = n - j$ with a spin-flip, $s = 1 \to s' = -1$, is

$$w^{\rm sf}_{Mn(n-j)}(\boldsymbol{k}) = \frac{\mu_0 e^2(\cos^2\theta + T_M^2)}{(1 + T_M^2)2\omega n_M\partial(\omega n_M)/\partial\omega} n_M^{2j}\sin^{2(j-1)}\theta$$
$$\times \frac{l!}{2^{j-1}(l-j+1)!\,[(j-1)!]^2}\left(\frac{B}{B_c}\right)^{1-j}\left(\frac{\omega}{m}\right)^{2j} 2\pi\delta(\varepsilon_n - \varepsilon'_{n-j} - \omega). \tag{6.2.10}$$

The probabilities corresponding to (6.2.9) and (6.2.10) for a positron may be obtained by replacing the polarization vector by it complex conjugate.

The dependence of the emissivity (6.2.9) on refractive index is $\propto n_N^{2j-1}$ for non-spin-flip transitions and $\propto n_N^{2j}$ for spin-flip transitions. These are the characteristic dependences on refractive index for 2^{2j}-electric multipole emission, and 2^{2j}-magnetic multipole emission, respectively.

6.2.2 *Spontaneous Gyromagnetic Emission in Vacuo*

An electron in a magnetic field emits gyromagnetic radiation spontaneously. The rate of transitions due to gyromagnetic emission is given by summing the probability (6.2.1) over the density of states of the emitted photon. For transverse waves in vacuo the resulting rate, after summing over the two states of polarization, is given by

$$R_{q'q} = \tfrac{1}{2}\mu_0 e^2 \int \frac{d^3\boldsymbol{k}}{(2\pi)^3} \frac{1}{\omega} \left[|\boldsymbol{\Gamma}^{++}_{q'q}(\boldsymbol{k})|^2 - |\boldsymbol{\kappa}\cdot\boldsymbol{\Gamma}^{++}_{q'q}(\boldsymbol{k})|^2 \right] 2\pi\,\delta(\varepsilon_n - \varepsilon'_{n'} - \omega), \tag{6.2.11}$$

with q and q' denoting p_z, n, s and p'_z, n', s', respectively, with ε_n, $\varepsilon'_{n'}$ denoting $\varepsilon_n(p_z)$, $\varepsilon_{n'}(p'_z)$, respectively, and with $p'_z = p_z - k_z$. The power radiated, $P_{q'q}$, is given by an analogous expression with the energy, ω, per photon included in the integrand:

$$P_{q'q} = \tfrac{1}{2}\mu_0 e^2 \int \frac{d^3\boldsymbol{k}}{(2\pi)^3} \left[|\boldsymbol{\Gamma}^{++}_{q'q}(\boldsymbol{k})|^2 - |\boldsymbol{\kappa}\cdot\boldsymbol{\Gamma}^{++}_{q'q}(\boldsymbol{k})|^2 \right] 2\pi\,\delta(\varepsilon_q - \varepsilon_{q'} - \omega). \tag{6.2.12}$$

Conservation of energy and parallel momentum imply that the energy and parallel momentum of the radiating particle change according to

$$\frac{d}{dt}\begin{pmatrix}\varepsilon_q\\ p_z\end{pmatrix} = -\sum_{q'} \tfrac{1}{2}\mu_0 e^2 \int \frac{d^3\boldsymbol{k}}{(2\pi)^3} \begin{pmatrix}1\\ \cos\theta\end{pmatrix} \times \left[|\boldsymbol{\Gamma}^{++}_{q'q}(\boldsymbol{k})|^2 - |\boldsymbol{\kappa}\cdot\boldsymbol{\Gamma}^{++}_{q'q}(\boldsymbol{k})|^2 \right] 2\pi\,\delta(\varepsilon_q - \varepsilon_{q'} - \omega). \tag{6.2.13}$$

In particular one has $d\varepsilon_q/dt = -\sum_{q'} P_{q'q}$.

In vacuo, the integral over $d^3\boldsymbol{k}$ in (6.2.11) or (6.2.12) may be rewritten as an integral over frequency and solid angle. To perform the ω-integral over the δ-function, it is convenient to transform to the frame $p_z = 0$. In this frame the resonant frequency is $\omega = \omega_{nn'}(\theta)$, with

$$\omega_{nn'}(\theta) = \frac{1}{\sin^2\theta} \left\{ \left[m^2 + 2eBn\right]^{1/2} - \left[m^2 + 2eB(n\cos^2\theta + n'\sin^2\theta)\right]^{1/2} \right\}. \tag{6.2.14}$$

The rate (6.2.11) becomes

$$R_{q'q} = \frac{\mu_0 e^2}{4\pi} \int_{-1}^{1} d\cos\theta \left[|\mathbf{\Gamma}_{q'q}^{++}(\boldsymbol{k})|^2 - |\boldsymbol{\kappa} \cdot \mathbf{\Gamma}_{q'q}^{++}(\boldsymbol{k})|^2 \right] \frac{\omega_{nn'}(\theta)[\varepsilon_n^0 - \omega_{nn'}(\theta)]}{\varepsilon_n^0 - \omega_{nn'}(\theta)\sin^2\theta}, \tag{6.2.15}$$

with $\varepsilon_n^0 = (m^2 + 2neB)^{1/2}$. The power (6.2.12) reduces to an analogous expression with an extra power of $\omega_{nn'}(\theta)$ in the integrand.

Lorentz Transformation to the Laboratory Frame

It is sometimes convenient to use the frame $p_z = 0$ to derive explicit expressions, and then to generalize these expressions to an arbitrary frame, $p_z \neq 0$, by making a Lorentz transformation.

Let quantities in the frame in which the particle has no parallel momentum be denoted by a tilde ($\tilde{p}_z = 0$), and let the Lorentz transformation to the laboratory frame, where the parallel momentum is p_z, have a velocity V and a Lorentz factor $\Gamma = (1 - V^2)^{-1/2}$. One has

$$\varepsilon_n(p_z) = \Gamma \varepsilon_n^0, \qquad p_z = \Gamma V \varepsilon_n^0. \tag{6.2.16}$$

In terms of the pitch angle, α, defined by

$$p_z = p\cos\alpha, \qquad p_n = p\sin\alpha, \qquad p = (p_z^2 + p_n^2)^{1/2}, \tag{6.2.17}$$

one has

$$\Gamma = \frac{1}{\sin\alpha}, \qquad V = \cos\alpha. \tag{6.2.18}$$

The frequency and angle of emission transform according to

$$\tilde{\omega} = \omega\,\frac{1 - \cos\alpha\cos\theta}{\sin\alpha}, \qquad \cos\tilde{\theta} = \frac{\cos\theta - \cos\alpha}{1 - \cos\alpha\cos\theta}. \tag{6.2.19}$$

Applying the Lorentz transformation to the expression (6.2.15) for $R_{q'q}$ and to the analogous expression for $P_{q'q}$ involves making the appropriate replacements on the right hand side, and noting that the rate, $R_{q'q}$, transforms as a frequency and that the power, $P_{q'q}$, is an invariant. Specifically, one has $R_{q'q} = \Gamma \tilde{R}_{q'q}$ with $\tilde{R}_{q'q}$ identified with the expression (6.2.15).

Spontaneous Cyclotron Emission in Vacuo

Consider spontaneous emission by an electron in vacuo, as described in the general case by (6.2.3)–(6.2.5). In the cyclotron limit, the transition rate (6.2.3) with (6.2.6) and (6.2.7) may be evaluated explicitly for transitions $n \to n' = n - j$ both without and with a spin flip. The rates for emission of photons polarized perpendicular and parallel to the projection of $\boldsymbol{B}$ onto the transverse plane may also be evaluated explicitly. Let $R_{n\to n-j}$ denote the rate summed over the states of polarization, and let $r_{n\to n-j}$ be the difference in the rates for perpendicular and parallel polarization, divided by the sum of these rates. One may interpret $r_{n\to n-j}$ as giving a measure of the degree of polarization, with positive values ($r_{n\to n-j} > 0$) favoring emission of perpendicularly polarized photons. The actual degree of polarization depends on the angle of emission.

For non-spin-flip transitions ($s' = s$), one finds

$$\begin{aligned} R^{\rm nsf}_{n\to n-j} &= \alpha_c m \,\frac{l!}{(l-j)!}\,\frac{2^{j+1}(j+1)!}{(j-1)!\,(2j+1)!}\left(\frac{\omega}{m}\right)^{2j-1}\left(\frac{B}{B_c}\right)^{2-j}, \\ r^{\rm nsf}_{n\to n-j} &= \frac{j}{j+1}, \end{aligned} \tag{6.2.20}$$

with $n = l + \frac{1}{2}(1+s)$ and where $\omega = \varepsilon_n - \varepsilon_{n-j}$ is the frequency of the emitted photons. The emission frequency at the jth harmonic is approximated by $\omega = j\Omega_e$, due to only the lowest order terms in B/B_c being retained in (6.2.20), with $(\varepsilon_n - \varepsilon_{n-j})/m = j(B/B_c)$ to this order.

For spin-flip transitions, $s = 1, s' = -1$, (6.2.20) is replaced by

$$R^{\rm sf}_{n\to n-j} = \frac{2R^{\rm nsf}_{n\to n-j}}{n-j}\left(\frac{\omega}{m}\right)^{2}\left(\frac{B}{B_c}\right)^{-1}, \qquad r^{\rm sf}_{n\to n-j} = -\frac{j}{j+1}, \tag{6.2.21}$$

with $l = n - 1$ in this case. The rate (6.2.21) for the spin-flip transition is smaller than the rates (6.2.20) for the non-spin-flip transition by $B/B_c \ll 1$. The rate for the reverse spin-flip transition, $s = -1, s' = 1$, is smaller than the rate for the direct spin-flip transition by a further factor of order $(B/B_c)^2$, and is neglected here.

The results (6.2.20) and (6.2.21) imply that for nonrelativistic electrons initially in a high Landau level in a field with $B/B_c \ll 1$, the branching ratio for jumps involving a jth-harmonic transition ($j \geq 2$) with no spin flip depends on the $(j+1)$th power of B/B_c, and a transition with spin flip ($s = 1, s' = -1$) is higher than that without spin flip by an extra factor of order B/B_c. Hence, the most probable sequence of transitions for $B/B_c \ll 1$ is that the electron jumps stepwise down the ladder to $l = 0$. That is, the most probable sequence of transitions is a sequence with $j = 1$ for each transition and with no spin flip. For an electron initially with $s = -1$, the level $l = 0$ is the ground state, $n = 0$. For an electron with $s = 1$, the likelihood of a spin-flip transition is low until it reaches $l = 0$ ($n = 1$ in this case), where it remains until it jumps to $n = 0$ by a spin-flip transition. The probability of a reverse spin flip ($s = -1, s' = 1$) is negligible in the cyclotron approximation.

In the cyclotron approximation, retaining only the electric-dipole transition, the rate of change in the energy and parallel momentum, cf. (6.2.5), reduce to

$$\frac{d\varepsilon_n}{dt} = -\frac{4\alpha_c m^2}{3}\left(\frac{B}{B_c}\right)^{3} l, \qquad \frac{dp_z}{dt} = 0, \tag{6.2.22}$$

with $n = l + \frac{1}{2}(1 + s)$. It follows from (6.2.22) with $\varepsilon_n = m + (p_n^2 + p_z^2)/2m$ that in the nonrelativistic limit the energy radiated comes from a decrease in p_n at fixed p_z.

For $j = 1$ and with the approximation $\omega = \Omega_e$, (6.2.20) and (6.2.21) gives (in ordinary units)

$$R^{\rm nsf}_{1\to 0} = \frac{4\alpha_c mc^2}{3\hbar}\left(\frac{B}{B_c}\right)^2, \qquad R^{\rm sf}_{1\to 0} = R^{\rm nsf}_{1\to 0}\frac{B}{B_c}. \tag{6.2.23}$$

For typical strong fields available in the laboratory, say several tesla, one has $B/B_c \sim 10^{-9}$, and the lifetime $(1/R^{\rm nsf}_{1\to 0})$ of an electron in the first excited state is tens of minutes. For typical pulsar fields, $B/B_c \sim 0.1$, the lifetime is extremely short, and all the electrons are expected to be in their ground states.

The foregoing discussion of spin-flip and non-spin-flip transitions assumes that the spin operator is the magnetic-moment operator. Suppose one chooses the helicity eigenfunctions (5.2.6) rather than the magnetic-moment eigenfunctions. For a nonrelativistic electron, the helicity eigenfunction $\sigma = 1$ is a mixture of the $s = 1$, $s = -1$ states in the ratio $\cos\alpha_n/2 : \sin\alpha_n/2$, where α_n is the counterpart of the classical pitch angle: $p_n/(p_n^2 + p_z^2)^{1/2} = \sin\alpha_n$, $p_z/(p_n^2 + p_z^2)^{1/2} = \cos\alpha_n$. As an electron radiates, n decreases, and hence p_n decreases, whereas p_z remains constant in the nonrelativistic approximation. Hence α_n decreases as a result of the gyromagnetic emission. Suppose that there is no spin-flip in s, and that in a sequence of electric-dipole transitions p_n decreases at fixed s. The linear relation between the σ- and s-states is a function of p_n/p_z and hence as p_n changes, the mixture of σ states changes. Such a change in σ should not be interpreted as a spin flip. 'Spin flip' should be used only to mean a change in the eigenvalue, s, of the magnetic-moment operator.

6.2.3 Cyclotron Emission and Absorption Coefficients

The gyromagnetic emission and absorption coefficients, (6.1.35) and (6.1.36) respectively, become cyclotron emission and absorption coefficients when the cyclotron approximation is made to them. Three separate approximations are involved. The approximation to the vertex function is straightforward, using (6.2.6) and (6.2.7). The simplest approximation to the resonance condition implies $p_z = m(\omega - j\Omega_e)/k_z$. With these approximations, (6.1.35) and (6.1.36) are replaced by

$$\begin{aligned}
\beta_M(\boldsymbol{k}) &= \frac{n_n(p_z)\,eB\mu_0 e^2 m}{(2\pi)^2(j^2\Omega_e^2 - \omega^2 + k_z^2)^{1/2}}\, R_M(\boldsymbol{k})\left|e^*_{M\mu}(\boldsymbol{k})[\Gamma^{++}_{q'q}(\boldsymbol{k})]^\mu\right|^2\Big|_{p_z=p_{zj}},\\
\gamma_M(\boldsymbol{k}) &= -\frac{[n_n(p_z) - n_n(p_z')]\,eB\mu_0 e^2 m}{(2\pi)^2(j^2\Omega_e^2 - \omega^2 + k_z^2)^{1/2}}\, R_M(\boldsymbol{k})\left|e^*_{M\mu}(\boldsymbol{k})[\Gamma^{++}_{q'q}(\boldsymbol{k})]^\mu\right|^2\Big|_{p_z=p_{zj}},
\end{aligned} \tag{6.2.24}$$

respectively, with $p_{zj} = m(\omega - j\Omega_e)/k_z$, $p'_{zj} = p_{zj} - k_z$. The quantum recoil is included simply by replacing p_{zj} by $p_{zj} = (m/k_z)[\omega - j\Omega_e + (\omega^2 - k_z^2)/2m]$.

Bi-Maxwellian Distribution

By way of illustration, consider the emission and absorption coefficients (6.2.24) for a bi-Maxwellian distribution, which is a nondegenerate, thermal-like distribution with different temperatures, $T_\parallel$ and $T_\perp$, parallel and perpendicular to the magnetic field, respectively. The occupation number for such a distribution is

$$n_n(p_z) = A\, e^{-na - p_z^2/2mT_\parallel}, \qquad A = \frac{(2\pi)^{3/2} n_e B_c}{Bm^3(T_\parallel/m)^{1/2}} \tanh(a/2), \tag{6.2.25}$$

with $a = (m/T_\perp)(B/B_c)$. The emission and absorption coefficients follow from (6.2.24) simply by interpreting $n_n(p_{zj})$, $n_n(p_{zj}) - n_n(p'_{zj})$, in terms of the distribution function (6.2.25). The emission coefficient then contains a factor $\exp[-m(\omega - j\Omega_e)^2/k_z^2 T_\parallel]$ that determines the line profile, at the jth harmonic. The Doppler width is of order $|k_z|(T_\parallel/m)^{1/2}$. In the denominator, one may make the approximation $(j^2\Omega_e^2 - \omega^2 + k_z^2)^{1/2} \approx |k_z|$.

Near threshold, both solutions $p^\pm_{nn'}/m \approx j\Omega_e k_z/(\omega^2 - k_z^2) \pm \omega g_{nn'}$, need to be included. The foregoing treatment breaks down for $k_z^2/2\omega \lesssim |\omega - j\Omega_e| \sim |k_z|(T_\parallel/m)^{1/2}$, which corresponds to a small range of angles about perpendicular propagation; emission for $k_z \to 0$ is possible only at $\omega < j\Omega_e$. In this range the two solutions may be approximated by $p^\pm_{nn'}/m \approx k_z/\omega \pm [2(j\Omega_e - \omega)/\omega]^{1/2}$. In the denominator, one may make the approximation $(j^2\Omega_e^2 - \omega^2 + k_z^2)^{1/2} \approx [2\omega(j\Omega_e - \omega)]^{1/2}$. The case where two nonrelativistic solutions need to be included is the counterpart of the case where the nonquantum resonance condition corresponds to a resonance ellipse that is approximately circular near the origin, as illustrated in Fig. 4.1.

6.3 Quantum Theory of Synchrotron Emission

The quantum theory of synchrotron emission differs from its nonquantum counterpart (§ 4.4) in relatively minor ways. The central assumptions are that the resonance condition is approximated by assuming that the particles are highly relativistic and that the transition involves very high harmonics such that the harmonic number, $n - n'$, is treated as a continuous variable. In the nonquantum theory the assumptions justify an Airy-integral approximation to the Bessel functions that appear in the theory, and in the quantum case then justify an Airy-integral approximation to the J-functions. The intrinsically quantum modifications to synchrotron emission discussed here are the quantum recoil, which is important when an emitted photon carries away a large fraction of the initial energy of the electron, quantum fluctuations in the orbit of the radiating electron, the effect of synchrotron emission on the spin of the electrons, and effects associated with superstrong fields.

6.3.1 Quantum Synchrotron Parameter

In the classical theory of synchrotron emission, the frequency dependence is described by a parameter $R = \omega/\omega_c$, with the characteristic frequency of synchrotron emission given by (4.4.2). The quantum treatment involves a parameter ξ introduced by writing the definition (4.4.2) in the form (ordinary units)

$$\omega_c = \frac{3\Omega_e \gamma^2 \sin\alpha}{2m} = \xi(mc^2/\hbar)\gamma_\perp, \qquad \xi = \frac{3B\gamma_\perp}{2B_c}. \tag{6.3.1}$$

It is convenient to develop the quantum theory of synchrotron emission in the frame in which the initial electron has no motion along the field lines, $p_z = 0$. The initial energy of the electron is then $\varepsilon_n = \varepsilon_n^0 = mc^2\gamma_\perp$; n is assumed large and continuous, and is re-expressed in terms of $\gamma_\perp$. The suppression due to quantum effects is significant when the final energy of the electron, $\varepsilon_{n'} \approx mc^2\gamma'_\perp$, is much smaller than the initial energy, due to the energy of the photon being comparable with this initial energy, $\hbar\omega \sim mc^2\gamma_\perp$. This is described by the parameter y, defined by

$$\xi y = \frac{\hbar\omega}{mc^2\gamma_\perp - \hbar\omega} = \frac{\hbar\omega}{mc^2\gamma'_\perp}, \tag{6.3.2}$$

such that the allowed range of frequencies corresponds to $0 < y < \infty$. The recoil implies that, in the frame $p_z = 0$, the final momentum is $p'_z = -\hbar k_z = -(\hbar\omega/c)\cos\theta$, which needs to be included in the detailed theory but which is unimportant in the definition (6.3.2). The suppression of synchrotron emission at high frequencies due to quantum effects is characterized by the product of parameters, ξy. To lowest order in the expansion in small parameters, ξy is related to the initial and final quantum numbers by

$$\frac{\sqrt{n}}{\sqrt{n'}} = 1 + \xi y, \tag{6.3.3}$$

such that one has

$$\frac{\hbar\omega}{mc^2\gamma'_\perp} = \frac{\sqrt{n}-\sqrt{n'}}{\sqrt{n'}} = \xi y, \qquad \frac{\hbar\omega}{mc^2\gamma_\perp} = \frac{\sqrt{n}-\sqrt{n'}}{\sqrt{n}} = \frac{\xi y}{1+\xi y}. \tag{6.3.4}$$

The classical limit applies for $\xi y \ll 1$, and the extreme quantum limit corresponds to $\xi y \gg 1$.

Sum over Final States

Reverting to natural units, the probability of synchrotron emission follows from the probability (6.2.1) for gyromagnetic emission (with $\epsilon = \epsilon' = 1$) when one assumes highly relativistic electrons and makes relevant approximations to the vertex function. For emission in vacuo one has $R_M(\boldsymbol{k}) \to 1/2$, $\omega_M(\boldsymbol{k}) \to \omega$, and there are two transverse states of polarization for the photons. Complete information on the polarization is retained by writing the probability as a (second-rank) polarization tensor which spans the two transverse components. In describing the emitting electron, the quantum numbers of the initial state are n, p_z, s. The parameter n is written in terms of the Lorentz factor, $\gamma_\perp = \varepsilon_n^0/m$, of the perpendicular motion, with $p_z = 0$ by choice of frame.

The rate at which transitions occur, which is the rate at which photons are emitted, is found by summing over the density of final states. For the electron this involves an integral over p_z', which is performed implicitly in using the probability (6.2.1), and sums over n', s'. The sum over n' is replaced by an integral which is performed over the δ-function in (6.2.1). One has

$$\int dn'\delta(\varepsilon_n - \varepsilon_{n'} - \omega) = \frac{\varepsilon_{n'}}{eB} = \frac{\varepsilon_n - \omega}{eB} = \frac{m\gamma_\perp}{eB}\,\frac{1}{1+\xi y}. \tag{6.3.5}$$

There are four possible transitions for the spin states, $s = \pm 1$, $s' = \pm 1$, and the transition rates for all four are different. The integral over the density of final states for the photon is written (natural units)

$$\int \frac{d^3\boldsymbol{k}}{(2\pi)^3} \to 2\pi \int_{-1}^{1} d\cos\theta \int_0^{m\gamma_\perp} \frac{d\omega\,\omega^3}{(2\pi)^3},$$

and the integral over ω may be rewritten using (6.3.2):

$$\int_0^{m\gamma_\perp} d\omega = \xi m\gamma_\perp \int_0^\infty \frac{dy}{(1+\xi y)^2}, \qquad \omega = m\gamma_\perp\,\frac{\xi y}{1+\xi y}. \tag{6.3.6}$$

6.3.2 Synchrotron Approximation

The synchrotron approximation in quantum theory is analogous to the nonquantum theory in that the harmonic number is assumed large and continuous.

Synchrotron Approximation to the Resonance Condition

As in the nonquantum case, important properties of synchrotron emission may be inferred from the resonance condition by appealing to known relativistic effects, simplified by choosing the frame $p_z = 0$. Rather than use the quantum gyroresonance condition in the standard form (6.1.7), it is more convenient to use the form $p_z = p^{\pm}_{nn'}$, given by (6.1.18), so that the resonance condition becomes $[p^{\pm}_{nn'}]^2 = 0$ in the frame $p_z = 0$. Assuming emission in vacuo, the resonance condition in this form implies

$$g^2_{nn'} = f^2_{nn'} \cos^2\theta, \tag{6.3.7}$$

with $f_{nn'}$, $g^2_{nn'}$ given by (6.1.16) and (6.1.17), respectively. The assumption of emission in vacuo implies $\omega^2 - k_z^2 = k_\perp^2$, so that one may write $\omega^2 - k_z^2 = 2eBx$. Then (6.1.16) and (6.1.17) give

$$f_{nn'} = \frac{(x_+ x_-)^{1/2} + x}{2x}, \qquad g^2_{nn'} \approx \frac{(x - x_-)(x - x_+)}{4x^2}, \qquad x_\pm = (\sqrt{n} \pm \sqrt{n'})^2, \tag{6.3.8}$$

where $(\varepsilon^0_n \pm \varepsilon^0_{n'})^2$ in (6.1.17) is expanded in $m^2/p_n^2 = 1/\gamma_\perp^2$, $m^2/p_{n'}^2 = 1/\gamma'^2_\perp$, and only the leading term is retained. The frame $p_z = 0$ corresponds to pitch angle $\alpha = \pi/2$, and the property that the emission is confined to a small range of angles about $\theta = \alpha$ implies that $\cos^2\theta \approx (\theta - \pi/2)^2$ is of order $1/\gamma_\perp^2$. This implies that the right hand side of (6.3.7) is of order $1/\gamma_\perp^2$, so that the left hand side must be of the same order. To lowest order this requires $x = x_-$. In the synchrotron approximation one sets $x = x_-$ except where the difference $x - x_-$ appears explicitly. Thus the lowest order resonance condition for synchrotron emission becomes $x = x_-$, or $\omega^2 = (p_n - p_{n'})^2$.

The difference $x - x_-$ is negative, and it is more convenient to express the difference in terms of the positive quantity $1 - x/x_-$. There are two contributions from the resonance condition in the form (6.3.7). One is from the exact form for $g^2_{nn'}$, given by (6.1.17), where $\omega^2 - k_z^2 - (\varepsilon^0_n - \varepsilon^0_{n'})^2$ is approximated by $2eB(x - x_-)$ to lowest order. Retaining the next order terms gives

$$\omega^2 - k_z^2 - (\varepsilon^0_n - \varepsilon^0_{n'})^2 \approx 2eB\left[x - x_-\left(1 - \frac{m^2}{p_n p_{n'}}\right)\right]. \tag{6.3.9}$$

The other contribution comes from the term proportional to $\cos^2\theta$ in (6.3.7). The resulting expansion gives

$$1 - \frac{x}{x_-} = \sqrt{\frac{n}{n'}}\frac{1}{\zeta^2}, \qquad \frac{1}{\zeta^2} = \frac{1}{\gamma_\perp^2} + \cos^2\theta. \tag{6.3.10}$$

The correction (6.3.10) appears explicitly in the Airy-integral approximation to the function $J^n_\nu(x)$. In an arbitrary frame, with $p_z \neq 0$, the parameter $\cos^2\theta$ in (6.3.10) may be reinterpreted as $(\theta - \alpha)^2$ in the frame in which the pitch angle is α.

Airy-Integral Approximation to $J^n_\nu(x)$

The synchrotron approximation to the functions $J^n_\nu(x)$ is based on an Airy-integral approximation [10, 19, 20]. Some details of a derivation of this approximation are summarized in Appendix J. In brief, the function $u = J^{n'}_{n-n'}(x)$ satisfies the differential equation

$$\frac{d^2(x^{1/2}u)}{dx^2} - \phi(x)\,(x^{1/2}u) = 0, \tag{6.3.11}$$

$$\phi(x) = \frac{(x - x_-)(x - x_+)}{4x^2}, \qquad x_\pm = \left(\sqrt{n + \tfrac{1}{2}} \pm \sqrt{n' + \tfrac{1}{2}}\right)^2. \tag{6.3.12}$$

In the synchrotron limit one has $n, n' \gg 1$, so that the terms $\frac{1}{2}$ are negligible, giving $\phi(x) \to (x - x_-)(x - x_+)/4x^2$, which implies $\phi(x) \to g_{nn'}$ in view of (6.1.20). Airy's differential equation and its relevant solution are

$$\frac{d^2w(z)}{dz^2} - zw(z) = 0, \qquad w(z) = \mathrm{Ai}\,(z) = \frac{1}{\pi}\left(\frac{z}{3}\right)^{1/2} K_{1/3}(\rho), \qquad \rho = \frac{2}{3}z^{3/2}. \tag{6.3.13}$$

The differential equation (6.3.11) is transformed into the differential equation (6.3.13) by writing

$$\rho = \tfrac{2}{3}z^{3/2} = \int_x^{x_-} dx'[\phi(x')]^{1/2}, \qquad \rho' = z^{1/2}z' = -[\phi(x)]^{1/2}, \tag{6.3.14}$$

where the prime denotes a derivative with respect to x, and with the second derivative, z'', neglected. In (6.3.14) one makes the approximation $\phi(x) \approx (x - x_-)(x_- - x_+)/4x_-^2$, and the integral is then elementary. One finds

$$\rho = \tfrac{2}{3}(nn')^{1/4}x_-^{1/2}\left(1 - \frac{x}{x_-}\right)^{3/2}, \qquad \rho' = -\frac{(nn')^{1/4}}{x_-^{1/2}}\left(1 - \frac{x}{x_-}\right)^{1/2}. \tag{6.3.15}$$

A normalization constant is determined by considering the limit $x \to 0$ [20]. The resulting Airy-integral approximation is

$$J^{n'}_{n-n'}(x) = \frac{1}{\pi\sqrt{3}}\left(1 - \frac{x}{x_-}\right)^{1/2} K_{1/3}(\rho), \tag{6.3.16}$$

which applies for $x < x_- = (\sqrt{n} - \sqrt{n'})^2$. The corresponding result for $dJ^{n'}_{n-n'}(x)/dx$ is also needed. This follows from (6.3.16) using the identity

$$\frac{d}{d\rho}[\rho^{1/3}K_{1/3}(\rho)] = -\rho^{1/3}K_{2/3}(\rho), \tag{6.3.17}$$

which implies

$$\frac{d}{dx} J^{n'}_{n-n'}(x) = -\rho' \frac{1}{\pi\sqrt{3}} \left(1 - \frac{x}{x_-}\right)^{1/2} K_{2/3}(\rho), \tag{6.3.18}$$

with ρ' given by (6.3.15).

Four specific J-functions are required to treat gyromagnetic emission: $J^n_{n-n'}(x)$, $J^{n-1}_{n-n'}(x)$, $J^n_{n-n'+1}(x)$, $J^{n-1}_{n-n'-1}(x)$. The other three are related to $J^{n'}_{n-n'}(x)$ using the properties of the J-functions given in Appendix J, specifically,

$$J^{n-1}_{n'-n}(x) = \frac{x}{(nn')^{1/2}} \left[\frac{n'+n-x}{2x} J^n_{n'-n}(x) - \frac{d}{dx} J^n_{n'-n}(x)\right],$$

$$J^n_{n'-n-1}(x) = \left(\frac{x}{n'}\right)^{1/2} \left[\frac{n'-n+x}{2x} J^n_{n'-n}(x) + \frac{d}{dx} J^n_{n'-n}(x)\right],$$

$$J^{n-1}_{n'-n+1}(x) = \left(\frac{x}{n}\right)^{1/2} \left[\frac{n'-n-x}{2x} J^n_{n'-n}(x) - \frac{d}{dx} J^n_{n'-n}(x)\right]. \tag{6.3.19}$$

In the Airy-integral approximation one sets $x = x_-$ except where $1 - x/x_-$ appears explicitly. Hence, the coefficients in (6.3.19) are approximated by

$$\frac{n'+n-x_-}{2(n'n)^{1/2}} = 1, \qquad \frac{n'-n+x_-}{2(n'x_-)^{1/2}} = -1, \qquad \frac{n'-n-x_-}{2(nx_-)^{1/2}} = -1. \tag{6.3.20}$$

Thus the functions $J^n_{n-n'}(x)$, $J^{n-1}_{n-n'}(x)$, $J^n_{n-n'+1}(x)$, $J^{n-1}_{n-n'-1}(x)$ are all equal in magnitude to a first approximation in the extreme relativistic limit, and only the differences resulting from the terms involving the derivatives in (6.3.19) need be retained in distinguishing between them. One finds

$$\begin{pmatrix} J^n_{n'-n} \\ J^{n-1}_{n'-n} \\ J^n_{n'-n-1} \\ J^{n-1}_{n'-n+1} \end{pmatrix} = \frac{(-)^{n'-n}(1+\xi y)^{1/2}}{\pi\sqrt{3}\,\zeta} \left[\begin{pmatrix} 1 \\ 1 \\ -1 \\ -1 \end{pmatrix} K_{1/3} + \begin{pmatrix} 0 \\ -\xi y \\ 1+\xi y \\ -1 \end{pmatrix} \frac{K_{2/3}}{\xi}\right], \tag{6.3.21}$$

where the argument x of the J-functions and $\rho = (y/2)(\gamma_\perp/\xi)^2$ of the Macdonald functions are omitted, with y defined by (6.3.1) and (6.3.3).

Approximation to the Vertex Function

The synchrotron approximation to the vertex function (5.4.18) involves the Airy-integral approximations to the J-functions and appropriate approximations to the parameters $a_\pm$, $a'_\pm$, $b_{\pm s}$, $b'_{\pm s'}$.

For $p_z = 0$, implying $p'_z = -k_z$, the parameters $a_\pm, a'_\pm$ that appear in the vertex function (5.4.18) reduce to

$$a_+ \approx 1, \qquad a'_+ \approx 1, \qquad a_- \approx 0, \qquad a'_- \approx -\tfrac{1}{2}\xi y \cos\theta. \tag{6.3.22}$$

The leading approximations to the other parameters that appear in the vertex function (5.4.18) are

$$b'_{s'} b_s \approx 1 + \frac{1}{2\gamma_\perp}[s + s'(1+\xi y)]. \tag{6.3.23}$$

Combining these various approximations, the synchrotron approximation to the space components of the vertex function (5.4.18) for $p_z = 0$ becomes

$$\begin{aligned}\boldsymbol{\Gamma}^{++}_{qq'}(\boldsymbol{k}) &= (-i)^{n-n'}\frac{(1+\xi y)^{1/2}}{2\pi\sqrt{3}\,\zeta}\Bigg\{\frac{1+s's}{2}\Bigg[2K_{1/3}(z),\\ &\qquad i\left(-\frac{s'\xi y}{\gamma_\perp}K_{1/3}(z) + (2+\xi y)\frac{K_{2/3}(z)}{\zeta}\right), -2\xi y\cos\theta\, K_{1/3}(z)\Bigg]\\ &\qquad + \frac{1-s's}{2}\Bigg[0, -is\xi y\cos\theta\, K_{1/3}(z), +\xi y\left(\frac{s'}{\gamma_\perp}K_{1/3}(z) - \frac{K_{2/3}(z)}{\zeta}\right)\Bigg]\Bigg\},\end{aligned} \tag{6.3.24}$$

where only the lowest order terms are retained in an expansion in $1/\gamma_\perp, 1/\zeta, \cos\theta$. Denoting the space components of (6.3.24), by superscripts x, y, z, the two transverse components correspond to $\Gamma^\perp_{s's} = [\Gamma^{++}_{qq'}]^y(\boldsymbol{k})$ and $\Gamma^\parallel_{s's} = \cos\theta[\Gamma^{++}_{qq'}]^x(\boldsymbol{k}) - \sin\theta[\Gamma^{++}_{qq'}]^z(\boldsymbol{k})$. Specifically, the transverse components are

$$\Gamma^{\perp,\parallel}_{s's} = (-i)^{n-n'}\frac{(1+\xi y)^{1/2}(1+t^2)}{2\pi\sqrt{3}\,\gamma_\perp^2}\,\gamma^{\perp,\parallel}_{s's}, \tag{6.3.25}$$

with $t = \gamma_\perp\cos\theta$, $\gamma_\perp/\zeta = 1+t^2$, and with

$$\begin{aligned}\gamma^\perp_{s's} &= i\Big\{\tfrac{1}{2}(1+s's)\big[-s'\xi y\, K_{1/3}(z) + (2+\xi y)(1+t^2)K_{2/3}(z)\big]\\ &\qquad -\tfrac{1}{2}(1-s's)\,\xi y\, t\, K_{1/3}(z)\Big\},\\ \gamma^\parallel_{s's} &= \tfrac{1}{2}(1+s's)\,(2+\xi y)\, t\, K_{1/3}(z)\\ &\qquad + \tfrac{1}{2}(1-s's)\,\xi y\,\big[s' K_{1/3}(z) - (1+t^2)K_{2/3}(z)\big].\end{aligned} \tag{6.3.26}$$

6.3.3 *Transition Rate for Synchrotron Emission*

Consider the rate of transitions, $R^{\perp,\parallel}_{s,s'}$, between an initial state with quantum numbers $n, s, p_z = 0$, and final states $n', s', p'_z = -k_z$ due to synchrotron emission of photons with $\perp$ and $\parallel$ polarizations. The initial state is described by the perpendicular Lorentz factor, $\gamma_\perp$, and the sum over n' is replaced by an integral. The rate follows by integrating the probability (6.1.1) over the density of final states for the photon. The integral over n' is performed over the δ-function, as in (6.3.5), and the integral over ω is replaced by an integral over y using (6.3.6). This gives transition rates

$$R^{\perp,\parallel}_{s's} = \frac{3\mu_0 e^2 m}{32\pi^3}\frac{B}{B_c}\int_0^\infty dt\,(1+t^2)^2\int_0^\infty \frac{dy\,y}{(1+\xi y)^3}\,|\gamma^{\perp,\parallel}_{s's}|^2, \tag{6.3.27}$$

with $\gamma^{\perp,\parallel}_{s's}(\boldsymbol{k})$ given by (6.3.26). The integrand in (6.3.27) is an even function of $\cos\theta$, and is dominated by the region $\cos\theta \lesssim 1/\gamma_\perp$, allowing the integral over $\cos\theta$ to be rewritten as two times the integral over $t = \gamma_\perp\cos\theta$ between 0 and ∞.

The power emitted is found by including an extra factor of $\omega = m\gamma_\perp^2\xi y/(1+\xi y)$ in the integrand in (6.3.27):

$$P^{\perp,\parallel}_{s's} = \frac{9\mu_0 e^2 m}{64\pi^3}\frac{B}{B_c}\,m\gamma_\perp^2\int_0^\infty dt\,(1+t^2)^2\int_0^\infty \frac{dy\,y^2}{(1+\xi y)^4}\,|\gamma^{\perp,\parallel}_{s's}|^2. \tag{6.3.28}$$

Integral Identities

The following integral identities involving Macdonald functions are required in (6.3.27) or (6.3.28) [19]:

$$\begin{aligned}
\int_0^\infty dt(1+t^2)^2[K_{2/3}(\rho)]^2 &= \frac{\pi}{2\sqrt{3}\,y}\left[\int_y^\infty dx\,K_{5/3}(x) + K_{2/3}(y)\right],\\
\int_0^\infty dt\,t^2(1+t^2)[K_{1/3}(\rho)]^2 &= \frac{\pi}{2\sqrt{3}\,y}\left[\int_y^\infty dx\,K_{5/3}(x) - K_{2/3}(y)\right],\\
\int_0^\infty dt(1+t^2)[K_{1/3}(\rho)]^2 &= \frac{\pi}{\sqrt{3}\,y}\int_y^\infty dx\,K_{1/3}(x),\\
\int_0^\infty dt(1+t^2)^{3/2}K_{1/3}(\rho)K_{2/3}(\rho) &= \frac{\pi}{\sqrt{3}\,y}K_{1/3}(y),
\end{aligned} \tag{6.3.29}$$

with $\rho = \frac{1}{2}y(1+t^2)^{3/2}$, and where the identity

$$\int_y^\infty dx\,K_{1/3}(x) = -\int_y^\infty dx\,K_{5/3}(x) + 2K_{2/3}(y), \tag{6.3.30}$$

is used.

Transition Rate and Power Emitted

The transition rate (6.3.27) becomes

$$R_{s's}^{\perp,\parallel} = \frac{\sqrt{3}\,\mu_0 e^2 m}{32\pi^2}\frac{B}{B_c}\int_0^\infty \frac{dy}{(1+\xi y)^3}\,G_{s's}^{\perp,\parallel}, \tag{6.3.31}$$

the power radiated (6.3.28) becomes

$$P_{s's}^{\perp,\parallel} = \frac{3\sqrt{3}\,\mu_0 e^2 m^2 \gamma_\perp^2}{64\pi^2}\frac{B}{B_c}\int_0^\infty \frac{dy\,y}{(1+\xi y)^4}\,G_{s's}^{\perp,\parallel}, \tag{6.3.32}$$

and the integral identities (6.3.29) imply

$$\begin{aligned}
G_{s's}^{\perp} &= \tfrac{1}{2}(1+s's)\,\left(1+\tfrac{1}{2}\xi y\right)^2\left[\int_y^\infty dx\,K_{5/3}(x) - K_{2/3}(y)\right] + \tfrac{1}{2}(1-s's)\,\frac{\xi^2 y^2}{4}\\
&\quad\times\left[\int_y^\infty dx\,K_{5/3}(x) + K_{2/3}(y) + 2\int_y^\infty dx\,K_{1/3}(x) + 2sK_{1/3}(y)\right],\\
G_{s's}^{\parallel} &= \tfrac{1}{2}(1+s's)\Big\{\left(1+\tfrac{1}{2}\xi y\right)^2\left[\int_y^\infty dx\,K_{5/3}(x) + K_{2/3}(y)\right]\\
&\quad+\tfrac{\xi^2 y^2}{2}\int_y^\infty dx\,K_{1/3}(x) - s\xi y(1+\frac{1}{2}\xi y)K_{1/3}(y)\Big\}\\
&\quad+\tfrac{1}{2}(1-s's)\,\frac{\xi^2 y^2}{4}\left[\int_y^\infty dx\,K_{5/3}(x) - K_{2/3}(y)\right].
\end{aligned} \tag{6.3.33}$$

The rate of transitions per unit frequency and the power per unit frequency follow from (6.3.31) and (6.3.32), respectively, by omitting the integral and using (6.3.6) to write $dy = d\omega\,(1+\xi y)/m\gamma_\perp\xi y$.

Power Emitted for Unpolarized Electrons

When the polarization of the electrons is of no interest, one averages over the initial spin, s, and sums over the final spin, s'. Denoting the resulting quantities by an overbar: $R_{s's}^{\perp,\parallel} \to \bar{R}^{\perp,\parallel}$, $P_{s's}^{\perp,\parallel} \to \bar{P}^{\perp,\parallel}$, $G_{s's}^{\perp,\parallel} \to \bar{G}^{\perp,\parallel}$. The spin-average of (6.3.33) gives

$$\bar{G}^{\perp,\parallel} = (1+\xi y)\left[\int_y^\infty dx\,K_{5/3}(x) + \frac{\xi^2 y^2}{1+\xi y}K_{2/3}(y)\right] \mp K_{2/3}(y). \tag{6.3.34}$$

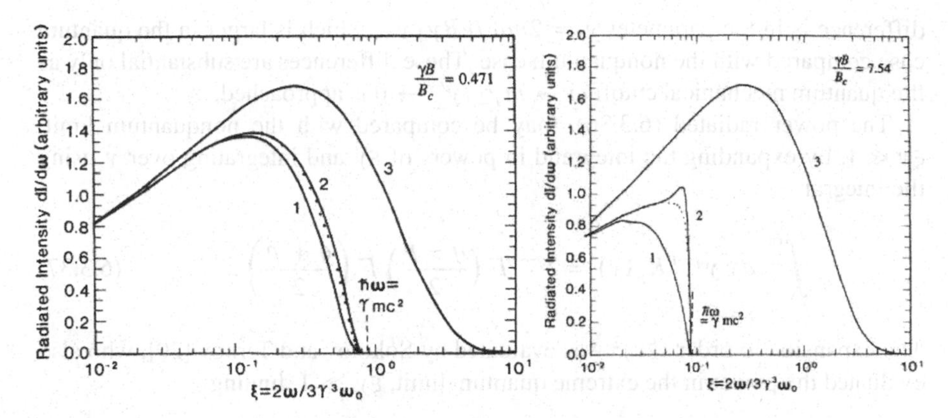

Fig. 6.3 Comparison of the quantum theory of synchrotron radiation for a spinless electron (*1*) and a spin-*half* electron (*2*) with the nonquantum theory (*3*) for two different values of the parameter $\gamma B/B_c$ (From [11], reprinted with permission AAS)

On summing over the two states of polarization of the photon, the spin-averages of (6.3.27) and (6.3.28) give the total transition rate and power radiated, respectively:

$$\bar{R} = \frac{\sqrt{3}\,\mu_0 e^2 m}{32\pi^2}\,\frac{B}{B_c}\int_0^\infty \frac{dy}{(1+\xi y)^2}\left[\int_y^\infty dx\,K_{5/3}(x) + \frac{\xi^2 y^2}{1+\xi y}K_{2/3}(y)\right], \tag{6.3.35}$$

$$\bar{P} = \frac{3\sqrt{3}\,\mu_0 e^2 m^2 \gamma_\perp^2}{64\pi^2}\,\frac{B}{B_c}\int_0^\infty \frac{dy\,y}{(1+\xi y)^3}\left[\int_y^\infty dx\,K_{5/3}(x) + \frac{\xi^2 y^2}{1+\xi y}K_{2/3}(y)\right]. \tag{6.3.36}$$

The final term in (6.3.35) or in (6.3.36) is associated with the spin, in the sense that this term is absent when the theory is developed for spinless particles [20]. Its effect is illustrated in Fig. 6.3, where three cases are compared [11]: the curve labeled 2 corresponds to the quantum theory of synchrotron emission by an unpolarized electron, the curve labeled 3 corresponds to the nonquantum theory, and the curve labeled 1 corresponds to a spinless particle. The dashed curve is calculated using a model based on the Klein-Nishina cross-section for scattering virtual photons [11]. Comparing the quantum, $\xi y \gtrsim 1$, and nonquantum, $\xi y \to 0$, limits there are three notable differences. The first is an overall suppression for the quantum case compared with the nonquantum limit, showing that the quantum mechanical suppression is large only when the particle loses a large fraction of its energy in the emission of a single photon. The second concerns the final term inside the square brackets in (6.3.36), which is absent for $\xi y \to 0$. The coefficient of this term is $\xi^2 y^2/(1+\xi y) = \omega^2/m^2\gamma_\perp\gamma_\perp'$, implying that the synchrotron spectrum differs substantially from its nonquantum counterpart for $\omega^2 \sim m^2\gamma_\perp\gamma_\perp'$. A third difference is in the parameter $y = 2m\omega/eB\gamma_\perp\gamma'_\perp$, which is larger in the quantum case compared with the nonquantum case. These differences are substantial only as the quantum mechanical cutoff $\omega \to m\gamma_\perp, \gamma'_\perp \to 0$ is approached.

The power radiated (6.3.36), may be compared with the nonquantum limit, $\xi y \ll 1$, by expanding the integrand in powers of ξy and integrating over y using the integral

$$\int_0^\infty dy\, y^{q-1} K_p(y) = 2^{q-1}\Gamma\left(\frac{q-p}{2}\right)\Gamma\left(\frac{q+p}{2}\right). \tag{6.3.37}$$

The expansion to order $(\xi y)^2$ was evaluated by Sokolov and Ternov [20], who also evaluated the power in the extreme quantum limit, $\xi y \gg 1$, finding

$$P = P_{\rm cl}\begin{cases} 1 - \dfrac{55\sqrt{3}}{24}\xi + \dfrac{56}{3}\xi^2 + \dfrac{8}{3}\xi^2 & \text{for } \xi \ll 1, \\ \dfrac{2^{8/3}\Gamma(2/3)}{9\xi^{4/3}} & \text{for } \xi \gg 1, \end{cases} \tag{6.3.38}$$

respectively, where $P_{\rm cl}$ is the power in the nonquantum limit. The final term, $8\xi^2/3$, arises from the final term in (6.3.36), and is the contribution from spin-flip transitions.

6.3.4 Change in the Spin During Synchrotron Emission

The rate of transitions is different for the different spin states. Summing over the two polarizations, and expanding the integrand to second order in ξy, evaluation of the resulting integrals using (6.3.37) gives [20]

$$R_{s's} = \frac{5\mu_0 e^2 m}{8\pi\sqrt{3}}\frac{B}{B_c}\left\{\frac{1+s's}{2}\left[1 - \frac{16\sqrt{3}}{45}\xi + \frac{25}{18}\xi^2 - s'\left(1 - \frac{20\sqrt{3}}{9}\xi\right)\frac{\xi}{5}\right] + \frac{1-s's}{2}\left(1 - s'\frac{8\sqrt{3}}{15}\right)\frac{\xi^2}{6}\right\}. \tag{6.3.39}$$

The final term in (6.3.39) describes the spin flips, and implies that the ratio of the rate of direct spin flips, $s = 1 \to s' = -1$, to the rate of reverse spin flips, $s = -1 \to s' = 1$, is $(15 + 8\sqrt{3})/(15 - 8\sqrt{3})$. It follows that after an arbitrarily long time, the occupation numbers for electrons in the states $s = \pm 1$ approach the ratio

$$\frac{n_-}{n_+} \to \frac{15 + 8\sqrt{3}}{15 - 8\sqrt{3}}. \tag{6.3.40}$$

This corresponds to 96 % of the electrons in the state $s = -1$. Sokolov and Ternov [19] showed that this limit is approached exponentially, $\propto \exp(-t/\tau)$, on a timescale (in SI units)

$$\tau_{\rm spin}^{-1} = \frac{5\sqrt{3}}{8}\frac{e^2 mc}{4\pi\epsilon_0\hbar^2}\gamma_\perp^2\left(\frac{B}{B_c}\right)^3 \approx 10^{27}\,s^{-1}\gamma_\perp^2\left(\frac{B}{B_c}\right)^3. \tag{6.3.41}$$

It is relevant to compare this time with the classical synchrotron half-lifetime, $\tau_\varepsilon = mc^2\gamma_\perp/P_{\rm cl}$ in ordinary units. One has

$$\frac{\tau_\varepsilon}{\tau_{\rm spin}} = \frac{15\sqrt{3}}{16}\gamma_\perp\left(\frac{B}{B_c}\right) = \frac{5\sqrt{3}}{8}\xi, \tag{6.3.42}$$

with ξ given by (6.3.1). It follows that for $\xi \gg 1$ the alignment of the spins in the state $s = -1$ occurs before the electron radiates away a significant fraction of it energy. Alignment of the spins does not have time to occur for $\xi \ll 1$.

6.3.5 Gyromagnetic Emission in Supercritical Fields

Gyromagnetic emission for $B/B_c \gtrsim 1$ is qualitatively different from both the cyclotron and synchrotron cases. For gyromagnetic emission to be possible the electron (or positron) must be in a state $n \geq 1$, and for $B/B_c \gg 1$, this implies that it is highly relativistic. However, the synchrotron approximation requires not only that the particles be highly relativistic but also that the quantum numbers n, n' be large enough to be treated as continuous variables. The latter conditions require $n \approx \gamma_\perp^2 B_c/2B \gg 1, n' \approx \gamma_\perp'^2 B_c/2B \gg 1$. However, for $B \gg B_c$, there is a range of $\gamma_\perp$ where the particles are highly relativistic for small n. In this case, gyromagnetic emission must be treated in terms of transitions, $n \to n'$, between discrete levels. In this sense the emission is more closely analogous to cyclotron emission than to synchrotron emission. However, an important simplifying approximation used in treating cyclotron emission does not apply: the argument, x, of the J-functions is not necessarily small. Explicit expressions for $J_\nu^n(x)$ for $n = 0, 1, 2, 3$ are written down in (A.1.47)–(A.1.50) in Appendix A.1.5. These functions are of the form $x^{\nu/2}e^{-x/2}$ times a polynomial in x of order n. In contrast with the cyclotron case, where the transition rates for transitions $n - n' > 1$ and those with a spin flip are smaller that the transition rate $n \to n' = n - 1$ without a spin flip by a power of B/B_c for $B/B_c \ll 1$, all these transition rates become comparable for $B/B_c \gtrsim 1$.

The exact value of the frequency of emission in vacuo for $p_z = 0$ in an arbitrary transition $n \to n'$ is $\omega_{nn'}(\theta)$, given by (6.2.14). In the cyclotron limit one expands the square roots in (6.2.14) assuming $2eB \ll m^2$, giving $\omega_{nn'}(\theta) \approx (n - n')\Omega_e$, $\Omega_e = eB/m$, In the relativistic case, neglecting the terms m^2, (6.2.14) gives

$$\omega_{nn'}(\theta) = \frac{(2eB)^{1/2}}{\sin^2\theta}\left[n^{1/2} - (n\cos^2\theta + n'\sin^2\theta)^{1/2}\right]. \tag{6.3.43}$$

In the synchrotron limit, relativistic beaming implies $\cos^2\theta \ll 1$ and (6.3.43) gives $\omega_{nn'}(\theta) \approx (2eB)^{1/2}(\sqrt{n} - \sqrt{n'})$, corresponding to the synchrotron approximation $x \approx x_-$. The synchrotron approximation is not valid for $n' \ll n$, specifically for $n' \sin^2\theta \ll n \cos^2\theta$, when (6.3.43) gives $\omega_{nn'}(\theta) \approx (2neB)^{1/2}/(1 + |\cos\theta|)$.

A specific case that has been treated in detail is for transitions from large n to $n' \ll n$ [5, 21, 28]. In this case, the emission is concentrated around $\theta = \pi/2$, and the frequency (6.2.14) becomes $\omega_{n0}(\pi/2) \approx (2eB)^{1/2}\sqrt{n}$, corresponding to a photon energy approximately equal to the initial electron energy. Such transitions have been treated using both the exact wave functions [28], and using a parabolic cylinder approximation to the J-functions [21]. The rate of transitions for $n' = 0$ is [21], in ordinary units,

$$R_{0n} = \frac{\alpha_c m^2 c^4}{2\hbar\varepsilon} \frac{B}{B_c} e^{-B_c/B}, \tag{6.3.44}$$

with the initial energy given by $\varepsilon \approx (2neB\hbar c^2)^{1/2}$. The inverse process in which a photon is absorbed by an electron at rest with excitation to $n \gg 1$ has a cross-section

$$\sigma_{n0} = \frac{2\pi^{3/2}\alpha_c\hbar}{m\omega} \left(\frac{B_c}{B}\right)^{1/2} e^{-B_c/B}. \tag{6.3.45}$$

The cross section, which has a maximum at $B = 2B_c$, can be substantially larger than the Thomson cross section.

Quantum Broadening

A more subtle quantum effect is related to spatial diffusion across the field lines. The description of such diffusion is gauge dependent: diffusion along the x-axis may be described using the Landau gauge (5.1.13) and radial diffusion away from a central field line may be described using the cylindrical gauge (5.1.14). In the Landau gauge, emission of a photon with given k_y implies a jump in the center of gyration of $\Delta x = \hbar k_y/eB$. The mean square displacement of the center of gyration increases at a rate $d\langle(\Delta x)^2\rangle/dt \sim \dot{N}_{\rm ph}(\hbar k_y/eB)^2$. In the cylindrical gauge, the radial position is described by a discrete radial quantum number, defined by (5.2.26), and an analogous dispersion occurs in this parameter. A semi-classical description is based on the quasilinear approximation, in which the motion is assumed to be a classical spiraling along a field line. This diffusive process is referred to as a quantum broadening of the classical orbit [16]. In the Landau gauge, assuming $k_y \approx \langle\omega\rangle/c$ with $\langle\omega\rangle$ the mean frequency of synchrotron emission near the peak of the classical spectrum, the gyrocenter jumps a fraction $\approx (B/B_c)\gamma$ of a gyroradius each time a photon is emitted. In the cylindrical gauge, the condition is that the emission of such a photon causes the radial quantum number to change by unity implies [16]

$$n \gtrsim (B_c/B)^5, \qquad \varepsilon \approx mc^2(B_c/B)^{5/2}. \tag{6.3.46}$$

This broadening effect has been confirmed experimentally [5, 20].

6.4 One-Photon Pair Creation

One-photon pair creation and annihilation are allowed in a magnetized vacuum [6, 10, 23, 24]. Pair creation and annihilation are related to gyromagnetic emission by crossing symmetries; they are described by the probability (6.1.1) for $\epsilon = -1$, $\epsilon' = 1$. The analogy between one-photon pair creation and gyromagnetic emission is relatively close. For a nonrelativistic pair the analogy is with cyclotron emission [27], and for an extremely relativistic pair the analogy is with synchrotron emission. The photon involved is necessarily of high energy, $\omega > 2m$, and the effect of an ambient plasma on the dispersion of the photons is usually negligible. Birefringence due to the magnetized vacuum can play a significant role, due to perpendicular and parallel polarized photons annihilating into pairs with different quantum numbers.

One-photon pair creation plays a central role in the theory of pulsar magnetospheres [22]. In this context, it was explored in detail during the 1980s, [1, 3, 4, 7, 9, 29, 32] (see [8] for a more recent review on this topic). Although early treatments of one-photon pair creation emphasized the ultrarelativistic limit, there is a strong argument that the nonrelativistic approximation is more relevant for pulsars. The reason is that the photons are assumed to be emitted nearly along curved field lines, at a very small initial angle, θ, with $\sin\theta$ increasing with distance due to this curvature. Pair creation becomes possible for given n, n' when the threshold $\omega \sin\theta > \varepsilon_n^0 + \varepsilon_{n'}^0$ is reached, and the absorption coefficient is singular at each threshold. Decay into a relativistic pair is possible only if the effective optical depth for absorption at low harmonics is small and that at high harmonics is of order unity or larger [27]. This condition is not satisfied for $B/B_c \gtrsim 0.1$ [27]. A different effect, that may prevent the photons reaching the range $\omega \sin\theta \gg 2m$ where they can decay into a relativistic pair, is that the photon may evolve into a bound pair (§ 6.5).

6.4.1 *Probability of Pair Creation and Decay*

The probability of pair creation or decay follows from (6.2.1) by setting $\epsilon = 1$, $\epsilon' = -1$:

$$w_{Mqq'}^{-+}(\boldsymbol{k}) = \frac{\mu_0 e^2 R_M(\boldsymbol{k})}{|\omega_M(\boldsymbol{k})|} \left|e_{M\mu}^*(\boldsymbol{k})[\Gamma_{q'q}^{-+}(\boldsymbol{k})]^\mu\right|^2 2\pi\delta\big(\varepsilon_q + \varepsilon_{q'} - \omega_M(\boldsymbol{k})\big). \quad (6.4.1)$$

As already remarked, in most cases of interest the wave dispersion can be neglected ($R_M(\boldsymbol{k}) \to \frac{1}{2}$, $\omega_M(\boldsymbol{k}) \to \omega = |\boldsymbol{k}|$) and the weak anisotropy limit applies with the polarization restricted to the transverse plane.

The absorption coefficient follows from the probability by summing over the final states. The sum over final states, q, q', includes sums over the discrete quantum numbers, n, s, n', s', and integrals over p_z, p_z'. As noted in connection with the definition (5.6.12) of the probability, the integral over $(VeB/2\pi)(dp_z'/2\pi)$

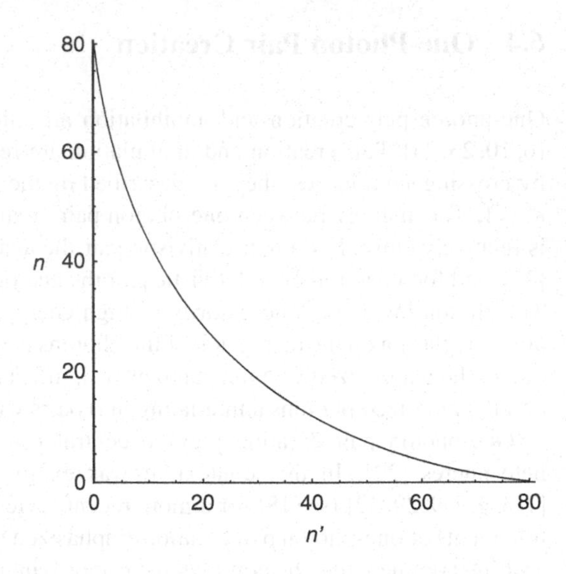

Fig. 6.4 The threshold condition for pair creation is plotted for $(\omega^2 - k_z^2)^{1/2} = 10\,m$, $B = 0.5\,B_c$. The allowed integral values of n, n' are below the curve, and the threshold for a given n, n' corresponds to the point lying on the curve, which is $x_+^2 = (\sqrt{n} + \sqrt{n'})^2 = 81$ in this case (After [4])

is performed over a δ-function, that is omitted from the probability, specifically over $(2\pi/VeB)\,2\pi\delta(p_z' + p_z - k_z)$ here. The factor V in the integral over $(VeB/2\pi)(dp_z/2\pi)$ is incorporated in the definition of the probability, and is omitted in the sum over final state q.

In treating pair creation it is often convenient to transform to the frame $k_z = 0$, in which the photon is propagating perpendicular to the magnetic field lines. This frame plays a role analogous to the frame with $p_z = 0$ in the case of gyromagnetic emission. In the frame $k_z = 0$, the parallel momenta of the electron and positron are equal and opposite. The resonance condition in the form (6.1.18) implies that the parallel momenta are given by $|p_{nn'}^{\pm}| = \omega g_{nn'}$. The threshold condition corresponds to $g_{nn'} = 0$, and pair creation is possible only above threshold, $\omega \geq \varepsilon_n^0 + \varepsilon_{n'}^0$. For given ω, B/B_c, this condition restricts the values of n, n'. For example, for $n' = 0$ the restriction implies $n \leq (\omega/m - 1)^2 B_c/2B$. The restriction is illustrated in Fig. 6.4, in an arbitrary frame ($\omega \to (\omega^2 - k_z^2)^{1/2}$), for values such that for $n' = 0$ the maximum is $n = 81$. A detailed investigation [4] leads to the following semi-quantitative conclusions for large $n + n'$: the most probable transitions are for n, n' near the boundary of the allowed region, the number of states n, n' within the boundary is of order $N_{\text{states}}(\omega, B) \approx 8\omega(\omega + 4m)(\omega - 2m)^2 B_c^2/3m^4 B^2$, and the photon energy tends to be shared roughly equally by the electron and positron ($n \sim n'$) for $B \ll B_c$, but unequally ($n \gg n'$ or $n' \gg n$) for $B \gtrsim B_c$.

Rate of Pair Creation

The rate of pair creation corresponds to the absorption coefficient (per unit time) for the photons. It may be obtained directly by summing the probability (6.4.1), or from the kinetic equation (6.1.30). In the latter case, provided that the number of pairs is well below the degeneracy limit, the occupation numbers of the pairs may be neglected in (6.1.30), which then reproduces the result obtained directly from (6.4.1). The resulting expression for the rate of production of pairs with quantum numbers $q = n, s$, $q' = n', s'$ through decay of a photon in mode M is

$$R^{M}_{q'q} = \frac{eB}{2\pi}\int\frac{dp_z}{2\pi}\,w^{-+}_{Mqq'}(\boldsymbol{k}). \tag{6.4.2}$$

The integral over p_z is carried out over the δ-function, and (6.4.2) gives

$$R^{M}_{q'q} = \frac{\mu_0 e^3 B}{\pi}\,\frac{R_M(\boldsymbol{k})}{\omega_M(\boldsymbol{k})}\,\frac{\left|e^*_{M\mu}(\boldsymbol{k})[\Gamma^{-+}_{q'q}(\boldsymbol{k})]^{\mu}\right|^2}{\omega^2_M(\boldsymbol{k}) - k_z^2}\,\frac{\varepsilon_{n'}\varepsilon_n}{g_{nn'}}, \tag{6.4.3}$$

where $p_z, p'_z, \varepsilon_n, \varepsilon_{n'}$ are to be evaluated at the $\pm$- solutions (6.1.18), (6.1.20), and with $g_{nn'}$ given by (6.1.17). An extra factor of 2 is included in (6.4.3) to take account of equal contributions from the two solutions, $p_z = p^{\pm}_{nn'}$, of the resonance condition.

The rate of pair production for photons in the mode M may also be interpreted as the absorption coefficient for waves in the mode M due to pair creation. The rate (6.4.3) may be calculated using the expression (3.1.27) for the absorption coefficient with the antihermitian part of the response tensor given by the general form (9.2.1), which includes contributions from both the vacuum and the electron gas. Only the contribution of the vacuum is included in (6.4.3).

One is usually justified in assuming that the wave dispersion is that of the vacuum ($\omega_M(\boldsymbol{k}) \to \omega = |\boldsymbol{k}|$, $R_M(\boldsymbol{k}) \to 1/2$), and that the two modes ($M \to \perp, \parallel$) of the birefringent vacuum have polarization vectors $\boldsymbol{e}^{\perp} = \boldsymbol{a}$, $\boldsymbol{e}^{\parallel} = \boldsymbol{t}$. The projections onto the polarization vectors is particularly simple in the frame in which the photon is propagating along the x-axis ($k_z = 0$, $\psi = 0$), when $\boldsymbol{e}^{\perp}$, $\boldsymbol{e}^{\parallel}$ are along the y- and z-axes, respectively. In this frame the only relevant components of the vertex function are

$$\Gamma^{\perp}_{q'q} = [\Gamma^{-+}_{q'q}(\boldsymbol{k})]^y, \qquad \Gamma^{\parallel}_{q'q} = [\Gamma^{-+}_{q'q}(\boldsymbol{k})]^z. \tag{6.4.4}$$

The rate (6.4.3) becomes

$$R^{\perp,\parallel}_{q'q} = \frac{\mu_0 e^2 m^2}{2\pi\omega^3}\,\frac{B}{B_c}\,\frac{\varepsilon_{n'}\varepsilon_n}{g_{nn'}}\,|\Gamma^{\perp,\parallel}_{q'q}|^2, \tag{6.4.5}$$

with $p'_z = -p_z$ for $k_z = 0$, and with $\Gamma^{\perp,\parallel}_{q'q}$ given by (6.4.4). It is usually convenient to calculate the transition rate in the frame $k_z = 0$ and to make a Lorentz transformation to an arbitrary frame with $k_z \neq 0$. The rate transforms like a frequency, as is apparent from the quantities on the right hand side of (6.4.3) which are invariants except for $\varepsilon'_{n'}, \varepsilon_n, g_{n'n}$ which all transform as frequencies. Hence, denoting a quantity in the frame $k_z = 0$ by a tilde, with $\tilde{R}^{\perp,\parallel}_{q'q}(\omega)$ given by (6.4.5), with the frequency dependence made explicit, the decay rate in an arbitrary frame is $R^{\perp,\parallel}_{q'q}(\omega) = \sin\theta\, \tilde{R}^{\perp,\parallel}_{q'q}(\omega\sin\theta)$.

When the dispersion is assumed to be that of the vacuum, the threshold condition $\omega^2 - k_z^2 > (\varepsilon_n^0 + \varepsilon_{n'}^0)^2$ becomes $k_\perp^2 > (\varepsilon_n^0 + \varepsilon_{n'}^0)^2$, and hence

$$x = \frac{k_\perp^2}{2eB} > \frac{(\varepsilon_n^0 + \varepsilon_{n'}^0)^2}{2eB}. \tag{6.4.6}$$

The lowest threshold at $n = n' = 0$, (6.4.6) requires $x > (B/B_c)^{-1}$. It follows that for weak fields, $B \ll B_c$, large x is required for pair creation, and that pair creation is compatible with small x only for supercritical fields, $B \gg B_c$.

6.4.2 Rate of Pair Production Near Threshold

The rate at which photons decay into pairs has singularities at the threshold, $\omega = \varepsilon_n^0 + \varepsilon_{n'}^0$ for creation of a pair with quantum numbers n, n' and $p_z = -p'_z = 0$ (in the frame $k_z = 0$). This is illustrated in Fig. 6.4. The (square-root) singularities correspond to the zeros of $g_{nn'}$ in (6.4.5), and may be regarded as the counterparts for pair creation of the quantum oscillations in gyromagnetic emission, illustrated in Fig. 6.2. Another notable feature in Fig. 6.4 is that the lowest threshold for the parallel and perpendicular polarizations are different. These features may be treated by approximating the rate by its form near each threshold, with the total absorption coefficient found by summing over all values of n, n' and s, s'.

The relevant components of the vertex function are

$$\Gamma^{\perp}_{q'q} = -is'(i)^{n-n'}(a'_{s'}a_{-s} + a'_{-s'}a_s)\,(b'_{-s'}b_s J^{n-1}_{n'-n+1} - s's b'_{s'}b_{-s} J^{n}_{n'-n-1}),$$

$$\Gamma^{\parallel}_{q'q} = (i)^{n-n'}(a'_{s'}a_s + a'_{-s'}a_{-s})\,(b'_{s'}b_s J^{n-1}_{n'-n} + s's b'_{-s'}b_{-s} J^{n}_{n'-n}),$$

$$a_\pm = \left(\frac{\varepsilon_n \pm \varepsilon_n^0}{2\varepsilon_n}\right)^{1/2}, \qquad b_\pm = \left(\frac{\varepsilon_n^0 \pm m}{2\varepsilon_n^0}\right)^{1/2}. \tag{6.4.7}$$

At threshold one has $p_z = -p'_z = 0$, and near threshold the small contributions from $p_z = -p'_z \neq 0$ are neglected. In this approximation, the factors involving the $a_\pm$ in (6.4.7) give

$$(a'_{s'}a_{-s} + a'_{-s'}a_s)^2 \approx \tfrac{1}{2}(1 - s's), \qquad (a'_{s'}a_s + a'_{-s'}a_{-s})^2 \approx \tfrac{1}{2}(1 + s's). \tag{6.4.8}$$

It follows that only $\Gamma^{\perp}_{q'q}$ contributes for $s = -s'$ and only $\Gamma^{\parallel}_{q'q}$ contributes for $s = s'$. The resulting approximations to the components (6.4.7) of the vertex function near threshold give

$$|\Gamma^{\perp}_{q'q}|^2 = \tfrac{1}{2}(1-s's)\,|b'_s b_s J^{n-1}_{n'-n+1} + b'_{-s} b_{-s} J^{n}_{n'-n-1}|^2,$$

$$|\Gamma^{\parallel}_{q'q}|^2 = \tfrac{1}{2}(1+s's)\,|b'_s b_s J^{n-1}_{n'-n} + b'_{-s} b_{-s} J^{n}_{n'-n}|^2, \tag{6.4.9}$$

Near the threshold $\omega = \varepsilon^0_n + \varepsilon^0_{n'}$, the expression (6.1.17) for $g_{nn'}$ in the denominator in (6.4.5) may be approximated by

$$g_{nn'} = \frac{1}{\omega^2}\left\{\varepsilon^0_n \varepsilon^0_{n'}\left[\omega^2 - (\varepsilon^0_n + \varepsilon^0_{n'})^2\right]\right\}^{1/2}. \tag{6.4.10}$$

An approximation to the rate (6.4.5) is found by setting ω equal to its threshold value, $\varepsilon^0_n + \varepsilon^0_{n'}$, except in the expression (6.4.10) for $g_{nn'}$. The argument of the J-functions is approximated by $x = (\varepsilon^0_n + \varepsilon^0_{n'})^2/2eB$.

Most interest is in the lowest thresholds, corresponding to $n = 0$ and $n' = 0, 1$. (There is symmetry in the interchange of n, n'.) In both cases, the only J-function required is $J^0_0(x) = e^{-x/2}$. The rate (6.4.5) gives, for $n = n' = 0$, (ordinary units)

$$R^{\parallel}_{0'0} = \frac{\alpha_c mc^2}{\hbar}\frac{B}{B_c}\frac{mc^2\, e^{-2B_c/B}}{[(\hbar\omega)^2 - (2mc^2)^2]^{1/2}}, \tag{6.4.11}$$

and for $n = 0, n' = 1$, (ordinary units)

$$R^{\perp}_{1'0} = \frac{\alpha_c mc^2}{\hbar}\frac{B}{B_c}\left(\frac{mc^2}{\varepsilon^0_1}\right)^{1/2}\frac{mc^2\, e^{-(B_c/B)(1+\varepsilon^0_1/mc^2)}}{[(\hbar\omega)^2 - (mc^2+\varepsilon^0_1)^2]^{1/2}}, \tag{6.4.12}$$

with $\varepsilon^0_1 = mc^2(1+2B/B_c)^{1/2}$.

6.4.3 *Creation of Relativistic Pairs*

When a photon with $\omega \sin\theta \gg 2m$ decays, the resulting pair is highly relativistic. As for synchrotron emission, in the extreme relativistic limit one treats n, n' as continuous variables and makes an Airy integral approximation to the J-functions.

Synchrotron-Like Approximation

A detailed treatment of pair creation when the pairs are relativistic is similar to the treatment of synchrotron emission. In the frame $k_z = 0$, the analysis is similar to that for synchrotron emission, with the important change being that in the expression (6.1.18) for $g^2_{nn'}$, one has $x \approx x_+$, rather than $x \approx x_-$, with $x = \omega^2/2eB$ and $x_\pm = (\sqrt{n} \pm \sqrt{n'})^2$. The relativistic approximation for pair creation involves setting $x = x_+$, except where the difference appears explicitly, when it is written in terms of $x/x_+ - 1$, replacing the sums over n, n' by integrals, and making the Airy-integral approximation to the J-functions.

It is convenient to change the variables of integration from n, n' to $g_{nn'}$, $f_{nn'}$ and thence to hyperbolic functions of variables u, w. The hyperbolic functions are introduced by writing $p_z = m \sinh u$ and the difference over the sum of the perpendicular momenta as $\tanh w$, specifically,

$$g_{nn'} = \frac{m}{\omega} \sinh u, \qquad \frac{\sqrt{n} - \sqrt{n'}}{\sqrt{n} + \sqrt{n'}} = \tanh w. \tag{6.4.13}$$

In the explicit expression (6.1.16) for $g^2_{nn'}$ with $k_z = 0$, one approximates $\varepsilon^0_n \approx p_n = \omega(n/x)^{1/2}$, $\varepsilon^0_{n'} \approx p_{n'} = \omega(n'/x)^{1/2}$, with $x \approx x_+$, except in

$$\omega^2 - (\varepsilon^0_n + \varepsilon^0_{n'})^2 \approx 2eBx_+ \left(\frac{x}{x_+} - 1 - \frac{4m^2}{\omega^2} \cosh^2 w \right), \tag{6.4.14}$$

which follows by expanding in m^2/p^2_n, $m^2/p^2_{n'}$. The resulting approximation to $g^2_{nn'}$ is

$$g^2_{nn'} \approx \frac{\sqrt{nn'}}{(\sqrt{n} + \sqrt{n'})^2} \left(\frac{x}{x_+} - 1 - \frac{4m^2}{\omega^2} \cosh^2 w \right). \tag{6.4.15}$$

The approximate form (6.4.15) implies

$$\frac{x}{x_+} - 1 \approx \frac{4m^2}{\omega^2} \cosh^2 u \cosh^2 w. \tag{6.4.16}$$

In the relativistic approximation to $f_{nn'} = (n - n' + x)/2x$ it suffices to set $x = x_+$, ignoring the corrections of order $(m/\omega)^2$:

$$f_{nn'} \approx \frac{\sqrt{n}}{\sqrt{n} + \sqrt{n'}} = \tfrac{1}{2}(1 + \tanh w), \qquad f_{n'n} \approx \frac{\sqrt{n'}}{\sqrt{n} + \sqrt{n'}} = \tfrac{1}{2}(1 - \tanh w). \tag{6.4.17}$$

A fraction $f_{nn'}$ of the photon energy goes into the electron, and the remaining fraction $f_{n'n} = 1 - f_{nn'}$ goes into the positron.

The integrals over n, n' may be rewritten as integrals over u, w. The Jacobian of the transformation implies

$$\frac{dn\, dn'}{g_{nn'}} = \frac{1}{2} \left(\frac{\omega}{m} \right)^3 \left(\frac{B_c}{B} \right)^2 du\, dw \frac{\cosh u}{\cosh^2 w}, \tag{6.4.18}$$

with the range of integration restricted to $u, w > 0$. The upper limits of integration, for n, n' or u, w, are finite and large, and a standard approximation is to extend them to infinity. Thus, after integrating over n, n', the rate (6.4.5) becomes

$$R_{s's}^{\perp,\parallel} = \frac{\mu_0 e^2 m}{\pi \omega^2} \int du\, dw \frac{\cosh u}{\cosh^2 w} \varepsilon_{n'} \varepsilon_n \, |\Gamma_{q'q}^{\perp,\parallel}|^2. \tag{6.4.19}$$

The result (6.4.19) depends on the spins of the electron and positron. The total rate of pair creation is found by summing over all four possible combinations $s', s = \pm 1$.

The restriction to $u > 0$ corresponds to $p_z > 0$, and the restriction to $w > 0$ corresponds to $n > n'$. There is a symmetry in the interchange of the electron and positron, such that the interpretation of the unprimed and primed quantum numbers corresponding to the electron or positron is arbitrary. The right hand side of (6.4.19) is an even function of u, and the contribution from negative u is taken into account by including an extra factor of two in (6.4.19). The assumption $w > 0$ for the unprimed particle implies that the particle with spin s has $p_n > p_{n'}$. Thus, in formulae that depend on the sign of $s \sinh w$, $s \sinh w > 0$ is interpreted as the particle with the larger perpendicular momentum having spin up and $s \sinh w < 0$ is interpreted as the particle with the larger perpendicular momentum having spin down.

Airy-Integral Approximation

The derivation of the Airy-integral approximation relevant to one-photon pair creation is analogous to the derivation of the Airy-integral (6.3.16) approximation for synchrotron emission. The main change is that the relevant zero of $\phi(x) = (x - x_-)(x - x_+)/4x^2$ changes from $x = x_-$ to $x = x_+$. The approximation for $x \approx x_+ = (\sqrt{n} + \sqrt{n'})^2$, with $x - x_+$ small and positive, leads to a Macdonald function $K_{1/3}(\rho)$ with argument

$$\rho = \tfrac{2}{3}[nn'x_+^2]^{1/4} \left(\frac{x}{x_+} - 1 \right)^{3/2}, \qquad \rho' = \left[\frac{\sqrt{n'n}}{x_+} \left(\frac{x}{x_+} - 1 \right) \right]^{1/2}. \tag{6.4.20}$$

The result (6.4.20) is the counterpart for pair creation of the result (6.3.15) for synchrotron emission.

The Airy-integral approximation to the J-functions for pair creation is

$$J_{n'-n}^{n}(x) = A_{n'n} \, K_{1/3}(\rho), \qquad A_{n'n} = \frac{(-)^{n'-n}}{\pi \sqrt{3}} \frac{2m}{\omega} \cosh u \cosh w, \tag{6.4.21}$$

where the argument, ρ, of the Macdonald functions is given by (6.4.20). It is convenient to introduce the parameter χ such that (6.4.20) becomes

$$\rho = \frac{2}{3\chi} \cosh^3 u \cosh^2 w, \qquad \chi = \frac{\omega}{2m} \frac{B}{B_c}. \tag{6.4.22}$$

The corresponding approximation to the derivative of the J-functions is derived using (6.3.17) and $\rho' = (m/\omega)\cosh u$:

$$\frac{d}{dx}J^n_{n'-n}(x) = -A_{n'n}\,\frac{m}{\omega}\cosh u\,K_{2/3}(\rho). \tag{6.4.23}$$

Approximations to the other J-functions that appear follow from (6.4.21) by using the relations (6.3.19). In place of (6.3.20), for $x \to x_+$ one finds

$$\frac{n'+n-x_+}{2(n'n)^{1/2}} = -1, \qquad \frac{n'-n+x_+}{2(n'x_+)^{1/2}} = 1, \qquad \frac{n'-n-x_+}{2(nx_+)^{1/2}} = -1. \tag{6.4.24}$$

In place of (6.3.21), for pair creation one finds

$$\begin{pmatrix} J^n_{n'-n} \\ J^{n-1}_{n'-n} \\ J^n_{n'-n-1} \\ J^{n-1}_{n'-n+1} \end{pmatrix} = A_{n'n}\left[\begin{pmatrix} 1 \\ -1 \\ 1 \\ -1 \end{pmatrix} K_{1/3}(\rho) + \frac{2m}{\omega}\cosh u \begin{pmatrix} 0 \\ a_+ + a_- \\ -a_+ \\ a_- \end{pmatrix} K_{2/3}(\rho)\right], \tag{6.4.25}$$

with $A_{n'n}$ given by (6.4.21), and with

$$\zeta_\pm = \frac{1}{1 \mp \tanh w} = \cosh w(\cosh w \pm \sinh w). \tag{6.4.26}$$

Relativistic Approximation to the Vertex Function

The 'relativistic' approximation, which is actually the large-(n, n') approximation, involves not only the Airy integral approximation to the J-functions, but also approximations to $a_\pm, a'_\pm$ and $b'_\pm, b_\pm$ in the components (6.4.7) of the vertex function. Only the leading terms, of zeroth and first order in m/ω, need be retained; these are

$$a_+ \approx 1, \qquad a'_+ \approx 1, \qquad a_- \approx \zeta_-\frac{m}{\omega}\sinh u, \qquad a'_- \approx -\zeta_+\frac{m}{\omega}\sinh u,$$
$$b_s b_{s'} \approx \frac{\omega}{2\cosh w}\left[1 + \frac{m}{\omega}(s\zeta_- + s'\zeta_+)\right], \tag{6.4.27}$$

with $\zeta_\pm$ given by (6.4.26).

The leading terms in the expansion of the components (6.4.7) of the vertex function follow from (6.4.25) and (6.4.27). They are of the form

$$(\varepsilon_n\varepsilon'_{n'})^{1/2}\Gamma^{\perp,\parallel}_{q'q} \approx \frac{(-i)^{n-n'}}{\pi\sqrt{3}}\frac{2m^2}{\omega}\cosh u\cosh w\,\gamma^{\perp,\parallel}_{s's}, \tag{6.4.28}$$

and explicit evaluation gives

$$\gamma^{\perp}_{s's} = i\left\{\tfrac{1}{2}(1+s's)\big[\sinh u \cosh w\, K_{1/3}\big]\right.$$
$$\left. +\tfrac{1}{2}(1-s's)\big[s\cosh w\, K_{1/3} + \cosh u \sinh w\, K_{2/3}\big]\right\}, \tag{6.4.29}$$

$$\gamma^{\parallel}_{s's} = \left\{\tfrac{1}{2}(1+s's)\big[-s\cosh w\, K_{1/3} + \cosh u \cosh w\, K_{2/3}\big]\right.$$
$$\left. +\tfrac{1}{2}(1-s's)\big[-\sinh u \sinh w\, K_{1/3}\big]\right\}, \tag{6.4.30}$$

where the argument of the functions $K_{1/3}(\rho)$, $K_{2/3}(\rho)$ is given by (6.4.22), specifically by $\rho = (4mB_c/3\omega B)\cosh^3 u \cosh^2 w$.

6.4.4 *Spin- and Polarization-Dependent Decay Rates*

The rate of photon decay for $\perp$- and $\parallel$-polarized photons propagating across the magnetic field reduces to

$$R^{\perp,\parallel}_{s's} = \frac{\mu_0 e^2 m^3}{3\pi^3\omega^2}\frac{B_c}{B}\int_0^\infty du\,\cosh^3 u \int_0^\infty dw\,|\gamma^{\perp,\parallel}_{s's}|^2. \tag{6.4.31}$$

The spin-dependent factors arise from the squares of the vertex components (6.4.9)–(6.4.30), respectively:

$$|\gamma^{\perp}_{s's}|^2 = \tfrac{1}{2}(1+s's)\sinh^2 u \cosh^2 w\, K^2_{1/3} + \tfrac{1}{2}(1-s's)\big[\cosh^2 w\, K^2_{1/3}$$
$$+ \cosh^2 u \sinh^2 w\, K^2_{2/3} + 2s\cosh u \cosh w \sinh w\, K_{1/3}K_{2/3}\big],$$
$$|\gamma^{\parallel}_{s's}|^2 = \tfrac{1}{2}(1+s's)\big[\cosh^2 w\, K^2_{1/3} + \cosh^2 u \cosh^2 w\, K^2_{2/3}$$
$$- 2s\cosh u \cosh^2 w\, K_{1/3}K_{2/3}\big] + \tfrac{1}{2}(1-s's)\sinh^2 u \sinh^2 w\, K^2_{1/3}. \tag{6.4.32}$$

The argument of the Macdonald functions is given by (6.4.22), specifically, $\rho = \frac{1}{2}y\cosh^3 u$, $y = (8m/3\omega)(B_c/B)\cosh^2 w$.

The integral over u in (6.4.31) with (6.4.32) may be performed using (6.3.29) with $t = \sinh u$ and $y = (8m/3\omega)(B_c/B)\cosh^2 w$. This gives

$$R^{\perp,\parallel}_{s's} = \frac{\mu_0 e^2 m^2}{8\pi^2\sqrt{3}\,\omega}\int_0^\infty \frac{dw}{\cosh^2 w}\, S^{\perp,\parallel}_{s's}, \tag{6.4.33}$$

with the spin- and polarization-dependence included in

$$S^{\perp}_{s's} = \tfrac{1}{2}(1+s's)\Big[\cosh^2 w\int_y^\infty dx\, K_{5/3}(x) - \cosh^2 w\, K_{2/3}(y)\Big]$$
$$+\tfrac{1}{2}(1-s's)\Big[-(\cosh^2 w+1)\int_y^\infty dx\, K_{5/3}(x) + (5\cosh^2 w-1)K_{2/3}(y)$$
$$+4s\cosh w\sinh w\, K_{1/3}(y)\Big],$$

$$S^{\parallel}_{s's} = \tfrac{1}{2}(1+s's)\Big[-\cosh^2 w\int_y^\infty dx\, K_{5/3}(x) + 5\cosh^2 w\, K_{2/3}(y)$$
$$-4s\cosh^2 w\, K_{1/3}(y)\Big]$$
$$+\tfrac{1}{2}(1-s's)\Big[\sinh^2 w\int_y^\infty dx\, K_{5/3}(x) - \sinh^2 w K_{2/3}(y)\Big]. \quad (6.4.34)$$

Spin-Summed Decay Rates

The total rate of decay is found by summing over the spins, $s', s = \pm 1$, of the electron and positron. For unpolarized photons one averages over the two polarizations. The resulting rates are denoted

$$R^{\perp,\parallel} = \sum_{s',s} R^{\perp,\parallel}_{s's}, \qquad \bar{R} = \tfrac{1}{2}(R^{\perp} + R^{\parallel}). \quad (6.4.35)$$

The sums over s', s are evaluated using (6.4.32):

$$\sum_{s',s}\gamma^{\perp}_{s's} = 2\big[\cosh^2 u\cosh^2 w\, K^2_{1/3} + \cosh^2 u\,\sinh^2 w\, K^2_{2/3}\big],$$
$$\sum_{s',s}\gamma^{\parallel}_{s's} = 2\big[(\cosh^2 w + \sinh^2 u\,\sinh^2 w)K^2_{1/3} + \cosh^2 u\,\cosh^2 w\, K^2_{2/3}\big]. \quad (6.4.36)$$

Alternatively, the decay rates summed over the spin states may also be derived more directly using (6.4.5). To lowest order in an expansion in m/ω the coefficients $\varepsilon_n\varepsilon_{n'} + m^2 \pm p_z^2$ and $p_n p_{n'}$ are equal, and the resulting combinations of J-functions, $(J^n_{n'-n-1} + J^{n-1}_{n'-n+1})^2$ and $(J^{n-1}_{n'-n} + J^n_{n'-n-1})^2$ respectively, are of second order in m/ω, and proportional to $K^2_{2/3}$. There are contributions of the same order from $(J^n_{n'-n-1} - J^{n-1}_{n'-n+1})^2$ and $(J^{n-1}_{n'-n} - J^n_{n'-n-1})^2$, which are proportional to $K^2_{1/3}$, with coefficients

$$\varepsilon_n\varepsilon_{n'} + m^2 \pm p_z^2 - p_n p_{n'} = m^2[1 \pm \sinh^2 u + \cosh^2 u\,(2\cosh^2 w - 1)], \quad (6.4.37)$$

respectively. The combinations (6.4.36) are reproduced. After integration over u, the spin-summed decay rate (6.4.35) becomes

$$R^{\perp,\parallel} = \frac{\mu_0 e^2 m^2}{4\pi^2\sqrt{3}\,\omega}\int_0^\infty \frac{dw}{\cosh^2 w}\left[-\int_y^\infty dx\, K_{5/3}(x) + (4\cosh^2 w \pm 1)K_{2/3}(y)\right], \tag{6.4.38}$$

with $y = (8m/3\omega)(B_c/B)\cosh^2 w$. An alternative derivation by Tsai and Erber [24] using Schwinger's proper time technique gives

$$R^{\perp,\parallel} = \frac{\mu_0 e^2 m^2}{4\pi^2\sqrt{3}\,\omega}\int_0^1 dv\left[\frac{9-v^2}{3(1-v^2)} \pm 1\right] K_{2/3}(y), \tag{6.4.39}$$

with $y = (8m/3\omega)(B_c/B)(1-v^2)^{-1}$. The equivalence of (6.4.38) and (6.4.39) may be shown by writing $v = \tanh w$ and using the identity

$$\int_0^\infty \frac{dw}{\cosh^2 w}\int_y^\infty dx\, K_{5/3}(x) = \int_0^\infty \frac{dw}{\cosh^2 w}\left[1 + \frac{4}{3}\sinh^2 w\right] K_{2/3}(y). \tag{6.4.40}$$

Asymptotic Approximation

The result (6.4.40) simplifies when the Macdonald functions are approximated by their asymptotic limit $K_\nu(y) \sim (\pi/2y)^{1/2}\exp(-y)$. This corresponds to the limit in which the energies of the electron and positron are close to $\omega/2$, and the integrals being dominated by the region $w \ll 1$, where $\tanh w$ is small. In this limit one may make the approximation $\sinh w \approx w$ and $\cosh w \approx 1$, except in the exponent $y \approx (4/3\chi)(1 + w^2)$. In this limit, (6.4.38) gives a well-known result (ordinary units)

$$\bar{R} = \tfrac{1}{2}(R^\perp + R^\parallel) = \alpha_c \frac{mc^2}{\hbar}\frac{B}{B_c}\frac{3\sqrt{3}}{16\sqrt{2}}\exp\left(-\frac{8mc^2 B_c}{3\hbar\omega B}\right), \tag{6.4.41}$$

which applies for $\hbar\omega/2mc^2(B/B_c) \ll 1$. An accurate approximation that includes (6.4.41) and applies more generally is [6]

$$\bar{R} = \alpha_c \frac{mc^2}{\hbar}\frac{B}{B_c}\frac{3\sqrt{3}}{16\sqrt{2}}\frac{2\chi}{\pi}K_{1/3}^2(\chi), \qquad \chi = \frac{4mc^2 B_c}{3\hbar\omega B}. \tag{6.4.42}$$

The approximate form (6.4.42) is a smooth function that ignores the saw-tooth variation characteristic of the exact result, as illustrated in Fig. 6.5. A comparison of the exact and approximate forms for a magnetic field characteristic of a pulsar [4] shows that the smooth approximation is a poor one near threshold, $\omega \approx 2m$, but provides a good approximation for $\omega \gg 2m$, $B \ll B_c$.

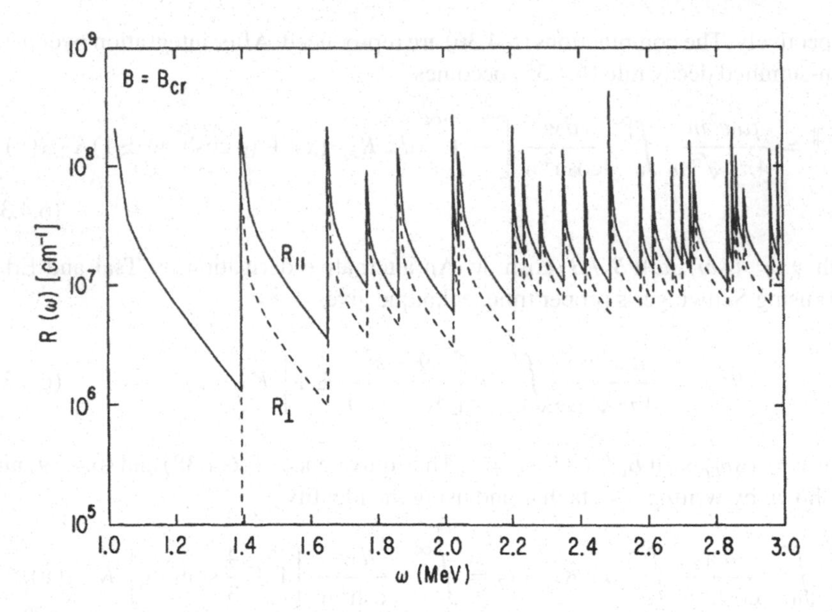

Fig. 6.5 The rate of one-photon pair creation per unit length for $\perp$ and $\parallel$ polarizations versus the (perpendicularly propagating) photon energy, for $B = B_c$ (From [4], reprinted with permission AAS)

The polarization- and spin-dependence is included in (6.4.33) with (6.4.34) and the leading terms that correspond to the approximation (6.4.41) give

$$R^{\perp}_{s's} = \tfrac{1}{2}(1 - s's)\,\frac{\bar{R}}{6}[1 + s\langle w\rangle], \qquad R^{\parallel}_{s's} = \tfrac{1}{2}(1 + s's)(1 - s)\,\frac{\bar{R}}{12}, \tag{6.4.43}$$

with $\langle w\rangle = (3\hbar\omega B/8\pi m B_c)^{1/2}$, and where terms $\langle w^2\rangle$ and higher are neglected. Thus, in this limit, $\perp$-polarized photons decay into pairs with opposite spins, with a small preference for the higher energy particle having spin up, and $\parallel$-polarized photons decay into pairs with the same spin with a strong preference for spin down. These spin preferences are consistent with the preferences found near the thresholds $n, n' = 0, 1$: the lowest threshold for $\parallel$-polarized photons is $n = n' = 0$, which corresponds to both particles having spin down ($s = s' = -1$), and the lowest threshold for $\perp$-polarized photons is $n = 0, n' = 1$ or $n = 1, n' = 0$, with $s' = -s$ the higher-energy particle having spin up.

6.4.5 *Energy Distribution of the Pairs*

The decay rates (6.4.38) contain information on the energy distribution between the electron and positron. Writing these energies as $\varepsilon_n \to \varepsilon$, $\varepsilon'_{n'} \to \omega - \varepsilon$, one has

$$4\cosh^2 w = \frac{\omega^2}{\varepsilon(\omega-\varepsilon)}, \qquad \frac{dw}{\cosh^2 w} = \frac{2d\varepsilon}{\omega}. \tag{6.4.44}$$

Let $dR^{\perp,\parallel}/d(\varepsilon)$ be the rate of decay with one of the particles in the energy range $d\varepsilon$. Then (6.4.38) implies

$$\frac{dR^{\perp,\parallel}}{d\varepsilon} = \frac{\mu_0 e^2 m^2}{2\pi^2\sqrt{3}\,\omega^2}\left[-\int_y^\infty dx\, K_{5/3}(x) + \left(\frac{\omega^2}{\varepsilon(\omega-\varepsilon)} \pm 1\right) K_{2/3}(y)\right],$$

$$y = \frac{2mB_c}{3\omega B}\frac{\omega^2}{\varepsilon(\omega-\varepsilon)}. \tag{6.4.45}$$

The distribution (6.4.45) averaged over the polarization of the photons has a functional form $\propto yK_{2/3}(y)$ that is well described by $y^{1/3}e^{-y}$, which increases $\propto y^{1/3}$ for $y \ll 1$, has a maximum at $y = 1/3$ and decreases exponentially for $y \gg 1$. Regarding y as a function of ε/ω for given $2mB_c/3\omega B$, it follows from (6.4.44) that the minimum value of y is for $\varepsilon = \frac{1}{2}\omega$ ($w = 0$), when the energies of the electron and positron are equal. For $2mB_c/3\omega B \gg 1$, one has $y \gg 1$ for all $w \geq 0$, and the exponential decrease of the Macdonald functions with y implies that the distribution is sharply peaked around the minimum of y. Hence, for $2mB_c/3\omega B \gg 1$ decay in which the energies of the electron and positron are nearly equal is strongly favored. For $2mB_c/3\omega B \ll 1$ the minimum value of y is small. The maximum of the Macdonald functions corresponds to $4\cosh^2 w = \omega B/2mB_c$. Hence, for $2mB_c/3\omega B \ll 1$, one of the two particles gains nearly all the photon energy, with the other having an energy $\varepsilon \approx 2mB_c/B \ll \omega$.

Lorentz Transformation to an Arbitrary Frame

The foregoing results for pair creation are derived in the frame in which the photon is propagating across the magnetic field lines ($k_z = 0$, $\sin\theta = 1$). Results for an arbitrary direction of propagation of the photon are obtained by making a Lorentz transformation to a frame with $k_z \neq 0$, $\sin\theta \neq 1$. The Lorentz transformation is closely analogous to that introduced above for synchrotron radiation, cf. (6.2.17). Specifically, the Lorentz factor for the transformation is $1/\sin\theta$, and the only changes to the foregoing formulae involve inclusion of factors of $\sin\theta$. The frequency, ω, and the Lorentz factor, γ, in the frame $k_z = 0$ are replaced by $\omega\sin\theta$ and $\gamma\sin\theta$, respectively, in an arbitrary frame. Suppose one denotes quantities in the frame $k_z = 0$ by tildes. On transforming from the frame $\theta = \pi/2$ to a frame with arbitrary θ, the rate $R^{\perp,\parallel}(\omega)$ is replaced by $\tilde{R}^{\perp,\parallel}(\tilde{\omega})$, with

$$R^{\perp,\parallel}(\omega) = \sin\theta\, \tilde{R}^{\perp,\parallel}(\omega\sin\theta). \tag{6.4.46}$$

In the analogous relation for $dR^{\perp,\parallel}/d\varepsilon$, multiplicative factors of $\sin\theta$ cancel, and one simply replaces ω and ε, by $\omega\sin\theta$ and $\varepsilon\sin\theta$, respectively, on the right hand side of (6.4.45). The relativistic approximation, $\omega \gg 2m$ in the frame $k_z = 0$, is replaced by $\omega \sin\theta \gg 2m$ in an arbitrary frame.

6.4.6 One-Photon Pair Annihilation

One-photon pair annihilation is the inverse of one-photon pair creation. In the absence of a magnetic field, a pair can decay only into two (or more) photons, and in the presence of a magnetic field, the one-photon and two-photon processes compete. Early estimates [3, 29] were based on pairs in their ground state, and suggested that the one-photon process dominates only for $B \gtrsim 10^9\,$T. Subsequently it was recognized that one-photon annihilation can dominate at lower B if the pair is in an excited state [1, 7, 8, 32].

The kinetic equation (6.1.30) includes both pair creation and decay, and it implies that the rate of increase of the photon occupation number due to pair annihilation is related in a simple way to the absorption coefficient for pair annihilation. Specifically, to obtain $\mathrm{D}N(\boldsymbol{k})/\mathrm{D}t$ for pair annihilation, one repeats the foregoing calculation for the decay rate for pair creation after first multiplying by the product of the occupation numbers, $n_s^+(n, p_z)n_{s'}^-(n', p_z')$, of the electrons and positrons.

A general expression for the rate of production of (linearly polarized) photons is

$$\frac{\mathrm{D}N^{\perp,\parallel}(\boldsymbol{k})}{\mathrm{D}t} = \sum_{n,n',\pm} \frac{\mu_0 e^3 B}{4\pi}\left[\frac{\varepsilon_{n'}\varepsilon_n |\Gamma_{q'q}^{\perp,\parallel}|^2 n_s^+(n, p_z)n_{s'}^-(n', k_z - p_z)}{\omega^3 \sin^2\theta\, g_{nn'}}\right]_\pm, \quad (6.4.47)$$

where $\Gamma_{q'q}^{\perp,\parallel}$ is defined by (6.4.4), and where $\pm$ indicates evaluation at the two solutions (6.1.18) of the resonance condition. The volume emissivity, $J(\omega,\theta)$, is related to $\mathrm{D}N((\boldsymbol{k})/\mathrm{D}t$ by

$$J^{\perp,\parallel}(\omega,\theta) = \frac{\omega^3}{(2\pi)^3}\frac{\mathrm{D}N^{\perp,\parallel}(\boldsymbol{k})}{\mathrm{D}t}, \quad (6.4.48)$$

with $k_\perp = \omega\sin\theta, k_z = \omega\cos\theta$.

For annihilation of nonrelativistic pairs, a given value of n and n' may be treated independently of any value. There is a threshold condition, $g_{nn'} = 0$, with $g_{nn'}$ given by (6.4.10) in this case, and the rate (6.4.47) of annihilation of a pair for given n, n' is singular at $g_{nn'} = 0$. As for creation of a pair, the annihilation of an electron and a positron with the same spin produces a $\parallel$-polarized photon, and annihilation of an electron and a positron with the opposite spins produces a $\perp$-polarized photon. For the two lowest thresholds, $n = 0 = n'$ and $n + n' = 1$, one finds

$$\frac{\mathrm{D}N^{\parallel}(\boldsymbol{k})}{\mathrm{D}t} = \sum_{\pm} \frac{\mu_0 e^3 B \sin\theta\, e^{-x}}{2\pi(\omega^2\sin^2\theta - 4m^2)^{1/2}}\left[n_-^+(0, p_z)n_-^-(0, k_z - p_z)\right]_\pm,$$

$$\frac{DN^{\perp}(\boldsymbol{k})}{Dt} = \sum_{\pm} \frac{\mu_0 e^3 B \sin\theta\, e^{-x}}{\pi[\omega^2 \sin^2\theta - 4m^2(1 + B/B_c)]^{1/2}}$$
$$\times \left[n_-^{\pm}(0, p_z) n_+^{\mp}(1, k_z - p_z) + n_+^{\pm}(1, p_z) n_-^{\mp}(0, k_z - p_z)\right]_{\pm}, \qquad (6.4.49)$$

where in the latter case the particle in the first excited state must have spin up, $s = 1, n = 1, s' = -1, n' = 0$ or $s = -1, n = 0, s' = 1, n' = 1$. For given n, n', ω, θ, the values $p^{\pm}_{nn'}$ are determined by (6.1.18), and the rate of annihilation is proportional to the occupation numbers of the electrons and positrons at $p_z = p^{\pm}_{nn'}$, $p'_z = k_z - p^{\pm}_{nn'}$. In the case $n = n' = 0$ the solutions (6.1.18) simplify to

$$p^{\pm}_{00'} = \frac{k_z}{2} \pm \frac{\omega}{2}\left(\frac{\omega^2 - k_z^2 - 4m^2}{\omega^2 - k_z^2}\right)^{1/2},$$

$$\varepsilon_0(p^{\pm}_{00'}) = \frac{\omega}{2} \pm \frac{k_z}{2}\left(\frac{\omega^2 - k_z^2 - 4m^2}{\omega^2 - k_z^2}\right)^{1/2}. \qquad (6.4.50)$$

In (6.4.49) vacuum dispersion is assumed in writing $\omega^2 - k_z^2 = \omega^2 \sin^2\theta$. When treating pair annihilation it is appropriate to regard the energies, $\varepsilon, \varepsilon'$, and momenta, p_z, p'_z, as given so that $\omega = \varepsilon + \varepsilon'$ and $k_z = p_z + p'_z$ are determined by energy and momentum conservation. This interpretation is implicit in (6.4.49). In particular, for pairs in the ground state the first of (6.4.49) may be rewritten

$$\frac{DN^{\parallel}(\boldsymbol{k})}{Dt} = \frac{\mu_0 e^3 B \sin\theta\, e^{-x}}{\pi[\omega^2 - k_z^2 - 4m^2]^{1/2}} n_-^{+}(0, p_z) n_-^{-}(0, p'_z),$$
$$\omega = (m^2 + p_z^2)^{1/2} + (m^2 + p_z'^2)^{1/2}, \qquad k_z = p_z + p'_z, \qquad (6.4.51)$$

with θ determined by writing $\sin^2\theta = (\omega^2 - k_z^2)/\omega^2$.

In the limit of ultrarelativistic pairs, the sum over n, n' and the factor $1/g_{nn'}$ are rewritten in terms of integrals over u, w, as in (6.4.18). To lowest order in the expansion in m/ω, n, n' are related to w by

$$n = \frac{\omega^2 \sin^2\theta}{8eB}(1 \pm \tanh w)^2, \qquad n' = \frac{\omega^2 \sin^2\theta}{8eB}(1 \mp \tanh w)^2. \qquad (6.4.52)$$

Evaluating the $\pm$-solutions (6.1.18) in the ultrarelativistic limit gives

$$p^{\pm}_{nn'} = \tfrac{1}{2}(1 + \tanh w)\omega \cos\theta \pm m \sinh u. \qquad (6.4.53)$$

The ultrarelativistic approximation involves expanding in $m/\omega \sin\theta$, and to lowest order one neglects the final term in (6.4.49) that contains the dependence on u. In this approximation, (6.4.52) and (6.4.49) imply that the pitch angles of the particles, $\alpha = \arctan[p^{\pm}_{nn'}/p_n]$, $p_n = (2neB)^{1/2}$, satisfy $\alpha = \theta$ to lowest order in $m/\omega \sin\theta$.

Hence, if one writes the occupation numbers in terms of polar coordinates, p, α, in momentum space, $n^{\pm}(n, p_z) \to n^{\pm}(p, \alpha)$, then the occupation numbers are to be evaluated at $p = p_n$, $p' = p_{n'}$, with n, n' given by (6.4.52), and at $\alpha = \theta$.

The final step is to integrate over u. Because the occupation numbers are independent of u to lowest order, the integral over u is identical to that performed in the derivation of (6.4.38). Hence, using (6.4.38), the emission of photons due to annihilation of ultrarelativistic pairs reduces to

$$\frac{\mathrm{D}N^{\perp,\parallel}(\boldsymbol{k})}{\mathrm{D}t} = \frac{\mu_0 e^2 m^2}{2\pi^2\sqrt{3}\,\omega} \int_0^\infty \frac{dw}{\cosh^2 w} \left[-\int_y^\infty dt K_{5/3}(t) + (4\cosh^2 w \pm 1) K_{2/3}(y) \right] [n^+(p_+, \theta) n^-(p_-, \theta) + n^+(p_-, \theta) n^-(p_+, \theta)], \tag{6.4.54}$$

with $p_{\pm} = \frac{1}{2}(1 \pm \tanh w)\,\omega \sin\theta$. The spin dependence is neglected in (6.4.54). The dependence on the spins of the electron and positron may be inferred from the discussion of the spin-dependence of pair creation. There is a preference for given initial spins to lead to either $\perp$- or $\parallel$-polarized photons, but there is no selection rule that forbids any of the possibilities.

6.5 Positronium in a Superstrong Magnetic Field

The properties of positronium in a strong magnetic field are discussed in this section. This discussion is motivated by the application to pulsars, where a photon can decay into a pair or evolve into a bound pair [17, 18].

6.5.1 *Qualitative Description of Positronium*

Positronium is familiar as a bound state of an electron and a positron for $B = 0$. In a weak magnetic field, the states of the positronium states are magnetically split, in a manner closely analogous to magnetic splitting of the states of hydrogen-like atoms. In this case, the magnetic field is treated as a weak perturbation compared with the Coulomb force. If the magnetic field is sufficiently strong ($B/B_c \gg \alpha_c^2$) the roles of the magnetic and Coulomb forces reverse. Thus, in the strong-B limit, to a first approximation the electron and positron are treated as free particles in a magnetic field, and the coupling due to the Coulomb force is treated as a perturbation.

Consider the dispersion curves of a photon and of positronium in which the energy, written ω or ε, is plotted as a function of perpendicular momentum, $k_\perp$ or $p_\perp$. For the photon the dispersion curve is approximately of the form $\omega \propto k_\perp$

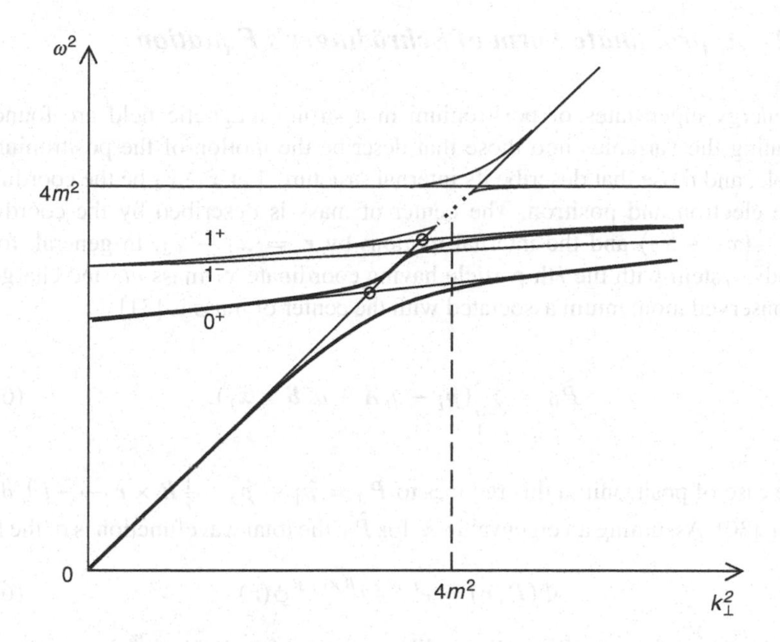

Fig. 6.6 The interference between photon dispersion curves and dispersion curves between bound states is shown schematically. In the absence of positronium states, the dispersion curves for photons with two transverse states of linear polarization would follow the *light line*, $\omega^2 \propto k_\perp^2$. Three positronium states, labeled 0^+, 1^- and 1^+ are indicated. The superscript denote parity, and photons with a particular polarization can evolve into positronium states a particular parity. The *circles* denote points where the dispersion curves would intersect, but they reconnect to form two mixed photon-positronium state denoted by the *solid lines*. For a photon to continue on the photon dispersion curve, it must tunnel across the gap between these two curves (After [17, 18])

for propagation nearly across the field line, $k_z^2 \ll k_\perp^2$. On a plot of ω^2 versus $k_\perp^2$, this dispersion curve forms a continuous light line. There are various states of positronium with slightly different binding energies, which depend weakly on a quantum number that can be interpreted as $k_\perp^2$. The "dispersion curves" for each bound state of positronium corresponds to an energy ε or ω just below 2 m, corresponding to an approximately horizontal line for each bound state. The photon and positronium dispersion curves would cross if they were independent. However, if the photon and positronium have the same quantum numbers then the two states interfere, causing the curves to deviate away from each other, as illustrated in Fig. 6.6. A photon following the dispersion curve evolves into positronium. A subsequent decay of positronium into a pair can occur, due to photo-dissociation say, and this constitutes an alternative source of free pairs to populate a pulsar magnetosphere. Quantum mechanical tunneling can prevent the conversion into positronium occurring, as illustrated in Fig. 6.6.

6.5.2 *Approximate Form of Schrödinger's Equation*

The energy eigenstates of positronium in a strong magnetic field are found by separating the variables into those that describe the motion of the positronium as a whole, and those that describe its internal structure. Let $\boldsymbol{x}_1, \boldsymbol{x}_2$ be the coordinates of the electron and positron. The center of mass is described by the coordinate $\boldsymbol{R} = \frac{1}{2}(\boldsymbol{x}_1 + \boldsymbol{x}_2)$ and the internal motions by $\boldsymbol{r} = \boldsymbol{x}_1 - \boldsymbol{x}_2$. In general, for an N-body system with the ith particle having coordinate $\boldsymbol{x}_i$ mass m_i and charge q_i, the conserved momentum associated with the center of mass is [31]

$$\hat{\boldsymbol{P}}_0 = \sum_{i=1}^{N} (\hat{\boldsymbol{p}}_i - q_i \boldsymbol{A} + q_i \boldsymbol{B} \times \boldsymbol{x}_i). \tag{6.5.1}$$

In the case of positronium this reduces to $\hat{\boldsymbol{P}}_0 = \hat{\boldsymbol{p}}_1 + \hat{\boldsymbol{p}}_2 - \frac{1}{2}\boldsymbol{B} \times \boldsymbol{r} \to -i\partial/\partial\boldsymbol{R} - \frac{1}{2}\boldsymbol{B}\times\boldsymbol{r}$ [30]. Assuming an eigenvalue $\boldsymbol{K}$ for $\hat{\boldsymbol{P}}_0$, the total wavefunction is of the form

$$\Phi(\boldsymbol{R}, \boldsymbol{r}) = e^{i(\boldsymbol{K}+\frac{1}{2}\boldsymbol{B}\times\boldsymbol{r})\cdot\boldsymbol{R}} \phi(\boldsymbol{r}), \tag{6.5.2}$$

with $\phi(\boldsymbol{r})$ satisfying a Schrödinger-like equation for the internal motions. The problem may be simplified by choosing the frame ($\boldsymbol{K} = 0$) in which the center of mass is at rest, but even in this frame a number of approximations need to be made to derive an appropriate approximate equation [17,18,30].

Analogy with the One-Dimensional Hydrogen Atom

In a zeroth approximation, in which the Coulomb force is neglected, the electron and positron are assumed to have energies $\varepsilon_n^0 = (m^2+2neB)^{1/2}$, $\varepsilon_{n'}^0 = (m^2+2n'eB)^{1/2}$, with $p_z = p_z' = 0$. With the Coulomb force included, let the total energy be E and let $\Delta\varepsilon = -(E - \varepsilon_n^0 - \varepsilon_{n'}^0)$ be the binding energy (defined to be positive for a bound state). Shabad and Usov [18] argued that an appropriate approximation to $\Delta\varepsilon$ is found by solving the equation

$$\left(\frac{1}{2M_{nn'}} \frac{d^2}{d(z_1 - z_2)^2} - V_0(z_1 - z_2)\right) \phi(z_1 - z_2) = \Delta\varepsilon\, \phi(z_1 - z_2) \tag{6.5.3}$$

with $M_{nn'} = \varepsilon_n^0 \varepsilon_{n'}^0 / (\varepsilon_n^0 + \varepsilon_{n'}^0)$. The derivation of (6.5.3) involves some detailed analysis, based on the Bethe-Salpeter equation [18], but it also requires further approximations to avoid difficulties with the ground state.

Physically, the main difficulty concerns the approximation of the Coulomb potential, $V(r) = -e^2/4\pi\varepsilon_0 r$, with $r = [(z_1 - z_2)^2 + (\boldsymbol{x}_1 - \boldsymbol{x}_2)_\perp^2]^{1/2}$. The form that appears in (6.5.3) may be understood by noting that the perpendicular motion is described by the gauge-dependent eigenvalues that describe the location of the centers of gyration of the electron and positron. Choosing the Landau gauge, these are p_{1y}, p_{2y}, with $P_y = p_{1y} + p_{2y}$ describing the relative separation of the centers of gyration. The corresponding approximation is $r = [(z_1 - z_2)^2 + P_y^2/(eB)^2]^{1/2}$. This leads to singular behavior for $P_y \to 0$, when the problem reduces to that of the one-dimensional hydrogen atom [12]. One way of avoiding the singularity at $P_y = 0$ is by including the term $1/eB$ in the potential [18]

$$V_0(z_1 - z_2) = -\frac{e^2}{4\pi\varepsilon_0} \frac{1}{|z_1 - z_2| + [1/eB + P_y^2/(eB)^2]^{1/2}}, \tag{6.5.4}$$

allowing the problem to be treated by analogy with the one-dimensional hydrogen atom [12]. A different way of avoiding this difficulty had been proposed earlier [30].

The eigenstates determined by (6.5.3) with (6.5.4) may be labeled by n, n' and two additional quantum numbers: a principal quantum number, $n_c = 0, 1, 2, \ldots$, analogous to that of the hydrogen atom, and a sign (parity), ± 1, depending on whether the wavefunctions are even or odd under $z_1 \leftrightarrow z_2$.

6.5.3 Bound States of Positronium

In a strong magnetic field, positronium is treated by starting with the exact eigenfunctions for the electron and positron and including the Coulomb interaction as a perturbation. The condition for this perturbation approach to be valid is that the radius of gyration of the electron or positron, $p_\perp/eB = (2n)^{1/2}/(eB)^{1/2}$, be small compared with the Bohr radius, $a = 4\pi\varepsilon_0/me^2$ of positronium. Requiring $a \gg 1/(eB)^{1/2}$ corresponds to $B \gg \alpha_c^2 B_c = 2.35 \times 10^5\,\mathrm{T}$, which is assumed to be satisfied for the purpose of discussion here. Let the unbound state of the electron and positron have quantum numbers n, n' such that the energy of the bound state is

$$\varepsilon_{nn'}(n_c, P_y) = \varepsilon_n^0 + \varepsilon_{n'}^0 - \Delta\varepsilon_{nn'}(n_c, P_y), \tag{6.5.5}$$

with the energy correction term, $\Delta\varepsilon_{nn'} = \Delta\varepsilon_{nn'}(n_c, P_y)$, given by

$$\Delta\varepsilon_{nn'} = \frac{\alpha_c^2 M_{nn'}}{2} \begin{cases} \left(2\ln\dfrac{r_{\mathrm{B}nn'}}{2[1/eB + P_y^2/(eB)^2]^{1/2}}\right)^2 & (n_c = 0), \\[2ex] \left[n_c + \left(\ln\dfrac{n_c r_{\mathrm{B}nn'}}{2[1/eB + P_y^2/(eB)^2]^{1/2}}\right)^{-1}\right]^{-2} & \text{(even)}, \\[2ex] \left(n_c + \dfrac{2[1/eB + P_y^2/(eB)^2]^{1/2}}{r_{\mathrm{B}nn'}}\right)^{-2} & \text{(odd)}, \end{cases} \tag{6.5.6}$$

where n_c is the principal quantum, and $r_{\mathrm{B}nn'} = 4\pi\varepsilon_0/e^2 M_{nn'}$ is the appropriate Bohr radius. The ground state has $n = n' = n_c = 0$ and is even. The wavefunction,

$\phi_{nn'n_c}(z_1 - z_2)$ for the excited states, $n_c > 0$, was determined by Loudon [12] in terms of Whittaker functions. In the following discussion only the wavefunction for the state $n_c = 0$ is required and this is approximated by

$$\phi_{nn'0}(z_1 - z_2) = \frac{1}{(r_{Bnn'}\Delta_{nn'})^{1/2}} \exp\left(-\frac{|z_1 - z_2|}{r_{Bnn'}\Delta_{nn'}}\right),$$

$$\Delta_{nn'} = \left(2\ln \frac{r_{Bnn'}}{2[1/eB + P_y^2/(eB)^2]^{1/2}}\right)^{-1}. \tag{6.5.7}$$

In the limit $\Delta_{nn'} \to 0$ the wavefunction becomes singular, $\phi_{nn'0}(z_1 - z_2) \to \delta(z_1 - z_2)$, but only for $n_c = 0$. It is this nearly singular case that is the most important when considering the coupling between the photon and positronium states. In the following discussion, only the case $n_c = 0$ need be retained explicitly. Thus, in discussing the interaction between photon and positronium states, the positronium states are a nearly singular state, $n_c = 0$, plus a set of excited bound states, $n_c = 1, 2, \ldots$, that are of only minor interest. These higher states merges into a continuum for $n_c \gg 1$ and join on smoothly to the continuum of unbound states corresponding to a nonrelativistic free pair.

6.5.4 *Evolution of Photons into Bound Pairs*

Pair formation in a pulsar magnetosphere is attributed to photons emitted nearly along a field line, $\sin\theta \approx 0$, having $k_\perp^2 = \omega^2 \sin^2\theta$ increase due to the angle, θ, increasing as the curve field line deviates away from the ray path.

One is free to make a Lorentz transformation to the local frame in which the photon is propagating perpendicular to the field lines, and only its perpendicular energy, $\omega\sin\theta$, is relevant in discussing photon splitting, one-photon pair decay and positronium formation. The possibility discussed here is that the photon simply evolves into positronium. The bound state energy of positronium corresponds to $\omega\sin\theta < 2m$, and the photon can transform into positronium before it reaches the threshold for decay into a free pair.

Interference Between Photon and Positronium States

In the absence of any interaction between them, the dispersion curves for photons and positronium intersect, at the points denoted by circles in Fig. 6.6. Interference between the states of a photon and positronium occurs where the energies of the photon and the positronium are equal at the same perpendicular momentum ($k_\perp$ and P_y, respectively). To treat this coupling, for given n, n', it suffices to estimate the contribution of the positronium state with $n_c = 0$ to the photon dispersion. The dispersion relation for $\|$-polarized photons near the intersection point and the positronium dispersion curve for $n = n' = n_c = 0$ may be approximated by an expression of the form [17]

$$\omega^2 - k_z^2 - k_\perp^2 = \kappa_\| (\omega^2 - k_z^2, k_\perp^2),$$

$$\kappa_\| (\omega^2 - k_z^2, k_\perp^2) = 4\alpha_c eB \frac{\epsilon_{00}(0, k_\perp) \, |\phi_{000}(0)|^2 \, e^{-k_\perp^2/2eB}}{\epsilon_{00}^2(0, k_\perp) - \omega^2 + k_z^2},$$

$$|\phi_{000}(0)|^2 = \frac{2}{r_{B00}} \ln \frac{r_{B00} eB}{2(k_\perp^2 + eB)^{1/2}}, \tag{6.5.8}$$

where only the wavefunction at $z_1 = z_2$ contributes. The term $\kappa_\|$ implies a coupling between the photon and positronium states. The form (6.5.8) corresponds to the dispersion of the mixed states being determined by the solutions

$$\omega^2 - k_z^2 = \tfrac{1}{2}\left[\varepsilon_{00}(0, k_\perp)\right]^2 + \tfrac{1}{2} k_\perp^2 \pm \tfrac{1}{2}\left(\left\{\left[\varepsilon_{00}(0, k_\perp)\right]^2 - k_\perp^2\right\}^2 + 4A(k_\perp)\right)^{1/2},$$

$$A(k_\perp) = \frac{4\alpha_c eB \varepsilon_{00}(0, k_\perp)}{a} \ln\left(\frac{a(eB)^{1/2}}{(1 + p_\perp^2/eB)^{1/2}}\right) \exp\left(-\frac{k_\perp^2}{2eB}\right), \tag{6.5.9}$$

with ω, k_z interpreted as ε, P_z for the positronium-like sections of the dispersion curves. The upper-frequency branch is positronium-like at low $k_\perp$ and photon-like at high $k_\perp$, and the lower-frequency branch is photon-like at low $k_\perp$ and positronium-like at high $k_\perp$, cf. Fig. 6.6. To treat the analogous interference for $\perp$-polarized photons one needs to carry out a similar analysis for the state with $n + n' = 1, n_c = 0$.

An implication of the mixing of the photon-positronium states is that a photon should evolve into positronium, as $k_\perp$ increases, before the threshold for creation of a free pair is reached. A $\|$-polarized photon evolves into positronium in its ground state, $n = n' = n_c = 0$, and a $\perp$-polarized photon evolves into positronium in the first excited state, $n + n' = 1, n_c = 0$. In the latter case, the electron or positron in the excited state, say $n = 1, n' = 0$, has $s = 1$ and it decays to its ground state, $n = 0$, through a spin-flip transition like a free particle, leaving the positronium in its ground state after emitting the cyclotron photon.

6.5.5 Tunneling Across the Intersection Point

The evolution of a photon into positronium as $k_\perp$ increases is an adiabatic process, and it occurs only if the adiabatic assumption is valid. Non-adiabatic behavior can result from any process that introduces an uncertainty into the definition of the wave or particle modes. One such effect is associated with the bending of the direction of propagation. The direction of propagation changes from along a straight line at an angle θ for the photon, to along the field line ($\theta = 0$) for the positronium. The adiabatic approximation requires that the rate of this change be small compared with the frequency of the wave.

Another effect that needs to be taken into account in the possible decay of positronium into two photons. Suppose that the positronium state (the upper of the mixed states) decays due to this process at a rate $\Gamma_{2\gamma}$. Then there is an uncertainty, $\Delta\varepsilon \sim \Gamma_{2\gamma}$, in the energy of the positronium state. A detailed calculation [30] gives

$$\Gamma_{2\gamma} \approx 8 \times 10^{12}\,\mathrm{s}^{-1} \left(\frac{B}{10^8\,T} \right). \tag{6.5.10}$$

The adiabatic assumption is valid only if the separation of the dispersion curves near the original intersection point is large compared with the net effect of all such broadenings.

In the application to pulsars, the two-photon decay of positronium is the most important broadening effect. Then the adiabatic condition is valid for $A(k_\perp) \gtrsim 2m\Gamma_{2\gamma}$, with $k_\perp \approx 2m$. Due to the exponential dependence of $A(k_\perp)$ on B/B_c, cf. (6.5.9), this condition is relatively insensitive to other details in the calculation. Photons evolve into positronium for $B \gtrsim 0.15\,B_c$, but not for $B \ll 0.15\,B_c$ [18, 25, 26].

References

1. M.G. Baring, Mon. Not. R. Astron. Soc. **235**, 51 (1988)
2. Y.G. Bezchastnov, G.G. Pavlov, Astrophys. Space Sci. **148**, 257 (1988)
3. J.K. Daugherty, R.W. Buzzard, Astrophys. J. **238**, 296 (1980)
4. J.K. Daugherty, A.K. Harding, Astrophys. J. **273**, 761 (1983)
5. O.F. Dorofeyev, A.V. Borisov, V.Ch. Zhukovsky, In: V.A. Bordovitsyn (ed.), *Synchrotron Radiation Theory and Its Development* (World Scientific, Singapore, 1999), p. 347
6. T. Erber, Rev. Mod. Phys. **38**, 626 (1966)
7. A.K. Harding, Astrophys. J. **300**, 167 (1986)
8. A.K. Harding, D. Lai, Rep. Prog. Phys. **69**, 2631 (2006)
9. H. Herold, H. Ruder, G. Wunner, Astron. Astrophys. **115**, 90 (1985)
10. N.P. Klepikov, Sov. Phys. JETP **26**, 19 (1954)
11. R. Lieu, W.I. Axford, Astrophys. J. **416**, 700 (1993)
12. R. Loudon, Am. J. Phys. **27**, 649 (1959)
13. D.B. Melrose, K. Russell, J. Phys. A **35**, 135 (2002)
14. G.G. Pavlov, Yu.A. Shibanov, D.G. Yakovlev, Astrophys. Space Sci. **73**, 33 (1980)
15. H. Pérez Rojas, A.E. Shabad, Ann. Phys. **138**, 1 (1982)
16. J. Schwinger, Phys. Rev. **75**, 898 (1949)
17. A.E. Shabad, V.V. Usov, Astrophys. Space Sci. **117**, 309 (1985)
18. A.E. Shabad, V.V. Usov, Astrophys. Space Sci. **128**, 377 (1986)
19. A.A. Sokolov, I.M. Ternov, *Synchrotron Radiation* (Pergamon Press, Oxford, 1968)
20. A.A. Sokolov, I.M. Ternov, *Radiation from Relativistic Electrons* (AIP, New York, 1986)
21. A.A. Sokolov, V.Ch. Zhukovskii, N.S. Nikitina, Phys. Lett. A **43**, 85 (1973)
22. P.A. Sturrock, Astrophys. J. **164**, 529 (1971)

23. J.S. Toll, Ph.D. thesis, Princeton University, 1952
24. W. Tsai, T. Erber, Phys. Rev. D **10**, 492 (1974)
25. V.V. Usov, D.B. Melrose, Aust. J. Phys. **48**, 571 (1985)
26. V.V. Usov, D.B. Melrose, Astrophys. J. **464**, 306 (1986)
27. J.I. Weise, D.B. Melrose, Mon. Not. R. Astron. Soc. **329**, 115 (2002)
28. D. White, Phys. Rev. D **9**, 868 (1974)
29. G. Wunner, Phys. Rev. Lett. **42**, 79 (1986)
30. G. Wunner, H. Herold, Astrophys. Space Sci. **63**, 503 (1979)
31. G. Wunner, H. Ruder, H. Herold, Astrophys. J. **247**, 374 (1981)
32. G. Wunner, J. Paez, H. Herold, H. Ruder, Astron. Astrophys. **170**, 179 (1986)

Chapter 7
Second Order Gyromagnetic Processes

Second order processes are described by Feynman diagrams with two electron-photon vertices. Such processes include Compton scattering, two-photon emission, two-photon pair creation and annihilation, Møller and Bhabha scattering, and the trident process. These processes have counterparts in the absence of a magnetic field, and an important motivation for the inclusion of the magnetic field was the discovery of radio and X-ray pulsars, and the recognition that the sources are strongly magnetized neutron stars [9]. The magnetic field modifies these processes in various ways, two of which are particularly notable: the fact that electrons quickly relax to the ground Landau state, $n = 0$, and the inclusion of gyromagnetic resonances.

General properties of Compton scattering and the related processes of two-photon emission and absorption are summarized in § 7.1, where some general formulae are derived by partially summing the probability over intermediate states. The special case of Compton scattering by an electron in the state $n = 0$ is discussed in § 7.2. A cyclotron-type approximation for the scattering is discussed in § 7.3, and the magnetized counterpart of inverse Compton scattering is discussed in § 7.3.3. Two-photon emission and two-photon pair annihilation and creation are discussed in § 7.4. Electron-ion and electron-electron scattering are discussed in § 7.5.

7.1 General Properties of Compton Scattering

The probability of Compton scattering includes all processes related to Compton scattering by crossing symmetries: Compton scattering by a positron, two-photon pair creation and annihilation, and double emission and absorption. The resonance condition for Compton scattering is used to derive results for the kinematics of these processes. Kinetic equations that describe the effects of Compton scattering on the distributions of photons and electrons are written down. Some analytic results

D. Melrose, *Quantum Plasmadynamics: Magnetized Plasmas*, Lecture Notes in Physics 854, DOI 10.1007/978-1-4614-4045-1_7,

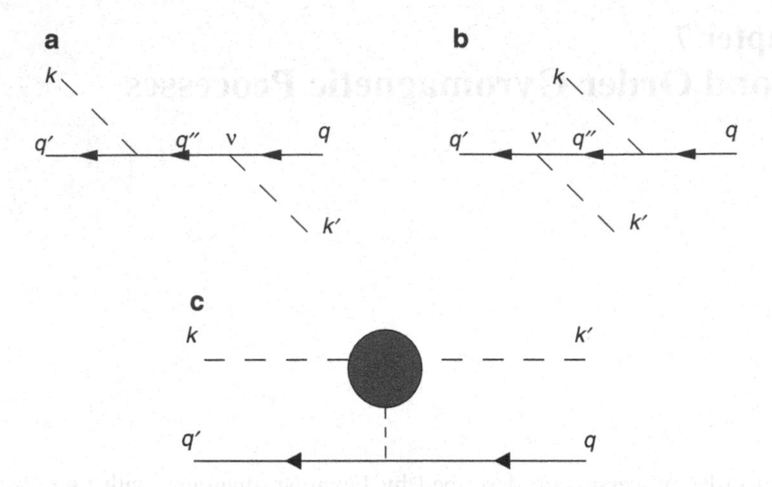

Fig. 7.1 The three diagrams that contributions Compton scattering in a magnetic field

relating to the evaluation of the probability are derived by partially performing the sum over an intermediate state, q'', specifically by performing the sums over s'', ϵ'' explicitly.

7.1.1 Probability of Compton Scattering

The Feynman diagrams for Compton scattering in a magnetic field are the same as those for Compton scattering in the absence of a magnetic field, except for the labeling of the states. In Fig. 7.1, the relevant Feynman diagrams are drawn for Compton scattering by an electron in an initial state $q = n, p_z, s$ and a final state $q' = n', p_z', s'$. Let the initial and final photons be in the modes M' and M, and have 4-momenta k' and k, respectively. Conservation of the parallel component of 4-momentum implies

$$\epsilon p_z - \epsilon' p_z' + k_z' - k_z = 0. \tag{7.1.1}$$

The condition (7.1.1) is implicit in the general expression for the probability of Compton scattering and related processes. The probability is written down in (5.6.13) in a generic form involving the transition matrix $\mathcal{T}_{\rm fi}$. On evaluating $\mathcal{T}_{\rm fi}$ using the vertex formalism, the probability becomes

$$w^{\epsilon'\epsilon}_{q'qMM'}(\boldsymbol{k},\boldsymbol{k}') = \frac{e^4 R_M(\boldsymbol{k}) R_{M'}(\boldsymbol{k}')}{|\varepsilon_0^2 \omega_M(\boldsymbol{k})\omega_{M'}(\boldsymbol{k}')|} \left| e^*_{M\mu}(\boldsymbol{k}) e_{M'\nu}(\boldsymbol{k}') [M^{\epsilon'\epsilon}_{q'q}(\boldsymbol{k},\boldsymbol{k}')]^{\mu\nu} \right|^2$$
$$\times\, 2\pi\delta[\epsilon\varepsilon_q - \epsilon'\varepsilon_{q'} - \omega_M(\boldsymbol{k}) + \omega_{M'}(\boldsymbol{k}')], \tag{7.1.2}$$

with $\varepsilon_q = \varepsilon_n(p_z)$, $\varepsilon_{q'} = \varepsilon_{n'}(p_z')$, usually written ε_n, $\varepsilon'_{n'}$, respectively. The 4-tensor in the numerator in (7.1.2) consists of three parts,

$$
\begin{aligned}
[M^{\epsilon'\epsilon}_{q'q}(\boldsymbol{k},\boldsymbol{k}')]^{\mu\nu} &= [M^{\epsilon'\epsilon}_{q'q}(\boldsymbol{k},\boldsymbol{k}')]^{\mu\nu}_1 + [M^{\epsilon'\epsilon}_{q'q}(\boldsymbol{k},\boldsymbol{k}')]^{\mu\nu}_2 + [M^{\epsilon'\epsilon}_{q'q}(\boldsymbol{k},\boldsymbol{k}')]^{\mu\nu}_{\mathrm{nl}}, \\
[M^{\epsilon'\epsilon}_{q'q}(\boldsymbol{k},\boldsymbol{k}')]^{\mu\nu}_1 &= \sum_{\epsilon'',q''} \frac{[\Gamma^{\epsilon'\epsilon''}_{q'q''}(\boldsymbol{k})]^{\mu}[\Gamma^{\epsilon''\epsilon}_{q''q}(-\boldsymbol{k}')]^{\nu}}{\epsilon\varepsilon_q - \epsilon''\varepsilon''_{q''} + \omega_{M'}(\boldsymbol{k}')} \exp\left(-i\frac{(\boldsymbol{k}\times\boldsymbol{k}')_z}{2eB}\right), \\
[M^{\epsilon'\epsilon}_{q'q}(\boldsymbol{k},\boldsymbol{k}')]^{\mu\nu}_2 &= \sum_{\epsilon'',q''} \frac{[\Gamma^{\epsilon'\epsilon''}_{q'q''}(-\boldsymbol{k}')]^{\nu}[\Gamma^{\epsilon''\epsilon}_{q''q}(\boldsymbol{k})]^{\mu}}{\epsilon\varepsilon_q - \epsilon''\varepsilon''_{q''} - \omega_{M}(\boldsymbol{k})} \exp\left(i\frac{(\boldsymbol{k}\times\boldsymbol{k}')_z}{2eB}\right), \\
[M^{\epsilon'\epsilon}_{q'q}(\boldsymbol{k},\boldsymbol{k}')]^{\mu\nu}_{\mathrm{nl}} &= \frac{2}{e}[\Gamma^{\epsilon'\epsilon}_{q'q}(\boldsymbol{k}-\boldsymbol{k}')]^{\theta} D_{\theta\eta}(k_M - k'_{M'}) \\
&\quad \times \Pi^{\eta\mu\nu}(k_M - k'_{M'}, -k_M, k'_{M'}),
\end{aligned}
\tag{7.1.3}
$$

which correspond to the three diagrams in Fig. 7.1. The term labeled nl describes nonlinear scattering, which is associated with the quadratic response tensor, $\Pi^{\eta\mu\nu}$. There are contributions from the medium and from the magnetized vacuum to $\Pi^{\eta\mu\nu}$, with the former typically the more important [17]. Nonlinear scattering in a thermal plasma is important for relatively long wavelengths, of order the Debye length, whereas quantum effects tend to be significant only for very much shorter wavelengths. As a consequence, when nonlinear scattering is important, the nonquantum theory suffices.

The case $\epsilon = \epsilon' = 1$ in (7.1.2) corresponds to Compton scattering by an electron. The various crossed processes correspond to changing the signs ϵ or ϵ' or changing the signs of $\boldsymbol{k}$ or $\boldsymbol{k}'$.

Kinematics of Compton Scattering

The resonance condition in the probability (7.1.2) with (7.1.1) can be reduced to the same form as the gyroresonance condition in (6.1.1). This is achieved by writing $\Omega = \omega - \omega'$, with $\omega = \omega_M(\boldsymbol{k})$, $\omega' = \omega_{M'}(\boldsymbol{k}')$, and $\boldsymbol{K} = \boldsymbol{k} - \boldsymbol{k}'$. The kinematics for gyromagnetic emission may then be used to describe the kinematics of Compton scattering. Specifically, the solutions of the gyroresonance condition found in § 6.1.2 translate trivially into corresponding solutions for Compton scattering and crossed processes.

In place of (6.1.16)–(6.1.18), and (6.1.20), one has

$$
F_{nn'} = \frac{(\varepsilon^0_n)^2 - (\varepsilon^0_{n'})^2 + \Omega^2 - K_z^2}{2(\Omega^2 - K_z^2)}, \qquad F_{n'n} = 1 - F_{nn'}, \tag{7.1.4}
$$

$$
G^2_{nn'} = G^2_{n'n} = \frac{[\Omega^2 - K_z^2 - (\varepsilon^0_n - \varepsilon^0_{n'})^2][\Omega^2 - K_z^2 - (\varepsilon^0_n + \varepsilon^0_{n'})^2]}{4(\Omega^2 - K_z^2)^2}, \tag{7.1.5}
$$

$$
p^{\pm}_{nn'} = K_z F_{nn'} \pm \Omega G_{nn'}, \qquad p'^{\pm}_{n'n} = K_z F_{n'n} \pm \Omega G_{n'n}, \tag{7.1.6}
$$

$$\varepsilon^{\pm}_{nn'} = \Omega F_{nn'} \pm K_z G_{nn'}, \qquad \varepsilon'^{\pm}_{n'n} = \Omega F_{n'n} \mp K_z G_{n'n}, \tag{7.1.7}$$

respectively. The two physically allowed regions correspond to $G^2_{nn'} \geq 0$, and are

$$\Omega^2 - K_z^2 \leq (\varepsilon^0_n - \varepsilon^0_{n'})^2, \qquad \Omega^2 - K_z^2 \geq (\varepsilon^0_n + \varepsilon^0_{n'})^2. \tag{7.1.8}$$

The inequality $\Omega^2 - K_z^2 \leq (\varepsilon^0_n - \varepsilon^0_{n'})^2$ applies to Compton scattering, with the equality being the threshold condition for Compton scattering at given n, n'. The region $\Omega^2 - K_z^2 \geq (\varepsilon^0_n + \varepsilon^0_{n'})^2$ corresponds to decay of a photon into another photon and a pair. For the crossed processes corresponding to $\omega', k'_z \to -\omega', -k'_z$, such that one has $\Omega = \omega + \omega'$, $K_z = k_z + k'_z$, the first of the regions (7.1.8) is where double emission is allowed, and the second region is where two-photon pair creation and decay are allowed.

7.1.2 Kinetic Equations for Compton Scattering

The kinetic equations for Compton scattering, two-photon emission and absorption and two-photon pair creation and decay are derived in the same way as the kinetic equations for the single-photon processes, cf. § 6.1.3.

For Compton scattering of photons in the mode M' into photons in the mode M by an electron, the kinetic equations for the photons are

$$\frac{\mathrm{D}N_M(\boldsymbol{k})}{\mathrm{D}t} = \sum_{q'q} \frac{eB}{2\pi} \int \frac{dp_z}{2\pi} \int \frac{d^3\boldsymbol{k}'}{(2\pi)^3} w^{++}_{q'qMM'}(\boldsymbol{k},\boldsymbol{k}')$$
$$\times \{n_q[1 + N_M(\boldsymbol{k})]N_{M'}(\boldsymbol{k}') - n_{q'} N_M(\boldsymbol{k})[1 + N_{M'}(\boldsymbol{k}')]\},$$
$$\frac{\mathrm{D}N_{M'}(\boldsymbol{k}')}{\mathrm{D}t} = -\sum_{q'q} \frac{eB}{2\pi} \int \frac{dp_z}{2\pi} \int \frac{d^3\boldsymbol{k}}{(2\pi)^3} w^{++}_{q'qMM'}(\boldsymbol{k},\boldsymbol{k}')$$
$$\times \{n_q[1 + N_M(\boldsymbol{k})]N_{M'}(\boldsymbol{k}') - n_{q'} N_M(\boldsymbol{k})[1 + N_{M'}(\boldsymbol{k}')]\}. \tag{7.1.9}$$

The kinetic equation for the particles is

$$\frac{dn_q}{dt} = \int \frac{d^3\boldsymbol{k}}{(2\pi)^3} \int \frac{d^3\boldsymbol{k}'}{(2\pi)^3} \Big(\sum_{q''} w^{++}_{qq''MM'}(\boldsymbol{k},\boldsymbol{k}')$$
$$\times \{n_{q''}[1 + N_M(\boldsymbol{k})]N_{M'}(\boldsymbol{k}') - n_q N_M(\boldsymbol{k})[1 + N_{M'}(\boldsymbol{k}')]\}$$
$$- \sum_{q'} w^{++}_{q'qMM'}(\boldsymbol{k},\boldsymbol{k}') \big) \{n_q[1 + N_M(\boldsymbol{k})]N_{M'}(\boldsymbol{k}') - n_{q'} N_M(\boldsymbol{k})[1 + N_{M'}(\boldsymbol{k}')]\} \Big), \tag{7.1.10}$$

where q' denotes the quantum numbers $n', s', p_z' = p_z + k_z' - k_z$ and q'' denotes $n'', s'', p_z'' = p_z - k_z' + k_z$.

Compton Cross Section

It is conventional to describe Compton scattering in terms of the scattering cross section. The relation between the cross section and the probability is discussed in §4.5.1, and given by (4.5.13), which translates into (ordinary units)

$$\Sigma_{MM'} = \int_0^\infty \frac{d\omega\,\omega^2 n_M^2}{(2\pi)^3\omega' v'_{gM'}} \frac{\partial(\omega n_M)}{\partial\omega} \frac{\partial\cos\theta}{\partial\cos\theta_r} \frac{\partial\cos\theta'}{\partial\cos\theta'_r} w^{++}_{q'qMM'}, \tag{7.1.11}$$

with the probability given by (7.1.2).

In the approximation in which the refractive indices are approximated by unity, and the ray and wave normal angles are not distinguished, the labels M, M' in (7.1.11) may be interpreted as describing (transverse) polarization states, and (7.1.11) reduces to

$$\Sigma_{MM'} = \frac{1}{c^4}\int_0^\infty \frac{d\omega\,\omega^2}{(2\pi)^3\omega'} w^{++}_{q'qMM'}. \tag{7.1.12}$$

On inserting the probability (7.1.2), the differential cross section (7.1.12) becomes

$$\Sigma_{MM'} = \frac{e^4\omega^2}{(4\pi\varepsilon_0)^2\omega'^2} \left|e^*_{M\mu} e_{M'\nu} \left[M^{\epsilon'\epsilon}_{q'q}\right]^{\mu\nu}\right|^2, \tag{7.1.13}$$

where $\omega - \omega' = \varepsilon_q - \varepsilon_{q'}$ is implicit. The total cross section is found by integrating (7.1.13) over the solid angle of the scattered photon.

7.1.3 Sum over Intermediate States: Vertex Formalism

Explicit evaluation of the probability (7.1.2) for Compton scattering involves carrying out the sum over intermediate states. The sum is over states labeled $q'' = \epsilon'', s'', n''$, and the sums over ϵ'', s'' can be performed explicitly.

The sums need to be performed for each of the two contributions labeled 1 and 2 to the matrix element in (7.1.3). The term labeled 1 corresponds to the photon $\boldsymbol{k}'$ being absorbed at the first vertex, labeled ν, and the photon $\boldsymbol{k}$ being emitted at the second vertex, labeled μ, so that the intermediate state has $\epsilon'' p_z'' = \epsilon p_z + k_z'$. The term labeled 2 corresponds to the sequence of absorption and emission being interchanged, and then the intermediate state has $\epsilon'' p_z'' = \epsilon p_z - k_z$. The sums over ϵ'', s'' can be performed explicitly by first rationalizing the denominators is these two terms, so that ϵ'' appears only in the numerator. Omitting the labels denoting the modes, the rationalization gives

$$\big[M^{\epsilon'\epsilon}_{q'q}(\boldsymbol{k},\boldsymbol{k}')\big]^{\mu\nu}_1 = \sum_{n''}\frac{\big[a^{\epsilon'\epsilon}_{q'q}(\boldsymbol{k},\boldsymbol{k}')\big]^{\mu\nu}_{1,n''}e^{i(xx')^{1/2}\sin(\psi-\psi')}}{2\epsilon(\varepsilon_q\omega'-p_zk'_z)+\omega'^2-k'^2_z-2n''eB},$$

$$\big[a^{\epsilon'\epsilon}_{q'q}(\boldsymbol{k},\boldsymbol{k}')\big]^{\mu\nu}_{1,n''} = \sum_{s'',\epsilon''}[\epsilon\varepsilon_q+\epsilon''\varepsilon''_{q''}+\omega']\,[\Gamma^{\epsilon'\epsilon''}_{q'q''}(\boldsymbol{k})]^{\mu}[\Gamma^{\epsilon\epsilon''}_{qq''}(\boldsymbol{k}')]^{*\nu}. \tag{7.1.14}$$

$$\big[M^{\epsilon'\epsilon}_{q'q}(\boldsymbol{k},\boldsymbol{k}')\big]^{\mu\nu}_2 = \sum_{n''}\frac{\big[a^{\epsilon'\epsilon}_{q'q}(\boldsymbol{k},\boldsymbol{k}')\big]^{\mu\nu}_{2,n''}e^{-i(xx')^{1/2}\sin(\psi-\psi')}}{-2\epsilon(\varepsilon_q\omega-p_zk_z)+\omega^2-k^2_z-2n''eB},$$

$$\big[a^{\epsilon'\epsilon}_{q'q}(\boldsymbol{k},\boldsymbol{k}')\big]^{\mu\nu}_{2,n''} = \sum_{s'',\epsilon''}[\epsilon\varepsilon_q+\epsilon''\varepsilon''_{q''}-\omega]\,[\Gamma^{\epsilon''\epsilon'}_{q''q'}(\boldsymbol{k}')]^{*\nu}[\Gamma^{\epsilon''\epsilon}_{q''q}(\boldsymbol{k})]^{\mu}. \tag{7.1.15}$$

The two tensors (7.1.14) and (7.1.15) are related by

$$\big[M^{\epsilon'\epsilon}_{q'q}(\boldsymbol{k},\boldsymbol{k}')\big]^{\mu\nu}_2 = \big[M^{\epsilon'\epsilon}_{q'q}(-\boldsymbol{k}',-\boldsymbol{k})\big]^{\nu\mu}_1,\quad \big[a^{\epsilon'\epsilon}_{q'q}(\boldsymbol{k},\boldsymbol{k}')\big]^{\mu\nu}_2 = \big[a^{\epsilon'\epsilon}_{q'q}(-\boldsymbol{k}',-\boldsymbol{k})\big]^{\nu\mu}_1. \tag{7.1.16}$$

On inserting the explicit form (5.4.18) for the vertex function, it is straightforward but tedious to perform the sums in (7.1.14) and (7.1.15) directly. It is convenient to write the result in the form

$$\big[a^{\epsilon'\epsilon}_{q'q}(\boldsymbol{k},\boldsymbol{k}')\big]^{\mu\nu}_{1,n''} = (ie^{-i\psi})^{n'}(-ie^{i\psi'})^{n}\,e^{-in''(\psi'-\psi)}\big[c^{\epsilon'\epsilon}_{q'q}(\boldsymbol{k},\boldsymbol{k}')\big]^{\mu\nu}_{n''},$$

$$\big[a^{\epsilon'\epsilon}_{q'q}(\boldsymbol{k},\boldsymbol{k}')\big]^{\mu\nu}_{2,n''} = (-ie^{-i\psi'})^{n'}(ie^{i\psi})^{n}\,e^{in''(\psi'-\psi)}\big[d^{\epsilon'\epsilon}_{q'q}(\boldsymbol{k},\boldsymbol{k}')\big]^{\mu\nu}_{n''}. \tag{7.1.17}$$

The components of $\big[c^{\epsilon'\epsilon}_{q'q}(\boldsymbol{k},\boldsymbol{k}')\big]^{\mu\nu}_{n''}$ and $\big[d^{\epsilon'\epsilon}_{q'q}(\boldsymbol{k},\boldsymbol{k}')\big]^{\mu\nu}_{n''}$ are written down in Tables 7.1 and 7.2, respectively. The components are related by the symmetry property (7.1.16).

7.1.4 Sum over Intermediate States: Ritus Method

The Ritus method (§5.5) provides an alternative way of evaluating the quantities $\big[M^{\epsilon'\epsilon}_{q'q}(\boldsymbol{k},\boldsymbol{k}')\big]^{\mu\nu}_{1,2}$ in the probability (7.1.2). To illustrate the use of this method, consider the evaluation of $\big[M^{\epsilon'\epsilon}_{q'q}(\boldsymbol{k},\boldsymbol{k}')\big]^{\mu\nu}_1$, which involves the product $[\Gamma^{\epsilon'\epsilon''}_{q'q''}(\boldsymbol{k})]^{\mu}[\Gamma^{\epsilon''\epsilon}_{q''q}(-\boldsymbol{k}')]^{\nu}$, and in the Ritus method this product becomes

$$V^2\bar{\varphi}^{\epsilon'}_{s'}(n',p'_z)\,\mathcal{J}^{\mu}_{n'n''}(\boldsymbol{k}_\perp)\,\phi^{\epsilon''}_{s''}(n'',p''_z)\,\bar{\phi}^{\epsilon''}_{s''}(n'',p''_z)\,\mathcal{J}^{\nu}_{n''n}(-\boldsymbol{k}'_\perp)\,\varphi^{\epsilon}_{s}(n,p_z).$$

Table 7.1 Table of tensor components of $[c^{\epsilon'\epsilon}_{q'q}(\boldsymbol{k},\boldsymbol{k}')]^{\mu\nu}_{n''}$, defined by (7.1.17), with $A^{\pm}_{q'q} = a'_{\epsilon's'}a_{\epsilon s} \pm a'_{-\epsilon's'}a_{-\epsilon s}$, $B^{\pm}_{q'q} = a'_{\epsilon's'}a_{-\epsilon s} \pm a'_{-\epsilon's'}a_{\epsilon s}$, $a'_{\pm} = [(\varepsilon'_{n'} \pm \varepsilon^0_{n'})/2\varepsilon_{n'})]^{1/2}$, $a_{\pm} = [(\varepsilon_n \pm \varepsilon^0_n)/2\varepsilon_n)]^{1/2}$, and with the remaining quantities defined by Table 7.3

$[c^{\epsilon'\epsilon}_{q'q}(\boldsymbol{k},\boldsymbol{k}')]^{00}_{n''}$	$[(\epsilon\varepsilon_q+\omega')A^{+}_{q'q}+(\epsilon p_z+k'_z)B^{+}_{q'q}]k'^{+}_{q'q}+A^{-}_{q'q}(mk'^{-}_{q'q}+p_{n''}l'^{+}_{q'q})$
$[c^{\epsilon'\epsilon}_{q'q}(\boldsymbol{k},\boldsymbol{k}')]^{01}_{n''}$	$[(\epsilon\varepsilon_q+\omega')A^{-}_{q'q}-(\epsilon p_z+k'_z)B^{-}_{q'q}]u'^{+}_{q'q}+A^{+}_{q'q}(mu'^{-}_{q'q}+p_{n''}t'^{+}_{q'q})$
$[c^{\epsilon'\epsilon}_{q'q}(\boldsymbol{k},\boldsymbol{k}')]^{10}_{n''}$	$[(\epsilon\varepsilon_q+\omega')A^{-}_{q'q}+(\epsilon p_z+k'_z)B^{-}_{q'q}]v'^{+}_{q'q}-A^{+}_{q'q}(mv'^{-}_{q'q}-p_{n''}r'^{+}_{q'q})$
$[c^{\epsilon'\epsilon}_{q'q}(\boldsymbol{k},\boldsymbol{k}')]^{02}_{n''}$	$-i\{[(\epsilon\varepsilon_q+\omega')A^{-}_{q'q}-(\epsilon p_z+k'_z)B^{-}_{q'q}]u^{-}_{q'q}+A^{+}_{q'q}(mu'^{+}_{q'q}-p_{n''}t'^{+}_{q'q})\}$
$[c^{\epsilon'\epsilon}_{q'q}(\boldsymbol{k},\boldsymbol{k}')]^{20}_{n''}$	$-i\{[(\epsilon\varepsilon_q+\omega')A^{-}_{q'q}+(\epsilon p_z+k'_z)B^{-}_{q'q}]v'^{-}_{q'q}-A^{+}_{q'q}(mv'^{+}_{q'q}-p_{n''}t'^{-}_{q'q})\}$
$[c^{\epsilon'\epsilon}_{q'q}(\boldsymbol{k},\boldsymbol{k}')]^{03}_{n''}$	$[(\epsilon\varepsilon_q+\omega')B^{+}_{q'q}+(\epsilon p_z+k'_z)A^{+}_{q'q}]k'^{+}_{q'q}+B^{-}_{q'q}(mk'^{-}_{q'q}+p_{n''}l'^{+}_{q'q})$
$[c^{\epsilon'\epsilon}_{q'q}(\boldsymbol{k},\boldsymbol{k}')]^{30}_{n''}$	$[(\epsilon\varepsilon_q+\omega')B^{+}_{q'q}+(\epsilon p_z+k'_z)A^{+}_{q'q}]k'^{+}_{q'q}-B^{-}_{q'q}(mk'^{-}_{q'q}+p_{n''}l'^{+}_{q'q})$
$[c^{\epsilon'\epsilon}_{q'q}(\boldsymbol{k},\boldsymbol{k}')]^{11}_{n''}$	$[(\epsilon\varepsilon_q+\omega')A^{+}_{q'q}-(\epsilon p_z+k'_z)B^{+}_{q'q}]p'^{+}_{q'q}-A^{-}_{q'q}(mp'^{-}_{q'q}-p_{n''}q'^{+}_{q'q})$
$[c^{\epsilon'\epsilon}_{q'q}(\boldsymbol{k},\boldsymbol{k}')]^{22}_{n''}$	$[(\epsilon\varepsilon_q+\omega')A^{+}_{q'q}-(\epsilon p_z+k'_z)B^{+}_{q'q}]p'^{+}_{q'q}-A^{-}_{q'q}(mp'^{-}_{q'q}+p_{n''}q'^{+}_{q'q})$
$[c^{\epsilon'\epsilon}_{q'q}(\boldsymbol{k},\boldsymbol{k}')]^{33}_{n''}$	$[(\epsilon\varepsilon_q+\omega')A^{+}_{q'q}+(\epsilon p_z+k'_z)B^{+}_{q'q}]k'^{+}_{q'q}-A^{-}_{q'q}(mk'^{-}_{q'q}+p_{n''}l'^{+}_{q'q})$
$[c^{\epsilon'\epsilon}_{q'q}(\boldsymbol{k},\boldsymbol{k}')]^{12}_{n''}$	$i\{[(\epsilon\varepsilon_q+\omega')A^{+}_{q'q}-(\epsilon p_z+k'_z)B^{+}_{q'q}]p'^{-}_{q'q}-A^{-}_{q'q}(mp'^{+}_{q'q}-p_{n''}q'^{-}_{q'q})$
$[c^{\epsilon'\epsilon}_{q'q}(\boldsymbol{k},\boldsymbol{k}')]^{21}_{n''}$	$-i\{[(\epsilon\varepsilon_q+\omega')A^{+}_{q'q}-(\epsilon p_z+k'_z)B^{+}_{q'q}]p'^{-}_{q'q}-A^{-}_{q'q}(mp'^{+}_{q'q}+p_{n''}q'^{-}_{q'q})\}$
$[c^{\epsilon'\epsilon}_{q'q}(\boldsymbol{k},\boldsymbol{k}')]^{13}_{n''}$	$[(\epsilon\varepsilon_q+\omega')B^{-}_{q'q}+(\epsilon p_z+k'_z)A^{-}_{q'q}]v'^{+}_{q'q}-B^{+}_{q'q}(mv'^{-}_{q'q}-p_{n''}r'^{+}_{q'q})\}$
$[c^{\epsilon'\epsilon}_{q'q}(\boldsymbol{k},\boldsymbol{k}')]^{31}_{n''}$	$[-(\epsilon\varepsilon_q+\omega')B^{-}_{q'q}+(\epsilon p_z+k'_z)A^{-}_{q'q}]u'^{+}_{q'q}+B^{+}_{q'q}(mu'^{-}_{q'q}+p_{n''}t'^{-}_{q'q})$
$[c^{\epsilon'\epsilon}_{q'q}(\boldsymbol{k},\boldsymbol{k}')]^{23}_{n''}$	$-i\{[(\epsilon\varepsilon_q+\omega')B^{-}_{q'q}+(\epsilon p_z+k'_z)A^{-}_{q'q}]v'^{-}_{q'q}-B^{+}_{q'q}(mv'^{+}_{q'q}-p_{n''}r'^{-}_{q'q})\}$
$[c^{\epsilon'\epsilon}_{q'q}(\boldsymbol{k},\boldsymbol{k}')]^{32}_{n''}$	$-i\{[-(\epsilon\varepsilon_q+\omega')B^{-}_{q'q}+(\epsilon p_z+k'_z)A^{-}_{q'q}]u'^{-}_{q'q}+B^{+}_{q'q}(mu'^{+}_{q'q}-p_{n''}t'^{-}_{q'q})\}$

The sums are over s'', ϵ'', n''. The sum over s'' is performed using (5.5.16), which gives

$$\sum_{s''}\phi^{\epsilon}_{s''}(n'',p''_z)\,\bar{\phi}^{\epsilon}_{s''}(n'',p''_z) = \frac{\not{P}^{\prime\prime\epsilon''}_{q''}+m}{\epsilon''\varepsilon_{q''}V},$$

with $[P^{\prime\prime\epsilon''}_{q''}]^{\mu} = (\epsilon''\varepsilon''_{q''}, 0, p_{n''}, \epsilon'' p''_z)$. In evaluating $[a^{\epsilon'\epsilon}_{q'q}(\boldsymbol{k},\boldsymbol{k}')]^{\mu\nu}_{1,n''}$, as given by (7.1.14), the sum over ϵ'' is performed using

$$\sum_{\epsilon''}[\epsilon\varepsilon_q+\epsilon''\varepsilon''_{q''}+\omega']\,\frac{\not{P}^{\prime\prime\epsilon''}_{q''}(p''_z)+m}{2\epsilon''\varepsilon_{q''}}$$

$$= (\epsilon\varepsilon_q+\omega')\gamma^0 - p_{n''}\gamma^2 - (\epsilon p_z+k'_z)\gamma^3 + m, \tag{7.1.18}$$

with $p''_z = \epsilon''(\epsilon p_z + k'_z)$, and where $\gamma^0, \gamma^2, \gamma^3$ are Dirac matrices. The final step is to evaluate the matrix elements of the resulting product of Dirac matrices for the projected states.

Table 7.2 As for Table 7.1 but for $[d^{\epsilon'\epsilon}_{q'q}(\boldsymbol{k},\boldsymbol{k}')]^{\mu\nu}_{n''}$

$[d^{\epsilon'\epsilon}_{q'q}(\boldsymbol{k},\boldsymbol{k}')]^{00}_{n''}$	$[(\epsilon\varepsilon_q-\omega)A^+_{q'q}+(\epsilon p_z-k_z)B^+_{q'q}]k^+_{q'q}+A^-_{q'q}(mk^-_{q'q}+p_{n''}l^+_{q'q})$
$[d^{\epsilon'\epsilon}_{q'q}(\boldsymbol{k},\boldsymbol{k}')]^{01}_{n''}$	$-[(\epsilon\varepsilon_q-\omega)A^-_{q'q}+(\epsilon p_z-k_z)B^-_{q'q}]v^+_{q'q}+A^+_{q'q}(mv^-_{q'q}-p_{n''}r^+_{q'q})$
$[d^{\epsilon'\epsilon}_{q'q}(\boldsymbol{k},\boldsymbol{k}')]^{10}_{n''}$	$-[(\epsilon\varepsilon_q-\omega)A^-_{q'q}-(\epsilon p_z-k_z)B^-_{q'q}]u^+_{q'q}-A^+_{q'q}(mu^-_{q'q}+p_{n''}t^+_{q'q})$
$[d^{\epsilon'\epsilon}_{q'q}(\boldsymbol{k},\boldsymbol{k}')]^{02}_{n''}$	$-i\{-[(\epsilon\varepsilon_q-\omega)A^-_{q'q}+(\epsilon p_z-k_z)B^-_{q'q}]v^-_{q'q}+A^+_{q'q}(mv^+_{q'q}-p_{n''}r^+_{q'q})\}$
$[d^{\epsilon'\epsilon}_{q'q}(\boldsymbol{k},\boldsymbol{k}')]^{20}_{n''}$	$i\{[(\epsilon\varepsilon_q-\omega)A^-_{q'q}-(\epsilon p_z-k_z)B^-_{q'q}]u^-_{q'q}+A^+_{q'q}(mu^+_{q'q}-p_{n''}t^-_{q'q})\}$
$[d^{\epsilon'\epsilon}_{q'q}(\boldsymbol{k},\boldsymbol{k}')]^{03}_{n''}$	$[(\epsilon\varepsilon_q-\omega)B^+_{q'q}+(\epsilon p_z-k_z)A^+_{q'q}]k^+_{q'q}-B^-_{q'q}(mk^-_{q'q}+p_{n''}l^+_{q'q})$
$[d^{\epsilon'\epsilon}_{q'q}(\boldsymbol{k},\boldsymbol{k}')]^{30}_{n''}$	$[(\epsilon\varepsilon_q-\omega)B^+_{q'q}+(\epsilon p_z-k_z)A^+_{q'q}]k^+_{q'q}+B^-_{q'q}(mk^-_{q'q}+p_{n''}l^+_{q'q})$
$[d^{\epsilon'\epsilon}_{q'q}(\boldsymbol{k},\boldsymbol{k}')]^{11}_{n''}$	$[(\epsilon\varepsilon_q-\omega)A^+_{q'q}-(\epsilon p_z-k_z)B^+_{q'q}]p^+_{q'q}-A^-_{q'q}(mp^-_{q'q}-p_{n''}q^+_{q'q})$
$[d^{\epsilon'\epsilon}_{q'q}(\boldsymbol{k},\boldsymbol{k}')]^{22}_{n''}$	$[(\epsilon\varepsilon_q-\omega)A^+_{q'q}-(\epsilon p_z-k_z)B^+_{q'q}]p^+_{q'q}-A^-_{q'q}(mp^-_{q'q}+p_{n''}q^+_{q'q})$
$[d^{\epsilon'\epsilon}_{q'q}(\boldsymbol{k},\boldsymbol{k}')]^{33}_{n''}$	$[(\epsilon\varepsilon_q-\omega)A^+_{q'q}+(\epsilon p_z-k_z)B^+_{q'q}]k^+_{q'q}-A^-_{q'q}(mk^-_{q'q}+p_{n''}l^+_{q'q})$
$[d^{\epsilon'\epsilon}_{q'q}(\boldsymbol{k},\boldsymbol{k}')]^{12}_{n''}$	$-i\{[(\epsilon\varepsilon_q-\omega)A^+_{q'q}-(\epsilon p_z-k_z)B^+_{q'q}]p^-_{q'q}-A^-_{q'q}(mp^+_{q'q}-p_{n''}q^-_{q'q})$
$[d^{\epsilon'\epsilon}_{q'q}(\boldsymbol{k},\boldsymbol{k}')]^{21}_{n''}$	$i\{[(\epsilon\varepsilon_q-\omega)A^+_{q'q}-(\epsilon p_z-k_z)B^+_{q'q}]p^-_{q'q}-A^-_{q'q}(mp^+_{q'q}+p_{n''}q^-_{q'q})\}$
$[d^{\epsilon'\epsilon}_{q'q}(\boldsymbol{k},\boldsymbol{k}')]^{13}_{n''}$	$[(\epsilon\varepsilon_q-\omega)B^-_{q'q}-(\epsilon p_z-k_z)A^-_{q'q}]u^+_{q'q}-B^+_{q'q}(mu^-_{q'q}+p_{n''}t^+_{q'q})\}$
$[d^{\epsilon'\epsilon}_{q'q}(\boldsymbol{k},\boldsymbol{k}')]^{31}_{n''}$	$[-(\epsilon\varepsilon_q-\omega)B^-_{q'q}-(\epsilon p_z-k_z)A^-_{q'q}]v^+_{q'q}+B^+_{q'q}(mv^-_{q'q}-p_{n''}r^-_{q'q})$
$[d^{\epsilon'\epsilon}_{q'q}(\boldsymbol{k},\boldsymbol{k}')]^{23}_{n''}$	$i\{[-(\epsilon\varepsilon_q-\omega)B^-_{q'q}+(\epsilon p_z-k_z)A^-_{q'q}]u^-_{q'q}+B^+_{q'q}(mu^+_{q'q}-p_{n''}t^-_{q'q})\}$
$[d^{\epsilon'\epsilon}_{q'q}(\boldsymbol{k},\boldsymbol{k}')]^{32}_{n''}$	$-i\{[-(\epsilon\varepsilon_q-\omega)B^-_{q'q}-(\epsilon p_z-k_z)A^-_{q'q}]v^-_{q'q}+B^+_{q'q}(mv^+_{q'q}-p_{n''}r^-_{q'q})\}$

For an arbitrary choice of spin operator the reduced wave functions have the form (5.5.12), so that $[\bar{\varphi}^{\epsilon'}_{s'}(n',p'_z)]_\pm$ become the row matrices $\bar{\varphi}'_+=(C_1'^*,0,-C_3'^*,0)$, $\bar{\varphi}'_-=(0,C_2'^*,0,-C_4'^*)$ and $[\varphi^\epsilon_s(n,p_z)]_\pm$ become the column matrices $\varphi_+=(C_1,0,C_3,0)$, $\varphi_-=(0,C_2,0,C_4)$. The matrix elements for the $\|$- and $\perp$-parts are nonzero only between $\bar{\varphi}'_\pm$, $\varphi_\pm$ and for $\bar{\varphi}'_\mp$, $\varphi_\pm$, respectively. The coefficient of $e^{i(\psi-\psi')}J^{n'-1}_{n''-n'-1}(x)\,J^{n-1}_{n''-n+1}(x')$ is the sum of $(\epsilon\varepsilon_q+\omega')$, $-(\epsilon p_z+k'_z)$, m times

$$\bar{\varphi}'_+\begin{pmatrix}\gamma^0\\ \gamma^3\\ 1\end{pmatrix}\varphi_+=\begin{pmatrix}C_1'^*C_1+C_3'^*C_3\\ C_1'^*C_3+C_3'^*C_1\\ C_1'^*C_1-C_3'^*C_3\end{pmatrix}=b'_{s'}b_s\begin{pmatrix}A^+_{q'q}\\ B^+_{q'q}\\ A^-_{q'q}\end{pmatrix},$$

respectively, and the coefficient of $e^{i(\psi+\psi')}J^{n'-1}_{n''-n'-1}(x)\,J^{n}_{n''-n+1}(x')$ is $-p_{n''}$ times

$$-\bar{\varphi}'_+\gamma^2\varphi_-=i(C_1'^*C_4+C_3'^*C_2)=-sb'_{s'}b_{-s}A^-_{q'q}.$$

The explicit forms for the combinations of C'^*s and Cs are for the magnetic-moment eigenfunctions (5.2.12), with $A^\pm_{q'q}$, $B^\pm_{q'q}$ defined in Table 7.1. These results

Table 7.3 Table of the combinations of J-functions that appear in Table 7.2, with $b'_{s'} = (\varepsilon^0_{n'} + s'm)^{1/2}$, $b_s = (\varepsilon^0_n + sm)^{1/2}$. The combinations of J-functions in Table 7.1 are modified, denoted by the prime, by (a) multiplication by $(-)^{n'-n}$, and (b) the interchanges $x \leftrightarrow x'$, $\psi \leftrightarrow \psi'$

$k^{\pm}_{q'q}$	$b'_{s'}b_s J^{n'-1}_{n''-n'}(x')J^{n-1}_{n''-n}(x) \pm s'sb'_{-s'}b_{-s}J^{n'}_{n''-n'}(x')J^{n}_{n''-n}(x)$
$l^{\pm}_{q'q}$	$sb'_{s'}b_{-s}J^{n'-1}_{n''-n'}(x')J^{n}_{n''-n}(x) \pm s'b'_{-s'}b_s J^{n'}_{n''-n'}(x')J^{n-1}_{n''-n}(x)$
$p^{\pm}_{q'q}$	$b'_{s'}b_s e^{i(\psi'-\psi)}J^{n'-1}_{n''-n'+1}(x')J^{n-1}_{n''-n+1}(x)$ $\pm s'sb'_{-s'}b_{-s}e^{-i(\psi'-\psi)}J^{n'}_{n''-n'-1}(x')J^{n}_{n''-n-1}(x)$
$q^{\pm}_{q'q}$	$sb'_{s'}b_{-s}e^{i(\psi'+\psi)}J^{n'-1}_{n''-n'+1}(x')J^{n}_{n''-n-1}(x)$ $\pm s'b'_{-s'}b_s e^{-i(\psi'+\psi)}J^{n'}_{n''-n'-1}(x')J^{n-1}_{n''-n+1}(x)$
$r^{\pm}_{q'q}$	$b'_{s'}b_s e^{i\psi'}J^{n'-1}_{n''-n'+1}(x')J^{n-1}_{n''-n}(x) \pm s'sb'_{-s'}b_{-s}e^{-i\psi'}J^{n'}_{n''-n'-1}(x')J^{n}_{n''-n}(x)$
$t^{\pm}_{q'q}$	$b'_{s'}b_s e^{-i\psi}J^{n'-1}_{n''-n'}(x')J^{n-1}_{n''-n+1}(x) \pm s'sb'_{-s'}b_{-s}e^{i\psi}J^{n'}_{n''-n'}(x')J^{n}_{n''-n-1}(x)$
$u^{\pm}_{q'q}$	$sb'_{s'}b_{-s}e^{i\psi}J^{n'-1}_{n''-n'}(x')J^{n}_{n''-n-1}(x) \pm s'b'_{-s'}b_s e^{-i\psi}J^{n'}_{n''-n'}(x')J^{n-1}_{n''-n+1}(x)$
$v^{\pm}_{q'q}$	$sb'_{s'}b_{-s}e^{i\psi'}J^{n'-1}_{n''-n'+1}(x')J^{n}_{n''-n}(x) \pm s'b'_{-s'}b_s e^{-i\psi'}J^{n'}_{n''-n'-1}(x')J^{n-1}_{n''-n}(x)$

reproduce half the terms in the entry for $[c^{\epsilon'\epsilon}_{q'q}(\boldsymbol{k},\boldsymbol{k}')]^{11}_{n''}$ in Table 7.1, with $p'^{\pm}_{q'q}, q'^{\pm}_{q'q}$ given by $\psi, x \leftrightarrow \psi', x'$ in the entries $p^{\pm}_{q'q}, q^{\pm}_{q'q}$ in Table 7.3. The remaining terms in $[c^{\epsilon'\epsilon}_{q'q}(\boldsymbol{k},\boldsymbol{k}')]^{11}_{n''}$ come from the other nonzero matrix elements, $\bar{\varphi}'_{+}(\gamma^0,\gamma^3,1)\varphi_{+}$, $\bar{\varphi}'_{-}\gamma^2\varphi_{+}$. Similar calculations reproduce the other μ, ν component in Tables 7.1 and 7.2.

In comparing the Ritus method and the vertex formalism, neither has an obvious major advantage over the other in detailed calculations. Both involve lengthy calculations. The Ritus method is closer to the conventional method used in the unmagnetized case to evaluate Feynman amplitudes and transition rates. In the unmagnetized case one is rarely interested in any spin-dependence, and one averages over the initial spins and sums over the final spins. This allows rules for evaluating transition rates to be formulated without requiring a specific choice of spin operator. The Ritus method allows one to develop an analogous spin-average approach in the magnetized case. However, the evaluation of the traces is considerably more cumbersome than in the unmagnetized case, as discussed in § 5.5.6.

7.2 Compton Scattering by an Electron with $n = 0$

The general form of the probability of Compton scattering in a magnetic field is too cumbersome to be of direct use in specific applications, and it needs to be replaced by simpler approximate forms and special cases. An important special case where the magnetized quantum theory is essential is for scattering by an electron in its ground Landau state $n = 0$. In a superstrong magnetic field, gryomagnetic losses occur at a very high rate: according to (6.2.20) this rate is of order (in ordinary units) $\alpha_c(mc^2/\hbar)(B/B_c)^2 \approx 10^{19}(B/B_c)^2\,s^{-1}$. The timescale for an electron to radiate away all its perpendicular energy can be so short (e.g., for B/B_c of order unity) that one expects most electrons to be in their ground state, $n = 0$. The case of Compton scattering by an electron in the state $n = 0$ is discussed in this section.

7.2.1 Scattering Probability for n = 0

The probability (7.1.2) for Compton scattering simplifies considerably for an electron with $n = 0$, when one has $\epsilon = \epsilon' = 1$ and $s = -1$. One is free to choose the frame $p_z = 0$, implying $p'_z = k'_z - k_z$, $\varepsilon_n = m$. The J-functions in Table 7.3 with upper index $n - 1$ are identically zero for $n = 0$, reducing the number of combinations that appear by one half. After omitting the labels M, M' referring to the modes, assuming $R_M, R_{M'}$ can both be approximated by 1/2, and neglecting nonlinear scattering, the probability (7.1.2) reduces to

$$w_{q'0}^{++}(\boldsymbol{k},\boldsymbol{k}') = \frac{e^4}{4\varepsilon_0^2|\omega\omega'|}\left|e_i^* e'_j\left[M_{q'0}^{++}(\boldsymbol{k},\boldsymbol{k}')\right]^{ij}\right|^2 2\pi\delta(\varepsilon_{q'} - m - \omega' + \omega), \quad (7.2.1)$$

with $\varepsilon_{q'} = [m^2 + 2n''eB + (k'_z - k_z)^2]^{1/2}$. The components of $\left[M_{q'0}^{++}(\boldsymbol{k},\boldsymbol{k}')\right]^{ij} = \left[M_{q'0}^{++}(\boldsymbol{k},\boldsymbol{k}')\right]_1^{ij} + \left[M_{q'0}^{++}(\boldsymbol{k},\boldsymbol{k}')\right]_2^{ij}$ follow from (7.1.14), (7.1.15), and (7.1.17), which give, for $n = 0$, $p_z = 0$,

$$\left[M_{q'0}^{++}(\boldsymbol{k},\boldsymbol{k}')\right]_1^{ij} = \sum_{n''} \frac{\left[c_{q'0}^{++}(\boldsymbol{k},\boldsymbol{k}')\right]_{n''}^{ij} e^{iF_1}}{2m\omega' + \omega'^2 - k_z'^2 - 2n''eB}, \quad (7.2.2)$$

with the phase factor $e^{iF_1} = (ie^{-i\psi})^{n'} e^{-in''(\psi'-\psi)} e^{i(xx')^{1/2}\sin(\psi-\psi')}$ and

$$\left[M_{q'0}^{++}(\boldsymbol{k},\boldsymbol{k}')\right]_2^{ij} = \sum_{n''} \frac{\left[d_{q'0}^{++}(\boldsymbol{k},\boldsymbol{k}')\right]_{n''}^{ij} e^{iF_2}}{-2m\omega + \omega^2 - k_z^2 - 2n''eB}, \quad (7.2.3)$$

with $e^{iF_2} = (-ie^{-i\psi'})^{n'} e^{in''(\psi'-\psi)} e^{-i(xx')^{1/2}\sin(\psi-\psi')}$. The components of $\left[c_{q'0}^{++}(\boldsymbol{k},\boldsymbol{k}')\right]_{n''}^{ij}$ and $\left[d_{q'0}^{++}(\boldsymbol{k},\boldsymbol{k}')\right]_{n''}^{ij}$ follow from Tables 7.1 and 7.2, respectively, with the combinations of J-functions in Table 7.3 simplifying to the forms given in Table 7.4.

The combinations of J-functions in Table 7.4 have relatively simple forms for small values of n', specifically $J_\nu^n(x)$ for $n = 0, 1, 2, 3$ are given by (A.1.47)–(A.1.50). For example, the exact form for $J_\nu^0(x)$ is

$$J_\nu^0(x) = \frac{e^{-x/2}x^{\nu/2}}{(\nu!)^{1/2}}, \quad (7.2.4)$$

Table 7.4 The entries in Table 7.3 reduce to the entries shown for $n = 0$, with $b'_\pm = [(\varepsilon^0_{n'} \pm m)/2\varepsilon^0_{n'}]^{1/2}$

$k^\pm_{q'0}$	$\mp s' b'_{-s'} J^{n'}_{n''-n'}(x') J^0_{n''}(x)$
$l^\pm_{q'0}$	$-b'_{s'} J^{n'-1}_{n''-n'}(x') J^0_{n''}(x)$
$p^\pm_{q'0}$	$\mp s' b'_{-s'} e^{-i(\psi'-\psi)} J^{n'}_{n''-n'-1}(x') J^0_{n''-1}(x)$
$q^\pm_{q'0}$	$-b'_{s'} e^{i(\psi'+\psi)} J^{n'-1}_{n''-n'+1}(x') J^0_{n''-1}(x)$
$r^\pm_{q'0}$	$\mp s' b'_{-s'} e^{-i\psi'} J^{n'}_{n''-n'-1}(x') J^0_{n''}(x)$
$t^\pm_{q'0}$	$\mp s' b'_{-s'} e^{i\psi} J^{n'}_{n''-n'}(x') J^0_{n''-1}(x)$
$u^\pm_{q'0}$	$-b'_{s'} e^{i\psi} J^{n'-1}_{n''-n'}(x') J^0_{n''-1}(x)$
$v^\pm_{q'0}$	$-b'_{s'} e^{i\psi'} J^{n'-1}_{n''-n'+1}(x') J^0_{n''}(x)$

Inserting (7.2.4) and the other relations into the entries in Table 7.4 leads to explicit forms for small values of n'.

7.2.2 *Transitions* $n = 0 \to n' \geq 0$

The possible final states, n', of the electron with $n = 0$ depend on the frequency, ω', of the unscattered photon. An electron can be excited to the state n' only if the photon energy exceeds the threshold for gyromagnetic absorption $0 \to n'$.

Transition Frequency

The frequency for gyromagnetic absorption $n = 0 \to n'$ is equal to that for the inverse transition of gyromagnetic emission $n' \to n = 0$, with the proviso that the electron have $p_z = 0$ in the state $n = 0$. The relevant frequency is determined by the zero of the denominator in (7.2.2), with n'' replaced by n'. This gives

$$\omega_{n'}(\theta') = \frac{m}{\sin^2\theta'}\left[\left(1 + \frac{B}{B_c} 2n' \sin^2\theta'\right)^{1/2} - 1\right]. \tag{7.2.5}$$

For $B/B_c \ll 1$, (7.2.5) reduces to $\omega_{n'}(\theta') \approx n'\Omega_e$, reproducing the familiar frequency of n'th harmonic gyromagnetic emission for $B/B_c \ll 1$. In a general frame, in which the electron with $n = 0$ has arbitrary p_z, (7.2.5) is replaced by

$$\omega_{n'}(\theta') = \frac{[(\varepsilon_0 - p_z\cos\theta')^2 + 2n'eB\sin^2\theta']^{1/2} - (\varepsilon_0 - p_z\cos\theta')}{\sin^2\theta'}, \tag{7.2.6}$$

with $\varepsilon_0 = (m^2 + p_z^2)^{1/2}$. The frequency (7.2.6) reduces to (7.2.5) for $p_z = 0$.

Scattering for $n = 0 \to n' = 0$

If the photon energy is below the threshold required for a transition $n = 0$ to $n' = 1$, then one has $n' = 0$. This applies for photons with $\omega' < \omega_1(\theta')$, with $\omega_1(\theta')$ given by (7.2.5) with $n' = 1$. This leads to a further simplification of the entries in Table 7.4, with $s' = -1$ and $l^{\pm}_{q'q} = q^{\pm}_{q'q} = u^{\pm}_{q'q} = v^{\pm}_{q'q} = 0$. A convenient notation for the resulting expressions is in term of the 3-vectors

$$\boldsymbol{e}_{\pm} = (1, \pm i, 0), \qquad \boldsymbol{b} = (0, 0, 1). \tag{7.2.7}$$

Using (7.2.4) and Tables 7.1, 7.2 and 7.4, one finds, for $n'' = 0$,

$$\begin{aligned}
\big[c^{++}_{0'0}(\boldsymbol{k}, \boldsymbol{k}')\big]^{ij}_{0} &= \frac{e^{-(x'+x)/2}}{[\varepsilon'_0(\varepsilon'_0 + m)]^{1/2}} \big[\omega'(\varepsilon'_0 + m) + k'_z(k'_z - k_z)\big]\, b^i b^j, \\
\big[d^{++}_{0'0}(\boldsymbol{k}, \boldsymbol{k}')\big]^{ij}_{0} &= \frac{e^{-(x'+x)/2}}{[\varepsilon'_0(\varepsilon'_0 + m)]^{1/2}} \big[-\omega(\varepsilon'_0 + m) - k_z(k'_z - k_z)\big]\, b^i b^j,
\end{aligned} \tag{7.2.8}$$

with $\varepsilon'_0 = [m^2 + (k'_z - k_z)^2]^{1/2}$ and, for $n'' \geq 1$,

$$\begin{aligned}
\big[c^{++}_{0'0}(\boldsymbol{k}, \boldsymbol{k}')\big]^{ij}_{n''} &= \frac{1}{[\varepsilon'_0(\varepsilon'_0 + m)]^{1/2}} \frac{e^{-(x'+x)/2}(x'x)^{n''-1}}{(n''-1)!} \\
&\quad \Big\{\big[\omega'(\varepsilon'_0 + m) + k'_z(k'_z - k_z)\big] e^{-i(\psi-\psi')}\, e^i_+ e^j_- \\
&\quad + \big[\omega'(\varepsilon'_0 + m) + k'_z(k'_z - k_z)\big]\, (x'x/n'')^{1/2}\, b^i b^j \\
&\quad + (k'_z - k_z)\big[k_\perp e^{i\psi'}\, e^i_+ b^j + k'_\perp e^{-i\psi}\, b^i e^j_-\big]\Big\},
\end{aligned} \tag{7.2.9}$$

$$\begin{aligned}
\big[d^{++}_{0'0}(\boldsymbol{k}, \boldsymbol{k}')\big]^{ij}_{n''} &= \frac{1}{[\varepsilon'_0(\varepsilon'_0 + m)]^{1/2}} \frac{e^{-(x'+x)/2}(x'x)^{n''-1}}{(n''-1)!} \\
&\quad \Big\{-\big[\omega(\varepsilon'_0 + m) - k_z(k'_z - k_z)\big] e^{i(\psi-\psi')}\, e^i_- e^j_+ \\
&\quad - \big[\omega(\varepsilon'_0 + m) + k_z(k'_z - k_z)\big]\, (x'x/n'')^{1/2}\, b^i b^j \\
&\quad - (k'_z - k_z)\big[k'_\perp e^{i\psi}\, e^i_- b^j + k_\perp e^{-i\psi'}\, b^i e^j_+\big]\Big\}.
\end{aligned} \tag{7.2.10}$$

The probability (7.2.1) for scattering $n = 0$ to $n' = 0$ in the frame $p_z = 0$ reduces to

$$\begin{aligned}
w_{0'0}(\boldsymbol{k}, \boldsymbol{k}') &= \frac{e^4}{4\varepsilon_0^2|\omega\omega'|} 2\pi\delta(\varepsilon'_{0'} - m - \omega' + \omega) \bigg| e^*_i e'_j \sum_{n''=0}^{\infty} \\
&\quad \bigg(\frac{\big[c^{++}_{0'0}(\boldsymbol{k}, \boldsymbol{k}')\big]^{ij}_{n''} e^{-i[n''(\psi'-\psi)+(xx')^{1/2}\sin(\psi-\psi')]}}{-2m\omega + \omega^2 - k_z^2 - 2n''eB} \\
&\quad + \frac{\big[d^{++}_{0'0}(\boldsymbol{k}', \boldsymbol{k})\big]^{ij}_{n''} e^{i[n''(\psi'-\psi)+(xx')^{1/2}\sin(\psi-\psi')]}}{2m\omega' + \omega'^2 - k'^2_z - 2n''eB}\bigg)\bigg|^2,
\end{aligned} \tag{7.2.11}$$

with $\varepsilon'_{0'} = [m^2 + (k'_z - k_z)^2]^{1/2}$. Further simplification of the probability (7.2.11) occurs for $x, x' \ll 1$, which corresponds to the cyclotron-like limit discussed in §7.3.

The resonance condition restricts the possible final frequencies, ω, for given ω'. For scattering by an electron with $p_z = 0$, the recoil implies $p'_z \neq 0$, and hence the energy of the final electron is greater than that of the initial electron. Energy conservation then requires $\omega < \omega'$. Using either the resonance condition in the δ-function in (7.2.11), or setting $p^{\pm}_{nn'} = 0$ and $n = n' = 0$ in (7.1.6), one finds that for $(k_z - k'_z)^2 \ll m^2$, the final frequency may be approximated by, in ordinary units,

$$\omega = \omega' \left[1 - \frac{\hbar\omega'}{mc^2} (\cos\theta - \cos\theta')^2 \right], \tag{7.2.12}$$

where the refractive index is assumed to be unity.

Scattering $n = 0 \to n' \geq 1$

When the frequency of the initial photon satisfies $\omega' > \omega_1(\theta')$, with $\omega_{n'}(\theta')$ given by (7.2.5), scattering can cause an electron in the ground state, $n = 0$, to jump to the first Landau state, $n' = 1$. This results in a scattered photon with frequency $\omega \approx \omega' - \omega_1(\theta')$. An independent subsequent process is possible: the electron in the first Landau state can radiate a cyclotron photon, jumping back to the ground state. This combination of scattering and emission may be regarded as a resonant form of photon splitting, in which the initial photon effectively splits into the final photon plus the emitted cyclotron photon. Similarly, for $\omega_{n_0+1}(\theta') > \omega' > \omega_{n_0}(\theta')$, the final electron can be in any of the states $n' = 0, 1, \ldots, n_0$. The electron can then jump back to the ground state through gyromagnetic emission involving any sequence of allowed transitions.

The probability of the scattering process is given by setting $n' = 1$ in (7.2.1), which does not simplify in any obvious way. A simplification that occurs for $n' = 0$ is that half the entries in Table 7.4 are identically zero, specifically $l^{\pm}_{q'q} = q^{\pm}_{q'q} = u^{\pm}_{q'q} = v^{\pm}_{q'q} = 0$, but for $n' \geq 1$ all eight entries in Table 7.4 are nonzero. As a consequence, the simplified forms (7.2.8)–(7.2.10) that involve only four 3-tensors formed from the vectors $\boldsymbol{e}_{\pm}, \boldsymbol{b}$ do not apply for $n' > 0$. Moreover, the two possible spin states, $s' = \pm 1$, require that one distinguish between transitions without spin flip, $s' = -1$, and transitions with spin flip, $s' = 1$. Significant simplification occurs if one has $x, x' \ll 1$, when a cyclotron-like approximation applies, as discussed in §7.3.

7.2.3 *Resonant Compton Scattering*

There are resonances in the probability (7.2.11) at the frequencies $\omega' = \omega_{n''}(\theta')$, where the denominator in the final term inside the modulus has zeros. The presence of a resonance implies enhanced scattering cross section. Near a resonance one may approximate the probability (7.2.11) by retaining only the term that diverges at the resonance. Such resonances in Compton scattering are associated with virtual transitions. A transition $n \to n'$ involves an intermediate state n'', such that the net transition may be regarded as a sum (over n'') of virtual transitions $n \to n'' \to n'$. A resonance occurs when the transition $n \to n''$ is allowed, that is, the virtual intermediate state becomes a real intermediate state. The net transition, $n \to n'$, becomes two sequential allowed transitions, gyromagnetic absorption $n \to n''$ and gyromagnetic emission $n'' \to n'$.

Consider the simplest case of resonant scattering: $n = 0, n' = 0, n'' = 1$. Formally, the probability (7.2.11) implies that the cross section becomes infinite at the resonance, but in practice the cross section remains finite. In particular, the enhancement factor is limited by the natural linewidth for gyromagnetic emission from the state $n'' = 1$ to $n' = 0$. In the immediate vicinity of the resonance at $n'' = 1$ one needs to take account of the virtual intermediate state becoming a real intermediate state. The conditions under which one should regard resonant scattering as an enhanced form of scattering and when one should regard it as a sequence of two independent processes, gyromagnetic absorption $n = 0 \to n'' = 1$ and gyromagnetic emission $n'' = 1 \to n' = 0$, depends on the relative magnitude of two frequencies. One of these frequencies is the mismatch, $\Delta\omega' = \omega' - \omega'_{n''}$, between the frequency of the initial photon and the resonant frequency, $\omega'_{n''}(\theta')$, determined by (7.2.5). The other is the natural linewidth of the emitted radiation, which is determined by the inverse of the lifetime of the excited state. The process should be regarded as scattering if the frequency mismatch exceeds the linewidth and as absorption and re-emission if the linewidth exceeds the frequency mismatch.

In the present case, the natural linewidth of the n''th Landau state is determined by the gyromagnetic decay rate, $\Gamma_{q''}$, for an electron in the state q'' (with q'' including n''). The limitation on the enhancement due to the n''th resonance may be determined by making the replacement $\varepsilon_{q''} \to \varepsilon_{q''} + i\Gamma_{q''}$ in (7.1.14), implying that the minimum value of the resonant denominator is $\approx \Gamma_{q''}$. For frequencies within $\Gamma_{q''}$ of the resonance, the intermediate state is real, and the resonant scattering must be treated as a sequence of absorptions and re-emissions. The occupation number of electrons in the excited states is then nonzero, and is determined by the ratio of the probability per unit time of an absorption event to that state divided by the decay rate from that state.

Resonant Scattering Versus Absorption Plus Emission

When the condition $\omega' = \omega_{n''}(\theta')$ is satisfied, the electron can absorb the photon, and jump from the state $n = 0$ to n''. Subsequent emission of a photon, with the electron jumping either back to $n' = 0$ or to some intermediate state $0 < n' < n''$, is equivalent to resonant scattering. It is useful to define an effective scattering probability that includes both limits.

The effective scattering probability is constructed by combining the probabilities per unit time for the two transitions, and dividing by the rate of decay of the intermediate state. This compound probability is

$$\bar{w}_{q'qMM'}(\boldsymbol{k},\boldsymbol{k}') = \sum_{q''} w_{q''qM'}(\boldsymbol{k}')\, w_{q'q''M}(\boldsymbol{k})/\Gamma_{q''}, \tag{7.2.13}$$

where the sum is over all possible resonances, and where $\Gamma_{q''}$ is the decay rate of the intermediate state. In the immediate vicinity of the resonance, specifically within a frequency less than $\Gamma_{q''}$ of the resonant frequency, the effective scattering probability is given by (7.2.13).

An interpolation between (7.2.13) in the core of the line, and (7.1.2) in the wings of the line is obtained by replacing the resonant factor by a Lorentzian line profile with the width determined by the lifetime of the intermediate state. Including the various crossed processes, an interpolation formula is [10, 14]

$$\begin{aligned}\mathrm{Res}\, w^{++}_{q'qMM'}(\boldsymbol{k},\boldsymbol{k}') &= \sum_{q''} \frac{e^4 R_M(\boldsymbol{k}) R_{M'}(\boldsymbol{k}')}{\varepsilon_0^2 |\omega_M(\boldsymbol{k})\omega_{M'}(\boldsymbol{k}')|} \left|e^*_{M\mu}(\boldsymbol{k})[\Gamma^{++'}_{q'q''}(\boldsymbol{k})]^{\mu}\right|^2 \\ &\times \left|e_{M'\nu}(\boldsymbol{k}')[\Gamma^{++}_{q''q}(\boldsymbol{k}')]^{*\nu}\right|^2 \frac{\pi\Gamma_{q''}/2}{[\varepsilon_q - \varepsilon_{q''} - \omega_M(\boldsymbol{k})]^2 + (\Gamma_{q''}/2)^2} \\ &\times 2\pi\delta[\varepsilon_q - \varepsilon_{q'} - \omega_M(\boldsymbol{k}) + \omega_{M'}(\boldsymbol{k}')].\end{aligned} \tag{7.2.14}$$

The form (7.2.14) reduces to (7.2.13) in the core of the line, $|\varepsilon_q - \varepsilon_{q''} - \omega_M(\boldsymbol{k})| \lesssim \Gamma_{q''}/2$, and it provides an approximation to resonant scattering in the wings of the line, $|\varepsilon_q - \varepsilon_{q''} - \omega_M(\boldsymbol{k})| \gtrsim \Gamma_{q''}/2$. In the far wings of the line, where resonant scattering is no longer dominant, the exact probability (7.1.2) needs to be used. An implication of (7.2.14) is that the maximum enhancement in resonant scattering is limited by the natural linewidth of the excited state. This quantum limitation tends to dominate over the nonquantum limitations, due to cold plasma and thermal effects at low densities in very strong magnetic fields.

7.3 Scattering in the Cyclotron Approximation

Compton scattering by a magnetized electron simplifies in the cyclotron-like approximation, defined as the limit in which the power series expansion of the J-functions converges rapidly. The cyclotron-like limit requires that the scattering electron be nonrelativistic.

7.3.1 Cyclotron-Like Approximation

Further simplification of the probability (7.2.11) occurs for $x', x \ll 1$, when only $J_0^0(x) \approx 1$ need be retained in (7.2.9) and (7.2.10). An associated approximation is that the kinetic energy of the final electron is negligible, $\varepsilon_0' \to m$, except where the difference $\varepsilon_0' - m$ appears explicitly. In this case (7.2.11) reduces to

$$w_{0'0}(\boldsymbol{k}, \boldsymbol{k}') = \frac{e^4}{4\varepsilon_0^2 m^2 |\omega\omega'|} \left| e_i^* e_j' \, g^{ij} \right|^2 2\pi\delta(\varepsilon_0' - m + \omega' - \omega), \tag{7.3.1}$$

with

$$\begin{aligned} g^{ij} &= \frac{[\omega'(\varepsilon_0' + m) + k_z'(k_z' - k_z)]e^{i(\psi'-\psi)} \, e_-^i e_+^j}{2m(\omega + \Omega_e) - \omega^2 + k_z^2} \\ &\quad + \frac{[\omega(\varepsilon_0' + m) - k_z(k_z' - k_z)]e^{-i(\psi'-\psi)} \, e_+^i e_-^j}{2m(\omega' - \Omega_e) + \omega'^2 - k_z'^2} \\ &\quad + \left[\frac{\omega(\varepsilon_0' + m) + k_z(k_z' - k_z)}{2m\omega' + \omega'^2 - k_z'^2} + \frac{\omega'(\varepsilon_0' + m) - k_z'(k_z' - k_z)}{2m\omega - \omega^2 + k_z^2} \right] b^i b^j. \end{aligned} \tag{7.3.2}$$

Only $n'' = 0, 1$ contribute in the sum over intermediate states, with $n'' = 1$ contributing to the term along $e_\pm^i e_\mp^j$, and $n'' = 0$ contributing to the term along $b^i b^j$.

The argument of the δ-function In (7.3.1) may be rationalized by writing

$$\begin{aligned} &\delta(\varepsilon_0' - m + \omega' - \omega) \\ &\quad = \frac{\varepsilon_0' + m + \omega' - \omega}{2m} \, \delta\left(\omega' - \omega - \frac{(\omega' - \omega)^2 - (k_z' - k_z)^2}{2m} \right), \end{aligned} \tag{7.3.3}$$

which includes the quantum recoil explicitly. When the quantum recoil is neglected, the right hand side of (7.3.3) reduces to $\delta(\omega' - \omega)$, and g^{ij} reduces to

$$g^{ij} = \frac{\omega}{\omega - \Omega_e} e^{-i(\psi'-\psi)} e_-^i e_+^j + \frac{\omega}{\omega + \Omega_e} e^{i(\psi'-\psi)} e_+^i e_-^j + 2b^i b^j. \tag{7.3.4}$$

The probability (7.3.1) then reduces to its nonquantum counterpart (4.5.15).

7.3.2 Scattering in the Birefringent Vacuum

In strong fields, where the electrons are predominantly in the state $n = 0$, the contribution of the plasma to the wave dispersion can usually be neglected in comparison with the contribution from the vacuum polarization. The relevant wave modes are usually those of the birefringent vacuum.

The scattering probabilities for the two modes of the birefringent vacuum are found by inserting the polarization vectors for these modes into (7.3.1). These are

$$\boldsymbol{e}_\perp = \boldsymbol{a} = (-\sin\psi, \cos\psi, 0), \qquad \boldsymbol{e}_\parallel = \boldsymbol{t} = (\cos\theta\cos\psi, \cos\theta\sin\psi, -\sin\theta), \tag{7.3.5}$$

with analogous expressions for the primed vectors. The projection of g^{ij} onto the specific polarizations (7.3.5) gives

$$\begin{pmatrix} g_{\perp\perp} \\ g_{\parallel\parallel} \\ g_{\perp\parallel} \\ g_{\parallel\perp} \end{pmatrix} = \begin{pmatrix} 0 \\ 2\sin\theta'\sin\theta \\ 0 \\ 0 \end{pmatrix} + \sum_{\pm} \frac{\omega}{\omega \pm \Omega_e} \begin{pmatrix} 1 \\ \cos\theta'\cos\theta \\ \pm i\cos\theta \\ \mp i\cos\theta' \end{pmatrix}. \tag{7.3.6}$$

On inserting (7.3.6) into (7.3.1) one may compare the rates of scattering for the different modes, $\perp\!\to\!\perp, \parallel$, $\parallel\!\to\!\perp, \parallel$. When the quantum recoil term is neglected in the resonance condition, implying $\omega = \omega'$, the cross sections for these scatterings are simply related to the Thomson cross section, $\sigma_T = (8\pi/3)(\mu_0 e^2/4\pi m)^2$. These cross-sections are [3]

$$\begin{pmatrix} \sigma_{\perp\perp}(\theta',\theta) \\ \sigma_{\parallel\parallel}(\theta',\theta) \\ \sigma_{\perp\parallel}(\theta',\theta) \\ \sigma_{\parallel\perp}(\theta',\theta) \end{pmatrix} = \frac{9\sigma_T}{7} \left\{ \begin{pmatrix} 0 \\ \sin^2\theta'\sin^2\theta \\ 0 \\ 0 \end{pmatrix} + \frac{1}{(1-Y^2)^2} \begin{pmatrix} 1 \\ \cos^2\theta'\cos^2\theta \\ Y^2\cos^2\theta \\ Y^2\cos\theta' \end{pmatrix} \right\}, \tag{7.3.7}$$

with $Y = \omega/\Omega_e$. The total cross section implied by (7.3.7) reduces to the Thomson cross section after averaging over the initial and summing over the final states polarization, averaging over $\cos\theta'$ and $\cos\theta$, and setting $Y \to 0$.

A notable feature of (7.3.7) is the singularity that occurs in each cross section for $Y = 1$. This corresponds to resonant Compton scattering, which is of specific interest in connection with secondary pair production in the polar cap region of pulsars [6, 7, 13, 20]. The scattering of thermal photons from the surface of the neutron star by a highly relativistic "primary" particle in the state $n = 0$ is greatly enhanced when the frequency of the thermal photon, transformed to the rest frame of the primary particle, is equal to the cyclotron frequency. The scattered photon is then of high energy in the pulsar frame.

7.3.3 Inverse Compton Scattering

Thomson scattering of photons by highly relativistic electrons is referred to as inverse Compton emission in the astrophysical literature. The quantum theory of scattering by highly relativistic electrons in a strong magnetic field has been discussed in detail only in one special case ($n = 0$, $x' = 0$) [8]. In this section, the scattering for $x' = 0$ is discussed, and some remarks are made about the generalization to $x' \neq 0$. A notable new feature for $x' \neq 0$ is that a resonance occurs in the cross section associated with the threshold for pair-creation.

7.3.4 Special Case $n = 0$, $x' = 0$

The simplifying assumption made in §7.3 involves the initial electron, which is assumed to be in the ground state $n = 0$. Another simplifying assumption is that the initial photon is propagating along the magnetic field. The argument, x', is then zero, and then $J^{n'}_{n''-n'}(x')$ is zero for $n'' \neq n'$ and unity for $n'' = n'$. Then the only J-functions that appear, $J^n_\nu(x)$, are the same as for gyromagnetic emission.

It has been argued [8] that the special case of parallel propagation is relevant to Compton scattering by a highly relativistic electron, $\gamma \gg 1$. The argument is that if one considers the scattering in the frame in which the electron is at rest, $p_z = 0$, the Lorentz transformation to this frame transforms the angle, θ', of propagation of nearly all photons into a small cone of angles, $\theta' \lesssim 1/\gamma$. The suggestion is that one may treat inverse Compton scattering in terms of the scattering of a parallel-propagating photon, $\theta' = 0$, by an electron with $p_z = 0$ [8], and that this is a valid approximation for $\gamma \gg 1$.

Consider the special case $n = 0$ and $x' = 0$. The cross section for inverse Compton scattering is shown in Fig. 7.2, where the final Landau quantum number is denoted $n' \to \ell$ and the cyclotron frequency is denoted $\Omega_e \to \omega_B$. The most obvious feature of the cross section is the singularity at the cyclotron frequency, which corresponds to resonant Compton scattering. The dependence of the cross section on the polarization of the initial and final photons is shown in Fig. 7.3, where the two linear polarizations are those of the natural modes of the birefringent vacuum.

However, the argument that inverse Compton scattering may be approximated by the case $x' = 0$, in the frame $p_z = 0$, ignores the fact that x' is an invariant. The assumption $x' \approx 0$ must apply in all frames. In particular, for $\theta' \neq 0$, one can transform to the frame in which $\sin\theta'$ is equal to unity, and in this frame $x' \ll 1$ requires $\omega'^2 \ll 2eB \ll 2m^2(B/B_c)$, or $\omega'/2m \ll (2B/B_c)^{1/2}$. The assumption $x' = k_\perp'^2/2eB \ll 1$ is overly restrictive in general.

PC-Induced Resonant Compton Scattering

A new feature of inverse Compton scattering that is excluded by the assumption $x' = 0$ is a pair-creation (PC) associated resonance (Weise JI, 2011, private communication). This occurs when the resonance condition passes through a threshold for PC. There is a close analogy between the threshold singularities in the gyromagnetic absorption (GA) and PC absorption coefficients, both of which occur

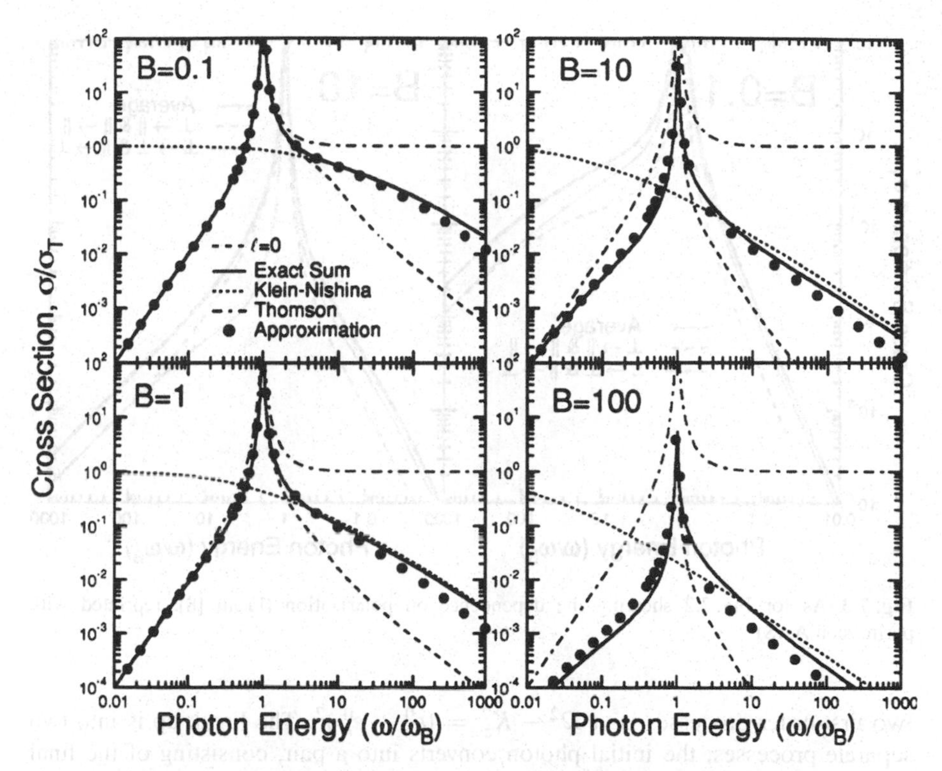

Fig. 7.2 The ratio of the Compton cross section to the Thomson cross section is plotted as a function of the incident photon energy (in units of the cyclotron energy) for $B = 0.1, 1, 10, 100$ in units of B_c. *Solid curve*: exact cross section; *dot-dashed curve*: nonrelativistic approximation; *dotted curve*: Klein-Nishina limit; *dashed curve*: case with only $\ell = 0$ (see text) (From [8], reprinted with permission AAS)

at $g_{nn'} = 0$, due to a factor $(g_{nn'})^{1/2}$ in the denominator. Similarly, the scattering cross section includes a factor $(G_{nn'})^{1/2}$ in the denominator, leading to a singularity at $G_{nn'} = 0$, with $G_{nn'}$ given by (7.1.5). The zero at $\Omega^2 - K_z^2 = (\varepsilon_n^0 - \varepsilon_{n'}^0)^2$ leads to resonant Compton scattering, which occurs at the cyclotron frequency in the case $n = 0$, $x' = 0$ in the frame $p_z = 0$. The new effect, PC-induced resonant Compton scattering, occurs at the threshold $\Omega^2 - K_z^2 = (\varepsilon_n^0 + \varepsilon_{n'}^0)^2$.

The existence of PC-induced resonant Compton scattering may be understood by analogy with resonant Compton scattering (RCS). One may interpret RCS in terms of the scattering process breaking up into two sequential real processes: gyromagnetic absorption of the initial photon by the initial electron into a real intermediate state consisting of an electron, and gyromagnetic emission by this intermediate electron to produce the final photon. The resonance occurs as the threshold for this break up into two first order processes is approached. In PC-induced resonant Compton scattering, the intermediate state is a virtual positron that becomes a real positron at the resonance. The threshold for this break-up into

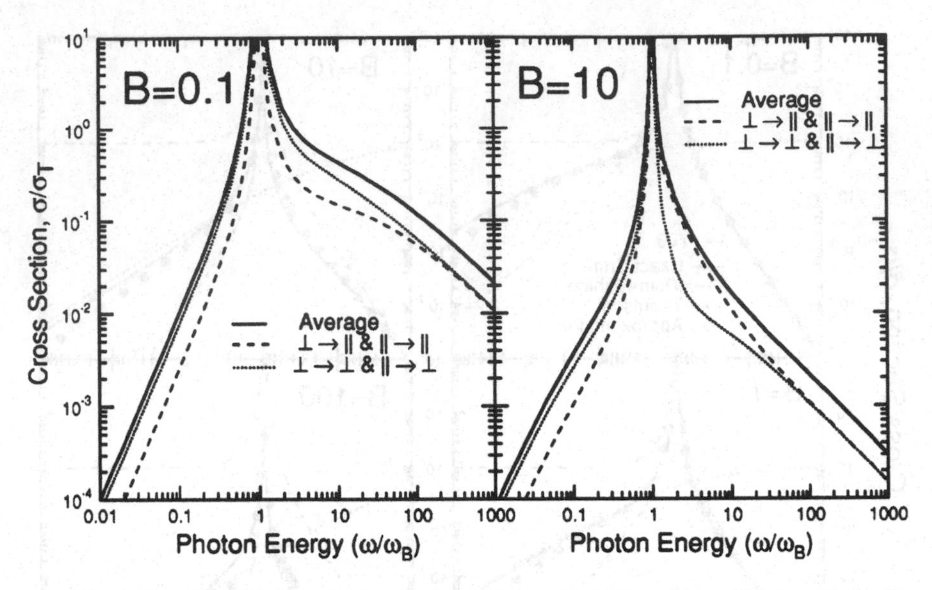

Fig. 7.3 As for Fig. 7.2 showing the dependence on polarization (From [8], reprinted with permission AAS)

two first-order processes is at $\Omega^2 - K_z^2 = (\varepsilon_n^0 + \varepsilon_{n'}^0)^2$. The break up is into two separate processes: the initial photon converts into a pair, consisting of the final electron and the intermediate positron; and the initial electron annihilates with the intermediate positron to produce the final photon. PC-induced resonant Compton scattering occurs when the beat, at $K^\mu = k^\mu - k'^\mu$, between the initial and final photons, satisfies a threshold condition for PC.

The PC-induced resonances do not occur for $x = 0$ or $x' = 0$. As with PC itself, near resonance one can choose any of three frames in which one of the beat disturbance, the initial photon or the final photon is propagating across the field lines, corresponding to $K_z = 0$, $k_z' = 0$ or $k_z = 0$, respectively. The perpendicular wavenumbers, $K_\perp, k'_\perp, k_\perp$, are invariant under the Lorentz transformations to any of these frames. The condition (6.4.6) implies that PC-induced resonance, at given n, n', requires $x' > (\varepsilon_n^0 + \varepsilon_{n'}^0)^2/2eB > 1$. This requirement implies that PC-induced resonance is likely to be of most interest in the limit of supercritical fields, $B \gg B_c$.

7.4 Two-Photon Processes

Processes in which there are two photons in the initial or final state and no photons in the other state include double emission and double absorption, two-photon pair creation and pair annihilation into two photons. Two-photon gyromagnetic emission is allowed under the same conditions as one-photon gyromagnetic emission, and because it is a higher order process it occurs at a rate of order α_c slower than the one-photon process. Similarly, two-photon pair creation and annihilation compete with one-photon pair creation and annihilation, respectively. These two-photon processes are treated here as crossed forms of Compton scattering, given by the probability (7.1.2). Investigations of these processes have been motivated by astrophysical applications [1, 2, 15, 16].

7.4.1 Kinetic Equations for Two-Photon Processes

Kinetic equations can be written down by considering the rate of transitions between different states, analogous to the derivation of the kinetic equations for gyromagnetic processes in § 6.1.3.

Kinetic Equations for Double Emission

Two-photon (or "double") emission is related to Compton scattering by making the crossing transformation $\boldsymbol{k}' \to -\boldsymbol{k}'$, with $\omega' \to -\omega'$, $\boldsymbol{e}' \to \boldsymbol{e}'^*$, in the probability (7.1.2). This converts the initial photon into a final photon, so that there are two final photons and no initial photon.

The kinetic equations for two-photon emission and absorption, between states q', q involving photons in the modes M, M', are

$$\frac{\mathrm{D}N_{M'}(\boldsymbol{k}')}{\mathrm{D}t} = \sum_{q'q} \frac{eB}{2\pi} \int \frac{dp_z}{2\pi} \int \frac{d^3\boldsymbol{k}}{(2\pi)^3} w^{++}_{q'qMM'}(\boldsymbol{k}, -\boldsymbol{k}')$$
$$\{n_q[1 + N_M(\boldsymbol{k})][1 + N_{M'}(\boldsymbol{k}')] - n_{q'} N_M(\boldsymbol{k}) N_{M'}(\boldsymbol{k}')\},$$
$$\frac{\mathrm{D}N_M(\boldsymbol{k})}{\mathrm{D}t} = \sum_{q'q} \frac{eB}{2\pi} \int \frac{dp_z}{2\pi} \int \frac{d^3\boldsymbol{k}'}{(2\pi)^3} w^{++}_{q'qMM'}(\boldsymbol{k}, -\boldsymbol{k}')$$
$$\{n_q[1 + N_M(\boldsymbol{k})][1 + N_{M'}(\boldsymbol{k}')] - n_{q'} N_M(\boldsymbol{k}) N_{M'}(\boldsymbol{k}')\},$$
$$\frac{dn_q}{dt} = \int \frac{d^3\boldsymbol{k}}{(2\pi)^3} \int \frac{d^3\boldsymbol{k}'}{(2\pi)^3} \Big[\sum_{q''} w^{++}_{qq''MM'}(\boldsymbol{k}, -\boldsymbol{k}')$$
$$\times\{n_{q''}[1 + N_M(\boldsymbol{k})][1 + N_{M'}(\boldsymbol{k}')] - n_q N_M(\boldsymbol{k}) N_{M'}(\boldsymbol{k}')\}$$
$$-\sum_{q'} w^{++}_{q'qMM'}(\boldsymbol{k}, -\boldsymbol{k}')\big)\{n_q[1 + N_M(\boldsymbol{k})][1 + N_{M'}(\boldsymbol{k}')]$$
$$-n_{q'} N_M(\boldsymbol{k}) N_{M'}(\boldsymbol{k}')\}\Big], \tag{7.4.1}$$

where q' denotes the quantum numbers $n', s', p'_z = p_z - k'_z - k_z$ and q'' denotes $n'', s'', p''_z = p_z + k'_z + k_z$. The term on the right hand side of each of (7.4.1) that is independent of $N_M(k), N_{M'}(k')$ describes the effect of spontaneous double emission. This term is intrinsically quantum mechanical.

Kinetic Equations for Two-Photon Pair Creation

The kinetic equations for the photons due to two-photon pair creation and annihilation are

$$\frac{\mathrm{D}N_M(\boldsymbol{k})}{\mathrm{D}t} = \sum_{qq'} \int \frac{d^3\boldsymbol{k}'}{(2\pi)^3} w^{+-}_{q'qMM'}(\boldsymbol{k}, -\boldsymbol{k}')\{n^+_q n^-_{q'}[1 + N_M(\boldsymbol{k}) + N_{M'}(\boldsymbol{k}')]$$
$$-N_M(\boldsymbol{k})N_{M'}(\boldsymbol{k}')(1 - n^+_q - n^-_{q'})\}, \qquad (7.4.2)$$

plus an analogous equation with primed and unprimed quantities interchanged. The corresponding kinetic equations for the electrons and positrons are

$$\frac{dn^+_q}{dt} = \int \frac{d^3\boldsymbol{k}}{(2\pi)^3} \int \frac{d^3\boldsymbol{k}'}{(2\pi)^3} \sum_{q'} w^{+-}_{q'qMM'}(\boldsymbol{k}, -\boldsymbol{k}')\{N_M(\boldsymbol{k})N_{M'}(\boldsymbol{k}')(1 - n^+_q - n^-_{q'})$$
$$-n^+_q n^-_{q'}[1 + N_M(\boldsymbol{k}) + N_{M'}(\boldsymbol{k}')]\},$$

$$\frac{dn^-_{q'}}{dt} = \int \frac{d^3\boldsymbol{k}}{(2\pi)^3} \int \frac{d^3\boldsymbol{k}'}{(2\pi)^3} \sum_{q} w^{+-}_{q'qMM'}(\boldsymbol{k}, -\boldsymbol{k}')[N_M(\boldsymbol{k})N_{M'}(\boldsymbol{k}')(1 - n^+_q - n^-_{q'})$$
$$-n^+_q n^-_{q'}[1 + N_M(\boldsymbol{k}) + N_{M'}(\boldsymbol{k}')]\}. \qquad (7.4.3)$$

In practice one usually considers pair annihilation in the absence of photons, $N_M(\boldsymbol{k}) = 0$, $N_{M'}(\boldsymbol{k}') = 0$, and pair creation in the absence of other particles, $n^+_q = n^-_{q'} = 0$.

7.4.2 *Double Cyclotron Emission*

Consider an electron in an initial state q corresponding to a Landau state n. The electron can jump to lower Landau states, $n' < n$, through gyromagnetic emission of a single photon. However, it can also make the jump emitting two photons simultaneously, with the sum of the frequencies of the two photons being equal to that of the single photon in one-photon emission. These two processes compete, and the ratio of the rates of transition determines the relative probability of the transition occurring due to the two processes. This ratio is referred to as a branching ratio.

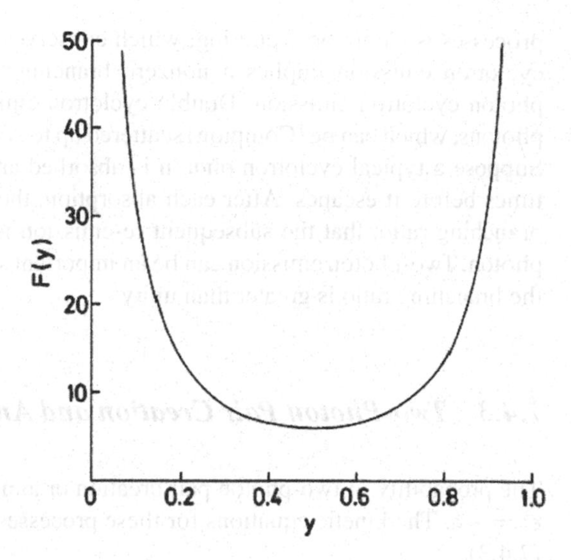

Fig. 7.4 The function $F(y)$ that describes the frequency dependence of double cyclotron emission is plotted as a function of $y = \omega/\Omega_e$ (After [14])

The transition from the first excited state, $n' = 1, s' = -1$, to the ground state, $n = 0$, in the nonrelativistic approximation, $k_z, k'_z \ll m, x, x' \ll 1$, averaged over polarizations and over angles [14] gives (ordinary units)

$$\langle w_{0'1}(\boldsymbol{k}, -\boldsymbol{k}')\rangle = \frac{(2\pi)^3\alpha_c^2 c^6}{45\Omega_e^4}\left(\frac{B}{B_c}\right)^3 \frac{F(y)}{y^2(1-y)^2} \times \delta\left[\omega + \omega' - \Omega_e - (k_z + k'_z)v_z + \frac{\hbar(k_z + k'_z)^2}{2m}\right], \tag{7.4.4}$$

with $y = \omega/\Omega_e$ and

$$F(y) = |y(1-y)|\left[7\{y^2 + (1-y)^2\} + 3\left\{\frac{y^4}{(2-y)^2} + \frac{(1-y)^4}{(1+y)^2}\right\} + 4\left(\frac{y^2}{(1-y)^2} + \frac{(1-y)^2}{y^2}\right) + 2\left(\frac{y^2}{1-y} + \frac{(1-y)^2}{y}\right) + 10\right]. \tag{7.4.5}$$

The function $F(y)$ is plotted in Fig. 7.4 as a function of $y = \omega/\Omega_e$, and (ignoring the quantum recoil) $y' = \omega'/\Omega_e = 1 - y$. The emission has an infra-red divergence at $y \to 0$ or $y \to 1$, corresponding to the one of the two photons having an arbitrarily low frequency. The function $F(y)$ is well approximated by $F(y) \approx 4[1 + y^3/(1-y) + (1-y)^3/y]$.

One application of double cyclotron emission is to X-ray pulsars, where the X-ray emission is from an accretion column containing hot, X-ray emitting gas. The column is optically thick at the cyclotron line, implying that cyclotron photons are emitted and absorbed many times before escaping. A problem in understanding emission in such a context is the ultimate source of photons: the dominant processes is Compton scattering, which conserves the number of photons. Double cyclotron emission implies a nonzero branching ratio for two-photon and one-photon cyclotron emission. Double cyclotron emission provides a source of soft photons, which can be (Compton) scattered up to X-ray frequencies before escaping. Suppose a typical cyclotron photon is absorbed and re-emitted (an average of) N times before it escapes. After each absorption, there is a probability, equal to this branching ratio, that the subsequent re-emission is two-photon, rather than single photon. Two-photon emission can be an important source of new photons if N times the branching ratio is greater than unity.

7.4.3 Two-Photon Pair Creation and Annihilation

The probability of two-photon pair creation or annihilation is given by (7.1.2) with $\epsilon' = -\epsilon$. The kinetic equations for these processes are written down in (7.4.2) and (7.4.3).

Two-Photon Pair Creation

It is helpful to distinguish two cases for two-photon pair creation: when the two photon have comparable energies, and when one photon is hard and the other is soft. In the former case, each photon must have an energy $\omega \gtrsim m$, requiring photons in the MeV range. In the latter case a photon with a higher energy can decay into a pair by interacting with a soft photon. The latter process is of interest in models for active galactic nuclei (AGN) where there is a simple way of estimating its importance. The cross section for the process is of order the Thomson cross section, σ_T. Suppose the temperature of the hot gas in the AGN is known, so that the number density, n_γ, of soft photons is known in a region of thickness L. Two-photon pair creation needs to be taken into account for $\sigma_T n_\gamma L \gtrsim 1$. Observation of hard photons from an AGN where this condition is satisfied implies that pairs are being created in the source.

Inclusion of a magnetic field in two-photon pair creation [11] is important for the application to magnetized neutron stars. The absorption coefficient for the photons involved in two-photon pair creation may be defined in an analogous way to that for one-photon pair creation, cf. (6.4.2). After integrating over the distribution of photons in the mode M', the absorption coefficient for photons in the mode M becomes

$$\gamma_M(\boldsymbol{k}) = \sum_{n,n',s,s'} \frac{eB}{2\pi} \int \frac{dp_z}{2\pi} \int \frac{d^3\boldsymbol{k}'}{(2\pi)^3}\, w^{+-}_{q'qMM'}(\boldsymbol{k},\boldsymbol{k}')\, N_{M'}(\boldsymbol{k}'). \tag{7.4.6}$$

For pair creation by a single distribution of photons, with $\omega \gtrsim m$, the labels M, M' may be interpreted as labeling the polarization of the photons, e.g., $M, M' = \perp, \parallel$

for the two modes of the birefringent vacuum. For $M = M'$, when the two photons are from the same distribution, an additional factor of $1/2$ needs to be included, e.g., in the definition of the probability, to avoid double counting. In the case of a hard photon decaying into a pair by interacting with a soft photon, the hard photons may be identified with the label M, in (7.4.6), and the soft photons with the label M'.

Qualitatively, many of the properties of two-photon pair creation can be understood in terms of the properties of one-photon pair creation and of Compton scattering. For nonrelativistic pairs, there is a sequence of thresholds corresponding to different values of n, n'. As for one-photon pair creation, the integral over p_z in (7.4.6) may be evaluated using the δ-function, to give an expression with a factor $G_{nn'}$ in the denominator, cf. (6.4.3). As for one-photon pair creation, the factor $1/G_{nn'}$ leads to square root singularities in the cross section, specifically at $(\omega + \omega')^2 - (k_z + k_z')^2 = (\varepsilon_n + \varepsilon'_{n'})^2$. A comparison of one-photon and two-photon pair creation in a mildly relativistic thermal plasma led to an estimate that the one-photon process dominates for $B/B_c \gtrsim 0.02$ [4].

Two-Photon Versus One-Photon Pair Annihilation

The annihilation of a pair into two-photons in a strong magnetic field competes with annihilation into one photon. Although the two-photon process is of higher order, the kinematic conditions for it to occur are less restrictive than for the one-photon process. The annihilation rate depends on the Landau levels of the electron and positron. For the simplest case when both are in their ground state, $n = n' = 0$, with energy ε in the center-of-mass frame, the cross section for annihilation into one photon is [19] (ordinary units)

$$\sigma_{1\gamma}(\varepsilon) = n_0^+(\varepsilon) n_0^-(\varepsilon) \frac{2\pi^2 \alpha_c \hbar c}{p\varepsilon B/B_c} \exp[-2(\varepsilon/mc^2)^2 (B/B_c)^{-1}], \tag{7.4.7}$$

where ε and p are the energy and momentum of the electron and positron, and $n_0^{\pm}(\varepsilon)$ are their occupation numbers. The energy of the photon into which they annihilate is 2ε. The exponential facto in (7.4.7) arises from $|J_0^0(x)|^2 = e^{-x}$ with $x = (2\varepsilon/mc^2)^2 (B_c/2B)$. This factor suppresses the one-photon process for small B/B_c, and there is no analogous suppression for annihilation into two-photons [5]. As a consequence of this suppression, it was estimated [19] that the two-photon process dominates for $B/B_c \lesssim 0.24$ and the one-photon process for $B/B_c \gtrsim 0.24$.

7.5 Electron–Ion and Electron–Electron Scattering

Particle-particle scattering is a second order process in QED. In the absence of a magnetic field, Mott scattering provides a simple model for electron-ion scattering in which the ion is replaced by a fixed charge at the origin; the neglect of the recoil of the ion can be justified as an approximation of lowest order in an expansion in the ratio of the masses of the electron and the ion. There is no such simplification for electron-electron scattering, called Møller scattering in QED. The generalization of these two processes to include a magnetized field are discussed in this section.

7.5.1 Collisional Excitation by a Classical Ion

In the discussion below emphasis is placed on collisionless processes. The general theory may also be used to treat the effects of collisions between particles. The simplest example is a collision between an electron and an ion, with the ion treated as an unmagnetized classical particle with charge Q and mass M. Two processes are of interest: non-radiative transitions and bremsstrahlung.

Collisional excitation or de-excitation of an electron or positron through interaction with an external field is described by the Feynman diagram for Mott scattering, cf. Fig. 7.3 of volume 1. The probability per unit time for an electron initially in the state q to be in the state q' after interacting with the ion is

$$w^{Q}_{q'q} = \lim_{T\to\infty} \frac{e^2}{T} \int \frac{d^4k}{(2\pi)^4} \left| A^{(Q)*}_{\mu(k)} [\Gamma^{\epsilon'\epsilon}_{q'q}(\boldsymbol{k})]^{\mu} \right|^2 2\pi\delta(\epsilon\varepsilon_q - \epsilon'\varepsilon_{q'} - \omega), \qquad (7.5.1)$$

where T is a normalization time. The 4-potential $A^{(Q)}(k)$ of the classical ion is

$$A^{(Q)\mu}(k) = e^{ikx_0} Q D^{\mu\nu}(k) u_\nu \, 2\pi\delta(kU), \qquad (7.5.2)$$

where U is the 4-velocity of the ion whose orbit is assumed to be of the from $x^\mu = x_0^\mu + U^\mu\tau$, with x_0^μ, U^μ constant and with τ the proper time. The δ-function in the classical form (7.5.2) appears squared in the probability (7.5.1), where one writes $[2\pi\delta(kU)]^2 = (T/\gamma)2\pi\delta(kU)$, where γ is the Lorentz factor of the particle. The normalization time, T, cancels with the factor $1/T$ in (7.5.1). The purely classical result can be generalized to a semi-classical form that takes the recoil of the ion into account. This generalization involves the replacement

$$\delta(kU) \to \frac{M}{E}\delta(E' - E - \omega), \qquad (7.5.3)$$

in (7.5.1), with $E = (M^2 + P^2)^{1/2}$, $\boldsymbol{P} = \Gamma M \boldsymbol{V}$, $\Gamma = 1/(1 - V^2)^{1/2}$, $E' = (M^2 + P'^2)^{1/2}$, $\boldsymbol{P}' = \boldsymbol{P} - \boldsymbol{k}$. One is free to choose the frame in which the ion is at rest, and then in the Coulomb gauge the 4-potential has only a time-component. This is

$$A^{(Q)0}(k) = e^{ikx_0} \frac{Q 2\pi\delta(\omega)}{4\pi\varepsilon_0|\boldsymbol{k}|^2 K^L(\omega, \boldsymbol{k})}, \qquad (7.5.4)$$

which is the electrostatic potential of the ion, with $K^L(\omega, \boldsymbol{k})$ the longitudinal part of the dielectric tensor. In this case the (7.5.1) gives

$$w^{Q}_{q'q} = \frac{e^2 Q^2}{(4\pi\varepsilon_0)^2} \int \frac{d^3\boldsymbol{k}}{(2\pi)^3} \frac{\left|[\Gamma^{\epsilon'\epsilon}_{q'q}(\boldsymbol{k})]^0\right|^2}{\{|\boldsymbol{k}|^2 K^L(0,\boldsymbol{k})\}^2} 2\pi\delta(\epsilon\varepsilon_q - \epsilon'\varepsilon_{q'}). \tag{7.5.5}$$

The δ-function in (7.5.5) can be satisfied only for $\epsilon' = \epsilon$, and then (7.5.5) describes collisional excitation or de-excitation of an electron ($\epsilon = 1$) or a positron ($\epsilon = -1$) as a result of the interaction with the ion.

The foregoing discussion of scattering of an electron by an ion involves a major simplifying assumption: the ion is treated as a classical particle, so that only its Coulomb field need be considered. A more rigorous treatment requires that the particle be treated quantum mechanically.

7.5.2 Electron–Electron (Møller) Scattering

The Feynman diagrams for electron-electron scattering are the same as in the absence of a magnetic field, cf. Fig. 7.10 of volume 1. It is convenient to denote the initial states of the two electrons by 1, 2 and the final states by 3, 4. Specifically, the label 1 denotes a set of quantum numbers $q_1 = (p_{1z}, p_{1y}, n_1, s_1)$, and similarly for 2, 3, 4. It is then convenient to describe the scattering as $1, 2 \to 3, 4$. The two different diagrams, for given initial states 1, 2, are related by the final states of the two electrons being interchanged, $3 \leftrightarrow 4$. The scattering amplitude for this process involves four vertex functions: for one diagram the vertex functions $[\Gamma^{++}_{q_1q_3}(\boldsymbol{k})]^\mu$, $[\Gamma^{++}_{q_2q_4}(-\boldsymbol{k})]^\nu$ appear, where k is the 4-momentum transfer. It is convenient to simplify the notation by writing these as $[\Gamma_{13}(\boldsymbol{k})]^\mu$, $[\Gamma_{24}(-\boldsymbol{k})]^\nu$, respectively. For the other diagram the vertex functions that appear are $[\Gamma_{14}(\boldsymbol{k}')]^\mu$, $[\Gamma_{23}(-\boldsymbol{k}')]^\nu$, where k' is the 4-momentum transferred. This simplified notation is also applied to the energies, $\varepsilon_{q_1}(p_{1z}) \to \varepsilon_1$, and so on.

One is usually not interested in the location of the centers of gyration of the electrons, and the average over the initial positions and the sum over the final positions are performed. The average transition rate then becomes

$$\begin{aligned}
\bar{w}_{i\to f} &= e^4 \frac{dp_{3z}}{2\pi}\frac{dp_{4z}}{2\pi} 2\pi\delta(p_{1z}+p_{2z}-p_{3z}-p_{4z})\, 2\pi\delta(\varepsilon_1+\varepsilon_2-\varepsilon_3-\varepsilon_4) \\
&\quad \times \Bigg[\int \frac{d^4k}{(2\pi)^4} 2\pi\delta(p_{1z}-p_{3z}+k_z)\, 2\pi\delta(\varepsilon_1-\varepsilon_3+\omega) \\
&\quad \times \left|[\Gamma_{31}(-\boldsymbol{k})]^\mu [\Gamma_{42}(\boldsymbol{k})]^\nu D_{\mu\nu}(k)\right|^2 \\
&\quad + \int \frac{d^4k}{(2\pi)^4} 2\pi\delta(p_{2z}-p_{3z}+k_z)\, 2\pi\delta(\varepsilon_2-\varepsilon_3+\omega)
\end{aligned}$$

$$
\begin{aligned}
&\times \left| [\Gamma_{32}(-\boldsymbol{k})]^{\mu} [\Gamma_{41}(\boldsymbol{k})]^{\nu} D_{\mu\nu}(k) \right|^2 \\
&-\frac{2\pi}{eB} \int \frac{d^4k}{(2\pi)^4} \int \frac{d^4k'}{(2\pi)^4} 2\pi\delta(p_{2z} - p_{3z} + k_z)\, 2\pi\delta(\varepsilon_2 - \varepsilon_3 + \omega) \\
&\times 2\pi\delta(p_{1z} - p_{3z} + k'_z)\, 2\pi\delta(\varepsilon_1 - \varepsilon_3 + \omega') \Big(e^{i(\boldsymbol{k}\times\boldsymbol{k}')_z/eB} \\
&\times \{[\Gamma_{32}(-\boldsymbol{k})]^{\mu} [\Gamma_{41}(\boldsymbol{k})]^{\nu} D_{\mu\nu}(k)\}^* [\Gamma_{31}(-\boldsymbol{k}')]^{\alpha} [\Gamma_{42}(\boldsymbol{k}')]^{\beta} D_{\alpha\beta}(k') + \text{c.c.} \Big) \Big],
\end{aligned}
\tag{7.5.6}
$$

where "c.c." denotes complex conjugate.

A case of particular interest is scattering of an electron initially in its ground state into an excited final state. This is relevant for cyclotron emission from the thermal plasmas in the polar cap of X-ray pulsars. A problem in understanding such cyclotron emission is the origin of the cyclotron photons. In such a strong magnetic field, the electrons all relax rapidly to their ground state, and the only possible source of cyclotron photons is some non-radiative process that excites the electron to a higher Landau state, when the resulting gyromagnetic emission produces cyclotron photons. An early calculation of the transition probability used the Johnson-Lippmann wavefunctions [12]. The transition rate calculated using the more appropriate magnetic-moment eigenfunctions [18] leads to the probability

$$
\begin{aligned}
\bar{w}_{i\to f} = {}& \frac{\alpha_c^2 2\pi}{eB} \frac{dp_{3z}}{2\pi} \frac{dp_{4z}}{2\pi} 2\pi\delta(p_{1z} + p_{2z} - p_{3z} - p_{4z})\, 2\pi\delta(\varepsilon_1 + \varepsilon_2 - \varepsilon_3 - \varepsilon_4) \\
&\times \frac{\varepsilon_1 + m}{2\varepsilon_1} \frac{\varepsilon_2 + m}{2\varepsilon_2} \frac{(\varepsilon_3 + \varepsilon_3^0)(\varepsilon_3^0 + m)}{4\varepsilon_3\varepsilon_3^0} \frac{\varepsilon_4 + m}{2\varepsilon_4} \frac{1}{n!} \\
&\left(\int_0^\infty dx \frac{x^n e^{-2x}}{(x^2 + Q^2)^2} (R_a^2 + R_b^2) - 2 \int_0^\infty dx \int_0^\infty dy \frac{(xy)^{n/2} e^{-(x+y)}}{(x^2 + Q^2)(y^2 + Q^2)} R_a R_b \right),
\end{aligned}
\tag{7.5.7}
$$

where the electrons labeled 1,2,4 are in their ground state and that label 3 is in its nth Landau state. In deriving (7.5.7), the photon propagator is assumed to have its vacuum form, $D^{\mu\nu}(k) \propto 1/k^2$. The functions R_a, R_b depend on the spin of the electron, with

$$
\begin{pmatrix} R_a \\ R_b \end{pmatrix}\Bigg|_{\text{nsf}} = \begin{pmatrix} A_{31}A_{42} - B_{31}B_{42} \\ A_{32}A_{41} - B_{32}B_{41} \end{pmatrix},
$$

$$
\begin{pmatrix} R_a \\ R_b \end{pmatrix}\Bigg|_{\text{sf}} = \frac{(2neB)^{1/2}}{\varepsilon_n^0 + m} \begin{pmatrix} B_{31}A_{42} - A_{31}B_{42} \\ B_{32}A_{41} - A_{32}B_{41} \end{pmatrix},
$$

$$A_{ab} = 1 + \frac{p_{za}p_{zb}}{(\varepsilon_a + \varepsilon_a^0)(\varepsilon_b + \varepsilon_b^0)}, \qquad B_{ab} = \frac{p_{za}}{\varepsilon_a + \varepsilon_a^0} + \frac{p_{zb}}{\varepsilon_b + \varepsilon_b^0}, \tag{7.5.8}$$

with $\varepsilon_a^0 = m$ for $a = 1, 2, 4$, and where non-spin-flip (nsf) transition corresponds to the 3-state having $s = -1$ and the spin-flip (sf) transition corresponds to the 3-state having $s = 1$.

References

1. S.G. Alexander, P. Meszaros, Astrophys. J. **372**, 554 (1991)
2. S.G. Alexander, P. Meszaros, Astrophys. J. **372**, 565 (1991)
3. R.D. Blandford, E.T. Scharlemann, Mon. Not. R. Astron. Soc. **174**, 59 (1976)
4. M.L. Burns, A.K. Harding, Astrophys. J. **285**, 747 (1984)
5. J.K. Daugherty, R.W. Bussard, Astrophys. J. **238**, 296 (1980)
6. J.K. Daugherty, A.K. Harding, Astrophys. J. **309**, 362 (1986)
7. C.D. Dermer, Astrophys. J. **360**, 197 (1990)
8. R.L.Gonthier, A.K. Harding, M.G. Baring, R.C. Costello, C.L. Mercer, Astrophys. J. **540**, 907 (2000)
9. A.K. Harding, D. Lai, Rep. Prog. Phys. **69**, 2631 (2006)
10. J.G. Kirk, D.B. Melrose, Astron. Astrophys. **156**, 277 (1986)
11. A.A. Kozlenkov, I.G. Mitrofanov, Sov. Phys. JETP **64**, 1173 (1987)
12. S.H. Langer, Phys. Rev. D **23**, 328 (1981)
13. Q. Luo, Astrophys. J. **468**, 338 (1996)
14. D.B. Melrose, J.G. Kirk, Astron. Astrophys. **156**, 268 (1986)
15. L. Semionova, D. Leahy, Phys. Rev. D **60**, 073011 (1999)
16. L. Semionova, D. Leahy, Astron. Astrophys. **373**, 272 (2001)
17. R.J. Stoneham, Opt. Acta **27**, 537 (1980)
18. M.C. Storey, D.B. Melrose, Aust J. Phys. **40**, 89 (1987)
19. G. Wunner, Phys. Rev. Lett. **42**, 79 (1979)
20. X.Y. Xia, G.J. Qiao, X.J. Wu, Y.Q. Hou, Astron. Astrophys. **152**, 93 (1985)

Chapter 8
Magnetized Vacuum

The magnetized vacuum has dispersive properties similar to a material medium. Its response may be described by a hierarchy of response tensors, which are functions of B/B_c. The linear response tensor, referred to as the vacuum polarization tensor, has a relatively simple form, first derived in the 1930s, that applies at frequencies, $\omega \ll 2m$, well below the pair-creation threshold. In this limit the linear and nonlinear response tensors may be derived from the Heisenberg-Euler Lagrangian, which includes a static electric field as well as a static magnetic field. When the low-frequency approximation is not made, the linear response tensor may be calculated from the Feynman amplitude for the bubble diagram. As in the unmagnetized case, this amplitude diverges, and it needs to be regularized. The magnetic field introduces no new divergences, so that the difference between the tensors for $B/B_c \neq 0$ and their limits for $B/B_c \to 0$ are necessarily divergence-free. The quadratic nonlinear response tensor of the magnetized vacuum is nonzero for $B/B_c \neq 0$, and it allows a three-wave interaction, referred to as photon splitting.

The linear response of the magnetized vacuum is derived from the amplitude of the bubble diagram in § 8.1. Schwinger's proper-time method is introduced in § 8.2, and used to derive a generalization of the Heisenberg-Euler Lagrangian. The inclusion of a homogeneous electrostatic field in the magnetized vacuum is discussed § 8.3. The properties of waves in the magnetized (birefringent) vacuum are derived in § 8.4. Photon splitting is discussed in § 8.5.

8.1 Linear Response of the Magnetized Vacuum

The linear response tensor for a magnetized vacuum is derived by regularizing the amplitude of the bubble diagram.

D. Melrose, *Quantum Plasmadynamics: Magnetized Plasmas*, Lecture Notes in Physics 854, DOI 10.1007/978-1-4614-4045-1_8,

8.1.1 Vacuum Polarization Tensor

There are several different methods of calculation of the linear response of the magnetized vacuum, referred to as the vacuum response tensor. The regularized form of the tensor can be written in term of three invariants, and the objective in calculating the vacuum response is to find forms for these invariants.

Methods of Calculation

A variety of different methods have been used to calculate the linear response tensor for a magnetized vacuum. These fall into three general classes: the Heisenberg-Euler approach, a dispersion-integral approach, and an S-matrix approach.

The properties of the two natural modes of the birefringent vacuum were first derived in the early 1930s [12, 34] using Dirac's model of a sea of filled negative energy states. The energy eigenstates of an electron in a magnetic field are different from those of an electron in the absence of a magnetic field. Summing over the energies of the electrons in the filled negative energy states gives an infinite result, but there is a finite energy difference, that depends on B/B_c, between the magnetized and the unmagnetized cases. This difference is identified with the Heisenberg-Euler Lagrangian [12], from which the wave properties can be calculated. This is one example of a more general feature of magnetized QED: infinities that exist in unmagnetized QED also exist in magnetized QED, but the differences between the magnetized and unmagnetized is finite and leads to observable effects. The singularities that occur in the Feynman amplitudes need to be removed by a regularization procedure. In this case the (negative) energy of the electrons in the Dirac sea is infinite, the correction to it for $B/B_c \neq 0$ is finite, and the regularization procedure is to subtract the infinity. The Heisenberg-Euler Lagrangian approach applies only in the low-frequency limit, below the threshold, $\omega = 2m$, for one-photon pair creation.

The second approach is based on a calculation of the absorption for one-photon pair creation in the magnetized vacuum. As discussed in § 6.4, one-photon pair creation can be treated as an absorption-like process in which a photon transforms into an electron-positron pair. The absorption coefficient for this process can be calculated using QED, specifically, using the relativistic quantum theory for gyromagnetic processes. Dispersion is related to absorption through the Kramers-Kronig relation, and the absorption coefficients for the two natural modes of the magnetized vacuum lead to expressions for relevant components of the linear response tensor [10, 18, 27]. This method avoids the singularities that appear in the Heisenberg-Euler and in the S-matrix approach. The method gives only two transverse components of the response tensor; the complete response tensor includes a third (longitudinal) component in a magnetized vacuum.

In the S-matrix approach the linear, quadratic and cubic response tensors are related to the Feynman amplitudes for the bubble, triangle and box diagrams, respectively. The amplitude for the bubble diagram diverges and requires regularization to derive the vacuum polarization tensor. Regularization is straightforward for the two components that may be derived using the dispersion-integral approach, as these must vanish in the limit $B/B_c \to 0$. However, the remaining component includes the divergence that exists in the unmagnetized limit, and simply subtracting the limit for $B/B_c \to 0$ does not suffice for this component.

Two superficially different forms of the polarization tensor for the magnetized vacuum are derived by using different forms for the electron propagator in calculating the amplitude for the bubble diagram. One derivation involves using the Géhéniau form (5.3.13) of the propagator for a magnetized electron. The other derivation is based on the vertex formalism, and involves a double sum over particles states. Physically, the response is interpreted in terms virtual electrons and positrons in the vacuum making transitions between the two states, summed over all states. A generalization of this method allows one to calculate the response of an electron gas by including real electrons and positrons, as described in detail in § 9.1.

Discarding Unacceptable Tensor Components

Regularization of the vacuum polarization tensor involves several steps, the first of which are to identify its tensorial form and to discard components that are not of this form. In the unmagnetized case, there is only one acceptable tensorial form, $\propto k^2 g^{\mu\nu} - k^\mu k^\nu$, and terms in the unregularized tensor that are not of this tensorial form are ignored, so that the vacuum response is described by a single invariant (the constant of proportionality). In the magnetized case there are two other acceptable tensorial forms, and terms that are not of one of the three forms are discarded.

The vacuum polarization tensor must be symmetric: an antisymmetric part would imply a gyrotropic part of the response, and charge-conjugation symmetry (interchange of electrons and positrons) implies this is zero. A symmetric 4-tensor has six independent components, and the charge-continuity and gauge-invariance conditions reduce the number of possible independent variables to 4. However, from the available 4-vector, k^μ, and 4-tensor, $f^{\mu\nu}$, one can construct only three independent 4-tensors that satisfy the charge-continuity and gauge-invariance conditions, and the vacuum response can be described by invariant components along these 4-tensors.

Invariant Components

A convenient choice of the three basis 4-tensors is the set introduced in (1.1.22). For $k^\mu = (\omega, k_\perp, 0, k_z)$ the first two of these basis 4-vectors are

$$b_1^\mu = k_G^\mu = (0, 0, k_\perp, 0), \quad b_2^\mu = k_D^\mu = \phi^{\mu\alpha} k_\alpha = (k_z, 0, 0, \omega). \tag{8.1.1}$$

The regularized vacuum polarization tensor can have invariant components corresponding to three 4-tensors, which may be chosen to be

$$f_0^{\mu\nu} = g^{\mu\nu} - \frac{k^\mu k^\nu}{k^2}, \qquad f_1^{\mu\nu} = \frac{k_G^\mu k_G^\nu}{k_G^2}, \qquad f_2^{\mu\nu} = \frac{k_D^\mu k_D^\nu}{k_D^2}, \tag{8.1.2}$$

with $k_G^2 = -k_\perp^2$, $k_D^2 = -(k^2)_\parallel$. Thus the physically relevant part of the tensor may be written in the form

$$\Pi^{\mu\nu}(k) = \sum_{i=0}^{2} \Pi_i(k) f_i^{\mu\nu}. \tag{8.1.3}$$

It is sometimes convenient to write the three invariants $\Pi_i(k)$ in terms of $\chi_i(k)$, defined by

$$\mu_0 \Pi_0(k) = \frac{\chi_0(k)}{k^2}, \qquad \mu_0 \Pi_1(k) = \frac{\chi_1(k)}{k_\perp^2}, \qquad \mu_0 \Pi_2(k) = \frac{\chi_2(k)}{\omega^2 - k_z^2}. \tag{8.1.4}$$

The vacuum response tensor then becomes

$$\mu_0 \Pi^{\mu\nu}(k) = \chi_0(k)(k^2 g^{\mu\nu} - k^\mu k^\nu) - \chi_1(k) k_G^\mu k_G^\nu - \chi_2(k) k_D^\mu k_D^\nu. \tag{8.1.5}$$

Unregularized Invariant Forms

The initial step in the regularization procedure is to discard the terms in the response tensor that do not correspond to components along the 4-tensors (8.1.2). Assuming the regularized tensor is of the form (8.1.3), the three invariants may be identified. For example, one can calculate the three invariants from $\Pi^{\mu\nu}(k)$ using

$$\begin{aligned} \Pi^\mu{}_\mu(k) &= 3\Pi_0(k) + \Pi_1(k) + \Pi_2(k), \\ k_G^\mu k_G^\nu \Pi_{\mu\nu}(k) &= -k_\perp^2 \Pi_1(k), \qquad k_D^\mu k_D^\nu \Pi_{\mu\nu}(k) = -(k^2)_\parallel \Pi_2(k). \end{aligned} \tag{8.1.6}$$

8.1.2 *Unregularized Tensor: Géhéniau Form*

The unregularized vacuum polarization tensor follows from the amplitude for the bubble diagram, and this amplitude involves electron propagators. Choosing the propagators in the Géhéniau form (5.3.12) leads to a coordinate form for the response tensor.

Response Tensor in Coordinate Space

The vacuum response tensor in coordinate space is obtained by the same procedure used to calculate the corresponding tensor in the unmagnetized case, cf. § 8.1 of volume 1. The Feynman amplitude of the bubble diagram gives

$$\Pi^{\mu\nu}(x-x') = -ie^2 \mathrm{Tr}\,[\gamma^\mu G(x,x')\gamma^\nu G(x',x)]. \tag{8.1.7}$$

The propagator in the Géhéniau form is given by (5.3.13), that is, $G(x,x') = -\phi(x,x')\Delta(x-x')$, where $\phi(x,x')$ is a phase factor that depends separately on x, x'. In (8.1.7) the phase factors cancel. Thus (8.1.7) reduces to

$$\Pi^{\mu\nu}(x-x') = -ie^2 \mathrm{Tr}\,[\gamma^\mu \Delta(x-x')\gamma^\nu \Delta(x'-x)], \tag{8.1.8}$$

with $\Delta(x)$ given by (5.3.14), which is written as an integral over λ of a function, $\mathcal{B}(\lambda, x)$, given by (5.3.15). The trace over the Dirac matrices in (8.1.8) reduces to

$$d^{\mu\nu}(\lambda,\lambda',x) = \mathrm{Tr}\,[\gamma^\mu \mathcal{B}(\lambda,x)\gamma^\nu \mathcal{B}(\lambda',-x)]. \tag{8.1.9}$$

The resulting expression for the unregularized response tensor is

$$\Pi^{\mu\nu}(x) = -i\frac{e^4B^2}{(4\pi)^4}\int_0^\infty \frac{d\lambda}{\lambda}\int_0^\infty \frac{d\lambda'}{\lambda'}\, d^{\mu\nu}(\lambda,\lambda',x)\exp\left[-i\frac{m^2}{2}\left(\frac{1}{\lambda}+\frac{1}{\lambda'}\right)\right]$$
$$\times\,\exp\left\{\frac{ieB}{4}(x^2+y^2)\left(\cot\frac{eB}{2\lambda}+\cot\frac{eB}{2\lambda'}\right)+\frac{i}{2}(\lambda+\lambda')(z^2-t^2)\right\}. \tag{8.1.10}$$

It is convenient to change the variables of integration in (8.1.10) to

$$\alpha = \frac{eB}{2}\left(\frac{1}{\lambda}+\frac{1}{\lambda'}\right), \quad \beta = \frac{eB}{2}\left(\frac{1}{\lambda}-\frac{1}{\lambda'}\right), \qquad \lambda = \frac{eB}{\alpha+\beta}, \quad \lambda' = \frac{eB}{\alpha-\beta}, \tag{8.1.11}$$

and to Fourier transform before evaluating the trace in (8.1.9).

Fourier Transform of the Unregularized Response Tensor

The Fourier transform of (8.1.10) involves integrals that can be reduced to the form

$$\int_{-\infty}^{\infty} d\eta\, e^{\pm ia\eta^2+ib\eta} = (1\pm i)\left(\frac{\pi}{2a}\right)^{1/2}\exp\left(\mp\frac{ib^2}{a}\right), \tag{8.1.12}$$

and derivatives of this integral. Writing

$$\left[\mathcal{J}(k), \mathcal{J}_\perp^\mu(k), \mathcal{J}_\parallel^\mu(k)\right] = \int d^4x \left[1, x_\perp^\mu, x_\parallel^\mu\right] e^{i[kx-a_\perp(x^2)_\perp - a_\parallel(x^2)_\parallel]}, \tag{8.1.13}$$

and using (8.1.12), one finds

$$\mathcal{J}(k) = \frac{i\pi^2}{a_\perp a_\parallel} \exp\left[\frac{i(k^2)_\perp}{4a_\perp} + \frac{i(k^2)_\parallel}{4a_\parallel}\right]. \tag{8.1.14}$$

With $\mathcal{J}^\mu_\perp(k) = -i\partial\mathcal{J}(k)/\partial k_{\perp\mu}$, $\mathcal{J}^\mu_\parallel(k) = -i\partial\mathcal{J}(k)/\partial k_{\parallel\mu}$, one finds

$$\mathcal{J}^\mu_\perp(k) = \left(k^\mu_\perp/2a_\perp\right)\mathcal{J}(k), \qquad \mathcal{J}^\mu_\parallel(k) = \left(k^\mu_\parallel/2a_\parallel\right)\mathcal{J}(k). \tag{8.1.15}$$

Integrals with higher powers of $x_\perp, x_\parallel$ in the integrand are evaluated in the same manner, corresponding to the replacements

$$x^\mu_\perp \to \frac{k^\mu_\perp}{2a_\perp}, \quad x^\mu_\parallel \to \frac{k^\mu_\parallel}{2a_\parallel}, \qquad a_\perp = -\frac{eB}{2}\frac{\sin\alpha}{\cos\alpha - \cos\beta}, \quad a_\parallel = \frac{eB\alpha}{\alpha^2 - \beta^2}. \tag{8.1.16}$$

The Fourier transform of (8.1.10) gives

$$\Pi^{\mu\nu}(k) = \frac{e^2m^2}{(2\pi)^2}\int_0^\infty \frac{d\alpha}{\alpha}\int_0^\alpha d\beta\, d^{\mu\nu}(k,\alpha,\beta)\exp\left[-i\frac{\alpha}{L} + i\frac{\alpha^2-\beta^2}{4\alpha L}\frac{k^2}{m^2} + \frac{i}{2L}\left(\frac{\alpha^2-\beta^2}{2\alpha} + \frac{\cos\alpha - \cos\beta}{\sin\alpha}\right)\frac{k_\perp^2}{m^2}\right], \tag{8.1.17}$$

with $L = B/B_c$, $B_c = m^2/e$.

The tensor $d^{\mu\nu}(k,\alpha,\beta)$ is defined by the replacements (8.1.11) and (8.1.16) in (8.1.9). Specifically, in (8.1.9) one makes the replacement

$$\frac{\mathcal{B}(\lambda,x)}{eB} \to \left[\frac{(\alpha+\beta)(\gamma k)_\parallel}{4a_\parallel} - \frac{k_\perp}{4a_\perp}\left(\gamma^1\cot\frac{\alpha+\beta}{2} + \gamma^2\right)\right] \left(\cot\frac{\alpha+\beta}{2} - i\Sigma_z\right), \tag{8.1.18}$$

with $(\gamma k)_\parallel = \gamma^0\omega - \gamma^3 k_z$, and with γ^1, γ^2 the components of γ along $e^\mu_1 = [0, \boldsymbol{k}_\perp/k_\perp]$, $e^\mu_2 = [0, \boldsymbol{b}\times\boldsymbol{\kappa}]$, respectively. The replacement for $\mathcal{B}(\lambda', -x)$ is given by (8.1.18) with $k \to -k$, $\beta \to -\beta$.

The regularized tensor must be nongyrotropic, and in a frame with $\boldsymbol{B}$ along the 3-axis and $\boldsymbol{k}$ in the 1–3 plane, this implies that one must discard the components d^{02}, d^{12}, d^{32} and d^{20}, d^{21}, d^{23}, where the arguments k, α, β are omitted. The only part of the tensor $d^{\mu\nu}(k,\alpha,\beta)$ in the integral (8.1.10) that can contribute is the part that is symmetric and is an even function of k and β. Hence one is to discard terms that do not satisfy

$$d^{\mu\nu}(k,\alpha,\beta) = d^{\nu\mu}(k,\alpha,\beta) = d^{\mu\nu}(-k,\alpha,\beta) = d^{\mu\nu}(k,\alpha,-\beta). \tag{8.1.19}$$

A straightforward way of removing the unacceptable terms is to construct the three invariant components (8.1.4).

Unregularized Invariant Components

The unregularized form for $\Pi_i(k)$ follows from (8.1.17), (8.1.3) and (8.1.6):

$$\Pi_i(k) = \frac{e^2m^2}{(2\pi)^2}\int_0^\infty \frac{d\alpha}{\alpha}\int_0^\alpha d\beta\, d_i(k,\alpha,\beta)\,\exp\left[-i\frac{\alpha}{L}+i\frac{\alpha^2-\beta^2}{4\alpha L}\frac{k^2}{m^2}\right.$$
$$\left.+\frac{i}{2L}\left(\frac{\alpha^2-\beta^2}{2\alpha}+\frac{\cos\alpha-\cos\beta}{\sin\alpha}\right)\frac{k_\perp^2}{m^2}\right], \tag{8.1.20}$$

with d_i defined by writing $d^{\mu\nu} = \sum_{i=0}^{2} d_i f_i^{\mu\nu}$. One finds

$$d_0(k,\alpha,\beta) = \frac{\cos\beta}{2\sin\alpha}\left(1-\frac{\beta\tan\beta}{\alpha\tan\alpha}\right)\frac{k^2}{m^2},$$
$$d_1(k,\alpha,\beta) = \left[\frac{\cos\alpha-\cos\beta}{\sin^3\alpha}+\frac{\cos\beta}{2\sin\alpha}\left(1-\frac{\beta\tan\beta}{\alpha\tan\alpha}\right)\right]\frac{k_\perp^2}{m^2},$$
$$d_2(k,\alpha,\beta) = \left[\frac{\alpha^2-\beta^2}{2\alpha^2\tan\alpha}-\frac{\cos\beta}{2\sin\alpha}\left(1-\frac{\beta\tan\beta}{\alpha\tan\alpha}\right)\right]\frac{\omega^2-k_z^2}{m^2}. \tag{8.1.21}$$

The integral over α in (8.1.20) diverges in the limit $\alpha \to 0$ for d_0, but not for d_1, d_2.

8.1.3 Regularization of the Vacuum Polarization Tensor

The regularization of the vacuum polarization tensor involves subtracting divergent terms from the invariant coefficients $\Pi_i(k)$ or $\chi_i(k)$.

Regularization of $\Pi_0(k)$

The unregularized expression for $\Pi_0(k)$ may be regularized by subtracting its zero-field limit from it, and adding the known regularized, zero-field vacuum polarization tensor to it [4, 5, 17, 18, 28].

In taking the limit $B \to 0$, it is helpful to incorporate $L = B/B_c$ into new variables of integration α/L, β/L, before taking the limit $L \to 0$. One finds, as expected, that $\Pi_1(k)$, $\Pi_2(k)$ are non-divergent and do not require regularization. The invariant $\Pi_0(k)$ is the coefficient of the part of the tensor $\propto k^2g^{\mu\nu} - k^\mu k^\nu$, and it must reduce to the unmagnetized vacuum polarization tensor for $B \to 0$. It is known that regularization in the unmagnetized case (§ 8.1.3 of volume 1) requires a double subtraction; specifically, regarding $\Pi_0(k)$ as a function of k^2, its regularized part is, cf. (8.1.9) of volume 1,

$$\mathrm{reg}\,\Pi_0(k^2) = \Pi_0(k^2) - \Pi_0(0) - k^2[\partial \Pi_0(k^2)/\partial k^2]_{k^2=0}.$$

Regularization of $\Pi_0(k)$ gives

$$\begin{aligned}\mathrm{reg}\,\Pi_0(k) = \frac{e^2k^2}{(2\pi)^2}\int_0^\infty \frac{d\alpha}{\alpha}\int_0^\alpha d\beta\, \exp\left(-i\frac{\alpha}{L} + i\frac{\alpha^2-\beta^2}{4\alpha L}\frac{k^2}{m^2}\right)\\ \times\left\{\frac{\cos\beta}{2\sin\alpha}\left(1-\frac{\beta\tan\beta}{\alpha\tan\alpha}\right)\exp\left[\frac{i}{2L}\left(\frac{\alpha^2-\beta^2}{2\alpha}+\frac{\cos\alpha-\cos\beta}{\sin\alpha}\right)\frac{k_\perp^2}{m^2}\right]\right.\\ \left. -\frac{1}{2\alpha}\left(1-\frac{\beta^2}{\alpha^2}\right)\right\}.\end{aligned} \tag{8.1.22}$$

The regularized, unmagnetized, vacuum polarization tensor needs to be added to (8.1.22) to obtain the full expression for reg $\Pi_0(k)$.

Weak-Field and Strong-Field Limits

In the weak field approximation, the power series expansions of the functions d_i in (8.1.21) converge rapidly. On retaining only the leading terms, the integrals to be evaluated are elementary, and one finds

$$\frac{\Pi_0(k)}{k^2} = -\frac{e^2}{90\pi^2}\frac{B^2}{B_c^2}, \quad \frac{\Pi_1(k)}{k_\perp^2} = \frac{4e^2}{45\pi^2}\frac{B^2}{B_c^2}, \quad \frac{\Pi_2(k)}{\omega^2-k_z^2} = \frac{7e^2}{45\pi^2}\frac{B^2}{B_c^2}. \tag{8.1.23}$$

In the strong field limit $B \gg B_c$, one finds that $\Pi_0(k)$, $\Pi_1(k)$ are much smaller than $\Pi_2(k)$, which may be approximated by

$$\begin{aligned}\Pi_2(k) = \frac{e^3B}{2\pi^2}e^{-k_\perp^2/2eB}\left\{-1\right.\\ \left. + \frac{4m^2}{\sqrt{[4m^2-(\omega^2-k_z^2)]^{1/2}(\omega^2-k_z^2)}}\arctan\sqrt{\frac{\omega^2-k_z^2}{4m^2-(\omega^2-k_z^2)}}\right\}.\end{aligned} \tag{8.1.24}$$

The result (8.1.24) simplifies to $\Pi_2(k) \approx (e^2/12\pi^2)(B/B_c)(\omega^2-k_z^2)$ for $4m^2 \gg \omega^2 - k_z^2, k_\perp^2 \ll 2eB$.

8.1.4 *Vacuum Polarization Tensor: Vertex Formalism*

An alternative procedure for calculating the vacuum polarization tensor is based on the amplitude of the bubble diagram derived using the vertex formalism. This approach is used in § 9.1 to write down the response tensor including the contributions from both the vacuum and an electron gas. The vacuum polarization tensor follows from the more general form (9.1.1) for the response tensor by setting the occupation numbers to zero.

Unregularized Tensor: Vertex Formalism

The expression for unregularized vacuum polarization tensor in the vertex formalism is

$$\Pi^{\mu\nu}(k) = -\frac{e^3 B}{2\pi} \sum_{\epsilon,q,\epsilon',q'} \int \frac{dp_z}{2\pi} \frac{\frac{1}{2}(\epsilon' - \epsilon)[\Gamma^{\epsilon'\epsilon}_{q'q}(k)]^{\mu}[\Gamma^{\epsilon'\epsilon}_{q'q}(k)]^{*\nu}}{\omega - \epsilon\varepsilon_q + \epsilon'\varepsilon_{q'} + i0}, \tag{8.1.25}$$

where the sets of quantum numbers q and q' denote p_z, n, s and p'_z, n', s', with $\epsilon' p'_z = \epsilon p_z - k_z$ implicit.

The sums over s, s' and ϵ', ϵ in (8.1.25) can be performed explicitly. The sum over s, s' is performed in § 5.4.4. The sum over the product of vertex function leads to the tensor $\left[C_{n'n}(\epsilon' p'_{\parallel}, \epsilon p_{\parallel}, k)\right]^{\mu\nu}$, defined by (5.4.23), and given explicitly by (5.4.24) and in Table 5.1. For the vacuum, only $\epsilon' = -\epsilon$ contributes, due to the factor $\frac{1}{2}(\epsilon' - \epsilon) = -\epsilon$ in (8.1.25).

The unregularized form of the response tensor is ill-defined, and the expressions obtained for it depend on the method of calculation. The regularized tensor must be independent of the method of calculation, but different methods can lead to superficially different forms. To illustrate this point, suppose one starts from (8.1.25), summed over the spins, and discards the gyrotropic terms and the terms that are inconsistent with the charge-continuity and gauge-invariance conditions. This leaves four different unregularized components (e.g., the 11, 22, 33 and 13 components). There are only three invariant components of the regularized form (8.1.3). One can choose different procedures to calculate the three invariants from the four components. One procedure is to note that the 22- and 33-components are proportional to $\Pi_0(k) + \Pi_1(k)$ and $\Pi_0(k) + \Pi_2(k)$, respectively, and one may use the trace of the 4-tensor to calculate $3\Pi_0(k) + \Pi_1(k) + \Pi_2(k)$. This procedure involves using the 00- and 03-components, but not the 13-component, and the result may be written in terms of the 11-, 22-, 33-components, e.g., as given in Table 5.1. Another procedure is to identify $\Pi^{13}(k)$ as $-\Pi_0(k)k_\perp k_z/k^2$. These two procedures lead to different results. The physical requirement is that the regularized tensor be well defined, and hence these two different procedure must lead to the same result after regularization.

By inspection of the entries in Table 5.1, the nongyrotropic components are symmetric under the interchange of the primed and unprimed variables, $\epsilon', n', \varepsilon'_{n'}, p'_z \leftrightarrow \epsilon, n, \varepsilon_n, p_z$. With the gyrotropic components discarded, one may make the dependence on ϵ explicit. A complication is that for $\epsilon = -\epsilon' = 1$ one has $p'_z = k_z - p_z$, and for $\epsilon = -\epsilon' = -1$ one has $p'_z = -k_z - p_z$; however, interchanging the primed and unprimed variables in the latter case, the relation $p'_z = k_z - p_z$ is restored, so that one $p'_z = k_z - p_z$ in both terms. This procedure ignores the 13-component, allowing the remaining nongyrotropic components in Table 5.1 to be written as a 4-tensor:

$$\begin{aligned}
\big[C_{n'n}(p'_\parallel, p_\parallel, k)\big]^{\mu\nu}_{\rm vac} &= -(p'^\mu_\parallel p^\nu_\parallel + p^\mu_\parallel p'^\nu_\parallel)\big[(J^{n-1}_{n'-n})^2 + (J^n_{n'-n})^2\big] \\
&\quad + g^{\mu\nu}_\parallel \{[(p'p)_\parallel + m^2]\big[(J^{n-1}_{n'-n})^2 + (J^n_{n'-n})^2\big] + 2p_{n'}p_n J^{n-1}_{n'-n} J^n_{n'-n}\} \\
&\quad + g^{\mu\nu}_\perp [(p'p)_\parallel + m^2]\big[(J^{n-1}_{n'-n+1})^2 + (J^n_{n'-n-1})^2\big] \\
&\quad + (e^\mu_1 e^\nu_1 - e^\mu_2 e^\nu_2) 2p_{n'}p_n J^{n-1}_{n'-n+1} J^n_{n'-n-1},
\end{aligned} \tag{8.1.26}$$

with $p'^\mu_\parallel = [\varepsilon'_{n'}, 0, 0, p'_z]$, $p^\mu_\parallel = [\varepsilon_n, 0, 0, p_z]$, and with $p'_z = k_z - p_z$. The vacuum contribution to (8.1.25) then becomes

$$\Pi^{\mu\nu}(k) = -\frac{e^3 B}{8\pi^2} \sum_{\epsilon, n', n} \int \frac{dp_z}{\varepsilon'_{n'}\varepsilon_n} \frac{\epsilon\big[C_{n'n}(p'_\parallel, p_\parallel, k)\big]^{\mu\nu}_{\rm vac}}{\omega - \epsilon(\varepsilon_n + \varepsilon'_{n'}) + i0}, \tag{8.1.27}$$

The sum over ϵ can be performed in (8.1.27) after multiplying numerator and denominator by $\omega + \epsilon(\varepsilon_n + \varepsilon'_{n'})$, but the form (8.1.27) is more useful in the following.

The alternative procedure for identifying $\Pi_0(k)$ is to choose the 13-component of (8.1.3)

$$\Pi^{13}(k) = -\frac{k_\perp k_z \Pi_0(k)}{k^2} = -\frac{e^3 B}{8\pi^2} \sum_{\epsilon, n', n} \int \frac{dp_z}{\varepsilon'_{n'}\varepsilon_n} \frac{\epsilon\big[C_{n'n}(p'_\parallel, p_\parallel, k)\big]^{13}}{\omega - \epsilon(\varepsilon_n + \varepsilon'_{n'}) + i0}, \tag{8.1.28}$$

with $\big[C_{n'n}(p'_\parallel, p_\parallel, k)\big]^{13}$ given by $\epsilon' = -\epsilon$ in the 13-entry in Table 5.1. It should be emphasized that (8.1.27) and (8.1.28) are meaningless in themselves, and are simply intermediate steps in the calculation of the unregularized invariant components $\Pi_i(k)$. One method of calculation of the invariants involves uses (8.1.27) and (8.1.6) to calculate all three invariants, and the alternative method uses (8.1.27) and (8.1.6) to calculate $\Pi_0(k) + \Pi_1(k)$ and $\Pi_0(k) + \Pi_2(k)$, and (8.1.28) to calculate $\Pi_0(k)$.

Unregularized Invariant Components: Vertex Formalism

One form of the unregularized invariant components $\Pi_i(k)$ follows from (8.1.27) and (8.1.6). Writing

$$\Pi_i(k) = -\frac{e^3 B}{8\pi^2} \sum_{\epsilon,n',n} \int \frac{dp_z}{\varepsilon'_{n'}\varepsilon_n} \frac{\epsilon\big[C_{n'n}(p'_\parallel, p_\parallel, k)\big]_i}{\omega - \epsilon(\varepsilon_n + \varepsilon'_{n'}) + i0}, \tag{8.1.29}$$

defines the (unregularized) invariants $C_i = \big[C_{n'n}(p'_\parallel, p_\parallel, k)\big]_i$ in the numerator.

One form for the invariants C_i follows by taking the trace of the 4-tensor and its projections onto $k_G^\mu k_G^\nu / k_G^2$ and $k_D^\mu k_D^\nu / k_D^2$. This procedure may be applied to (8.1.26), resulting in $3C_0 + C_1 + C_2$, $C_0 + C_1$ and $C_0 + C_2$ respectively. In evaluating the projection onto $k_D^\mu k_D^\nu / k_D^2$ one uses the identity

$$(k_D p')(k_D p) = (p'k)_\parallel (pk)_\parallel - (k^2)_\parallel (p'p)_\parallel. \tag{8.1.30}$$

The (unregularized) invariant components are

$$\begin{aligned}
C_0 &= \left[-\frac{2(p'k)_\parallel (pk)_\parallel}{(k^2)_\parallel} + (p'p)_\parallel + m^2\right]\left[(J^{n-1}_{n'-n})^2 + (J^n_{n'-n})^2\right] \\
&\quad + [(p'p)_\parallel + m^2]\left[(J^{n-1}_{n'-n+1})^2 + (J^n_{n'-n-1})^2\right] \\
&\quad + 2p_{n'}p_n\left[J^{n-1}_{n'-n}J^n_{n'-n} + J^{n-1}_{n'-n+1}J^n_{n'-n-1}\right], \\
C_1 &= -C_0 + [(p'p)_\parallel + m^2]\left[(J^{n-1}_{n'-n+1})^2 + (J^n_{n'-n-1})^2\right] \\
&\quad - 2p_{n'}p_n J^{n-1}_{n'-n+1}J^n_{n'-n-1}, \\
C_2 &= -C_0 + \left[\frac{2(p'k)_\parallel (pk)_\parallel}{(k^2)_\parallel} - (p'p)_\parallel + m^2\right]\left[(J^{n-1}_{n'-n})^2 + (J^n_{n'-n})^2\right] \\
&\quad + 2p_{n'}p_n J^{n-1}_{n'-n}J^n_{n'-n}.
\end{aligned} \tag{8.1.31}$$

The alternative method for calculating $\Pi_0(k)$ involves using (8.1.28), which gives

$$\begin{aligned}
C_0 = \frac{\epsilon k^2}{k_\perp k_z}\{&p_{n'}p_z\left[J^{n-1}_{n'-n}J^{n-1}_{n'-n+1} + J^n_{n'-n}J^n_{n'-n-1}\right] \\
&- p'_z p_n\left[J^n_{n'-n}J^{n-1}_{n'-n+1} + J^{n-1}_{n'-n}J^n_{n'-n-1}\right]\}.
\end{aligned} \tag{8.1.32}$$

The expression (8.1.32) is an alternative to the expression for C_0 in (8.1.31); the remaining two invariants are unaffected.

The final step in the calculation is to regularize the invariants (8.1.27) with (8.1.31). This may be achieved by identifying the antihermitian part of the tensor and using a Kramers-Kronig relation to construct the hermitian part of the regularized tensor from it.

8.1.5 Antihermitian Part of the Vacuum Polarization Tensor

The antihermitian part of the vacuum response tensor describes dissipation in the vacuum. The only allowed dissipative process is one-photon pair creation. Besides describing this process, the anithermitian part is also useful in regularizing the hermitian part, using a method due to Toll [27].

The vacuum response tensor is symmetric, and hence its antihermitian part is its imaginary part. The imaginary part of the Géhéniau form (8.1.17) is obtained by replacing the exponential function according to $e^{iy} \to i \sin y$. The interpretation of this imaginary part is not obvious in the Géhéniau form. In contrast, the interpretation of the antihermitian part of the response tensor (8.1.25) in the vertex formalism, is straightforward in terms of one-photon pair creation.

Antihermitian Part: Vertex Formalism

The antihermitian part of the vacuum response tensor may be described by the imaginary parts of its invariant components. The imaginary parts of the components (8.1.29) are

$$\mathrm{Im}\, \Pi_i(k) = \frac{e^3 B}{8\pi} \sum_{\epsilon,n',n} \int \frac{dp_z}{\varepsilon'_{n'}\varepsilon_n} \epsilon\big[C_{n'n}(p'_\parallel, p_\parallel, k)\big]_i\, \delta[\omega - \epsilon(\varepsilon_n + \varepsilon'_{n'})]. \quad (8.1.33)$$

The integral over p_z may be performed over the δ-function, with $\epsilon = 1$ contributing for $\omega > 0$ and $\epsilon = -1$ contributing for $\omega < 0$. There are two solutions of the resonance condition, labeled $\pm$ and given by (6.1.16)–(6.1.19). This gives

$$\mathrm{Im}\, \Pi_i(k) = \frac{e^3 B}{8\pi} \sum_{n',n,\pm} \frac{\big[C_{n'n}(p'_\pm, p_\pm, k)\big]_i}{(k^2)_\parallel g_{nn'}}, \quad (8.1.34)$$

with $g_{nn'}$ defined by (6.1.17) and with the resonant values of p_z, p'_z and $\varepsilon_n, \varepsilon'_{n'}$ given by (6.1.18) and (6.1.19), respectively. The resonant values of the invariants in (8.1.31) are the same for the $\pm$-solutions with

$$[(pk)_\parallel]_\pm = (pk)_{nn'} = (k^2)_\parallel f_{nn'} = \frac{1}{2}[p_n^2 - p_{n'}^2 + k^2)_\parallel],$$

$$[(p'k)_\parallel]_\pm = (k^2)_\parallel - [(pk)_\parallel]_\pm, \qquad [(p'p)_\parallel]_\pm = [(pk)_\parallel]_\pm - (\varepsilon_n^0)^2. \quad (8.1.35)$$

The antihermitian part of the vacuum response tensor may be used to calculate the absorption coefficient for one-photon pair creation. The absorption coefficient is equivalent to the rate of pair production, which is calculated in § 6.4.1 from the probability for pair production, leading to (6.4.3). The absorption coefficient may calculated from the general expression (3.1.27) for the absorption coefficient in terms of the antihermitian part of the response tensor. The latter procedure, using (8.1.34) in (8.1.3), is summed over the spins, and does not includes information on the spins of the electron and positron that is retained in (6.4.3).

Regularization Using Kramers-Kronig Relation

The complete response tensor may be constructed from its antihermitian part using the Kramers-Kronig relation

$$\Pi^{\mu\nu}(\omega,\boldsymbol{k}) = -\frac{i}{\pi}\int_{-\infty}^{\infty}\frac{d\omega'}{\omega'-\omega+i0}\,\Pi^{A\mu\nu}(\omega',\boldsymbol{k}). \tag{8.1.36}$$

A regularization procedure is to construct the antihermitian part of the unregularized form of the tensor, and use (8.1.36) to construct the regularized tensor from its antihermitian part. This method applies in wider contexts, and is referred to as the dispersion-integral method in § 9.2.1, where it is applied to the response of an electron gas.

The dispersion-integral method for calculating the vacuum polarization tensor [10, 18, 27] is equivalent here to inserting (8.1.34) into (8.1.36). The result may be rewritten as a p_z-integral by considering the antihermitian part of (8.1.37) and reversing the step between (8.1.34) and (8.1.33). This gives

$$\mathrm{Reg}\,\Pi_i(k) = -\frac{e^3B}{8\pi^2}\sum_{\epsilon,n',n}\left[C_{n'n}(p'_\pm,p_\pm,k)\right]_i\int\frac{dp_z}{\varepsilon'_{n'}\varepsilon_n}\,\frac{\epsilon}{\omega-\epsilon(\varepsilon_n+\varepsilon'_{n'})+i0}. \tag{8.1.37}$$

The important change between (8.1.29) and (8.1.37) is that the numerator is evaluated at the resonant values, so that it no longer depends on p_z and is taken outside the integral. This removes terms in the numerator in (8.1.29) that cause the integral to diverge at the limits $|p_z|\to\infty$, thereby regularizing these integrals. Thus a regularized form of the vacuum response tensor is given by (8.1.37) in (8.1.3) with $\left[C_{n'n}(p'_\pm,p_\pm,k)\right]_i$ identified by replacing $(pk)_\parallel, (p'k)_\parallel, (p'p)_\parallel$ in the expressions (8.1.31) for C_i by the resonant values (8.1.35). One has

$$\left[C_{n'n}(p'_\pm,p_\pm,k)\right]_0 = \frac{2(pk)^2_{nn'}}{(k^2)_\parallel}\left[(J^{n-1}_{n'-n})^2+(J^{n}_{n'-n})^2\right]$$
$$-\frac{1}{2}\left[(p_n^2+p_{n'}^2-k_\perp^2\right]\left[(J^{n-1}_{n'-n+1})^2+(J^{n}_{n'-n-1})^2\right]$$

$$
\begin{aligned}
&\qquad +2p_{n'}p_n J_{n'-n+1}^{n-1} J_{n'-n-1}^{n}, \\
&\left[C_{n'n}(p'_{\pm}, p_{\pm}, k)\right]_1 = -C_0 - \frac{1}{2}[p_n^2 + p_{n'}^2 - (k^2)_{\parallel}][(J_{n'-n+1}^{n-1})^2 + (J_{n'-n-1}^{n})^2] \\
&\qquad -2p_{n'}p_n J_{n'-n+1}^{n-1} J_{n'-n-1}^{n}, \\
&\left[C_{n'n}(p'_{\pm}, p_{\pm}, k)\right]_2 = -C_0 + \left[\frac{2(pk)_{nn'}^2}{(k^2)_{\parallel}} + 2(\varepsilon_n^0)^2 - \frac{1}{2}[p_n^2 + p_{n'}^2 - (k^2)_{\parallel}]\right] \\
&\qquad \times[(J_{n'-n}^{n-1})^2 + (J_{n'-n}^{n})^2] + 2p_{n'}p_n J_{n'-n}^{n-1} J_{n'-n}^{n},
\end{aligned} \tag{8.1.38}
$$

where the expression for $\left[C_{n'n}(p'_{\pm}, p_{\pm}, k)\right]_0$ is simplified using the identity (A.1.36).

The alternative method of calculation of C_0 is based on (8.1.32). The resonant values are $p_z = k_z - p'_z = \omega f_{nn'} \pm \omega g_{nn'}$. The $\pm$ terms cancel due to the identifies (A.1.35). The remaining terms give

$$
\begin{aligned}
\left[C_{n'n}(p'_{\pm}, p_{\pm}, k)\right]_0 &= \frac{\epsilon k^2}{k_\perp}\{p_{n'} f_{nn'}[J_{n'-n}^{n-1} J_{n'-n+1}^{n-1} + J_{n'-n}^{n} J_{n'-n-1}^{n}] \\
&\quad - p_n f_{n'n}[J_{n'-n}^{n} J_{n'-n+1}^{n-1} + J_{n'-n}^{n-1} J_{n'-n-1}^{n}]\},
\end{aligned} \tag{8.1.39}
$$

with $f_{n'n} = (1 - f_{nn'})$, $f_{nn'} = [p_n^2 - p_{n'}^2 + (k^2)_{\parallel}]/2(k^2)_{\parallel}$. Using the identities (A.1.35), (8.1.39) may be rewritten as

$$
\begin{aligned}
\left[C_{n'n}(p'_{\pm}, p_{\pm}, k)\right]_0 &= \epsilon\frac{k^2}{(k^2)_{\parallel}}\left\{\frac{1}{2}[p_n^2 - p_{n'}^2 + (k^2)_{\parallel}][(J_{n'-n+1}^{n-1})^2 + (J_{n'-n-1}^{n})^2] \right.\\
&\quad \left. + \frac{p_n}{k_\perp}(p_n^2 - p_{n'}^2)[J_{n'-n}^{n} J_{n'-n+1}^{n-1} + J_{n'-n}^{n-1} J_{n'-n-1}^{n}]\right\}.
\end{aligned} \tag{8.1.40}
$$

8.1.6 Vacuum Polarization: Limiting Cases

The vacuum polarization tensor simplifies in the long-wavelength limit and in the strong-B limit.

Long-Wavelength Limit

The long-wavelength limit corresponds to $k_\perp \to 0$, $k_z \to 0$. In the limit $k_\perp \to 0$, the argument, $x = k_\perp^2/2eB$, of the J-functions tends to zero, and this is referred to as the small-x approximation in § 9.4.2. The sums over n', n simplify, with the only contributions being from $n' = n, n \pm 1$. Useful expressions can be derived in the low-frequency limits, $\omega^2 \ll 4m^2$ and $B \ll B_c$, when the sum over n may be performed using the Euler-Maclaurin summation formula. The result gives [18]

$$\operatorname{Re} \Pi_0(k) = -\frac{e^2\omega^4}{60\pi^2 m^2} - \frac{e^2\omega^2}{90\pi^2}\left(\frac{B}{B_c}\right)^2 + \frac{e^2\omega^2}{105\pi^2}\left(\frac{B}{B_c}\right)^4 + \cdots,$$

$$\operatorname{Re} \Pi_2(k) = \frac{7e^2\omega^2}{180\pi^2}\left(\frac{B}{B_c}\right)^2 - \frac{13e^2\omega^2}{630\pi^2}\left(\frac{B}{B_c}\right)^4 + \cdots, \tag{8.1.41}$$

with $\operatorname{Re} \Pi_1(k) = 0$ in this approximation.

Strong-*B* Limit

In the strong field limit, $B \gg B_c$, for $k_\perp^2 \ll m^2$, $\omega^2 - k_z^2 \ll m^2$, the sum over n', n is dominated by $n = n' = 0$. Retaining only this contribution, one finds that $\operatorname{Re} \Pi_0(k)$ and $\operatorname{Re} \Pi_1(k)$ are negligible in comparison with $\operatorname{Re} \Pi_2(k)$, which reduces to [18]

$$\frac{\operatorname{Re} \Pi_2(k)}{\omega^2 - k_z^2} = \frac{e^2}{12\pi^2}\frac{B}{B_c}. \tag{8.1.42}$$

The result (8.1.42) reproduces the $\omega^2 - k_z^2 \ll 4m^2$ limit of (8.1.24).

8.1.7 *Wave Modes of the Magnetized Vacuum*

The properties of waves in the magnetized vacuum may be found by identifying the relevant wave equation and solving it. The photon propagator is identified as the Green function for this wave equation, and the wave modes may be identified from the poles of the photons propagator. The latter procedure is adopted here.

Photon Propagator in the Magnetized Vacuum

In the homogeneous wave equation, $\Lambda^{\mu\nu}(k)A_\nu(k) = 0$, one has

$$\Lambda^{\mu\nu}(k)A_\nu(k) = 0,$$

$$\Lambda^{\mu\nu}(k) = [1 + \chi_0](k^2 g^{\mu\nu} - k^\mu k^\nu) - \chi_1 k_G^\mu k_G^\nu - \chi_2 k_D^\mu k_D^\nu, \tag{8.1.43}$$

where the dependence of the χ_i on k is implicit.The wave properties are determined by the solutions of (8.1.43). A direct covariant method involves constructing the invariants $t^{(1)}, t^{(2)}, t^{(3)}$ that appear in the invariant dispersion equation (3.1.4).

A simpler approach in this case is to construct the photon propagator, and to identify the wave modes from its poles [24, 25]. The photon propagator is defined as a solution of

$$\Lambda^{\mu}{}_{\nu}(k) D^{\nu\rho}(k) = \mu_0 (g^{\nu\rho} - k^{\nu} k^{\rho}/k^2). \tag{8.1.44}$$

One may construct $D^{\mu\nu}(k)$ by assuming it has components along the tensors (8.1.2), inserting this form in (8.1.44), and solving for the coefficients. This gives

$$D^{\mu\nu} = \frac{\mu_0}{[1+\chi_0]k^2} \left\{ g^{\mu\nu} - \frac{k^\mu k^\nu}{k^2} + \frac{\chi_1 k_G^\mu k_G^\nu}{[1+\chi_0]k^2 + \chi_1 k_\perp^2} + \frac{\chi_2 k_D^\mu k_D^\nu}{[1+\chi_0]k^2 + \chi_2 (k^2)_\parallel} \right\}, \tag{8.1.45}$$

where the argument k is omitted.

Wave Modes in Invariant Form

The poles in the photon propagator determine the dispersion relations for the modes, and the polarization vector for a particular mode is determined by the tensor in the corresponding numerator. The photon propagator is gauge dependent, and the form (8.1.45) corresponds to the Lorenz gauge, implying that polarization vectors identified from (8.1.45) are in the Lorenz gauge.

The invariant dispersion relations for the $\perp$-mode and the $\parallel$-mode are

$$\begin{aligned} (1+\chi_0)k^2 + \chi_2 k_\perp^2 &= 0, & e_\perp^\mu &\propto k_G^\mu, \\ (1+\chi_0)k^2 + \chi_2 (k^2)_\parallel &= 0, & e_\parallel^\mu &\propto k_D^\mu, \end{aligned} \tag{8.1.46}$$

respectively. The polarization 4-vectors in the Lorenz gauge need to be transformed to the temporal gauge and normalized to find the corresponding polarization 3-vectors. For the $\perp$-mode the polarization 3-vector, $\mathbf{e}_\perp = \mathbf{a}$, is perpendicular to both $\mathbf{k}$ and $\mathbf{B}$; it has no longitudinal component. For the $\parallel$-mode the polarization 4-vector $\propto k_D^\mu$ in the Lorenz gauge implies $\mathbf{e} \propto (k_\perp k_z, 0, -\omega^2 + k_z^2)$ in the temporal gauge. This polarization vector has both longitudinal and transverse components, and writing it in the form $\mathbf{e} = (L\boldsymbol{\kappa} + T\mathbf{t})/(1+L^2+T^2)^{1/2}$, the ratio of the longitudinal and transverse parts is $L/T = (1-n^2)\cot\theta$, where $k^2 = \omega^2(1-n^2)$ defines the refractive index, n. When the dispersion is weak, $n^2 \approx 1$, the longitudinal component is small, and the polarization vector is nearly transverse, with $\mathbf{e}_\parallel \approx \mathbf{t}$.

3-Tensor Form for the Vacuum Response

There are two different 3-tensor descriptions of the response of the magnetized vacuum. The more traditional description is in terms of separate electric and magnetic responses, involving an electric susceptibility tensor and a magnetic susceptibility tensor. When an electrostatic field is included, one also needs additional magneto-electric susceptibility tensors. This form of the response is based on a multipole expansion, with the electric response described by electric moments per unit volume and the magnetic response by magnetic moments per unit volume. This formalism, which is discussed briefly following (8.3.21) below, is not used here. The other 3-tensor form is in terms of an equivalent susceptibility tensor or equivalent dielectric tensor, as used in conventional plasma dispersion theory. After Fourier transforming, one cannot distinguish electric and magnetic effects uniquely, and both are combined in a single **k**-dependent response 3-tensor. The equivalent dielectric tensor $\mathbf{K}(\omega,\mathbf{k}) = \mathbf{1} + \boldsymbol{\chi}(\omega,\mathbf{k})$, differs from the equivalent susceptibility tensor, $\boldsymbol{\chi}(\omega,\mathbf{k})$, by the unit tensor, and $\boldsymbol{\chi}(\omega,\mathbf{k})$ is equal to $\boldsymbol{\Pi}(\omega,\mathbf{k})/\varepsilon_0\omega^2$, where the ij component of $\boldsymbol{\Pi}(\omega,\mathbf{k})$ is identified as $-\Pi^{ij}(k)$.

The equivalent susceptibility tensor for the magnetized vacuum is

$$\boldsymbol{\chi}(\omega,\mathbf{k}) = \begin{pmatrix} \chi_0(1-n_z^2) & 0 & -\chi_0 n_\perp n_z \\ 0 & \chi_0(1-n^2)+\chi_1 n_\perp^2 & 0 \\ -\chi_0 n_\perp n_z & 0 & \chi_0(1-n_\perp^2)+\chi_2 \end{pmatrix}, \tag{8.1.47}$$

with $n_\perp = k_\perp/\omega = n\sin\theta$, $n_z = k_z/\omega = n\cos\theta$, and where the dependence of χ_i on $k^\mu = (\omega,\mathbf{k})$ is implicit. The wave equation in 3-tensor form can be written as the matrix equation

$$\begin{pmatrix} (1-n_z^2)(1+\chi_0) & 0 & -(1+\chi_0)n_\perp n_z \\ 0 & (1-n^2)(1+\chi_0)+\chi_1 n_\perp^2 & 0 \\ -(1+\chi_0)n_\perp n_z & 0 & (1-n_\perp^2)(1+\chi_0)+\chi_2 \end{pmatrix} \begin{pmatrix} e_x \\ e_y \\ e_z \end{pmatrix} = 0. \tag{8.1.48}$$

The solutions of (8.1.48) reproduce the two modes (8.1.46). These solutions are written down explicitly and discussed in § 8.4, cf. (8.4.9) and (8.4.10).

8.2 Schwinger's Proper-Time Method

Schwinger [23] developed what is known as the proper-time method for treating QED processes in a static electromagnetic field. Two related applications the proper-time method are discussed in this section: Schwinger's generalization of the Heisenberg-Euler Lagrangian [12], and spontaneous pair creation in an electromagnetic wrench.

8.2.1 Proper-Time Method

In the proper-time method the propagator, $G(x,x')$, is regarded as the matrix element, $\langle x'|\,\hat{G}\,|x\rangle$, of an operator, $\hat{G}$, between space-time states, $|x\rangle$, and $|x'\rangle$. The essence of the proper-time method is to identify an equivalent dynamical system that determines how the position and momentum operators, x and Π, respectively, evolve as a function of a variable, s, identified as the proper time.

Proper-Time Method: Propagator

The electron propagator may be defined as a solution of the inhomogeneous Dirac equation. It satisfies the differential equation

$$(i\partial\!\!\!/ - eA\!\!\!/(x) + m)G(x, x') = \delta^4(x - x'). \tag{8.2.1}$$

Writing $G(x, x') = \langle x'|\,\hat{G}\,|x\rangle$, and noting $\langle x'|\,|x\rangle = \delta^4(x - x')$, the operator $\hat{G}$ satisfies

$$(\Pi\!\!\!/ - m)G = 1, \qquad \Pi = p + eA, \tag{8.2.2}$$

where the hats on operators are omitted for simplicity. The solution of (8.2.2) is written

$$\hat{G} = \frac{\Pi\!\!\!/ + m}{\Pi\!\!\!/^2 - m^2} = i(\Pi\!\!\!/ + m)\int_0^\infty ds\, \exp\left[i(\Pi\!\!\!/^2 - m^2)s\right]. \tag{8.2.3}$$

The parameter s is (8.2.3) is interpreted as a proper time.

Proper-Time Method: Hamilton's Equations

In the proper-time method, the state $|x\rangle$ is regarded as an eigenfunction of the space-time operator, $\hat{x}$. An equivalent dynamical system is introduced to determine how the eigenstates $|x(0)''\rangle$ and $\langle x(s)'|$ of the operator $x(s)$ evolve with s. Let the Hamiltonian of this equivalent dynamical system be $\mathcal{H}$, which is a function of the coordinate operator, x, and the momentum operator, Π, as well as of the parameter s. Assuming that s plays the role of a time, the evolution of the states in the Schrödinger picture leads to the requirement

$$\langle x'|\exp[-i\mathcal{H}s]\,|x''\rangle = \langle x(s)'|x(0)''\rangle. \tag{8.2.4}$$

In effect, the operator $\exp[-i\mathcal{H}s]$ corresponds to the S-matrix for the system, with $t \to s$, $t_0 \to 0$, $H_\mathrm{I} \to \mathcal{H}$. The notation in (8.2.4) reflects the fact that $x(s)$ is to be identified as x'' for $s = 0$ and as x' for arbitrary s.

The Hamiltonian may be written down by inspection of (8.2.3):

$$\mathcal{H} = \Pi\!\!\!/^2 = \Pi^2 + eS^{\mu\nu}F_{\mu\nu}, \qquad \Pi^\mu = i[\partial^\mu - ieA^\mu(x)]. \tag{8.2.5}$$

For a uniform static electromagnetic field, $F^{\mu\nu}$, Hamilton's equations for the operators x and Π give

$$\frac{dx^\mu}{ds} = -i[x^\mu, \mathcal{H}] = 2\Pi^\mu, \qquad \frac{d\Pi^\mu}{ds} = -i[\Pi^\mu, \mathcal{H}] = 2eF^{\mu\nu}\Pi_\nu, \tag{8.2.6}$$

respectively. Schwinger wrote these equations in the symbolic form $dx/ds = 2\Pi$, $d\Pi/ds = 2eF\Pi$, and wrote down the solutions:

$$\Pi(s) = e^{2eFs}\,\Pi(0), \qquad x(s) - x(0) = \frac{e^{2eFs} - 1}{eF}\,\Pi(0). \tag{8.2.7}$$

The commutation relations $[x^\mu(0), \Pi^\nu(0)] = -ig^{\mu\nu}$ are written symbolically as $[x(0), \Pi(0)] = -i$. The commutation relation of the second of (8.2.7) with $x(0)$ implies

$$[x(s), x(0)] = i\,\frac{e^{2eFs} - 1}{eF}. \tag{8.2.8}$$

The essential step in linking the proper-time method to the propagator is to note that the factor involving $\not{\Pi}^2 \to \mathcal{H}$ in the integrand in (8.2.3) may be written as a matrix element of the position operator at different proper times:

$$\begin{aligned} G(x', x'') &= i(\not{\Pi} + m)\int_0^\infty ds\, e^{-im^2 s}\,\langle x(s)'|x(0)''\rangle \\ &= i\int_0^\infty ds\, e^{-im^2 s}\,[\langle x(s)'|\not{\Pi}(s)|x(0)''\rangle + m\langle x(s)'|x(0)''\rangle]. \end{aligned} \tag{8.2.9}$$

The evaluation of the propagator is thereby reduced to the evaluation of the matrix element in the integrand of (8.2.9). This is determined by

$$\begin{aligned} \frac{d\,\langle x(s)'|x(0)''\rangle}{ds} &= \langle x(s)'|\mathcal{H}|x(0)''\rangle, \\ \langle x(s)'|\Pi_\mu(s)|x(0)''\rangle &= [i\partial''_\mu - eA_\mu(x')]\langle x(s)'|x(0)''\rangle. \end{aligned} \tag{8.2.10}$$

The relevant solutions are

$$\begin{aligned} \langle x(s)'|x(0)''\rangle &= -i\,\frac{\phi(x', x'')}{16\pi^2 s^2}\, e^{-L(s)}\, e^{\frac{1}{4}i(x'-x'')eF\,[\coth(eFs)](x'-x'')}\, e^{-ieSF}, \\ \langle x(s)'|\Pi(s)|x(0)''\rangle &= \frac{1}{2}[eF\coth(eFs) - eF]\,(x' - x'')\,\langle x(s)'|x(0)''\rangle, \\ L(s) &= \frac{1}{2}\mathrm{tr}\,\ln\left[\frac{\sinh eFs}{eFs}\right], \end{aligned} \tag{8.2.11}$$

with $\phi(x', x'')$ given by (5.3.4), $SF = S^{\mu\nu}F_{\mu\nu}$, and where tr denotes the trace over the 4-tensor indices, with $\mathrm{tr}\,F = 0$ due to $F^{\mu\nu}$ being antisymmetric.

8.2.2 Propagator in an Electromagnetic Field

The form (8.2.9) with (8.2.11) for the propagator depends on the electromagnetic field, F, in a manner that can be made explicit by expressing the result in terms of the invariants S, P, defined by (1.1.5).

Propagator in a Magnetostatic Field

The case of a magnetostatic field corresponds to $\boldsymbol{E} = 0$, implying $S = -B^2/2$, $P = 0$. To show that Schwinger's method reproduces the result (5.3.12), it is convenient to start from the first form of (8.2.9).

In a magnetostatic field the Maxwell tensor is of the form

$$F^{\mu\nu} = Bf^{\mu\nu}, \qquad g_{\perp}^{\mu\nu} = -f^{\mu}{}_{\alpha}f^{\alpha\nu}, \qquad g_{\parallel}^{\mu\nu} = g^{\mu\nu} - g_{\perp}^{\mu\nu}. \tag{8.2.12}$$

The quantity $L(s)$ in (8.2.11) is determined by the eigenvalues of $F^{\mu\nu}$: two of the four eigenvalues are zero and the other two are $\pm iB$. After summing over the contributions from each of these eigenvalues, one has $e^{-L(s)} = eBs/\sin(eBs)$. The next factor in (8.2.11) involves a 4-tensor $eFs\coth(eFs)$, which is defined formally in terms of the expansion in a power series in the 4-tensor eFs. In the case of a magnetostatic field, the Maxwell tensor and its powers are confined to the two-dimensional x-y space, whose metric tensor is $g_{\perp}^{\mu\nu}$. The invariant $xeFs\,[\coth(eFs)]x$ separates into a trivial part, $x_{\parallel}^2$, and a part $(xeFs\,[\coth(eFs)]x)_{\perp}$ that is evaluated by summing over the eigenvalues $F \to \pm iB$. Thus one finds

$$xeFs\,[\coth(eFs)]x = x_{\parallel}^2 + x_{\perp}^2\, eBs\,\cot(eBs). \tag{8.2.13}$$

One uses (8.2.13) in (8.2.11) with $x \to x' - x''$ and $(x'-x'')_{\parallel}^2 = (t'-t'')^2 - (z'-z'')^2$, $(x'-x'')_{\perp}^2 = -(x'-x'')^2 - (y'-y'')^2$. The final factor in (8.2.11) is $\exp(-ieSF)$, with $S^{\mu\nu}F_{\mu\nu} = -\boldsymbol{\Sigma}\cdot\boldsymbol{B}$, that is $S^{\mu\nu}F_{\mu\nu} = -\Sigma_z B$ for $\boldsymbol{B}$ along the z axis. Hence one has

$$\exp[-ieSFs] = \exp[ie\Sigma_z Bs] = \cos(eBs) + i\Sigma_z\sin(eBs). \tag{8.2.14}$$

Combining these factors, (8.2.11) reproduces the propagator (5.3.12) with $1/2\lambda$ interpreted as Schwinger's proper time, s.

Propagator for an Electromagnetic Wrench

The generalization to an arbitrary static electromagnetic field, that is, to an electromagnetic wrench, involves assuming the invariant P to be nonzero. In general the eigenvalues, F', of the electromagnetic field, $F^{\mu}{}_{\nu}$, satisfy

$$\det[F^{\mu}{}_{\nu} - F'\delta^{\mu}_{\nu}] = F'^4 - 2SF'^2 - P^2 = 0. \tag{8.2.15}$$

The four eigenvalues are $F' = \pm f_{\pm}$, with $f_{\pm}^2 = S \pm (S^2 + P^2)^{1/2}$ and hence

$$f_{\pm} = \frac{1}{2^{1/2}}\left[(S + iP)^{1/2} \pm (S - iP)^{1/2}\right]. \tag{8.2.16}$$

In the general case of an electromagnetic wrench, one has

$$\begin{aligned}
e^{-L(s)} &= \frac{(es)^2 P}{\mathrm{Im}\{\cosh(esX)\}},\\
xeF\,[\coth(eFs)]x &= x_{\parallel}^2 eE\coth(eEs) + x_{\perp}^2 eB\cot(eBs),\\
\mathrm{tr}\{\exp[-ieSFs]\} &= 4\mathrm{Re}\{\cosh(esX)\},
\end{aligned} \tag{8.2.17}$$

where tr denotes the trace over the 4-tensor indices, and where one has

$$X^2 = (\boldsymbol{B} + i\boldsymbol{E})^2 = -2S + 2iP, \tag{8.2.18}$$

with S, P defined by (1.1.5). The final result for the propagator is

$$\begin{aligned}
G(x' - x'') &= \phi(x', x'')\,\Delta(x' - x''),\\
\Delta(x) &= \frac{-e^2EB}{16\pi^2}\int_0^{\infty} ds\,\frac{e^{-im^2 s}}{\sinh(eEs)\sin(2Bs)}\\
&\quad\times\left[\frac{eE(\gamma x)_{\parallel}\,e^{-i\sigma eBs}}{2\sinh(eEs)} + \frac{eB(\gamma x)_{\perp}\,e^{-\alpha eEs}}{2\sin(eBs)} + m\,e^{-\alpha eEs}\,e^{-i\sigma eBs}\right]\\
&\quad\times\exp\left[-\frac{ieEx_{\parallel}^2}{4\tanh(eEs)} - \frac{ieBx_{\perp}^2}{4\tan(eBs)}\right].
\end{aligned} \tag{8.2.19}$$

The gauge dependent factor, $\phi(x', x'')$, is defined by (5.3.4), and for the choice

$$A^{\mu}(x) = (0, 0, Bx, Et), \tag{8.2.20}$$

it has the explicit form

$$\phi(x', x'') = e^{ie[E(t'+t'')(z'-z'')-B(x'+x'')(y'-y'')]/2}. \tag{8.2.21}$$

8.2.3 *Weisskopf's Lagrangian*

Before using the proper-time method to generalize the Heisenberg-Euler Lagrangian, it is appropriate to consider the original derivation in the 1930s, by Weisskopf [34] using Dirac's hole theory. In Dirac's hole theory the vacuum consists of empty positive-energy states and filled negative-energy states, with a positron identified as a hole in the negative-energy sea.

Effect of a Magnetic Field on the Dirac Sea

The presence of a static electromagnetic field, $F_0^{\mu\nu}$, modifies the energy of the virtual particles in the Dirac sea. It follows that the vacuum energy density for $F_0^{\mu\nu} \neq 0$ is different from that for $F_0^{\mu\nu} = 0$. Suppose that the additional vacuum energy is $U_V(F_0)$, with $U_V(0) = 0$ for $F_0 = 0$. The change in the Lagrangian density is numerically equal to minus this energy density. The nonlinear corrections to the Lagrangian are included in

$$\mathcal{L}_I = -U_V(F_0), \tag{8.2.22}$$

and these nonlinear terms give the Heisenberg-Euler Lagrangian.

The vacuum energy consists of the contributions from all the filled negative energy states. The energy eigenvalues are $\pm(m^2 + p_z^2 + 2neB)^{1/2}$, with the state $n = 0$ being nondegenerate and the states $n \geq 1$ being doubly degenerate. The vacuum energy for the filled negative energy states is identified as

$$U_V = -\frac{eB}{2\pi}\int \frac{dp_z}{2\pi}\sum_{n=0}^{\infty} a_n(m^2 + p_z^2 + 2neB)^{1/2} \tag{8.2.23}$$

where the degeneracy factor is $a_0 = 1$, $a_n = 2$ for $n \geq 1$.

Euler-Maclaurin Summation Formula

The sum over n in (8.2.23) is performed using the Euler-Maclaurin summation formula

$$\sum_{n=0}^{\infty} F(nb) = \frac{1}{2}[F(0) + F(\infty)] + \frac{1}{b}\int_0^{\infty} dx\, F(x) + \sum_{k=1}^{\infty} \frac{B_{2k}b^{2k-1}}{(2k)!}\left[F^{(2k-1)}(\infty) - F^{(2k-1)}(0)\right], \tag{8.2.24}$$

where the B_n are the Bernoulli numbers, with $B_2 = 1/6$, $B_4 = -1/30$, and with $F^{(n)}(x)$ denoting the nth derivative of $F(x)$. In this way one finds

$$U_V = -\frac{1}{2\pi}\int \frac{dp_z}{2\pi}\left[\int_0^\infty dx\,(m^2+p_z^2+x)^{1/2} - \sum_{k=1}^{\infty}\frac{B_{2k}(2eB)^{2k}}{(2k)!}\left.\frac{d^{(2k-1)}(m^2+p_z^2+x)^{1/2}}{dx^{2k-1}}\right|_{x\to 0}\right]. \tag{8.2.25}$$

Regularization

The vacuum energy is divergent and needs to be regularized. Regularization of (8.2.25) requires that the divergent term independent of B be omitted, and that the term proportional to B^2 be replaced by the known correct value for the Lagrangian of the Maxwell field. The nonlinear terms that remain are

$$\operatorname{reg} U_V = \frac{1}{2\pi}\int \frac{dp_z}{2\pi}\sum_{k=2}^{\infty}\frac{B_{2k}b^{2k-1}}{(2k)!}\left.\frac{d^{(2k-1)}(m^2+p_z^2+x)^{1/2}}{dx^{2k-1}}\right|_{x\to 0}. \tag{8.2.26}$$

The integral over p_z gives

$$\int \frac{dp_z}{2\pi}\left.\frac{d^{2k-1}(m^2+p_z^2+x)^{1/2}}{dx^{2k-1}}\right|_{x\to 0} = \frac{\Gamma(2k-2)}{4\pi m^{4(k-1)}}. \tag{8.2.27}$$

The integral representation of the Γ-function,

$$\Gamma(n) = \int_0^\infty d\eta\,\eta^{n-1}\,e^{-\eta}, \tag{8.2.28}$$

allows one to rewrite (8.2.26) with (8.2.27) in the form

$$\operatorname{reg} U_V = \frac{m^4}{8\pi^2}\int d\eta\,\frac{e^{-\eta}}{\eta^3}\sum_{k=2}^{\infty}\frac{2^{2k}B_{2k}}{(2k)!}\left(\frac{\eta eB}{m^2}\right)^{2k}. \tag{8.2.29}$$

It is convenient to replace the variable of integration according to $\eta \to m^2\eta$. After evaluating the sum in (8.2.29) explicitly, one obtains

$$\mathcal{L}_I = -\frac{1}{8\pi^2}\int \frac{d\eta}{\eta^3}\,e^{-m^2\eta}\left[\eta eB\,\coth(\eta eB) - 1 - \frac{1}{3}\eta^2 e^2 B^2\right]. \tag{8.2.30}$$

The result (8.2.30) is for a magnetostatic field.

Weisskopf [34] also evaluated the Lagrangian including an electrostatic field by assuming the electric field to be spatially periodic and allowing the wavelength to go to infinity. In the limit where the spatial period of the field is arbitrarily long, the result corresponds to (8.2.30) with $B \to iE$.

8.2.4 Generalization of Heisenberg-Euler Lagrangian

Schwinger [23] used the proper-time method to rederive and generalize the Heisenberg-Euler Lagrangian.

Lagrangian Density

The Heisenberg-Euler Lagrangian is calculated by considering the energy density in the vacuum using the proper-time method. This energy associated with a variation, δA, in the electromagnetic field is

$$\delta W^{(1)} = \int d^4x\, \delta A_\mu(x)\, \langle J^\mu(x)\rangle = \delta \int d^4x\, \mathcal{L}_I(x), \tag{8.2.31}$$

where $\mathcal{L}_I(x)$ is the nonlinear correction to the Lagrangian and $\langle J^\mu(x)\rangle$ is the vacuum expectation value of the current density.

The calculation of $\langle J^\mu(x)\rangle$ using the proper-time method involves first writing it in the form

$$\langle J^\mu(x)\rangle = ie\, \mathrm{tr}\{\gamma^\mu\, \langle x|G|x\rangle\}, \tag{8.2.32}$$

where G is the operator introduced in (8.2.2). An operator δA is also introduced such that its diagonal matrix elements give the variation in the electromagnetic field:

$$\langle x|\delta A^\mu|x'\rangle = \delta^4(x - x')\, \delta A^\mu(x). \tag{8.2.33}$$

Then using (8.2.3) and the identity

$$- e\delta A = \delta(\not{\Pi} + m), \tag{8.2.34}$$

the variation (8.2.31) is written in the form

$$\delta W^{(1)} = ie\mathrm{Tr}\{\gamma\, \delta A\, G\} = \delta\left[\frac{i}{2}\int_0^\infty \frac{ds}{s}\, e^{-im^2 s}\, e^{i\not{\Pi}^2 s}\right], \tag{8.2.35}$$

where Tr denotes summation over the diagonal terms with respect to both the Dirac spinor indices and the space-time coordinates. To within an additive constant, the Lagrangian density is identified as

$$\mathcal{L}^{(1)}(x) = \frac{i}{2}\int_0^\infty \frac{ds}{s}\, e^{-im^2 s}\, \langle x|e^{i\not{\Pi}^2 s}|x\rangle. \tag{8.2.36}$$

The proper-time method gives $\langle x|\exp(i\Pi^2 s)|x\rangle = \langle x(s)|x(0)\rangle$, and an explicit expression for $\mathcal{L}^{(1)}(x)$ follows from (8.2.11):

$$\mathcal{L}^{(1)}(x) = \frac{1}{32\pi^2}\int_0^\infty \frac{ds}{s^3}\, e^{-im^2 s} e^{-L(s)}\, \mathrm{tr}\{e^{ieSFs}\}. \tag{8.2.37}$$

The total Lagrangian is obtained by subtracting the divergent term for $F = 0$, and also the terms of first order in F^2, and adding the Lagrangian, S/μ_0, for the Maxwell field. The final result for the Lagrangian is

$$\mathcal{L}(x) = \frac{S}{\mu_0} - \frac{1}{8\pi^2}\int_0^\infty \frac{ds}{s^3}\, e^{-im^2 s}\left[(es)^2\, P\, \frac{\mathrm{Re}\,\cosh(esX)}{\mathrm{Im}\,\cosh(esX)} - 1 + \frac{2(es)^2 S}{3}\right], \tag{8.2.38}$$

with $X^2 = (\boldsymbol{B} + i\boldsymbol{E})^2$, as in (8.2.17). The result (8.2.38) is Schwinger's generalization of the Heisenberg-Euler Lagrangian.

Expansion for Weak Fields

The lowest order terms in the expansion of (8.2.38) in powers of the fields are

$$\begin{aligned}\mathcal{L} = \frac{S}{\mu_0} &+ \frac{e^4}{360\pi^2 m^4}(4S^2 + 7P^2) + \frac{e^6}{630\pi^2 m^8}(8S^3 + 13SP^2)\\ &+ \frac{e^8}{945\pi^2 m^{12}}(48S^4 + 88S^2P^2 + 19P^4) + \cdots .\end{aligned} \tag{8.2.39}$$

The nonlinear terms may be rewritten in terms of the fine structure constant, $\alpha_c = e^2/4\pi\varepsilon_0 m = \mu_0 e^2/4\pi m$, and the critical magnetic field, $B_c = m^2/e$. For a magnetostatic field, the nonlinear terms are proportional to α_c/μ_0 times $(B/B_c)^4$, $(B/B_c)^6$, $(B/B_c)^8$, and so on.

8.3 Vacuum in an Electromagnetic Wrench

Inclusion of an electrostatic field, as well as a magnetostatic field, leads to an electromagnetic field referred to as an electromagnetic wrench. In this section, the vacuum polarization tensor for an electromagnetic wrench is evaluated using the proper-time method. At low frequencies, $\omega \ll m$, this reduces to the result derived using the Heisenberg-Euler Lagrangian. The vacuum in an electromagnetic wrench is unstable to spontaneous pair creation, described by an antihermitian part of the response tensor. The nonlinear response tensors for an electromagnetic wrench are also discussed in this section.

8.3.1 Response Tensor for an Electromagnetic Wrench

A derivation of the vacuum response tensor for an electromagnetic wrench is analogous to the derivation in § 8.1 for a magnetostatic field with the propagator given by the proper-time method, (8.2.19) with $S \neq 0$, $P \neq 0$, in place of the Géhéniau form (5.3.13), which corresponds to $S < 0$, $P = 0$. It is possible to choose a set of basis 4-vectors that generalize the set (8.1.1).

The Maxwell 4-tensor and its dual for an electromagnetic wrench may be written

$$F^{\mu\nu} = Bf^{\mu\nu} + E\phi^{\mu\nu} \qquad \mathcal{F}^{\mu\nu} = -Ef^{\mu\nu} + B\phi^{\mu\nu} \tag{8.3.1}$$

respectively, where $f^{\mu\nu}$, $\phi^{\mu\nu}$ are as defined for $E = 0$ by (1.1.6) and (1.1.7) respectively. The resulting (unnormalized) basis 4-vectors, $\tilde{b}_i^\mu$ say, are

$$\tilde{b}_1^\mu = k_\alpha F^{\alpha\mu} = Bb_1^\mu + Eb_2^\mu, \qquad \tilde{b}_2^\mu = k_\alpha \mathcal{F}^{\alpha\mu} = -Eb_1^\mu + Bb_2^\mu, \tag{8.3.2}$$

Unlike the magnetostatic case, the response tensor for an electromagnetic wrench is not diagonal for any choice of real basis 4-vectors. The nonzero off-diagonal terms are along $\tilde{f}_3^{\mu\nu}$, with

$$\tilde{f}_1^{\mu\nu} = \frac{\tilde{b}_1^\mu \tilde{b}_1^\nu}{\tilde{b}_1^2}, \qquad \tilde{f}_2^{\mu\nu} = \frac{\tilde{b}_2^\mu \tilde{b}_2^\nu}{\tilde{b}_2^2}, \qquad \tilde{f}_3^{\mu\nu} = \frac{\tilde{b}_1^\mu \tilde{b}_2^\nu + \tilde{b}_2^\mu \tilde{b}_1^\nu}{2\tilde{b}_1\tilde{b}_2}. \tag{8.3.3}$$

The response tensor is of the form, cf. (8.1.3),

$$\Pi^{\mu\nu}(k) = \sum_{i=0}^{3} \tilde{\Pi}_i(k) \tilde{f}_i^{\mu\nu}, \tag{8.3.4}$$

with $\tilde{f}_0^{\mu\nu} = f_0^{\mu\nu}$ defined by (8.1.3). One finds

$$\begin{aligned}
\tilde{\Pi}_0(k) &= \Phi_0 k^2, \\
\tilde{\Pi}_1(k) &= \frac{B^2 k_\perp^2 + E^2(\omega^2 - k_z^2)}{(B^2 + E^2)^2} (B^2\Phi_1 + E^2\Phi_2 + 2EB\Phi_3), \\
\tilde{\Pi}_2(k) &= \frac{B^2(\omega^2 - k_z^2) + E^2 k_\perp^2}{(B^2 + E^2)^2} (E^2\Phi_1 + B^2\Phi_2 - 2EB\Phi_3), \\
\tilde{\Pi}_3(k) &= \frac{2EB[(B^2 - E^2)\Phi_3 - EB(\Phi_1 - \Phi_2)]}{(B^2 + E^2)^2} k^2,
\end{aligned} \tag{8.3.5}$$

with the quantities Φ_i defined in terms of an integral over functions ϕ_i:

$$\Phi_i = \frac{e^4 EB}{(2\pi)^2} \int_0^\infty d\alpha \int_0^\alpha d\beta \, \frac{\phi_i e^{-im^2\alpha}}{\sin(eB\alpha)\,\sinh(eE\alpha)}$$
$$\exp\left\{ i \left[\frac{\cos(eB\alpha) - \cos(eB\beta)}{2eB\,\sin(eB\alpha)} k_\perp^2 + \frac{\cosh(eE\alpha) - \cosh(eE\beta)}{2eE\,\sinh(eE\alpha)} (\omega^2 - k_z^2) \right] \right\}, \tag{8.3.6}$$

$$\phi_0 = \frac{\cos(eB\beta)\cosh(eE\beta)}{2}\left[1 - \frac{\tan(eB\beta)\tanh(eE\beta)}{\tan(eB\alpha)\tanh(eE\alpha)}\right],$$

$$\phi_1 = \cosh(eE\alpha)\,\frac{\cos(eB\alpha) - \cos(eB\beta)}{\sin^2(eB\alpha)} + \phi_0,$$

$$\phi_2 = \cos(eB\alpha)\,\frac{\cosh(eE\alpha) - \cosh(eE\beta)}{\sinh^2(eE\alpha)} - \phi_0,$$

$$\phi_3 = -\frac{1}{2}\left\{ \frac{[1 - \cos(eB\alpha)\cos(eB\beta)][1 - \cosh(eE\alpha)\cosh(eE\beta)]}{\sin(eB\alpha)\sinh(eE\alpha)} + \sin(eB\beta)\sinh(eE\beta) \right\}. \tag{8.3.7}$$

The results (8.3.1)–(8.3.7) apply to an arbitrary electromagnetic wrench, and include the special cases of a magnetostatic field ($B \neq 0$, $E = 0$) and an electrostatic field ($B = 0$, $E \neq 0$).

8.3.2 Response Tensors for $\omega \ll m$

In the low-frequency limit, the linear an nonlinear response of the vacuum with an electromagnetic wrench can be derived from the Heisenberg-Euler Lagrangian. One writes $F^{\mu\nu} = F_0^{\mu\nu} + F_1^{\mu\nu}$, where $F_0^{\mu\nu}$ corresponds to an electromagnetic wrench, and $F_1^{\mu\nu}$ describes a test field to which the vacuum responds.

Interaction Energy

The existence of linear and nonlinear responses implies a nonzero interaction energy $\int d^4x\, J(x)\delta A(x)$, where $J(x)$ is the 4-current induced by a perturbation, $\delta A(x)$, in the electromagnetic field.

After Fourier transforming, the weak-turbulence expansion of the current allows the linear and nonlinear response tensors to be identified:

$$J^\mu(k) = \Pi^{\mu\nu}(k)A_\nu(k) + \sum_{n=2}^{\infty} \int d\lambda^{(n)}\, \Pi^{(n)\mu\nu_1\dots\nu_n}(-k, k_1, \dots, k_n) A_{\nu_1}(k_1)\dots A_{\nu_n}(k_n), \tag{8.3.8}$$

where the convolution integral is defined by (1.1.32). One then has

$$\int d^4x\, J(x)\delta A(x) = \delta\Bigg[\frac{1}{2}\int \frac{d^4k}{(2\pi)^4} J^\mu(k)\, A_\mu(-k) A_\nu(k) + \sum_{n=2}^{\infty}\frac{1}{n+1}\int d\lambda^{(n)}\, \Pi^{(n)\nu_0\nu_1\dots\nu_n}(k_0, k_1, \dots, k_n) \times A_{\nu_0}(k_0)A_{\nu_1}(k_1)\dots A_{\nu_n}(k_n)\Bigg], \tag{8.3.9}$$

where the factor $1/(n+1)$ arises from the dependence of the nth term on the right hand side on the $(n+1)$th power of the wave amplitude, and the symmetry of the response tensors under arbitrary permutations of the suffices $0, \dots, n$.

The interaction energy is also related to the Lagrangian of the field,

$$\int d^4x\, J^\mu(x)\delta A_\mu(x) = \delta \int d^4x\, \mathcal{L}(x), \tag{8.3.10}$$

with $\mathcal{L}(x) \to \mathcal{L}_I$ identified as the Heisenberg-Euler Lagrangian (8.2.38). This leads to the identifications

$$\mu_0 \Pi^{\mu\nu}(k) = k_\alpha k_\beta \frac{\partial^2 \mathcal{L}_I}{\partial F_{\alpha\mu}\partial F_{\beta\nu}}, \tag{8.3.11}$$

for the linear response tensor, and

$$\Pi^{\nu_0\nu_1\dots\nu_n}(k_0, k_1, \dots, k_n) = \frac{(-i)^{n+1}}{n!} k_0^{\alpha_0} k_1^{\alpha_1} \dots k_n^{\alpha_n} \frac{\partial^{n+1}\mathcal{L}_I}{\partial F^{\alpha_0}{}_{\nu_0} F^{\alpha_1}{}_{\nu_1} \dots F^{\alpha_n}{}_{\nu_n}}, \tag{8.3.12}$$

for the nonlinear response tensors. The linear and nonlinear response tensors derived in this way apply only at low frequencies and long wavelengths, which corresponds to $|\omega_i| \ll m$, $|\boldsymbol{k}_i| \ll m$ for all $i = 0, \dots n$.

Derivatives of the Heisenberg-Euler Lagrangian

The differentiation involved in (8.3.11) and (8.3.12) may be carried out by first noting that the Heisenberg-Euler Lagrangian depends on $F_{\mu\nu}$ only through the invariants

$$S = -\frac{1}{4}F^{\mu\nu}F_{\mu\nu} = \frac{1}{2}(\boldsymbol{E}^2 - \boldsymbol{B}^2), \qquad P = -\frac{1}{4}F^{\mu\nu}\mathcal{F}_{\mu\nu} = \boldsymbol{E}\cdot\boldsymbol{B}. \tag{8.3.13}$$

The derivatives are

$$\frac{\partial S}{\partial F_{\mu\nu}} = -F^{\mu\nu}, \qquad \frac{\partial P}{\partial F_{\mu\nu}} = \mathcal{F}^{\mu\nu}. \tag{8.3.14}$$

One has

$$\frac{\partial F^{\alpha\beta}}{\partial F_{\mu\nu}} = g^{\mu\alpha}g^{\nu\beta} - g^{\mu\beta}g^{\nu\alpha}, \qquad \frac{\partial \mathcal{F}^{\alpha\beta}}{\partial F_{\mu\nu}} = \epsilon^{\mu\nu\alpha\beta}, \tag{8.3.15}$$

where the definition $\mathcal{F}^{\mu\nu} = \frac{1}{2}\epsilon^{\mu\nu\alpha\beta}F_{\alpha\beta}$, of the dual is used, with the permutation symbol $\epsilon^{\mu\nu\alpha\beta}$ completely antisymmetric with $\epsilon^{0123} = 1$.

Susceptibility 4-Tensor for the Vacuum

There are two different descriptions of the linear response of the vacuum, and the one used here is in terms of the linear response tensor, $\Pi^{\mu\nu}(k)$. The alternative (and older) description is in terms of electric and magnetic susceptibility 3-tensors; a magneto-electric susceptibility 3-tensor is also required in general. These 3-tensors may be combined into a single fourth rank susceptibility 4-tensor. The derivation from the Heisenberg-Euler Lagrangian leads to this tensor directly.

The Maxwell tensor, $F^{\mu\nu}$, is assumed to consist of two parts:

$$F^{\mu\nu} = F_0^{\mu\nu} + \delta F^{\mu\nu}, \tag{8.3.16}$$

where $F_0^{\mu\nu}$ is the static field and

$$\delta F^{\mu\nu}(x) = -i\int\frac{d^4k}{(2\pi)^4}\,e^{ikx}\left[k^{\mu}A^{\nu}(k) - k^{\nu}A^{\mu}(k)\right], \tag{8.3.17}$$

describes a test field. One then expands in powers of $\delta F^{\mu\nu}(x)$. The term linear in $\delta F^{\mu\nu}(x)$ allows one to identify the fourth-rank susceptibility 4-tensor,

$$M^{\mu\nu} = \chi^{\mu\nu}{}_{\rho\sigma}\,\delta F^{\rho\sigma}, \qquad \chi^{\mu\nu\rho\sigma} = \frac{1}{2}\frac{\partial^2\mathcal{L}_I}{\partial F_{\mu\nu}\partial F_{\rho\sigma}}. \tag{8.3.18}$$

The linear response tensor is related to the susceptibility 4-tensor through (8.3.11):

$$\mu_0\Pi^{\mu\nu}(k) = 2\,k_{\alpha}k_{\beta}\chi^{\alpha\mu\beta\nu}(k) = k_{\alpha}k_{\beta}\frac{\partial^2\mathcal{L}_I}{\partial F_{\alpha\mu}\partial F_{\beta\nu}}. \tag{8.3.19}$$

Expansion of the Linear Response Tensor

The general expression for the linear response tensor may be expanded in powers of the background electromagnetic field, $F^{\mu\nu}$, where the subscript 0 is now redundant. This gives

$$\mu_0\Pi^{\mu\nu}(k) = \frac{e^2}{45\pi B_c^2}\left[4S(k^2g^{\mu\nu} - k^\mu k^\nu) + 4k_\alpha F^{\alpha\mu}k_\beta F^{\beta\nu} + 7k_\alpha \mathcal{F}^{\alpha\mu}k_\beta \mathcal{F}^{\beta\nu}\right]$$

$$+ \frac{2\alpha_c}{315\pi B_c^4}\Big[(24S^2 + 13P^2)(k^2g^{\mu\nu} - k^\mu k^\nu) + 48S\, k_\alpha F^{\alpha\mu}k_\beta F^{\beta\nu},$$

$$+ 26S\, k_\alpha \mathcal{F}^{\alpha\mu}k_\beta \mathcal{F}^{\beta\nu} - 26P\left(k_\alpha F^{\alpha\mu}k_\beta \mathcal{F}^{\beta\nu} + k_\alpha \mathcal{F}^{\alpha\mu}k_\beta F^{\beta\nu}\right)\Big]. \tag{8.3.20}$$

In a frame in which the electric and magnetic fields are parallel and along the 3-axis and $\boldsymbol{k}$ is in the 1–3 plane, the leading terms in the response tensor (8.3.20) have the form

$$\mu_0\Pi^{\mu\nu} = \frac{e^2}{45\pi B_c^2}\left[2(B^2-E^2)\begin{pmatrix} -k_\perp^2 - k_z^2 & -\omega k_\perp & 0 & -\omega k_z \\ -\omega k_\perp & -\omega^2 + k_z^2 & 0 & -k_\perp k_z \\ 0 & 0 & -k^2 & 0 \\ -\omega k_z & -k_\perp k_z & 0 & -\omega^2 + k_\perp^2 \end{pmatrix}\right.$$
$$+ 4\begin{pmatrix} k_z^2E^2 & 0 & k_\perp k_z EB & -\omega k_z E^2 \\ 0 & 0 & 0 & 0 \\ k_\perp k_z EB & 0 & k_\perp^2 B^2 & -\omega k_\perp EB \\ -\omega k_z E^2 & 0 & -\omega k_\perp EB & \omega^2 E^2 \end{pmatrix}$$
$$\left. + 7\begin{pmatrix} k_z^2B^2 & 0 & -k_\perp k_z EB & -\omega k_z B^2 \\ 0 & 0 & 0 & 0 \\ -k_\perp k_z EB & 0 & k_\perp^2 E^2 & \omega k_\perp EB \\ -\omega k_z B^2 & 0 & \omega k_\perp EB & \omega^2 B^2 \end{pmatrix}\right]. \tag{8.3.21}$$

The terms in the 3-tensor components of (8.3.21) include not only an electric response and magnetic response, but also a magneto-electric response, which is the component along $f_3^{\mu\nu}$, cf. (8.3.3). With the response written in terms of the polarization, $\boldsymbol{P}$, and the magnetization, $\boldsymbol{M}$, one has

$$\boldsymbol{P}/\varepsilon_0 = \boldsymbol{\chi}^{(e)}\cdot\boldsymbol{E} + \boldsymbol{\chi}^{(em)}\cdot\boldsymbol{B},$$
$$\mu_0\boldsymbol{M} = \boldsymbol{\chi}^{(m)}\cdot\boldsymbol{B} + \boldsymbol{\chi}^{(me)}\cdot\boldsymbol{E}, \tag{8.3.22}$$

where the χ are the susceptibility 3-tensors. The terms proportional to ω^2 are included in $\chi^{(e)}$, the terms independent of ω^2 are included in $\chi^{(m)}$ and the terms proportional to ω are included in $\chi^{(em)}$ and $\chi^{(me)}$. The magneto-electric response of the vacuum is nonzero in an electromagnetic wrench, when there is a nonzero component of $\boldsymbol{E}$ along $\boldsymbol{B}$.

In the absence of an electrostatic field ($E = 0$), (8.3.20) describes the response of a magnetized vacuum. In this case, comparison with (8.1.1) leads to the identifications

$$k_\alpha F_0^{\alpha\mu} k_\beta F_0^{\beta\nu} = -k_\perp^2 B^2 f_1^{\mu\nu}, \qquad k_\alpha \mathcal{F}_0^{\alpha\mu} k_\beta \mathcal{F}_0^{\beta\nu} = -(\omega^2 - k_z^2) B^2 f_2^{\mu\nu}. \tag{8.3.23}$$

Then (8.3.20) is of the form (8.1.3), leading to the identifications

$$\frac{\mu_0 \Pi_0(k)}{k^2} = -\frac{2\alpha_c B^2}{45\pi B_c^2}, \qquad \frac{\mu_0 \Pi_1(k)}{k_\perp^2} = \frac{4\alpha_c B^2}{45\pi B_c^2}, \qquad \frac{\mu_0 \Pi_2(k)}{\omega^2 - k_z^2} = \frac{7\alpha_c B^2}{45\pi B_c^2}, \tag{8.3.24}$$

which is equivalent to (8.1.23).

8.3.3 Nonlinear Response Tensors for $\omega \ll m$

The derivation from the Heisenberg-Euler Lagrangian provides expressions for the quadratic and cubic response tensor purpose that are valid at low frequencies, $\omega \ll m$.

Quadratic Nonlinear Response Tensor

The term $n = 2$ in (8.3.12) gives

$$\Pi^{\mu\nu\rho}(k_0, k_1, k_2) = \frac{i}{2} k_0^\alpha k_1^\beta k_2^\gamma \frac{\partial^3 \mathcal{L}_I}{\partial F^\alpha{}_\mu \partial F^\beta{}_\nu \partial F^\gamma{}_\rho}. \tag{8.3.25}$$

Inserting the explicit form for the Heisenberg-Euler Lagrangian and carrying out the indicated operations gives

$$\begin{aligned}
\mu_0 \Pi^{\mu\nu\rho}(k_0, k_1, k_2) = {} & \frac{i\alpha_c}{90\pi B_c^2}\big[4(k_0 k_1\, g^{\mu\nu} - k_1^\mu k_0^\nu)\big) k_{2\gamma} F^{\gamma\rho} \\
& \qquad + 7 k_{0\alpha} k_{1\beta} \epsilon^{\mu\alpha\nu\beta} k_{2\gamma} \mathcal{F}^{\gamma\rho} + \text{perm.}\big] \\
& + \frac{i\alpha_c}{315\pi B_c^4}\big[- \big(16 k_{0\alpha} F^{\alpha\mu} k_{1\beta} F^{\beta\nu} + 26 k_{0\alpha} \mathcal{F}^{\alpha\mu} k_{1\beta} \mathcal{F}^{\beta\nu}\big) k_{2\gamma} F^{\gamma\rho} \\
& \qquad - \big(k_0 k_1\, g^{\mu\nu} - k_1^\mu k_0^\nu)\big)\big(48 S\, k_{2\gamma} F^{\gamma\rho} - 26 P\, k_{2\gamma} \mathcal{F}^{\gamma\rho}\big) \\
& \qquad + 26 k_{0\alpha} k_{1\beta} \epsilon^{\mu\alpha\nu\beta}\big(S\, k_{2\gamma} \mathcal{F}^{\gamma\rho} + P\, k_{2\gamma} F^{\gamma\rho}\big) + \text{perm.}\big],
\end{aligned} \tag{8.3.26}$$

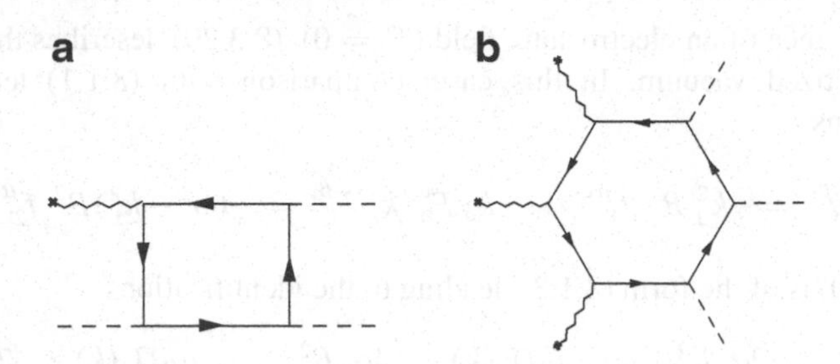

Fig. 8.1 (**a**) The box diagram, and (**b**) the hexagon diagram, whose Feynman amplitudes are used to calculate the transition rate for photon scattering to first and third order in B, respectively. Three of the vertices are associated with the three photons, and the other one and three, respectively with B

where "+perm." indicates two additional sets of terms obtained from those written, one by making the interchanges $k_0, \mu \leftrightarrow k_2, \rho$ and the other by making the interchanges $k_1, \nu \leftrightarrow k_2, \rho$.

The two terms in (8.3.26) are proportional to the first power, F, and the third power, F^3, of the external field. These arise from the Feynman amplitudes for the box and hexagon diagrams, respectively, as illustrated in Fig. 8.1. For the box diagram one of the external lines is associated with the static electromagnetic field, and the other three vertices with the three waves involved; for the hexagon diagram, three of the vertices are associated with the static electromagnetic field, leading to the dependencies as F and F^3, respectively.

Cubic Response Tensor

The same procedure applied to the cubic response leads to a leading term that is independent of the external field:

$$\mu_0 \Pi^{\mu\nu\rho\sigma}(k_0, k_1, k_2, k_2) = \frac{\alpha_c}{270\pi B_c^2}\big[4\big(k_0 k_1\, g^{\mu\nu} - k_1^\mu k_0^\nu\big)\big(k_2 k_3\, g^{\rho\sigma} - k_3^\rho k_2^\sigma\big)$$
$$+ 7 k_{0\alpha} k_{1\beta} \epsilon^{\mu\alpha\nu\beta} k_{2\alpha} k_{3\beta} \epsilon^{\rho\alpha\sigma\beta} + \text{perm.}\big], \qquad (8.3.27)$$

where "+perm." corresponds to two other pairs of terms obtained from the pair written by the interchanges $k_1, \nu \leftrightarrow k_2, \rho$ and $k_1, \nu \leftrightarrow k_3, \sigma$.

8.3.4 *Spontaneous Pair Creation*

An electrostatic field and an electromagnetic wrench are intrinsically unstable to decay into pairs. A simple physical model is for a static electric field, E, along the x axis, with potential $\Phi = -Ex$. Consider a virtual pair at $x = 0$. Quantum mechanical tunneling implies a nonzero probability of the electron and positron appearing as real particles after tunneling a distance such the $e\Phi$ exceeds the rest energy m, that is, with a separation $>2m/eE$.

The probability of spontaneous pair creation is identified by noting that probability amplitude for a system whose action integral is A is $\exp(iA)$ ($\exp(iA/\hbar)$ in ordinary units), so that the probability of the system remaining in its initial state decays as $|\exp(iA)|^2 = \exp(-2\text{Im}\{A\})$. Assuming the probability of decay is small, it is $2\text{Im}\{A\} \ll 1$. With $A = \int d^4x\, \mathcal{L}(x)$, the probability of spontaneous pair creation per unit time and per unit volume is identified as

$$w_{\text{SPC}} = 2\text{Im}\{\mathcal{L}(x)\}, \tag{8.3.28}$$

where SPC denotes spontaneous pair creation, with $\mathcal{L}(x)$ identified as the Heisenberg-Euler Lagrangian.

The imaginary part of the Heisenberg-Euler Lagrangian, in the form (8.2.38), arises from the term in the integrand involving the function $\cosh(esX)$. On writing $X = a + ib$, one has

$$(es)^2\, P\,\frac{\text{Re}\{\cosh(esX)\}}{\text{Im}\{\cosh(esX)\}} = (eas)(ebs)\frac{\cosh(eas)\cos(ebs)}{\sinh(eas)\sin(ebs)}. \tag{8.3.29}$$

The singularities of the function (8.3.29) occur where $(ebs)\cot(ebs) = \infty$, the solutions of which are $s = s_n$, with

$$s_n = \frac{\pi n}{eb}, \qquad n = 1, 2, \ldots. \tag{8.3.30}$$

The imaginary parts are determined by applying the Landau prescription to each of these singularities. This gives

$$w_{\text{SPC}} = \frac{e^2}{4\pi^2}ab\sum_{n=1}^{\infty}\frac{1}{n}\exp\left(-n\pi\frac{m^2}{eb}\right)\coth\left(n\pi\frac{a}{b}\right). \tag{8.3.31}$$

Further evaluation of (8.3.31) requires explicit values for a, b, with $X = a + ib$. From the definition (8.2.18), one has $X^2 = (\boldsymbol{B} - i\boldsymbol{E})^2 = B^2 - E^2 + 2i\,\boldsymbol{E}\cdot\boldsymbol{B}$. In the case of an electrostatic field, $\boldsymbol{B} = 0$, one may choose $a = 0, b = E$. Then (8.3.31) gives [23]

$$w_{\text{SPC}} = \frac{e^2E^2}{4\pi^2}\sum_{n=1}^{\infty}\frac{1}{n^2}\exp\left(-n\pi\frac{m^2}{eE}\right). \tag{8.3.32}$$

For an electromagnetic wrench, choosing the frame in which the electric and magnetic fields are parallel, one has $a = B, b = E$. Then (8.3.31) gives [8]

$$w_{\rm SPC} = \frac{e^2 EB}{4\pi^2} \sum_{n=1}^{\infty} \frac{1}{n} \exp\left(-n\pi \frac{m^2}{eE}\right) \coth\left(n\pi \frac{B}{E}\right). \tag{8.3.33}$$

The probability $w_{\rm SPC}$ is zero for a magnetostatic field, which does not decay, and nonzero for an electrostatic field or an electromagnetic wrench, which do decay.

The probability (8.3.32) involves an exponential factor with exponent $\propto -E_c/E$, with $E_c = m^2/e$ the critical electric field. The critical electric field is the electric counterpart of the critical magnetic field B_c (5.1.28); in SI units its value is

$$E_c = \frac{m^2 c^3}{\hbar e} = 1.3 \times 10^{18}\,{\rm V\,m^{-1}}. \tag{8.3.34}$$

In virtually all contexts one has $E \ll E_c$ and the derivation of (8.3.31) is valid only if this condition is satisfied. An exception where spontaneous pair creation is of interest and where this condition is not satisfied is for a model of bare strange stars (composed of u, d and s quarks) where there is a surface layer with an electric field $\gg E_c$ [2, 30]; in this case the surface layer contains degenerate electrons with a sufficiently highly relativistic Fermi energy that strongly suppresses spontaneous pair creation through the Pauli exclusion principle.

Although the Heisenberg-Euler Lagrangian allows one to determine the rate of decay of an electric field, it does not determine the spectrum of the pairs produced. For an electrostatic field, the spectrum of the pairs is an ill-defined concept because each electron and positron is subject to ongoing linear acceleration. For an electromagnetic wrench, the solution of Dirac's equation factorizes into a part describing the motion in the t-z plane and a part describing the motion in the x-y plane, where $\boldsymbol{E}$ and $\boldsymbol{B}$ are assumed along the z axis. The solutions in the x-y plane may be described in terms of the Landau states. Each pair produced must have the same eigenvalues as the vacuum, which requires that the electron and positron be in the same Landau state with the same spin. However, the approach based on the Heisenberg-Euler Lagrangian provides no information on the distribution of the pairs in these Landau states. (Note that integer n in the sum in (8.3.33) is not the Landau quantum number.)

8.4 Waves in Strongly Magnetized Vacuum

The magnetized vacuum is birefringent. For weak fields, $B \ll B_c$, the properties of the two natural wave modes are well known, having been derived in the 1930s [12, 34]. These properties are rederived here and generalized to include the case where the magnetic field is not necessarily weak. The high-frequency limit ($\omega \gg m$) and the generalization to an electromagnetic wrench are also discussed.

8.4.1 *Weak-Field, Weak-Dispersion Limit*

The properties of waves in a birefringent vacuum can be derived from the Heisenberg-Euler Lagrangian for low-frequencies, $\omega \ll m$, and weak fields, $B \ll B_c$. The dispersion relations may be written in covariant form, $k^2 = k_\pm^2$, or in terms of the refractive indices, $|\mathbf{k}|^2/\omega^2 = n_\pm^2$, with $k_\pm^2 = \omega^2(1 - n_\pm^2)$. The polarization vectors of the two modes are transverse, in the plane spanned by the vectors $\boldsymbol{t}, \boldsymbol{a}$.

The $\perp$ and $\parallel$ Modes

The usual labelling of the vacuum wave modes is as the $\perp$ and $\parallel$ modes, but unfortunately there are two different conventions. The more widely used convention is to define these in terms of the direction of the electric vector in the wave relative to the projection of the magnetostatic field on the plane orthogonal to $\boldsymbol{k}$, with $\boldsymbol{t} = \boldsymbol{e}_\parallel$ and $\boldsymbol{a} = \boldsymbol{e}_\perp$. The alternative convention [1] is to define the $\perp$ and $\parallel$ modes in terms of the direction of the magnetic vector in the wave relative to the projection of the magnetostatic field on the plane orthogonal to $\boldsymbol{k}$ and this interchanges the labelling $\perp \leftrightarrow \parallel$ compared with the more widely used convention. The $\parallel$ and $\perp$ modes were denoted the two- and three-modes in the Russian literature [24].

Susceptibilities for the Magnetized Vacuum

The response of the vacuum can be described by the fourth rank susceptibility 4-tensor defined by the first of (8.3.18), viz. $M^{\mu\nu} = \chi^{\mu\nu}{}_{\rho\sigma}\,\delta F^{\rho\sigma}$. The nonzero components for a magnetostatic field can be written in term of electric and magnetic susceptibilities that are different perpendicular and parallel to the magnetostatic field. These may be written as

$$\chi_\perp^{(e)} = -\frac{2\alpha_c}{45\pi}\frac{B^2}{B_c^2}, \quad \chi_\parallel^{(e)} = \frac{\alpha_c}{9\pi}\frac{B^2}{B_c^2}, \quad \chi_\perp^{(m)} = \frac{2\alpha_c}{45\pi}\frac{B^2}{B_c^2}, \quad \chi_\parallel^{(m)} = \frac{2\alpha_c}{15\pi}\frac{B^2}{B_c^2}, \tag{8.4.1}$$

with $\alpha_c = e^2/4\pi\varepsilon_0 \approx 1/137$ the fine structure constant, and $B_c = m_e^2/e \approx 4.4 \times 10^9\,T$ the critical magnetic field. The nonzero components of the fourth rank 4-response susceptibility tensor are

$$\chi^{0101} = \chi^{0202} = -\frac{1}{2}\chi_\perp^{(e)}, \qquad \chi^{0303} = -\frac{1}{2}\chi_\parallel^{(e)},$$
$$\chi^{1313} = \chi^{2323} = -\frac{1}{2}\chi_\perp^{(m)}, \qquad \chi^{1212} = -\frac{1}{2}\chi_\parallel^{(m)}, \tag{8.4.2}$$

with the other nonzero terms determined the antisymmetry in both the first-written and the second-written pair of indices. The construction of $\Pi^{\mu\nu}(k)$ follows from (8.3.18).

The simplest approximation to the wave modes corresponds to the weak-anisotropy approximation, where only the transverse components $t^{\mu\nu}(k) = \mu_0 \Pi^{\mu\nu}(k)$ of the response tensor contribute. The components for $\mu, \nu = a, t$ are

$$t^{aa} = -\omega^2 \chi_\perp^{(e)} - |\mathbf{k}|^2 \left(\chi_\parallel^{(m)} \sin^2\theta + \chi_\perp^{(m)} \cos^2\theta \right), \qquad t^{at} = 0 = t^{ta},$$
$$t^{tt} = -\omega^2 \left(\chi_\parallel^{(e)} \sin^2\theta + \chi_\perp^{(e)} \cos^2\theta \right) - |\mathbf{k}|^2 \chi_\perp^{(m)}. \tag{8.4.3}$$

The properties of waves in the magnetized vacuum follow by inserting (8.4.3) into (3.6.6) and (3.6.7).

Approximate Dispersion Relations

To lowest order in B^2/B_c^2, the resulting dispersion relations are

$$k_\pm^2 = -\frac{(11 \pm 3)\alpha_c}{90\pi} \frac{B^2}{B_c^2} \omega^2 \sin^2\theta, \qquad n_\pm^2 = 1 + \frac{(11 \pm 3)\alpha_c}{90\pi} \frac{B^2}{B_c^2} \sin^2\theta, \tag{8.4.4}$$

and the polarization 4-vectors are

$$e_+^\mu = t^\mu = (0, \cos\theta, 0, -\sin\theta), \qquad e_-^\mu = a^\mu = (0, 0, 1, 0), \tag{8.4.5}$$

where the $\boldsymbol{k}$ is assumed to be in the 1–3 plane. The polarization vectors (8.4.5) are referred to as $\parallel$ and $\perp$ polarizations, respectively.

The next highest order terms in the expansion in B^2/B_c^2 in the dispersion relations (8.4.4) were given by Herold et al. [13]:

$$n_\parallel^2 = 1 + \frac{\alpha_c}{\pi} \left(\frac{7}{45\pi} \frac{B^2}{B_c^2} - \frac{26}{315} \frac{B^4}{B_c^4} + \frac{176}{945} \frac{B^6}{B_c^6} + \frac{52}{945} \frac{\omega^2 \sin^2\theta}{\Omega_e^2} \frac{B^6}{B_c^6} \right) \sin^2\theta,$$

$$n_\perp^2 = 1 + + \frac{\alpha_c}{\pi} \left(\frac{4}{45\pi} \frac{B^2}{B_c^2} - \frac{16}{105} \frac{B^4}{B_c^4} + \frac{64}{105} \frac{B^6}{B_c^6} + \frac{4}{135} \frac{\omega^2 \sin^2\theta}{\Omega_e^2} \frac{B^6}{B_c^6} \right) \sin^2\theta,$$
$$\tag{8.4.6}$$

where the leading frequency-dependent terms are retained, and where the $+, -$ modes are relabeled as the $\parallel, \perp$ modes, respectively.

Waves in an Electromagnetic Wrench

The wave properties (8.4.4) and (8.4.5) can be derived from the leading term in the expansion of the Heisenberg-Euler Lagrangian in B/B_c for $E = 0$. Various generalizations follow by relaxing these assumptions.

The inclusion of an electrostatic field with a component along the magnetostatic field (an electromagnetic wrench) modifies both the dispersion relations and the polarization vectors compared with the case $\boldsymbol{E} \cdot \boldsymbol{B} = 0$. In place of (8.4.4) one has

$$n_{\pm}^2 = 1 + \frac{(11 \pm 3)\alpha_c}{90\pi} \frac{(E^2 + B^2)}{B_c^2} \sin^2\theta, \tag{8.4.7}$$

and in place of (8.4.5) one has

$$e_{+}^{\mu} = \frac{(0, B\cos\theta, -E, B\sin\theta)}{(E^2 + B^2)^{1/2}}, \qquad e_{-}^{\mu} = \frac{(0, E\cos\theta, B, -E\sin\theta)}{(E^2 + B^2)^{1/2}}. \tag{8.4.8}$$

The results (8.4.7) and (8.4.8) include the special case $E \neq 0, B = 0$. The refractive indices for the two modes are the same as for the case $E = 0$, $B \neq 0$ with B replaced by E, but with the polarization vectors of the two modes interchanged.

8.4.2 Vacuum Wave Modes: General Case

In the general case, when the weak field, $B \ll B_c$, and low-frequency, $\omega \ll m$, approximations are not made, the properties of the wave modes may determined by the wave equation (8.1.43), with the regularized vacuum response tensor given by the forms derived in § 8.1. In the weak-dispersion limit, one may approximate the χ_i by evaluating them for $k^2 = 0$, and this reproduces the results derived using the Heisenberg-Euler Lagrangian. The generalization to $B \gtrsim B_c$ without assuming weak dispersion involves retaining the exact forms for the $\chi_i(k)$.

General Dispersion Relations

The dispersion equation in 3-tensor form (8.1.48) may be solved for the wave properties. Changing notation by writing $n_\perp = n\sin\theta$, $n_z = n\cos\theta$, and labeling the two modes as $\perp$ and $\parallel$, the resulting wave dispersion relations are

$$n_\perp^2 = \frac{1 + \chi_0}{1 + \chi_0 - \chi_1 \sin^2\theta}, \qquad n_\parallel^2 = \frac{1 + \chi_0 + \chi_2}{1 + \chi_0 + \chi_2 \cos^2\theta}. \tag{8.4.9}$$

The dispersion relations (8.4.9) are explicit only in the approximation where the k-dependence of the $\chi_i(k)$ is neglected. More generally, when the k-dependence is included, (8.4.9) are implicit equations for $n_\perp, n_\parallel$, respectively.

The polarization 3-vectors for the two modes are

$$\boldsymbol{e}_\perp = \boldsymbol{a}, \qquad \boldsymbol{e}_\parallel = \frac{((1+\chi_0+\chi_2)\cos\theta, 0, -(1+\chi_0)\sin\theta)}{[(1+\chi_0+\chi_2)^2\cos^2\theta + (1+\chi_0)^2\sin^2\theta]^{1/2}}. \tag{8.4.10}$$

The $\perp$-mode is strictly transverse, with $\boldsymbol{e}_\perp = \boldsymbol{a} = (0,1,0)$. The $\parallel$ mode has a longitudinal component, which is exhibited by writing

$$\boldsymbol{e}_\parallel = \frac{K_\parallel \boldsymbol{\kappa} + T_\parallel \boldsymbol{t}}{(K_\parallel^2 + T_\parallel^2)^{1/2}}, \qquad K_\parallel = \chi_2 \cos\theta \sin\theta, \quad T_\parallel = 1 + \chi_0 + \chi_2 \cos^2\theta. \tag{8.4.11}$$

The results in the remainder of this section follow from (8.4.9) and (8.4.10).

Weak-Field, Weak-Dispersion

First consider application of (8.4.9) and (8.4.10) to the weak-dispersion, weak-field limit. In the weak-field limit the χ_i are small, of order B^2/B_c^2, and retaining only the leading order terms in an expansion in B^2/B_c^2 gives

$$\chi_0 = -\frac{2\alpha_c}{45\pi}\frac{B^2}{B_c^2}, \qquad \chi_1 = \frac{4\alpha_c}{45\pi}\frac{B^2}{B_c^2}, \qquad \chi_2 = \frac{7\alpha_c}{45\pi}\frac{B^2}{B_c^2}. \tag{8.4.12}$$

To lowest order in an expansion in the χ_i the dispersion relations (8.4.9) are independent of χ_0 and reduce to $n_{\perp,\parallel}^2 = 1 + \chi_{1,2}\sin^2\theta$. On inserting the values (8.4.12) into (8.4.9) the wave properties (8.4.4) and (8.4.5) are reproduced.

Strong-Field, Weak-Dispersion

The strong-field limit corresponds to $B \gtrsim B_c$. In this case, $\Pi_0(k)$ and $\Pi_1(k)$ are small, being of order $\exp(-B/B_c)$, and $\Pi_2(k)$ is approximated by (8.1.24). For $4m^2 \gg \omega^2 - k_z^2$, $k_\perp^2 \ll 2eB$ (8.1.24) reduces to (8.1.42), giving $\chi_2 \approx (\mu_0 e^2/12\pi^2)(B/B_c)$, and one has

$$n_\perp^2 = 1 + \frac{\alpha_c}{3\pi}\left[1 - \frac{3\ln(2B/B_c)}{2B/B_c}\right]\sin^2\theta,$$

$$n_\parallel^2 = \begin{cases} 1 + \dfrac{\alpha_c}{3\pi}\dfrac{B}{B_c}\sin^2\theta & \text{for } B \ll (3\pi/e^2)B_c, \\ \dfrac{1}{\cos^2\theta} & \text{for } B \gg (3\pi/e^2)B_c. \end{cases} \tag{8.4.13}$$

The terms retained in (8.4.13) reproduce the leading terms in an expansion derived by Tsai and Erber [29], cf. also [14]. In the limit $B \gg (3\pi/e^2)B_c = 2.5 \times 10^{13}\,T$, the modes defined by (8.4.13) may be interpreted as MHD modes with Alfvén speed $v_A \gg 1$. The $\perp$ and $\parallel$ modes are counterparts of the magnetoacoustic and Alfvén modes, respectively.

High-Frequency, Weak-Field Limit

The foregoing results for the wave dispersion apply for frequencies well below the pair-creation threshold, $\omega^2 \sin^2\theta \ll 4m^2$. The response tensor in the form (8.3.4)–(8.3.7) applies both below and above this threshold and so may be used to derive the wave properties at higher frequencies.

The dispersion at sufficiently high frequencies necessarily approaches the (unmagnetized) vacuum value $k^2 = 0$, implying that weak-dispersion limit applies at sufficiently high frequency. Denoting the values in the weak-dispersion limit by an asterisk, the ϕ_i that appear in (8.3.7) reduce to

$$\chi_i^* = -\frac{e^4}{4\pi}\int_0^1 d\eta\,(1-\eta^2)\,\phi_i^*$$

$$\times \int_0^\infty d\alpha\alpha \cos\left(m^2\alpha + \frac{(1-\eta^2)^2\alpha^3 e^2(E^2+B^2)\omega^2\sin^2\theta}{48}\right)$$

$$\phi_1^* = \left(1-\frac{\eta^2}{3}\right)E^2 + \left(\frac{1}{2}+\frac{\eta^2}{6}\right)B^2,$$

$$\phi_2^* = \left(\frac{1}{2}+\frac{\eta^2}{6}\right)E^2 + \left(1-\frac{\eta^2}{3}\right)B^2, \qquad \phi_3^* = -\frac{1-\eta^2}{2}EB, \tag{8.4.14}$$

with $\eta = \beta/\alpha$. The α-integral in (8.4.14) is performed in terms of the derivative, $\mathrm{Gi}'(z)$, of an Airy function

$$\mathrm{Gi}\,(z) = \frac{1}{\pi}\int_0^\infty dt\,\sin\left(zt + \frac{1}{3}t^3\right). \tag{8.4.15}$$

The integral in (8.4.14) gives

$$\chi_i^* = \frac{e^4}{(2\xi^2)^{2/3}m^4}\int_0^1 d\eta\,\frac{\phi_i^*}{(1-\eta^2)^{1/3}}\,\mathrm{Gi}'\,([2/\xi(1-\eta^2)]^{2/3}),$$

$$\xi = \frac{\omega(E^2+B^2)^{1/2}\sin\theta}{2mB_c}. \tag{8.4.16}$$

The limiting forms of Gi′ are

$$\mathrm{Gi}'(z) = -\frac{1}{\pi z^2} \quad \text{for } z \gg 1, \qquad \mathrm{Gi}'(z) = \frac{\Gamma(2/3)}{2\pi 3^{1/3}} \quad \text{for } z \ll 1. \tag{8.4.17}$$

On using (8.4.17) to approximate (8.4.16), the upper limit applies for $\xi \ll 1$ and leads to the wave properties that reproduce those derived using the weak-field limit of the Heisenberg-Euler Lagrangian.

The lower limit in (8.4.17) applies for $\xi \gg 1$ and leads to

$$n_\perp^2 = 1 - 2A\frac{(E^2 + B^2)}{B_c^2}\sin^2\theta, \qquad n_\parallel^2 = 1 - 3A\frac{(E^2 + B^2)}{B_c^2}\sin^2\theta, \tag{8.4.18}$$

with $A = 3[\Gamma(2/3)]^2\alpha_c/7\pi^{1/2}\Gamma(1/6)(2\xi^2)^{2/3}$. The polarization vectors are the same as in (8.4.8).

High Frequency, $n(\omega) < 1$

According to (8.4.18) the refractive indices at high frequencies are less than unity, whereas they are greater than unity at low frequencies. The limit $k^2 \to 0$ for $\omega \to \infty$ must be approached from above, $k^2 > 0$. Put another way, the difference between the refractive index and unity for each mode must change sign, from positive to negative, with increasing ω, and approach zero from negative values. This is implied by the sum rule

$$\int_0^\infty d\omega\,[n(\omega) - 1] = 0, \tag{8.4.19}$$

which is well known for an isotropic system [3]. This sum rule requires a negative contribution, $n_{\perp,\parallel}(\omega) < 1$, at high frequency to balance the positive contribution, $n_{\perp,\parallel}(\omega) > 1$, at low frequency. The refractive indices depend on ω only in the combination $\lambda = (3/2)(\omega/m)(B/B_c)$; they increase monotonically with λ for $\lambda < 1.2$, decrease monotonically for $1.2\lambda \le 24$, crossing unity, and then increase towards the limit (8.4.18).

The form (8.4.16) describes the real part of χ_i^*, and the corresponding imaginary part describes damping due to one-photon absorption. The real and imaginary parts are related by the Kramers-Kronig relations. The imaginary part may be identified by noting that the real and imaginary parts of the Airy functions satisfy the Kramers-Kronig relations: the imaginary part is obtained from (8.4.16) by the replacement $\mathrm{Gi}'(z) \to -i\,\mathrm{Ai}'(z)$.

8.4.3 Vacuum Plus Cold Electron Gas

The combination of the magnetized vacuum and a cold electron gas has interesting properties, due to waves in the vacuum alone having refractive index greater than unity and being linearly polarized, and waves in the electron gas alone having refractive index less than unity and being approximately circularly polarized. The so-called vacuum resonance is a feature of the combined system.

Combined Susceptibility Tensor

The combined effect of dispersion in a cold electron gas and the magnetized vacuum is of particular interest in connection with X-ray pulsars, where the plasma parameters satisfy $\omega_p \ll \Omega_e \ll m$, and the frequencies of interest, $\omega \lesssim \Omega_e$, include the cyclotron resonance and its low harmonics [1, 9, 15, 16, 20–22]. The response tensor for such a plasma is the sum of contributions from the plasma and the magnetized vacuum. Qualitatively, at high frequencies, $\omega \gg \Omega_e$, the difference between the refractive indices and unity for the modes of the plasma alone is negative and decreases in magnitude with increasing $\omega \gg \Omega_e$, whereas the contribution from the magnetized vacuum is positive and is independent of ω for $\omega \ll m$. The natural modes of the combined system at $\Omega_e \ll \omega \ll m$ must be plasma-like at lower frequencies where the negative contribution dominates and vacuum-like at higher frequencies where the positive contribution dominates.

The susceptibility tensor for the vacuum plus cold electron gas is the sum of the separate susceptibility tensors for the vacuum and the cold electron gas. It is convenient to use 3-tensor notation in this case. The vacuum contribution is given by (8.1.47), and the susceptibility for a cold electron gas follows from (1.2.29) with (1.2.36),

$$\chi_c(\omega) = \begin{pmatrix} S-1 & -iD & 0 \\ iD & S-1 & 0 \\ 0 & 0 & P-1 \end{pmatrix}, \tag{8.4.20}$$

with

$$S-1 = -\frac{X}{1-Y^2}, \qquad D = -\frac{\epsilon XY}{1-Y^2}, \qquad P-1 = -X, \tag{8.4.21}$$

with $X = \omega_p^2/\omega^2$, $Y = \Omega_e/\omega$, and with $\epsilon = 1$ for an electron gas, and $\epsilon = 0$ for a pure pair plasma.

Dispersion Equation for Combined System

For an electron gas, $D \neq 0$, the combined plasma plus vacuum is gyrotropic, so that the natural modes are elliptically polarized. This general case may be treated as a modified form of a cold electron gas, with the χ_i assumed independent of k. The dispersion equation reduces to the cold-plasma form (3.2.2) and (3.2.4),

$$An^4 - Bn^2 + C = 0, \qquad n^2 = n_\pm^2 = \frac{B \pm F}{2A}, \qquad F = (B^2 - 4AC)^{1/2}. \tag{8.4.22}$$

The generalization of the coefficients (3.2.3) to include the vacuum contribution gives

$$A = (1 + \chi_0)(1 + \chi_0 - \chi_1 \sin^2\theta)][P' \cos^2\theta + S' \sin^2\theta],$$

$$B = (1 + \chi_0)(S'^2 - D^2)\sin^2\theta + P'S'[(1 + \chi_0)(1 + \cos^2\theta) - \chi_1 \sin^2\theta],$$

$$C = P'(S'^2 - D^2), \tag{8.4.23}$$

with

$$S' = S + \chi_0. \qquad P' = P + \chi_0 + \chi_2. \tag{8.4.24}$$

An explicit expression for F^2 is

$$F^2 = [P'S'(1 + \chi_0 - \chi_1) - (1 + \chi_0)(S'^2 - D^2)]^2 \sin^4\theta$$
$$+4(1 + \chi_0)^2 P'^2 D^2 (1 + \chi_0 - \chi_1 \sin^2\theta)\cos^2\theta. \tag{8.4.25}$$

The results (8.4.22)–(8.4.25) reduces to the expression given by (3.2.7) for $\chi_i \to 0$, and they give the vacuum modes (8.4.9) for $P, S \to 1$, $D \to 0$.

The transverse parts of the polarization of the modes become elliptical when the plasma is included, and may be described by the axial ratio, T, of the polarization ellipse. As in a cold plasma, T satisfies a quadratic equation: the generalization of (3.2.6) is

$$T^2 - 2bT - 1 = 0, \qquad b = \frac{[P'S'(1 + \chi_0 - \chi_1) - (1 + \chi_0)(S'^2 - D^2)]\sin^2\theta}{2(1 + \chi_0)P'D\cos\theta}, \tag{8.4.26}$$

and the solutions, $T = T_\pm$, that generalize (3.2.7) are

$$T_\pm = b \pm (1 + b^2)^{1/2}. \tag{8.4.27}$$

These become linear polarizations, $T_\pm \to 0, \infty$, for $b \to \infty$ and circular polarizations, $T = \pm 1$, to $b \to 0$.

8.4.4 Vacuum Resonance

The combination of plasma dispersion and vacuum polarization leads to an intrinsically new effect, referred to as the vacuum resonance [9, 20, 21]. This corresponds to the parameter b in (8.4.26) passing through zero, at which point the modes are circularly polarized. In the cold electron gas alone, the modes are elliptically polarized in general, and are only circularly polarized for parallel propagation.

On inserting the expressions (8.4.21) in (8.4.26), assuming that X and the χ_i are all much smaller than unity, and expanding in these quantities, one finds

$$b = -\frac{(1-Y^2)\sin^2\theta}{2XY\cos\theta}\left[\frac{XY^2}{1-Y^2} + 3\delta\right], \tag{8.4.28}$$

where (8.4.12) is used to write $\chi_2 - \chi_1 = 3\delta$ with $\delta = (\alpha_c/45\pi)(B^2/B_c^2)$. One can have $b = 0$ for $X \ll 1$ and $Y^2 > 1$, when the two terms inside the square brackets in (8.4.28) cancel each other. This cancelation occurs at an electron number density $n_e = n_c$ with (in ordinary units)

$$n_c = \frac{1}{60\pi^2}\left(\frac{mc^2}{\hbar}\right)^3 \frac{|1-Y^2|}{Y^4}\left(\frac{B}{B_c}\right)^4, \tag{8.4.29}$$

with $Y^2 > 1$.

The vacuum resonance is of interest in connection with hard photons escaping from a pulsar [31]. If the photon is emitted in a region where the magnetic field is strong enough for the dispersion to be dominated by the vacuum polarization, as it propagates away from the star B decreases, and it can enter a region where the dispersion due to the plasma dominates.

Interpretation of the Vacuum Resonance

In interpreting the vacuum resonance, there are two separate effects: crossing of the dispersion curves and the vanishing of the linear polarization. Both effects occur at $F = 0$, implying $n_+^2 = n_-^2$ and $|T_\pm| = 1$. The significance of the vacuum resonance is associated with the polarization of the modes, which changes from nearly linear to nearly circular as the resonance is crossed. This may be illustrated by considering a case where there is a mode crossing, $n_+^2 = n_-^2$, but no change from linear to circular polarization. Such a case is the vacuum plus a pure pair plasma.

For a pure pair plasma, when one has $D = 0$, so that F is a perfect square. The $-$ solution implies $T_- = 0$, so that the polarization vector corresponds to the $\perp$ mode with the $+$ mode implying $T_+ = \infty$ corresponding to the transverse part of the polarization of the $\parallel$ mode. For $D = 0$, the refractive indices for the two modes become

$$n_+^2 = \frac{(S+\chi_0)(P+\chi_0+\chi_2)}{(1+\chi_0)[(P+\chi_0+\chi_2)\cos^2\theta + (S+\chi_0)\sin^2\theta]},$$
$$n_-^2 = \frac{S+\chi_0}{1+\chi_0-\chi_1\sin^2\theta}. \tag{8.4.30}$$

The $\pm$ modes in (8.4.30) reduce to the $\parallel$ mode (8.4.9), and $\perp$ mode (8.4.10), respectively, in the absence of the plasma, $S, P \to 1$; they reduce to the o and x modes for a pure pair plasma, given by (3.3.15) and (3.3.14), respectively, when the vacuum contribution is neglected, $\chi_i \to 0$. In the approximation in which X and the χ_i are assumed small, the refractive indices of the two modes are approximated by

$$n_+^2 - 1 \approx -\frac{XY^2}{1-Y^2}\cos^2\theta - X\sin^2\theta + 7\delta\sin^2\theta, \qquad n_-^2 - 1 \approx -\frac{XY^2}{1-Y^2} + 4\delta\sin^2\theta. \tag{8.4.31}$$

In this case the two refractive index curves pass through unity at different frequencies that both depend on θ; the two curves also cross each other at another point, specifically at $X/(1-Y^2) = 3\delta$, independent of θ. However, there is no effect on the polarization, with the + mode being a combination of $\|$-transverse and longitudinal polarization, and the − mode having $\perp$ polarization.

The new feature in the vacuum resonance is that the contributions to the linear polarization from the vacuum and from the plasma cancel, resulting in circular polarization. As the case of a pure pair plasma illustrates, the refractive indices of the two modes can pass through unity (as functions of either ω or θ) without affecting the polarization.

8.5 Photon Splitting

Photon splitting is the decay of a single photon into two photons, and it relies on the quadratic nonlinear response of the magnetized vacuum. Photon splitting is of interest in connection with emission from pulsars [1, 6, 11].

8.5.1 Photon Splitting as a Three-Wave Interaction

Photon splitting is a three-wave process, and the probability for photon splitting in vacuo is formally the same as that for three-wave interactions in a plasma (§ 5.7.1 of volume 1), with the nonlinear response tensor interpreted as that of the magnetized vacuum.

Quadratic Nonlinearity of the Vacuum

The quadratic nonlinear response tensor of the vacuum may be calculated in three different ways. At low frequencies, $\omega \ll m$, the nonlinearity can be treated using the Heisenberg-Euler Lagrangian; the quadratic nonlinear response is derived in this way in § 8.3.3. A general form of this approach involves calculating the Heisenberg-Euler Lagrangian using the proper-time method. Using the resulting form, (8.2.38), allows one to calculate the splitting for arbitrarily strong fields; it also allows one to include an electric field (electromagnetic wrench) [32]. For weak fields, and alternative approach involves treating the virtual electrons in vacuum as unmagnetized, and calculating the interaction involving three photons from the amplitude of a Feynman diagram. The lowest order diagram is a box, with three of the four vertices corresponding to the three photons, and the fourth to a zero-frequency field identified as the background field, **B**. Higher order terms in B are included by considering higher order diagrams with the additional vertices also associated with **B**. Only diagrams with an even number of sides contribute to the response of the unmagnetized vacuum, and hence the next lowest order contribution is $\propto B^3$ from the hexagon diagram, cf. Fig. 8.1. The contribution from the hexagon diagram dominates, for $B/B_c \ll 1$, leading to a transition amplitude $\propto (eB)^3$ and hence a transition rate $\propto \alpha_c^3 B^6$.

An exact treatment of photon splitting involves calculating the quadratic nonlinear response tensor for the magnetized vacuum without assuming low frequencies (or weak fields). As with the linear response tensor for the magnetized vacuum, § 8.1, there are two general approaches to this calculation, involving using different forms for the electron propagator. Using the Géhéniau form (5.3.13) for the electron propagator leads to an expression for the nonlinear response tensor that involves integrals over elementary functions [26]. This approach is restricted to the vacuum, and does not allow the inclusion of real electrons. An alternative approach, as for the linear response § 8.1.4, involves using the vertex formalism, which allows the theory to be developed in momentum space rather than coordinate space. This allows one to include the response of real electrons using the same formalism. The theory of photon splitting may the be extended to include the effect of the plasma, but this is significant only under extreme conditions [7], and is not discussed here.

Kinematics of Photon Splitting

The birefringence of the magnetized vacuum implies six possible three-wave interactions involving the two modes. To see this, let a specific splitting be denoted by $M, P, Q = \parallel, \perp$, denoting the $\parallel$- and $\perp$-modes. Due to the crossing symmetry of the probability of the three-wave interaction, the probability for $M \to P + Q$ also describes $P \to M + Q$ and $Q \to M + P$ simply by reversing the signs of the relevant wave 4-vectors. There are four different three-wave interactions, corresponding to $MPQ = \perp\perp\perp; \perp\perp\parallel; \perp\parallel\parallel; \parallel\parallel\parallel$. Only the first and third of these are allowed, with the other two being zero because the projection of the polarization vectors onto the quadratic nonlinear response tensor gives zero, sometimes attributed to CP invariance, as discussed below.

Conservation of 4-momentum in a photon splitting process requires that the three-wave matching condition on the wave 4-vectors, $k = k' + k''$, be satisfied. This condition is

$$\omega = \omega' + \omega'', \qquad \mathbf{k} = \mathbf{k}' + \mathbf{k}''. \tag{8.5.1}$$

A splitting is kinematically forbidden if the dispersion relations do not allow (8.5.1) to be satisfied for that specific splitting.

Consider the three-wave matching conditions (8.5.1) for waves in a magnetized vacuum in the weak-dispersion limit. To lowest order in B/B_c, the refractive indices are approximated by unity. The conditions (8.5.1) are satisfied only if the three photons are collinear. To show this, note that when one has $k^2 = 0$, $k'^2 = 0$, $k''^2 = 0$, the condition $k^2 = (k' + k'')^2$ requires $k'k'' = 0$, that is, $\omega'\omega'' = \boldsymbol{k}' \cdot \boldsymbol{k}''$. Then $\omega' = |\boldsymbol{k}'|$, $\omega'' = |\boldsymbol{k}''|$ requires that to satisfy $\omega'\omega'' = \boldsymbol{k}' \cdot \boldsymbol{k}''$ the angle between $\boldsymbol{k}'$, $\boldsymbol{k}''$ must be zero. A similar consideration of $k''^2 = (k - k')^2$ implies that the angle between $\boldsymbol{k}$, $\boldsymbol{k}'$ is π. It follows that the three photons are collinear.

Assuming that the three photons are collinear, one is free to make a Lorentz transformation such that the initial photon is propagating perpendicular to the magnetic field. (An exception is when the initial photon is propagating along $\boldsymbol{B}$, but then photon splitting is forbidden.) Specifically, for a photon propagating in an arbitrary direction a boost along $\boldsymbol{B}$ can be made such that in the new frame the photon is propagating perpendicular to $\boldsymbol{B}$. Then the two final photons are propagating perpendicular to $\boldsymbol{B}$ in the same direction as the initial photon. The perpendicular wavenumbers, $k_\perp = \omega \sin\theta$, $k'_\perp = \omega' \sin\theta'$, $k''_\perp = \omega'' \sin\theta''$, and the magnetic field, $\boldsymbol{B}$, are unchanged as a result of this Lorentz transformation. It follows that in the laboratory frame, the collinearity condition (for weak dispersion) reduces to $\sin\theta = \sin\theta' = \sin\theta''$. Choosing the frame in which all three photons are propagating perpendicular to $\boldsymbol{B}$ to treat the splitting, the matrix element is expressed in terms of the variables in the laboratory frame simply by interpreting the wavenumbers according to $k_\perp = \omega \sin\theta$, $k'_\perp = \omega' \sin\theta$, $k''_\perp = \omega'' \sin\theta$.

Probability for Photon Splitting

The probability for photon splitting is formally identical to the probability for a three-wave interaction, § 5.7.1 of volume 1. The probability is given by (5.7.4) with (5.7.5) of volume 1, viz.

$$w_{MPQ}(-k, k', k'') = 4\mu_0^3 \frac{R_M(k)R_P(k')R_Q(k'')}{\omega_M(k)\omega_P(k')\omega_Q(k'')} \times |\Pi^{MPQ}(-k, k', k'')|^2 (2\pi)^4 \delta^4(k_M - k'_P - k''_Q),$$

$$\Pi^{MPQ}(-k, k', k'') = e_M^{*\mu}(k) e_P^{\nu}(k') e_Q^{\rho}(k'') \Pi^{(2)}{}_{\mu\nu\rho}(-k_M, k'_P, k''_Q). \qquad (8.5.2)$$

The probability (8.5.2) applies when all three wave modes are different; if the modes P, Q are the same, the factor 4 is replaced by 2.

In photon splitting, the modes are identified as those of the magnetized vacuum, and the quadratic response tensor, $\Pi^{(2)}{}_{\mu\nu\rho}(k_0, k_1, k_2)$, is identified as that of the magnetized vacuum with $k_0 = -k$, $k_1 = k'$, $k_2 = k''$. For photon splitting $M \to P + Q$ it is convenient to rewrite (8.5.2) as

$$w_{MPQ}(-k,k',k'') = \mu_0 \frac{|M^{MPQ}(-k,k',k'')|^2}{4\omega\omega'\omega''}(2\pi)^4\delta^4(k-k'-k''),$$
$$M^{MPQ}(-k,k',k'') = e_M^{*\mu} e_P^{\nu} e_Q^{\rho} \mu_0 \Pi^{(2)}{}_{\mu\nu\rho}(-k,k',k''), \tag{8.5.3}$$

where $k^\mu = k_M^\mu$, $k'^\nu = k_P'^\nu$, $k''^\rho = k_Q''^\rho$ implicitly satisfy the relevant dispersion relation, and where the ratio of electric to total energy for each of the modes is approximated by $1/2$. The quantity $M^{MPQ}(-k,k',k'')$ is referred to as the matrix element for the process $M \to P + Q$.

8.5.2 *Three-Wave Interactions in the Vacuum*

In treating photon splitting in the magnetized vacuum, the photons are identified as in the $\parallel$ and $\perp$ modes of the magnetized vacuum. In the weak-dispersion limit, where all three photons are collinear, the invariants kk', kk'', $k'k''$ all vanish, and the two directions in the transverse plane, $b_1^\mu \propto k_\alpha F^{\alpha\mu}$, $b_2^\mu \propto k_\alpha \mathcal{F}^{\alpha\mu}$, are the same for all three photons. These properties lead to the matrix elements for three of the possible six splittings being zero in the weak-field, weak-dispersion limit. The matrix element is nonzero only if there is an even number of photons with $\parallel$-polarization and an odd number with $\perp$-polarization. One interpretation of this constraint is in terms of CP-invariance [1].

CP Invariance in the Weak-Field Limit

The selection rule on the modes for photon splitting arises as follows. The 4-vectors $(e_\parallel)_\mu = b_1^\mu \propto k_\alpha F^{\alpha\mu}$ and $(e_\perp)_\mu = b_2^\mu \propto k_\alpha \mathcal{F}^{\alpha\mu}$ appear in the quadratic response tensor (8.3.26) and in the projection of it onto the relevant polarization vectors in the matrix element (8.5.2). Consider the terms in (8.3.26) that are linear in the strength of the external field. The terms shown explicitly are $4(k_0k_1\, g^{\mu\nu} - k_1^\mu k_0^\nu)k_{2\gamma}F^{\gamma\rho}$ and $7k_{0\alpha}k_{1\beta}\epsilon^{\mu\alpha\nu\beta}k_{2\gamma}\mathcal{F}^{\gamma\rho}$. We have $k_0 \to -k$, $k_1 \to k'$, $k_2 \to k''$ here. The projection of the polarization vectors onto these terms gives zero. Term by term, this arises as follows. The term proportional to $g^{\mu\nu}$ gives zero due to $kk' = 0$; the term proportional to $k^\mu k'^\nu$ gives zero because the waves are transverse; and the projection onto $k_\alpha k'_\beta \epsilon^{\mu\alpha\nu\beta}$ vanishes because k, k' are parallel. Hence, the lowest order contribution comes from the terms proportional to the cube of the external magnetic field in (8.3.26).

It is simplest to calculate the matrix element in the frame in which the three photons are propagating perpendicular to the magnetic field. In this frame, the polarization vectors of the $\perp$ and $\parallel$ modes are along the 2 and 3 axes, respectively. Hence the only components of the quadratic response tensor that contribute are $\Pi^{(2)\mu\nu\rho}$ with $\mu,\nu,\rho = 2,3$. By inspection of (8.3.26) one finds that there is a nonzero contribution only if all three of μ, ν, ρ are equal to 3, or if one is equal to 3 and the other two to 2. The terms written explicitly in (8.3.26) have coefficients 16 and 26, and when the symmetries are taken into account, the matrix elements for these two cases are in the ratio 48 to 26. The fact that the matrix elements are nonzero only if an even number of the photons have $\parallel$ polarization is a selection rule for photon splitting.

Adler [1] gave an argument for the selection rule based on the invariance of QED under the parity, P, and charge-conjugation, C, transformations to explain why half the possible splitting are forbidden. The invariance of QED under C implies that only even powers of the electromagnetic field appear, and hence that the theory is invariant under $\boldsymbol{B} \to -\boldsymbol{B}$, $\boldsymbol{E} \to -\boldsymbol{E}$. The P transformation reverses the signs of vectors but not of axial vectors, implying the QED is invariant under $\boldsymbol{k} \to -\boldsymbol{k}$, $\boldsymbol{B} \to \boldsymbol{B}$, $\boldsymbol{E} \to -\boldsymbol{E}$. Combining these, CP invariance requires that the theory be unchanged by the replacements

$$\text{CP-transformation:} \quad \boldsymbol{k} \to -\boldsymbol{k}, \quad \boldsymbol{B} \to -\boldsymbol{B}, \quad \boldsymbol{E} \to \boldsymbol{E}. \tag{8.5.4}$$

Applying the CP transformation to the wave modes (8.3.13), one finds $\boldsymbol{e}_\perp \to +\boldsymbol{e}_\perp$, $\boldsymbol{e}_2 \to -\boldsymbol{e}_2$, so that the $\perp$ mode has eigenvalue $\eta_{\rm CP} = +1$ and $\parallel$ mode has eigenvalue $\eta_{\rm CP} = -1$ under this transformation.

In photon splitting, CP must be conserved. Hence, the selection rule is satisfied if and only if the number of photons of mode 2 is even. Thus the allowed decays (when only this selection rule is taken into account) are $\perp\to\perp, \perp$ and $\parallel\to\perp, \parallel$. An unrelated constraint forbids $\perp\to\perp, \perp$ at low frequencies and weak fields.

8.5.3 Decay Rates in the Weak-Field Approximation

The decay rate is given by integrating the relevant probability over the phase space of the final photons:

$$R_{MPQ} = \int \frac{d^3\boldsymbol{k}'}{(2\pi)^3} \frac{d^3\boldsymbol{k}''}{(2\pi)^3} \, w_{MPQ}(-k, k', k''). \tag{8.5.5}$$

On noting the dependence of the probability (8.5.3) on k, k', k'', the integral that needs to be evaluated is

$$J(\omega) = \int \frac{d^3\boldsymbol{k}'}{(2\pi)^3} \frac{d^3\boldsymbol{k}''}{(2\pi)^3} \, \omega\omega'\omega'' \, (2\pi)^4 \delta^4(k - k' - k''). \tag{8.5.6}$$

It is straightforward to evaluate $J(\omega)$ in the weak-dispersion limit, $k^2 = 0$, $k'^2 = 0$, $k''^2 = 0$. After performing the $\boldsymbol{k}''$-integral over $\delta^3(\boldsymbol{k} - \boldsymbol{k}' - \boldsymbol{k}'')$ and using the dispersion relations to write $|\boldsymbol{k}| = \omega$, $|\boldsymbol{k}'| = \omega'$, $|\boldsymbol{k}''| = \omega''$, one has

$$J(\omega) = \int_0^\omega \frac{d\omega'\omega'^2}{2\pi} \int_{-1}^1 d\cos\chi\, \omega\omega'\omega''\, \delta(\omega - \omega' - \omega''), \tag{8.5.7}$$

with $\omega'' = (\omega^2 + \omega'^2 - 2\omega\omega'\cos\chi)^{1/2}$, where χ is the angle between $\boldsymbol{k}, \boldsymbol{k}'$. The $\cos\chi$-integral is performed over the remaining δ-function, and the ω'-integral is then elementary, giving

$$J(\omega) = \frac{\omega^5}{60\pi}. \tag{8.5.8}$$

In the approximation in which the dispersion is neglected, the decay rate for photon splitting has a simple form for $B/B_c \ll 1$, $\omega \ll m$. This is

$$R_{MPQ} = \frac{\alpha_c^3 m}{60\pi^2} \left(\frac{B\sin\theta}{B_c}\right)^6 \left(\frac{\omega}{m}\right)^5 |M_{MPQ}(B)|^2, \tag{8.5.9}$$

with $M, P, Q = \perp, \parallel$. The nonzero transitions for $B/B_c \ll 1$ have

$$M_{\perp\perp\perp}(0) = \frac{48}{315}, \qquad M_{\perp\parallel\parallel}(0) = \frac{26}{315}. \tag{8.5.10}$$

The other transition rates are equal to 0 in this approximation. These results are derived in the approximation in which the refractive indices are all assumed equal to unity, and when $n_\perp - 1 \neq 0$ is taken into account, the splitting $\perp\to\perp,\perp$ is also forbidden.

Inclusion of Weak Dispersion

The weak-dispersion approximation needs to be complemented with a detailed consideration of the implications of the actual dispersion. In the absence of dispersion, the three photons are collinear, and hence one has $\cos\chi = 1$ in (8.5.7). The inclusion of the small differences of the refractive indices from unity changes the angles between the three photons slightly. If this change tends to reduce $\cos\chi$ then the process is allowed and (8.1.21) is a good approximation to the integral $J(\omega)$. However, if this change tends to increase $\cos\chi$, then $\cos\chi$ becomes nonphysical, and the three-wave matching cannot be satisfied. Such splittings are kinematically forbidden and in such cases (8.1.21) is to be replaced by $J(\omega) = 0$.

In the weak-field low-frequency limit, for the two modes one has $k = \omega(1 + \delta n_{\perp,\parallel})$ with $\delta n_\parallel = (7/4)\delta n_\perp > \delta n_\perp$, where (8.3.13) is used. Splitting is forbidden for $|\boldsymbol{k}| > |\boldsymbol{k}'| + |\boldsymbol{k}''|$, but not for $|\boldsymbol{k}| < |\boldsymbol{k}'| + |\boldsymbol{k}''|$, when a small angle, ψ say, between $\boldsymbol{k}'$ and $\boldsymbol{k}''$ allows $\boldsymbol{k} = \boldsymbol{k}' + \boldsymbol{k}''$ to be satisfied. This condition forbids the splittings $\parallel\to\perp,\perp$ and $\parallel\to\perp,\parallel$. For the splittings $\perp\to\perp,\perp$ and $\parallel\to\parallel,\parallel$ the $\omega \to 0$ limits of $\delta n_{\perp,\parallel}$ are inadequate to determine whether three-wave matching is possible. Adler [1] considered the frequency dependence of the refractive indices and argued that in an expansion in $\omega/m \ll 1$, the lowest order nonzero contribution is $\propto \omega^2$ and is positive for both modes. When this dependence is taken into account, one finds $|\boldsymbol{k}| - |\boldsymbol{k}'| + |\boldsymbol{k}''| \propto n''(0)[\omega'^3 + \omega''^3 - (\omega' + \omega'')^3 < 0$, with $n''(\omega) = d^2n(\omega)/d\omega^2$, which applies for both $\perp \to \perp, \perp$ and $\| \to \|, \|$. It follows that both the splitting $\perp \to \perp, \perp$ and $\| \to \|, \|$ are forbidden in the low frequency regime. The only allowed splitting is $\perp \to \|, \|$.

Low-Frequencies and Arbitrary Field Strengths

Assuming that the low-frequency approximation remains valid, one may relax the weak-field approximation by deriving the wave properties from the full Heisenberg-Euler Lagrangian, rather that using the weak-field expansion in the form (8.3.26). The nonzero matrix elements are [26]:

$$M_{\perp\|\|}(B) = \left(\frac{B}{B_c}\right)^4 \int_0^\infty \frac{ds}{s} e^{-sB_c/B} \left\{ \left(-\frac{3}{4s} + \frac{s}{6}\right) + \left(\frac{1}{4} + \frac{s^2}{6}\right) \frac{1}{\sinh^2 s} + \frac{s}{2} \frac{\cosh s}{\sinh^3 s} \right\} \tag{8.5.11}$$

$$M_{\perp\perp\perp}(B) = \left(\frac{B}{B_c}\right)^4 \int_0^\infty \frac{ds}{s} e^{-sB_c/B} \left\{ \frac{3}{4s} \frac{\cosh s}{\sinh s} + \left(\frac{3}{4} - s^2\right) \frac{\cosh^2 s}{\sinh^2 s} - \frac{3s^2}{2} \frac{\cosh^2 s}{\sinh^2 s} \right\}. \tag{8.5.12}$$

To obtain the weak-field limit of (8.5.12), one expands the hyperbolic functions for $s \ll 1$; low order terms cancel, and carrying out the integral for the leading nonzero term gives the result (8.5.10).

8.5.4 S-Matrix Approach

An alternative treatment of photon splitting is based on the S-matrix approach [19]. Formally, this approach is equivalent to treating the splitting using (8.5.3) with the quadratic response tensor identified as the vacuum contributions to the form (9.7.14) with (9.7.15). The S-matrix elements involve sums over quantum numbers. The sum over the spins is straightforward, and the integral over p_z, although tedious, can be performed using elementary methods. The sum over n remains, and its evaluation contains some subtleties [33].

The S-matrix contains terms with odd powers of the frequencies, and only the terms cubic in the frequencies (proportional to $\omega\omega'\omega''$) is considered here. After summing over the spins and integrating over p_z, these terms are of the form

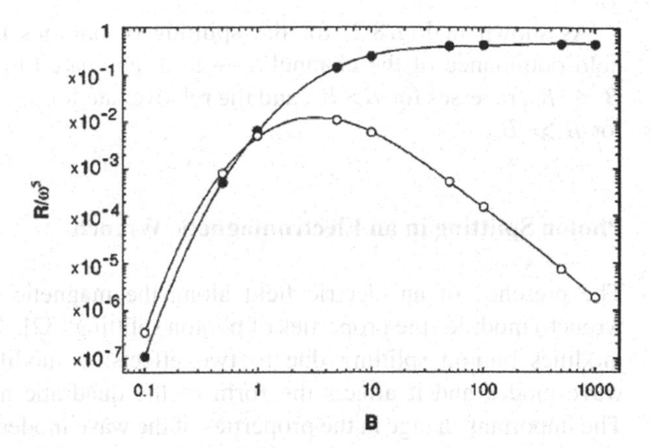

Fig. 8.2 The rates of splitting of a $\perp$ photon into two $\perp$ photons (*light circles*) and into two $\parallel$ photons (*dark circles*) are plotted on a log-log scale (*tick marks* represent powers of 10) as functions of B/B_c (From [33], reprinted with permission AIP)

$$S_{fi} = \frac{4\pi^2(4\pi\alpha_c)^{3/2}(\omega\omega'\omega'')^{1/2}}{(2V)^{3/2}}\delta(k_x - k'_x - k''_x)\delta(\omega - \omega' - \omega'')\sum_{n=0}^{\infty} T(n, B), \tag{8.5.13}$$

where the photons are collinear along the x axis. The $T(n, B)$ do not depend on the frequencies, and the remaining factors do not depend on B. Explicit expressions for the $T(n, B) \to T_{MPQ}(n, B)$ for the two independent splittings, $MPQ = \perp\parallel\parallel$, $MPQ = \perp\perp\perp$, are relatively cumbersome [33], and are not written down here; they involve only powers that depend on n, B, and logarithms involving ratios of $\varepsilon_n^0 = m(1 + 2nB/B_c)^{1/2}$ with different values of n.

The decay rate is

$$R_{MPQ} = \frac{1}{4\,60\pi^2}\frac{\alpha_c^3 m}{} \left(\frac{\omega}{m}\right)^5 \left|\sum_{n=0}^{\infty} T_{MPQ}(n)\right|^2. \tag{8.5.14}$$

In practice, the sum over n must be terminated as some $n_{\max}$. The Euler-Maclaurin summation formula is used to include the contributions from $n > n_{\max}$:

$$\sum_{n=0}^{\infty} T(n) = T(0) + \cdots + T(i-1) + \frac{1}{2}T(i) + \int_i^{\infty} dn\, T(n) - \frac{1}{12}T'(i) + \frac{1}{720}T'''(i) + \cdots, \tag{8.5.15}$$

where primes denote derivatives. The choices of $i = n_{\max}$ and of the number of derivatives of $T(i)$ to be retained are made by trial and error, such that increasing the numbers does not lead to significant improvement beyond some predetermined accuracy. The sum over n converges increasingly rapidly with increasing B, allowing the sum to be truncated at small values for $B \gtrsim B_c$.

As shown in Fig. 8.2, for the splitting of photons in the $\perp$ mode, the weak-field dominance of the channel $\perp\to\perp,\perp$ compared to the channel $\perp\to\|,\|$ for $B \ll B_c$, reverses for $B \gtrsim B_c$, and the relative rate for $\perp\to\perp,\perp$ becomes negligible for $B \gg B_c$.

Photon Splitting in an Electromagnetic Wrench

The presence of an electric field along the magnetic field (an electromagnetic wrench) modifies the properties of photon splitting [32]. The inclusion of $E/B \neq 0$ modifies photon splitting due to two effects: it modifies the properties of the wave modes and it affects the form of the quadratic nonlinear response tensor. The important change in the properties of the wave modes is on the polarization: the identification as the $\perp$-mode and $\|$-mode, respectively, is no longer appropriate, with the polarizations (8.4.8) of both modes being mixtures of $\perp$- and $\|$-polarizations. The quadratic nonlinear response tensor includes additional terms that depend on $E/B \neq 0$, and these need to be included in the matrix elements for the various possible splittings.

The inclusion of $E/B \neq 0$ has only a small effect on splitting for $E/B \ll 1$ and $B/B_c \ll 1$. The components of an exact expression for the quadratic nonlinear response tensor are evaluated by expanding in $E/B \ll 1$ for arbitrary B/B_c, and used to calculate the matrix elements for the various splittings [32]. The most notable effect is that for $B/B_c \gg 1$, the splitting $+ \to ++$, which reduces to the forbidden $\|\to\|\|$ for $E/B = 0$, becomes of comparable strength to the other splittings for a sufficiently strong B/B_c, that depends on the value of E/B. The argument based on CP invariance remains valid, but it does not exclude this splitting. Specifically, the matrix element for a nonzero splitting can depend only even powers of E, and the additional splitting occurs because an additional component of the polarization vector proportional to E combines with an additional term proportional to E in the quadratic nonlinear response tensor to give an intrinsically new contribution that satisfies this requirement.

References

1. S.L. Adler, Ann. Phys. **67**, 599 (1971)
2. C. Alcock, E. Farhi, A. Olinto, Astrophys. J. **310**, 261 (1986)
3. M. Altarelli, D.L. Dexter, H.M. Nussenzveig, D.Y. Smith, Phys. Rev. B **6**, 4502 (1972)
4. P. Bakshi, R.A Cover, G., Kalman, Phys. Rev. D **14**, 2532 (1976)
5. I.A. Batalin, A.E. Shabad, Sov. Phys. JETP **33**, 483 (1971)
6. Z. Bialynicka-Birula, I. Bialynicki-Birula, Phys. Rev. D **2**, 2341 (1970)
7. T. Bulik, Acta Astron. **48**, 695 (1998)
8. J.K Daugherty, I. Lerche, Phys. Rev. D **14**, 340 (1976)
9. Yu.N. Gnedin, G.G. Pavlov, Yu.A. Shibanov, JETP Lett. **27**, 305 (1978)
10. T. Erber, Rev. Mod. Phys. **38**, 626 (1966)

11. A.K. Harding, D. Lai, Rep. Prog. Phys. **69**, 2631 (2006)
12. W. Heisenberg, H. Euler, Z. Phys. **98**, 714 (1936)
13. H. Herold, H. Ruder, G. Wunner, Plasma Phys. **23**, 755 (1981)
14. J.S. Heyl, L. Hernquist, J. Phys. A **30**, 6485 (1997)
15. J.G. Kirk, Plasma Phys. **22**, 639 (1980)
16. J.G. Kirk, N.F. Cramer, Aust J. Phys. **38**, 715 (1985)
17. D.B. Melrose, R.J. Stoneham, Nuovo Cim. **32A**, 435 (1976)
18. D.B. Melrose, R.J. Stoneham, J. Phys. A **10**, 1211 (1977)
19. M. Mentzel, D. Berg, G. Wunner, Phys. Rev. D **50**, 1125 (1994)
20. P. Mészáros, J. Ventura, Phys. Rev. Lett. **41**, 1544 (1978)
21. P. Mészáros, J. Ventura, Phys. Rev. D **19**, 3565 (1979)
22. F. Özel, Astrophys. J. **563**, 276 (2001)
23. J. Schwinger, Phys. Rev. **82** , 664 (1951)
24. A.E. Shabad, Ann. Phys. **90**, 166 (1975)
25. A.E. Shabad, *Polarization of the Vacuum and a Quantum Relativistic Gas in an External Field* (Nova Science, New York, 1991)
26. R.J. Stoneham, J. Phys. A **12**, 2187 (1979)
27. J.S. Toll, PhD Thesis, Princeton University, 1952
28. W.-Y. Tsai, Phys. Rev. D **10**, 2699 (1974)
29. W.-Y. Tsai, T. Erber, Phys. Rev. D **12**, 1132 (1975)
30. V.V. Usov, Phys. Rev. D **70**, 067301 (2004)
31. C. Wang, D. Lai, Mon. Not. Roy. Astron. Soc. **377**, 1095 (2007)
32. J.I. Weise, Phys. Rev. D **69**, 105017 (2004)
33. J.I. Weise, M.G. Baring, D.B. Melrose, Phys. Rev. D **57**, 5526 (1998)
34. V. Weisskopf, Mat. Fys. Medd. Dan. Vid. Selsk. **14**, Nr.6 (1936)

Chapter 9
Response of a Magnetized Electron Gas

Dispersion in a magnetized, quantum electron gas was first discussed in the nonrelativistic case in the 1960s and 1970s [2, 6, 7, 15, 24]. Extension to the relativistic case, with one early exception [20], was carried out mainly in the 1980s [3, 5, 8, 13] and continues to the present [11, 14, 22]. An exact expression for the linear response tensor includes the following quantum effects: degeneracy, the quantization of the Landau states, the spin, the quantum recoil and dispersion associated with one-photon pair creation. It involves sums over two sets of quantum numbers, denoted $q = \epsilon, n, p_z, s$ and $q' = \epsilon', n', p_z', s'$ here. This leads to a cumbersome form, and only a few special cases, notably parallel propagation, have been explored in any detail.

The approach adopted in this chapter is to start from the general form, involving sums over q, q', and first perform the sums over s, s' and ϵ, ϵ'; this is done in § 9.1. In § 9.2, it is shown that the resulting expression for the response tensor may be evaluated in terms of a single relativistic plasma dispersion functions (RPDF), evaluated at the resonant values of $p_\parallel^\mu$. The response of a thermal electron gas is discussed in § 9.3, where the RPDF is evaluated in some particular electron distributions. Special and limiting cases of the linear response tensor are discussed in § 9.4. Some specific results for wave dispersion in relativistic quantum magnetized plasmas are presented in § 9.5. Spin-dependence is included in § 9.6. Nonlinear response tensors for a relativistic magnetized quantum plasma are written down in § 9.7.

9.1 Response of a Magnetized Electron Gas

General expressions for the linear response tensor for a magnetized Dirac electron gas are written down and discussed in this section. The sums over spin states, s, s', and electron and positron states, ϵ, ϵ', are carried out explicitly. The neglect of quantum effects is then discussed, and the reduction to nonquantum forms is outlined.

D. Melrose, *Quantum Plasmadynamics: Magnetized Plasmas*, Lecture Notes in Physics 854, DOI 10.1007/978-1-4614-4045-1_9,

9.1.1 Calculation of the Response Tensor

There are several alternative methods for calculating the response of a relativistic quantum electron gas in the absence of a magnetic field, and these may be adapted to apply in a magnetic field. Three methods are discussed in § 8.4 of volume 1: the forward-scattering method, the Wigner-matrix method and the density-matrix approach. From a quantum mechanical viewpoint, these three methods may be interpreted in terms of different pictures, where a picture involves the choice of how to include time dependence in quantum mechanics. The forward-scattering method is based on the S-matrix approach, also called the Green function approach, which involves adopting the interaction picture. In the interaction picture the wave functions evolve due to the interaction Hamiltonian, and the operators evolve due to the unperturbed Hamiltonian. The Wigner function in nonrelativistic quantum mechanics is defined in terms of the outer product of the Schrödinger wave function with its adjoint. The evolution of the Wigner function is determined by the Schrödinger equation. The Wigner matrix in the relativistic case is a generalization of the nonrelativistic Wigner function; it is a 4×4 Dirac matrix defined in terms of the outer product of the Dirac wave function and its adjoint. The evolution of the Wigner matrix is determined by the Dirac equation, and this corresponds to choosing the Schrödinger picture. The density matrix is defined as an operator constructed from the state function (a vector in a Hilbert space, usually written as a ket) and its adjoint (a bra). The density-matrix method is based on the Heisenberg picture, in which all the time dependence is in the operators, with the density-matrix interpreted as an operator. Expressions for the response tensor for a nonrelativistic magnetized quantum electron gas were derived in the 1960s using both the density-matrix method [15, 24] and the Wigner-function method [7]. The S-matrix approach is adopted here, but it should be emphasized that the three methods are formally equivalent, and the choice of method has no effect on the result.

In the S-matrix approach, the linear response tensor for a relativistic quantum electron gas in a magnetic field is derived from the Feynman amplitude for the bubble diagram. As discussed in § 8.1 of volume 1, the amplitude for the bubble diagram gives the vacuum polarization tensor. By replacing the electron propagators in the bubble diagram by their statistical average over the electron gas, the calculation of the vacuum polarization tensor also includes the response tensor for the electron gas.

9.1.2 Vertex Form of $\Pi^{\mu\nu}(k)$

The general expression for the linear response tensor, derived from the amplitude for the bubble diagram using the vertex formalism, is referred to as the vertex form. This form is

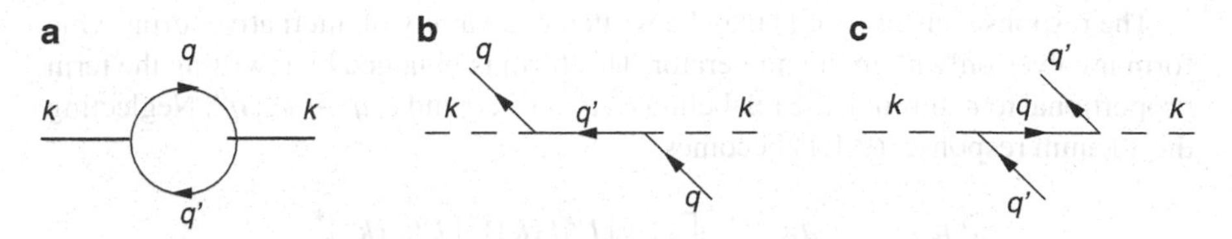

Fig. 9.1 Cuts in the *upper* and *lower* electron lines in the Feynman bubble diagram (**a**) produce the forward scattering diagrams (**b**) and (**c**), respectively

$$\Pi^{\mu\nu}(k) = -\frac{e^3 B}{2\pi} \sum_{\epsilon,q,\epsilon',q'} \int \frac{dp_z}{2\pi} \int \frac{dp'_z}{2\pi} 2\pi\delta(\epsilon' p'_z - \epsilon p_z + k_z)$$
$$\times \frac{\frac{1}{2}(\epsilon' - \epsilon) + \epsilon n_q^{\epsilon} - \epsilon' n_{q'}^{\epsilon'}}{\omega - \epsilon\varepsilon_q + \epsilon'\varepsilon_{q'}} [\Gamma_{q'q}^{\epsilon'\epsilon}(\boldsymbol{k})]^{\mu} [\Gamma_{qq'}^{\epsilon\epsilon'}(-\boldsymbol{k})]^{\nu}, \tag{9.1.1}$$

with $[\Gamma_{qq'}^{\epsilon\epsilon'}(-\boldsymbol{k})]^{\nu} = [\Gamma_{q'q}^{\epsilon'\epsilon}(\boldsymbol{k})]^{*\nu}$. The resonance condition is to be imposed by making the replacement $\omega \to \omega + i0$ in the denominator.

The response tensor (9.1.1) includes the response of the magnetized vacuum, which corresponds to the term $\frac{1}{2}(\epsilon' - \epsilon)$ in the numerator. Retaining only the vacuum terms gives an unregularized form of the vacuum polarization tensor, and a regularization procedure needs to be used to derive the (physically relevant) vacuum response tensor, as discussed in § 8.1. The terms in (9.1.1) that are proportional to an occupation number describe the response of the electron gas, and these do not need to be regularized. In the following, the vacuum terms are neglected in deriving various forms of the response tensor for an electron gas. Formally, these various forms also apply to the vacuum: from (9.1.1) it can be seen that response tensor for an electron gas becomes the (unregularized) response tensor for the vacuum by making the replacement $n_q^{\epsilon} \to -\frac{1}{2}\epsilon$. No physical significance is attached to this replacement. It is used below to note various forms for the unregularized vacuum response tensor whose regularization is discussed in § 8.1.

To illustrate the interpretation of the sums in (9.1.1), consider the contribution from the terms with occupation number n_q^{ϵ}. This includes electrons ($\epsilon = 1$) and positrons ($\epsilon = -1$) in the state q, which denotes n, s, p_z collectively. For electrons, the contribution involves the electron absorbing and re-emitting (or emitting and re-absorbing) a wave quantum, with the electron returning to the state q. The sums over ϵ, q are over all the electrons and positron states that are occupied, in the sense that n_q^{ϵ} is nonzero. Between the absorption and re-emission the electron or positron is in a virtual state with quantum numbers q'. The sum over ϵ', q' is over all possible virtual states, which includes both virtual electron states ($\epsilon' = 1$) and virtual positron states ($\epsilon' = -1$) irrespective of whether the real state ϵ, q corresponds to an electron or a positron. The Feynman diagrams Fig. 9.1b, c correspond to virtual states with $\epsilon' = \pm\epsilon$, respectively.

The response tensor (9.1.1) may be written in a variety of alternative forms. One form involves only n^ϵ_q in the numerator. This form is obtained by rewriting the term proportional to ϵ' through the relabeling $\epsilon', q' \to \epsilon, q$ and $\epsilon, q \to \epsilon'', q''$. Neglecting the vacuum response, (9.1.1) becomes

$$\Pi^{\mu\nu}(k) = -\frac{e^3 B}{2\pi}\sum_{\epsilon,q}\int\frac{dp_z}{2\pi}\,\epsilon n^\epsilon_q\Bigg[\sum_{\epsilon',q'}\frac{[\Gamma^{\epsilon'\epsilon}_{q'q}(\boldsymbol{k})]^\mu[\Gamma^{\epsilon'\epsilon}_{q'q}(\boldsymbol{k})]^{*\nu}}{\omega-\epsilon\varepsilon_q+\epsilon'\varepsilon_{q'}}$$
$$-\sum_{\epsilon'',q''}\frac{[\Gamma^{\epsilon\epsilon''}_{qq''}(\boldsymbol{k})]^\mu[\Gamma^{\epsilon\epsilon''}_{qq''}(\boldsymbol{k})]^{*\nu}}{\omega-\epsilon''\varepsilon_{q''}+\epsilon\varepsilon_q}\Bigg], \tag{9.1.2}$$

where $\epsilon' p'_z = \epsilon p_z - k_z$, $\epsilon'' p''_z = \epsilon p_z + k_z$ are implicit.

The dependence of the specific quantum numbers, $q = n, s, p_z$ can be made explicit by writing $n^\epsilon_q \to n^\epsilon_{ns}(p_z)$. The energy eigenstates, $\varepsilon_q \to \varepsilon_n(p_z) = [m^2 + p_z^2 + 2neB]^{1/2}$, are independent of the spin, and it is convenient to use the notation

$$\varepsilon_n(p_z) \to \varepsilon_n, \qquad \varepsilon_{n'}(p'_z) \to \varepsilon'_{n'}. \tag{9.1.3}$$

For a spin-independent electron gas, the occupation numbers for $s = \pm 1$ are equal, and are written as $n^\epsilon_n(p_z)$.

Explicit expressions for the vertex function in (9.1.1) depend on the choice of spin operator. As discussed in § 5.4, the choice of the magnetic-moment operator is the most appropriate, leading to the form (5.4.18), viz.

$$[\Gamma^{\epsilon'\epsilon}_{q'q}(\boldsymbol{k})]^\mu = (ie^{i\psi})^{n-n'}\big(A^{\epsilon'\epsilon(+)}_{q'q}\,J^{(0)}_{q'q}(\boldsymbol{k}),$$
$$-A^{\epsilon'\epsilon(-)}_{q'q}\,J^{(+)}_{q'q}(\boldsymbol{k}), -iA^{\epsilon'\epsilon(-)}_{q'q}\,J^{(-)}_{q'q}(\boldsymbol{k}), B^{\epsilon'\epsilon}_{q'q}\,J^{(0)}_{q'q}(\boldsymbol{k})\big),$$
$$A^{\epsilon'\epsilon(\pm)}_{q'q} = a'_{\epsilon's'}a_{\epsilon s} \pm a'_{-\epsilon's'}a_{-\epsilon s}, \qquad B^{\epsilon'\epsilon}_{q'q} = a'_{\epsilon's'}a_{-\epsilon s} + a'_{-\epsilon's'}a_{\epsilon s},$$
$$J^{(0)}_{q'q}(\boldsymbol{k}) = b'_{s'}b_s J^{n-1}_{n'-n}(x) + s'sb'_{-s'}b_{-s}J^n_{n'-n}(x),$$
$$J^{(\pm)}_{q'q}(\boldsymbol{k}) = s'b'_{-s'}b_s e^{-i\psi}J^{n-1}_{n'-n+1}(x) \pm sb'_{s'}b_{-s}e^{i\psi}J^n_{n'-n-1}(x),$$
$$a_\pm = P_\pm\left(\frac{\varepsilon_n \pm \varepsilon^0_n}{2\varepsilon_n}\right)^{1/2}, \quad b_s = \left(\frac{\varepsilon^0_n + sm}{2\varepsilon^0_n}\right)^{1/2}, \tag{9.1.4}$$

with $x = k_\perp^2/2eB$, and $P_\pm = \frac{1}{2}(1+P) \pm \frac{1}{2}(1-P)$, $P = p_z/|p_z|$.

9.1.3 Summed Form of $\Pi^{\mu\nu}(k)$

The sums over s, s' and ϵ, ϵ' in (9.1.2) can be performed explicitly for an arbitrary distribution. The dependence on s, s' and ϵ, ϵ' is explicit except for the occupation number, $n^{\epsilon}_{n,s}(p_z)$. The occupation number may be separated into parts that are even and odd under $s \to -s$ and $\epsilon \to -\epsilon$; this separates $n^{\epsilon}_{n,s}(p_z)$ into four terms, proportional to $1, s, \epsilon, s\epsilon$, respectively, times averages over s and ϵ. For a spin-independent distribution there is no dependence on s, and the terms proportional to 1 and ϵ are half the sum and difference, respectively, between the occupation numbers for electrons and positrons. The effect of spin-dependence, in $n^{\epsilon}_{n,s}(p_z)$, is discussed in § 9.6.

Sum over s, s'

In a spin-independent electron gas, the sums over s', s in (9.1.1) define the tensor (5.4.23), viz.

$$[C_{n'n}(\epsilon' p'_{\parallel}, \epsilon p_{\parallel}, k)]^{\mu\nu} = 2\epsilon'\epsilon\varepsilon'_{n'}\varepsilon_n \sum_{s',s} [\Gamma^{\epsilon'\epsilon}_{q'q}(\boldsymbol{k})]^{\mu} [\Gamma^{\epsilon'\epsilon}_{q'q}(\boldsymbol{k})]^{*\nu}. \tag{9.1.5}$$

Explicit forms for the tensor are given in (5.4.24) and in Table 5.1.

The response tensor (9.1.1) for a spin-independent electron gas becomes

$$\Pi^{\mu\nu}(k) = -\frac{e^3 B}{2\pi} \sum_{\epsilon,n,\epsilon',n'} \left[\int \frac{dp_z}{2\pi} n^{\epsilon}_n(p_z) \frac{\epsilon'[C_{n'n}(P'_{\parallel}, P_{\parallel}, k)]^{\mu\nu}}{2\varepsilon'_{n'}\varepsilon_n(\omega - \epsilon\varepsilon_n + \epsilon'\varepsilon'_{n'})} \right.$$
$$\left. - \int \frac{dp'_z}{2\pi} n^{\epsilon'}_{n'}(p'_z) \frac{\epsilon[C_{n'n}(P'_{\parallel}, P_{\parallel}, k)]^{\mu\nu}}{2\varepsilon'_{n'}\varepsilon_n(\omega - \epsilon\varepsilon_n + \epsilon'\varepsilon'_{n'})} \right], \tag{9.1.6}$$

with $P'^{\mu}_{\parallel} = [\epsilon'\varepsilon'_{n'}, 0, 0, \epsilon p_z - k_z]$, $P^{\mu}_{\parallel} = [\epsilon\varepsilon_n, 0, 0, \epsilon p_z]$ in the first integral, and $P'^{\mu}_{\parallel} = [\epsilon'\varepsilon'_{n'}, 0, 0, \epsilon' p'_z]$, $P^{\mu}_{\parallel} = [\epsilon\varepsilon_n, 0, 0, \epsilon' p'_z + k_z]$ in the second integral. The symmetry property (5.4.24) allows one to make the interchange $\epsilon', n', p'_z \leftrightarrow \epsilon, n, p_z$, so that the second integral in (9.1.6) can be written in terms of a p_z-integral over the occupation number $n^{\epsilon}_n(p_z)$ and combined with the first integral.

Sum over ϵ' or ϵ

The sums over ϵ' and ϵ can be performed for the terms in (9.1.6) proportional to $n^{\epsilon}_n(p_z)$ and $n^{\epsilon'}_{n'}(p'_z)$, respectively. The remaining sum (over ϵ and ϵ', respectively) can be performed after separating into the parts even and odd in the interchange of electrons and positrons.

The sums over ϵ', ϵ in the two terms, respectively, in (9.1.6) give

$$\sum_{\epsilon'} \frac{\epsilon'[C_{n'n}(\epsilon' p'_\parallel, \epsilon p_\parallel, k)]^{\mu\nu}}{2\varepsilon'_{n'}\varepsilon_n(\omega - \epsilon\varepsilon_n + \epsilon'\varepsilon'_{n'})} = -\frac{[C_{n'n}(\epsilon p_\parallel - k_\parallel, \epsilon p_\parallel, k)]^{\mu\nu}}{\varepsilon_n[(\omega - \epsilon\varepsilon_n)^2 - \varepsilon'^2_{n'}]},$$

$$\sum_{\epsilon} \frac{\epsilon[C_{n'n}(\epsilon' p'_\parallel, \epsilon p_\parallel, k)]^{\mu\nu}}{2\varepsilon'_{n'}\varepsilon_n(\omega - \epsilon\varepsilon_n + \epsilon'\varepsilon'_{n'})} = \frac{[C_{n'n}(\epsilon' p'_\parallel, \epsilon' p'_\parallel + k_\parallel, k)]^{\mu\nu}}{\varepsilon'_{n'}[(\omega + \epsilon'\varepsilon'_{n'})^2 - \varepsilon_n^2]}. \tag{9.1.7}$$

The denominators on the right hand sides of (9.1.7) can be written

$$(\omega - \epsilon\varepsilon_n)^2 - \varepsilon'^2_{n'} = -2[\epsilon(pk)_\parallel - (pk)_{nn'}],$$
$$(\omega + \epsilon'\varepsilon'_{n'})^2 - \varepsilon_n^2 = 2[\epsilon'(p'k)_\parallel + (pk)_{n'n}], \tag{9.1.8}$$

respectively, with $(pk)_\parallel = \varepsilon_n\omega - p_z k_z$, $(p'k)_\parallel = \varepsilon'_{n'}\omega - p'_z k_z$, and with

$$(pk)_{nn'} = (k^2)_\parallel f_{nn'} = \tfrac{1}{2}(k^2)_\parallel + \tfrac{1}{2}p_n^2 - \tfrac{1}{2}p_{n'}^2. \tag{9.1.9}$$

The form (9.1.6) for the response tensor then becomes

$$\Pi^{\mu\nu}(k) = -\frac{e^3 B}{2\pi}\sum_{n,n'}\left[\int \frac{dp_z}{2\pi}\sum_{\epsilon} n_n^\epsilon(p_z)\frac{[C_{n'n}(\epsilon p_\parallel - k_\parallel, \epsilon p_\parallel, k)]^{\mu\nu}}{2\varepsilon_n[\epsilon(pk)_\parallel - (pk)_{nn'}]}\right.$$
$$\left. - \int \frac{dp'_z}{2\pi}\sum_{\epsilon'} n_{n'}^{\epsilon'}(p'_z)\frac{[C_{n'n}(\epsilon' p'_\parallel, \epsilon' p'_\parallel + k_\parallel, k)]^{\mu\nu}}{2\varepsilon'_{n'}[\epsilon'(p'k)_\parallel + (pk)_{n'n}]}\right]. \tag{9.1.10}$$

An alternative derivation of the hermitian part starts from the form (9.1.2), rather than from (9.1.1). After summing over the spins, in place of (9.1.6), this gives

$$\Pi^{\mu\nu}(k) = -\frac{e^3 B}{2\pi}\sum_{\epsilon,n}\int \frac{dp_z}{2\pi} n_n^\epsilon(p_z)\left\{\sum_{\epsilon',n'}\frac{\epsilon'[C_{n'n}(\epsilon' p'_\parallel, \epsilon p_\parallel, k)]^{\mu\nu}}{2\varepsilon'_{n'}\varepsilon_n(\omega - \epsilon\varepsilon_n + \epsilon'\varepsilon'_{n'})}\right.$$
$$\left. - \sum_{\epsilon'',n''}\frac{\epsilon''[C_{nn''}(\epsilon p_\parallel, \epsilon'' p''_\parallel, k)]^{\mu\nu}}{2\varepsilon_n\varepsilon''_{n''}(\omega - \epsilon''\varepsilon''_{n''} + \epsilon\varepsilon_n)}\right\}, \tag{9.1.11}$$

with $\epsilon'' p''^{\mu}_\parallel = [\epsilon''\varepsilon''_{n''}, 0, 0, \epsilon p_z + k_z]$, $\varepsilon''_{n''} = \varepsilon_{n''}(\epsilon p_z + k_z)$. On performing the sums over ϵ' and ϵ'', and relabeling n'' as n', (9.1.11) gives

$$\Pi^{\mu\nu}(k) = -\frac{e^3 B}{2\pi} \sum_{\epsilon,n,n'} \int \frac{dp_z}{2\pi} \frac{n_n^\epsilon(p_z)}{2\varepsilon_n} \left\{ \frac{[C_{n'n}(\epsilon p_\parallel - k_\parallel, \epsilon p_\parallel, k)]^{\mu\nu}}{\epsilon(pk)_\parallel - (pk)_{nn'}} - \frac{[C_{nn'}(\epsilon p_\parallel, \epsilon p_\parallel + k_\parallel, k)]^{\mu\nu}}{\epsilon(pk)_\parallel + (pk)_{nn'}} \right\}. \tag{9.1.12}$$

The results (9.1.10) and (9.1.12) are equivalent, as may be shown by interchanging primed and unprimed quantities in the second integral in (9.1.10).

The second integrals in either (9.1.10) and (9.1.12) can be rewritten so that the denominator is the same as in the first integral. This is achieved by making the replacement $\epsilon \to -\epsilon$ in the second term in (9.1.12). The two terms can then be combined, but the significance of ϵ as the sign of the charge is then lost. When this procedure is used below, the sign ϵ is relabeled as η.

9.1.4 Nongyrotropic and Gyrotropic Parts of $\Pi^{\mu\nu}(k)$

The remaining sum in (9.1.12) that can be performed explicitly is that over ϵ. To make the dependence on ϵ explicit, one separates the occupation number into the sum and difference under the interchange of electrons and positrons, $\epsilon \to -\epsilon$. This is achieved by writing

$$n_n^\epsilon(p_z) = \tfrac{1}{2}\left[n_n^{\rm sum}(p_z) + \epsilon n_n^{\rm diff}(p_z)\right]. \tag{9.1.13}$$

This is equivalent to separating the response tensor into two parts, referred to here as the nongyrotropic and gyrotropic parts, respectively. The nongyrotropic part is independent of the sign of the charge, so that electrons and positrons contribute to it with the same sign, and the gyrotropic part depends on the sign of the charge, so that electrons and positrons contribute to it with the opposite sign. For an appropriate choice of basis vectors, $\Pi^{\mu\nu}(k)$ separates into nongyrotropic and gyrotropic components which satisfy different symmetry relations, as implied by the Onsager relations.

Onsager Relations

The Onsager relations, cf. § 1.4 of volume 1, imply relations between components of the response tensor under the transformation $\boldsymbol{B} \to -\boldsymbol{B}$, which is equivalent to changing the sense of gyration of particles, and hence to changing the sign of the charge. A relevant form of the Onsager relations is

$$\Pi^{00}(\omega, -\boldsymbol{k})|_{-\boldsymbol{B}} = \Pi^{00}(\omega, \boldsymbol{k})|_{\boldsymbol{B}}, \qquad \Pi^{0i}(\omega, -\boldsymbol{k})|_{-\boldsymbol{B}} = -\Pi^{i0}(\omega, \boldsymbol{k})|_{\boldsymbol{B}},$$
$$\Pi^{ij}(\omega, -\boldsymbol{k})|_{-\boldsymbol{B}} = \Pi^{ji}(\omega, \boldsymbol{k})|_{\boldsymbol{B}}. \tag{9.1.14}$$

With $\boldsymbol{B}$ along the 3-axis, and with $\boldsymbol{k}$ defining the vector $e_1^\mu = [0, \boldsymbol{k}_\perp/k_\perp]$ and with $e_2^\mu = -f^{\mu\nu} e_{1\nu} = [0, \boldsymbol{k}_G/k_\perp]$, the gyrotropic terms are the 02-, 12- and 23-components, and their negatives, the 20-, 21- and 32-components, respectively. The remaining terms are the nongryotropic terms. The nongryotropic and gyrotropic parts of the $\Pi^{\mu\nu}(k)$ are symmetric and antisymmetric, respectively.

Summed Form of $\Pi^{\mu\nu}(k)$

The nongyrotropic and gyrotropic parts are identified as those proportional to $n_n^{\rm sum}(p_z)$ and $n_n^{\rm diff}(p_z)$, respectively, in

$$\begin{aligned}
n_n^\epsilon(p_z)\big[C_{n'n}(\epsilon p_\parallel - k_\parallel, \epsilon p_\parallel, k)\big]^{\mu\nu} + n_n^{-\epsilon}(p_z)\big[C_{nn'}(-\epsilon p_\parallel, -\epsilon p_\parallel + k_\parallel, k)\big]^{\mu\nu} \\
= n_n^{\rm sum}(p_z)\big[N_{n'n}(\epsilon p_\parallel, k)\big]^{\mu\nu} + \epsilon n_n^{\rm diff}(p_z)\big[G_{n'n}(\epsilon p_\parallel, k)\big]^{\mu\nu}. \qquad (9.1.15)
\end{aligned}$$

This definition corresponds to

$$\begin{aligned}
\big[C_{n'n}(\epsilon p_\parallel - k_\parallel, \epsilon p_\parallel, k)\big]^{\mu\nu} &= \big[N_{n'n}(\epsilon p_\parallel, k)\big]^{\mu\nu} + \big[G_{n'n}(\epsilon p_\parallel, k)\big]^{\mu\nu}, \\
\big[C_{nn'}(-\epsilon p_\parallel, -\epsilon p_\parallel + k_\parallel, k)\big]^{\mu\nu} &= \big[N_{n'n}(\epsilon p_\parallel, k)\big]^{\mu\nu} - \big[G_{n'n}(\epsilon p_\parallel, k)\big]^{\mu\nu}. \qquad (9.1.16)
\end{aligned}$$

Explicit expressions for these two parts follow by inspection of the expression (5.4.24) for $\big[C_{n'n}(P'_\parallel, P_\parallel, k)\big]^{\mu\nu}$. The nongyrotropic part is

$$\begin{aligned}
[N_{n'n}(\epsilon p_\parallel, k)]^{\mu\nu} = &\Big\{2p_\parallel^\mu p_\parallel^\nu - \epsilon(p_\parallel^\mu k_\parallel^\nu + k_\parallel^\mu p_\parallel^\nu) - g_\parallel^{\mu\nu}[p_n^2 - \epsilon(pk)_\parallel]\Big\}\,[(J_{n'-n}^{n-1})^2 + (J_{n'-n}^n)^2] \\
&-\{g_\perp^{\mu\nu}[p_n^2 - \epsilon(pk)_\parallel] + \epsilon(p_\parallel^\mu k_\perp^\nu + p_\parallel^\nu k_\perp^\mu)\}[(J_{n'-n+1}^{n-1})^2 + (J_{n'-n-1}^n)^2] \\
&+2p_{n'}p_n g_\parallel^{\mu\nu} J_{n'-n}^{n-1} J_{n'-n}^n + 2p_{n'}p_n(e_1^\mu e_1^\nu - e_2^\mu e_2^\nu) J_{n'-n+1}^{n-1} J_{n'-n-1}^n \\
&-p_n\{(2\epsilon p_\parallel^\mu - k_\parallel^\mu)e_1^\nu + (2\epsilon p_\parallel^\nu - k_\parallel^\nu)e_1^\mu\}[J_{n'-n}^{n-1} J_{n'-n-1}^n + J_{n'-n}^n J_{n'-n+1}^{n-1}],
\end{aligned}$$
(9.1.17)

and the gyrotropic part is

$$\begin{aligned}
[G_{n'n}(\epsilon p_\parallel, k)]^{\mu\nu} = &\, i\{f^{\mu\nu}[p_n^2 - \epsilon(pk)_\parallel] + \epsilon(p_\parallel^\mu k_G^\nu - p_\parallel^\nu k_G^\mu)\}[(J_{n'-n+1}^{n-1})^2 - (J_{n'-n-1}^n)^2] \\
&-ip_n\{(2\epsilon p_\parallel^\mu - k_\parallel^\mu)e_2^\nu - (2\epsilon p_\parallel^\nu - k_\parallel^\nu)e_2^\mu\}[J_{n'-n}^{n-1} J_{n'-n-1}^n - J_{n'-n}^n J_{n'-n+1}^{n-1}],
\end{aligned}$$
(9.1.18)

where one has $k_\perp^\mu = k_\perp e_1^\mu$, $k_G^\mu = k_\perp e_2^\mu$.

Sum over η

In the resulting expression for the response tensor, the sign ϵ has lost its meaning as a label for electrons and positrons, and it is convenient to relabel it as $\eta = \pm 1$. With this change, the summed form of $\Pi^{\mu\nu}(k)$ becomes

$$\Pi^{\mu\nu}(k) = -\frac{e^3 B}{8\pi^2}\sum_{\eta,n,n'}\int \frac{dp_z}{\varepsilon_n}\,\frac{1}{\eta(pk)_\parallel - (pk)_{nn'}}$$
$$\times \left\{n_n^{\rm sum}(p_z)\big[N_{n'n}(\eta p_\parallel, k)\big]^{\mu\nu} + \eta n_n^{\rm diff}(p_z)\big[G_{n'n}(\eta p_\parallel, k)\big]^{\mu\nu}\right\}. \quad (9.1.19)$$

All the dependence on η in (9.1.19) is explicit, allowing the sum over η to be performed, giving

$$\Pi^{\mu\nu}(k) = -\frac{e^3 B}{8\pi^2}\sum_{n,n'}\int \frac{dp_z}{\varepsilon_n}\,\frac{\big[B_{nn'}(p_\parallel, k)\big]^{\mu\nu}}{(pk)_\parallel^2 - (pk)_{nn'}^2}, \quad (9.1.20)$$

with the tensor in the numerator given by

$$[B_{nn'}(p_\parallel, k)]^{\mu\nu} = \sum_{\eta=\pm1}[\eta(pk)_\parallel + (pk)_{nn'}]$$
$$\times\left\{n_n^{\rm sum}(p_z)\big[N_{n'n}(\eta p_\parallel, k)\big]^{\mu\nu} + \eta n_n^{\rm diff}(p_z)\big[G_{n'n}(\eta p_\parallel, k)\big]^{\mu\nu}\right\}. \quad (9.1.21)$$

The remaining sums in (9.1.20) are over n', n, and cannot be performed explicitly in general.

Further evaluation of (9.1.20) or (9.1.21) involves performing the integral over p_z, which can be expressed in terms of a relativistic plasma dispersion function (RPDF) for given $n_n^{\rm sum}(p_z)$ or $n_n^{\rm diff}(p_z)$. One complication is the dependence on p_z in the numerator. Some progress in removing this dependence is made by writing $p_\parallel = [(pk)_\parallel k^\mu - (pk)_D k_D^\mu]/(k^2)_\parallel$, and replacing $(pk)_\parallel$ everywhere in the numerator by its resonant value $(pk)_{nn'}$, with the extra nondispersive terms being evaluated separately. However, the terms involving $(pk)_D = \varepsilon_n k_z - p_z\omega$ cannot be eliminated in any obvious way. An alternative way of evaluating the p_z-integral in terms of RPDFs is developed in § 9.2.

9.1.5 Response Tensor: Ritus Method

For a spin-independent electron gas, the Ritus method results in an expression for $\Pi^{\mu\nu}(k)$ that has some analogies to the corresponding result in the unmagnetized case, which is given in Eq. (8.1.2) of volume 1. The result for the unmagnetized case can be written in the form

$$\Pi^{\mu\nu}(k) = ie^2\int\frac{d^4P}{(2\pi)^4}\frac{d^4P'}{(2\pi)^4}(2\pi)^4(P'-P+k)\,{\rm Tr}\,[\gamma^\mu\bar G(P)\gamma^\nu\bar G(P')], \quad (9.1.22)$$

where the statistical average of the propagator, $G(P) = (\not{P} + m)/(P^2 - m^2 + i0)$, can be written

$$\bar{G}(P) = \sum_{\epsilon=\pm1} \frac{\epsilon\not{p} + m}{2\epsilon\varepsilon} \left[\wp\frac{1}{E - \epsilon\varepsilon} - i\epsilon[1 - 2n^{\epsilon}(\boldsymbol{p})]\pi\delta(E - \epsilon\varepsilon) \right], \qquad (9.1.23)$$

with $E = P^0$ and $\not{p} = \gamma^0\varepsilon - \boldsymbol{\gamma}\cdot\boldsymbol{p}$.

The form of the response tensor implied by the Ritus method for the magnetized case is analogous to (9.1.22) in that it can be written as

$$\Pi^{\mu\nu}(k) = \sum_{n',n} \frac{ie^3 B}{2\pi} \int \frac{dE\, dE'}{2\pi\ 2\pi} 2\pi(E' - E + \omega) \int \frac{dP_z\, dP_z'}{2\pi\ 2\pi} 2\pi\delta(P_z' - P_z + k_z)$$

$$\times \mathrm{Tr}\left[\mathcal{J}^{\mu}_{n'n}(\boldsymbol{k}_\perp)\, \bar{G}_n(E, P_z) \mathcal{J}^{\nu}_{nn'}(-\boldsymbol{k}_\perp)\, \bar{G}_{n'}(E', P_z')\right], \qquad (9.1.24)$$

where the statistical average of the propagator $G_n(E, P_z) = (\not{P}_n + m)/(E^2 - \varepsilon_n^2)$ is given by

$$\bar{G}_n(E, P_z) = \sum_{\epsilon=\pm1} \frac{\not{P}_n^{\epsilon} + m}{2\epsilon\varepsilon_n} \left[\wp\frac{1}{E - \epsilon\varepsilon_n} - i\epsilon[1 - 2n_n^{\epsilon}(p_z)]\pi\delta(E - \epsilon\varepsilon_n) \right], \qquad (9.1.25)$$

with $P_z = \epsilon p_z, \not{P}_n^{\epsilon} = \gamma^0\epsilon\varepsilon_n - \gamma^2 p_n - \gamma^3\epsilon p_z$. The vertex function in the unmagnetized case is replaced, in the magnetized case, by (5.5.26), viz.

$$\mathcal{J}^{\mu}_{n'n}(\boldsymbol{k}_\perp) = (-ie^{-i\psi})^{n'-n}\{[J^{n-1}_{n'-n}(x)\,\mathcal{P}_+ + J^{n}_{n'-n}(x)\,\mathcal{P}_-]\gamma^{\mu}_{\parallel}$$

$$+ [-ie^{-i\psi}J^{n-1}_{n'-n+1}(x)\,\mathcal{P}_- + ie^{i\psi}J^{n}_{n'-n-1}(x)\,\mathcal{P}_+]\gamma^{\mu}_{\perp}\}. \qquad (9.1.26)$$

Further evaluation of the response tensor in the form (9.1.24) involves taking the trace over the Dirac matrices. The result is the same as that derived above by summing over the spins directly. The direct method has the advantage that it can also be applied to a spin-dependent electron gas (§ 9.6), whereas the Ritus method applies only to a spin-independent electron gas.

9.1.6 *Neglect of Quantum Effects*

Reduction of the relativistic quantum form for the response tensor to the non-quantum limit involves identifying the quantum effects and then neglecting them. Several independent approximations need to be made in neglecting quantum effects, and these can be made in different orders, leading to various intermediate (partly quantum) approximate forms. The specific approximations that need to be made are the large-n limit, the Bessel-function approximation to the J-functions, the neglect of the quantum recoil, the neglect of the spin, and the neglect of dispersion associated with pair creation.

Large-n limit: The large-n limit corresponds to cases where the important contribution to the response tensor is due to electrons in Landau states, $n \gg 1$. The explicit assumption is that n and n' are large, with $n-n'$ much smaller than either. In the large-n limit motion perpendicular to the magnetic field is quasi-classical, with $\hbar n \to p_\perp^2/2eB$, corresponding to $p_n \to p_\perp$. Thus the ratio $a/n = (n-n')eB\hbar/p_\perp^2$ is first order in $\hbar$. The sum over n' for fixed n is replaced by a sum over $a = n - n'$. The number a is identified as the harmonic number in a classical treatment. The sum over n in any specific form of the response tensor is replaced by an integral over $p_\perp$, specifically, by the integral over $dp_\perp p_\perp/eB\hbar$. The occupation number, $n_n^\epsilon(p_z)$, is reinterpreted as a quasi-classical occupation number, $n^\epsilon(p_\perp, p_z)$, that is a function of the continuous variable $p_\perp = (2neB\hbar)^{1/2}$.

Bessel-function approximation: In the large-n limit, the J-functions can be expanded in Bessel functions. In the nonquantum limit only the leading term is retained in this expansion, which gives $J^n_{n-n'}(x) = J_a(z)$, with $a = n - n'$, and where the relation between the arguments $x = \hbar k_\perp^2/2eB$ and $z = k_\perp p_\perp/eB$ involves n through $p_\perp = (2neB\hbar)^{1/2}$. Quantum corrections to this leading term are included in (A.1.54)–(A.1.56).

Retaining only the leading terms in $\hbar$ gives the approximations (6.1.11) and (6.1.12), viz.

$$[\Gamma^{\epsilon\epsilon}_{q'q}]^\mu = \frac{(ie^{i\psi})^{-a}}{\gamma} U^\mu(a,k),$$
$$\left[N_{n'n}(\epsilon p_\parallel, k)\right]^{\mu\nu} + \left[G_{n'n}(\epsilon p_\parallel, k)\right]^{\mu\nu} = 4m^2 U^\mu(a,k) U^{*\nu}(a,k), \quad (9.1.27)$$

with $a = n - n'$, where $U^\mu(a,k)$ is given by (2.1.28) with (2.1.29), and where the Lorentz factor is interpreted as $\gamma = \varepsilon_n/mc^2$ with $\varepsilon_n^2 = m^2c^4 + p_n^2c^2 + p_z^2c^2$, $p_n = p_\perp$. Note that the sign ϵ is implicit on the right hand side of (9.1.27), and appears explicitly in (2.1.29).

Quantum recoil: The resonant denominator in the classical case is $(ku)_\parallel - a\Omega_0$. The denominators that appear in the general form (9.1.12) are $\eta(pk)_\parallel \pm (pk)_{nn'}$, which may be rewritten using

$$(pk)_\parallel = \varepsilon_n\omega - p_z k_z \to m(ku)_\parallel, \qquad (pk)_{nn'} = aeB + \tfrac{1}{2}\hbar(k^2)_\parallel, \qquad (9.1.28)$$

with $\varepsilon_n \to \gamma mc^2$ in the large-n limit. The term $\frac{1}{2}\hbar(k^2)_\parallel$ is identified as the quantum recoil, and it is neglected in the resonant denominator in the nonquantum limit. With this neglect, $\eta(pk)_\parallel \pm (pk)_{nn'}$ reproduces the classical resonant denominator in the large-n limit. The sum over $n' = n - a$ for fixed n in the quantum case can be written as a sum over a, the restriction $n' \geq 0$ on a is ignored in the nonquantum limit, and the sum is extended to $-\infty < a < \infty$.

Spin: Spin can be neglected in two different ways in a relativistic quantum treatment. The procedure used here is to include the spin, by using Dirac's equation, and assume the occupation number is independent of spin, and average over the spin states. The other procedure is to assume scalar particles, described by the Klein–Gordon equation rather than the Dirac equation, so that the spin is identically zero. These procedures lead to different results for $\Pi^{\mu\nu}(k)$, with the difference being of order $\hbar^2$. Spin is an intrinsically quantum effect and, even in a spin-dependent electron gas, the spin introduces a correction of order $\hbar$ to the response tensor. There is a context in which the spin cannot be neglected: when the magnetic properties of materials are important: paramagnetism and ferromagnetism are directly related to the spin of the electron. Spin-dependent plasmas are discussed in § 9.6.

One-photon pair creation: Dispersion due to one-photon pair creation, which has no nonquantum counterpart, corresponds to the denominator in (9.1.1) with $\epsilon' = -\epsilon$, specifically, $\hbar\omega \pm (\varepsilon_q + \varepsilon_{q'})$. The nonquantum limit for these terms corresponds $\hbar\omega \to 0$. Although these terms do not contribute to the dispersion in the nonquantum limit, they need to be retained to reproduce the nondispersive part correctly.

9.1.7 Nonquantum Limit of $\Pi^{\mu\nu}(k)$

In Chap. 2, the response tensor in the nonquantum limit is derived in two different ways, referred to as the forward-scattering and Vlasov methods, leading to the superficially different forms (2.3.10) and (2.3.29), respectively. It is of interest to derive both these forms from the relativistic quantum form for the response tensor.

Forward-Scattering Form

The form (9.1.19) for the response tensor reproduces the forward-scattering form (2.3.10) in the nonquantum limit. The reduction of (9.1.19) to this limit involves neglecting the quantum recoil term, taking the limits $\hbar \to 0$, $n \to \infty$, and making the replacement $\hbar n \to p_\perp^2/2eB$. The vertex functions are replaced by their nonquantum counterparts through (9.1.27). Only the contribution of the electrons needs to be considered explicitly, with positrons contributing with the same sign to the nongyrotropic part and with the opposite sign to the gyrotropic part. The derivation is straightforward, and reproduces (2.3.10) in the form

$$\Pi^{\mu\nu}(k) = -e^2\sum_a \int \frac{d^3\boldsymbol{p}}{(2\pi)^3}\frac{n(p_\perp, p_z)}{\varepsilon}\Bigg\{g_\parallel^{\mu\nu}J_a^2 - \frac{k_\parallel^\mu U^{*\nu}(a,k) + k_\parallel^\nu U^\mu(a,k)}{(ku)_\parallel - a\Omega_0}J_a$$
$$+\frac{(k^2)_\parallel U^\mu(a,k)U^{*\nu}(a,k)}{[(ku)_\parallel - a\Omega_0]^2}$$

$$+\frac{[(ku)_\parallel - a\Omega_0]^2}{[(ku)_\parallel - a\Omega_0]^2 - \Omega_0^2}\left[g_\perp^{\mu\nu} J_a^2 - \frac{k_\perp^\mu U^{*\nu}(a,k) + k_\perp^\nu U^\mu(a,k)}{(ku)_\parallel - a\Omega_0} J_a \right.$$

$$\left. + \frac{(k^2)_\perp U^\mu(a,k) U^{*\nu}(a,k)}{[(ku)_\parallel - a\Omega_0]^2}\right]$$

$$\left. - \frac{i\bar{\epsilon}\Omega_0[(ku)_\parallel - a\Omega_0]}{[(ku)_\parallel - a\Omega_0]^2 - \Omega_0^2}\left[f^{\mu\nu} J_a^2 + \frac{k_G^\mu U^{*\nu}(a,k) - k_G^\nu U^\mu(a,k)}{(ku)_\parallel - a\Omega_0} J_a\right]\right\}, \quad (9.1.29)$$

with $J_a = J_a(k_\perp u_\perp/\Omega_0)$ and with $\bar{\epsilon} = n^{\rm diff}(p)/n^{\rm sum}(p)$ ranging between $\bar{\epsilon} = 1$ for an electron gas to $\bar{\epsilon} = -1$ for a positron gas.

Vlasov Form

The reduction of (9.1.1), with $\epsilon = \epsilon' = 1$, to its nonquantum counterpart in the Vlasov form (2.3.29) involves the following steps. First, take the large-n limit and replace the vertex functions by their nonquantum counterparts using (9.1.27). Second, replace the resonant denominator by its nonquantum counterpart, by neglecting the quantum recoil. Third, expand the difference between the occupation numbers in (9.1.1) using the differential operator $\hat{D}$, introduced in (6.1.8). One has

$$n_{n'}^+(p_z') = n_{n-a}^+(p_z - k_z) = \left(1 - \hat{D} + \tfrac{1}{2}\hat{D}^2 + \cdots\right) n_n^+(p_z). \quad (9.1.30)$$

Only up to first order is retained in (9.1.30). Then $n_n^+(p_z)$ is replaced by the classical distribution function, $n_n^+(p_z) \to n^+(p_\perp, p_z)$. A related step is to replace the sum over n by the integral over $dp_\perp\, p_\perp/eB$. The occupation number, $n^+(p_\perp, p_z)$, is proportional to the classical distribution function. The constant of proportionality is identified by noting that the number density of electrons is

$$n^+ = 4\pi \int_0^\infty \frac{dp_\perp\, p_\perp}{(2\pi\hbar)^2} \int_{-\infty}^\infty \frac{dp_z}{2\pi\hbar} n^+(p_\perp, p_z) = \int \frac{d^4p}{(2\pi\hbar)^4} \gamma F(p), \quad (9.1.31)$$

where a factor 2 arises from the sum over spins ($a_n = 2$). The sum over n' is replaced by a sum over a, and $n_n^+(p_z)$ is rewritten as $n^+(p_\perp, p_z)$. This gives

$$n_n^+(p_z) - n_{n'}^+(p_z') = \hbar\left[\frac{aeB}{p_\perp}\frac{\partial}{\partial p_\perp} + k_z \frac{\partial}{\partial p_z}\right] n^+(p_\perp, p_z). \quad (9.1.32)$$

The resulting approximation to the response tensor reproduces the term involving the sum over harmonic number in (2.3.29) which, when rewritten in the notation used in this section, becomes

$$\Pi^{\mu\nu}(k) = \frac{4\pi e^2}{c} \int_0^\infty \frac{dp_\perp\, p_\perp}{(2\pi\hbar)^2} \int_{-\infty}^{\infty} \frac{dp_z}{2\pi\hbar} \left[\frac{1}{m} \left(p_\parallel^\mu p_\parallel^\nu \frac{1}{p_\perp} \frac{\partial}{\partial p_\perp} - b^\mu b^\nu\, p_z \frac{\partial}{\partial p_z} \right) \right.$$
$$\left. - \sum_{a=-\infty}^{\infty} \frac{U^\mu(a,k) U^{*\nu}(a,k)}{(ku)_\parallel - a\Omega_0} \left(\frac{m(ku)_\parallel}{u_\perp} \frac{\partial}{\partial p_\perp} + k_z \frac{\partial}{\partial p_z} \right) \right] n^+(p_\perp, p_z), \tag{9.1.33}$$

with $u_\parallel^\mu = p_\parallel^\mu / m$.

9.2 Relativistic Plasma Dispersion Functions

The response tensor for a relativistic magnetized quantum electron gas involves only one integral, over p_z, and this integral can be written in terms of a single relativistic plasma dispersion function (RPDF). The RPDF chosen here, denoted as $\mathcal{Z}_n^\epsilon(t_0)$, depends on the occupation number $n_n^\epsilon(p_z)$, which is arbitrary. Some explicit forms for $\mathcal{Z}_n^\epsilon(t_0)$ are evaluated in § 9.3 for thermal distributions.

9.2.1 Dispersion-Integral Method

A method used by Toll [21] to calculate the regularized form of the vacuum polarization tensor (§ 8.1.5) has a counterpart for an electron gas. The idea is to calculate the dissipative part of the response and relate the dispersive part to it through a dispersion integral, which is a Kramers–Kronig relation in the present context. This idea was used by Silin [17] to calculate the response tensor for an isotropic relativistic quantum electron gas (§ 4.4.4 of volume 1). In the wider context of quantum field theory, the same idea was introduced by Cutkovsky [4], and it is referred to here as the dispersion-integral method. In the present context, the idea that underlies these methods is that one can construct the hermitian part of the response tensor from the antihermitian part, and that one can calculate the antihermitian part by considering only resonant interactions, as in the evaluation of the absorption coefficient for gyromagnetic absorption (6.1.36) and one-photon pair creation (6.4.3).

Antihermitian Part of the Response Tensor

A general form for the antihermitian part of the response tensor may be derived from (9.1.1). Using the Plemelj formula (2.1.17), one finds

$$\Pi^{A\mu\nu}(k) = i\pi \tfrac{e^3 B}{2\pi} \sum_{\epsilon,q,\epsilon',q'} \int \frac{dp_z}{2\pi} \left[\tfrac{1}{2}(\epsilon' - \epsilon) + \epsilon n_q^{\epsilon} - \epsilon' n_{q'}^{\epsilon'}\right]$$

$$\times\left[\Gamma_{q'q}^{\epsilon'\epsilon}(\boldsymbol{k})\right]^{\mu}\left[\Gamma_{qq'}^{\epsilon\epsilon'}(-\boldsymbol{k})\right]^{\nu} \delta(\omega - \epsilon\varepsilon_q + \epsilon'\varepsilon_{q'}). \tag{9.2.1}$$

The sum over ϵ, ϵ' involves four terms. The term $\epsilon = \epsilon' = 1$ is associated with gyromagnetic absorption by electrons. In this case the primed state is identified as the electron state before absorption, and the unprimed state as the electron state after absorption. The term $\epsilon = \epsilon' = -1$ is associated with gyromagnetic absorption by positrons, with the unprimed state the initial state and the primed state the final state in this case. The terms $\epsilon \neq \epsilon'$ are associated with dispersion due to one-photon pair creation. One-photon pair creation exists in the vacuum, and the terms in (9.2.1) that are independent of the occupation numbers give the antihermitian part of the vacuum response tensor. The presence of electrons or positrons suppresses the vacuum contribution. Specifically, for $\epsilon = -\epsilon' = \pm 1$ the factor in square brackets is $\mp[1 - n_q^+ - n_{q'}^-]$, and it implies that one-photon pair creation is suppressed when the state into which the electron or the positron would be created is occupied, in accord with the Pauli exclusion principle.

The antihermitian part of the response tensor (9.2.1) becomes, for a spin-independent electron gas,

$$\Pi^{A\mu\nu}(k) = i\pi \frac{e^3 B}{4\pi^2} \sum_{\epsilon,n,\epsilon',n',\pm} \frac{\tfrac{1}{2}(\epsilon' - \epsilon) + \epsilon n_n^{\epsilon}(\epsilon p_{nn'}^{\pm}) - \epsilon' n_{n'}^{\epsilon'}(\epsilon' p'_{z\pm})}{2(\omega^2 - k_z^2)\, g_{nn'}}$$

$$\times\left[C_{n'n}(p'_{\parallel\pm}, p_{\parallel\pm}, k)\right]^{\mu\nu}, \tag{9.2.2}$$

with the resonant values given by (6.1.16)–(6.1.18). Including $\hbar$ and c explicitly, one has (ordinary units)

$$p_{nn'}^{\pm} = \hbar k_z f_{nn'} \pm \hbar\omega g_{nn'}/c, \qquad \varepsilon_{nn'}^{\pm} = \hbar\omega f_{nn'} \pm \hbar k_z c g_{nn'},$$

$$f_{nn'} = \frac{(\varepsilon_n^0)^2 - (\varepsilon_{n'}^0)^2 + \hbar^2(\omega^2 - k_z^2 c^2)}{2\hbar^2(\omega^2 - k_z^2 c^2)},$$

$$g_{nn'}^2 = \frac{\left[\hbar^2(\omega^2 - k_z^2 c^2) - (\varepsilon_n^0 - \varepsilon_{n'}^0)^2\right]\left[\hbar^2(\omega^2 - k_z^2 c^2) - (\varepsilon_n^0 + \varepsilon_{n'}^0)^2\right]}{[2\hbar^2(\omega^2 - k_z^2 c^2)]^2}. \tag{9.2.3}$$

The resonant values can be expressed in terms of the 4-vector $p_{\parallel\pm}^{\mu} = (\varepsilon_{nn'}^{\pm}/c, 0, 0, p_{nn'}^{\pm})$ with

$$p_{\parallel\pm}^{\mu} = \hbar(k_{\parallel}^{\mu} f_{nn'} \pm k_D^{\mu} g_{nn'}), \qquad p'^{\mu}_{\parallel\pm} = p_{\parallel\pm}^{\mu} - \hbar k_{\parallel}^{\mu}, \tag{9.2.4}$$

with $k_\parallel^\mu = (\omega/c, 0, 0, k_z)$, $k_D^\mu = (k_z, 0, 0, \omega/c)$. The $\pm$ subscripts of the vertex functions in (9.2.1) indicate that they are to be evaluated at these resonant values.

Kramers–Kronig Relations

The Kramers–Kronig relations follow from the requirement that the response be causal, which implies

$$\Pi^{\mu\nu}(\omega, \boldsymbol{k}) = i \int_{-\infty}^{\infty} \frac{d\omega'}{2\pi} \frac{\Pi^{\mu\nu}(\omega', \boldsymbol{k})}{\omega - \omega' + i0}. \tag{9.2.5}$$

The causal condition imposed in deriving (9.2.5) expresses the requirement that cause precede effect. A stronger condition is that this must apply in all inertial frames. In the magnetized case, only Lorentz transformations along the direction of the magnetic field are relevant. The stronger condition then corresponds to (9.2.5) being replaced by

$$\Pi^{\mu\nu}(\omega, \boldsymbol{k}_\perp, k_z) = i \int_{-\infty}^{\infty} \frac{d\omega'}{2\pi} \frac{\Pi^{\mu\nu}(\omega', \boldsymbol{k}_\perp, k_z + \beta_0[\omega - \omega'])}{\omega - \omega' + i0}, \tag{9.2.6}$$

which applies for all $\beta_0^2 < 1$. The stronger condition (9.2.6) has various implications but these have not been used in plasma dispersion theory. An example is the identity found by differentiating (9.2.6) with respect to β_0 and setting $\beta_0 = 0$. This implies that the ω-integral of $\partial \Pi^{\mu\nu}(k)/\partial k_z$ is zero.

The Kramers–Kronig relations follow from (9.2.5) by separating into hermitian and antihermitian parts, which are related by a Hilbert transform. The hermitian part of (9.2.5) gives

$$\Pi^{\mathrm{H}\mu\nu}(\omega, \boldsymbol{k}) = \Pi^{\mu\nu}_{\mathrm{ND}}(\omega, \boldsymbol{k}) - \frac{i}{\pi} \wp \int_{-\infty}^{\infty} \frac{d\omega'}{\omega' - \omega} \Pi^{A\mu\nu}(\omega', \boldsymbol{k}), \tag{9.2.7}$$

where there is an undetermined nondispersive part, $\Pi^{\mu\nu}_{\mathrm{ND}}(\omega, \boldsymbol{k})$, whose Hilbert transform is zero. Use of (9.2.7) is the basis for one method of regularizing the vacuum polarization tensor [21]: first evaluate the integrals in the vacuum contribution in (9.2.1), and then insert this into (9.2.7) to find the hermitian part of the vacuum response tensor.

The relation (9.2.5) implies that the dispersive part of the response, described by $\Pi^{\mathrm{H}\mu\nu}(\omega, \boldsymbol{k})$, can be determined from the resonant part, described by $\Pi^{\mathrm{A}\mu\nu}(\omega, \boldsymbol{k})$. On inserting the antihermitian part (9.2.1) into (9.2.7), the dependence of $\Pi^{A\mu\nu}(\omega', \boldsymbol{k})$ on ω' is through the dependence on $p_{nn'}^{\pm}$, on $\omega \to \omega'$. The integral over ω' in (9.2.7) may be rewritten as an integral over p_z, with a resonant denominator $p_z - p_{nn'}^{\pm}$, suggesting a form

$$\Pi^{\mathrm{H}\mu\nu}(\omega,\boldsymbol{k})=\Pi^{\mu\nu}_{\mathrm{ND}}(\omega,\boldsymbol{k})-\sum_{\pm}\frac{i}{\pi}\wp\int_{-\infty}^{\infty}\frac{dp_z}{p_z-p^{\pm}_{nn'}}\,\Pi^{A\mu\nu}(p_z,\boldsymbol{k}),\qquad(9.2.8)$$

with $\Pi^{A\mu\nu}(p_z,\boldsymbol{k})$, given by (9.2.1). As in Silin's method (§ 4.4.4 of volume 1), the nondispersive (ND) part is undetermined by (9.2.8), and the only general constraints on it are that it involve only functions whose Hilbert transform is zero, and that it satisfy the charge-continuity and gauge-invariance conditions.

9.2.2 *Evaluation of Dispersion Integrals*

The foregoing arguments suggest that the dispersive part of the response tensor can be expressed in a form in which the numerator is evaluated at the resonant values and taken outside the integral over p_z, which then corresponds to a RPDF. The following systematic procedure leads to this result.

Rationalized Resonant Denominator

Before any sums are performed, there are four resonant denominators in the response tensor (9.1.1), specifically, $\omega-\epsilon\varepsilon_n+\epsilon'\varepsilon'_{n'}$ with $\epsilon,\epsilon'=\pm1$. The product of all four resonant denominators is a quadratic function of p_z:

$$\frac{1}{D(\omega,\varepsilon_n,\varepsilon'_{n'})}=-\frac{1}{4(k^2)_\parallel(\epsilon p_z-p^{+}_{nn'})(\epsilon p_z-p^{-}_{nn'})},\qquad(9.2.9)$$

with $D(\omega,\varepsilon_n,\varepsilon'_{n'})$ defined by (6.1.13).

After rationalizing the denominator in this way, the response tensor in the form (9.1.19) gives

$$\Pi^{\mu\nu}(k)=\frac{e^3B}{16\pi^2(k^2)_\parallel}\sum_{\eta,n,n'}\int\frac{dp_z}{\varepsilon_n}\frac{\eta(\varepsilon_n\omega+p_zk_z)+(pk)_{nn'}}{(\eta p_z-p^{+}_{nn'})(\eta p_z-p^{-}_{nn'})}$$
$$\times\{n^{\mathrm{sum}}_n(p_z)[N_{n'n}(\eta p_\parallel,k)]^{\mu\nu}+\eta n^{\mathrm{diff}}_n(p_z)[G_{n'n}(\eta p_\parallel,k)]^{\mu\nu}\},\qquad(9.2.10)$$

with $p^{\pm}_{nn'}$ implicit functions of n',n through (9.2.3) and with $(pk)_{nn'}=\frac{1}{2}(k^2)_\parallel+\frac{1}{2}p_n^2-\frac{1}{2}p_{n'}^2$. The sign ϵ is relabeled as η.

The denominator in (9.2.10) can be written as a sum of two denominators with poles at $\eta p_z=p^{\pm}_{nn'}$:

$$\frac{1}{(\eta p_z-p^{+}_{nn'})(\eta p_z-p^{-}_{nn'})}=\frac{1}{2\omega g_{nn'}}\sum_{\pm}\frac{\pm1}{\eta p_z-p^{\pm}_{nn'}}.\qquad(9.2.11)$$

This reduces the p_z-integral to a standard dispersive form with a resonant denominator $\eta p_z - p^{\pm}_{nn'}$. The response tensor (9.2.10) involves sums over two independent signs: $\eta = \pm 1$ and the $\pm$ in the sum in (9.2.11).

Reduction to a Single Dispersive Integral

All the integrals that appear in the response tensor (9.2.10) with (9.2.11) can be evaluated in terms of a single dispersive function, written here as $K^{\epsilon}_n(\eta p_{\parallel\pm})$, with the argument denoting dependence on $p^{\mu}_{\parallel\pm} = (\varepsilon^{\pm}_{nn'}, 0, 0, p^{\pm}_{nn'})$, given by (9.2.4). The single dispersive integral is defined by writing

$$\int dp_z \frac{n^{\epsilon}_n(p_z)}{\varepsilon_n} \frac{\eta(\varepsilon_n \omega + p_z k_z) + (pk)_{nn'}}{(\eta p_z - p^{+}_{nn'})(\eta p_z - p^{-}_{nn'})} = \frac{1}{2g_{nn'}} \sum_{\pm} \pm K^{\epsilon}_n(\eta p_{\parallel\pm}), \tag{9.2.12}$$

$$K^{\epsilon}_n(\eta p_{\parallel\pm}) = \int dp_z \frac{n^{\epsilon}_n(p_z)}{\varepsilon_n} \frac{\eta\varepsilon_n + \varepsilon^{\pm}_{nn'}}{\eta p_z - p^{\pm}_{nn'}}. \tag{9.2.13}$$

The p_z-integral in the response tensor (9.2.10) with (9.2.11) includes the integral (9.2.12) and integrals of the form (9.2.12) with additional terms, $\eta p^{\mu}_{\parallel}$ and $p^{\mu}_{\parallel} p^{\nu}_{\parallel}$ in the integrand. Omitting the superscript ϵ on $n^{\epsilon}_n(p_z)$, the integrals reduce to

$$H^{(a,b)}_{nn'} = \int dp_z \frac{n_n(p_z)}{\varepsilon_n} \frac{\eta(\varepsilon_n \omega + p_z k_z) + (\omega^2 - k_z^2) f_{nn'}}{(\eta p_z - p^{+}_{nn'})(\eta p_z - p^{-}_{nn'})} (\eta\varepsilon_n)^a (\eta p_z)^b, \tag{9.2.14}$$

which may be evaluated in terms of (9.2.13) with additional nondispersive contributions:

$$H^{(a,b)}_{nn'} = \left[H^{(a,b)}_{nn'}\right]_{\rm ND} + \frac{1}{2g_{nn'}} \sum_{\pm} \pm K_n(\eta p_{\parallel\pm})(\varepsilon^{\pm}_{nn'})^a (p^{\pm}_{nn'})^b. \tag{9.2.15}$$

The response tensor (9.2.10) with (9.2.11) reduces to

$$\Pi^{\mu\nu}(k) = \Pi^{\mu\nu}_{\rm ND}(k) + \frac{e^3 B}{32\pi^2 (k^2)_{\parallel} g_{nn'}} \sum_{\eta,n,n',\pm} \pm\{[N_{n'n}(p^{\pm}_{nn'}, k)]^{\mu\nu} K^{\rm sum}_n(\eta p_{\parallel\pm})$$
$$+\eta[G_{n'n}(p^{\pm}_{nn'}, k)]^{\mu\nu} K^{\rm diff}_n(\eta p_{\parallel\pm})\}, \tag{9.2.16}$$

where $\Pi^{\mu\nu}_{\rm ND}(k)$ is the nondispersive part, and with

$$\begin{bmatrix} K^{\rm sum}_n(\eta p_{\parallel\pm}) \\ K^{\rm diff}_n(\eta p_{\parallel\pm}) \end{bmatrix} = \begin{bmatrix} K^{+}_n(\eta p_{\parallel\pm}) + K^{-}_n(\eta p_{\parallel\pm}) \\ K^{+}_n(\eta p_{\parallel\pm}) - K^{-}_n(\eta p_{\parallel\pm}) \end{bmatrix}. \tag{9.2.17}$$

The result (9.2.16) with (9.2.17) achieves the objective of writing the response tensor in a form in which the numerator is evaluated at the resonant values and taken outside the integral, leaving a p_z-integral that defines a RPDF, plus a nondispersive part that can be identified explicitly.

9.2.3 Nondispersive Part

The nondispersive (ND) part, $\Pi^{\mu\nu}_{\rm ND}(k)$, results from terms in the numerator of the p_z-integral that cancel with the resonant denominator. The ND terms in the response tensor arise from the ND terms in integrals of the form (9.2.14) with $a, b = 0, 1, 2$. However, note that the ND part is not uniquely defined: formally, it is defined only by the requirement that its Hilbert transform be zero. It follows that the explicit form of the ND part can depend on the definition of the RPDF. The ND part in (9.2.17) applies specifically to the RPDF defined by (9.2.13). Additional ND terms may appear on introducing a different RPDF.

Explicit evaluation of the ND terms in (9.2.15) gives

$$[H^{(a,b)}_{nn'}]_{\rm ND} = \int dp_z \, \frac{n_n(p_z)}{\varepsilon_n} \, h^{(a,b)}_{nn'},$$

$$h^{(0,0)}_{nn'} = 0, \quad h^{(1,0)}_{nn'} = \omega, \quad h^{(0,1)}_{nn'} = k_z, \quad h^{(1,1)}_{nn'} = \eta(\varepsilon_n k_z + p_z\omega) + 2\omega k_z \, f_{nn'},$$

$$h^{(0,2)}_{nn'} = h^{(2,0)}_{nn'} = \eta(\varepsilon_n\omega + p_z k_z) + (\omega^2 + k_z^2) f_{nn'}. \tag{9.2.18}$$

On inserting these expressions into (9.2.10), the sum over $\eta = \pm 1$ gives zero for terms in the integrand proportional to η, and only the terms independent of η in (9.2.18) contribute to the ND part of the response tensor.

For the ND terms, the sum over n' may be performed explicitly. The relevant sums over the J-functions were derived by Sokolov and Ternov [18], and are written down in (A.1.42):

$$\sum_{n'=0}^{\infty} J^{n'}_{n-n'}(x) J^{n'}_{n''-n'}(x) = \delta_{nn''}, \qquad \sum_{n'=0}^{\infty} (n'-n)[J^{n'}_{n-n'}(x)]^2 = x, \tag{9.2.19}$$

The resulting sum over n is proportional to the proper number density,

$$\bar{n}_{\rm pr} = \sum_{\epsilon=\pm} \sum_{n=0} n^{\epsilon}_{n\rm pr}, \tag{9.2.20}$$

where the proper number density in the nth Landau state is

$$n^{\epsilon}_{n\mathrm{pr}} = \frac{eBm}{2\pi} a_n \int \frac{dp_z}{2\pi} \frac{n^{\epsilon}_n(p_z)}{\varepsilon_n} = \frac{eBm}{2\pi^2} a_n \int_{-1}^{1} dt \, \frac{1}{1-t^2} \, n^{\epsilon}_n(p_z). \tag{9.2.21}$$

The resulting expression is [11]

$$\Pi^{\mu\nu}_{\mathrm{ND}}(k) = -\frac{e^2 \bar{n}_{\mathrm{pr}}}{m} \left[g^{\mu\nu}_{\perp} - \frac{k^{\mu}_{\parallel} k^{\nu}_{\perp} + k^{\nu}_{\parallel} k^{\mu}_{\perp}}{(k^2)_{\parallel}} - \frac{(k^{\mu}_{\parallel} k^{\nu}_{\parallel} + k^{\mu}_{D} k^{\nu}_{D}) k^2_{\perp}}{[(k^2)_{\parallel}]^2} \right], \tag{9.2.22}$$

with $\bar{n}_{\mathrm{pr}}$ the sum of the proper number densities for electrons and positrons. The ND part (9.2.22) applies specifically for the RPDF (9.2.13), and it is of little physical interest in itself.

9.2.4 Plasma Dispersion Function $\mathcal{Z}^{\epsilon}_n(t_0)$

The dispersive part of the response tensor, (9.2.16), involves the RPDF $K^{\epsilon}_n(\eta p_{\parallel\pm})$, given by (9.2.13). The following steps allow this to be re-expressed in terms of a more convenient RPDF, denoted $\mathcal{Z}^{\epsilon}_n(t_0)$.

Evaluation of Dispersion Integrals

Evaluation of the p_z-integral in (9.2.13) is complicated by the square-root dependence that appears in $\varepsilon_n = [(\varepsilon^0_n)^2 + p_z^2]^{1/2}$, with $\varepsilon^0_n = (m^2 + p_n^2)^{1/2}$. One may eliminate square roots and evaluate the integral by rewriting the p_z-integral in terms of a variable t defined by

$$\frac{p_z}{\varepsilon^0_n} = \frac{2t}{1-t^2}, \qquad \frac{\varepsilon_n}{\varepsilon^0_n} = \frac{1+t^2}{1-t^2}, \qquad dp_z = 2\varepsilon^0_n dt \, \frac{1+t^2}{(1-t^2)^2}. \tag{9.2.23}$$

The resonant values $\eta p_z = p^{\pm}_{nn'}$ may be written

$$p^{\pm}_{nn'} = \varepsilon^0_n \frac{2t^{\pm}_{nn'}}{1-[t^{\pm}_{nn'}]^2}, \qquad \varepsilon^{\pm}_{nn'} = \varepsilon^0_n \frac{1+[t^{\pm}_{nn'}]^2}{1-[t^{\pm}_{nn'}]^2}, \tag{9.2.24}$$

where $\eta t = t^{\pm}_{nn'}$ is one of the solutions of the quadratic equation $2\eta t/(1-t^2) = p^{\pm}_{nn'}/\varepsilon^0_n$. The two solutions are $\eta t = t^{\pm}_{nn'}, -1/t^{\pm}_{nn'}$, with

$$t^{\pm}_{nn'} = \frac{\varepsilon^{\pm}_{nn'} - \varepsilon^0_n}{p^{\pm}_{nn'}}, \qquad -\frac{1}{t^{\pm}_{nn'}} = -\frac{\varepsilon^{\pm}_{nn'} + \varepsilon^0_n}{p^{\pm}_{nn'}}. \tag{9.2.25}$$

The combination that appears in the integrand in (9.2.10) with (9.2.11) becomes

$$\frac{\eta(\varepsilon_n\omega + p_z k_z) + (pk)_{nn'}}{(\eta p_z - p^+_{nn'})(\eta p_z - p^-_{nn'})} = \sum_{\pm} \pm \frac{1}{p^+_{nn'} - p^-_{nn'}} \left[\frac{\eta\varepsilon_n + \varepsilon^\pm_{nn'}}{\eta p_z - p^\pm_{nn'}} + k_z \right], \qquad (9.2.26)$$

with the final term inside the square brackets summing to zero. On rewriting the integral in (9.2.13) in terms of t one finds

$$K^\epsilon_n(\eta p_{\parallel\pm}) = \int dt\, n^\epsilon_n(p_z) \left[\frac{2t}{1-t^2} + \frac{\eta+1}{\eta t - t^\pm_{nn'}} + \frac{\eta-1}{\eta t + 1/t^\pm_{nn'}} \right]. \qquad (9.2.27)$$

The dispersive contribution arises from the resonances, with that at $t = t^\pm_{nn'}$ associated with $\eta = +1$, and that at $t = -1/t^\pm_{nn'}$ associated with $\eta = -1$. The term involving $2t/(1-t^2)$ in (9.2.27) gives a nondispersive (ND) contribution, and there are analogous ND terms when additional powers of ε_n and p_z are included in the integrand.

Definition of $\mathcal{Z}^\epsilon_n(t_0)$

It follows from (9.2.27) that the dispersive contribution can be evaluated in terms of a RPDF defined by

$$\mathcal{Z}^\epsilon_n(t_0) = \int_{-1}^{1} dt\, \frac{n^\epsilon_n(p_z)}{t - t_0}, \qquad \eta\mathcal{Z}^\epsilon_n(\eta t_0) = \int_{-1}^{1} dt\, \frac{n^\epsilon_n(p_z)}{\eta t - t_0}, \qquad (9.2.28)$$

where the second form is equivalent to the first. The resonances in (9.2.27) correspond to $t_0 = t^\pm_{nn'}, -1/t^\pm_{nn'}$, and it follows from (9.2.27) that the RPDF (9.2.28) appears with all four arguments. The dispersive part of (9.2.27) becomes

$$K^\epsilon_n(\eta p_{\parallel\pm}) = (\eta+1)\mathcal{Z}^\epsilon_n(t^\pm_{nn'}) - (\eta-1)\mathcal{Z}^\epsilon_n(1/t^\pm_{nn'}). \qquad (9.2.29)$$

The dispersive part of (9.2.15) becomes

$$H^{(a,b)}_{nn'} = \frac{1}{2g_{nn'}} \sum_{\pm} \pm \left[(\eta+1)\mathcal{Z}^\epsilon_n(t^\pm_{nn'}) - (\eta-1)\mathcal{Z}^\epsilon_n(1/t^\pm_{nn'})\right] (\varepsilon^\pm_{nn'})^a (p^\pm_{nn'})^b. \qquad (9.2.30)$$

This confirms the result, suggested in (9.2.8) based on the causal condition, that the terms in the numerator that involve powers of ε_n and p_z can be evaluated at the resonant values and taken outside the dispersion integral, in evaluating the dispersive part.

The combinations of electron and positron contributions appear in the response tensor through

$$\begin{bmatrix} \mathcal{Z}_n^{\rm sum}(t_0) \\ \mathcal{Z}_n^{\rm diff}(t_0) \end{bmatrix} = \begin{bmatrix} \mathcal{Z}_n^+(t_0) + \mathcal{Z}_n^-(t_0) \\ \mathcal{Z}_n^+(t_0) - \mathcal{Z}_n^-(t_0) \end{bmatrix}. \tag{9.2.31}$$

The dependence on the sign η is now explicit, and the sum over η gives

$$\sum_{\eta=\pm1} \begin{pmatrix} 1 \\ \eta \end{pmatrix} [(\eta+1)\mathcal{Z}_n^\epsilon(t_{nn'}^\pm) - (\eta-1)\mathcal{Z}_n^\epsilon(1/t_{nn'}^\pm)] = 2 \begin{pmatrix} \mathcal{Z}_n^\epsilon(t_{nn'}^\pm) + \mathcal{Z}_n^\epsilon(1/t_{nn'}^\pm) \\ \mathcal{Z}_n^\epsilon(t_{nn'}^\pm) - \mathcal{Z}_n^\epsilon(1/t_{nn'}^\pm) \end{pmatrix}. \tag{9.2.32}$$

It follows that the upper combination appears in the nongyrotropic part, which involves $\mathcal{Z}_n^{\rm sum}(t_0)$, and the lower combination appears in the gyrotropic part, which involves $\mathcal{Z}_n^{\rm diff}(t_0)$.

9.2.5 RPDF Form of $\Pi^{\mu\nu}(k)$

The expression for the response tensor that results from (9.2.16) is referred to as the RPDF form of $\Pi^{\mu\nu}(k)$. The resulting form is

$$\begin{aligned} \Pi^{\mu\nu}(k) = \Pi_{\rm ND}^{\mu\nu}(k) &+ \frac{e^3 B}{16\pi^2 (k^2)_\parallel} \\ &\times \sum_{n,n',\pm} \frac{\pm 1}{g_{nn'}} \Big\{ \big[N_{n'n}(p_{nn'}^\pm, k)\big]^{\mu\nu} [\mathcal{Z}_n^{\rm sum}(t_{nn'}^\pm) + \mathcal{Z}_n^{\rm sum}(1/t_{nn'}^\pm)] \\ &+ \big[G_{n'n}(p_{nn'}^\pm, k)\big]^{\mu\nu} [\mathcal{Z}_n^{\rm diff}(t_{nn'}^\pm) - \mathcal{Z}_n^{\rm diff}(1/t_{nn'}^\pm)] \Big\}. \end{aligned} \tag{9.2.33}$$

To perform the sum over $\pm$ one needs to separate the tensors into contributions that are even and odd, by writing

$$\begin{bmatrix} [N_{n'n}(k)]_\pm^{\mu\nu} \\ [G_{n'n}(k)]_\pm^{\mu\nu} \end{bmatrix} = \frac{1}{2} \begin{bmatrix} [N_{n'n}(p_{nn'}^+, k)]^{\mu\nu} \pm [N_{n'n}(p_{nn'}^-, k)]^{\mu\nu} \\ [G_{n'n}(p_{nn'}^+, k)]^{\mu\nu} \pm [G_{n'n}(p_{nn'}^-, k)]^{\mu\nu} \end{bmatrix}, \tag{9.2.34}$$

Explicit forms for the tensors defined by (9.2.34) follow by inserting $\epsilon p_\parallel^\mu \to p_{\parallel\pm}^\mu$, given by (9.2.4), into (9.1.17) and (9.1.18). This gives

$$\begin{aligned} [N_{n'n}(k)]_+^{\mu\nu} = &\left[2k_\parallel^\mu k_\parallel^\nu f_{nn'}(f_{nn'}-1) + 2k_D^\mu k_D^\nu g_{nn'}^2 + \frac{1}{2}(k^2)_\parallel g_\parallel^{\mu\nu} \right] \\ &\times \left[\left(J_{n'-n}^{n-1}\right)^2 + \left(J_{n'-n}^n\right)^2 \right] \\ &- \left[\frac{1}{2}k_\perp^2 g_\parallel^{\mu\nu} + \frac{1}{2}(p_n^2 + p_{n'}^2 - (k^2)_\parallel) g_\perp^{\mu\nu} + f_{nn'}(k_\parallel^\mu k_\perp^\nu + k_\parallel^\nu k_\perp^\mu) \right] \\ &\times [(J_{n'-n+1}^{n-1})^2 + (J_{n'-n-1}^n)^2] \\ &- p_n(2f_{nn'}-1)(k_\parallel^\mu e_1^\nu + k_\parallel^\nu e_1^\mu)[J_{n'-n}^{n-1} J_{n'-n-1}^n + J_{n'-n}^n J_{n'-n+1}^{n-1}] \end{aligned}$$

$$
\begin{aligned}
&\quad + 2p_{n'}p_n(e_1^\mu e_1^\nu - e_2^\mu e_2^\nu)J_{n'-n+1}^{n-1}J_{n'-n-1}^{n},\\
[N_{n'n}(k)]_-^{\mu\nu} &= g_{nn'}\{(2f_{nn'}-1)(k_\parallel^\mu k_D^\nu + k_\parallel^\nu k_D^\mu)[(J_{n'-n}^{n-1})^2 + (J_{n'-n}^{n})^2]\\
&\quad - 2p_n(k_D^\mu e_1^\nu + k_D^\nu e_1^\mu)[J_{n'-n}^{n-1}J_{n'-n-1}^{n} + J_{n'-n}^{n}J_{n'-n+1}^{n-1}]\\
&\quad - (k_D^\mu k_\perp^\nu + k_D^\nu k_\perp^\mu)[(J_{n'-n+1}^{n-1})^2 + (J_{n'-n-1}^{n})^2]\},\\
[G_{n'n}(k)]_+^{\mu\nu} &= i\left\{f^{\mu\nu}\frac{1}{2}\left(p_n^2 + p_{n'}^2 - (k^2)_\parallel\right)[(J_{n'-n+1}^{n-1})^2 - (J_{n'-n-1}^{n})^2]\right.\\
&\quad - p_n(2f_{nn'}-1)(k_\parallel^\mu e_2^\nu - k_\parallel^\nu e_2^\mu)[J_{n'-n}^{n-1}J_{n'-n-1}^{n} - J_{n'-n}^{n}J_{n'-n+1}^{n-1}]\\
&\quad \left.+ f_{nn'}(k_\parallel^\mu k_G^\nu - k_\parallel^\nu k_G^\mu)[(J_{n'-n+1}^{n-1})^2 - (J_{n'-n-1}^{n})^2]\right\},\\
[G_{n'n}(k)]_-^{\mu\nu} &= i g_{nn'}\{-2p_n(k_D^\mu e_2^\nu - k_D^\nu e_2^\mu)[J_{n'-n}^{n-1}J_{n'-n-1}^{n} - J_{n'-n}^{n}J_{n'-n+1}^{n-1}]\\
&\quad + (k_D^\mu k_G^\nu - k_D^\nu k_G^\mu)[(J_{n'-n+1}^{n-1})^2 - (J_{n'-n-1}^{n})^2]\}.
\end{aligned}
\tag{9.2.35}
$$

The term involving $g_{nn'}^2$ can be rewritten in terms of $f_{nn'}^2$ using

$$
g_{nn'}^2 = f_{nn'}^2 - \frac{(\varepsilon_n^0)^2}{\omega^2 - k_z^2}.
\tag{9.2.36}
$$

The tensors (9.2.35) apply for $g_{nn'}^2 > 0$; for $g_{nn'}^2 < 0$, corresponding to the dissipation-free region where the resonance is not allowed, one makes the replacement $g_{nn'} \to i g'_{nn'}$ with $g'^2_{nn'} = -g_{nn'}^2$.

Four Combined RPDFs

Four different combinations of the RPDF appear. These are

$$
\begin{bmatrix}
\mathcal{Z}_{nn'}^{(1)}(\omega,k_z)\\
\mathcal{Z}_{nn'}^{(2)}(\omega,k_z)\\
\mathcal{Z}_{nn'}^{(3)}(\omega,k_z)\\
\mathcal{Z}_{nn'}^{(4)}(\omega,k_z)
\end{bmatrix}
=
\begin{bmatrix}
\mathcal{Z}_n^{\rm sum}(t_{nn'}^+) + \mathcal{Z}_n^{\rm sum}(1/t_{nn'}^+) - \mathcal{Z}_n^{\rm sum}(t_{nn'}^-) - \mathcal{Z}_n^{\rm sum}(1/t_{nn'}^-)\\
\mathcal{Z}_n^{\rm sum}(t_{nn'}^+) + \mathcal{Z}_n^{\rm sum}(1/t_{nn'}^+) + \mathcal{Z}_n^{\rm sum}(t_{nn'}^-) + \mathcal{Z}_n^{\rm sum}(1/t_{nn'}^-)\\
\mathcal{Z}_n^{\rm diff}(t_{nn'}^+) - \mathcal{Z}_n^{\rm diff}(1/t_{nn'}^+) - \mathcal{Z}_n^{\rm diff}(t_{nn'}^-) + \mathcal{Z}_n^{\rm diff}(1/t_{nn'}^-)\\
\mathcal{Z}_n^{\rm diff}(t_{nn'}^+) - \mathcal{Z}_n^{\rm diff}(1/t_{nn'}^+) + \mathcal{Z}_n^{\rm diff}(t_{nn'}^-) - \mathcal{Z}_n^{\rm diff}(1/t_{nn'}^-)
\end{bmatrix}.
\tag{9.2.37}
$$

The resulting general form for the response tensor for an arbitrary spin-independent distribution is

$$
\begin{aligned}
\Pi^{\mu\nu}(k) &= \Pi_{\rm ND}^{\mu\nu}(k) + \frac{e^3B}{8\pi^2(k^2)_\parallel}\sum_{n,n'}\frac{1}{g_{nn'}}\{[N_{n'n}(k)]_+^{\mu\nu}\mathcal{Z}_{nn'}^{(1)}(\omega,k_z) + [N_{n'n}(k)]_-^{\mu\nu}\mathcal{Z}_{nn'}^{(2)}(\omega,k_z)\\
&\quad + [G_{n'n}(k)]_+^{\mu\nu}\mathcal{Z}_{nn'}^{(3)}(\omega,k_z) + [G_{n'n}(k)]_-^{\mu\nu}\mathcal{Z}_{nn'}^{(4)}(\omega,k_z)\}.
\end{aligned}
\tag{9.2.38}
$$

The general form (9.2.38) with (9.2.35) applies when the resonances are allowed, $g_{nn'}^2 > 0$, and the dissipation is then described by the antihermitian part that arises form the imaginary parts of the RPDFs. In the dissipation-free region (for given $n, n', (k^2)_\parallel$), $g_{nn'}$ is imaginary, the $\pm$ solutions $(p_{nn'}^\pm, \varepsilon_{nn'}^\pm, t_{nn'}^\pm)$ are complex conjugates of each other, the RPDF has real and imaginary parts and the imaginary contributions sum to zero, implying that the antihermitian part is zero, and that the response tensor (9.2.38) with (9.2.35) is hermitian in the dissipation-free region.

9.2.6 Imaginary Parts of RPDFs

The antihermitian part of the form (9.2.38) for the response tensor includes all dissipative processes (for $g_{nn'}$ and hence $t_{nn'}^\pm$ real). Dissipation is described by the imaginary parts of the RPDFs obtained by imposing the Landau prescription; the sign of the imaginary part needs to be deduced from the dependence of $t_{nn'}^\pm$ on ω. The antihermitian part is evaluated in a more straightforward manner in (9.2.2), and comparison of this result with the imaginary parts of the RPDFs allow one to identify the signs of imaginary parts of the RPDFs indirectly. One finds

$$\mathrm{Im}\mathcal{Z}_n^\epsilon(t_{nn'}^\pm) = i\pi\, n_n^\epsilon(p_{nn'}^\pm), \qquad \mathrm{Im}\mathcal{Z}_n^\epsilon(1/t_{nn'}^\pm) = i\pi\, n_n^\epsilon(-p_{nn'}^\pm). \tag{9.2.39}$$

The result (9.2.39) also follows from the definition (9.2.28) of the RPDFs by writing $t_{nn'}^\pm \to t_{nn'}^\pm + i0$ and using the Plemelj formula.

The imaginary parts of the RPDFs (9.2.37) become

$$\begin{bmatrix} \mathrm{Im}\mathcal{Z}_{nn'}^{(1)}(\omega,k_z) \\ \mathrm{Im}\mathcal{Z}_{nn'}^{(2)}(\omega,k_z) \\ \mathrm{Im}\mathcal{Z}_{nn'}^{(3)}(\omega,k_z) \\ \mathrm{Im}\mathcal{Z}_{nn'}^{(4)}(\omega,k_z) \end{bmatrix} = i\pi \begin{bmatrix} n_n^{\mathrm{sum}}(p_{nn'}^+) + n_n^{\mathrm{sum}}(-p_{nn'}^+) - n_n^{\mathrm{sum}}(p_{nn'}^-) - n_n^{\mathrm{sum}}(-p_{nn'}^-) \\ n_n^{\mathrm{sum}}(p_{nn'}^+) + n_n^{\mathrm{sum}}(-p_{nn'}^+) + n_n^{\mathrm{sum}}(p_{nn'}^-) + n_n^{\mathrm{sum}}(-p_{nn'}^-) \\ n_n^{\mathrm{diff}}(p_{nn'}^+) - n_n^{\mathrm{diff}}(-p_{nn'}^+) - n_n^{\mathrm{diff}}(p_{nn'}^-) + n_n^{\mathrm{diff}}(-p_{nn'}^-) \\ n_n^{\mathrm{diff}}(p_{nn'}^+) - n_n^{\mathrm{diff}}(-p_{nn'}^+) + n_n^{\mathrm{diff}}(p_{nn'}^-) - n_n^{\mathrm{diff}}(-p_{nn'}^-) \end{bmatrix}. \tag{9.2.40}$$

In interpreting (9.2.40) it is helpful to note that there are sums over n, n', and one may relabel these, effectively making the interchange $n \leftrightarrow n'$. Under this interchange one has

$$p_{nn'}^\pm = k_z f_{nn'} \pm \omega g_{nn'} \leftrightarrow k_z f_{n'n} \pm \omega g_{n'n} = k_z(1 - f_{nn'}) \pm \omega g_{nn'} = k_z - p_{nn'}^\mp.$$

This allows one to rewrite the right hand side of (9.2.40) under the interchange $n \leftrightarrow n'$. The nongyrotropic part of the tensor does not change sign under this interchange, so that the first two lines in (9.2.40) become

$$i\pi\left[\begin{array}{c}-n_{n'}^{\rm sum}(p_{nn'}^{+}-k_z)+n_n^{\rm sum}(p_{nn'}^{+})-n_{n'}^{\rm sum}(-p_{nn'}^{-}-k_z)+n_n^{\rm sum}(-p_{nn'}^{-})\\ n_n^{\rm sum}(p_{nn'}^{+})+n_{n'}^{\rm sum}(k_z-p_{nn'}^{+})+n_n^{\rm sum}(-p_{nn'}^{-})+n_{n'}^{\rm sum}(-k_z+p_{nn'}^{-})\end{array}\right],$$

respectively. The first of these two lines may be interpreted in terms of gyromagnetic absorption (GA) from state $n', p_{nn'}^{+} - k_z$ to state $n, p_{nn'}^{+}$ plus GA from state $n', -p_{nn'}^{-} - k_z$ to state $n, -p_{nn'}^{-}$. The second line may be interpreted in terms of pair creation (PC) into electron and positron states $n, p_{nn'}^{+}$ and $n', k_z - p_{nn'}^{+}$ plus PC into states $n, -p_{nn'}^{-}$ and $n', k_z + p_{nn'}^{-}$. The presence of electrons or positrons suppresses PC, compared to its value in vacuo, and the sign of the imaginary part is opposite to that for GA. The gyrotropic part of the tensor changes sign under the interchange $n \leftrightarrow n'$, and when the additional change in sign is taken into account, the third line in (9.2.40) reduces to the same combination as the first line, and the fourth line to the same combinations as the second line. Thus the first and third lines in (9.2.40) correspond to GA and the second and fourth lines to PC.

The contribution of the vacuum to the dissipation may be included by making the replacements $n_n^{\rm sum}(p_z) \to -1$, $n_n^{\rm diff}(p_z) \to 0$. Then only $\mathrm{Im}\mathcal{Z}_{nn'}^{(2)}(\omega, k_z)$ contributes. Combining the contributions of the vacuum and the electron gas is equivalent to replacing $n_n^{\rm sum}(p_z) \to n_n^{\rm sum}(p_z) - 1$ in $\mathcal{Z}_{nn'}^{(2)}(\omega, k_z)$, and leaving the other three RPDFs unchanged.

Dissipation-Free Region

The dissipation-free region is defined as the range of $(k^2)_\parallel = \omega^2 - k_z^2$ between the maximum value for gyromagnetic absorption and the minimum value to pair creation. From (6.1.17) this range is, for given n, n',

$$(\varepsilon_n^0 - \varepsilon_{n'}^0)^2 < (k^2)_\parallel < (\varepsilon_n^0 + \varepsilon_{n'}^0)^2. \tag{9.2.41}$$

This condition can also be written as $g_{nn'}^2 < 0$ or $(\varepsilon_n^0)^2 > (k^2)_\parallel f_{nn'}^2$. The contribution to the dispersion from electrons in the nth Landau state involves a sum over different virtual levels n' and, depending on ω, k_z and n, can correspond to the range where GA is allowed, to the dissipation-free range or to the range where PC is allowed.

In the dissipation-free region, $g_{nn'}$ is imaginary, it is convenient to write $g_{nn'} = i g'_{nn'}$, with $g'_{nn'} = [(\varepsilon_n^0)^2/(k^2)_\parallel - f_{nn'}^2]^{1/2}$, and the $\pm$-solution become complex conjugates of each other. Using the form (9.2.25), one may write

$$t_{nn'}^{\pm} = R_{nn'} \exp[\pm i\phi_{nn'}], \tag{9.2.42}$$

$$R_{nn'} = \left(\frac{\omega\varepsilon_n^0 - (k^2)_\parallel f_{nn'}}{\omega\varepsilon_n^0 + (k^2)_\parallel f_{nn'}}\right)^{1/2}, \qquad \phi_{nn'} = \arctan\left(\frac{(k^2)_\parallel g'_{nn'}}{k_z\varepsilon_n^0}\right). \tag{9.2.43}$$

Then the definition (9.2.28) of the RPDF gives

$$\mathcal{Z}_n^\epsilon(t_{nn'}^\pm) = \int_{-1}^{1} dt\, n_n^\epsilon(p_z) \frac{t - R_{nn'}\cos\phi_{nn'} \pm i R_{nn'}\sin\phi_{nn'}}{t^2 - 2tR_{nn'}\cos\phi_{nn'} + R_{nn'}^2}, \tag{9.2.44}$$

allowing the real and imaginary parts to be identified directly.

9.3 Magnetized Thermal Distributions

The response tensor for a thermal distribution of particles plays a central role in the theory of dispersion in plasmas. In relativistic quantum theory, a thermal distribution of electrons is a Fermi–Dirac (FD) distribution. Two limiting cases of a FD distribution are the nondegenerate and completely degenerate cases. In this section, the RPDF defined by (9.2.28) is evaluated explicitly in these limiting cases. The nonrelativistic and cold-plasma limits are also discussed.

9.3.1 Fermi–Dirac Distribution for Magnetized Electrons

In the magnetized case, the occupation number for electrons ($\epsilon = +1$) and positrons ($\epsilon = -1$) in a FD distribution is

$$n_n^\epsilon(p_z) = \frac{1}{e^{(\varepsilon_n - \mu^\epsilon)/T} + 1}, \tag{9.3.1}$$

where T is the temperature (in energy units), and where $\mu^\pm$ are the chemical potentials for the electrons and positrons, respectively. In thermal equilibrium the chemical potentials satisfy $\mu^+ + \mu^- = 0$. Suppose that the number density of free electrons is specified; this is the difference between the number densities of electrons and positrons, $n^+ - n^-$. In equilibrium the electron gas consists of this fixed number of free electrons plus a number of pairs that is determined by the condition for thermal equilibrium. The fixed number $n^+ - n^-$ is (SI units)

$$n^+ - n^- = \sum_{n=0}^{\infty} a_n (n_n^+ - n_n^-), \qquad n_n^\epsilon = \frac{eB}{2\pi\hbar} \int \frac{dp_z}{2\pi\hbar} n_n^\epsilon(p_z), \tag{9.3.2}$$

with $a_0 = 1$, $a_n = 2$ for $n \geq 1$. For the FD distribution (9.3.1), the number densities are functions of the chemical potential $\mu_+ = -\mu_-$ and the temperature, T. A pure pair plasma corresponds to $n^+ = n^-$, requiring $\mu_+ = \mu_-$, and hence, $\mu_+ = 0$. The chemical potential has no simple physical interpretation, except in the completely degenerate limit, $T \to 0$, when it corresponds to the Fermi energy; there are no positrons in this limit.

In terms of integrals over t, defined by (9.2.23), the number densities in (9.3.2) become (SI units)

$$n_n^\epsilon = \frac{eB\varepsilon_n^0}{2\pi^2\hbar^2 c} a_n \int_{-1}^{1} dt\, \frac{1+t^2}{(1-t^2)^2}\, n_n^\epsilon(p_z), \tag{9.3.3}$$

with $n_n^\epsilon(p_z)$ written as a function of t using (9.2.23).

The proper number density for electrons or positrons is defined by (9.2.21). For a FD distribution $n_n^\epsilon(p_z)$ is given by (9.3.1) with ε_n written as a function of t using (9.2.23).

9.3.2 Completely Degenerate and Nondegenerate Limits

The completely degenerate limit corresponds to $T \to 0$ in the FD distribution (9.3.1). A degenerate distribution in the absence of a magnetic field corresponds to all the states being filled for $\varepsilon < \varepsilon_F$, and empty for $\varepsilon > \varepsilon_F$, where the Fermi energy, ε_F, is equal to the chemical potential. Similarly, for a degenerate distribution in the presence of a magnetic field, the Fermi energy is equal to the chemical potential, and the occupation number is unity for each Landau state with energy less than the Fermi energy. This corresponds to all the states being filled for $|p_z| < p_{nF}$, where the Fermi momentum for the nth level is

$$p_{nF} = [\varepsilon_F^2 - (\varepsilon_n^0)^2]^{1/2} = (\varepsilon_F^2 - m^2 - 2neB)^{1/2}. \tag{9.3.4}$$

For given ε_F, the Landau states are filled up to a maximum n_F such that p_{nF} is real for $n \le n_F$ and imaginary for $n > n_F$. The levels are occupied for $n \le n_F$ and $-p_{nF} < p_z < p_{nF}$ and empty for $|p_z| > p_{nF}$, $n \le n_F$, and for all $n > n_F$. For $\varepsilon_F^2 < m^2 + 2eB$ only the ground state, $n = 0$, is occupied.

There are no positrons present in the completely degenerate limit. The number density of electrons is (SI units)

$$n_e = \sum_{n=0}^{n_F} a_n \frac{eB}{2\pi\hbar} \int_{-p_{nF}}^{p_{nF}} \frac{dp_z}{2\pi\hbar} = \sum_{0}^{n_F} a_n \frac{eB\, p_{nF}}{2\pi^2\hbar^2}, \tag{9.3.5}$$

with $a_0 = 1$, $a_n = 2$ for $n \ge 1$, with $p_{nF} = (\varepsilon_F^2/c^2 - m^2c^2 - 2neB\hbar)^{1/2}$. The density increases monotonically with ε_F at fixed B, due to the increase in p_{nF} for $n \le n_F$, and due to n_F increasing in a stepwise manner, with $\partial n_e/\partial\varepsilon_F$ singular at the threshold for each stepwise increase in n_F. The proper number density is (SI units)

$$n_{\rm pr} = \sum_{n=0}^{n_F} a_n \frac{eB}{2\pi\hbar} \int_{-p_{nF}}^{p_{nF}} \frac{dp_z}{2\pi\hbar} \frac{mc^2}{\varepsilon_n} = \sum_{n=0}^{n_F} a_n \frac{eBmc}{4\pi^2\hbar^2} \ln\left(\frac{\varepsilon_F + p_{nF}c}{\varepsilon_F - p_{nF}c}\right). \tag{9.3.6}$$

In the nondegenerate limit, the chemical potential, $\mu^+ - mc^2$, is large and negative. The unit term in the denominator in (9.3.1) is then negligible, and the resulting distribution reduces to a sum of 1D Jüttner distributions, with one such distribution for each n. The condition for the nondegenerate limit to apply is that the occupation number be much less than unity. The occupation number decreases with increasing n, and the condition for degeneracy can be applied separately to each Landau state. For an unmagnetized thermal distribution, the condition for nondegeneracy is that the number of electrons per cubic de Broglie wavelength be much less than unity. In the magnetized case, the condition for electrons in the nth Landau state to be nondegenerate is that the number of electrons in this level in a volume determined by the product of B/B_c times the Compton wavelength squared and the de Broglie wavelength be much less than unity.

In the nondegenerate limit, it is convenient to incorporate the chemical potential into a normalization constant, A^ϵ, by writing

$$n_n^\epsilon(p_z) = A^\epsilon \exp\left[-\rho\left(1 + p_z^2/m^2c^2 + 2nB/B_c\right)^{1/2}\right], \quad A^\epsilon = \exp\left(\rho\mu^\pm/mc^2\right), \tag{9.3.7}$$

with $\rho = mc^2/T$ the inverse temperature in units of the rest energy. The number densities, $n^\pm$, are given by (SI units)

$$\frac{n^\pm}{A^\pm} = \sum_{n=0}^{\infty} a_n \frac{eB}{2\pi\hbar} \int_{-\infty}^{\infty} \frac{dp_z}{2\pi\hbar}\, e^{-\rho\varepsilon_n/mc^2}, \tag{9.3.8}$$

with $\varepsilon_n/mc^2 = (1 + p_z^2/m^2c^2 + 2nB/B_c)^{1/2}$. By changing the variable of integration to χ, with $\sinh\chi = p_z/mc(1 + 2nB/B_c)^{1/2}$, the integral in (9.3.8) reduces to an integral representation of a Macdonald function:

$$K_\nu(x) = \frac{(x/2)^\nu \Gamma(\frac{1}{2})}{\Gamma(\nu + \frac{1}{2})} \int_0^\infty d\chi\, \sinh^{2\nu}\chi\, e^{-x\cosh\chi}. \tag{9.3.9}$$

The number density in a nondegenerate, relativistic, thermal electron gas is given by (SI units)

$$\frac{n^\pm}{A^\pm} = \sum_{n=0}^{\infty} a_n \frac{eB\varepsilon_n^0}{2\pi^2\hbar^2 c} K_1(\rho\varepsilon_n^0/mc^2), \tag{9.3.10}$$

with $\varepsilon_n^0/mc^2 = (1 + 2nB/B_c)^{1/2}$. The proper number density is given by (SI units)

$$\frac{n_{\rm pr}^\pm}{A^\pm} = \sum_{n=0}^{\infty} a_n \frac{eBmc}{2\pi^2\hbar^2} K_0(\rho\varepsilon_n^0/mc^2). \tag{9.3.11}$$

The result (9.3.10) simplifies in the nonrelativistic limit when the asymptotic expansion, $K_\nu(x) = (\pi/2x)^{1/2}e^{-x}$ applies, provided that one also has $B \ll B_c$, such that the approximation $(1 + 2nB/B_c)^{1/2} \approx 1 + nB/B_c$ applies. The sum over

n in (9.3.10) gives

$$\sum_{n=0}^{\infty} a_n e^{-\rho n B/B_c} = \frac{1}{\tanh(\rho B/2B_c)}. \tag{9.3.12}$$

In the ultrarelativistic limit, $\rho \ll 1$, the Macdonald functions in (9.3.10) may be approximated, for $\rho(2nB/B_c)^{1/2} \ll 1$, by the leading term in the power series expansion, $K_1(x) \approx 1/x$. For $\rho \ll 1$ and $\rho(2B/B_c)^{1/2} \gg 1$ nearly all the particles are in the ground state, $n = 0$, so that the electron gas is essentially one dimensional.

9.3.3 RPDFs: Completely Degenerate Limit

The RPDF $\mathcal{Z}_n^{\epsilon}(t_0)$, defined by (9.2.28), reduces to a logarithm for a completely degenerate distribution. The absence of positrons in this limit implies that $n_n^{\rm sum}(p_z)$ and $n_n^{\rm diff}(p_z)$ are equal, so that the superscripts are superfluous. Let t_{nF} be defined by $p_{nF} = \varepsilon_n^0 2t_{nF}/(1-t_{nF}^2)$. Then (9.2.28) gives, for the real part,

$$\mathcal{Z}_n(t_0) = \int_{-t_{nF}}^{t_{nF}} \frac{dt}{t-t_0} = \ln\left|\frac{t_{nF}-t_0}{t_{nF}+t_0}\right|, \tag{9.3.13}$$

where t_0 is assumed real. The imaginary part is given by (9.2.39). The four RPDFs (9.2.37) are all logarithmic functions (for $t_{nn'}^{\pm}$ real), and it is convenient to write

$$\mathcal{Z}_{nn'}^{(N)}(\omega, k_z) = \ln \Lambda_{nn'}^{(N)}, \tag{9.3.14}$$

with $N = 1, 2, 3, 4$. Explicit evaluation gives

$$\Lambda_{nn'}^{(1)} = \left|\frac{t_{nF}-t_{nn'}^+}{t_{nF}+t_{nn'}^+}\,\frac{t_{nF}-1/t_{nn'}^+}{t_{nF}+1/t_{nn'}^+}\,\frac{t_{nF}+t_{nn'}^-}{t_{nF}-t_{nn'}^-}\,\frac{t_{nF}+1/t_{nn'}^-}{t_{nF}-1/t_{nn'}^-}\right|,$$

$$\Lambda_{nn'}^{(2)} = \left|\frac{t_{nF}-t_{nn'}^+}{t_{nF}+t_{nn'}^+}\,\frac{t_{nF}-1/t_{nn'}^+}{t_{nF}+1/t_{nn'}^+}\,\frac{t_{nF}-t_{nn'}^-}{t_{nF}+t_{nn'}^-}\,\frac{t_{nF}-1/t_{nn'}^-}{t_{nF}+1/t_{nn'}^-}\right|,$$

$$\Lambda_{nn'}^{(3)} = \left|\frac{t_{nF}-t_{nn'}^+}{t_{nF}+t_{nn'}^+}\,\frac{t_{nF}+1/t_{nn'}^+}{t_{nF}-1/t_{nn'}^+}\,\frac{t_{nF}+t_{nn'}^-}{t_{nF}-t_{nn'}^-}\,\frac{t_{nF}-1/t_{nn'}^-}{t_{nF}+1/t_{nn'}^-}\right|,$$

$$\Lambda_{nn'}^{(4)} = \left|\frac{t_{nF}-t_{nn'}^+}{t_{nF}+t_{nn'}^+}\,\frac{t_{nF}+1/t_{nn'}^+}{t_{nF}-1/t_{nn'}^+}\,\frac{t_{nF}-t_{nn'}^-}{t_{nF}+t_{nn'}^-}\,\frac{t_{nF}+1/t_{nn'}^-}{t_{nF}-1/t_{nn'}^-}\right|. \tag{9.3.15}$$

Alternatively, (9.3.15) may be written in the form

$$\Lambda^{(1)}_{nn'} = \left| \frac{\varepsilon_F p^+_{nn'} - p_{nF}\varepsilon^+_{nn'}}{\varepsilon_F p^+_{nn'} + p_{nF}\varepsilon^+_{nn'}} \frac{\varepsilon_F p^-_{nn'} + p_{nF}\varepsilon^-_{nn'}}{\varepsilon_F p^-_{nn'} - p_{nF}\varepsilon^-_{nn'}} \right|,$$

$$\Lambda^{(2)}_{nn'} = \left| \frac{\varepsilon_F p^+_{nn'} - p_{nF}\varepsilon^+_{nn'}}{\varepsilon_F p^+_{nn'} + p_{nF}\varepsilon^+_{nn'}} \frac{\varepsilon_F p^-_{nn'} - p_{nF}\varepsilon^-_{nn'}}{\varepsilon_F p^-_{nn'} + p_{nF}\varepsilon^-_{nn'}} \right|,$$

$$\Lambda^{(3)}_{nn'} = \left| \frac{p_{nF} - p^+_{nn'}}{p_{nF} + p^+_{nn'}} \frac{p_{nF} + p^-_{nn'}}{p_{nF} - p^-_{nn'}} \right|, \quad \Lambda^{(4)}_{nn'} = \left| \frac{p_{nF} - p^+_{nn'}}{p_{nF} + p^+_{nn'}} \frac{p_{nF} - p^-_{nn'}}{p_{nF} + p^-_{nn'}} \right|. \quad (9.3.16)$$

The resonant values $p^\pm_{nn'}$, $\varepsilon^\pm_{nn'}$ are given in terms of $f_{nn'}$, $g_{nn'}$ by (9.2.3), and rewriting (9.3.16) in terms of these quantities gives

$$\Lambda^{(1)}_{nn'} = \left| \frac{k_z^2(\varepsilon_n^0)^4 - (k^2)_\parallel^2 (p_{nF} f_{nn'} - \varepsilon_F g_{nn'})^2}{k_z^2(\varepsilon_n^0)^4 - (k^2)_\parallel^2 (p_{nF} f_{nn'} + \varepsilon_F g_{nn'})^2} \right|,$$

$$\Lambda^{(2)}_{nn'} = \left| \frac{(\omega\varepsilon_F - k_z p_{nF})^2 - [(k^2)_\parallel f_{nn'}]^2}{(\omega\varepsilon_F + k_z p_{nF})^2 - [(k^2)_\parallel f_{nn'}]^2} \right|,$$

$$\Lambda^{(3)}_{nn'} = \left| \frac{k_z^2 f^2_{nn'} - (p_{nF} - \omega g_{nn'})^2}{k_z^2 f^2_{nn'} - (p_{nF} + \omega g_{nn'})^2} \right|, \qquad \Lambda^{(4)}_{nn'} = \left| \frac{(p_{nF} - k_z f_{nn'})^2 - \omega^2 g^2_{nn'}}{(p_{nF} + k_z f_{nn'})^2 - \omega^2 g^2_{nn'}} \right|.$$
(9.3.17)

The logarithmic form for the RPDF assumes that $g_{nn'}$ is real, so that $t^\pm_{nn'}$ are real. The logarithmic functions need to be analytically continued into the dissipation-free region, determined by (9.2.41), where $t^\pm_{nn'}$ are complex conjugates of each other. One may write $t^\pm_{nn'} = R_{nn'} \exp[\pm i\phi_{nn'}]$, and evaluate the real and imaginary parts of the RPDF evaluated using (9.2.44). However, it is simpler to use the logarithmic form (9.3.13) with the argument of the logarithm written

$$\frac{t_{nF} - t^\pm_{nn'}}{t_{nF} + t^\pm_{nn'}} = \frac{t_{nF} - R_{nn'}\cos\phi_{nn'} \mp iR_{nn'}\sin\phi_{nn'}}{t_{nF} + R_{nn'}\cos\phi_{nn'} \pm iR_{nn'}\sin\phi_{nn'}} = \frac{R^-_{Fnn'}\exp[\mp i\phi^-_{Fnn'}]}{R^+_{Fnn'}\exp[\pm i\phi^+_{Fnn'}]},$$
(9.3.18)

with

$$R^\pm_{Fnn'} = [t^2_{nF} \pm t_{nF} R_{nn'}\cos\phi_{nn'} + R^2_{nn'}]^{1/2},$$

$$\phi^\pm_{Fnn'} = \arctan\left(\frac{R_{nn'}\sin\phi_{nn'}}{t_{nF} \pm R_{nn'}\cos\phi_{nn'}}\right), \quad (9.3.19)$$

with $R_{nn'}$ and $\phi_{nn'}$ given by (9.2.43). The combinations of the $\pm$ solutions that appear in (9.3.15) then give

$$\ln\left(\frac{t_{nF}-t^+_{nn'}}{t_{nF}+t^+_{nn'}}\,\frac{t_{nF}+t^-_{nn'}}{t_{nF}-t^-_{nn'}}\right) = -i\,[\phi^+_{Fnn'}+\phi^-_{Fnn'}],$$

$$\ln\left(\frac{t_{nF}-t^+_{nn'}}{t_{nF}+t^+_{nn'}}\,\frac{t_{nF}-t^-_{nn'}}{t_{nF}+t^-_{nn'}}\right) = 2\ln\left(\frac{R^-_{Fnn'}}{R^+_{Fnn'}}\right). \tag{9.3.20}$$

For the analogous combinations with $t^{\pm}_{nn'} \to 1/t^{\pm}_{nn'}$ that appear in (9.3.15) one has analogous expressions with $R^{\pm}_{Fnn'} \to R'^{\pm}_{Fnn'}$, $\phi^{\pm}_{Fnn'} \to \phi'^{\pm}_{Fnn'}$ on the right hand side of (9.3.20), where $R'^{\pm}_{Fnn'}$, $\phi'^{\pm}_{Fnn'}$ may be defined analogous to (9.3.19) with $R_{nn'} \to 1/R_{nn'}$, $\phi_{nn'} \to -\phi_{nn'}$. With $g_{nn'} \to i g'_{nn'}$ imaginary in the dissipation-free region, the response tensor in the form (9.2.38), with $g_{nn'} \to i g'_{nn'}$ in (9.2.35), is hermitian.

An explicit expression for the response tensor in the completely degenerate limit follows by inserting the RPDFs (9.3.14) with (9.3.16) into the dispersive part of (9.2.38) with (9.2.35). In the ND part (9.2.22) of the response tensor, the proper number density is given explicitly by (9.3.6).

9.3.4 Dissipation in a Completely Degenerate Electron Gas

Two forms of (collisionless) dissipation are possible in a completely degenerate electron gas: gyromagnetic absorption (GA) by electrons and pair creation (PC). Both GA and PC are resonant processes, and for given n, n', ω, k_z, the resonant values of p_z and ε_n are $p^{\pm}_{nn'}$ and $\varepsilon^{\pm}_{nn'}$, respectively. These correspond to a resonant 4-momentum $p^{\mu}_{nn'} = (\varepsilon^{\pm}_{nn'}, 0, 0, p^{\pm}_{nn'}) = f_{nn'}k^{\mu}_{\parallel} + g_{nn'}k^{\mu}_{D}$. If the initial state for GA corresponds to $p'^{\mu}_{n'n} = p^{\mu}_{nn'} - k^{\mu}_{\parallel}$, the final state corresponds to $p^{\mu}_{nn'}$. The resonant values of p_z for PC are $p^{\mu}_{nn'}$ and $k_z - p^{\mu}_{nn'}$, corresponding to the electron and positron. These resonant values must be real for GA or PC to be allowed, and this requires $g^2_{nn'} \geq 0$.

In the completely degenerate case, for given n, n', ω, k_z, the requirements on GA that the initial state (n') be occupied and the final state (n) be unoccupied, imply $|p^{\pm}_{nn'} - k_z| < p_{n'F}$, $\varepsilon^0_{n'} \leq \varepsilon^{\pm}_{nn'} - \omega < \varepsilon_F$, $n' < n_F$, and either $|p^{\pm}_{nn'}| > p_{nF}$, $\varepsilon^{\pm}_{nn'} > \varepsilon_F$, $n < n_F$ or $n > n_F$. A qualitative change occurs at $|k_z| = 2p_{nF}$ (or $|k_z| = 2p_{n'F}$): for $|k_z| < 2p_{nF}$ both initial and final states can be occupied (so that there is no dissipation), and for $|k_z| > 2p_{nF}$ at most one of the states can be occupied. The imaginary parts of the RPDFs change abruptly by $\pm i\pi$ when ω, k_z is such that one of these inequalities, for given n, n', is replaced by an equality. The values where these abrupt changes occur correspond to the boundaries of the allowed regions. It is informative to identify these boundaries for two different choices of independent variable: for k_z and ω as functions of given $\omega^2 - k_z^2$, and for ω as a function of given k_z.

Given $n, n', \omega^2 - k_z^2$ implies that $f_{nn'}$ and $g_{nn'}$ are specified, and one can solve $p^{\pm}_{nn'} = \pm p_{nF}, \varepsilon^{\pm}_{nn'} = \varepsilon_F$ for k_z and ω. The solutions are

$$k_{F\pm} = \frac{\omega^2 - k_z^2}{(\varepsilon_n^0)^2}\,(\varepsilon_F g_{nn'} \pm p_{nF} f_{nn'}), \qquad \omega_{F\pm} = \frac{\omega^2 - k_z^2}{(\varepsilon_n^0)^2}\,(\varepsilon_F f_{nn'} \pm p_{nF} g_{nn'}), \tag{9.3.21}$$

with $\omega_{F\pm}^2 - k_{F\pm}^2 = \omega^2 - k_z^2$. The condition $g^2_{nn'} > 0$, requires $\omega^2 - k_z^2 < (\varepsilon_n^0 - \varepsilon_{n'}^0)^2$ for GA and $\omega^2 - k_z^2 > (\varepsilon_n^0 + \varepsilon_{n'}^0)^2$ for PC. The threshold for given $n, n', \omega^2 - k_z^2$ corresponds to $g^2_{nn'} = 0$, when the $\pm$ solutions coincide. Let the threshold values be denoted $k_{F+} = -k_{F-} = k_{F0}$ and $\omega_{F+} = \omega_{F-} = \omega_{F0}$. They are given by

$$k_{F0} = \frac{p_{nF}}{\varepsilon_n^0}\,|\omega^2 - k_z^2|^{1/2}. \qquad \omega_{F0} = \frac{\varepsilon_F}{\varepsilon_n^0}\,|\omega^2 - k_z^2|^{1/2}. \tag{9.3.22}$$

The solutions $k_{F\pm}$ or $\omega_{F\pm}$ diverge from each other as $g^2_{nn'}$ increases, and the regions between the $\pm$ curves define the allowed regions of absorption for given n, n', ε_F. Examples of the regions in α–k_z space, $\alpha = -(\omega^2 - k_z^2)$, were plotted by Pérez Rojas and Shabad [13]. The allowed regions for different n, n' can overlap.

An alternative choice of independent variables corresponds to given n, n', k_z. One can solve for ω by eliminating $g_{nn'}$ between either $p^{\pm}_{nn'} = p_{nF}$ or $p^{\pm}_{nn'} = -p_{nF}$, and $\varepsilon^{\pm}_{nn'} = \varepsilon_F$, giving the quadratic equation

$$\omega^2 - 2\omega\varepsilon_F \pm 2k_z p_{nF} + 2(n - n')eB - k_z^2 = 0. \tag{9.3.23}$$

The solutions are [22]

$$\begin{aligned}\omega_{G1,2} &= \left|\,\left(\varepsilon_F^2 - 2(n - n')eB \pm 2p_{nF}k_z + k_z^2\right)^{1/2} - \varepsilon_F\right|,\\ \omega_{P1,2} &= \left(\varepsilon_F^2 - 2(n - n')eB \pm 2p_{nF}k_z + k_z^2\right)^{1/2} + \varepsilon_F,\end{aligned} \tag{9.3.24}$$

which apply to GA and PC, respectively. An example of such curves is shown in Fig. 9.2 for the specific case $n = n' = 1$, $p_{1F} = 2.2m$, $B = B_c$. For $n - n' = 0$, the frequency ω_{G2} has a local maximum at $k_z = p_{nF}$ and is zero at $k_z = 2p_{nF}$; the frequency ω_{P2} has a local minimum at $k_z = p_{nF}$. For $n - n' \neq 0$, including the dependence on n, n' explicitly, the solutions $\omega_{G2}(n, n')$, $\omega_{P2}(n, n')$ for $k_z < 2p_{nF}$ continue as $\omega_{G2}(n', n)$, $\omega_{P2}(n', n)$ for $k_z > 2p_{nF}$.

9.3.5 RPDFs: Nondegenerate Limit

In the nondegenerate limit, the FD distribution reduces to a 1D Jüttner distribution. The RPDF (9.2.28) becomes

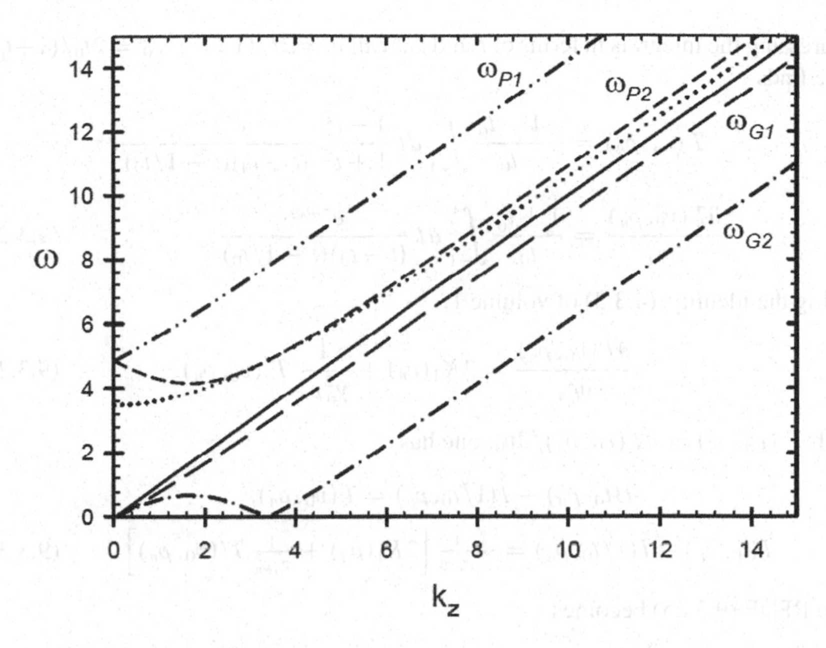

Fig. 9.2 The boundaries of the different regions in the $\omega - k_z$ plane for $p_{nF} = 2.2m$, $B = B_c$ and $n = n' = 1$ [22]. The *solid line* is $\omega = k_z$, the *dotted line* is the threshold for PC, $\omega^2 - k_z^2 = (2\varepsilon_1^0)^2$. The *dashed-double-dotted* and *short-dashed lines*, $\omega = \omega_{P1}$ and $\omega = \omega_{P2}$, respectively. The *long-dashed* and *dashed-dotted lines*, are $\omega = \omega_{G1}$ and $\omega = \omega_{G2}$, respectively (From [22], reprinted with permission AIP)

$$\mathcal{Z}_n^\epsilon(t_0) = \int_{-1}^{1} dt \, \frac{n_n^\epsilon(p_z)}{t - t_0} = A_n^\epsilon I(t_0, \rho_n). \tag{9.3.25}$$

with A_n^ϵ defined by (9.3.7), and where the RPDF

$$I(t_0, \rho_n) = \int_{-1}^{1} \frac{dt}{t - t_0} \exp\left[-\rho_n \frac{1 + t^2}{1 - t^2}\right], \tag{9.3.26}$$

is introduced, with $\rho_n = \rho\varepsilon_n^0/m$. The function (9.3.26) may be evaluated in terms of the RPDF (§ 4.4.3 of volume 1) used to describe dispersion for a nonquantum, unmagnetized Jüttner distribution. This RPDF is defined by the integral (2.4.29), in the form

$$T(v_0, \rho_n) = \int_{-1}^{1} \frac{dv}{v - v_0} \exp(-\rho_n \gamma), \tag{9.3.27}$$

with $\gamma = (1 - v^2)^{-1/2}$. The properties of $T(v, \rho_n)$ are summarized in § 4.4.3 of volume 1.

The combinations of the RPDF that appear in (9.2.37) are $I(t_0, \rho_n) \pm I(1/t_0, \rho_n)$ with $t_0 = t_{nn'}^+$ and $t_0 = t_{nn'}^-$. These may be related to $T(v_0, \rho_n)$ and $\partial T(v_0, \rho_n)\partial\rho_n$ by expressing the integrals in terms of t and t_0, with $v = 2t/(1+t^2)$, $v_0 = 2t_0/(1+t_0^2)$. One finds

$$T(v_0, \rho_n) = -\frac{1+t_0^2}{t_0}\int_{-1}^{1} dt\, \frac{1-t^2}{1+t^2}\, \frac{e^{-\rho_n \gamma}}{(t-t_0)(t-1/t_0)},$$

$$\frac{\partial T(v_0, \rho_n)}{\partial \rho_n} = \frac{1+t_0^2}{t_0}\int_{-1}^{1} dt\, \frac{e^{-\rho_n \gamma}}{(t-t_0)(t-1/t_0)}. \tag{9.3.28}$$

Using the identity, (4.3.3) of volume 1,

$$v_0 \frac{\partial T(v_0, \rho_n)}{\partial \rho_n} = 2K_1(\rho_n) + \frac{1}{\gamma_0^2 \rho_n}\, T'(v_0, \rho_n), \tag{9.3.29}$$

with $T'(v_0, \rho_n) = \partial T(v_0, \rho_n)/\partial v_0$, one has

$$I(t_0, \rho_n) + I(1/t_0, \rho_n) = T(v_0, \rho_n),$$

$$I(t_0, \rho_n) - I(1/t_0, \rho_n) = -\tfrac{1}{\gamma_0 v_0}\left[2K_1(\rho_n) + \tfrac{1}{\gamma_0^2 \rho_n}\, T'(v_0, \rho_n)\right]. \tag{9.3.30}$$

The RPDF (9.3.25) becomes

$$\mathcal{Z}_n^{\epsilon}(t_0) = \frac{A_n^{\epsilon}}{2}\left[-\frac{(1-v_0^2)^{1/2}}{v_0}\left(2K_1(\rho_n) + \frac{(1-v_0^2)}{\rho_n} T'(v_0, \rho_n)\right) + T(v_0, \rho_n)\right]. \tag{9.3.31}$$

The four RPDFs (9.2.37), for either electrons or positrons, become

$$\begin{bmatrix} \mathcal{Z}_{nn'}^{(1)}(\omega, k_z) \\ \mathcal{Z}_{nn'}^{(2)}(\omega, k_z) \\ \mathcal{Z}_{nn'}^{(3)}(\omega, k_z) \\ \mathcal{Z}_{nn'}^{(4)}(\omega, k_z) \end{bmatrix} = A_n^{\epsilon} \sum_{\pm} \begin{bmatrix} \pm T(v_\pm, \rho_n) \\ T(v_\pm, \rho_n) \\ \mp[2K_1(\rho_n)/\gamma_\pm v_\pm + T'(v_\pm, \rho_n)/\gamma_\pm^2 v_\pm \rho_n] \\ -[2K_1(\rho_n)/\gamma_\pm v_\pm + T'(v_\pm, \rho_n)/\gamma_\pm^2 v_\pm \rho_n] \end{bmatrix}. \tag{9.3.32}$$

The specific functions that appear in the arguments of $T(v_\pm, \rho_n)$, $T'(v_\pm, \rho_n)$ are

$$v_\pm = \frac{2t_{nn'}^{\pm}}{1+[t_{nn'}^{\pm}]^2} = \frac{k_z f_{nn'} \pm \omega g_{nn'}}{\omega f_{nn'} \pm k_z g_{nn'}}, \tag{9.3.33}$$

with $f_{nn'}$, $g_{nn'}$ defined by (6.1.16) and (6.1.17), respectively.

The effects of partial degeneracy can be included through the expansion of the FD distributions (9.3.1):

$$n_n^{\epsilon}(p_z) = \sum_{N=1}^{\infty} (-)^{N+1} e^{-N(\varepsilon_n - \mu^{\epsilon})/T}. \tag{9.3.34}$$

The normalization constant, A_n^ϵ, and the RPDFs can all be expanded using the sum (9.3.34). Each term in the expansion (9.3.34) has the same functional form as a Jüttner distribution, and hence may be evaluated using the foregoing formulae, essentially by replacing ρ_n by $N\rho_n$. This leads to an expression for the RPDF for a FD distribution as an infinite sum. The expansion must converge when the degeneracy is sufficiently weak, but there is no obvious way of determining the range of convergence.

9.3.6 *Nonrelativistic Distributions*

The integrals that appear for a relativistic FD distribution cannot be evaluated in terms of known functions. For a nonrelativistic FD distribution, the relevant integrals can be evaluated in terms of polylogarithms.

In the nonrelativistic case, the occupation number (9.3.1) is approximated by (ordinary units)

$$n_n(p_z) = \frac{1}{e^{(\varepsilon_n^0 + p_z^2c^2/2\varepsilon_n^0 - \mu)/T} + 1}, \tag{9.3.35}$$

where only the electrons are considered and the superscript $+$ is omitted. The distribution (9.3.35) can be interpreted as a 1D FD distribution of nonrelativistic particles with effective mass ε_n^0/c^2 and chemical potential $\mu - \varepsilon_n^0$.

In the unmagnetized nonrelativistic case the normalization of the FD distribution may be expressed in terms of a polylogarithm function, defined by

$$-\mathrm{Li}_{s+1}(-\xi) = \frac{1}{\Gamma(s+1)} \int_0^\infty dt \, \frac{t^s}{\xi^{-1}e^t + 1}. \tag{9.3.36}$$

The power series expansion,

$$\mathrm{Li}_n(\xi) = \sum_{k=1}^{\infty} \frac{\xi^k}{k^n}, \tag{9.3.37}$$

is an alternative definition. The expansion converges rapidly for small ξ, which corresponds to the nondegenerate limit. The leading term in the asymptotic expansion of the polylogarithm, for large ξ, is

$$-\lim_{\xi\to\infty} \mathrm{Li}_s(-\xi)\Gamma(s+1) = (\ln \xi)^s. \tag{9.3.38}$$

In this limit one may interpret $\ln \xi = T_F/T \gg 1$ as the ratio of the Fermi temperature to actual temperature.

The normalization for a strong-B nonrelativistic magnetized FD distribution involves (9.3.36) with $s = -1/2$. One finds (SI units)

$$n_e = \sum_{n=0}^{\infty} a_n \frac{eB}{2\pi\hbar} \int \frac{dp_z}{2\pi\hbar} n_n(p_z) = -\sum_{n=0}^{\infty} a_n \frac{eB\varepsilon_n^0}{4\pi\hbar)^2 c} \left(\frac{2\pi T}{\varepsilon_n^0}\right)^{1/2} \mathrm{Li}_{1/2}(-\xi_n), \tag{9.3.39}$$

with $\xi_n = \exp[(\mu - \varepsilon_n^0)/T]$, and where $\Gamma(\frac{1}{2}) = \pi^{1/2}$ is used. The nondegenerate limit corresponds to $\xi_n \ll 1$, when one may approximate $\mathrm{Li}_{1/2}(-\xi_n)$ by the leading term, $-\xi_n$, in the power series (9.3.37), and then (9.3.39) reproduces the result for a 1D Maxwellian distribution. In the completely degenerate limit, using (9.3.38) reproduces the number density (9.3.5).

In the nondegenerate limit, the FD distribution (9.3.35) reduces to a 1D Maxwellian distribution. The nonrelativistic plasma dispersion function, $Z(y)$, defined by (2.5.5), may be used to approximate $\mathcal{Z}_n(t_0)$ in this case. The nonrelativistic approximation, $|p_z| \ll \varepsilon_n^0/c$, corresponds to $2|t| \ll 1$, allowing the t-integral in (9.2.28) to be extended to $-\infty < t < \infty$. One then finds

$$\mathcal{Z}_n(t_0) = \int_{-\infty}^{\infty} \frac{dp_z\, n_n^\epsilon(p_z)}{p_z - 2\varepsilon_n^0 t_0} = \pi^{1/2}\xi_n\, Z(y_{n0}), \tag{9.3.40}$$

with $y_{n0} = (2\varepsilon_n^0 t_0/T)^{1/2}$, $\xi_n = \exp[(\mu - \varepsilon_n^0)/T]$. In the completely degenerate limit, one has

$$\mathcal{Z}_n(t_0) = \ln\left|\frac{p_{nF} - 2\varepsilon_n^0 t_0}{p_{nF} + 2\varepsilon_n^0 t_0}\right|. \tag{9.3.41}$$

A more general plasma dispersion function for partial degeneracy must reduce to (9.3.40) and (9.3.41) in the nondegenerate and completely degenerate limits, respectively.

A generalization of the plasma dispersion function $Z(y)$ for an unmagnetized, nonrelativistic FD distribution may be defined by Melrose and Mushtaq [9]

$$Z(y,\xi) = \frac{2}{\pi^{1/2}} \int_0^{\infty} \frac{dt\, t}{e^{t^2} + \xi} \ln\frac{t-y}{t+y} = \frac{1}{\pi^{1/2}} \int_{-\infty}^{\infty} \frac{dt}{t-y} \ln(1+\xi e^{-t^2}). \tag{9.3.42}$$

The function $\phi(y,\xi) = -yZ(y,\xi)$ is shown in Fig. 9.3 for several different values ranging from the nondegenerate limit, $\xi \to 0$, to completely degenerate limit, $\xi \to \infty$. Near the completely degenerate limit, where $\ln\xi$ is large, the real part of the plasma dispersion function may be approximated by [9]

$$\phi(y,\xi) = -yZ(y,\xi) = \frac{4y^2}{\sqrt{\pi}(\ln\xi)^{3/2}}\left(1 - \frac{y^2}{3\ln\xi} + \cdots\right). \tag{9.3.43}$$

In the case of partial degeneracy, the RPDF (9.2.28) becomes,

$$\mathcal{Z}_n(t_0) = \pi^{1/2}\xi_n\, Z(y_{n0}, \xi_n), \tag{9.3.44}$$

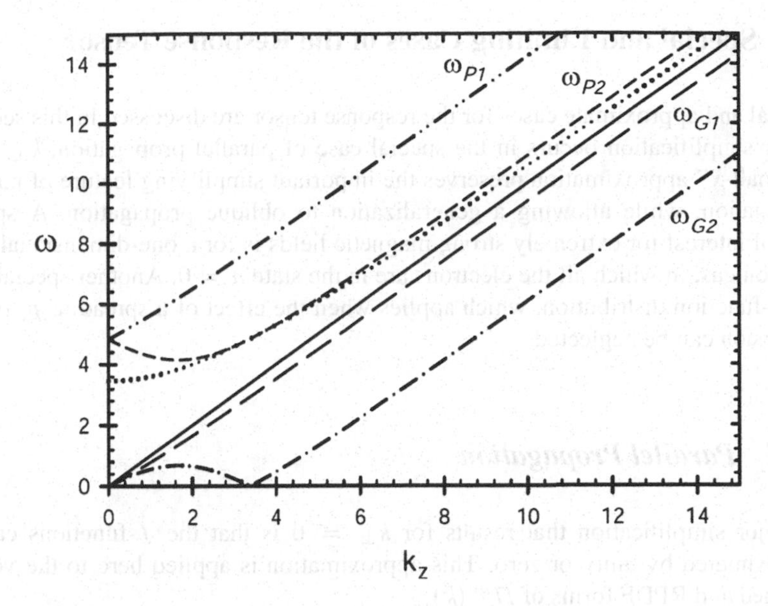

Fig. 9.3 (**a**) The function $\phi(y,\xi) = -yZ(y,\xi)$ is plotted versus y for $\xi = 0, 0.1, 1, 10$. (**b**) A rescaled form of $\phi'(y',\xi) = \xi\phi(y,\xi)/(\ln\xi)^{3/2}$ is plotted versus $y' = y/(\ln\xi)$ for $\xi = 10, 100, 10^4, \infty$. The curves for $\xi = 10$ are the same, apart from the scaling. The curve for $\xi = 0$ corresponds to the plasma dispersion function in the nondegenerate limit, and that for $\xi = \infty$ corresponds to a logarithmic function in the completely degenerate limit (From [9], reprinted with permission AIP)

with y_{n0} defined following (9.3.40). The result (9.3.44) applies to a magnetized, nonrelativistic FD distribution. The generalization to an arbitrary magnetized FD distribution does not appear to have a representation in terms of known functions.

The RPDF appears in the response tensor with arguments $t_0 = t^{\pm}_{nn'}, 1/t^{\pm}_{nn'}$, and the only important contributions come from values that correspond to $|t_0| \ll 1$. Suppose either $|t^{\pm}_{nn'}|$ is much smaller than unity; then $1/|t^{\pm}_{nn'}|$ is much larger than unity, and gives a small contribution to the dispersion that can be neglected. On the other hand, if either $|t^{\pm}_{nn'}|$ is much larger than unity, then $1/|t^{\pm}_{nn'}|$ is much smaller than unity, and its contribution to the dispersion needs to be retained while that from $|t^{\pm}_{nn'}|$ is neglected. In these two cases, the resonant value in the denominator in (9.3.40) are $t^{\pm}_{nn'} \approx p^{\pm}_{nn'}c/2\varepsilon^0_n$ for $|t^{\pm}_{nn'}| \ll 1$, and $-1/t^{\pm}_{nn'} \approx p^{\pm}_{nn'}c/2\varepsilon^0_n$ for $|t^{\pm}_{nn'}| \gg 1$. The RPDFs that appear in the response tensor are then approximated by

$$[\mathcal{Z}^{\epsilon}_n(t^{\pm}_{nn'}),\ \mathcal{Z}^{\epsilon}_n(1/t^{\pm}_{nn'})] \approx \begin{cases} [\mathcal{Z}^{\epsilon}_n(p^{\pm}_{nn'}c/2\varepsilon^0_n),\ 0] & |t^{\pm}_{nn'}| \ll 1, \\ [0,\ \mathcal{Z}^{\epsilon}_n(-p^{\pm}_{nn'}c/2\varepsilon^0_n)] & |t^{\pm}_{nn'}| \gg 1. \end{cases} \tag{9.3.45}$$

9.4 Special and Limiting Cases of the Response Tensor

Special and approximate cases for the response tensor are discussed in this section. Major simplification occurs in the special case of parallel propagation, $k_\perp = 0$. A "small-x" approximation preserves the important simplifying feature of parallel propagation, while allowing a generalization to oblique propagation. A special case of interest for extremely strong magnetic fields is for a one-dimensional (1D) electron gas, in which all the electrons are in the state $n = 0$. Another special case is a δ-function distribution, which applies when the effect of a spread in p_z on the dispersion can be neglected.

9.4.1 *Parallel Propagation*

A major simplification that results for $k_\perp = 0$ is that the J-functions can be approximated by unity or zero. This approximation is applied here to the vertex, summed and RPDF forms of $\Pi^{\mu\nu}(k)$.

Vertex Form of $\Pi^{\mu\nu}(k)$: Parallel Propagation

Parallel propagation corresponds to $k_\perp = 0$, implying that the argument, $x = \hbar k_\perp^2/2eB$, of the J-functions is zero. It follows from the definition (A.1.25) that for $x = 0$ the J-functions reduce to $J_0^n(0) = 1$, $J_\nu^n(0) = 0$ for $\nu \neq 0$. Specifically, one has

$$J^{n-1}_{n'-n}(0) = \delta_{n',n}, \qquad J^{n}_{n'-n}(0) = \delta_{n',n},$$
$$J^{n-1}_{n'-n+1}(0) = \delta_{n',n-1}, \qquad J^{n}_{n'-n-1}(0) = \delta_{n',n+1}. \tag{9.4.1}$$

With the approximation (9.4.1), the vertex form (9.1.1) of $\Pi^{\mu\nu}(k)$ becomes

$$\Pi^{\mu\nu}(k) = -\frac{e^3Bc^2}{2\pi\hbar}\sum_{\epsilon,q,\epsilon',q'}\int\frac{dp_z}{2\pi\hbar}\int\frac{dp'_z}{2\pi\hbar}\,2\pi\hbar\,\delta(\epsilon' p'_z - \epsilon p_z + \hbar k_z)$$
$$\times\frac{\frac{1}{2}(\epsilon'-\epsilon) + \epsilon n^\epsilon_q - \epsilon' n^{\epsilon'}_{q'}}{\hbar\omega - \epsilon\varepsilon_q + \epsilon'\varepsilon_{q'}}\,[\Gamma^{\epsilon'\epsilon}_{q'q}(\boldsymbol{k})]^\mu_\parallel[\Gamma^{\epsilon\epsilon'}_{qq'}(-\boldsymbol{k})]^\nu_\parallel, \tag{9.4.2}$$

where the vertex functions (9.1.4) simplify through

$$J^{(0)}_{q'q}(\boldsymbol{k}) = (b'_{s'}b_s + s'sb'_{-s'}b_{-s})\delta_{n',n},$$
$$J^{(\pm)}_{q'q}(\boldsymbol{k}) = s'b'_{-s'}b_s e^{-i\psi}\delta_{n',n-1} \pm sb'_{s'}b_{-s}e^{i\psi}\delta_{n',n+1}. \tag{9.4.3}$$

The two terms in (9.4.3) appear in the 03- and 12-components of the vertex function, respectively. It follows that only $n' = n$ contributes to the $\mu, \nu = 0, 3$ components of $\Pi^{\mu\nu}(k)$, that only $n' = n \pm 1$ contribute to the $\mu, \nu = 1, 2$ components, and that the 01-, 02-, 13- and 23-components are zero.

Summed Form of $\Pi^{\mu\nu}(k)$: Parallel Propagation

For parallel propagation the summed form (9.1.19) of $\Pi^{\mu\nu}(k)$ simplifies due to

$$\begin{aligned}
[N_{n'n}(\eta p_\parallel, k)]_\parallel^{\mu\nu} &= [2p_\parallel^\mu p_\parallel^\nu - \eta\hbar(p_\parallel^\mu k_\parallel^\nu + p_\parallel^\nu k_\parallel^\mu) + \eta\hbar(pk)_\parallel g_\parallel^{\mu\nu}]\, a_n \delta_{n',n} \\
&\quad -[p_n^2 - \eta\hbar(pk)_\parallel]\, g_\perp^{\mu\nu}(\delta_{n',n-1} + \delta_{n',n+1}), \\
[G_{n'n}(\eta p_\parallel, k)]_\parallel^{\mu\nu} &= i[p_n^2 - \eta\hbar(pk)_\parallel]\, f^{\mu\nu}(\delta_{n',n-1} - \delta_{n',n+1}),
\end{aligned} \tag{9.4.4}$$

with $a_0 = 1$, $a_n = 2$ for $n > 0$. The summed form (9.1.19) reduces to

$$\begin{aligned}
\Pi^{\mu\nu}(k) &= -\frac{e^3 Bc^2}{\pi\hbar}\sum_n a_n \int \frac{dp_z}{2\pi\hbar}\frac{1}{\varepsilon_n}\Bigg\{ \\
&\quad n_n^{\mathrm{sum}}(p_z)\frac{(k^2)_\parallel p_\parallel^\mu p_\parallel^\nu - (pk)_\parallel(p_\parallel^\mu k_\parallel^\nu + p_\parallel^\nu k_\parallel^\mu) + (pk)_\parallel^2 g_\parallel^{\mu\nu}}{(pk)_\parallel^2 - [\frac{1}{2}\hbar(k^2)_\parallel]^2} \\
&\quad +\frac{1}{2}\sum_{\eta,\pm}\frac{[\eta(pk)_\parallel - 2neB]\{n_n^{\mathrm{sum}}(p_z)g_\perp^{\mu\nu} \pm i\eta n_n^{\mathrm{diff}}(p_z)f^{\mu\nu}\}}{\eta(pk)_\parallel - \frac{1}{2}\hbar(k^2)_\parallel \pm eB}\Bigg\}.
\end{aligned} \tag{9.4.5}$$

The two terms inside the curly brackets correspond to separate parallel (confined to the 0–3 plane and perpendicular 1–2 plane) parts of the response tensor.

The nondispersive (ND) part of (9.4.5) follows by writing $\eta(pk)_\parallel/[\eta(pk)_\parallel - (pk)_{nn'}] = 1 + (pk)_{nn'}/[\eta(pk)_\parallel - (pk)_{nn'}]$, with $(pk)_{nn'} = \frac{1}{2}\hbar(k^2)_\parallel \mp eB$ here, and identifying it with the integral over the unit term. This gives

$$\Pi_{\mathrm{ND}}^{\mu\nu}(k) = -\frac{e^3 Bc^2}{4\pi^2\hbar^2}\sum_n a_n \int \frac{dp_z}{\varepsilon_n} n_n^{\mathrm{sum}}(p_z) g_\perp^{\mu\nu}, \tag{9.4.6}$$

which corresponds to the general results (9.2.22) with (9.2.21) for $k_\perp = 0$.

In (9.4.5) the parallel-components are already summed over η. The sums over $\eta, \pm$ in the perpendicular components in (9.4.5) may be performed using

$$\frac{1}{2}\sum_{\eta,\pm}\frac{1}{\eta a - b \pm c}\begin{pmatrix}1\\ \eta\\ \pm 1\\ \pm\eta\end{pmatrix} = \frac{2}{D}\begin{pmatrix}b(a^2 - b^2 + c^2)\\ a(a^2 - b^2 - c^2)\\ -c(a^2 + b^2 - c^2)\\ -2abc\end{pmatrix},$$

$$D = (a^2 + b^2 - c^2)^2 - 4a^2b^2, \quad a = (pk)_\parallel, \quad b = \frac{1}{2}\hbar(k^2)_\parallel, \quad c = eB. \tag{9.4.7}$$

The resulting form is cumbersome and is not written down explicitly here.

There are effectively six resonant denominators in (9.4.5): at $\eta(pk)_\parallel = \frac{1}{2}\hbar(k^2)_\parallel, \frac{1}{2}\hbar(k^2)_\parallel \pm eB$, with $\eta = \pm 1$. The term $\frac{1}{2}\hbar(k^2)_\parallel$ is attributed to the quantum recoil. When the recoil is neglected, there are only three different denominators, $(pk)_\parallel, (pk)_\parallel \pm eB$. In the nonquantum limit, these three denominators reduce to m times ku, $ku \pm \Omega_0$.

RPDF Form of $\Pi^{\mu\nu}(k)$: Parallel Propagation

The RPDF form (9.2.38) of $\Pi^{\mu\nu}(k)$ simplifies for $k_\perp = 0$ due to

$$\begin{aligned}
\left[N_{n'n}(k)\right]_+^{\mu\nu} &= -\frac{(\varepsilon_n^0)^2}{\hbar^2(k^2)_\parallel c^2} k_D^\mu k_D^\nu \, a_n \delta_{a,0} \\
&\quad - \left(\frac{(2n-a)eB}{\hbar} - \frac{1}{2}(k^2)_\parallel\right) g_\perp^{\mu\nu} \left[\delta_{a,1} + \delta_{a,-1}\right], \\
\left[G_{n'n}(k)\right]_+^{\mu\nu} &= i\left(\frac{(2n-a)eB}{\hbar} - \frac{1}{2}(k^2)_\parallel\right) f^{\mu\nu} \left[\delta_{a,1} - \delta_{a,-1}\right],
\end{aligned} \tag{9.4.8}$$

with $\left[N_{n'n}(k)\right]_-^{\mu\nu} = \left[G_{n'n}(k)\right]_-^{\mu\nu} = 0$. The RPDFs $\mathcal{Z}_{nn'}^{(N)}(\omega, k_z)$, $N = 1, 2, 3, 4$, given by (9.2.37), are independent of $k_\perp$.

The RPDF form (9.2.38) becomes

$$\begin{aligned}
\Pi^{\mu\nu}(k) &= \Pi_{\rm ND}^{\mu\nu}(k) + \frac{e^3Bc}{8\pi^2\hbar^2(k^2)_\parallel} \\
&\quad \times \sum_{n,n'} \frac{1}{g_{nn'}} \left\{\left[N_{n'n}(k)\right]_+^{\mu\nu} \mathcal{Z}_{nn'}^{(1)}(\omega, k_z) + \left[G_{n'n}(k)\right]_+^{\mu\nu} \mathcal{Z}_{nn'}^{(3)}(\omega, k_z)\right\},
\end{aligned} \tag{9.4.9}$$

with the tensors in the integrand given by (9.4.8), and with the ND part given by (9.4.6). The RPDFs in (9.4.9) with (9.4.8) have $n - n' = a = 0, \pm 1$, and for these values the arguments, $t_{nn'}^\pm$, $1/t_{nn'}^\pm$, of the RPDFs are given by (ordinary units)

$$t_{nn'}^\pm = \frac{\omega f_{nn'} \pm k_z c g_{nn'} - \varepsilon_n^o/\hbar}{\hbar_z c f_{nn'} \pm \omega g_{nn'}}. \tag{9.4.10}$$

Response 3-Tensors: Parallel Propagation

For parallel propagation there are only three independent components of the response tensor. These are $\Pi^{11}(k) = \Pi^{22}(k)$, $\Pi^{12}(k) = -\Pi^{21}(k)$ and $\Pi^{33}(k)$, with $\Pi^{13}(k) = 0 = \Pi^{23}(k)$. The form (9.4.5) of the response tensor for an arbitrary distribution gives, for $k_\perp = 0$,

$$\Pi^{33}(k) = -\frac{e^3 B\omega^2}{2\pi^2\hbar^2 c^2}\sum_n a_n(\varepsilon_n^0)^2 \int \frac{dp_z}{\varepsilon_n}\frac{n_n^{\rm sum}(p_z)}{(pk)_\parallel^2 - [\frac{1}{2}\hbar(k^2)_\parallel]^2}, \tag{9.4.11}$$

$$\begin{bmatrix}\Pi^{11}(k)\\ \Pi^{12}(k)\end{bmatrix} = \frac{e^2\bar{n}_{\rm pr}}{m}\begin{bmatrix}1\\0\end{bmatrix} - \frac{e^3 Bc^2}{2\pi^2\hbar^2}\sum_n \int \frac{dp_z}{\varepsilon_n}$$

$$\times\frac{1}{2}\sum_{\eta,\pm}\frac{2(n\pm 1)eB - \frac{1}{2}\hbar(k^2)_\parallel}{\eta(pk)_\parallel - \frac{1}{2}\hbar(k^2)_\parallel \pm eB}\begin{bmatrix}n_n^{\rm sum}(p_z)\\ \pm i\eta n_n^{\rm diff}(p_z)\end{bmatrix}, \tag{9.4.12}$$

with $(pk)_\parallel = \varepsilon_n\omega/c^2 - p_z k_z$, $(k^2)_\parallel = \omega^2/c^2 - k_z^2$. The sums over η and $\pm$ in the expressions (9.4.12) can be performed using (9.4.7). Note that although the term proportional to n appears to be of lower order in $\hbar$ than the other terms, this is not the case when the sums are performed; in the nonquantum limit, this term sums to zero.

The form (9.4.9) of the response tensor in terms of RPDFs gives, for the parallel and perpendicular parts,

$$\Pi^{33}(k) = -\frac{e^3 Bc}{8\pi^2\hbar^2(k^2)_\parallel}\sum_n \frac{(\varepsilon_n^0)^2\omega^2}{\hbar^2(k^2)_\parallel c^4}\frac{\mathcal{Z}_{nn}^{(1)}(\omega,k_z)}{g_{nn}}, \tag{9.4.13}$$

$$\begin{bmatrix}\Pi^{11}(k)\\ \Pi^{12}(k)\end{bmatrix} = \frac{e^2\bar{n}_{\rm pr}}{m}\begin{bmatrix}1\\0\end{bmatrix} + \frac{e^3 Bc}{8\pi^2\hbar^3(k^2)_\parallel}\sum_n\sum_{a=\pm1}\frac{1}{g_{n\,n-a}}$$

$$\times[(2n-a)eB - \frac{1}{2}\hbar(k^2)_\parallel]\begin{bmatrix}\mathcal{Z}_{n\,n-a}^{(1)}(\omega,k_z)\\ ia\mathcal{Z}_{n\,n-a}^{(3)}(\omega,k_z)\end{bmatrix}, \tag{9.4.14}$$

respectively. Only $n - n' = a = 0$ contributes to the parallel part (33-component), and $a = \pm 1$ contribute to the perpendicular part.

The corresponding nonzero components of the dielectric tensor are

$$K^1{}_1 = K^2{}_2 = 1 - \frac{\Pi^{11}}{\varepsilon_0\omega^2}\quad K^3{}_3 = 1 - \frac{\Pi^{33}}{\varepsilon_0\omega^2},\quad K^1{}_2 = -K^2{}_1 = -\frac{\Pi^{12}}{\varepsilon_0\omega^2}, \tag{9.4.15}$$

where arguments are omitted.

9.4.2 Small-x Approximation

The small-x approximation corresponds to assuming that the argument, $x = \hbar k_\perp^2/2eB$, of the J-functions is small, and retaining terms of low order in x. This leads to a generalization of the results for parallel propagation, to include terms corresponding to oblique propagation. The small-x approximation may be regarded as a quantum counterpart of the small-gyroradius approximation for "magnetized" particles in the nonquantum case. The gyroradius, R, is a classical concept, which applies only in the limit of large n, such that $p_n = (2neB\hbar)^{1/2} \to p_\perp$ can be regarded as a continuous variable. One then has $nx = (\frac{1}{2}k_\perp R)^2$. Thus the condition $nx \ll 1$ is equivalent to the classical small-gyroradius approximation, $k_\perp R \ll 1$, for $n \gg 1$. The small-x approximation applies for any n, and it is an intrinsically quantum approximation for small n.

One may write x in the form

$$x = \tfrac{1}{2}\left(\frac{\hbar k_\perp}{mc}\right)^2 \left(\frac{B}{B_c}\right)^{-1}. \tag{9.4.16}$$

The small-x approximation is satisfied either for sufficiently small $k_\perp$ for any B or for sufficiently strong B for any $k_\perp$. More specifically, for photons with refractive index approximately equal to unity, implying $k_\perp \approx (\omega/c)\sin\theta$, the condition $x \ll 1$ is satisfied sufficiently near parallel propagation, $\sin^2\theta \ll 1$, for sufficiently soft photons, $\hbar\omega \ll mc^2(B_c/B)^{1/2}$, or for a sufficiently strong field; for a supercritical field, $B \gg B_c$, one can have $x \ll 1$ even for photons with energy of order an MeV.

Care is required in approximating the J-functions in the small-x approximation. There is a term $\propto k_\perp^2$ that contributes even in the nonquantum limit, cf. (9.1.29). This term is present in the $\mu, \nu = 0, 3$ components, and to reproduce it, one needs to retain nonzero contributions from $J^{n-1}_{n'-n}(x)$, $J^{n}_{n'-n}(x)$ for $n' - n = \pm 1$. For small x, the expansion (A.1.51) in x, implies that the lowest order corrections are of order $x^{1/2} \propto k_\perp$. The term $\propto k_\perp^2$ in the nonquantum limit arises from

$$J^{n-1}_{1}(x) = -J^{n}_{-1}(x) \approx (nx)^{1/2} = \frac{k_\perp p_n}{2eB}. \tag{9.4.17}$$

9.4.3 One-Dimensional (1D) Electron Gas

The case of a 1D electrons gas is of interest in connection with superstrong magnetic fields, particularly pulsar fields, where the time scale for electrons to radiate away their perpendicular energy is so short that effectively all the electrons (and positrons) are in the ground Landau state, $n = 0$. In a more general context, where states $n > 0$ are also populated, the ground state is non-degenerate and needs to be treated separately. The result for a 1D electron distribution applies to the contribution from $n = 0$ for an arbitrary electron gas, where the contributions from $n \geq 1$ can be treated by summing over the spin states.

Summed Form of $\Pi^{\mu\nu}(k)$: 1D Electron Gas

An expression for the response tensor for a 1D electron gas follows by setting $n = 0$ in (9.1.19) and performing the sum over η. This gives

$$\Pi^{\mu\nu}(k) = -\frac{e^3 B}{(2\pi\hbar)^2}\sum_{n'}\int \frac{dp_z}{\varepsilon_0(p_z)}\frac{[B_{n'}(p_\parallel, k)]^{\mu\nu}}{(pk)_\parallel^2 - [\frac{1}{2}\hbar(k^2)_\parallel]^2 - n'eB]^2}, \tag{9.4.18}$$

with $\varepsilon_0(p_z) = (m^2c^4 + p_z^2c^2)^{1/2}$. For $n = 0$ the J-functions with superscript $n-1$ are identically zero, and $[B_{n'}(p_\parallel, k)]^{\mu\nu}$ reduces to

$$\begin{aligned}[B_{n'}(p_\parallel, k)]^{\mu\nu} &= n_0^{\rm sum}(p_z)[(k^2)_\parallel p_\parallel^\mu p_\parallel^\nu - (pk)_\parallel(p_\parallel^\mu k_\parallel^\nu + p_\parallel^\nu k_\parallel^\mu) + (pk)_\parallel^2 g_\parallel^{\mu\nu}](J_{n'}^0)^2 \\ &+ n_0^{\rm sum}(p_z)[(pk)_\parallel^2 g_\perp^{\mu\nu} - (pk)_\parallel(p_\parallel^\mu k_\perp^\nu + p_\parallel^\nu k_\perp^\mu) - k_\perp^2 p_\parallel^\mu p_\parallel^\nu](J_{n'-1}^0)^2 \\ &+ i n_0^{\rm diff}(p_z)[\tfrac{1}{2}\hbar(k^2)_\parallel]^2 - n'eB]\,[(pk)_\parallel f^{\mu\nu} - (p_\parallel^\mu k_G^\nu - p_\parallel^\nu k_G^\mu)](J_{n'-1}^0)^2.\end{aligned} \tag{9.4.19}$$

A given term in the sum over n' in (9.4.18) describes the contribution due to virtual transitions $n = 0 \to n' \to n = 0$. The state $n = 0$ has a unique spin, $s = -1$, and the spin obviously does not change for $n' = 0$, For $n' > 0$ the sum over spins is already performed, so that the contribution from n' includes the effect of both non-spin-flip transitions, $s' = -1$, and spin-flip transitions, $s' = +1$. The sum over η is also performed in deriving (9.4.18), implying that both virtual electron and virtual positron states are included.

The J-functions that appear in (9.4.19) have simple values, $(J_{n'}^0)^2 = x^{n'}e^{-x}/n'!$ for $n' \geq 0$, with $(J_{n'-1}^0)$ identically zero for $n' = 0$. The contribution from increasing n' decreases rapidly with increasing n' for $x \ll 1$, when only the lowest values of n' need be retained. The small-x approximation may not apply for transitions from large n' to $n = 0$ and, as discussed in connection with (6.3.44) for real transitions, virtual transitions between large n' and $n = 0$ can be significant at high frequencies.

RPDF Form of $\Pi^{\mu\nu}(k)$: 1D Electron Gas

The form (9.2.38) for the response tensor for a 1D distribution is obtained by setting $n = 0$ in (9.2.3), giving

$$f_{0n'} = \tfrac{1}{2} - \frac{n'eB}{\hbar(k^2)_\parallel}, \qquad g_{0n'}^2 = f_{0n'}^2 - \frac{m^2c^2}{\hbar^2(k^2)_\parallel}. \tag{9.4.20}$$

The tensors (9.2.35) in the numerator of (9.2.38) become

$$
\begin{aligned}
[N_{n'0}(k)]_{+}^{\mu\nu} &= 2\left\{k_{\parallel}^{\mu}k_{\parallel}^{\nu}\left(f_{0n'}-\tfrac{1}{2}\right)^{2}+k_{D}^{\mu}k_{D}^{\nu}\left[\left(f_{0n'}^{2}-\tfrac{1}{4}\right)-m^{2}c^{2}/\hbar^{2}(k^{2})_{\parallel}\right]\right\}\left(J_{n'}^{0}\right)^{2} \\
&\quad -\left[\tfrac{1}{2}k_{\perp}^{2}g_{\parallel}^{\mu\nu}+-\tfrac{1}{2}(k^{2})_{\parallel}f_{0n'}g_{\perp}^{\mu\nu}+f_{0n'}\left(k_{\parallel}^{\mu}k_{\perp}^{\nu}+k_{\parallel}^{\nu}k_{\perp}^{\mu}\right)\left(J_{n'-1}^{0}\right)^{2},\right. \\
[N_{n'0}(k)]_{-}^{\mu\nu} &= g_{0n'}\left\{2\left(f_{0n'}-\tfrac{1}{2}\right)\left(k_{\parallel}^{\mu}k_{D}^{\nu}+k_{\parallel}^{\nu}k_{D}^{\mu}\right)\left(J_{n'}^{0}\right)^{2}\right. \\
&\quad \left.-(k_{D}^{\mu}k_{\perp}^{\nu}+k_{D}^{\nu}k_{\perp}^{\mu})[(J_{n'-1}^{0})^{2}]\right\}, \\
[G_{n'0}(k)]_{+}^{\mu\nu} &= i\left\{\tfrac{1}{2}(k^{2})_{\parallel}f_{0n'}f^{\mu\nu}+f_{0n'}\left(k_{\parallel}^{\mu}k_{G}^{\nu}-k_{\parallel}^{\nu}k_{G}^{\mu}\right)\right\}\left(J_{n'-1}^{0}\right)^{2}, \\
[G_{n'0}(k)]_{-}^{\mu\nu} &= i g_{0n'}\left(k_{D}^{\mu}k_{G}^{\nu}-k_{D}^{\nu}k_{G}^{\mu}\right)\left(J_{n'-1}^{0}\right)^{2}.
\end{aligned}
\tag{9.4.21}
$$

The remarks made above concerning reproducing the term $\propto k_{\perp}^{2}$ in the nonquantum limit can be made specific in this case: one needs to retain the contribution from $n'=1$ to the term $\propto (J_{n'}^{0})^{2}$ to reproduce the nonquantum limit.

9.4.4 δ-Function Distribution Function

A "$\delta(p_z)$-model" is defined by assuming that the p_z-distribution is a δ-function distribution for each Landau state. This corresponds to

$$
n_{n}^{\epsilon}(p_z)=a_{n}^{\epsilon}\,\delta(p_z), \qquad n_{n}^{\epsilon}=\frac{eB}{(2\pi\hbar)^{2}}\,a_{n}a_{n}^{\epsilon},
\tag{9.4.22}
$$

where n_{n}^{ϵ} is the number density of particles in the nth Landau state.

RPDFs for $\delta(p_z)$-Model

For the distribution (9.4.22), the RPDF defined by (9.2.28) becomes

$$
\mathcal{Z}_{n}^{\epsilon}(t_0)=-\frac{a_{n}^{\epsilon}c}{2\varepsilon_{n}^{0}}\,\frac{1}{t_0}.
\tag{9.4.23}
$$

The RPDFs defined by (9.2.37) become

$$
\begin{bmatrix}
\mathcal{Z}_{nn'}^{(1)}(\omega,k_z)\\
\mathcal{Z}_{nn'}^{(2)}(\omega,k_z)\\
\mathcal{Z}_{nn'}^{(3)}(\omega,k_z)\\
\mathcal{Z}_{nn'}^{(4)}(\omega,k_z)
\end{bmatrix}
=\frac{2a_{n}^{\epsilon}c}{\varepsilon_{n}^{0}}\,\frac{1}{p_{nn'}^{+}p_{nn'}^{-}}
\begin{bmatrix}
\hbar^{2}(k^{2})_{\parallel}f_{nn'}g_{nn'}\\
-\omega k_z(\varepsilon_{n}^{0})^{2}/(k^{2})_{\parallel}c^{3}\\
\hbar\omega\varepsilon_{n}^{0}g_{nn'}/c^{2}\\
-\hbar k_z\varepsilon_{n}^{0}f_{nn'}/c
\end{bmatrix}.
\tag{9.4.24}
$$

There are poles in these RPDFs at the solutions of $p^+_{nn'}p^-_{nn'} = 0$ for ω. Using (9.2.36), one has

$$p^+_{nn'}p^-_{nn'} = \frac{(\varepsilon_n^0)^2}{(k^2)_\parallel c^4}\left[\omega - a\Omega_{0n} - \frac{\hbar(k^2)_\parallel c^2}{2\varepsilon_n^0}\right]\left[\omega + a\Omega_{0n} + \frac{\hbar(k^2)_\parallel c^2}{2\varepsilon_n^0}\right], \tag{9.4.25}$$

with $a = n - n'$ and $\Omega_{0n} = eBc^2/\varepsilon_n^0$. Inserting these approximations into (9.2.38) gives a general expression for the response tensor for a cold-electron gas for arbitrary B/B_c.

Validity of the $\delta(p_z)$-Model

The conditions under which a δ-function model for the distribution function is a valid approximation are relatively well understood in the nonquantum limit, where assuming a δ-function in kinetic theory is equivalent to making the cold-plasma approximation. For example, the dispersion associated with the plasma dispersion function, $Z(y)$ defined by (2.5.5) with $y \propto \omega$, for a thermal distribution is replaced by a pole, $\propto 1/y$, in the cold limit $T \to 0$, when a Maxwellian distribution reduces to a δ-function distribution. A pole in a cold-plasma model corresponds to a resonant frequency; when thermal effects are included Doppler broadening removes the singularity; thermal motions effectively smear out the resonance over a Doppler width. Nevertheless, a pole is a useful approximation to the plasma dispersion for frequencies more than a few Doppler widths away from the resonant frequency. This smearing effect due to thermal motions is present in a quantum treatment and, as in the nonquantum case, it does not invalidate the δ-function model as an approximation.

There are two inconsistencies that need to be considered when applying the δ-function model to a quantum treatment of an electron gas.

One inconsistency is that a δ-function is formally incompatible with Fermi statistics: the occupation number in any state cannot exceed unity, and a δ-function distribution clearly violates this. This is unimportant in the nondegenerate case, but it can invalidate the δ-function approximation for a degenerate distribution. It is straightforward to compare the RPDFs for a δ-function and for a completely degenerate distribution, for which the occupation number is unity for $|p_z| < p_{nF}$, $n < n_F$, and zero otherwise. This involves comparing the RPDF (9.3.13), which is $\mathcal{Z}_n(t_0) = \ln|(t_{nF} - t_0)/(t_{nF} + t_0)|$, with the RPDF $\propto -1/t_0$ for the δ-function model. The comparison is illustrated in Fig. 9.4. The RPDF for the degenerate distribution has logarithmic singularities at $x = \pm 1$, corresponding to $t_0 = \pm t_{nF}$. For $|t_0| \gg t_{nF}$, the wings of the function $1/t_0$ approach the asymptotes of the logarithmic functions. In this limit, the $\delta(p_z)$-model may be a useful approximation, but it is clearly invalid for $t_0 \lesssim t_{nF}$.

The second inconsistency is associated with the quantum recoil. Dispersion in a quantum plasma can be attributed to virtual transitions between two states, summed

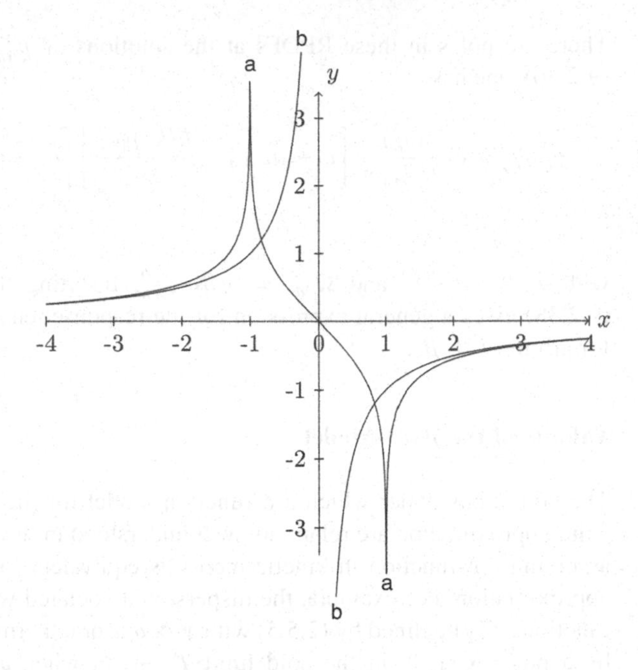

Fig. 9.4 The functions (**a**) $y = \frac{1}{2}\ln|(1-x)/(1+x)|$ and (**b**) $y = -1/x$ are compared; they correspond to RPDFs for completely degenerate and $\delta(p_z)$-model distributions, and have two logarithmic singularities and a pole, respectively

over all pairs of states for which the (virtual) transition is possible. The response tensor involves the difference between the occupation numbers of these two states. These two states have parallel momenta that differ by k_z, due to the quantum recoil, being p_z and $p_z - k_z$ say. The $\delta(p_z)$-model can give a contribution for $p_z = 0$ or for $p_z = k_z$, but not to both simultaneously. Making the $\delta(p_z)$-approximation, in (9.1.19) for example, excludes the possibility of taking resonances at p_z and $p_z - k_z$ into account simultaneously in the numerator. This inconsistency leads to seemingly unavoidable difficulties in including the quantum recoil in a self-consistent manner in a δ-function model. There is no logical difficulty if one neglects the recoil before assuming a $\delta(p_z)$-model.

The RPDF for a distribution $\propto \delta(p_z)$ has poles at $p^+_{nn'}, p^-_{nn'} = 0$, given by (9.4.25), and these include quantum recoil terms. However, the $\delta(p_z)$-model is not valid near these poles.

9.5 Wave Dispersion: Parallel, Degenerate Case

The properties of dispersion in a relativistic quantum electron gas have been explored only in a few special cases. The discussion in this section is restricted to examples discussed in the literature for parallel propagation ($k_\perp = 0$) in a completely degenerate electron gas ($T \to 0$) [3, 13, 14, 22]. After some remarks on the static response, the emphasis in this section is on unique features of wave dispersion in a magnetized relativistic degenerate electron gas. Even in the special case $k_\perp = 0$, interesting new features appear, notably additional wave modes, referred to as gyromagnetic absorption modes and pair modes.

9.5.1 Static Response

The static electric response is associated with screening of a test charge, and includes contributions from both the electron gas and the vacuum. The effect of the vacuum polarization on screening has been shown to becomes highly anisotropic in a superstrong magnetic field [16], with the anisotropy due to the different dependence on $k_\perp$ and k_z. The static electric response for a relativistic degenerate magnetized electron gas is discussed below for $k_\perp = 0$, following [3]. The static magnetic response can be described by the magnetic susceptibility tensor, which has contributions from Landau diamagnetism and Pauli spin paramagnetism. The static magnetic response of a relativistic degenerate electron gas can also be derived from the general expression for $\Pi^{\mu\nu}(k)$ [3].

The static limit corresponds to $\omega \to 0$. For an isotropic electron gas, it is known that the longitudinal (electric) susceptibility approaches $-\omega_p^2/\omega^2$ when the limit $|\mathbf{k}| \to 0$ is taken before $\omega \to 0$, and approaches $1/|\mathbf{k}|^2\lambda_0^2$ when the limit $\omega \to 0$ is taken before $|\mathbf{k}| \to 0$, with λ_0 the Debye length in a thermal electron gas and the Thomas-Fermi length in a degenerate electron gas, which exhibits Friedel oscillations (§ 9.5.2 of volume 1). For the static magnetic response it is known that in the weak-field limit, Landau diamagnetism and Pauli spin paramagnetism contribute to the magnetic susceptibility in the ratio $-1:3$ (§ 9.5.3 of volume 1).

For $k_\perp = 0$, the static longitudinal electric susceptibility and the static magnetic susceptibility are given by

$$\chi^{(e)}(k_z) = \lim_{\omega\to 0} \frac{\Pi^{33}(k)}{\varepsilon_0\omega^2}, \qquad \chi^{(m)}(k_z) = -\lim_{\omega\to 0} \frac{\mu_0\Pi^{22}(k)}{k_z^2}. \tag{9.5.1}$$

The relevant forms for $\Pi^{33}(k)$ and $\Pi^{22}(k) = \Pi^{11}(k)$ are given by (9.4.13) and (9.4.14), respectively. The static electric response follows from $k_\perp = 0$, $\omega \to 0$ in (9.4.9) with (9.4.8), giving

$$\chi^{(e)}(k_z) = \sum_{n=0}^{n_F} a_n \frac{e^3 B(\varepsilon_n^0)^2}{4\pi^2\varepsilon_0\hbar^4 k_z^4 c^3} \frac{\mathcal{Z}_{nn}^{(1)}(0,k_z)}{g_{nn}}, \tag{9.5.2}$$

with $g_{nn} = [(\varepsilon_n^0)^2/\hbar^2c^2 + k_z^2/4]^{1/2}/|k_z|$. The result (9.5.2) applies for an arbitrary distribution of electrons. For a completely degenerate distribution in the limit $\omega \to 0$, it is convenient to use the form (9.3.16) for $n' \neq n$ and the form (9.3.17) for $n' = n$. The latter gives

$$\mathcal{Z}_{nn}^{(1)}(0,k_z) = \ln\left|\frac{(\varepsilon_n^0)^4 - \hbar^2k_z^2c^2(\frac{1}{2}p_{nF}c - \varepsilon_F g_{nn})^2}{(\varepsilon_n^0)^4 - \hbar^2k_z^2c^2(\frac{1}{2}p_{nF}c + \varepsilon_F g_{nn})^2}\right|. \tag{9.5.3}$$

As in the unmagnetized case, the screening involves a logarithmic function of the wavenumber. There are logarithmic singularities at the values of k_z where the argument in the logarithm in (9.5.3) passes through zero or infinity. The values of k_z

at which the logarithmic singularities occur are related to the limit $\omega \to 0$ of the longitudinal GA modes discussed below. However, the physical significance of these results is limited by the restriction to $k_\perp = 0$.

The static magnetic response for a completely degenerate relativistic distribution was derived for $k_\perp = 0$ in a similar manner in reference [3]. This restriction needs to be relaxed in a more general discussion.

9.5.2 Dispersion Equation for Parallel Propagation

For parallel propagation, the dispersion equation, $\lambda(k) = 0$, factors into a dispersion equation for longitudinal waves,

$$\omega^2 - \Pi^{33}(k)/\varepsilon_0 = 0, \tag{9.5.4}$$

and dispersion equations for transverse waves with right-hand and left-hand circular polarization,

$$\omega^2 - k_z^2c^2 - \Pi^{\pm}(k)/\varepsilon_0 = 0, \qquad \Pi^{\pm}(k) = \Pi^{11}(k) \pm i\Pi^{12}(k). \tag{9.5.5}$$

Explicit expressions for the components of the response tensor for a completely degenerate distribution follow from (9.4.13) and (9.4.14). Only $a = 0$ contributes for longitudinal dispersion relation, and in this case it is convenient to use the form (9.3.17) for the RPDF,

$$\Pi^{33}(k) = \frac{e^3B\omega^2}{4\pi^2\hbar^4(k^2)_\parallel^2c^3}\sum_{n=0}^{n_F}\frac{(\varepsilon_n^0)^2}{g_{nn}}\ln\left|\frac{4k_z^2(\varepsilon_n^0)^4 - (k^2)_\parallel^2(p_{nF} - 2\varepsilon_F g_{nn})^2}{4k_z^2(\varepsilon_n^0)^4 - (k^2)_\parallel^2(p_{nF} + 2\varepsilon_F g_{nn})^2}\right|, \tag{9.5.6}$$

with $f_{nn} = \frac{1}{2}$ and $g_{nn} = \frac{1}{2}[1 - 4(\varepsilon_n^0)^2/\hbar^2(k^2)_\parallel c^2]^{1/2}$. Only $a = \pm 1$ contribute to the transverse dispersion relations, and (9.4.14) with (9.3.16) gives

$$\begin{aligned}\Pi^{\pm}(k) = \frac{e^2\bar{n}_{\rm pr}}{m} - \frac{e^3Bc}{4\pi^2\hbar^3(k^2)_\parallel}\sum_{n=0}^{n_F}\sum_{a=\pm1}\frac{(2n-a)eB - \frac{1}{2}\hbar(k^2)_\parallel}{g_{n\,n-a}} \\ \times\left\{\ln\left|\frac{k_z^2(\varepsilon_n^0)^4 - (k^2)_\parallel^2(p_{nF}f_{n\,n-a} - \varepsilon_F g_{n\,n-a})^2}{k_z^2(\varepsilon_n^0)^4 - (k^2)_\parallel^2(p_{nF}f_{n\,n-a} + \varepsilon_F g_{n\,n-a})^2}\right|\right. \\ \left.\pm a\ln\left|\frac{k_z^2f_{n\,n-a}^2 - (p_{nF} - \omega g_{n\,n-a})^2}{k_z^2f_{n\,n-a}^2 - (p_{nF} + \omega g_{n\,n-a})^2}\right|\right\},\end{aligned} \tag{9.5.7}$$

$$f_{n\,n-a} = \frac{aeB}{\hbar(k^2)_\parallel} + \frac{1}{2}, \qquad g_{n\,n-a}^2 = f_{n\,n-a}^2 - \frac{(\varepsilon_n^0)^2}{\hbar^2(k^2)_\parallel c^2}, \tag{9.5.8}$$

with only $a = -1$ allowed for $n = 0$. In principle, the response tensor $\Pi^{\mu\nu}(k)$ also includes the contributions of both the electron gas and the magnetized vacuum [14]. The combination of the vacuum polarization and dispersion in the cold plasma modes is discussed in § 8.4.3; the vacuum polarization is neglected here.

The components (9.5.6) and (9.5.7) of the response tensor apply in the dissipative region, $g^2_{n\,n-a} > 0$, where the RPDFs for a completely degenerate distribution are logarithmic functions. These RPDFs have logarithmic singularities at values where the argument of the logarithm has a zero in the numerator or a zero in the denominator. These singularities correspond to the values at which the antihermitian part of the response tensor changes discontinuously by $\pm i\pi$, as discussed in § 9.3.4. For a zero in the numerator of the argument of the logarithm, the real part of the logarithm approaches $-\infty$ on either side of the singularity. For a zero in the denominator, the real part approaches $+\infty$ from either side. A logarithmic singularity is qualitatively different from a pole, where the function changes sign by jumping from $\mp\infty$ to $\pm\infty$ as a pole is crossed. Typically, associated with a given singularity, the dispersion equation has either a doublet of closely separated solutions on either side of the singularity, or no solution, depending on the sign of the singularity.

9.5.3 *Longitudinal Modes*

The solutions of the longitudinal dispersion relation (9.5.4) for a completely degenerate electron gas include a Langmuir-type mode and GA and PC modes whose existence depends on intrinsically quantum effects. Both the GA and PC modes are associated with logarithmic singularities in the response function, for $(k^2)_\parallel < 0$ and $(k^2)_\parallel > 0$, respectively.

Langmuir-Type Mode

In an isotropic thermal plasma, the familiar Langmuir mode has an approximate dispersion relation $\omega^2 = \omega_p^2 + 3|\mathbf{k}|^2 V_e^2$, and when relativistic effects are included, the cutoff frequency is decreased to below ω_p and the peak is increased to above ω_p (Figs. 9.2 and 9.7 of volume 1). A counterpart of this mode exists for parallel propagation in the magnetized relativistic quantum degenerate case.

The cutoff frequencies satisfy (9.5.4) with $k_z = 0$ (as well as $k_\perp = 0$). With $g_{nn} = ig'_{nn}$ imaginary for $\omega^2 < k_z^2c^2$, the RPDF becomes an arctangent rather than a logarithmic function, as given by (9.3.20) with (9.3.19). The small-angle approximation to the arctangent applies for $p_{nF}c \ll 2\varepsilon_F g'_{nn}$, and then one has

$$\frac{Z_{nn}^{(1)}(\omega, 0)}{g_{nn}} = \frac{2p_{nF}c}{\varepsilon_F}\,\frac{\hbar^2\omega^2}{(\varepsilon_n^0)^2 + \hbar^2\omega^2/4}. \tag{9.5.9}$$

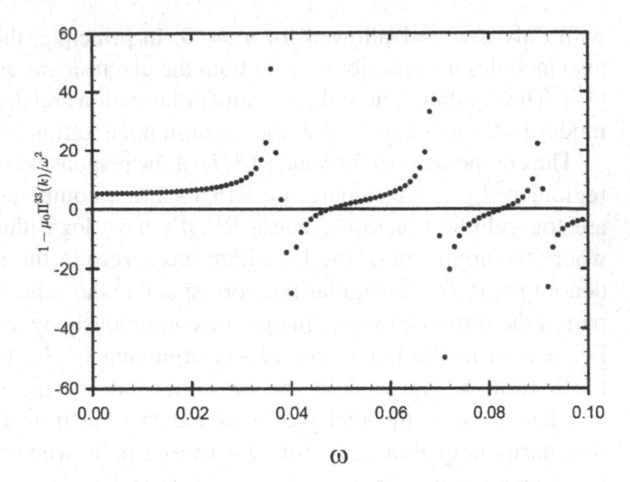

Fig. 9.5 The real part of $1 - \Pi^{33}(k)/\varepsilon_0\omega^2 = 0$, is plotted as a function of ω/m; the points are calculated for $p_{nF} = 2.2m$, $B = B_c$, $k_z = 0.1m$. A solution of the longitudinal dispersion relation (9.5.4) occurs where the curve passes through zero (From [22], reprinted with permission AIP)

With the further approximation $\hbar^2\omega^2/4 \ll (\varepsilon_n^0)^2$, (9.5.4) with (9.5.6) implies the cutoff frequency

$$\omega_c^2 = \frac{e^3Bc^2}{2\pi^2\varepsilon_0\hbar^2}\sum_{n=0}^{n_F}\frac{a_n p_{nF}}{\varepsilon_F}. \tag{9.5.10}$$

The frequency at which the dispersion curve for the longitudinal mode crosses the light line follows from (9.5.4) with (9.5.6) by taking the limit $(k^2)_\parallel \to 0$. This gives

$$\omega_0^2 = \frac{e^3Bc^2}{2\pi^2\varepsilon_0\hbar^2}\sum_{n=0}^{n_F}\frac{a_n p_{nF}\varepsilon_F}{(\varepsilon_n^0)^2}. \tag{9.5.11}$$

The inequality $\varepsilon_F > \varepsilon_n^0$ implies $\omega_0 > \omega_c$, implying that ω increases as a function of k_z over the range $0 < k_z < \omega_0/c$, as for a Langmuir wave in a nonrelativistic plasma. Similarly, in a highly relativistic plasma ω_0^2/ω_c^2 is large, related to the Lorentz factor ε_F/mc^2, as in the unmagnetized case.

Longitudinal GA Modes

For $k_z^2c^2 > \omega^2$ there are logarithm singularities in $\Pi^{33}(k)$ at the frequencies $\omega_{G1,2}$, given by (9.3.24) with $n' = n$, that is, at $\omega = |(\varepsilon_F^2 - 2p_{nF}\hbar k_z c^2 + \hbar^2k_z^2c^2)^{1/2} - \varepsilon_F|/\hbar$ and at $\omega = [(\varepsilon_F^2 + 2p_{nF}\hbar k_z c^2 + \hbar^2k_z^2c^2)^{1/2} - \varepsilon_F]/\hbar$ for $k_z > 0$. As illustrated in Fig. 9.5, $1 - \Pi^{33}(k)/\varepsilon_0\omega^2$ increases from a positive value for each n, starting with n_F, to $+\infty$ at the lower of these two frequencies, then decreases, crossing zero, and continuing to $-\infty$ at the higher of the two frequencies, and then increasing and crossing zero. Thus there is a doublet of solutions for each n either side of the higher of these frequencies. The dispersion relations for the longitudinal GA modes at given n can be approximated by Weise [22]

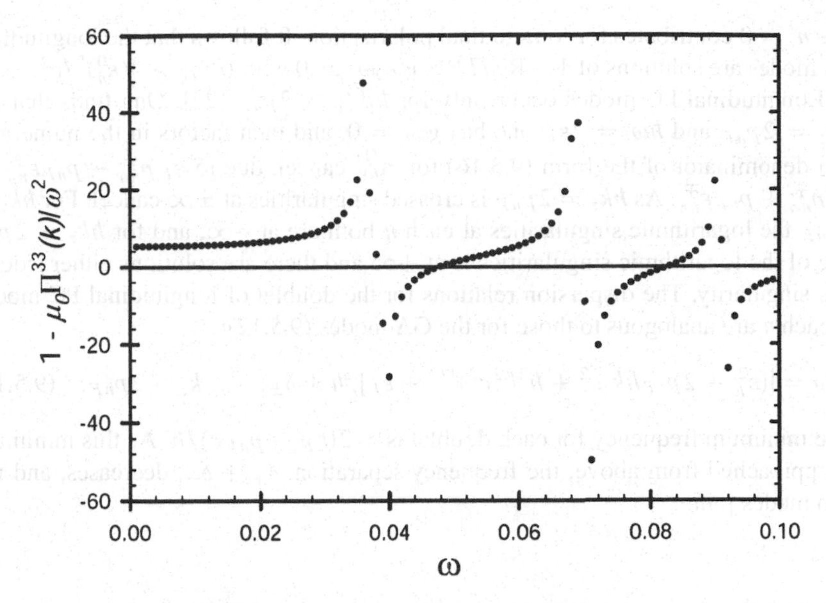

Fig. 9.6 Dispersion relations for longitudinal modes plotted for the same parameters as in Fig. 9.5 (From [22], reprinted with permission AIP)

$$\omega = [(\varepsilon_F^2 + 2p_{nF}\hbar k_z c^2 + \hbar^2 k_z^2 c^2)^{1/2} - \varepsilon_F]/\hbar \pm \delta_\pm, \tag{9.5.12}$$

where $\delta_\pm$ are the frequency differences between the singularity at $-\infty$ and the two zeros on either side of it. It is evident from Fig. 9.5 that δ_- is of order half the frequency separation between the singularities, and is smaller than δ_+. The two branches of the doublets are not resolved in Fig. 9.5.

The frequencies of each doublet of GA modes decrease with increasing n. The dispersion curves are illustrated in Fig. 9.6, where the frequency separation between the doublets is too small to be resolved. The highest frequency GA mode is for $n = 0$, and in this case there is a single mode, rather than a doublet. This mode joins onto the Langmuir-type mode at the light line [22].

The GA modes (with the exception of the Langmuir-type mode) do not have a cutoff frequency. The dispersion curves begin at $\omega = 0$ with $k_z \neq 0$,

Longitudinal PC Modes

PC modes in a magnetized electron gas were identified by Pulsifer and Kalman [14], who noted an analogy with Cooper pairs in a solid-state plasma. As with the GA modes, there can be PC modes associated with different n, n'. In a completely degenerate electron gas, there are logarithmic singularities associated with given n, n' provided that ω, k_z are in the range where PC is allowed, that is, for $(k^2)_\parallel > (\varepsilon_n^0 + \varepsilon_{n'}^0)^2$. For parallel propagation only $n - n' = 0, \pm 1$ contribute, and only

$n - n' = 0$ contributes for longitudinal polarization. It follows that the longitudinal PC modes are solutions of $1 - \mathrm{Re}\,\Pi^{33}(k)/\varepsilon_0\omega^2 = 0$ with $(k^2)_\parallel > 4(\varepsilon_n^0)^2/c^2$.

Longitudinal PC modes occur only for $\hbar|k_z| > 2p_{nF}$ [22]. One finds that for $\hbar k_z = 2p_{nF}$ and $\hbar\omega = 2\varepsilon_F$ one has $g_{nn} = 0$, and then factors in the numerator and denominator of the form (9.3.16) for $\Lambda_{nn}^{(1)}$ cancel, due to $\varepsilon_F p_{nn}^\pm - p_{nF}\varepsilon_{nn}^\pm = \varepsilon_F p_{nn}^\mp + p_{nF}\varepsilon_{nn}^\mp$. As $\hbar k_z = 2p_{nF}$ is crossed singularities at $\pm\infty$ cancel. For $\hbar k_z < 2p_{nF}$ the logarithmic singularities at each n both are at $-\infty$, and for $\hbar k_z > 2p_{nF}$ one of the logarithmic singularities is at $+\infty$ and there are solutions either side of this singularity. The dispersion relations for the doublet of longitudinal PC modes at each n are analogous to those for the GA modes (9.5.12):

$$\omega = [(\varepsilon_F^2 + 2p_{nF}\hbar k_z c^2 + \hbar^2 k_z^2 c^2)^{1/2} + \varepsilon_F]/\hbar \pm \delta_\pm, \qquad k_z > 2p_{nF}. \quad (9.5.13)$$

The minimum frequency for each doublet is $> 2(\varepsilon_F + p_{nF}c)/\hbar$. As this minimum is approached from above, the frequency separation, $\delta_+ + \delta_-$, decreases, and the two modes join.

9.5.4 Transverse Modes

Transverse waves (for $k_\perp = 0$) in a degenerate magnetized electron gas include modes that are generalizations of the cold plasma modes, absorption-edge modes, GA modes and PC modes. The GA and PC modes arise from logarithmic singularities in a similar manner to the longitudinal modes. For given n, GA is allowed for $\hbar^2\omega^2 < (\varepsilon_n^0 - \varepsilon_{n-a}^0)^2 + \hbar^2 k_z^2 c^2$, and PC is allowed for $\hbar^2\omega^2 > (\varepsilon_n^0 + \varepsilon_{n-a}^0)^2 + \hbar^2 k_z^2 c^2$, with $a = \pm 1$.

Cutoff Frequencies

Near the cutoff frequencies, the transverse modes of a relativistic degenerate electron gas reduce to counterparts of the modes of a cold classical electron gas. The transverse dispersion equation (9.5.5) for a cold classical electron gas gives, by using (1.2.29) with (1.2.38),

$$\omega^2 - k_z^2 c^2 - \frac{\omega_p^2 \omega}{\omega \mp \Omega_e} = 0, \qquad (9.5.14)$$

with the upper (lower) sign corresponding to right (left) hand polarization. The cutoff frequencies are the solutions of (9.5.14) for $k_z = 0$, and these are

$$\omega_{c\pm} = \frac{1}{2}\left[4\omega_p^2 + \Omega_e^2\right]^{1/2} \pm \frac{1}{2}\Omega_e. \qquad (9.5.15)$$

The upper and lower cutoffs are for the magnetoionic x mode and z mode, respectively. The z mode joins on continuously to the o mode at $\omega = \omega_p$ for parallel propagation. The $\pm$ solutions in (9.5.15) correspond to right and left hand polarizations in the sense that electron gyration is right hand.

The generalization of (9.5.15) to the relativistic degenerate case involves setting $k_z \to 0$ in $\Pi^{\pm}(k)$ and solving (9.5.5) for ω^2:

$$\omega^2 = \omega_{p0}^2 - \frac{e^3 B c^3}{\varepsilon_0 2\pi^2 \hbar^3 \omega^2} \sum_n \sum_{a=\pm 1} \frac{(2n-a)eB - \frac{1}{2}\hbar\omega^2/c^2}{g_{n\,n-a}}$$
$$\times \left\{ \ln \left| \frac{p_{nF} c f_{n\,n-a} - \varepsilon_F g_{n\,n-a}}{p_{nF} c f_{n\,n-a} + \varepsilon_F g_{n\,n-a}} \right| \pm a \ln \left| \frac{p_{nF} c - \hbar\omega g_{n\,n-a}}{p_{nF} c + \hbar\omega g_{n\,n-a}} \right| \right\}, \tag{9.5.16}$$

where $\omega_{p0}^2 = e^2 \bar{n}_{\rm pr}/\varepsilon_0 m$ is the proper plasma frequency. The leading terms in the limit in which $f_{n\,n-a}$, $g_{n\,n-a}$ are large in magnitude give

$$f_{n\,n-a} = a\frac{\Omega_{0n}\varepsilon_n^0}{\hbar\omega^2} + \frac{1}{2} \approx a\frac{\Omega_{0n}\varepsilon_n^0}{\hbar\omega^2}, \qquad g_{n\,n-a}^2 = -f_{n\,n-a}^2 \left(\frac{\omega^2 - \Omega_{0n}^2}{\Omega_{0n}^2} \right), \tag{9.5.17}$$

with $\Omega_{0n} = eBc^2/\varepsilon_n^0$. Formally, the logarithmic functions in (9.5.16) are replaced by arctangents for $g_{n\,n-a}^2 < 0$, but no error is introduced by retaining the logarithmic forms when $g_{n\,n-a}$ is imaginary.

Simplification of the logarithmic factors occurs for $p_{nF} c \ll \varepsilon_F, \varepsilon_n^0$. Retaining only the leading term in an expansion in p_{nF}, (9.5.16) gives

$$\omega^2 = \omega_{p0}^2 \mp \frac{e^3 B}{\varepsilon_0 2\pi^2 \hbar^2} \sum_n a_n p_{nF} \frac{\omega \pm \Omega_F}{\omega^2 - \Omega_{0n}^2}, \tag{9.5.18}$$

with $\Omega_F = eBc^2/\varepsilon_F$. In the nonrelativistic limit, $\Omega_{0n}, \Omega_F \to \Omega_e$, (9.5.18) reduces to the cold plasma form, (9.5.14) with $k_z = 0$, that leads to the solutions (9.5.15). In the general case, the cutoffs are given by the solutions of (9.5.16) with (9.5.17).

The dispersion curves for the $\pm$ modes are qualitatively similar to the cold plasma limit. These curves are plotted in Figs. 9.7 and 9.8, but the scale is such that the difference between these curves and the light line is not resolved. As in the cold-plasma limit, the curves begin at the cutoff frequencies for $k_z = 0$ and asymptote to the light line for large k_z.

GA Edge Modes

There is an additional class of transverse modes that start at (or near) cutoff frequencies. These modes are near the absorption edge, where $g_{nn'}$ passes through zero. The allowed regions for GA correspond to $(k^2)_\parallel < (\varepsilon_n^0 - \varepsilon_{n'}^0)^2/\hbar^2 c^2$, and the

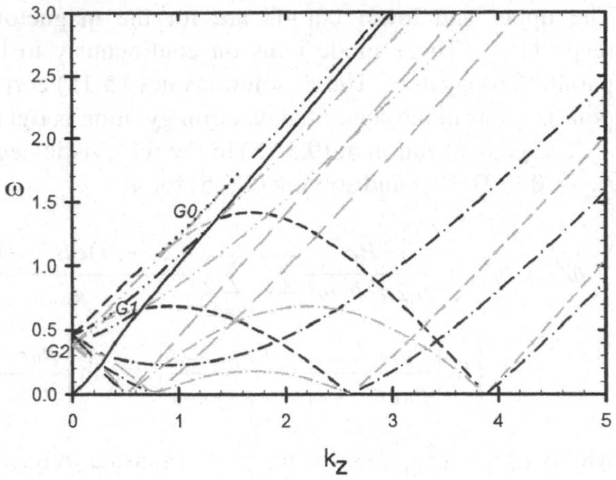

Fig. 9.7 Dispersion relations for transverse modes with *right hand* polarization plotted for the same parameters as in Fig. 9.5 (From [22], reprinted with permission AIP)

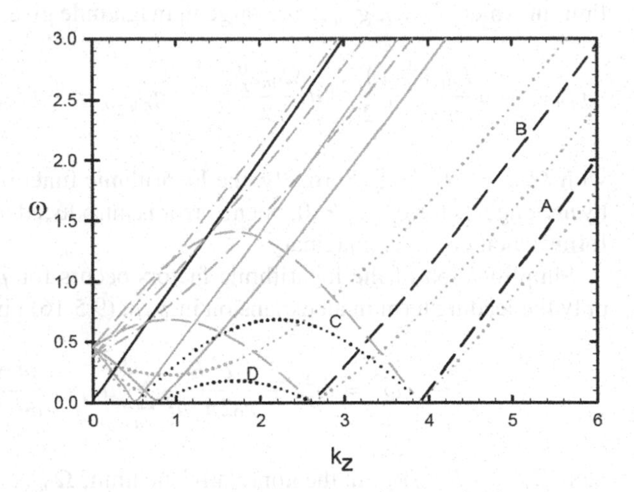

Fig. 9.8 Dispersion relations for transverse modes with *left hand* polarization plotted for the same parameters as in Fig. 9.5 (From [22], reprinted with permission AIP)

allowed regions for PC correspond to $(k^2)_\parallel > (\varepsilon_n^0 + \varepsilon_{n'}^0)^2/\hbar^2c^2$. At an absorption edges, $g_{nn'} = 0$ implies $p_{nn'}^+ = p_{nn'}^- = k_z f_{nn'}$, $\varepsilon_{nn'}^+ = -\varepsilon_{nn'}^- = \omega f_{nn'}$. Zeros in the arguments of the RPDF (9.3.16) for a completely degenerate distribution occur at $\pm p_{nF} = \hbar k_z \varepsilon_n^0/(\varepsilon_n^0 - \varepsilon_{n'}^0)$. Consider a given frequency that is slightly larger than $(\varepsilon_n^0 - \varepsilon_{n'}^0)/\hbar$, with $n' = n - 1$. For $k_z = 0$, $g_{nn'}$ is imaginary, and (this particular) GA is not allowed. As k_z increases at constant ω, one can cross the absorption edge and enter the region where GA is allowed and the RPDFs have logarithmic singularities. This allows a doublet of modes to exist, provided the singularities have the appropriate sign, which needs to be considered separately for each case. By varying ω and looking for these solutions as a function of k_z, one can identify the dispersion relations for these absorption edge modes.

Absorption edge modes occur only for + polarization, and there is one such mode for each of $n = 0, 1, \ldots, n_F$. These doublet modes exist only over

$p_{n+1\,F}(\varepsilon^0_{n+1}-\varepsilon^0_n)/\varepsilon^0_{n+1} < \hbar k_z < p_{nF}(\varepsilon^0_{n+1}-\varepsilon^0_n)/\varepsilon^0_{n+1}$ for $1 \leq n \leq n_F - 1$ and $\hbar k_z < p_{nF}(1-\varepsilon^0_{n+1}/\varepsilon^0_n)$ for $n = n_F$. These modes, as with the GA and PC modes, are doublets, with the separation between the doublet being unresolved on the scale of the figure. The dispersion curves are shown in Fig. 9.7 as the dotted curves labeled $G0, G1, G2$, with $n_F = 2$ in this case. The range over which these modes exist is only a short portion of what appears to be a continuous curve, changing from dashed to dotted to dashed-double-dotted. Only the dotted portions are the absorption edge modes, and the two arms of the doublet join at the ends of the dotted portions. The dotted portion is obvious for $G0$, but is barely resolved for $G2$.

Transverse GA Modes

As for the longitudinal modes, the transverse GA modes are associated with logarithmic singularities, and are doublets. There are two classes of GA modes, determined by solutions of (9.5.5) with the + and − signs, respectively. These correspond to right and left hand polarizations, and it is in convenient to label them as the GA± modes, respectively.

The GA modes occur near logarithmic singularities, which occur at frequencies $\omega_{G1,2}$, given by (9.3.24). It is convenient to rewrite these as

$$\omega_{G1,2}(n,a) = |(\varepsilon_F^2 + 2aeB\hbar c^2 \pm 2p_{nF}\hbar k_z c^2 + \hbar^2 k_z^2 c^2)^{1/2} - \varepsilon_F|/\hbar, \quad (9.5.19)$$

with $a = n - n' = \pm 1$ here. A particular logarithmic singularity then results in a GA mode doublet only if it has the appropriate sign to lead to a solution of (9.5.5).

The dispersion relations for the GA± modes are illustrated in Figs. 9.7 and 9.8, respectively. The parameters chosen are $p_F = 2.2\,mc$ ($\varepsilon_F = 2.4\,mc^2$) and $B = B_c$, and these imply $\varepsilon^0_n = (1+2n)^{1/2}mc^2$ and hence $n_F = 2$. The allowed Landau states, n and $n' = n - a$ with $a = \pm 1$, are $n = 0$, $n' = 1$; $n = 1$, $n' = 0, 2$; $n = 2$, $n' = 1, 3$. The solid black lines in Figs. 9.7 and 9.8 represent the modes $\omega_\pm$, and the dashed and dotted dark lines represent the GA± modes, respectively. Each GA± mode is a doublet and the separation between the two modes in each doublet is unresolved in the figures. The gray lines correspond to the frequencies (9.5.19), where logarithmic singularities occur. The frequencies $\omega_{G1}(n,a)$, $\omega_{G2}(n,a)$ are equal for $k_z = 0$, and they initially increase and decrease, respectively, with increasing k_z. There is an extremum of $\omega_{G2}(n,a)$ at $\hbar k_z = p_{nF}$ ($p_{2F} = 0.9$, $p_{1F} = 1.7$, $p_{0F} = 2.2$) and this is a minimum ($n = 2$) or a maximum ($n = 1, 0$). A maximum occurs if $\omega_{G2}(n,a)$ passes through zero, with the modulus sign in (9.5.19) requiring that the curve reflects at the line $\omega = 0$; the zeros occur at $\hbar k_z = p_{nF} - p_{n+1\,F}$.

For the GA− modes in Fig. 9.8, the dotted lines labelled C and D effectively overlie the curves showing $\omega_{G2}(n,1)$ in (9.5.19) for $n = 0$ and $n = 1$, respectively. The long-dashed lines labelled A and B effectively overlie the curves showing $\omega_{G2}(n,-1)$ for $n = 0$ and $n = 1$, respectively. These modes all occur below the light line.

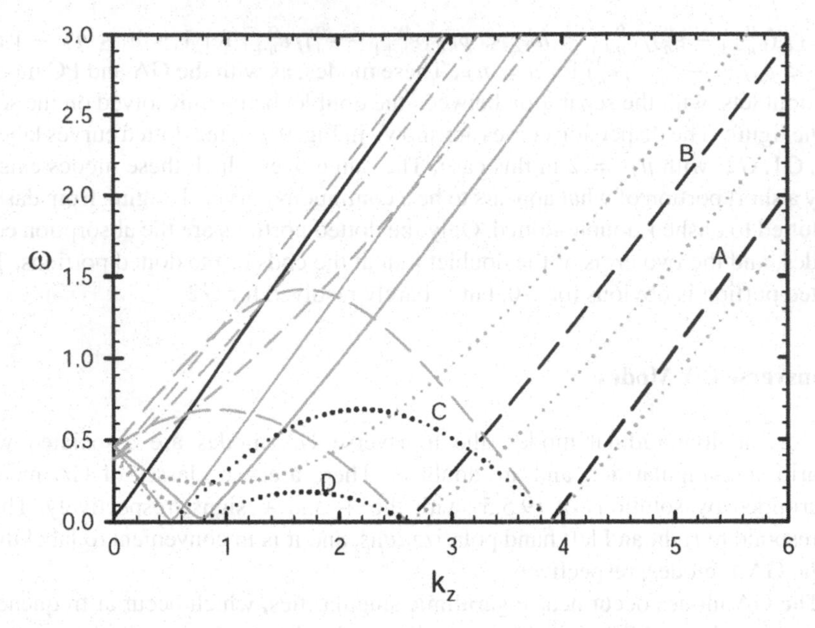

Fig. 9.9 Transverse PC modes for the same parameters as in Fig. 9.5 (From [22], reprinted with permission AIP)

The GA+ modes, shown by the dashed dark lines in Fig. 9.7, occur both above and below the light line. The modes are doublets which are unresolved and are well-approximated the frequency $\omega_{G1,2}(n,a)$ corresponding to the relevant logarithmic singularity. A given curve $\omega_{G2}(n,a)$ can correspond to more than one mode, with stop bands separating the modes. An example is the curve labeled $G0$, which is above the light line and extends to infinity; this curve corresponds to three separate modes, with no solution near $k_z = p_{n+1\,F}(\varepsilon^0_{n+1} - \varepsilon^0_n)/\varepsilon^0_{n+1}$ and $k_z = p_{nF}(\varepsilon^0_{n+1} - \varepsilon^0_n)/\varepsilon^0_{n+1}$.

Transverse PC Modes

Transverse PC modes are doublets associated with logarithmic singularities. The relevant singularities occur at the frequencies ω_{P2}, given by (9.3.24), and rewritten here as

$$\omega_{P1,2}(n,a) = \left[\left(\varepsilon_F^2 + 2aeB\hbar c^2 \pm 2p_{nF}\hbar k_z c^2 + \hbar^2 k_z^2 c^2\right)^{1/2} + \varepsilon_F\right]/\hbar. \quad (9.5.20)$$

The singularities correspond to modes only over restricted ranges of k_z. The frequencies (9.5.20) are plotted as dashed gray curves in Fig. 9.8. There are zeros of $\omega_{P2}(n,a)$, which occur at $\hbar k_z = p_{nF} + p_{n-a\,F}$ ($k_z = 2.6, 3.9$).

The dispersion curves for the PC modes are shown as the dark portions of the curves in Fig. 9.9. The dotted lines correspond to $\omega_{P2}(n,1)$ for the − mode, with the frequencies increasing with $n = 0, 1, 2$. The dashed lines correspond to $\omega_{P2}(n, -1)$ for the $+$ mode, with the frequencies increasing with $n = 1, 2$. The gray portions denote the PC threshold where no modes exist.

9.5.5 *Discussion of GA and PC Modes*

The GA and PC modes, which are uniquely relativistic quantum plasma phenomena, have been investigated only under very restrictive conditions [3, 22], in particular, for parallel propagation and in the absence of dissipation. The following qualitative remarks concern the implications of relaxing these restrictions.

Dissipation of the $\pm$ modes, and of the GA-edge, GA and PC modes, occurs at a given $n, n' = n - a$ if the dispersion curve is in region where the absorption is allowed. The GA-edge, GA and PC modes are associated with specific values of n, a, and absorption can be associated with this n, a, or at any other n, a that is allowed. GA is restricted to $(k^2)_\parallel < 0$ and PC to $(k^2)_\parallel > (\varepsilon_n^0)^2$: PC is forbidden for GA modes, and GA is forbidden for PC modes. The GA and PC modes are associated with specific logarithmic singularities, and these occur only when the absorption is allowed, $g^2_{n\,n-a} > 0$. For these modes, the absorption associated with the values of n, a that define the mode can never be zero. Formally, absorption can be treated by including the antihermitian part of the response tensor in the dispersion equation, (9.5.4) or (9.5.5), and searching for complex solutions for ω. In the absence of absorption, the solutions are doublets. On including absorption, these solutions can be interpreted as modes only if the imaginary part of the frequency is less than the doublet separation.

On relaxing the assumption of parallel propagation, the dispersion equation no longer factorizes into three equations, for longitudinal and oppositely circularly polarized transverse modes. Solutions of the general dispersion equation then have polarizations that have both longitudinal and transverse components, with the transverse component being elliptical in general. Although this greatly increases the algebraic complexity, an important feature does not change. The location of the logarithmic singularities, which determine the GA and PC modes, is independent of $k_\perp$. Hence, for $k_\perp \neq 0$, one expects there to be GA and PC modes with dispersion curves in ω–k_z space similar to the dispersion curves for the modes identified above. These dispersion curves are well approximated by $\omega_{G1,2}(n, a)$ and $\omega_{P1,2}(n, a)$, with $a = 0, \pm 1$ and $0 \leq n \leq n_F$. The restrictions on the allowed ranges of k_z are also independent of $k_\perp$, and hence one expects these dispersion curves to be closely analogous to those shown in Figs. 9.6–9.9.

The physical implications of the existence of these modes has not been explored in detail. In particular, the PC modes can exist only under exotic conditions, and can be excited and dissipated only through PC. An interesting additional effect, that has not be discussed in this context, is the existence of bound (positronium) states § 6.5, which should introduce a qualitatively different singularity into the dispersion equation.

9.6 Response of a Spin-Polarized Electron Gas

The results for $\Pi^{\mu\nu}(k)$ derived in § 9.1 and § 9.2 apply to an unpolarized electron gas. In a spin-polarized electron gas the average (over the electron gas) value of the spin is nonzero. In a magnetic field, the only spin eigenstates that do not precess are those of the magnetic-moment operator. A net spin then corresponds to a net magnetization of the electron gas. A nonzero magnetization leads to an additional contribution to the response tensor, $\Pi^{\mu\nu}_{\rm m}(k)$, that describes the spin-dependent part of the response. In this section, this contribution to the response tensor is calculated assuming that the occupation number depends on the spin, s, with $s = \pm 1$ labeling the eigenstates of the magnetic-moment operator.

9.6.1 Spin-Dependent Occupation Number

In general, the occupation number, $n^\epsilon_q \to n^\epsilon_{n,s}(p_z)$, is different for the two spin states, $s = \pm 1$. Introducing the sum and difference

$$\bar{n}^\epsilon_n(p_z)=\tfrac{1}{2}\left[n^\epsilon_{n,+}(p_z)+n^\epsilon_{n,-}(p_z)\right], \quad \Delta n^\epsilon_n(p_z)=\tfrac{1}{2}\left[n^\epsilon_{n,+}(p_z)-n^\epsilon_{n,-}(p_z)\right], \tag{9.6.1}$$

the dependence on s is made explicit by writing

$$n^\epsilon_{ns}(p_z) = \bar{n}^\epsilon_n(p_z) + s\Delta n^\epsilon_n(p_z). \tag{9.6.2}$$

A 'spin-independent' electron gas means $\Delta n^\epsilon_n(p_z) = 0$. An exception is for $n = 0$, for which there is a unique spin state, $s = -1$, implying that $n^\epsilon_{0+}(p_z)$ is identically zero.

The magnetic-moment eigenvalues for an electron or positron are $\frac{1}{2}g\mu_B\epsilon s$, where $g = 2.00232\ldots$ is the gyromagnetic ratio, and $\mu_B = e/\hbar 2m$ is the Bohr magneton. A spin-dependent electron gas has a nonzero magnetization, $\boldsymbol{M}$, directed along the magnetic field. The contributions of the electrons and positrons in the nth Landau state to the number density, N^ϵ_n, and magnetization, M^ϵ_n, are given by

$$\begin{pmatrix} N^\epsilon_n \\ M^\epsilon_n \end{pmatrix} = \frac{eB}{2\pi}\int\frac{dp_z}{2\pi}\begin{pmatrix} 2\bar{n}^\epsilon_n(p_z) \\ g\mu_B\epsilon\Delta n^\epsilon_n(p_z) \end{pmatrix}, \tag{9.6.3}$$

respectively. Thus the spin-independent part of the response tensor is proportional to N^ϵ_n and the spin-dependent part is proportional to M^ϵ_n.

9.6.2 General Forms for $\Pi_{\rm m}^{\mu\nu}(k)$

The derivation of the spin-dependent part of the response tensor involves a straightforward generalization of the derivations in § 9.1 and § 9.2.

Sum over s, s'

On inserting (9.6.2) into (9.1.1), there are three different sums over s, s':

$$\frac{1}{2\epsilon'\epsilon\varepsilon'_{n'}\varepsilon_n}\begin{pmatrix}[C_{n'n}(\epsilon' p'_\parallel, \epsilon p_\parallel, k)]^{\mu\nu}\\ [D_{n'n}(\epsilon' p'_\parallel, \epsilon p_\parallel, k)]^{\mu\nu}\\ [D'_{n'n}(\epsilon' p'_\parallel, \epsilon p_\parallel, k)]^{\mu\nu}\end{pmatrix} = \sum_{s',s}\begin{pmatrix}1\\ s\\ s'\end{pmatrix}[\Gamma^{\epsilon'\epsilon}_{q'q}(\boldsymbol{k})]^\mu[\Gamma^{\epsilon'\epsilon}_{q'q}(\boldsymbol{k})]^{*\nu}. \tag{9.6.4}$$

The first sum applies to the spin-independent part, and is given by (5.4.24). The other two sums in (9.6.4) contribute only to the spin-dependent part. Explicit evaluation gives

$$\begin{aligned}\Big[D_{n'n}\big(P'_\parallel, P_\parallel, k\big)\Big]^{\mu\nu} = \frac{m}{\varepsilon_n^0}\Big(&\Big\{P'^\mu_\parallel P^\nu_\parallel + P'^\nu_\parallel P^\mu_\parallel - \Big[(P'P)_\parallel - \big(\varepsilon_n^0\big)^2\Big]g^{\mu\nu}_\parallel\Big\}\,[(J^{n-1}_{n'-n})^2 - (J^n_{n'-n})^2]\\ &-\Big\{[(P'P)_\parallel - (\varepsilon_n^0)^2]\,g^{\mu\nu}_\perp + P^\mu_\parallel k^\nu_\perp + P^\nu_\parallel k^\mu_\perp\Big\}](J^{n-1}_{n'-n+1})^2 - (J^n_{n'-n-1})^2]\\ &+i\Big\{[(P'P)_\parallel - (\varepsilon_n^0)^2]f^{\mu\nu} + P^\mu_\parallel k^\nu_G - P^\nu_\parallel k^\mu_G][(J^{n-1}_{n'-n+1})^2 + (J^n_{n'-n-1})^2\Big\}\Big),\end{aligned} \tag{9.6.5}$$

with the other tensor implied by (9.6.5) and the symmetry property

$$[D'_{n'n}(P'_\parallel, P_\parallel, k)]^{\mu\nu} = [D_{nn'}(-P_\parallel, -P'_\parallel, k)]^{\nu\mu}. \tag{9.6.6}$$

Summed Form of $\Pi_{\rm m}^{\mu\nu}(k)$

The spin-dependent contribution to the response tensor in the form (9.1.10) is

$$\begin{aligned}\Pi_{\rm m}^{\mu\nu}(k) = -\frac{e^3 B}{2\pi}\sum_{n,n'}\Bigg[&\int\frac{dp_z}{2\pi}\sum_\epsilon \epsilon\Delta n_n^\epsilon(p_z)\frac{\big[D_{n'n}(\epsilon p_\parallel - k_\parallel, \epsilon p_\parallel, k)\big]^{\mu\nu}}{2\varepsilon_n[\epsilon(pk)_\parallel - (pk)_{nn'}]}\\ &-\int\frac{dp'_z}{2\pi}\sum_{\epsilon'}\epsilon'\Delta n_{n'}^{\epsilon'}(p'_z)\frac{\big[D'_{n'n}(\epsilon' p'_\parallel, \epsilon' p'_\parallel + k_\parallel, k)\big]^{\mu\nu}}{2\varepsilon'_{n'}[\epsilon'(p'k)_\parallel - (p'k)_{nn'}]}\Bigg].\end{aligned} \tag{9.6.7}$$

As in the spin-independent case, one may rewrite (9.6.7) in a variety of ways. By making the replacements $n', p'_z \to n, p_z$ and $\epsilon' \to -\epsilon$ in the second integral in

(9.6.7), one derives a form analogous to the summed form (9.1.19):

$$\Pi_{\rm m}^{\mu\nu}(k) = -\frac{e^3 B}{8\pi^2}\sum_{\eta,n,n'}\int\frac{dp_z}{\varepsilon_n}\frac{1}{\eta(pk)_\parallel - (pk)_{nn'}}$$
$$\times\{\eta\Delta n_n^{\rm sum}(p_z)[\Delta N_{n'n}(\eta p_\parallel,k)]^{\mu\nu} + \Delta n_n^{\rm diff}(p_z)[\Delta G_{n'n}(\eta p_\parallel,k)]^{\mu\nu}\}, \tag{9.6.8}$$

where the sign ϵ is replaced by η, which can no longer be interpreted as labeling electron and positron states. In place of (9.1.16) one has

$$[D_{n'n}(\eta p_\parallel - k_\parallel, \eta p_\parallel, k)]^{\mu\nu} = [\Delta N_{n'n}(\eta p_\parallel,k)]^{\mu\nu} + [\Delta G_{n'n}(\eta p_\parallel,k)]^{\mu\nu}, \tag{9.6.9}$$

with the numerator in the second term becoming $[D_{n'n}(\eta p_\parallel - k_\parallel, \eta p_\parallel, k)]^{\nu\mu}$, which is equal to $[\Delta N_{n'n}(\eta p_\parallel,k)]^{\mu\nu} - [\Delta G_{n'n}(\eta p_\parallel,k)]^{\mu\nu}$. Explicit forms for the tensors in the numerator of (9.6.8) follow from (9.6.5), (9.6.6), and (9.6.9):

$$[\Delta N_{n'n}(\eta p_\parallel,k)]^{\mu\nu} = \frac{m}{\varepsilon_n^0}$$
$$\times\Big\{[2p_\parallel^\mu p_\parallel^\nu - \eta(p_\parallel^\mu k_\parallel^\nu + k_\parallel^\mu p_\parallel^\nu) + \eta(pk)_\parallel g_\parallel^{\mu\nu}][(J_{n'-n}^{n-1})^2 - (J_{n'-n}^{n})^2]$$
$$+[\eta(pk)_\parallel g_\perp^{\mu\nu} - \eta p_\parallel^\mu k_\perp^\nu - \eta p_\parallel^\nu k_\perp^\mu][(J_{n'-n+1}^{n-1})^2 - (J_{n'-n-1}^{n})^2]\Big\},$$
$$[\Delta G_{n'n}(\eta p_\parallel,k)]^{\mu\nu} = -i\frac{m}{\varepsilon_n^0}\{[\eta(pk)_\parallel f^{\mu\nu} - \eta p_\parallel^\mu k_G^\nu + \eta p_\parallel^\nu k_G^\mu]$$
$$\times[(J_{n'-n+1}^{n-1})^2 + (J_{n'-n-1}^{n})^2]\}. \tag{9.6.10}$$

Spin-Dependent Response in Terms of RPDFs

The spin-dependent part, $\Pi_{\rm m}^{\mu\nu}(k)$, of the response tensor can be rewritten in terms of the RPDFs introduces in § 9.2. The counterpart of (9.2.10) is

$$\Pi_{\rm m}^{\mu\nu}(k) = \frac{e^3 B}{16\pi^2 (k^2)_\parallel}\sum_{\eta,n,n'}\int\frac{dp_z}{\varepsilon_n}\frac{\eta(\varepsilon_n\omega + p_z k_z) + (pk)_{nn'}}{(\eta p_z - p_{nn'}^+)(\eta p_z - p_{nn'}^-)}$$
$$\times\{\eta\Delta n_n^{\rm sum}(p_z)[\Delta N_{n'n}(\eta p_\parallel,k)]^{\mu\nu}$$
$$+\Delta n_n^{\rm diff}(p_z)[\Delta G_{n'n}(\eta p_\parallel,k)]^{\mu\nu}\}, \tag{9.6.11}$$

As for the spin-independent part, (9.6.11) can be evaluated in terms of a single RPDF, defined by analogy with (9.2.13) as

$$\Delta K_n^\epsilon(\eta p_{\parallel\pm}) = \int dp_z\,\frac{\Delta n_n^\epsilon(p_z)}{\varepsilon_n}\frac{\eta\varepsilon_n + \varepsilon_{nn'}^\pm}{\eta p_z - p_{nn'}^\pm}. \tag{9.6.12}$$

This RPDF can be expressed in terms of RPDFs defined by analogy with (9.2.28),

$$\Delta Z_n^\epsilon(t_0) = \int_{-1}^{1} dt\, \frac{\Delta n_n^\epsilon(p_z)}{t - t_0}, \tag{9.6.13}$$

giving, in place of (9.2.29),

$$K_n^\epsilon(\eta p_{\parallel\pm}) = (\eta + 1) Z_n^\epsilon(t_{nn'}^\pm) - (\eta - 1) Z_n^\epsilon(1/t_{nn'}^\pm). \tag{9.6.14}$$

The extra sign η, in the integrand of (9.6.11) compared with (9.2.10), reverses the signs of the terms with argument $1/t_{nn'}^\pm$ in the counterpart of (9.2.37).

The resulting expression for the spin-dependent part of the response tensor is

$$\begin{aligned} \Pi_{\rm m}^{\mu\nu}(k) = {} & \frac{e^3 B}{8\pi^2 (k^2)_\parallel} \sum_{n,n'} \frac{1}{g_{nn'}} \\ & \times \Big\{ [\Delta N_{n'n}(k)]_+^{\mu\nu} Z_{nn'}^{(1)}(\omega, k_z) + [\Delta N_{n'n}(k)]_-^{\mu\nu} Z_{nn'}^{(2)}(\omega, k_z) \\ & + [\Delta G_{n'n}(k)]_+^{\mu\nu} Z_{nn'}^{(3)}(\omega, k_z) + [\Delta G_{n'n}(k)]_-^{\mu\nu} Z_{nn'}^{(4)}(\omega, k_z) \Big\}, \end{aligned} \tag{9.6.15}$$

where there is no spin-dependent contribution to the ND part. The RPDFs are

$$\begin{bmatrix} \Delta Z_{nn'}^{(1)}(\omega, k_z) \\ \Delta Z_{nn'}^{(2)}(\omega, k_z) \\ \Delta Z_{nn'}^{(3)}(\omega, k_z) \\ \Delta Z_{nn'}^{(4)}(\omega, k_z) \end{bmatrix} = \begin{bmatrix} \Delta Z_n^{\rm sum}(t_{nn'}^+) - \Delta Z_n^{\rm sum}(1/t_{nn'}^+) - \Delta Z_n^{\rm sum}(t_{nn'}^-) + \Delta Z_n^{\rm sum}(1/t_{nn'}^-) \\ \Delta Z_n^{\rm sum}(t_{nn'}^+) - \Delta Z_n^{\rm sum}(1/t_{nn'}^+) + \Delta Z_n^{\rm sum}(t_{nn'}^-) - \Delta Z_n^{\rm sum}(1/t_{nn'}^-) \\ \Delta Z_n^{\rm diff}(t_{nn'}^+) + \Delta Z_n^{\rm diff}(1/t_{nn'}^+) - \Delta Z_n^{\rm diff}(t_{nn'}^-) - \Delta Z_n^{\rm diff}(1/t_{nn'}^-) \\ \Delta Z_n^{\rm diff}(t_{nn'}^+) + \Delta Z_n^{\rm diff}(1/t_{nn'}^+) + \Delta Z_n^{\rm diff}(t_{nn'}^-) + \Delta Z_n^{\rm diff}(1/t_{nn'}^-) \end{bmatrix}. \tag{9.6.16}$$

The tensors labeled $\pm$ are the parts of (9.6.10) that are evaluated at $\eta p_\parallel^\mu = p_{\parallel\pm}^\mu$, and separated into the parts that are even and odd, respectively, under the interchange of the $\pm$ solutions, as for the spin-independent tensors (9.2.34). Explicit forms for these tensors are

$$\begin{aligned} [\Delta N_{n'n}(k)]_+^{\mu\nu} = {} & \frac{m}{\varepsilon_n^0} [(k_\parallel^\mu k_\parallel^\nu + k_D^\mu k_D^\nu) f_{nn'}(2 f_{nn'} - 1) - 2 k_D^\mu k_D^\nu (\varepsilon_n^0)^2/(k^2)_\parallel] \\ & \times [(J_{n'-n}^{n-1})^2 - (J_{n'-n}^{n})^2] \\ & + f_{nn'} \{(k^2)_\parallel g_\perp^{\mu\nu} - k_\parallel^\mu k_\perp^\nu - k_\parallel^\nu k_\perp^\mu\} [(J_{n'-n+1}^{n-1})^2 - (J_{n'-n-1}^{n})^2], \\ [\Delta N_{n'n}(k)]_-^{\mu\nu} = {} & \frac{m}{\varepsilon_n^0} g_{nn'} \{(2 f_{nn'} - 1)(k_\parallel^\mu k_D^\nu + k_\parallel^\nu k_D^\mu) [(J_{n'-n}^{n-1})^2 - (J_{n'-n}^{n})^2] \\ & - (k_D^\mu k_\perp^\nu + k_D^\nu k_\perp^\mu) [(J_{n'-n+1}^{n-1})^2 - (J_{n'-n-1}^{n})^2], \}, \end{aligned}$$

$$[\Delta G_{n'n}(k)]^{\mu\nu}_{+} = -i\frac{m}{\varepsilon^0_n} f_{nn'}\{(k^2)_\parallel f^{\mu\nu} - k^\mu_\parallel k^\nu_G + k^\nu_\parallel k^\mu_G\}$$
$$\times[(J^{n-1}_{n'-n+1})^2 + (J^n_{n'-n-1})^2],$$
$$[\Delta G_{n'n}(k)]^{\mu\nu}_{-} = i\frac{m}{\varepsilon^0_n} g_{nn'}\{k^\mu_D k^\nu_G - k^\nu_D k^\mu_G\}[(J^{n-1}_{n'-n+1})^2 + (J^n_{n'-n-1})^2]. \quad (9.6.17)$$

The term proportional to $(J^{n-1}_{n'-n})^2 - (J^n_{n'-n})^2$ can be rewritten using the identity (A.1.37), giving

$$(2f_{nn'} - 1)[(J^{n-1}_{n'-n})^2 - (J^n_{n'-n})^2] = \frac{-k_\perp^2}{(k^2)_\parallel}[(J^{n-1}_{n'-n+1})^2 - (J^n_{n'-n-1})^2] \quad (9.6.18)$$

This replacement is convenient in the small-x approximation, where the left hand side of (9.6.18) vanishes to lowest order in x, requiring that higher order terms be retained. Use of (9.6.18) gives the relevant limiting form directly.

The combination of the spin-independent part (9.2.38) and the spin-dependent part (9.6.15) gives a general expression for the response tensor for an arbitrary (magnetized relativistic quantum) electron gas.

9.6.3 Small-x Approximation to $\Pi^{\mu\nu}_{\mathrm{m}}(k)$

The small-x approximation is discussed in § 9.4.2. The small-x approximation to the spin-dependent contribution (9.6.16) to the response tensor involves only the tensors (9.6.18), and it allows one to derive an approximate form for $\Pi^{\mu\nu}_{\mathrm{m}}(k)$ that is of relatively wide validity. In the small-x approximation the J-functions, in (9.6.10) or (9.6.17), are approximated by the leading term in the expansion in x. Only $n' - n = \pm 1$ need be retained, with $n' = n$ and $|n' - n| \geq 2$ not contributing to lowest order in an expansion in $\hbar$.

The terms in (9.6.18) that involve $(J^{n-1}_{n'-n+1})^2 \pm (J^n_{n'-n-1})^2$ are approximated by $\delta_{n',n-1} \pm \delta_{n',n+1}$. The term that involves $(J^{n-1}_{n'-n})^2 - (J^n_{n'-n})^2$ gives a contribution of order x for $n' - n = \pm 1$,

$$[J^{n-1}_{n'-n}(x)]^2 - [J^n_{n'-n}(x)]^2 \approx -x(\delta_{n',n-1} + \delta_{n',n+1}). \quad (9.6.19)$$

One may derive (9.6.19) either by using the expansion (A.1.51), or by first applying the identity (A.1.37), in the form

$$f_{nn'}[(J^{n-1}_{n'-n})^2 - (J^n_{n'-n})^2] = \frac{-k_\perp^2}{2(k^2)_\parallel}[(J^{n-1}_{n'-n+1})^2 - (J^n_{n'-n-1})^2]$$
$$+\frac{1}{2}[(J^{n-1}_{n'-n})^2 - (J^n_{n'-n})^2], \quad (9.6.20)$$

to (9.6.17). The resulting term $\propto k_\perp^2$ is of the same order in $\hbar$ as the other leading terms in (9.6.17). The small-x approximation to (9.6.17) then gives

$$
\begin{aligned}
[\Delta N_{n'n}(k)]_+^{\mu\nu} &= \frac{m}{\varepsilon_n^0}(-k_\perp^2)\frac{k_\parallel^\mu k_\parallel^\nu + k_D^\mu k_D^\nu}{(k^2)_\parallel}(\delta_{a,1}+\delta_{a,-1}) \\
&\quad + \frac{m}{\varepsilon_n^0}\left[aeB + \frac{1}{2}(k^2)_\parallel\right]\left[g_\perp^{\mu\nu} - \frac{k_\parallel^\mu k_\perp^\nu + k_\parallel^\nu k_\perp^\mu}{(k^2)_\parallel}\right](\delta_{a,1}-\delta_{a,-1}), \\
[\Delta N_{n'n}(k)]_-^{\mu\nu} &= -\frac{m}{\varepsilon_n^0} g_{nn'}(k_D^\mu k_\perp^\nu + k_D^\nu k_\perp^\mu)(\delta_{a,1}-\delta_{a,-1}), \\
[\Delta G_{n'n}(k)]_+^{\mu\nu} &= -i\frac{m}{\varepsilon_n^0}\left[aeB + \frac{1}{2}(k^2)_\parallel\right]\left[f^{\mu\nu} - \frac{k_\parallel^\mu k_G^\nu - k_\parallel^\nu k_G^\mu}{(k^2)_\parallel}\right](\delta_{a,1}+\delta_{a,-1}), \\
[\Delta G_{n'n}(k)]_-^{\mu\nu} &= i\frac{m}{\varepsilon_n^0} g_{nn'}(k_D^\mu k_G^\nu - k_D^\nu k_G^\mu)(\delta_{a,1}+\delta_{a,-1}).
\end{aligned}
\tag{9.6.21}
$$

The small-x approximation does not affect the RPDFs.

9.6.4 Reduction to the Cold-Plasma Limit

Some physical implications of spin dependence in an electron gas are discussed in the context of a quasi-classical treatment of the spin § 1.5, and the comments here are restricted to two aspects: the validity of the cold-plasma model and the validity of the quasi-classical treatment of spin. In brief, there are unresolved logical inconsistencies between the spin-dependent relativistic quantum result and quasi-classical models for the inclusion of spin.

A specific form for the response tensor for a spin-dependent cold electron gas is derived in § 1.5.4 using a covariant quasi-classical model (the BMT model) for the spin. It is of interest to compare the resulting form (1.5.20) for $\Pi_{\mathrm{m}}^{\mu\nu}(k)$ for a cold electron gas [10], with the relativistic quantum results. The relevant approximations to the relativistic quantum result include the small-x approximation (9.6.21), the neglect of the quantum recoil, and the cold-plasma approximation. However, the relativistic quantum forms (9.6.8) or (9.6.15) do not reproduce the quasi-classical result (1.5.20) in any obvious way. In the spin-independent case, the known cold-plasma limit is reproduced. In the spin-dependent case, the leading terms in the numerator can all be regarded as quantum corrections, and the cold-plasma limits of either (9.6.8) with (9.6.10) or (9.6.15) with (9.6.16) and (9.6.21) lead to different results.

The failure of the relativistic quantum expression to reproduce the quasi-classical result suggests that the quasi-classical theory for spin is inadequate in this context. In a quasi-classical theory the spin is represented by a vector s, and no distinction is made between different spin operators. This is consistent with the Schrödinger-Pauli theory, in which the spin and the dynamics are independent, but it is not consistent with Dirac's theory. This difference can be illustrated by the role that the magnetic-moment tensor plays in the two theories. In the quasi-classical theory, the magnetic-moment tensor (1.5.13), viz. $m^{\mu\nu} = -\frac{1}{2}g\mu_B\epsilon^{\mu\nu\alpha\beta}s_\alpha u_\beta$, is identified with the spin-dependent part of the response of the medium, whose induced 4-magnetization is $M^{\mu\nu} = n_e m^{\mu\nu}$, which is used to identify $\Pi_{\rm m}^{\mu\nu}(k)$ in the derivation of (1.5.20). No distinction is made between different spin operators in a quasi-classical model: the existence of $\mathbf{s}$ is simply postulated, as it also is in the Schrödinger-Pauli theory.

There is no uniquely defined spin operator in Dirac's theory, and there are many possible choices of different spin operators. The forms (9.6.8) and (9.6.15) are derived under the explicit assumption that s is the eigenvalue of the magnetic-moment operator, as discussed in § 5.2.2. In the relativistic quantum theory, the magnetic-moment operator for the electron is constructed so that it commutes with the Dirac Hamiltonian. This operator is decomposed of the electric dipole and magnetic dipole operators. These operators are written down in the unmagnetized case in § 10.1.2 of volume 1, and the magnetic momentum operator in a magnetic field is written down in (5.2.8), viz. $\hat{\boldsymbol{\mu}} = m\boldsymbol{\Sigma} - i\boldsymbol{\gamma}\times\hat{\boldsymbol{p}}$, where $\boldsymbol{\Sigma}$ and $\boldsymbol{\gamma}$ are Dirac matrices written in 3-vector forms. In the presence of a static electromagnetic field, the operator $\hat{\boldsymbol{\mu}}$ evolves according to

$$\frac{d\hat{\boldsymbol{\mu}}}{dt} = e\gamma^0\boldsymbol{\Sigma}\times\boldsymbol{B} - ie\boldsymbol{\gamma}\times\boldsymbol{E}, \tag{9.6.22}$$

implying that for $\boldsymbol{E} = 0$ the components perpendicular $\mathbf{B}$ precess. The choice of the component of $\hat{\boldsymbol{\mu}}$ along the $\boldsymbol{B}$ as the spin operator leads to eigenvalues that do not precess. In contrast, in the quasi-classical theory, the spin is identified as a quasi-classical vector $\mathbf{s}$ with no distinction being made (or seemingly being possible) between the possible spin operators it supposedly describes.

The validity of the quasi-classical models in deriving the response of a spin-dependent electron gas needs to be confirmed by showing that they are valid limits of the relativistic quantum results. At the time of writing, this is an unresolved problem.

9.7 Nonlinear Response Tensors

As with the linear response, the nonlinear responses are of interest for both an electron gas and for the magnetized vacuum. The method used in § 9.1 to calculate the response of an electron gas generalizes in a straightforward manner to the hierarchy of nonlinear response tensors, and allows one to treat both the vacuum and the electron gas in the same way. This approach involves an infinite sum over the Landau quantum number. An alternative approach, using the Géhéniau form for the electron propagator, leads to closed form expressions, with no infinite sum, but applies only to the vacuum response.

The general forms for the nonlinear response tensors derive in this section are cumbersome, and for most purposes simpler forms suffice. The response of the magnetized vacuum at low frequencies can be treated using the Heisenberg-Euler Lagrangian, and the contribution from an electron gas can be treated assuming the cold plasma model, which applies when the phase speeds of the wave are much higher than the thermal speed of the particles.

9.7.1 Closed Loop Diagrams

In QPD, the $(n+1)$th rank response tensor is derived from the amplitude for a closed loop diagram with $n+1$ sides, with $n=1, n=2, n=3$ corresponding to the linear, quadratic and cubic responses, respectively.

The Feynman amplitude for an n-sided electron loop follows from the nth order term in the expansion of S-matrix with no contractions:

$$\hat{S}^{(n)} = -\frac{(-e)^n}{n!}\int d^4x_1 \int d^4x_2 \dots \int d^4x_n\, \gamma^{\nu_1} G(x_1-x_2)\gamma^{\nu_2} \dots \gamma^{\nu_n} G(x_n-x_1)$$
$$\times : \hat{A}_{\nu_1}(x_1)\,\hat{A}_{\nu_2}(x_2)\dots\hat{A}_{\nu_n}(x_n) : . \tag{9.7.1}$$

The term with $n=2$ involves two propagators, and leads to the linear response, the term with $n=3$ involves three propagators and leads to the quadratic response, and so on. The quadratic response tensor, in coordinate space, is identified from the integrand of (9.7.1) for $n=3$, as

$$\Pi^{\mu\nu\rho}(x,x',x'') = \frac{1}{2}\left[\Pi_1^{\mu\nu\rho}(x,x',x'') + \Pi_1^{\mu\rho\nu}(x,x'',x')\right],$$
$$\Pi_1^{\mu\nu\rho}(x,x',x'') = ie^3\mathrm{Tr}\left[\gamma^\mu G(x,x')\gamma^\nu G(x',x'')\gamma^\rho G(x'',x)\right]. \tag{9.7.2}$$

The two contributions come from two triangle diagrams with different ordering of the three vertices. The quadratic response of the magnetized vacuum follows directly from (9.7.2) by inserting the Géhéniau form for the electron propagator, and Fourier transforming [19]. The response of the vacuum plus that of an electron gas may be obtained from (9.7.2) using the vertex formalism, and replacing the propagators by their statistical averages over the electron gas. Both approaches are discussed below.

9.7.2 Quadratic Response Tensor for the Vacuum

The quadratic response tensor for the magnetized vacuum was derived by Stoneham [19] in connection with photon splitting § 8.5. Inserting the electron propagator in the form (5.3.15) into (9.7.2) gives

$$\Pi_1^{\mu\nu\rho}(x,x',x'') = i\,e^3 e^{ie(x-x')_\alpha(x''-x)_\beta F^{\alpha\beta}} \times \mathrm{Tr}\left[\gamma^\mu \Delta(x,x')\gamma^\nu \Delta(x',x'')\gamma^\rho \Delta(x'',x)\right]. \tag{9.7.3}$$

There is freedom to locate the space-time origin, and this may be used to write (9.7.3) in terms only of the differences between the space-time points. Specifically, if one chooses the origin at $x = 0$, then (9.7.3) simplifies to

$$\begin{aligned}\Pi_1^{\mu\nu\rho}(x',x'') = -i\,&\frac{e^3(eB)^3}{(16\pi^2)^3}\,e^{-iex'_\alpha x''_\beta F^{\alpha\beta}}\int_0^\infty \frac{ds}{s}\int_0^\infty \frac{ds'}{s'}\int_0^\infty \frac{ds''}{s''}\\ &D_1^{\mu\nu\rho}(x',x'')\frac{e^{-im^2(s+s'+s'')}}{\sin(eBs)\sin(eBs')\sin(eBs'')}\\ &\times\exp\left[-\frac{i}{4}\left(\frac{eBx'^2_\perp}{\tan(eBs)}+\frac{eB(x'-x'')^2_\perp}{\tan(eBs')}\right.\right.\\ &\left.\left.+\frac{eBx''^2_\perp}{\tan(eBs'')}+\frac{x'^2_\parallel}{s}+\frac{(x'-x'')^2_\parallel}{s'}+\frac{x''^2_\parallel}{s''}\right)\right],\end{aligned} \tag{9.7.4}$$

$$\begin{aligned}D_1^{\mu\nu\rho}(x',x'') = \mathrm{Tr}\Bigg\{&\gamma^\mu\left[\exp(-i\Sigma eBs)\left(m-\frac{(\gamma x')_\parallel}{2s}\right)-\frac{eB(\gamma x')_\perp}{2\sin(eBs)}\right]\\ &\times\gamma^\nu\left[\exp(-i\Sigma eBs')\left(m+\frac{(\gamma(x'-x''))_\parallel}{2s'}\right)+\frac{eB(\gamma(x'-x''))_\perp}{2\sin(eBs')}\right]\\ &\times\gamma^\rho\left[\exp(-i\Sigma eBs')\left(m+\frac{(\gamma x'')_\parallel}{2s''}\right)+\frac{eB(\gamma x'')_\perp}{2\sin(eBs'')}\right]\Bigg\},\end{aligned} \tag{9.7.5}$$

The quadratic response tensor in momentum space then follows from

$$\begin{aligned}\Pi^{\mu\nu\rho}(-k,k',k'') &= \Pi_1^{\mu\nu\rho}(-k,k',k'') + \Pi_1^{\mu\rho\nu}(-k,k'',k'),\\ \Pi_1^{\mu\nu\rho}(-k,k',k'') &= \int d^4x\,d^4x'\,e^{-ik'x'-ik''x''}\,\Pi_1^{\mu\nu\rho}(x',x''),\end{aligned} \tag{9.7.6}$$

with $k = k' + k''$.

The Fourier transform may be performed following [1], to find

$$\begin{aligned}\Pi_1^{\mu\nu\rho}(-k,k',k'') = i\,&\frac{e^4 B}{16\pi^2}\int_0^\infty ds\int_0^\infty ds'\int_0^\infty ds''\,D_1^{\mu\nu\rho}(-k,k',k'')\\ &\times\frac{e^{-im^2(s+s'+s'')}}{(s+s'+s'')\sin[eB(s+s'+s'')]}\exp\left[i\left(\frac{ss'k^2_\parallel+ss''k'^2_\parallel+ss'k''^2_\parallel}{s+s'+s''}\right)\right]\end{aligned}$$

$$
\times \exp\left[i\left(\frac{C(s)S(s')S(s'')k_\perp^2 + S(s)C(s')S(s'')k_\perp'^2 + S(s)S(s')C(s'')k_\perp''^2}{eB\sin[eB(s+s'+s'')]}\right)\right]
$$
$$
\times \exp\left[2i\frac{S(s)S(s')S(s'')\,k_\alpha k'_\beta f^{\alpha\beta}}{eB\sin[eB(s+s'+s'')]}\right)\Bigg],
$$
$$
S(s) = \sin(eBs), \qquad C(s) = \cos(eBs), \qquad f^{\alpha\beta} = F^{\alpha\beta}/B. \tag{9.7.7}
$$

The resulting expression is

$$
D_1^{\mu\nu\rho}(-k,k',k'') = \int d^4x\, d^4x'\, e^{-ik'x'-ik''x''}\, e^{-\frac{1}{2}ieB(x'y''-x''y')}\, D_1(x',x''). \tag{9.7.8}
$$

It is helpful to write $k \to -k_0$, $k' \to k_1$, $k'' \to k_2$, so that one has $k_0+k_1+k_2 = 0$, and to make the dependence on s, s', s'' explicit, writing $s \to s_0$, $s' \to s_1$, $s'' \to s_2$. Then one has

$$
D_1^{\mu\nu\rho}(k_0,k_1,k_2;s_0,s_1,s_2) = D_1^{\nu\rho\mu}(k_1,k_2,k_0;s_1,s_2,s_0). \tag{9.7.9}
$$

However, even when all the symmetries are taken into account, the result remains cumbersome [19], and is not written down here. For most purposes, the low-frequency response suffices, and this may be derived much more simply by using the Heisenberg–Euler Lagrangian.

9.7.3 Nonlinear Responses: The Vertex Formalism

Using the vertex formalism allows one to include the contribution of the electron gas and the vacuum together. In the unmagnetized case one can identify the nth order nonlinear response tensor in momentum space from (9.7.1) simply by expressing the propagators in terms of their Fourier transforms and then carrying out the space-time integrals, which all give δ-functions. The n δ-functions ensure conservation of 4-momentum at the n vertices, and one can carry out $n-1$ of the 4-momentum integrals over these δ-functions, leaving a single δ-function, which expresses conservation of the external 4-momentum, plus the integral over the undetermined loop 4-momentum. In the magnetized case this procedure is not possible because the Fourier transform of the propagator does not exist. This difficulty can be overcome by writing (9.7.1) in terms of vertex functions whose Fourier transform does exist. The energy and the parallel momentum are conserved at each vertex, and are treated in the same manner as any of the components of 4-momentum in the unmagnetized case.

The propagators in (9.7.1) are interpreted as statistical averages over the electron gas, as in (5.5.22). The n space integrals in (9.7.1) are then trivial, giving n δ-functions. Similarly, the n time-integrals in (9.7.1) give n δ-functions. The n

propagators have both real (principal value) and imaginary parts, and the statistical average affects only the latter. The only physical contributions are those corresponding to the imaginary part of one propagator times the real parts of all the others, and there is one such contribution for each line in the diagram. (The term with all principal value parts gives zero, and terms with more than one resonant part are nonphysical due to the use of the Feynman propagator.)

In this way (9.7.1) reduces to

$$\hat{S}^{(n)} = \frac{(ie)^n}{n!} \int \frac{d^4k_n}{(2\pi)^4} \cdots \frac{d^4k_1}{(2\pi)^4}\, L^{\nu_n\cdots\nu_1}(-k_1,\ldots,-k_n)\, \hat{A}_{\nu_n}(k_n)\, \ldots\, \hat{A}_{\nu_1}(k_1),$$

$$L^{\nu_n\cdots\nu_1}(k_1,\ldots,k_n) = \sum_{q_1,\ldots,q_n} \gamma^{\epsilon_1\epsilon_n}_{q_1q_n}(\boldsymbol{k}_n)\ldots\gamma^{\epsilon_2\epsilon_1}_{q_2q_1}(\boldsymbol{k}_1)\, 2\pi\delta(\omega_1+\omega_2+\cdots\omega_n)$$

$$\times \int \frac{dE}{2\pi} \sum_{i=1}^{n} \frac{-i\epsilon_i(1-2n^{\epsilon}_{q_i})\,\pi\delta(E_i-\epsilon_i\varepsilon_{q_i})}{\prod^{n}_{r\neq i}[E_r-\epsilon_r\varepsilon_{q_r}]}. \tag{9.7.10}$$

This procedure leads only to the nondissipative part of the response tensor. The causal condition needs to be imposed separately to obtain the dissipative part. The causal condition in included by the Landau prescription of giving the frequencies in $E_r - E_{r-1} + \omega_r = 0$ an infinitesimal positive imaginary part. The resonant parts of the nonlinear response tensors are usually ignored, and only the principal value parts of the E-integral at each of the poles is retained in (9.7.10).

The specific form of the vertex functions $[\gamma^{\epsilon'\epsilon}_{q'q}(\boldsymbol{k})]^{\mu}$ in (9.7.10) depends on the choice of gauge. The gauge-dependent factors can be included in a factor $d^{\epsilon'\epsilon}_{q'q}(\boldsymbol{k})$ multiplying a gauge-independent vertex function $[\Gamma^{\epsilon'\epsilon}_{q'q}(\boldsymbol{k})]^{\mu}$, as in (5.4.2), viz.

$$[\gamma^{\epsilon'\epsilon}_{q'q}(\boldsymbol{k})]^{\mu} = d^{\epsilon'\epsilon}_{q'q}(\boldsymbol{k})\, [\Gamma^{\epsilon'\epsilon}_{q'q}(\boldsymbol{k})]^{\mu}. \tag{9.7.11}$$

For the Landau gauge, the gauge-dependent factor is given by (5.4.9). There are n integrals over the $p_{1y}\ldots p_{ny}$ and n integrals over the $p_{1z}\ldots p_{nz}$, and $n-1$ of each of these are performed over the δ-functions in the gauge-dependent factors. The resulting expression, which is independent of the choice of gauge, is

$$L^{\nu_n\cdots\nu_1}(k_1,\ldots,k_n) = (2\pi)^4\delta^4(k_1+\cdots+k_n)\,\exp\left[-\sum_{j<i}\frac{(\boldsymbol{k}_i\times\boldsymbol{k}_j)_z}{2eB}\right]$$

$$\times \sum_{q_1,\ldots,q_n} \frac{eB}{2\pi}\int\frac{dE}{2\pi}\int\frac{dp_z}{2\pi}\, \Gamma^{\epsilon_1\epsilon_n}_{q_1q_n}(\boldsymbol{k}_n)\ldots\Gamma^{\epsilon_2\epsilon_1}_{q_2q_1}(\boldsymbol{k}_1)$$

$$\times \sum_{i=1}^{n} \frac{-i\epsilon_i(1-2n^{\epsilon}_{q_i})\,\pi\delta(E_i-\epsilon_i\varepsilon_{q_i})}{\prod^{n}_{r\neq i}[E_r-\epsilon_r\varepsilon_{q_r}]}, \tag{9.7.12}$$

where q_r denotes the numbers $n_r, s_r, (p_z)_r$, with $r = 1, \ldots, n$. The $(p_z)_r$ are related by $(p_z)_r - (p_z)_{r-1} + (k_z)_r = 0$, and the p_z-integral is over the undetermined additive constant in each of the $(p_z)_r$. The exponential factor in (9.7.12) arises from the sum rule, (5.4.15), obeyed by the gauge-dependent factors.

The final step in the formal identification of the nonlinear response tensors involves the identification of the interaction energy in terms of S-matrix elements with the interaction energy implied by the nonlinear response. This corresponds to the identification

$$-i\Pi^{\nu_0\nu_1\ldots\nu_n}(k_0, k_1, \ldots, k_n)(2\pi)^4\delta^4(k_1 + \cdots + k_n)$$
$$= \frac{(ie)^{n+1}}{n!}\sum_P P^{1\ldots n}_{1'\ldots n'} L^{\nu_{n'}\cdots\nu_{1'}\nu_0}(k_{n'}, \ldots, k_{1'}, k_0), \tag{9.7.13}$$

where the sum is over all permutations of $1' \ldots n'$ amongst $1 \ldots n$.

Applying (9.7.13) to the linear case $n = 1$ gives the expression (9.1.1) for the linear response tensor. The notation in (9.7.13) corresponds to a linear response tensor $\Pi^{\mu\nu}(k_0, k_1)$ with two argument that satisfy $k_0 + k_1 = 0$. One writes $k_1 \to k$, $k_0 \to -k$ and $\Pi^{\mu\nu}(k_0, k_1) \to \Pi^{\mu\nu}(k)$. The contribution of the linear response from the vacuum needs to be regularized, but this is unnecessary for the quadratic response tensor.

9.7.4 *Quadratic Response Tensor*

The resulting expression for the quadratic nonlinear response tensor has two related contributions (from the two triangle diagrams with different order of the vertices),

$$\Pi^{\mu\nu\rho}(k_1, k_2, k_3) = \Pi_1^{\mu\nu\rho}(k_1, k_2, k_3) + \Pi_1^{\mu\nu\rho}(k_1, k_3, k_2),$$

$$\Pi_1^{\mu\nu\rho}(k_1, k_2, k_3) = -\frac{e^3B}{4\pi}\int\frac{dp_z}{2\pi}\exp\left[-\sum_{j<i}\frac{(\boldsymbol{k}_i \times \boldsymbol{k}_j)_z}{2eB}\right]$$
$$\times \sum_{q_1,q_2,q_3}\left[\frac{\frac{1}{2}\epsilon_1(1 - 2n^{\epsilon_1}_{q_1})}{(\omega_1 + \epsilon_1\varepsilon_{q_1} - \epsilon_3\varepsilon_{q_3})(\omega_2 + \epsilon_2\varepsilon_{q_2} - \epsilon_1\varepsilon_{q_1})}\right.$$
$$+\frac{\frac{1}{2}\epsilon_2(1 - 2n^{\epsilon_2}_{q_2})}{(\omega_2 + \epsilon_2\varepsilon_{q_2} - \epsilon_1\varepsilon_{q_1})(\omega_3 + \epsilon_3\varepsilon_{q_3} - \epsilon_2\varepsilon_{q_2})}$$
$$\left.+\frac{\frac{1}{2}\epsilon_3(1 - 2n^{\epsilon_3}_{q_3})}{(\omega_3 + \epsilon_3\varepsilon_{q_3} - \epsilon_2\varepsilon_{q_2})(\omega_1 + \epsilon_1\varepsilon_{q_1} - \epsilon_3\varepsilon_{q_3})}\right]$$
$$\times[\Gamma^{\epsilon_3\epsilon_1}_{q_3q_1}(-\boldsymbol{k}_1)]^\mu[\Gamma^{\epsilon_1\epsilon_2}_{q_1q_2}(-\boldsymbol{k}_2)]^\nu[\Gamma^{\epsilon_2\epsilon_3}_{q_2q_3}(-\boldsymbol{k}_3)]^\rho. \tag{9.7.14}$$

The contribution of the electron gas arises from the terms proportional to the occupation numbers. In contrast with the linear response, where the sum and the difference of the contribution of the electrons and positrons appear in the nongyrotropic and gyrotropic parts of the response tensor, respectively, for the quadratic response the situation is reversed, and the difference between the contribution of the electrons and positrons appears in the nongyrotropic terms. The gyrotropic terms, which vanish $\propto B$ for $B \to 0$, contain the sum of the contributions of the electrons and positrons.

The terms in (9.7.14) that are independent of the occupation numbers, n_q, correspond to the quadratic response of the vacuum. The contribution of the vacuum to the quadratic response tensor does not need to be regularized.

Further the evaluation of the quadratic response tensor for a magnetized vacuum using the form (9.7.14), involves performing the sum over spins, and the sums over the signs ϵ_i. The full result is cumbersome; some particular combinations of components were written down by Mentzel et al. [12] and Weise et al. [23], in connection with what is called the S-matrix treatment of photon splitting in vacuo. The forms (9.7.14) and (9.7.15) are the most convenient when the sums over the Landau quantum numbers, $n_1, \ldots$, converge rapidly, which is the case when the arguments of the J-functions are small. This is the small-x condition, which here requires $k_{i\perp}^2/2eB \ll 1$ for all wavenumbers, $\boldsymbol{k}_i$. For photon splitting this requires $\omega_i \sin\theta_i \ll m(2B/B_c)^{1/2}$ for all three photons. Thus the form (9.7.14) is convenient for treating photon splitting for sufficiently low frequencies or in sufficiently strong fields, $B \gtrsim B_c$.

9.7.5 Cubic Response Tensor

The cubic response tensor follows from (9.7.13) with $n = 3$:

$$
\begin{aligned}
\Pi^{\mu\nu_2\nu_3\nu_4}(k_1,k_2,k_3,k_4) &= \sum_P P^{2\,3\,4}_{2'3'4'}\,\Pi_1^{\mu\nu_{2'}\nu_{3'}\nu_{4'}}(k_1,k_{2'},k_{3'},k_{4'}),\\
\Pi_1^{\mu\nu\rho\sigma}(k_1,k_2,k_3,k_4) &= \frac{e^3B}{12\pi}\int\frac{dp_z}{2\pi}\\
&\quad\times\exp\left[-\sum_{j<i}\frac{(\boldsymbol{k}_i\times\boldsymbol{k}_j)_z}{2eB}\right]\sum_{q_1,q_2,q_3,q_4}\left[\sum_{r=1}^{4}\frac{\frac{1}{2}\epsilon_r(1-2n_{q_r}^{\epsilon_r})}{d_r d_{r+1} d_{r+2}}\right]\\
&\quad\times[\gamma_{q_4q_1}^{\epsilon_4\epsilon_1}(-\boldsymbol{k}_1)]^\mu[\gamma_{q_1q_2}^{\epsilon_1\epsilon_2}(-\boldsymbol{k}_2)]^\nu[\gamma_{q_2q_3}^{\epsilon_2\epsilon_3}(-\boldsymbol{k}_3)]^\rho[\gamma_{q_3q_4}^{\epsilon_3\epsilon_4}(-\boldsymbol{k}_4)]^\sigma,\\
d_r &= \omega_r+\epsilon_r\varepsilon_{q_r}-\epsilon_{r-1}\varepsilon_{q_{r-1}},
\end{aligned}
\tag{9.7.15}
$$

with $r + 4 \equiv r$. The response of the medium arises from the terms proportional to the occupation numbers; as for the linear response, the contributions of the electrons and positrons add for the nongyrotropic terms.

The terms in (9.7.15) that are independent of the occupation numbers, n_q, correspond to the cubic response of the vacuum. As with the cubic response tensor for $B \to 0$, the divergent parts are removed by discarding the part that does not satisfy the charge-continuity and gauge-invariance conditions. This may be achieved by projecting (9.7.15) onto the set of basis vectors (8.1.2).

References

1. N.N. Bogoliubov, D.V. Shirkov, *The Theory of Quantized Fields* (Interscience, New York, 1959)
2. V. Canuto, J. Ventura, Astrophys. Space Sci. **18**, 104 (1972)
3. R.A. Cover, G. Kalman, P. Bakshi, Phys. Rev. D **20** 3015 (1979)
4. R.E. Cutkovsky, J. Math. Phys. **1**, 429 (1960)
5. A.E. Delsant, N.E. Frankel, Ann. Phys. **125**, 135 (1980)
6. M.P. Greene, H.J. Lee, J.J. Quinn, S. Rodriguez, Phys. Rev. **177**, 1019 (1969)
7. D.C. Kelly, Phys. Rev. **134**, A641 (1964)
8. V. Kowalenko, N.E. Frankel, K.C. Hines, Phys. Rep. **126**, 109 (1985)
9. D.B. Melrose, A. Mushtaq, Phys. Plasmas **17**, 122103 (2010)
10. D.B. Melrose, A. Mushtaq, Phys. Rev. E **83**, 056404 (2011)
11. D.B. Melrose, J.I. Weise, J. Phys. A **42**, 345502 (2009)
12. M. Mentzel, D. Berg, G. Wunner, Phys. Rev. D **50**, 1125 (1964)
13. H. Pérez Rojas, A.E. Shabad, Ann. Phys. **138**, 1 (1982)
14. P. Pulsifer, G. Kalman, Phys. Rev. A **45**, 5820 (1992)
15. J.J. Quinn, S. Rodriguez, Phys. Rev. **128**, 2487 (1962)
16. A.E. Shabad, V.V. Usov, Phys. Rev. Lett. **98**, 180403 (2007)
17. V.P. Silin, Sov. Phys. JETP **11**, 1136 (1960)
18. A.A. Sokolov, I.M. Ternov, *Synchrotron Radiation* (Pergamon Press, Oxford, 1968)
19. R.J. Stoneham, J. Phys. A **12**, 2187 (1979)
20. G.N. Svetozarova, V.N. Tsytovich, Izv. Vuzov. Radiofiz. **5**, 658 (1962)
21. J.S. Toll, Ph.D. thesis Princeton University, 1952
22. J.I. Weise, Phys. Rev. E **78**, 046408 (2008)
23. J.I. Weise, M.G. Baring, D.B. Melrose, Phys. Rev. D **57**, 5526 (1998)
24. P.S. Zyryanov, V.P. Kalashnikov, Sov. Phys. JETP **14**, 799 (1962)

Appendix A
Special Functions

A.1 Bessel Functions and J-Functions

The properties of Bessel function are summarized in standard references [1, 3, 8].

A.1.1 Ordinary Bessel Functions

Ordinary Bessel functions, $J_\nu(z)$, have the following properties.
Differential equation:

$$z^2 J_\nu''(z) + z J_\nu'(z) + (z^2 - \nu^2) J_\nu(z) = 0, \tag{A.1.1}$$

where a prime denotes differentiation with respect to z.
Power series:

$$J_\nu(z) = \sum_{k=0}^{\infty} \frac{(-1)^k}{k!\Gamma(k+\nu+1)} (z/2)^{2k+\nu}. \tag{A.1.2}$$

Recursion relations:

$$J_{\nu-1}(z) + J_{\nu+1}(z) = 2\frac{\nu}{z} J_\nu(z), \tag{A.1.3}$$

$$J_{\nu-1}(z) - J_{\nu+1}(z) = 2J_\nu'(z). \tag{A.1.4}$$

Generating function:

$$e^{iz\sin\phi} = \sum_{n=-\infty}^{\infty} e^{in\phi} J_n(z). \tag{A.1.5}$$

D. Melrose, *Quantum Plasmadynamics: Magnetized Plasmas*, Lecture Notes in Physics 854, DOI 10.1007/978-1-4614-4045-1,

Sum rules:

$$\sum_{n=-\infty}^{\infty} J_n^2(z) = 1, \qquad \sum_{n=-\infty}^{\infty} nJ_n^2(z) = 0, \qquad \sum_{n=-\infty}^{\infty} J_n(z)J_n'(z) = 0,$$

$$\sum_{n=-\infty}^{\infty} n^2 J_n^2(z) = \frac{1}{2}z^2, \qquad \sum_{n=-\infty}^{\infty} J_n'^2(z) = \frac{1}{2}. \tag{A.1.6}$$

A.1.2 Modified Bessel Functions $I_\nu(z)$

Differential equation:

$$I_\nu''(z) + \frac{1}{z}I_\nu'(z) - \left(1 + \frac{\nu^2}{z^2}\right) I_\nu(z) = 0, \tag{A.1.7}$$

Power series:

$$I_\nu(z) = \sum_{k=0}^{\infty} \frac{1}{k!\Gamma(k+\nu+1)} (z/2)^{2k+\nu}. \tag{A.1.8}$$

Recursion relations:

$$I_{\nu-1}(z) - I_{\nu+1}(z) = 2(\nu/z)I_s(z),$$

$$I_{\nu-1}(z) + I_{\nu+1}(z) = 2I_\nu'(z). \tag{A.1.9}$$

Generating function:

$$e^{z\cos\phi} = \sum_{s=-\infty}^{\infty} I_s(z)e^{\pm is\phi}, \tag{A.1.10}$$

A.1.3 Macdonald Functions $K_\nu(z)$

Differential equation:

$$\frac{d^2}{dz^2}K_\nu(z) + \frac{1}{z}\frac{d}{dz}K_\nu(z) - \left(1 + \frac{\nu^2}{z^2}\right) K_\nu(z) = 0, \tag{A.1.11}$$

Recursion relations:

$$K_{\nu-1}(z) - K_{\nu+1}(z) = -2(\nu/z)K_\nu(z),$$
$$K_{\nu-1}(z) + K_{\nu+1}(z) = -2K'_\nu(z). \tag{A.1.12}$$

The recursion relations imply $K_{-\nu}(z) = K_\nu(z)$ and

$$\frac{1}{z}\frac{d}{dz}[z^{\pm\nu}K_\nu(z)] = -z^{\pm\nu-1}K_{\nu\mp1}(z). \tag{A.1.13}$$

Expansion of $K_\nu(z)$ for small z is

$$K_\nu(z) \approx 2^{\nu-1}\Gamma(\nu)\, z^{-\nu}. \tag{A.1.14}$$

The asymptotic expansion for large z is

$$K_\nu(z) = \left(\frac{\pi}{2z}\right)^{1/2} e^{-z}\left(1 + \frac{4\nu^2-1}{8z} + \frac{(4\nu^2-1)(4\nu^2-9)}{128z^2} + \cdots\right). \tag{A.1.15}$$

Integral representation:

$$K_\nu(x) = \frac{(x/2)^\nu \Gamma(\frac{1}{2})}{\Gamma(\nu+\frac{1}{2})}\int_0^\infty d\chi\, \sinh^{2\nu}\chi\, e^{-x\cosh\chi} \tag{A.1.16}$$

The Gamma function satisfies

$$\Gamma(x+1) = x\Gamma(x), \qquad \Gamma(1) = 1, \qquad \Gamma\left(\tfrac{1}{2}\right) = \pi^{1/2}. \tag{A.1.17}$$

The integral (A.1.16) also applies when ν is negative, and then $K_{-\nu}(x) = K_\nu(x)$ implies

$$K_\nu(x) = \frac{(x/2)^{-\nu}\Gamma(\nu+\frac{1}{2})\cos\pi\nu}{\Gamma(\frac{1}{2})}\int_0^\infty d\chi\, \frac{e^{-x\cosh\chi}}{\sinh^{2\nu}\chi}, \tag{A.1.18}$$

$$\Gamma\left(\tfrac{1}{2}+\nu\right)\Gamma\left(\tfrac{1}{2}-\nu\right) = \frac{\pi}{\cos\pi\nu}. \tag{A.1.19}$$

An integral identity due to Schwinger is

$$\int_0^\infty d\xi\, \xi^2 K_\mu^2(\xi) = \frac{\pi^2(1-4\mu^2)}{32\cos\pi\mu}. \tag{A.1.20}$$

A.1.4 Airy Functions

The two Airy functions that appear are defined by

$$\mathrm{Ai}\,(z) = \frac{1}{\pi}\int_0^\infty dt\,\cos\left(zt + \tfrac{1}{3}t^3\right), \qquad \mathrm{Gi}\,(z) = \frac{1}{\pi}\int_0^\infty dt\,\sin\left(zt + \tfrac{1}{3}t^3\right). \tag{A.1.21}$$

For $z > 0$ one has

$$\mathrm{Ai}\,(z) = \frac{1}{\pi}\left(\frac{z}{3}\right)^{1/2} K_{1/3}(\zeta), \qquad \mathrm{Ai}'\,(z) = -\frac{z}{\pi\sqrt{3}}K_{2/3}(\zeta), \tag{A.1.22}$$

with $\zeta = 2z^{3/2}/3$.

The approximations available for Gi (z) are for large and small z. The leading terms in the asymptotic expansion for $z \gg 1$ are [5]

$$\mathrm{Gi}\,(z) \sim \frac{1}{\pi}\left(\frac{1}{z} + \frac{2}{z^4} + \cdots\right), \qquad \mathrm{Gi}'\,(z) \sim \frac{1}{\pi}\left(-\frac{1}{z^2} + \cdots\right),$$

$$\int^z dz'\,\mathrm{Gi}\,(z') \sim \frac{1}{\pi}\left(\ln z + \frac{2C + \ln 3}{3} - \frac{2}{3z^3} + \cdots\right), \tag{A.1.23}$$

where $C = 0.577\cdots$ is Euler's constant. The expansion for $z \ll 1$ gives

$$\mathrm{Gi}\,(z) = \frac{1}{\pi}\left[\frac{3^{1/3}}{2}\Gamma(4/3) + \frac{3^{2/3}}{4}\Gamma(5/3)\,z - \frac{z^2}{2} + \cdots\right],$$

$$\mathrm{Gi}\,(0) = 0.205, \quad \mathrm{Gi}'\,(0) = 0.149. \tag{A.1.24}$$

Rothman [5] found that the asymptotic expansion is accurate for $z \gtrsim 8$ and tabulated the functions for lower z.

A.1.5 J-Functions

Definition

The J-functions used here are defined by, for $\nu \geq 0$,

$$J_\nu^n(x) = \left(\frac{n!}{(n+\nu)!}\right)^{1/2} e^{-x/2} x^{\nu/2} L_n^\nu(x). \tag{A.1.25}$$

By requiring $J_\nu^n(x) = (-)^\nu J_{-\nu}^{n+\nu}(x)$, for $\nu < 0$ one has

$$J_\nu^n(x) = (-)^\nu \left(\frac{(n-|\nu|)!}{n!}\right)^{1/2} e^{-x/2} x^{|\nu|/2} L_n^{|\nu|}(x), \tag{A.1.26}$$

with $L_n^\nu(x)$ the generalized Laguerre polynomial, defined by

$$L_n^\nu(x) = \frac{e^x x^{-\nu}}{n!} \frac{d^n}{dx^n}(e^{-x} x^{n+\nu}) = \sum_{k=0}^{n} \frac{(n+\nu)!\,(-x)^k}{(n-k)!(k+\nu)!k!}. \tag{A.1.27}$$

Sokolov and Ternov Function

The function defined by Sokolov and Ternov [6, 7] is related to (A.1.25) by

$$I_{n,n'}(x) = J_{n-n'}^{n'}(x). \tag{A.1.28}$$

Recursion Relations

The J-functions satisfy recursion relations

$$x^{1/2} J_{\nu+1}^{n-1}(x) = (n+\nu)^{1/2} J_\nu^{n-1}(x) - n^{1/2} J_\nu^n(x), \tag{A.1.29}$$

$$x^{1/2} J_{\nu-1}^{n}(x) = -n^{1/2} J_\nu^{n-1}(x) + (n+\nu)^{1/2} J_\nu^n(x), \tag{A.1.30}$$

and also

$$\nu J_\nu^{n-1}(x) = x^{1/2}\big[(n+\nu)^{1/2} J_{\nu+1}^{n-1}(x) + n^{1/2} J_{\nu-1}^{n}(x)\big], \tag{A.1.31}$$

$$\nu J_\nu^{n}(x) = x^{1/2}\big[n^{1/2} J_{\nu+1}^{n-1}(x) + (n+\nu)^{1/2} J_{\nu-1}^{n}(x)\big]. \tag{A.1.32}$$

A further pair of relations that is similar to the recursion relations for Bessel functions is

$$(x+\nu) J_\nu^n(x) = [x(n+\nu)]^{1/2} J_{\nu-1}^n(x) + [x(n+\nu+1)]^{1/2} J_{\nu+1}^n(x), \tag{A.1.33}$$

$$2x \frac{d}{dx} J_\nu^n(x) = [x(n+\nu)]^{1/2} J_{\nu-1}^n(x) - [x(n+\nu+1)]^{1/2} J_{\nu+1}^n(x). \tag{A.1.34}$$

Relations Involving J-Functions

With $\nu = n - n'$ $p_n = (2neB)^{1/2}$, $x = k_\perp^2/2eB$, relations (A.1.33) and (A.1.34) become

$$p_{n'} J^n_{n'-n}(x) = p_n J^{n-1}_{n'-n}(x) + k_\perp J^n_{n'-n-1}(x),$$
$$p_{n'} J^{n-1}_{n'-n}(x) = p_n J^n_{n'-n}(x) + k_\perp J^{n-1}_{n'-n+1}(x). \tag{A.1.35}$$

The following identities result from squares of the relations (A.1.35):

$$(p^2_{n'} + p^2_n)[(J^{n-1}_{n'-n})^2 + (J^n_{n'-n})^2] - 4p_{n'}p_n J^{n-1}_{n'-n} J^n_{n'-n}$$
$$= k^2_\perp[(J^{n-1}_{n'-n+1})^2 + (J^n_{n'-n-1})^2], \tag{A.1.36}$$

$$(p^2_{n'} - p^2_n)[(J^{n-1}_{n'-n})^2 - (J^n_{n'-n})^2] = k^2_\perp[(J^{n-1}_{n'-n+1})^2 - (J^n_{n'-n-1})^2], \tag{A.1.37}$$

$$(p^2_{n'} + p^2_n)[(J^{n-1}_{n'-n})^2 - (J^n_{n'-n})^2] = 2p_n k_\perp[J^{n-1}_{n'-n} J^n_{n'-n-1} - J^n_{n'-n} J^{n-1}_{n'-n+1}]$$
$$+k^2_\perp[(J^{n-1}_{n'-n+1})^2 - (J^n_{n'-n-1})^2], \tag{A.1.38}$$

$$(p^2_{n'} - p^2_n)[(J^{n-1}_{n'-n})^2 + (J^n_{n'-n})^2] = 2p_n k_\perp[J^{n-1}_{n'-n} J^n_{n'-n-1} + J^n_{n'-n} J^{n-1}_{n'-n+1}]$$
$$+k^2_\perp[(J^{n-1}_{n'-n+1})^2 + (J^n_{n'-n-1})^2]. \tag{A.1.39}$$

In evaluating the response tensor in the summed form (9.1.20) some tensorial components are multiplied by $(pk)_{nn'} = \frac{1}{2}[(k^2)_\parallel + p^2_n - p^2_{n'}]$, and (A.1.37), (A.1.39) allow one to rewrite some of the terms that are multiplied by $p^2_{n'} - p^2_n$. Other terms that are multiplied by $p^2_{n'} - p^2_n$ can be rewritten using

$$(p^2_{n'} - p^2_n) J^{n-1}_{n'-n} = k_\perp[p_n J^n_{n'-n-1} + p_{n'} J^{n-1}_{n'-n+1}], \tag{A.1.40}$$

$$(p^2_{n'} - p^2_n) J^n_{n'-n} = k_\perp[p_n J^{n-1}_{n'-n+1} + p_{n'} J^n_{n'-n-1}]. \tag{A.1.41}$$

The remaining terms that are multiplied by $p^2_{n'} - p^2_n$ involve the square and products of $J^{n-1}_{n'-n+1}$, $J^n_{n'-n-1}$, and these can be rewritten by first expressing these in terms of $J^{n-1}_{n'-n}$, $J^n_{n'-n}$ using (A.1.36)–(A.1.39), but no major simplifications occur.

Sum Rules

The sum rules

$$\sum_{n'=0}^{\infty} J^{n'}_{n-n'}(x) J^{n'}_{n''-n'}(x) = \delta^{nn''}, \tag{A.1.42}$$

$$\sum_{n'=0}^{\infty} (n'-n)[J^{n'}_{n-n'}(x)]^2 = x, \tag{A.1.43}$$

were derived by Quinn and Rodriguez [4] and Sokolov and Ternov [6].

Orthogonality Relation

$$\int_0^\infty dx\, J_\nu^n(x) J_\nu^{n'}(x) = \delta^{nn'}. \tag{A.1.44}$$

Integral Identities

$$\int_0^\infty dx x^{1/2} [J_\nu^n(x)]^2 = (n+\nu+1)^{1/2}\left(1+\frac{n+\frac{1}{2}}{4(n+\nu+1)}\right), \tag{A.1.45}$$

$$\int_0^\infty dx\, x\, [J_\nu^n(x)]^2 = 2n+\nu+\tfrac{3}{2}, \tag{A.1.46}$$

Particular Values

For $\nu \geq 0$, one has

$$J_\nu^0(x) = (-)^\nu J_{-\nu}^{\nu+1}(x) = \frac{x^{\nu/2}e^{-x/2}}{(\nu!)^{1/2}}, \tag{A.1.47}$$

$$J_\nu^1(x) = (-)^\nu J_{-\nu}^{\nu+1}(x) = \frac{x^{\nu/2}e^{-x/2}}{((\nu+1)!)^{1/2}}(\nu+1-x), \tag{A.1.48}$$

$$J_\nu^2(x) = (-)^\nu J_{-\nu}^{\nu+2}(x) = \frac{x^{\nu/2}e^{-x/2}}{(2!(\nu+2)!)^{1/2}} \times[(\nu+1)(\nu+2) - 2(\nu+2)x + x^2], \tag{A.1.49}$$

$$J_\nu^3(x) = (-)^\nu J_{-\nu}^{\nu+3}(x) = \frac{x^{\nu/2}e^{-x/2}}{(3!(\nu+3)!)^{1/2}}[(\nu+1)(\nu+2)(\nu+3) -3(\nu+2)(\nu+3)x + 3(\nu+3)x^2 - x^3]. \tag{A.1.50}$$

Expansion in x

For $x \ll 1$, the J-functions may be approximated by the leading term in their expansion in powers of x:

$$J_{n'-n}^n(x) = \left(\frac{n'!}{n!}\right)^{1/2} \frac{x^{(n'-n)/2}}{(n'-n)!}\left[1 - \frac{n'+n+1}{2(n'-n+1)}x + \cdots\right], \tag{A.1.51}$$

which applies for $n' \geq n$. The limit $x \to 0$ gives

$$J_0^n(0) = 1, \qquad J_\nu^n(0) = 0 \quad \text{for } \nu \neq 0. \tag{A.1.52}$$

Approximation by Bessel Functions

The expansion of the J-functions in terms of Bessel functions,

$$J_\nu^n\left(\frac{z^2}{4n}\right) = \left[\frac{(n+\nu)!}{n!n^\nu}\right]^{1/2} \sum_{a=0}^{\infty} b_a \left(\frac{z}{2n}\right)^a J_{\nu+a}(z),$$

$$b_0 = 1, \quad b_1 = -\tfrac{1}{2}(\nu+1), \quad b_2 = \tfrac{1}{8}(\nu+1)(\nu+2),$$

$$(a+1)b_{a+1} = -\tfrac{1}{2}(\nu+1)b_a + \tfrac{1}{4}(\nu+a)b_{a-1} - \tfrac{1}{4}nb_{a-2}, \tag{A.1.53}$$

converges rapidly for sufficiently large n.

In taking the nonquantum limit, one takes the limit $\hbar \to 0$, with $n \to \infty$ so that $p_n = (2neB\hbar)^{1/2} \to p_\perp$ remains finite; the ratio $a/n = (n-n')/n$ is regarded as of order $\hbar$. To first order in $\hbar$ one has

$$J_{n-n'}^n(x) = J_a(z) - \tfrac{1}{2}(a+1)\frac{\hbar k_\perp}{p_\perp} J_{a+1}(z). \tag{A.1.54}$$

The J-functions with upper index $n-1$ and n differ at first order in $\hbar$:

$$J_{n'-n}^{n-1}(x) - J_{n-n'}^n(x) = -\frac{\hbar k_\perp}{p_\perp} J_a'(z). \tag{A.1.55}$$

Related identities (with arguments x and z omitted) are

$$(J_{n'-n}^{n-1})^2 + (J_{n'-n}^n)^2 = J_a^2 - \frac{2a\hbar k_\perp}{p_\perp} J_a' J_a,$$

$$J_{n'-n}^{n-1} J_{n'-n}^n = J_a^2 - \frac{a\hbar k_\perp}{p_\perp} J_a' J_a,$$

$$(J_{n'-n+1}^{n-1})^2 + (J_{n'-n-1}^n)^2 = \sum_{\eta=\pm 1} J_{a-\eta}^2 \left(1 + \eta \frac{a(a-\eta)eB}{p_\perp^2}\right) + \frac{2a\hbar k_\perp}{p_\perp} J_a' J_a,$$

$$J_{n'-n+1}^{n-1} J_{n'-n-1}^n = J_{a+1} J_{a-1} \left(1 + \frac{aeB}{p_\perp^2}\right) - \frac{a\hbar k_\perp}{p_\perp} J_a' J_a,$$

$$(J_{n'-n}^{n-1})^2 - (J_{n'-n}^n)^2 = -\frac{2a\hbar k_\perp}{p_\perp} J_a' J_a,$$

$$(J^{n-1}_{n'-n+1})^2 - (J^{n}_{n'-n-1})^2 = \sum_{\eta=\pm 1} \eta J^2_{a-\eta}\left(1 + \eta\frac{a(a-\eta)eB}{p^2_\perp}\right) + \frac{a}{n}J^2_a. \tag{A.1.56}$$

A.2 Relativistic Plasma Dispersion Functions

A.2.1 *Relativistic Thermal Function $T(z, \rho)$*

The function $T(z, \rho)$, defined by (2.4.29), has alternative integral representations:

$$\begin{aligned} T(z,\rho) &= -\rho \int_0^\infty d\chi \,\sinh\chi\, e^{-\rho\cosh\chi} \ln\left(\frac{z+\tanh\chi}{z-\tanh\chi}\right) \\ &= 2z \int_0^\infty d\chi \frac{e^{-\rho\cosh\chi}}{(1-z^2)\cosh^2\chi - 1} \\ &= -\frac{2\rho}{1-z^2}\int^z d\zeta \frac{K_1(\rho R)}{R}, \end{aligned} \tag{A.2.1}$$

with $R = [(1-\zeta^2)(1-z^2)]^{1/2}$.

The function $T(z, \rho)$ satisfies the partial differential equations [2]:

$$(1-z^2)\frac{\partial^2}{\partial\rho^2}T(z,\rho) = 2zK_0(\rho) + T(z,\rho), \tag{A.2.2}$$

$$\begin{aligned} z(1-z^2)^3\, T''(z,\rho) - (1-z^2)^2(1+2z^2)\, T'(z,\rho) - \rho^2 z^3 T(z,\rho) \\ = 2z^2\rho^2 K_0(\rho) + 2(1-z^2)\rho K_1(\rho), \end{aligned} \tag{A.2.3}$$

$$z\frac{\partial}{\partial\rho}T(z,\rho) = 2K_1(\rho) + \frac{(1-z^2)}{\rho}T'(z,\rho), \tag{A.2.4}$$

with $T'(z,\rho) = \partial T(z,\rho)/\partial z$, $T''(z,\rho) = \partial^2 T(z,\rho)/\partial z^2$.

A.2.2 *Trubnikov Functions*

Trubnikov functions are defined by

$$t^n_\nu(z,\rho) = (k\tilde{u})^{n+1}\int_0^\infty d\xi\, \xi^n \frac{K_\nu(r(\xi))}{r^\nu(\xi)}, \tag{A.2.5}$$

with $r(\xi)$ given by (2.4.10), and where the power of $k\tilde{u}$ is included so that the integral is dimensionless. They satisfy the recursion relations

$$t_{\nu+1}^{n+1}(z,\rho) = \frac{i\rho z^2}{1-z^2}\, t_{\nu+1}^{n}(z,\rho) + \frac{z^2}{1-z^2} \begin{cases} \dfrac{K_\nu(\rho)}{\rho^\nu} & \text{for } n = 0, \\ n t_\nu^{n-1}(z,\rho) & \text{for } n > 0, \end{cases} \tag{A.2.6}$$

$$\frac{\partial t_\nu^n(z,\rho)}{\partial \rho} = -i\rho\, t_{\nu+1}^{n}(z,\rho) - i t_{\nu+1}^{n+1}(z,\rho). \tag{A.2.7}$$

Two further identities are

$$t_{\nu+1}^{n}(z,\rho) = -\frac{1-z^2}{\rho}\frac{\partial t_\nu^n(z,\rho)}{\partial \rho} + \frac{iz^2}{\rho} \begin{cases} \dfrac{K_\nu(\rho)}{\rho^\nu} & \text{for } n = 0, \\ n\, t_\nu^{n-1}(z,\rho) & \text{for } n > 0, \end{cases} \tag{A.2.8}$$

$$t_{\nu+1}^{n+2}(z,\rho) = z^3\frac{\partial t_\nu^n(z,\rho)}{\partial z}. \tag{A.2.9}$$

The relation to $T(z,\rho)$ follows from

$$t_0^0(z,\rho) = \frac{iz}{2}\frac{\partial T(z,\rho)}{\partial \rho} = \frac{i}{2}\left[2K_1(\rho) + \frac{(1-z^2)}{\rho}\, T'(z,\rho)\right], \tag{A.2.10}$$

$$t_1^0(z,\rho) = -\frac{iz}{2\rho}\, T(z,\rho). \tag{A.2.11}$$

The functions for higher n are generated from these using (A.2.6).

A.2.3 Shkarofsky and Dnestrovskii Functions

The generalized Shkarofsky functions are defined by (2.5.28) for real q, integer $r \geq 0$ and complex z, a with $\mathrm{Im}\,(z-a) > 0$ by

$$\begin{aligned} \mathcal{F}_{q,r}(z,a) &= -i\int_0^\infty dt\, \frac{(it)^r}{(1-it)^q}\exp\left[izt - \frac{at^2}{1-it}\right] \\ &= -ie^{-a}\int_0^\infty dt\, \frac{(it)^r}{(1-it)^q}\exp\left[i(z-a)t + \frac{a}{1-it}\right]. \end{aligned} \tag{A.2.12}$$

The definition is extended to $\mathrm{Im}\,(z-a) < 0$ by analytic continuation. Generalized Dnestrovskii functions are defined by (2.5.34), viz. $F_{q,r}(z) = \mathcal{F}_{q,r}(z,0)$. The usual Shkarofsky functions, $\mathcal{F}_q(z,a) = \mathcal{F}_{q,0}(z,a)$, and Dnestrovskii functions, $F_q(z) = F_{q,0}(z)$, are the special cases $r = 0$.

The Shkarofsky functions and the Dnestrovskii functions are related by an expansion in modified Bessel functions:

$$\mathcal{F}_q(z,a) = \sum_{s=-\infty}^{\infty} e^{-2a}\, I_s(2a)\, F_{q-s}(z). \qquad \text{(A.2.13)}$$

Recursion Relations and Differential Equations

Recursion relations satisfied by the Shkarofsky functions are

$$a\mathcal{F}_{q-2}(z,a) = 1 + (a-z)\mathcal{F}_q(z,a) - q\mathcal{F}_{q+1}(z,a), \qquad \text{(A.2.14)}$$

$$\mathcal{F}_q'(z,a) = \mathcal{F}_q(z,a) - \mathcal{F}_{q-1}(z,a), \qquad \text{(A.2.15)}$$

$$\mathcal{F}_q''(z,a) = \mathcal{F}_q(z,a) - 2\mathcal{F}_{q-1}(z,a) + \mathcal{F}_{q-2}(z,a), \qquad \text{(A.2.16)}$$

where a prime denotes a derivative with respect to z. Eliminating $\mathcal{F}_{q-1}(z,a)$ and $\mathcal{F}_{q-2}(z,a)$ between these gives a second order differential equation satisfied by the Shkarofsky functions:

$$(a-z)\mathcal{F}_q''(z,a) - [2(a-z)-q-2]\mathcal{F}_q'(z,u) - (z+q-2)\mathcal{F}_q(z,a) + 1 = 0. \quad \text{(A.2.17)}$$

Recursion relations for the Dnestrovskii functions follow from (A.2.14) and (A.2.15) for $a = 0$:

$$(q-1)F_q(z) = 1 - zF_{q-1}(z), \qquad \text{(A.2.18)}$$

$$F_q'(z) = F_q(z) - F_{q-1}(z). \qquad \text{(A.2.19)}$$

Eliminating $F_{q-1}(z)$ between these gives a first order differential equation satisfied by the Dnestrovskii functions:

$$zF_q'(z) = (z+q-1)F_q(z) - 1. \qquad \text{(A.2.20)}$$

The function $F_q(z)$ also satisfies (A.2.17) with $a = 0$. Equation (A.2.19) integrates to give

$$F_q(z) = z^{q-1}e^z\Gamma(1-q,z), \qquad \Gamma(q,z) = \int_z^{\infty} d\zeta\, \zeta^{q-1}e^{-\zeta}, \qquad \text{(A.2.21)}$$

where $\Gamma(q,z)$ is the incomplete gamma function.

Limiting Cases

The expansion of the Dnestrovskii functions for small arguments z follows from (A.2.21) and the relevant expansion of the incomplete gamma function:

$$F_q(z) = z^{q-1} e^z \Gamma(1-q) - \sum_j^{\infty} \frac{z^j \Gamma(1-q)}{\Gamma(j+q-1) j!}$$

$$= z^{q-1} e^z \Gamma(1-q) - e^z \sum_j^{\infty} \frac{(-z)^j \Gamma(1-q)}{\Gamma(j+2-q)}. \tag{A.2.22}$$

For real, positive z there is an expansion in generalized Laguerre polynomials:

$$F_q(z) = \sum_{j=0}^{\infty} \frac{L_j^{(1-q)}(z)}{j+1}. \tag{A.2.23}$$

For large argument, $|z| \gg 1$, the limit

$$F_q(z) \sim \sum_{j=0}^{\infty} (-1)^j z^{-1-j} \Gamma(q+j) \tag{A.2.24}$$

applies for $\arg(z) < 3\pi/2$.

Half-Integer q

In evaluating (2.5.27) in terms of Shkarofsky functions, the function and its derivative with $q = 5/2$ appear. The expansion (2.5.38) then leads to Dnestrovskii functions with half-integer q. For q a positive half-integer, the Dnestrovskii functions are expressible in terms of the plasma dispersion function

$$Z(y) = \pi^{-1/2} \int_{-\infty}^{\infty} dt \, \frac{e^{-t^2}}{t-y} = -\frac{\phi(y)}{y} + i\pi^{1/2} e^{-z^2}, \tag{A.2.25}$$

The relevant form is

$$\Gamma(q) F_q(z) = \sum_{j=0}^{q-3/2} (-z)^j \Gamma(q-1-j) + \pi^{1/2} (-z)^{q-3/2} [i z^{1/2} e^z Z(i z^{1/2})]. \tag{A.2.26}$$

Expansions for small and large arguments are

$$\Gamma(q) F_q(z) = \begin{cases} \sum_{j=0}^{\infty} (-z)^j \Gamma(q-1-j) - i\pi(-z)^{q-1} e^z & \text{for } |z|^2 \ll 1, \\ -\sum_{j=0}^{\infty} \Gamma(q+j)(-z)^{-1-j} - i\sigma\pi(-z)^{q-1} e^z & \text{for } |z| \gg 1, \end{cases} \tag{A.2.27}$$

with $\sigma = 0$ for $\arg z < \pi$, $\sigma = 1$ for $\arg z = \pi$ and $\sigma = 2$ for $\pi < \arg z < 2\pi$.

A.3 Dirac Algebra

In this section some results associated with the properties of Dirac matrices are summarized.

A.3.1 Definitions and the Standard Representation

The Dirac matrices are defined to satisfy

$$\gamma^\mu \gamma^\nu + \gamma^\nu \gamma^\mu = 2g^{\mu\nu}, \tag{A.3.1}$$

where the unit Dirac matrix is implicit on the right hand side. The Dirac Hamitonian is

$$\hat{H} = \alpha \cdot \hat{p} + \beta m, \qquad \alpha = \gamma^0 \gamma, \quad \beta = \gamma^0. \tag{A.3.2}$$

The requirement that the Dirac Hamiltonian be self-adjoint implies

$$(\gamma^\mu)^\dagger = \gamma^0 \gamma^\mu \gamma^0. \tag{A.3.3}$$

Standard Representation

The specific choice for the Dirac matrices used here is referred to as the standard representation. It corresponds to

$$\gamma^0 = \begin{pmatrix} 1 & 0 & 0 & 0 \\ 0 & 1 & 0 & 0 \\ 0 & 0 & -1 & 0 \\ 0 & 0 & 0 & -1 \end{pmatrix}, \qquad \gamma^1 = \begin{pmatrix} 0 & 0 & 0 & 1 \\ 0 & 0 & 1 & 0 \\ 0 & -1 & 0 & 0 \\ -1 & 0 & 0 & 0 \end{pmatrix},$$

$$\gamma^2 = \begin{pmatrix} 0 & 0 & 0 & -i \\ 0 & 0 & i & 0 \\ 0 & i & 0 & 0 \\ -i & 0 & 0 & 0 \end{pmatrix}, \qquad \gamma^3 = \begin{pmatrix} 0 & 0 & 1 & 0 \\ 0 & 0 & 0 & -1 \\ -1 & 0 & 0 & 0 \\ 0 & 1 & 0 & 0 \end{pmatrix}. \tag{A.3.4}$$

A convenient way of writing these and other 4×4 matrices is in terms of block matrices. Let $\mathbf{0}$ and $\mathbf{1}$ be the null and unit 2×2 matrices. One writes

$$\boldsymbol{\Sigma} = \begin{pmatrix} \boldsymbol{\sigma} & \mathbf{0} \\ \mathbf{0} & \boldsymbol{\sigma} \end{pmatrix}, \qquad \rho_x = \begin{pmatrix} \mathbf{0} & \mathbf{1} \\ \mathbf{1} & \mathbf{0} \end{pmatrix},$$

$$\rho_y = \begin{pmatrix} \mathbf{0} & -i\mathbf{1} \\ i\mathbf{1} & \mathbf{0} \end{pmatrix}, \quad \rho_z = \begin{pmatrix} \mathbf{1} & \mathbf{0} \\ \mathbf{0} & -\mathbf{1} \end{pmatrix}, \tag{A.3.5}$$

where the 2×2 matrices

$$\sigma_x = \begin{pmatrix} 0 & 1 \\ 1 & 0 \end{pmatrix}, \quad \sigma_y = \begin{pmatrix} 0 & -i \\ i & 0 \end{pmatrix}, \quad \sigma_z = \begin{pmatrix} 1 & 0 \\ 0 & -1 \end{pmatrix}, \tag{A.3.6}$$

are the usual Pauli matrices. In this representation one has

$$\gamma^\mu = [\rho_z, i\rho_y \boldsymbol{\Sigma}], \qquad \boldsymbol{\alpha} = \rho_x \boldsymbol{\sigma}, \qquad \beta = \rho_z. \tag{A.3.7}$$

Dirac Matrices $\sigma^{\mu\nu}$ and γ^5

Two additional Dirac matrices that play an important role in the theory are

$$\sigma^{\mu\nu} = \tfrac{1}{2}[\gamma^\mu, \gamma^\nu], \tag{A.3.8}$$

which plays the role of a spin angular momentum, and

$$\gamma^5 = -i\gamma^0\gamma^1\gamma^2\gamma^3, \tag{A.3.9}$$

which satisfies the relations

$$\gamma^\mu\gamma^5 + \gamma^5\gamma^\mu = 0, \qquad (\gamma^5)^2 = 1, \qquad (\gamma^5)^\dagger = \gamma^5. \tag{A.3.10}$$

One also has

$$\gamma^\mu\gamma^\nu\gamma^\rho\gamma^\sigma\gamma^5 = -i\epsilon^{\mu\nu\rho\sigma}. \tag{A.3.11}$$

In the standard representation one has $\gamma^5 = -\rho_x$. The spin 4-tensor $\sigma^{\mu\nu}$, defined by (A.3.8), has components

$$\sigma^{\mu\nu} = \begin{pmatrix} 0 & \alpha_x & \alpha_y & \alpha_z \\ -\alpha_x & 0 & -i\sigma_z & i\sigma_y \\ -\alpha_y & i\sigma_z & 0 & -i\sigma_x \\ -\alpha_z & -i\sigma_y & i\sigma_x & 0 \end{pmatrix}. \tag{A.3.12}$$

A.3.2 *Basic Set of Dirac Matrices*

There are 16 independent 4×4 matrices and for the Dirac matrices it is sometimes convenient to choose a set of 16 basis vectors. A specific choice of 16 independent matrices is the set

$$\gamma^A = \left[1,\ \gamma^\mu,\ i\sigma^{\mu\nu},\ i\gamma^\mu\gamma^5,\ \gamma^5\right]. \tag{A.3.13}$$

This choice involves a scalar and a pseudo scalar $(1, \gamma^5)$, a 4-vector and a pseudo 4-vector $(\gamma^\mu, i\gamma^\mu\gamma^5)$ and an antisymmetric second rank 4-tensor $(\sigma^{\mu\nu})$. These have 1, 1, 4, 4, and 6 components, respectively. This set is chosen such that the analogous set, γ_A with indices down, $\gamma_A = [1,\ \gamma_\mu, i\sigma_{\mu\nu}, i\gamma_\mu\gamma^5, \gamma^5]$ satisfy

$$\gamma^A\gamma_A = 1 \quad \text{(no sum)}, \qquad \gamma^A\gamma_B = \delta^A_B. \tag{A.3.14}$$

The expansion of an arbitrary Dirac matrix, O say, in this basis gives

$$O = \sum_A c_A\gamma^A, \qquad c_A = \tfrac{1}{4}\mathrm{Tr}\,[\gamma_A O]. \tag{A.3.15}$$

Traces of Products of γ-Matrices

The traces of products of γ-matrices are important in detailed calculations in QED. Consider

$$T^{\alpha_1\alpha_2\ldots\alpha_n} = \mathrm{Tr}\left(\gamma^{\alpha_1}\gamma^{\alpha_2}\ldots\gamma^{\alpha_n}\right). \tag{A.3.16}$$

The trace of γ^μ is zero, as are the traces of $\sigma^{\mu\nu}$, $\gamma^\mu\gamma^5$ and γ^5. The trace of a product of an odd number of γ-matrices is also zero: $T^{\alpha_1\alpha_2\ldots\alpha_n} = 0$ for n odd. The trace of a product of two γ-matrices is nonzero. This trace is evaluated as follows. First the invariance of the trace of a product of matrices under cyclic permutations of the matrices implies $T^{\mu\nu} = T^{\nu\mu}$. The trace of (5.1.1) implies $T^{\mu\nu} = 4g^{\mu\nu}$, where the factor of 4 arising from the trace of the unit 4×4 matrix. Using the invariance of the trace under cyclic permutations and (5.1.1) allows one to evaluate the traces (A.3.16) for all even n. One finds

$$T^{\mu\nu} = 4g^{\mu\nu}, \qquad T^{\mu\nu\rho\sigma} = 4\left[g^{\mu\nu}g^{\rho\sigma} - g^{\mu\rho}g^{\nu\sigma} + g^{\mu\sigma}g^{\nu\rho}\right], \tag{A.3.17}$$

$$T^{\mu\nu\rho\sigma\alpha\beta} = 4\left[g^{\mu\nu}T^{\rho\sigma\alpha\beta} - g^{\mu\rho}T^{\nu\sigma\alpha\beta} + g^{\mu\sigma}T^{\nu\rho\alpha\beta} - g^{\mu\alpha}T^{\nu\rho\sigma\alpha} + g^{\mu\beta}T^{\nu\rho\sigma\alpha}\right], \tag{A.3.18}$$

and so on.

References

1. M. Abramowitz, I.A. Stegun, *Handbook of Mathematical Functions* (Dover, New York, 1965)
2. B.B. Godfrey, B.S. Newberger, K.A. Taggart, IEEE Trans. Plasma Phys. **PS-3**, 60, 68 (1975)
3. I.S. Gradshteyn, I.M. Ryzhik, *Table of Integrals, Series, and Products* (Academic, New York, 1965)

4. J.J. Quinn, S. Rodriguez, Phys. Rev. **128**, 2487 (1962)
5. M. Rotham, Q. J. Mech. Appl. Math. **7**, 379 (1954)
6. A.A. Sokolov, I.M. Ternov, *Synchrotron Radiation* (Pergamon Press, Oxford, 1968)
7. A.A. Sokolov, I.M. Ternov, *Radiation from Relativistic Electrons* (AIP, New York, 1986)
8. G.N. Watson, *A Treatise on the Theory of Bessel Functions* (Cambridge University Press, Cambridge, 1944)

Index

D. Melrose, *Quantum Plasmadynamics: Magnetized Plasmas*, Lecture Notes in Physics 854, DOI 10.1007/978-1-4614-4045-1,

Q

R

S